ELECTRICITY PRICING

Engineering Principles
and Methodologies

POWER ENGINEERING

Series Editor
H. Lee Willis
Quanta Technology
Raleigh, North Carolina

Advisory Editor
Muhammad H. Rashid
University of West Florida
Pensacola, Florida

1. Power Distribution Planning Reference Book, *H. Lee Willis*
2. Transmission Network Protection: Theory and Practice, *Y. G. Paithankar*
3. Electrical Insulation in Power Systems, *N. H. Malik, A. A. Al-Arainy, and M. I. Qureshi*
4. Electrical Power Equipment Maintenance and Testing, *Paul Gill*
5. Protective Relaying: Principles and Applications, Second Edition, *J. Lewis Blackburn*
6. Understanding Electric Utilities and De-Regulation, *Lorrin Philipson and H. Lee Willis*
7. Electrical Power Cable Engineering, *William A. Thue*
8. Electric Systems, Dynamics, and Stability with Artificial Intelligence Applications, *James A. Momoh and Mohamed E. El-Hawary*
9. Insulation Coordination for Power Systems, *Andrew R. Hileman*
10. Distributed Power Generation: Planning and Evaluation, *H. Lee Willis and Walter G. Scott*
11. Electric Power System Applications of Optimization, *James A. Momoh*
12. Aging Power Delivery Infrastructures, *H. Lee Willis, Gregory V. Welch, and Randall R. Schrieber*
13. Restructured Electrical Power Systems: Operation, Trading, and Volatility, *Mohammad Shahidehpour and Muwaffaq Alomoush*
14. Electric Power Distribution Reliability, *Richard E. Brown*
15. Computer-Aided Power System Analysis, *Ramasamy Natarajan*

16. Power System Analysis: Short-Circuit Load Flow and Harmonics, *J. C. Das*
17. Power Transformers: Principles and Applications, *John J. Winders, Jr.*
18. Spatial Electric Load Forecasting: Second Edition, Revised and Expanded, *H. Lee Willis*
19. Dielectrics in Electric Fields, *Gorur G. Raju*
20. Protection Devices and Systems for High-Voltage Applications, *Vladimir Gurevich*
21. Electrical Power Cable Engineering, Second Edition, *William Thue*
22. Vehicular Electric Power Systems: Land, Sea, Air, and Space Vehicles, *Ali Emadi, Mehrdad Ehsani, and John Miller*
23. Power Distribution Planning Reference Book, Second Edition, *H. Lee Willis*
24. Power System State Estimation: Theory and Implementation, *Ali Abur*
25. Transformer Engineering: Design and Practice, *S. V. Kulkarni and S. A. Khaparde*
26. Power System Capacitors, *Ramasamy Natarajan*
27. Understanding Electric Utilities and De-regulation: Second Edition, *Lorrin Philipson and H. Lee Willis*
28. Control and Automation of Electric Power Distribution Systems, *James Northcote-Green and Robert G. Wilson*
29. Protective Relaying for Power Generation Systems, *Donald Reimert*
30. Protective Relaying: Principles and Applications, Third Edition, *J. Lewis Blackburn and Thomas J. Domin*
31. Electric Power Distribution Reliability, Second Edition, *Richard E. Brown*
32. Electrical Power Equipment Maintenance and Testing, Second Edition, *Paul Gill*
33. Electricity Pricing: Engineering Principles and Methodologies , *Lawrence J. Vogt*

ELECTRICITY PRICING

Engineering Principles and Methodologies

Lawrence J. Vogt, P.E.

CRC Press
Taylor & Francis Group
Boca Raton   London   New York

CRC Press is an imprint of the
Taylor & Francis Group, an **informa** business

CRC Press
Taylor & Francis Group
6000 Broken Sound Parkway NW, Suite 300
Boca Raton, FL 33487-2742

© 2009 by Taylor and Francis Group, LLC
CRC Press is an imprint of Taylor & Francis Group, an Informa business

No claim to original U.S. Government works

Printed in the United States of America on acid-free paper
10 9 8 7 6 5 4 3 2 1

International Standard Book Number: 978-0-8247-2753-6 (Hardback)

This book contains information obtained from authentic and highly regarded sources. Reasonable efforts have been made to publish reliable data and information, but the author and publisher cannot assume responsibility for the validity of all materials or the consequences of their use. The authors and publishers have attempted to trace the copyright holders of all material reproduced in this publication and apologize to copyright holders if permission to publish in this form has not been obtained. If any copyright material has not been acknowledged please write and let us know so we may rectify in any future reprint.

Except as permitted under U.S. Copyright Law, no part of this book may be reprinted, reproduced, transmitted, or utilized in any form by any electronic, mechanical, or other means, now known or hereafter invented, including photocopying, microfilming, and recording, or in any information storage or retrieval system, without written permission from the publishers.

For permission to photocopy or use material electronically from this work, please access www.copyright.com (http://www.copyright.com/) or contact the Copyright Clearance Center, Inc. (CCC), 222 Rosewood Drive, Danvers, MA 01923, 978-750-8400. CCC is a not-for-profit organization that provides licenses and registration for a variety of users. For organizations that have been granted a photocopy license by the CCC, a separate system of payment has been arranged.

Trademark Notice: Product or corporate names may be trademarks or registered trademarks, and are used only for identification and explanation without intent to infringe.

Library of Congress Cataloging-in-Publication Data

Vogt, Lawrence J.
   Electricity pricing : engineering principles and methodologies / Lawrence J. Vogt.
     p. cm. -- (Power engineering ; 33)
   Includes bibliographical references and index.
   ISBN 978-0-8247-2753-6
   1. Electric utilities--Prices. I. Title. II. Series.

HD9685.A2V64 2009
333.793'231--dc22
                                                                                         2009029821

Visit the Taylor & Francis Web site at
http://www.taylorandfrancis.com

and the CRC Press Web site at
http://www.crcpress.com

To Deanna

Contents

Foreword	xiii
Preface	xv
About the Author	xix

Chapter 1
Overview of Electricity Pricing

1.1	Introduction	1
1.2	Objectives of the Electric Tariff	7
1.3	Traditional Price Regulation	10
1.4	Impacts of Industry Restructuring on Electricity Pricing	12
1.5	The Role of Engineering in Electricity Pricing	18
1.6	Chapter Organization	20

Chapter 2
Prerequisite Concepts

2.1	Introduction	23
2.2	The Cardinal Laws of Electricity	24
2.3	Fundamental Load Characteristics	36
2.4	Elements of Utility Costs and Pricing	50
2.5	Summary	59

Chapter 3
Electric End Uses

3.1	Introduction	61
3.2	Customer End-Use Needs	62
3.3	End-Use Value	89
3.4	Customer End-Use Research	91
3.5	Summary	104

Chapter 4
Customers: Electric Service Requirements

4.1	Introduction	105
4.2	Customer Electricity Requirements	106
4.3	Nonphysical Service Requirements	116
4.4	The Customer as Energy Manager	120
4.5	Customer-Owned Power Systems	124
4.6	Wholesale and Compound Retail Customers	132
4.6	Summary	134

Chapter 5
Power Delivery Systems: Transmission and Distribution

5.1	Introduction	135
5.2	Power Distribution	142
5.3	Power Transmission	173
5.4	Summary	190

Chapter 6
Electric Power Production

6.1	Introduction	193
6.2	Power Production Technologies	195
6.3	Classifications of Generation	214
6.4	Generation Operation and Control	219
6.5	Summary	225

Chapter 7
Revenue Metering and Billing

7.1	Introduction	227
7.2	Metering of Electrical Energy	229
7.3	Metering of Demand	237
7.4	Special Metering Considerations	244
7.5	Meter Data Acquisition, Processing, and Billing	252
7.6	Summary	262

Chapter 8
The Rate Schedule: Pricing Methods and Risk

8.1	Introduction	263
8.2	Qualifying Service Provisions	267
8.3	Rate Structure Forms	271
8.4	Attendant Billing Provisions	309
8.5	Summary	331

Contents   xi

Chapter 9
The Mathematics of Rates

9.1	Introduction	335
9.2	Graphical Representation of Rates	337
9.3	Formulary Representation of Rates	352
9.4	Rate Comparison Methods	364
9.5	Summary	387

Chapter 10
Billing Data Statistics and Applications

10.1	Introduction	389
10.2	Characteristics of Billing Data	391
10.3	Bill Frequency Analyses	404
10.4	Bill Frequency Revenue Analyses	407
10.5	Revenue Forecasting Methodologies	421
10.6	Summary	440

Chapter 11
Cost-of-Service Methodology

11.1	Introduction	443
11.2	Cost-of-Service Study Overview	445
11.3	Cost Functionalization	454
11.4	Cost Classification	494
11.5	Cost Assignment	530
11.6	Cost-of-Service Study Results	562
11.7	Summary	570

Chapter 12
Translating Costs to Prices

12.1	Introduction	571
12.2	Coincidence Factor—Load Factor Relationship	576
12.3	Cost Curve Development	587
12.4	Rate Design Methodology	597
12.5	Guiding Principles of Ratemaking	610
12.6	Summary	611

Appendix A
Unit Conversions   613

Appendix B
Uniform System of Electric Accounts   617

**Appendix C
Industry Classification Systems** 639

**Appendix D
Internet Resources** 651

Bibliography 655

Index 661

Foreword

My first exposure to electricity pricing came eight years into my career as a power engineer. In 1977, now a registered PE with past tours of duty in several departments, I was appointed Lead Transmission Planning Engineer at Houston Lighting and Power Company. I felt that in T&D planning I had found my true calling: the nearly intractable combinatorial nature of T&D planning, the myriad interactions among often poorly measured factors, and the uncertainty inherent in dealing with the future, created an irresistible challenge: here was a game one could never win entirely, but doing the best one could do was both valuable to the company and technically intoxicating to me.

Not so appealing was an onerous duty that came with my new job. As the junior member of the company's leadership team, it was expected that I "lead" the next rate case cost-of-service (COS) study. Older co-workers warned me that turning down the assignment would ruin any chance of further promotion, and that "leading" meant I would actually do the tiresome data collection, tedious analysis and lengthy report writing myself, while spending interminable hours in meetings with lawyers, economists and others not imbued with the crystal-clear thinking that came only with an Engineering degree and a PE license.

But the COS study I did over the next many months was one of the most fascinating projects of my career. First, it was indisputably valuable, for COS was the foundation of the company's pricing; if you don't get the money right, you don't get anything right. Second, like T&D planning, it required a complicated, multi-faceted analysis that included every element in the system and every part of the customer base in a consistent, traceable process from beginning to end. And finally, in its own way it had to deal with factors and assumptions that were fuzzy and often not precisely measureable.

There were many long and boring meetings, but while the economists and lawyers I met often seemed to have been raised on another planet, I grew to realize they were clearly expert in important areas then completely alien to me. Uncharacteristically keeping my mouth shut about the numerous inconsistencies and shortcomings I saw in the existing method, and doing the job entirely as prescribed by upper management, in due course I completed the COS study and moved on.

But I was sorry it was over. I sensed I had been exposed to something fundamental, and that there was much to learn. I started car pooling with one of the company's younger, brighter rate economists. I let it be known that I was eager to do *all* future COS studies. I told my bosses I could make valuable improvements next time without realizing that that was, to some, about the last thing they wanted to hear. Looking back, this was without doubt the beginning of HL&P's and my joint realization that I belonged somewhere else. About a year later I moved to Westinghouse (later ABB) Advanced Systems Technology, where I made a career of trying to improve planning methodologies and was privileged to work with many of the brightest people in our industry, among them Ralph Masiello, Jim Burke, Damir Novosel, Hahn Tram, Richard Brown . . . and Larry Vogt.

I first met Larry in the early 1980s and I was immediately impressed because he was a Rate *Engineer*. I was even more impressed because he had a solid T&D engineering background and not only recognized the shortcomings I had seen in COS and rate engineering methods, but understood how to fix them and make further improvements. Eventually Larry joined me at Westinghouse and I was lucky enough to collaborate with and learn from him for a decade, before diverging professional and personal paths took us our separate ways.

Electricity Pricing: Engineering Principles and Methodologies is Larry at his best: detail-oriented but mindful of the big picture, theoretically correct but grounded in no-nonsense practical considerations, and crystal clear throughout. This is by far the best book I have seen on electricity pricing, and one of the best in power engineering, period. Of course, I'm particularly partial to Chapter 11, Cost of Service Methodology, a terrifically comprehensive, 128-page treatment of that foundation of pricing that still intrigues me after thirty years. Take into consideration its very accessible tutorials on the basics, and its wonderful discussions on rate theory, rate mathematics, translating cost to prices and the like, and there is no doubt this book is a necessity for any modern electric utility engineer.

As both a close personal friend of Larry and the editor of Marcel Dekker's Power Engineering Series, I am proud to include *Electricity Pricing: Engineering Principles and Methodologies* in that group of books. But more important, this book comes at just the right time. The advent of Smart Grid means that our industry is about to undergo a revolution in how homeowners and businesses can purchase and manage their power. Pricing will be more complicated and intricate than ever. *Electricity Pricing: Engineering Principles and Methodologies* provides a solid and accessible resource for those who have to "get the money right" in this new industry environment. I will use this book a great deal.

Lee Willis
Cary NC

Preface

On September 4, 1882, Thomas Edison started up his 100 kilowatt coal-fired Pearl Street dynamo to demonstrate that central station generation and distribution of electricity along the streets of lower Manhattan in New York could make city electric lighting a reality. Customers perceived great value in the use of electric illumination over the flame. Demand for incandescent lighting grew. Billing for electric service soon followed – electricity pricing was born.

It took some time to develop a functional electromechanical watt-hour meter for measuring a customer's energy usage. As electric lighting of buildings was the first practical end-use device for small consumers, electricity prices, or rates, were often based on a simple count of bulbs or even rooms within the home or business (typical energy consumption could be estimated reasonably).

In the early days of the commercialization of electricity, unit prices were relatively high. As new consumer end-use devices flourished, more and more generators and electric lines were built to meet the need. The construction of large capacity, central station generators offered economies of scale, which put downward pressure on electricity prices. Domestic service rates were structured and priced to encourage customer acceptance of new appliances that would forever change our lifestyles for the better. Contemporary electric rates and their associated metering requirements have become much more sophisticated than those of the original Pearl Street Station era.

Today, electricity is essential to providing a significant amount of society's energy requirements. No viable substitutes for electrical energy exist for many end-use applications. Consider the need for electricity for a much broader array of lighting purposes and far more complex applications in computer systems and manufacturing. Even so, electricity competes with alternative energy sources to serve some end uses, such as space heating, although even a natural gas-fired furnace requires electricity in order to operate.

Our dependency upon energy in today's economy has made electricity vital for maintaining and advancing the quality of life. As such, society demands an abundant, dependable, and economical supply of electricity. Utility engineers have responded to this charge by planning, designing, building, and operating electric power systems to ensure the universal availability of reliable electric service at a reasonable cost. Governmental authorities at both the state and federal levels have also responded by establishing more stringent service criteria and regulating electricity pricing structures.

Traditional price regulation treats electricity as a cost-based service. As a result, engineering doctrine has been used extensively to develop and support the ratemaking process through a cost-of-service approach. For example, the costs of customer-shared facilities, such as central generating stations, obviously cannot be directly assessed to individual customers. Therefore, a logical methodology for apportioning the shared costs of electricity production and delivery to different consumer segments is necessary. The engineering solution is to allocate capital investments and operating expenses on a cost causation criteria that models the technical characteristics of the power system.

Today's electric industry environment is a mix of traditional regulated services and market-based services. In some locations, utilities have been restructured into separate competitive energy service providers and monopolistic power delivery service providers. It follows that restructuring has transformed electricity pricing in those jurisdictions from an integrated cost-based structure to a functionally unbundled configuration of prices that are governed accordingly by both energy market forces and conventional price regulation. Such transitions have yielded opportunities and challenges for electricity purchasers and suppliers alike.

Today's electric industry is also facing increasing costs which in turn put upward pressure on the prices for electricity. Due in part to cost volatility, the need for more frequent rate revisions is once again a front and center utility issue. Recovering costs while meeting the requirements of customers is a paramount challenge for the rate designer. In essence, electricity pricing is a risk management process which takes into account the needs of all stakeholders.

Electricity pricing is a multi-disciplined endeavor. Effective utility rate design requires the melding of information, methodologies, and techniques from accounting, economics, engineering, finance, and marketing. Engineering is often referred to as the *art of applied science*. Utility pricing or ratemaking can be thought of as the *art of applied business technologies*. This book examines the comprehensive scope of ratemaking, while focusing on the underlying engineering and technical principles of electricity pricing.

Rate engineering is a unique skill that ultimately is mastered not in an academic curriculum but by means of on-the-job longevity. One must be engaged in the ratemaking process to fully comprehend the discipline. While general ratemaking concepts are widely published, rather few "how to do it" resources exist for aiding the electricity pricing practitioner. Typically, such details have been passed along through task succession channels. This book consolidates and preserves valuable analysis techniques, realistic examples, and lessons that have been gained through on the job experience and research, and which are not widely documented in contemporary literature.

Preface

Because of the multi-discipline aspects of ratemaking, this book is intended for a broad audience. Engineers involved in ratemaking will not only learn about the role that engineering doctrine play in electric rate analysis and design, but they will also learn about how other business functions collaborate in the overall electricity pricing process. Conversely, ratemaking professionals from other business disciplines will learn about the significance of engineering applications that are key to successful electricity pricing.

Electricity Pricing: Engineering Principles and Methodologies has a dual purpose. It serves as a textbook for those who are new to the discipline and who are interested in learning about the technical principles and methodologies of the ratemaking profession. It also serves as a comprehensive reference book for the experienced practitioner.

I have always held a great admiration for the founders of electricity pricing – iconic names such as Dr. John Hopkinson and Mr. Arthur Wright. I have also learned much about rates from the later writings of Mr. Constantine Bary, Dr. James Bonbright, and Mr. Russell Caywood. But I have also developed a deeper knowledge and appreciation of electricity pricing from having worked for distinguished rate experts whose careers have overlapped with my generation, including: Mr. Don Gimbel at Public Service Indiana, Mr. Clarence Grund and Mr. Robbie Norman at Southern Company Services, Mr. Marty Blake and Mr. Steve Seelye at Louisville Gas and Electric Company, and Mr. Jim Patton at Mississippi Power Company. I also pay tribute to Dr. David Conner, my thesis advisor at the University of Louisville, and Mr. Lee Willis, who I worked for at ABB Power T&D Company. They have been longtime mentors, and their inspiration helped me to discover my desire to be a published writer. I also have a great appreciation for Professor Ralph Powell of Penn State University and Mr. Cass Bielski of EEI's Rate and Regulatory Affairs Committee, who have both provided me with exceptional opportunities to teach the rate discipline. All of these career experiences and influences have helped me to produce this book.

Lawrence J. (Larry) Vogt

R8maker@ljvogt.com

About the Author

Lawrence J. (Larry) Vogt is the Manager of Rates for Mississippi Power Company, a Southern Company, where he is responsible for managing the research, development, and implementation of cost-of-service studies and rate designs. He is also responsible for administration of the Company's electric tariff and for supporting other regulatory matters of the Company. He previously worked for ABB Power T&D Company as Manager of the Distribution Technologies Center and for Southern Company Services as Principal Engineer – Rates & Regulation. He has also held various pricing, marketing, and engineering positions at Public Service Company of Indiana, Inc. (now known as Duke Energy – Indiana) and Louisville Gas & Electric Company. Larry has over 36 years of experience in the electric utility business in the areas of pricing, cost analysis, commercial and industrial marketing, demand-side management, and distribution engineering and planning.

Larry is Mississippi Power Company's representative on the EEI Rate & Regulatory Affairs Committee, and he also serves as a principal instructor in the Committee-sponsored E-Forum Rate College. In addition, he serves as an Adjunct Professor in Penn State University's International Power Engineering Program. He has also conducted a number of industry workshops under the sponsorships of the University of South Alabama and the Electric League of Indiana, Inc. As an instructor, he has taught numerous courses both domestically and internationally on the topics of rate design, cost-of-service methodology, spatial load forecasting, distribution system planning, and the impacts of demand-side management on the transmission and distribution system. His audiences have included engineers, accountants, attorneys, and other professionals from utilities, regulatory agencies, and business.

Larry is a graduate of the University of Louisville where he earned Bachelor of Science and Master of Engineering degrees in the field of electrical engineering. He is a member of AEE, IEEE, and NSPE, and he is a registered Professional Engineer in several states. He has served as an expert witness on a variety of rate and regulatory issues. Larry is the coauthor of several technical papers as well as the textbook *Electrical Energy Management*, Lexington Books (1977).

1
Overview of Electricity Pricing

1.1 INTRODUCTION

Electricity provides consumers with the ability to realize numerous conveniences in their everyday lives. One does not buy electricity as an end product. One buys electricity as an input which is then utilized to produce consumable goods and services or to enjoy domestic comforts. Compared to most other everyday products, electricity is elusive because it does not possess a physical form. Electricity is energy, and it is readily available to the consumer in a most unique way.

Flip a switch and an electric lamp is immediately illuminated. The "order" is simultaneously placed by the consumer and filled by the supplier. Electricity arrives on a just-in-time delivery basis. No practical applications of mass electricity storage exist today; therefore, electricity must be produced and transmitted to the point of use upon demand. Electricity travels fast at about 186,000 miles per hour. The continuous, high speed transportation of this energy requires a reliable power production and delivery system to ensure that the demands of all consumers are met continuously.

A massive infrastructure of electric utility systems has been developed. Electricity is generated at central stations and then transmitted to consumers via a complex network of wires which are operated at a variety of voltages. The basic organization of the electrical power system is shown in Figure 1.1. Three major functions are identified: production, transmission, and distribution.

The production plants are facilities where other forms of energy are converted into electrical energy by means of a generator. The relatively low voltage output of the generator is transformed to higher voltages in order to effectively transmit power to consumers, many of whom are located far away from the plants. The transmission system operates at high voltage levels in order to move large amounts of power over long distances to load centers (i.e., towns and regions within large cities). At the load centers, the high level voltages are transformed into medium level voltages for the distribution or routing of power within different locales. At various service points along the distribution system, the medium voltages are transformed to low voltages at which many electrical devices, or loads, operate.

As indicated in Figure 1.1, customers may take electric service at any functional level of the power system depending on their specific needs. In general, large commercial and industrial customers receive service at higher voltages than smaller customers. Residential customers are predominantly served at the lowest voltage level.

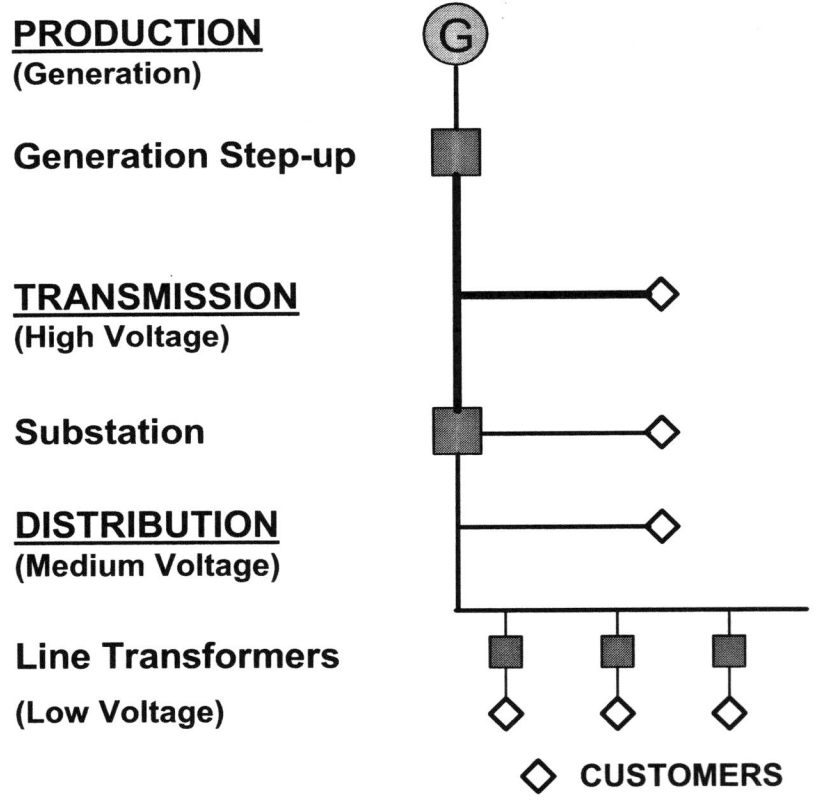

Figure 1.1 Functional organization of the electric power system. The major functions represent different levels of operating voltage.

Overview of Electricity Pricing

In addition to the principal functions of producing and delivering energy, electric service includes a number of attendant services. The power system provides energy to customers having rather dynamic requirements. The energy needs of customers vary significantly in correspondence to the diverse operations of a variety of loads composed of electrical equipment and appliances. Taken together, the load requirements imposed on the power system fluctuate significantly from moment to moment. The power system is designed and operated to provide the capacity necessary to adequately serve the load in this fashion. Thus, the ongoing process of balancing generation with dynamic loads represents one example of an ancillary service which is provided as a responsibility of the system operators. Another example of an ancillary service is reactive power, which is required for system voltage control and to meet the power factor needs of customers.

Other nonelectric attendant services are also included with the rendering of electric service. Metering, billing, and customer accounting are functions which are necessary in order to properly invoice customers and maintain the records of customers' usage and payment histories. Additional utility functions include customer service representatives and education programs which deal with such topics as electrical safety and energy usage and management methods.

Types of Customers

Electricity sales are generally categorized along the lines of *customer classes*. Figure 1.2 illustrates the customer segment groups which are typically used to delineate customers by class of service. A utility may provide service to both *retail* and *wholesale* customers. Customers in the retail sector are ultimate users of electricity. Lighting service customers are actually residential, commercial, public authority, and industrial types of customers. Each of these primary classes may separately purchase outdoor lighting service for safety and security reasons. Cities, towns, and communities require lighting service for illumination of streets, highways, and interchanges. Lighting service can consist of only electricity, or it can also include the actual fixtures, lamps, and the associated maintenance responsibility. Customers in the wholesale sector purchase electricity from a utility for resale to other entities or to their own ultimate customers.

Territorial or *native customers* represent a utility's retail customers and its wholesale requirements customers, or municipals and cooperatives. Wholesale customers may be classified as taking either *full requirements service* or *partial requirements service*. Partial requirements customers have some generation resources which must be supplemented with wholesale electricity purchases in order to meet the needs of their ultimate customers. Traditionally, other wholesale customers consisted of adjacent, interconnected utilities that established interchange agreements for the basic purpose of economy and emergency purchases and to augment generation construction schedules. But with contemporary changes in the wholesale sector, a number of new buyers and sellers have appeared, including power marketers, brokers, and customer aggregators.

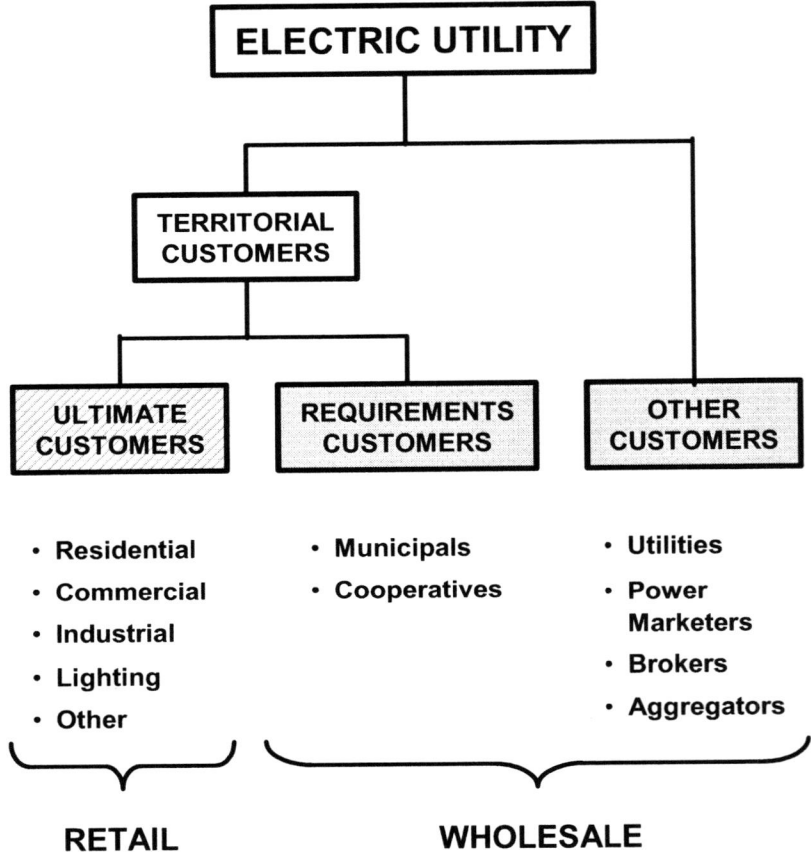

Figure 1.2 Classifications of utility customers. An example of an investor-owned utility's customers. Retail service is provided to ultimate customers (end users). Wholesale service is provided to other utilities and service providers for resale purposes. Requirements service customers may own production facilities which meet a portion of their load.

Key Electricity Pricing Concepts

Pricing of electric service must take into account both the production and delivery of electrical energy as well as the attendant services required to adequately render service. While similar to other products and services in many respects, electricity possesses its own exceptional characteristics. As a result, the process of pricing electricity is both complex and unique.

Overview of Electricity Pricing

To simply compute a single price per kilowatt-hour (kWh) to be applied to all kWh usage, regardless of customer type, would grossly ignore the distinct differences that exist in the ways in which customers utilize electricity. By carefully assessing these distinctions, a number of factors which justify price differentials can be recognized. For example, much of the power system is jointly shared by all customers. All customers share in the use of generation resources. Furthermore, all customers share in the use of the transmission system. But customers taking service at the transmission level do not utilize the utility's distribution system. Transmission customers have their own transformation facilities and distribution systems; thus, the prices for customers served at a transmission voltage should not include any cost recovery for the utility's distribution facilities. Meanwhile, customers served at primary and secondary voltage levels would share in the costs of the primary distribution system. This observation leads to the rationale for pricing which is differentiated by voltage level.

Temporal variations in the way in which electricity is utilized lead to another significant pricing differential. The power system experiences periods during which the combined loading of all of the customers is much higher than at other times. This *on-peak* period vs. *off-peak* period situation is quite prevalent throughout the industry in terms of both seasonal and daily variations. During peak loading conditions, such as a summer afternoon in climate zones when customers are using air conditioning extensively, a major portion of the power system's capability is utilized. The system is designed and built to carry such loads adequately (since storage is not possible). As a result, when the load falls off at night, the system's capability becomes underutilized. Simply stated, more costs must be incurred to serve peak period loads than to serve off-peak loads. This observation leads to the rationale for pricing which is differentiated by time.

A distinction in pricing can also be raised due to spatial variations of the system. Consider, for example, an area that is experiencing an accelerated level of growth compared to an adjacent area that has been "built out," possibly for some time. Such a situation often occurs within the same city. The high growth area requires significant distribution system upgrades in addition to completely new facilities, including perhaps a new substation. In contrast, customers located in the older area continue to be served by fully operational distribution system facilities of an earlier vintage, which were constructed to serve the electric loads of the day. An inspection of the costs relative to these two areas would reveal a higher cost to serve customers located in the fast growing area.[1] This observation leads to the rationale for pricing which is differentiated by location.

Another pricing distinction occurs relative to the configuration of local facilities. Obviously, customers with high peak loads require higher capacity service

---

[1] The premise for this example is that two different distribution systems can create a geographical cost differential. One might also wish to argue that only those customers located in the high growth area are potentially creating a need for additional transmission and generation facilities.

equipment than do smaller customers. But some customers (e.g., those with large motor loads) have a requirement for three-phase power while many other customers only need single-phase power. In comparison, three-phase service requires more conductors, transformers, and protective devices than single-phase service. Metering cost is also greater for three-phase service. Even at the same overall capacity requirement, three-phase service typically would be more expensive than single-phase service. This observation leads to the rationale for pricing which is differentiated by character of service.

A final observation relative to pricing distinctions can be recognized from the type of construction. Specifically, service provided from underground transmission and/or distribution facilities may require a greater investment than service provided from overhead facilities (except for higher density loads). On the other hand, underground facilities are less exposed to the types of physical hazards faced by overhead facilities; thus, underground distribution may offer an advantage with respect to reliability. In general, the overhead method of delivering electricity often represents the least-cost approach.

The common theme uncovered in each of the preceding discussion points is that, within an electric power system, certain situations often occur which lead to the existence of cost-of-service differentials between individual customers (even within the same customer segment) and between dissimilar groups of customers. Recognition of the more significant cost-of-service differentials through pricing structures is necessary to ensure reasonable equity between customers while balancing the inherent risks assumed by each of the parties. In short, costs should be more accurately attributed to those who require that those costs be incurred.

To adequately capture the cost-related attributes of electricity through pricing, while preserving equity among customers, three individual pricing components are recognized.

- **Customer Component** - the costs associated with a customer's connectivity to the power system[2]
- **Demand Component** - the costs associated with a customer's maximum load requirements
- **Energy Component** - the costs associated with a customer's requirements for a volume of electricity, i.e., kWh

These elementary pricing components represent *unit prices* or *rates* and are combined and modified in a variety of fashions to form the basic billing mechanism or *rate structure*. The form of rate structure is varied for different customer segments in order to reflect the cost distinctions of each of the groups. Typically, simplistic rate structures are utilized for small residential and commercial custom-

---

[2] This reference to connection costs applies to the recurring costs (investment and expenses) of an established customer and not to the "hook up" fee that made be charged initially to establish a customer's account.

ers while more complex rate structures are used for large commercial and industrial customers. Taken together, the rate structures of all of the customer classes represent the nucleus of the electric tariff.

1.2 OBJECTIVES OF THE ELECTRIC TARIFF

The *tariff* is a comprehensive document which contains all of the prices and contract features under which electric service is rendered.[3] In addition to the basic rate structures for the various customer classes, a number of supplementary stipulations and clauses are required. These auxiliary provisions are needed to define such aspects as the qualifications for service, the character of service, and the term of service. Furthermore, a variety of additional billing clauses are needed to price specific service issues, such as power factor, which are not always addressed within the basic rate structure. To accomplish this objective, a *rate schedule* is established for each classification of service.

Rate schedules may be end-use specific, customer class specific, or general in nature. General service rate schedules apply across the commercial and industrial customer classes. Examples of common electric service rate schedule applications are shown in Figure 1.3.

In addition to the rate schedules, the tariff specifies the *general terms and conditions* for service which are rules and regulations that apply commonly to all classifications of customers. Some examples of such stipulations would be restrictions in the use of the service, billing and metering methods and requirements, and line extension policies.

The Stakeholders of Electricity Pricing

While many aspects of a utility and its operations are rather obscure to those outside of the business, the tariff receives a lot of attention from many bearings. The electric tariff has the objective of satisfying the interests of three principal stakeholder groups: customers, the utility, and jurisdictional regulators.

Historically, customers have been served exclusively by a local utility which typically operates in a specified franchise area. No other suppliers have been allowed to compete for customers within these restricted service territories. All new customers locating within a franchise area are required to take electrical service only from the incumbent utility. This utility-customer relationship is a classical example of a monopoly operation. Consequently, pricing of electrical services have fallen under the scrutiny of local and national governmental agencies which serve as utility regulators.

---

[3] Utilities may have both a formal retail tariff for ultimate use customers and a formal wholesale tariff for requirements services customers. Standard and special contracts are used for other retail and wholesale transactions. These contracts contain many of the same types of features found in the published tariffs.

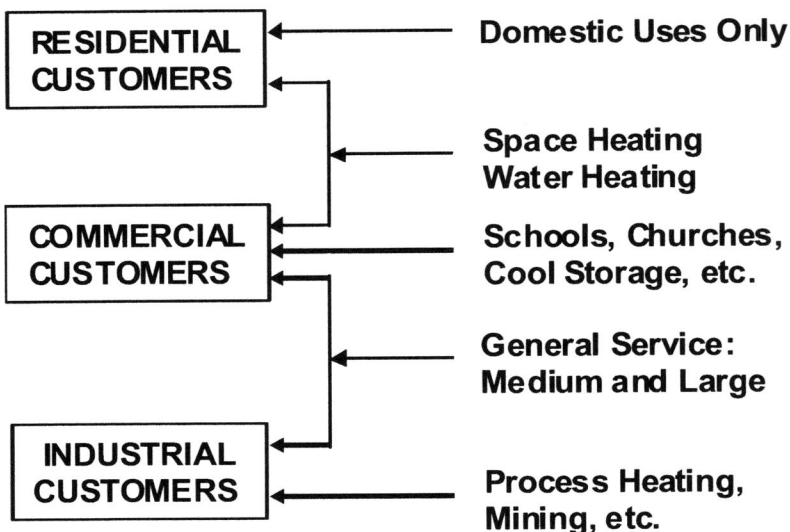

Figure 1.3 Examples of retail electric service rate schedule applications. Rate schedules may be customer class specific or general in nature. General service rates apply across two or more customer classes. Special rates may apply only to specific customer end-use applications.

Customers have come to view electricity as an essential public service. Like any other product or service, customers seek value from electricity and expect a price level that is commensurate with that perceived value. Value perceptions vary widely across the customer classes and even among customers within the same class because customers have unique needs relative to their energy usage. Consequently, social and economic issues arise in the regulatory arena as price structures and revenue levels are formulated. The regulatory process usually requires a formal rate review procedure and public hearings with expert testimony before final rates are approved.

Customers also seek price options because of a basic human desire to have choice, particularly within the constraints imposed by the monopoly structure of the utility industry. The variations in customer needs and characteristics serve as another impetus for employing pricing differentiation. Utilities have responded by offering alternative pricing methodologies so that customers can select the best rate schedule under which to receive service. For example, customers having a fair amount of operational flexibility could benefit through load shifting under time-differentiated prices as opposed to a uniform price for all kWh usage. Customers who lack such flexibility would pay higher prices under the time-of-use rate schedule.

Overview of Electricity Pricing

Although utilities are considered by customers to be providers of a public service, most electric utilities operate with a well-focused business perspective. Investor-owned utilities must earn a profit in order to compensate stockholders with a reasonable return on their investment and to pay interest on bonds and other debt. Municipal utilities, which are owned by cities, raise all of their capital through the issuance of bonds and other forms of debt financing. Likewise cooperative utilities also borrow funds to support their operations. Regardless of the means of financing, each of these utility entities must recover revenues through rates that will adequately provide compensation for operating expenses and service the cost of capital.

As with other businesses, utilities must ensure customer satisfaction in terms of both pricing and customer service. Despite the exclusivity of the local franchise, a customer's desire to add load and use more electricity may very well be influenced by the treatment from the incumbent utility. In some cases, customers may have the ability to meet at least some of their energy needs through alternative resources, such as natural gas or renewable resources. Growth of the business is just as important for a utility as it is for any other enterprise.

One offset to the monopoly aspect of the industry is that utilities have an *obligation to serve*. The incumbent utility must provide service for all customers within its franchise area. This responsibility includes connecting customers to the distribution system and planning sufficient resources to meet customers' needs through construction of new generation and/or procurement of energy from other suppliers. Under the obligation to serve criteria, the utility accepts much of the risks of serving customers that may not fulfill their contracts for service.

Regulation is also intended to serve as an offset to the monopoly situation. Regulatory agencies are charged with the responsibility to balance the interests of both the customers and the utility. Regulation strives to keep prices as low as possible for consumers while allowing the utility to earn a fair return on its investment. To achieve this balance, regulators generally operate from a position that prices should be based upon costs. Thus, a detailed *cost-of-service study* is a key factor to developing tariff prices that perform efficiently and ensure a reasonable level of equity between all parties.

But the cost-of-service study is based on an allocation of the costs of facilities which are largely shared by all customer classes. For example, a typical distribution feeder serves hundreds of customers ranging from neighborhoods of homes to areas of commercial and industrial buildings. A number of methodologies for allocating the costs of joint use facilities have been supported by the different stakeholders. Adoption of an allocation methodology is challenging since the interested parties often have opposing perspectives. The selection of one cost allocation methodology is often preferential to one customer class, e.g., residential, at the expense of another, e.g., industrial. Application of a different allocation technique could easily reverse the direction of advantage between the two classes. When taking into account the considerations of all stakeholders, the difficult question to answer is: Which methodology is (most) correct?

1.3 TRADITIONAL PRICE REGULATION

Price regulation of the electric utility industry has been in effect for many years. The need for regulation developed primarily because of the monopoly aspects of the electricity business. Having multiple utilities compete for customers in the same geographic area was deemed to be inefficient and costly. By providing a single utility with the exclusive rights to serve a defined geographical area, the duplication of facilities would be prevented. Only one company's pole line would run along the streets and alleys. The economies of scale should provide customers with the best prices. Consequently, regulation of this monopoly would guarantee quality service and price protection for all consumers.

In the United States, two basic forms of governmental regulation exists. At the national level, the Federal Energy Regulatory Commission (FERC) has regulatory jurisdiction over the pricing of all wholesale electric services.[4] Thus, prices for many of the interchanges of electricity between utilities must be reviewed and approved by the FERC.[5] Such transactions also include the use of a utility's transmission system to transmit power in support of a nonassociated electricity sale, a process referred to as *wheeling*.

The FERC consists of five Commissioners who are appointed by the President, subject to approval by the Senate. The President designates one of these Commissioners to serve as the chairman. The Commissioners serve staggered five year terms. The FERC is supported by a staff which includes an Office of Electric Power Regulation. Although the FERC is a regulatory agency, one of its primary objectives is to encourage the development of competitive electricity markets.

The regulation of prices for retail electric sales to ultimate consumers is the responsibility of the states. Each state, and the District of Columbia, has established a "Public Service Commission" (PSC) typically consisting of three or more Commissioners. The Commissioners are chosen for office either through governmental appointment or by the electorate, depending on the specific laws of each state. The term served by a state Commissioner varies from state to state. Similar to the FERC, the state commissions are supported by staffs which help to conduct the regulatory process.

Not all of the state commissions have exclusive jurisdiction over all utilities within their states. In some states, municipal utilities and rural electric cooperatives are exempt from state regulation. In such cases, municipal electric boards provide a form of "self regulation." Likewise, the customers of electric coopera-

---

[4] The FERC was created in 1977 by the Department of Energy Organization Act and is the successor to the Federal Power Commission (FPC) which was originally established in 1935. In addition to regulating wholesale electric services, the FERC is charged with other jurisdictional responsibilities, such as the regulation of wholesale natural gas prices.

[5] The FERC is now allowing more market-based transactions to take place. Previously, the prices for these sales were set by the FERC through formal proceedings.

tives, who are also cooperative members, elect a board of directors having the responsibilities of managing the utility's operation and setting prices for electric service. In some states, the Commissions have only partial jurisdiction over municipals and cooperatives, where, for example, their regulatory authority may be restricted to approval of pricing structures but not revenue levels.

Regardless of the form, regulation has long served as a means of overseeing the ratemaking process. Decades ago, the utility industry began to grow rapidly, and additions of large central generating stations provided economies of scale which resulted in declining electricity prices. But the accelerated demands for electricity of the 1970s, followed by the high inflation rates of the 1980s, ensured a considerable increase in rate case activity and an upswing in prices. In the early 1990s, as inflation slowed and construction programs matured, rate increases started to moderate considerably. In some cases, prices actually started to decline somewhat. With a trend toward a restructured industry, regional prices would likely converge as electric service providers and the regulatory processes work toward achieving higher efficiencies and lower costs.

Competition in the Regulated Environment

Although the electric utility industry is a monopoly and consequently is subject to regulation to ensure price protection of customers, elements of competition have always existed. With very few exceptions, the society has a requirement for electricity. At a minimum, customers need some electric lighting capability and convenience outlets in order to operate basic appliances and equipment. But a number of other end-use devices can be operated using alternative fuels. Distribution of natural is common in urban areas where residential customers can choose between electric and gas for space heating, water heating, cooking, and clothes drying. In rural areas where gas distribution systems are not available, customers often utilize propane to fuel some or all of these domestic end uses. In addition to natural gas and propane, commercial and industrial customers have the additional option of using coal as an energy source for heating and operating thermal processes. Fuel oil is yet another alternative for some customers.

Some customer operations require a large amount of steam which is often produced in gas or coal-fired boilers. Any excess steam that is available can sometimes be utilized to generate electricity on site. Alternatively, a combustion turbine can be used to simultaneously generate electricity and produce process steam via recovery of its exhaust heat. Such processes are known as *cogeneration*, and the customer can use the electrical output to supplement or replace the electric service otherwise provided by the utility. The customer may also be capable of producing electricity at a volume which exceeds the needs of the site. By interconnecting with the local utility, the customer may be able to sell excess power to the grid.

Another element of competition that exists within the conventional regulated utility environment is business development in both the commercial and industrial sectors. The price of electricity can be a key input to the siting of new manufacturing plants and other business facilities.[6] Utilities compete on a regional, national, and even global basis in an effort to attract new customers. In Georgia, for example, the regulatory process has allowed the in-state electric utilities to compete with each other to serve any new customer having a connected load over 900 kW, regardless of location within the state.

Customers having multiple plant sites often look at the local operating costs of each of their facilities when deciding where production expansions should be made. This internal rivalry for capital brings about a secondary competitive pressure on local utilities to offer price concessions or special contracts to support these customers' efforts to expand. Sometimes multi-site operations can shift production between plants in order to optimize regional electric costs. Aside from attracting new customers or load additions, utilities are faced with the concern for retaining customers and thus sales. Economic conditions may sometimes threaten a plant's viability. Mergers and acquisitions between businesses can impact work sites as these "new" corporations realign their operations.

To meet such competitive challenges, utilities have developed strategies, including flexible pricing methods, to support their local customers and to attract new customers. Many utilities have established formal *economic development* programs, often in concert with local commerce officials, in order to promote the local advantages. The regulatory bodies have had to ensure that any unconventional pricing approaches that meet the needs of a single customer or a restricted customer segment would not create a condition of undue discrimination between customers. Customers as a whole should benefit from special rate schedules or contracts through long-term lower costs.

In one fashion or another, electric utilities have been competing for sales and for customers in order to maintain and grow their businesses. Although competing is not new to the electric utility business, contemporary industry restructuring efforts which allow for further competition are beginning to alter the face of utility regulation and thus the methods of electricity pricing.

1.4 IMPACTS OF INDUSTRY RESTRUCTURING ON ELECTRICITY PRICING

The electric utility business around the world has been undergoing a dramatic transformation. In the late 1900s, the utility system in Great Britain was restructured into three separate entities: a competitive generation function, a national transmission grid, and discrete distribution operating companies. Other countries

---

[6] Electricity represents only one item of cost that must be considered when siting a new plant or expanding an existing facility. Other factors such as taxes and the labor pool may be equally if not more important. The price of electricity is more of a critical item for businesses which are energy intensive.

Overview of Electricity Pricing

in Europe and elsewhere have also implemented electric utility restructuring in some form or fashion. Some nations have gone through the process of transforming their utility systems, which were formerly owned and operated by the government, into private companies. In some cases, a competitive generation business and a separate power delivery business have been established.

In the United States, the first seeds of competition were sown as early as 1978 when the Public Utility Regulatory Policies Act (PURPA) was enacted. Under PURPA, a utility was required to purchase energy from Qualifying Facilities (QFs) at its (i.e., the utility's) avoided costs of energy and capacity.[7] Then in 1992, the Energy Policy Act (EPAct) mandated the nondiscriminatory access to transmission systems for wholesale transactions. EPAct required that no utility could charge another transmission user more than it charges itself for the use of its transmission system. In 1996, the FERC issued its landmark orders 888 and 889 which required utilities to file open access transmission services tariffs and to publish electronically their prices on an Open Access Same-Time Information System (OASIS). The EPAct of 2005 repealed the Public Utility Holding Company Act (PUHCA) of 1935, which was originally established to place utility holding companies under federal control in order to ensure efficient utility system operation. PUHCA required related holding companies to be vertically integrated.

Prior to the implementation of the first legislatively-directed competition proceeding, utilities began positioning themselves for change. Foremost, to be price competitive, a business must manage its costs aggressively, and many utilities today are scrutinizing their expenditures and practices microscopically. Regulation of a monopoly is intended to promote cost control, but regulation alone does not induce a business in quite the same way as a competitive marketplace. Regulation is effective because of carefully crafted rules and procedures. Competition is effective because it simply and directly challenges a company's will and ability to survive. Whether or not the entire electric utility business is ever formally restructured, it functions today much more like an unregulated industry.

The terms *restructured* or *reregulated*, as opposed to *deregulated*, is a much better descriptor of the impending changes in the utility industry, because only a portion of the business actually undergoes deregulation, at least at this time. Competition in electricity generation, or energy supply, is the major target for regulatory reform. This effort results in the disaggregation of the vertically integrated utility structure into a generation business unit which stands separate from the transmission and distribution, or *wires*, business unit.[8]

---

[7] QFs are nonutility cogenerators and small power producers which either simultaneously produce electricity and thermal energy (e.g., steam) using the same fuel source or utilize a renewable energy resource as a fuel source based on FERC prescribed efficiency requirements. Other FERC established operating and ownership criteria apply for QF status.

[8] Some utilities have divested themselves of one of their business units and focus on operating as either an energy supplier or a wires company.

Under this model, customers are allowed to purchase electricity from a supplier of choice, which includes other utilities, nonutility generators, power marketers and brokers, and customer aggregators, as well as the incumbent utility. But regulation would continue to be involved in order to ensure that all suppliers meet standards of certification and performance as well as operate in a scrupulous manner. Otherwise, customers exercising a choice of suppliers assume a risk since the obligation to serve criteria is diminished in a competitive market. The power delivery system which transports electricity from the generation suppliers to the customers would remain regulated. Avoiding the duplication of transmission and distribution facilities is just as prudent in a restructured industry as it is in a fully regulated industry. Thus, the fundamental charge of the regulated wires business is the *obligation to connect*.

Another aspect of the utility business that may be viewed as a target for deregulation is revenue metering and billing. While traditionally considered to be a part of the distribution function, energy suppliers may want the option to meter and/or bill their customers for the electricity they provide. Others feel that these customer services can be provided competitively by a third party.

Attempts at national legislation in the United States, which would require the country as a whole to proceed with restructuring, have been unsuccessful thus far. Consequently, restructuring efforts have only progressed on a state-by-state basis. Generally, both the state legislative body and the state regulatory commission must enact law and rule changes before restructuring could be legally permitted in a given state. Some states are currently engaged in full implementation of restructuring. Other states may continue to study the many issues and impacts of restructuring as they continue to employ more traditional regulatory methods.[9]

Restructuring Alters the Pricing Structure

Conventional bundled rate structures provide little if any delineation of the costs of production, transmission, or distribution. The cost-of-service study used for supporting ratemaking does begin with discrete cost information that corresponds with the major functional areas of the power system.[10] But, the roll-up, or bundling, of these cost elements into the basic customer, demand, and energy components obscures much of this functional detail.

The actual rate design process provides an additional level of filtering of cost information since the final approved unit prices rarely align exactly with the unit costs derived from a cost-of-service study. For example, the cost-of-service study computes separate customer, demand, and energy cost components for the

---

[9] In the early 2000s, U.S restructuring activity has slowed considerably, and some states that had implemented restructuring have since reverted back to a more regulated approach.

[10] The cost-of-service analysis must deal with a considerable amount of plant and expense related cost elements, which must be allocated ultimately to the major functions of production, transmission, distribution, and customer services.

residential class. However, the most common practice for structuring residential rates is to establish a stand-alone customer charge and then combine the total demand and energy cost recovery under a kWh-based rate structure. However, a further distortion is created as a portion of the customer-related costs are typically recovered by mean of the kWh charge structure.

Bundled wholesale prices are approved by the FERC. Bundled retail prices are approved by each of the state PSCs. The functional disaggregation or *unbundling* of the vertically integrated electric utility naturally leads to the unbundling of prices. Cost of service and rate design must be redefined. With restructuring, the pricing of retail sales of electricity, which traditionally falls exclusively within each state commission's jurisdiction, is redistributed in three ways.

- **Distribution prices** are regulated by the state PSC. The rates for this portion of the business continue to be cost based within allowed limits of profitability (as governed by the PSC authorized return on equity). Furthermore, distribution rates might also be subject to performance measures to ensure quality service at reasonable prices.
- **Transmission prices** are regulated by the FERC. The rates for this portion of the business continue to be based on the costs of service within allowed limits of profitability to ensure nondiscriminatory access by all users.
- **Generation prices** are established by the marketplace. Rates are based on supply and demand; thus, the low cost producers realize a higher level of profitability than high cost producers. The highest cost producers would not be viable sellers and thus would realize the risk consequences of operating in a market-based environment.

Table 1.1 presents a striking example of how price structures can change as they are unbundled for a competitive generation market. Pre- and post-restructuring rate schedules for the same class of residential electric service are compared. As illustrated, the charges under competition, listed in the right-hand column, are itemized in detail by functional level, i.e., distribution, transmission, and generation. The state PSC approves the prices for the distribution portion of the rate structure. In the pre-restructuring rate schedule, listed in the left-hand column, some of the distribution system costs would have been collected in the Customer Charge, while the remainder of the distribution costs would have been collected through the Energy Charge. The new Distribution Charges are divided into two components. The fixed component represents costs that are constant each month regardless of the level of energy usage, which is not unlike the Customer Charge. At a minimum, these costs would include a meter, a service line, and basic account-related expenses, such as meter reading and billing. The variable component represents those distribution costs that are considered to be a function of usage, i.e., as a result of a load requirement, and these costs are being recovered by means of a kWh rate in this example rate design.

Table 1.1 An Example of Bundled and Unbundled Base Rate Structures

DOMESTIC SERVICE
SCHEDULE 10

Bundled Rate Structure:	**Unbundled Rate Structure:**	
Effective 5/1/97	Effective 1/1/99	
CUSTOMER CHARGE: $5.00	**DISTRIBUTION CHARGES:**	
ENERGY CHARGE:	FIXED DISTRIBUTION CHARGE	$4.73
All kWh 6.155 ¢/kWh		
	VARIABLE DISTRIBUTION CHARGE	
	All kWh	1.699 ¢/kWh
	TRANSMISSION CHARGES:	
	All kWh	0.303 ¢/kWh
	Ancillary Services (all kWh)	
	Scheduling, System Control & Dispatch	0.000 ¢/kWh
	Energy Imbalance	0.000 ¢/kWh
	Reactive and Voltage Control	0.019 ¢/kWh
	Regulation and Frequency Response	0.020 ¢/kWh
	Spinning Reserve	0.054 ¢/kWh
	Supplemental Reserve	0.048 ¢/kWh
	The transmission charges are based on the Company's Pro Forma Open Access Transmission Tariff which will change from time to time and is subject to Federal Energy Regulatory Commission approval.	
	COMPETITIVE TRANSITION CHARGE:	
	All kWh	0.654 ¢/kWh
	GENERATION CHARGE:	
	All kWh	3.221 ¢/kWh
	The generation charge applies only to Customers receiving Provider of Last Resort Service from the Company. The generation charge does not apply to Customers obtaining Competitive Energy Supply.	

Source: West Penn Power Company residential electric service rate schedules prior to and after the implementation of restructuring in Pennsylvania.

Once a retail rate schedule is unbundled, the FERC-approved transmission tariff rates become the prices for the transmission portion of retail electric service. In essence, the state PSC simply adopts the FERC transmission prices. Since the FERC typically establishes the transmission service rate as a charge per kW, it must be converted to an equivalent charge per kWh for small customers, such as residential, whose usage is commonly measured with a simple watt-hour type of meter. The six transmission ancillary services prescribed by the FERC may also be itemized as part of this new rate schedule, as shown in Table 1.1.

The charge for generation is applicable only to those customers who are not served by an alternative energy supplier. Included in this group are customers who either are not eligible or do not choose to purchase electricity from an alternative provider, customers whose alternative supplier fails to deliver, and customers who choose to leave by agreeing to purchase electricity from an alternative supplier but later return to the incumbent utility for generation service. The state PSC approves this default customer group generation charge. Customers who do successfully secure alternative suppliers would pay their specific charges for electricity in lieu of the incumbent utility's generation charge. But customers who buy electricity from an alternative supplier must also pay the incumbent utility's transmission and distribution prices for power delivery service.

The Competitive Transition Charge (CTC) is a mechanism designed to recover certain costs that arise as a result of implementing a competitive form of restructuring. The CTC includes PSC approved charges for *stranded cost* recovery and other costs which must be incurred to develop the methods and systems necessary to invoke a competitive utility structure.[11] The CTC is a nonbypassable charge that must be paid by all customers. CTCs are temporary as they are needed only to bridge the period that lies between the traditional regulated utility operation and a mature restructured industry.

While the example may not serve as the only blueprint for an unbundled rate structure design, it certainly highlights many of the electricity pricing issues that must be dealt with when implementing restructuring. In addition to the rate design aspects of unbundling, the billing of electric service also becomes much more complicated. With restructuring, customers are exposed to many more pricing and billing details than in the past, and many questions arise with respect to changes in the industry. The unbundled commercial and industrial rate structures are much more complex than the illustrated residential rate. An adequate amount of customer education in all sectors is crucial in order for any type of restructuring activity to produce benefits. Electricity pricing is complex, and many customers could have great difficulty in understanding because electricity is so unlike most other everyday goods and services.

---

[11] Stranded costs result when the fixed costs of an asset, installed under traditional rate regulation, cannot be fully recovered in a competitive marketplace. Generation facilities having costs higher than the prevailing market price of electricity represent an example of how stranded costs can occur.

1.5 THE ROLE OF ENGINEERING IN ELECTRICITY PRICING

The early literature on electric utilities and the pricing of electricity makes reference to the *rate engineer*. In the late 1800s, distinguished engineers, including Dr. John Hopkinson and Mr. Arthur Wright of England, debated the scientific aspects of power system cost analysis and rate design. The technical issues they raised and the methodologies they pioneered have long served as underlying principles of electricity pricing. Even today, their legacy rate structure designs represent the cornerstones of both electric and gas utility pricing practice. Today, and into the future, the findings of the original ratemaking architects continue to guide the contemporary utility pricing professional.

Since that early era, electricity pricing concepts and techniques have evolved considerably. Not strictly an engineering solution, effective electricity pricing must take into account a number of other issues, such as economic theory and applications. In the 1960s, Dr. James C. Bonbright, a professor of economics, gained notoriety with his marginal cost approach to ratemaking. In addition, Dr. Bonbright outlined a list of eight key criteria for a sound rate structure which have been widely cited as "Bonbright's Big Eight." These criteria relate to basic business objectives, economic efficiency, and customer fairness, and thus they complement the pure scientific aspects of electricity pricing. Bonbright's rate criteria continue to serve as a key checklist for effective rate design.

Electricity pricing is based on a balance of viewpoints. The electrical engineer who practices electricity pricing must understand a cross-section of key utility business concepts, methods, and technologies in addition to his or her chosen discipline. Pricing is also about accounting, finance, marketing, regulatory law, and customer service. Likewise, the economist and business professional, who practices electricity pricing, must develop a fundamental grasp of the technical workings of the electric power system and its customers in order to produce a balanced and thorough pricing mechanism. Overall, the engineering perspective represents a significant input to the electricity pricing decision-making process.

The engineering aspects of pricing are important for several reasons. First, electricity is technically unique compared to any other products or services. Electricity flows according to the laws of physics which, in a large network, means that the loading of equipment is constantly changing as customers' energy requirements vary on demand. The complexity of the power system necessitates the use of extensive and sophisticated monitoring and control equipment to ensure the adequacy, reliability, and quality of electric service in a real-time sense for all customers. Modeling the dynamic flow of electricity from the production facilities to the customers is a complex problem. In addition, electricity has a complicating component known as reactive power which is required for the operation of certain customer devices, such as induction motors, and is necessary to transport electricity through the system.

Recovering the costs of the power system in terms of its asset structure and operations is a fundamental objective of electricity pricing that requires an engineering perspective. The cost-of-service study is a model based on utility accounting and financial data and which relies heavily on a variety of engineering principles to appropriately assign electric service costs to customers. Allocating these costs properly to customer classes requires that all of the utility's investments and expenditures, along with the customers' load and energy requirements, are attributed to functional voltage levels because customers are served at these different voltages. The cost-of-service model must include a reasonable facsimile of the power flows and electrical losses that are experienced throughout the system.

Complete cost data and customer information required to produce the cost-of-service study and design effective rate structures are not always readily available. Plant investment and expenses are recorded in accounting ledgers which typically do not align with the level of detail necessary to conduct adequate cost studies. Engineering analyses of account information are thus needed to achieve an acceptable resolution of the elements of cost of service. The pricing function critically relies on the correct recognition of the drivers of the customer, demand, and energy cost components.

Likewise, customer load and energy requirements data often fall short of providing an adequate level of detail for costing and pricing. The relatively simple metering methods that are typically used to capture customer usage data for billing purposes simply fail to record key ratemaking requirements.[12] The use of statistically-based sampling techniques and engineering studies is needed to reasonably estimate such factors as the temporal nature of energy usage, the relationships between energy usage and peak demands, and other key load characteristics.

A number of mathematical analysis and graphing techniques are used to facilitate the rate design process and to conduct special rate studies, such as revenue forecasting. These methods provide a clear insight to the way that complex pricing structures behave. They also help to identify and mitigate areas of risk due to the lack of missing input data, assumptions, or uncertainty as to how customers will respond to various price structure options. Special computer applications serve as tools which are utilized for rate design and analysis. These programs are commercially available from vendors and industry trade groups. But rate engineers also have ample opportunity to develop ad hoc analyses and reporting methods which support the electricity pricing decision-making process.

In total, the engineering discipline is essential to understanding the technical nature of electricity and power system operations, how and why costs are incurred to provide electric service to customers, and the types of pricing structures which best fit with customer operations and requirements. Engineering is often referred to as the "art of applied science." This assertion is well exemplified in the strategic field of electricity pricing.

---

[12] For example, meters used for the majority of residential and small commercial customers typically provide only accumulated kWh readings and not time-of-use or demand-related data. The cost of metering increases somewhat substantially when such other attributes of electric use must be measured.

1.6 CHAPTER ORGANIZATION

This book is structured in a progressive fashion. Each chapter generally builds upon the previous chapters' topics. For the most part, the chapters are grouped into major subject areas. Chapters 2, 3, and 4 are about understanding electricity as a service product, the customer's electric power usage characteristics, and the customer's service requirements. Chapter 2, Prerequisite Concepts, reviews key technical principles of electricity and the behavior of electrical loads within the power system, both of which are fundamental to pricing electricity. Knowledge of costs is critical for ensuring profitable pricing. Therefore, Chapter 2 also presents an overview of how the costs of electricity are viewed from various perspectives in order to support the objectives of pricing, and it provides an introduction to the basic concepts and approaches used to establish electricity prices. Chapter 3, Electric End Uses, investigates some of the essential underlying principles of electricity pricing by investigating electricity utilization requirements from the perspective of the customer's end-use needs. End-use requirements are evaluated for each of the major customer classes, and end-use research is discussed. Chapter 4, Customers: Electric Service Requirements, discusses the customers' overall electricity requirements in unison with the utility's perspective of rendering service. The discussion continues with customer initiatives for procuring electricity and managing its utilization.

Chapters 5, 6, and 7 focus on the power system infrastructure. Chapter 5, Power Delivery Systems: Transmission and Distribution, presents an overview of the power delivery system's transmission and distribution functions. An understanding of the components of these systems as well as many of the associated engineering design principles is key to conducting quality cost of service analyses and designing rates which recognize pricing differentials, such as those previously discussed. Chapter 6, Electric Power Production, discusses power generation in terms of technologies employed, classifications of generation, and system operations. Chapter 7, Revenue Metering and Billing, discusses the functions of metering and billing for electric service. A variety of metering techniques are discussed in light of capturing the variety of customer usage data and information that is necessary to develop rates that are sophisticated enough to meet the objectives of electricity pricing.

Chapter 8, The Rate Schedule: Pricing Methods and Risk, explores the building blocks of electricity pricing and investigates the various ways in which these elementary pricing components are combined to produce practical rate schedules. Alternative methodologies for structuring various pricing mechanisms are also discussed and evaluated. The discussion ties the concepts of electric service pricing to the characteristics and requirements of both the power system and customer usage. The aspect of rate design risk is discussed and illustrated.

Chapters 9 and 10 present analytical tools that are used for electricity pricing. Chapter 9, The Mathematics of Rates, delves into the mathematical analysis of rate structures and billing provisions. Methods for developing various graphical

Overview of Electricity Pricing

representations of rates are also presented. The results of such analyses are not only key to the task of rate design but also provide valuable information that is useful for applying and administering rates. Chapter 10, Billing Data Statistics and Applications, presents various billing information analysis techniques that are essential for the rate design process, whose success relies on quality database preparation. Methods of forecasting rate revenues and billing determinants are developed and compared.

Chapters 11 and 12 bring all of the forgoing technical discussion and analytical methods and techniques together to describe and illustrate the cost-of-service methodology and the ultimate translation of costs to workable rates. By sheer page volume, Chapter 11, Cost-of-Service Methodology, indicates the complexity of the cost analysis process needed to support electricity pricing. A number of engineering analysis methods are blended with the accounting and financial aspects of ratemaking in order to produce results that serve as a basis for developing fair and equitable pricing structures for customers. Finally, Chapter 12, Translating Costs to Prices, unites cost-of-service and load research data resources in an investigation of cost-based rate design principles and practices.

2
Prerequisite Concepts

2.1 INTRODUCTION

Understanding what electricity is and how it behaves is an important aspect of electricity pricing. While electricity does have similarities to other products, it is more dynamic than most goods, which are produced, packaged, shipped, inventoried, and ultimately sold to consumers over a period of time. Electricity must be produced immediately and delivered upon demand. Electricity is somewhat mysterious because, while the production plants and the power delivery systems are prominently visible everywhere, the actual product itself is invisible. A number of electrical phenomena are not directly comparable to the nature of any other products, including other utility services.

Similarities and dissimilarities aside, the characteristics of electricity and the various ways in which customers utilize it largely determine its cost, and cost is a primary determinant of electricity prices, particularly within the regulated functions of the industry. Cost also determines whether or not an energy supplier can be profitable when competing against market prices for electricity.

A basic understanding of accounting and finance is necessary for evaluating a utility's cost structure and earnings position. The revenues that are collected from customers through the sale of electricity must recover the costs of electric operations adequately and provide for a suitable financial return for the utility's or service provider's investors.

Thus, electricity pricing requires a comprehensive knowledge of both the electric industry and the manner in which customers utilize electricity. Although such vital points are discussed throughout the book, it is best to begin with a foundation composed of a few key technical and business-related concepts. The following sections review the most fundamental aspects of electricity and the electric power business.

2.2 THE CARDINAL LAWS OF ELECTRICITY

Electricity is defined as "electric current [which is] used or regarded as a source of power."[1] Electric *current*, measured in amperes or amps, is the flow of free electrons that occurs in a conductor when electrical pressure, or *voltage*, is applied.[2] Good conductors are made of materials which have an abundance of free electrons, while poor conductors have a deficiency of free electrons. Poor conductors are good insulators. Electrical wires and devices are most often composed of copper or aluminum, since these metals have a high conductivity and are economical. Common insulators include a variety of materials such as rubber and porcelain.

As current flows through a material, it is met by an opposition referred to as *resistance*, which is measured in ohms (Ω). Insulators have a high resistance to current flow, while good conductors have a relatively low resistance. Since some level of resistance is present in all conductors, electrical energy must be expended to overcome the opposition and allow current to flow.

The flow of electricity in a conductor is somewhat similar to the flow of water in a pipe. When a head pressure is applied, water will flow at a proportional rate (e.g., gallons per minute). As the pressure is increased or decreased, the rate of water flow will also increase or decrease. The friction of the water on the inside of the pipe restricts the flow of water in a manner analogous to the way in which a conductor's resistance restricts the flow of electrical current. Under increasing pressure, the rate of water flow that can be transported will ultimately reach a maximum, and further pressure would burst the pipe. Similarly, as a conductor's capacity for current flow is exceeded, it will first anneal and then burn through.

Ohm's Law

The most fundamental rule of electricity is Ohm's Law. Ohm's Law states that voltage is directly proportional to current by

$$V = i \times R \tag{2.1a}$$

or

$$i = \frac{V}{R} \tag{2.1b}$$

where V = voltage in volts
  i = current in amps
  R = resistance in ohms

---

[1] American Heritage Dictionary, 2nd College Edition, s.v. "electricity."

[2] By convention, electric current is deemed to flow in the opposite direction to the flow of electrons.

Prerequisite Concepts

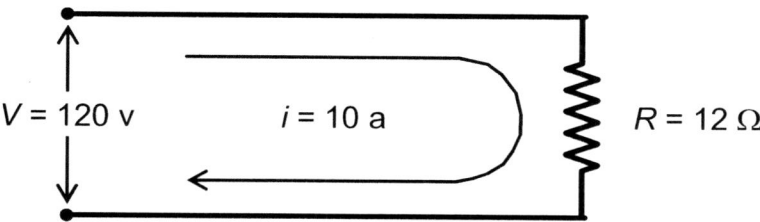

Figure 2.1 A fundamental electric circuit. A voltage (V) is applied across the terminals of a resistor (R) thereby causing a current (i) to flow.

If a voltage is applied across the terminals of a circuit containing a resistor (i.e., a circuit component which is designed to limit the flow of electricity), the resulting current passing through the resistor would be indirectly proportional to its resistance, as indicated by Equation 2.1b. For example, if 120 volts were applied across the terminals of a device having a resistance of 12 Ω, the current flow would be 10 amps, as shown in Figure 2.1. If the voltage was held constant at 120 volts while the resistance was doubled to 24 Ω, the resulting current flow would be cut in half to 5 amps. Likewise, if the value of the resistance was cut in half, the current flow would double.

The Ohm's Law relationship is valid in both *direct current* (DC) and *alternating current* (AC) circuit applications.[3] Unlike DC circuits where the polarity of the voltage is constant, the applied voltage in AC circuits continually reverses polarity at a given frequency, f, such as 50 or 60 times per second, as shown in Figure 2.2. Frequency is measured in Hertz (Hz). Power system AC voltages and currents are sinusoidal in form but may also be represented by phasors that rotate counterclockwise with one revolution being equal to one cycle. Each time the polarity of the AC voltage changes, the current flow in the circuit reverses direction. Thus, if an AC voltage is applied to the circuit illustrated in Figure 2.1, the current would flow clockwise (as depicted) when the voltage is positive and counterclockwise when the voltage is negative. In a circuit containing only resistance, the directional change of the current occurs at the exact instance when the voltage is at zero (i.e., at the x-axis). Thus, in a purely resistive circuit, the AC voltage and current operate in phase with each other.

In AC circuits, *reactance* may be present in addition to resistance. Similar to resistance, reactance also offers an opposition to the flow of electrical current. Reactance consists of two components: *inductive reactance* and *capacitive reactance*. Inductive reactance is a characteristic of an electrical circuit element known as an inductor while capacitive reactance is attributed to a circuit element known as a capacitor.

---

[3] Most DC applications occur at the end-use level. Modern power systems typically operate on an AC basis although DC is used in some cases for long transmission lines.

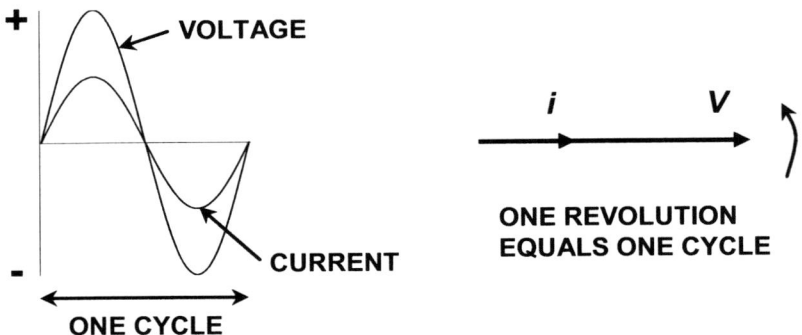

Figure 2.2 **AC voltages and currents.** AC voltages and currents can be recognized as periodic waveforms or phasors where the direction of phasor rotation is counterclockwise. In a resistive circuit, the current (*i*) is in phase with the voltage (*V*). During the positive portion of the voltage cycle, the current flows in one direction. The direction of current flow is reversed during the negative portion of the voltage cycle.

Inductors and Capacitors

An inductor is simply a coil of conductor wire, which produces a magnetic field as electrical current flows through it.[4] Often a coil is wound around an iron core in order to intensify its magnetic field thereby increasing its inductance rating, which is measured in henrys. Motors and transformers are common examples of magnetic field applications that are based on the principle of inductance.

The relationship between the voltage and current of an inductor is much different than for a resistor. The voltage across an inductor is zero when the current it is carrying is constant.[5] In an AC circuit, the current is constant only at the point where its time rate of change is zero. This situation occurs when the current flow has reached its maximum value (in either the positive or negative direction), as can be seen by the graph in Figure 2.3. Conversely, the voltage across an inductor is at its maximum positive or negative value when the time rate of change of the current is greatest, which is when the current flow changes direction (i.e., crosses the *x*-axis). Thus, as illustrated by both the sinusoidal waveform plot and the rotational vector plot in Figure 2.3, the current through the inductor lags the voltage across the inductor by 90°.

---

[4] A weak but detectable magnetic field is created around a straight wire conductor as current flows through it. Winding a conductor into a coil increases the magnetic field substantially. The inductance of the coil is directly proportional to the square of the number of turns in the coil.

[5] In a DC circuit with a constant current source, the inductor acts as a short circuit.

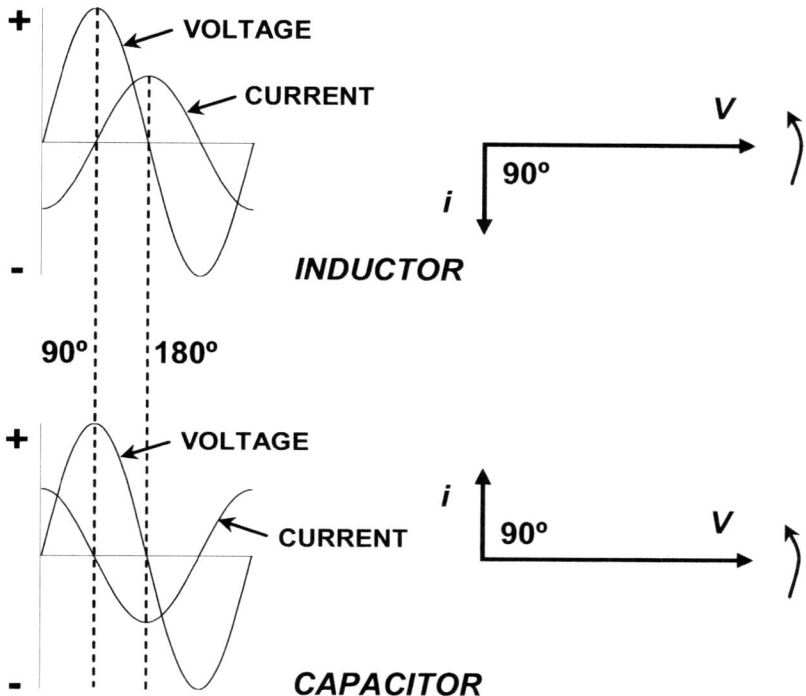

Figure 2.3 Reactive device impacts on AC voltages and currents. In a purely inductive circuit, the current lags the voltage by 90°. In a purely capacitive circuit, the current leads the voltage by 90°. With either device, when voltage is at a maximum, the current is at zero, and when the current is at a maximum, the voltage is at zero.

A capacitor is a circuit device composed of two conductive surfaces, or plates, which are separated by an insulator or dielectric medium. The surface area of the plates and the insulation medium between the two plates determine its capacitance rating, which is measured in farads. When a voltage is applied to the terminals of a capacitor, current will flow thereby charging the plates of the capacitor and creating an electrostatic field across the insulating material. Essentially, electrons are collected on one plate and extracted from the other plate until the voltage differential between the plates is equal to the applied voltage. Electrons do not actually flow through the capacitor because of the dielectric medium. The appearance of current flow through a capacitive device is referred to as the *displacement current*. When fully charged by a DC source, the voltage across the capacitor becomes constant, the current flow ceases, and the capacitor acts like an open circuit.

In an AC circuit, the voltage across a capacitor is constant when its time rate of change is zero (i.e., when the voltage is at its maximum positive or negative value). At these points, the current flow through the capacitor becomes zero. Conversely, the current flow through the capacitor is at its maximum value (in either the positive or negative direction) when the time rate of change of the voltage is greatest, which is when it crosses the *x*-axis. The voltage and current relationship for a capacitor is opposite to the relationship of the voltage and current for an inductor. As shown in Figure 2.3, the current through a capacitor leads the voltage across the capacitor by 90°.

Impedance
All electrical load devices are composed of one or more of the three fundamental circuit elements of resistance, inductance, and capacitance.[6] Figure 2.4 provides an illustration of a circuit with a resistance (R), an inductive reactance (X_L), and a capacitive reactance (X_C), in series connection to which a voltage source is applied. The same current flows through each of the circuit components, and a drop in voltage occurs across each component. The voltage drop across the resistor is in phase with the current. However, the voltage drop across the inductor leads the current, while the voltage drop across the capacitor lags the current. The total of the voltage drops across the three components is equal to the source voltage of 120 volts, but the voltages are complex and can not be summed algebraically.

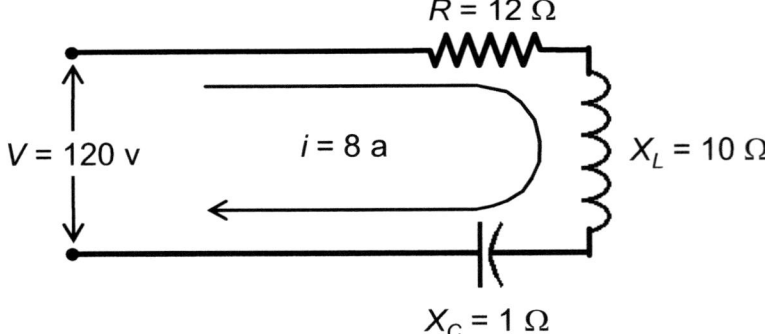

Figure 2.4 A series circuit. A circuit containing the elements of resistance, inductive reactance, and capacitive reactance connected in series. Impedance is the total opposition to the flow of current provided by the combination of these three circuit components.

---

[6] Furthermore, inductors are not purely inductive and capacitors are not purely capacitive. Both of these devices contain a small amount of internal resistance; therefore, the displacement angle between voltage and current is slightly less than 90°.

Prerequisite Concepts

The combined effect of resistance, inductive reactance, and capacitive reactance is *impedance*, which represents the total opposition to the flow of current. For series circuits, such as shown in Figure 2.4, impedance is defined by

$$Z = R + jX_L - jX_C \qquad (2.2a)$$

or

$$Z = \sqrt{R^2 + (X_L - X_C)^2} \qquad (2.2b)$$

where Z = impedance in ohms
R = resistance in ohms
X_L = inductive reactance in ohms
X_C = capacitive reactance in ohms

The resistance and reactance components are in quadrature with each other (i.e., angularly separated from each other by 90°) and therefore are not algebraically additive.[7] Instead, these components must be added vectorially.[8]

The series circuit illustrated in Figure 2.4 consists of a resistance of 12 Ω, an inductive reactance of 10 Ω, and a capacitive reactance of 1 Ω. Thus, from Equation 2.2b, the impedance for this circuit is found to be

$$Z = \sqrt{12^2 + (10-1)^2} \ = \ 15 \ \Omega.$$

In AC circuits, the Ohm's Law formula expressed in Equation 2.1b is modified by simply substituting impedance (Z) in place of resistance (R) as expressed by

$$i = \frac{V}{Z} \qquad (2.3)$$

Thus, if an AC voltage of 120 volts is applied across the combined circuit elements shown in Figure 2.4, a current of 120 volts ÷ 15 Ω = 8 amps will flow through the circuit.

---

[7] Resistance lies along the real axis (x) while reactance lies along the imaginary axis (y). In circuit analysis, the imaginary operator $\sqrt{-1}$ is represented by j rather than i, the mathematical convention, since i is used to represent current in electrical circuits.

[8] When impedance components are connected in parallel, their reciprocal quantities must be used. Thus, for parallel circuits, Equation 2.2a becomes $1/Z = 1/R - j(1/X_L) + j(1/X_C)$.

The net effect of the inductive and capacitive reactance components of the impedance in a circuit determines if the current flowing through the circuit leads or lags the voltage applied to the circuit and by what angle. Since the reactance component of the circuit in Figure 2.4 is predominantly inductive ($X_L - X_C = 10 - 1 = 9\,\Omega$), the current lags the applied voltage. In this example, the angle at which the current lags the voltage is not 90°, as with the voltage and current relationship associated with an inductor, because a resistor is also present in the circuit. The angle between the voltage and current for a circuit is determined by

$$\theta = \tan^{-1}\left(\frac{X}{R}\right) \qquad (2.4)$$

where θ = angle between R and X in degrees
X = net reactance $X_L - X_C$ in ohms
R = resistance in ohms

When $X_L - X_C$ is positive, the angle θ is positive, and the current lags the voltage. When $X_L - X_C$ is negative, the angle θ is negative, and the current leads the voltage. For the example circuit of Figure 2.4, where $R = 12\,\Omega$ and $X = 9\,\Omega$, the current lags the voltage by 36.87°.

If the inductive and capacitive reactance in a circuit nets out to zero, then the circuit impedance is simply equal to the resistance, and the circuit is said to be in resonance. While the voltage across an inductor or a capacitor in a resonant circuit would be out of phase with the current, the voltage applied to the circuit would be in phase with the current.

Impedance is a phasor (although it does not rotate like voltage and current phasors), and it can be expressed either in rectangular form, as in Equation 2.2a, or in polar form as

$$Z\angle\theta$$

where the magnitude of Z is determined by Equation 2.2b and θ is determined by Equation 2.4. Thus, for the example circuit shown in Figure 2.4, the impedance can be expressed as either $12 + j9\,\Omega$ or $15\angle 36.87°\,\Omega$.

AC Power

Alternating voltages and currents are sinusoidal in nature, and thus their values vary as a function of time. The voltages and currents reach maximum values in both the positive and negative directions, and they are zero when they cross the x-axis. But, as shown in Figure 2.5, the average value of a sinusoidal waveform over a complete cycle is zero.

Prerequisite Concepts

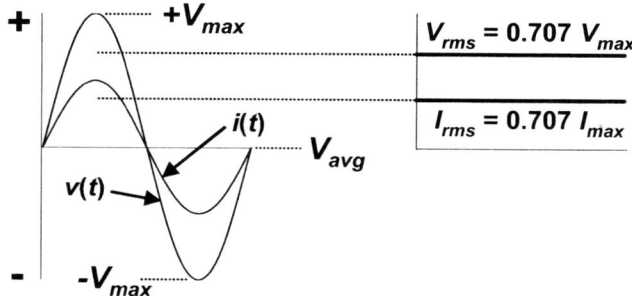

Figure 2.5 **Effective voltages and currents.** The average value of either voltage or current over a full cycle is zero since both waveforms are symmetrical about the *x*-axis. The effective, or rms, values of time-varying voltage and current are related to an AC source's capability to supply power to a resistive type of load.

Electrical applications generally do not specify voltages or currents in terms of either their maximum or instantaneous values. A low voltage commonly used for convenience outlets in homes and businesses is specified as 120 volts AC (the magnitude of the applied voltage in the example circuits in Figures 2.1 and 2.4). The 120 volt value represents the *effective voltage*.

The effective voltage (or current) for any waveform[9] is known as the root-mean-square (rms) value, which is determined by

$$V_{rms} = \sqrt{\frac{1}{T}\int_0^T v(t)^2 \, dt} \qquad (2.5)$$

where V_{rms} =   the effective voltage
            $v(t)$ =   instantaneous voltage
            T =   the duration of one cycle in seconds

For sinusoidal waveform applications, such as electric AC power systems, the instantaneous voltage can be described by $v(t) = V_{max} \times \sin \omega t$, where $\omega = 2\pi f$. Thus, by substituting $V_{max} \times \sin \omega t$ for $v(t)$ in Equation (2.5), the relationship between the effective voltage and the maximum voltage for a waveform having a frequency *f* of 60 cycles per second, or 60 Hertz (Hz), is found to be

---

[9] Equation 2.5 is applicable to both sinusoidal and nonsinusoidal waveforms.

$$V_{rms} = \frac{V_{max}}{\sqrt{2}} = 0.707 \times V_{max} \qquad (2.6a)$$

Likewise, for the current

$$I_{rms} = 0.707 \times I_{max} \qquad (2.6b)$$

Recall that in Figure 2.1 a voltage source of 120 volts is applied to the terminals of a 12 Ω resistor. A battery or other DC source, rated at 120 volts, would cause a current flow of 10 amps in this circuit. An AC voltage with a maximum value of 169.7 volts would also cause a current flow of 10 amps since the rms value of the applied voltage would be equal to 120 volts. The 10 amps that flows as a result of the AC voltage source represents the rms value of the current. Thus, the maximum value of current would actually be 11.3 amps. The rms values of AC voltage and current have an equivalent heating effect in the resistor as the DC voltage and current. Under either scenario, the power delivered to the load would be the same.

Apparent Power

The product of voltage and current is *apparent power*, and it is expressed in units of volt-amperes (VA). Thus, apparent power is determined by

$$S = V \times i \qquad (2.7)$$

where S = power in volt-amperes
  V = voltage in volts (i.e., the rms value)
  i = current in amps

The load (i.e., the resistive and reactive circuit elements) illustrated in Figure 2.4 has a combined impedance of 15 Ω. A voltage of 120 volts is applied across the load thereby causing a current flow of 8 amps. Therefore, the total power requirement of the load is 120 volts × 8 amps = 960 VA.[10]

Apparent power is composed of two components: *real power* and *reactive power*. Real power is associated with resistive loads and reactive power is associated with reactance loads. As with the resistance and reactance components of impedance discussed above, the real and reactive components of apparent power are in quadrature with each other and, therefore, must be added vectorially.

---

[10] In a DC application, the product of voltage and current is more correctly defined to have the unit of watts since power is supplied to only a resistive load; whereas in an AC application, power is fed to both the resistive and reactive components of a load.

Prerequisite Concepts

Real and Reactive Power Components

Real power is the component associated with converting electrical energy into another form of energy, i.e., the performance of work. For example, real power consumed by a motor will create shaft torque which then provides mechanical power to a mechanical load. The basic unit of real power is the *watt* (W).

The real power component is determined by

$$P = V \times i \times \cos\theta \qquad (2.8)$$

where P = power in watts
V = voltage in volts
i = current in amps
θ = angle between voltage and current

Referring again to the load device in Figure 2.4, the real power delivered to the load is 120 volts × 8 amps × cos (36.87°) = 768 W. The angle θ is the impedance relationship between resistance and reactance and thus the angle between the voltage and the current. The quantity $i \times \cos\theta$ represents the component of the current that is in phase with the voltage, as shown in Figure 2.6. The real power is consumed only by the resistance portion of the load.

Alternatively, by substituting $V = i \times R$ from Equation 2.1 into Equation 2.7, while replacing S with P, the real power can also determined by

$$P = i^2 \times R \qquad (2.9)$$

Again for the circuit illustrated in Figure 2.4, the real power delivered to the load is equal to (8 amps)² × 12 Ω = 768 W.

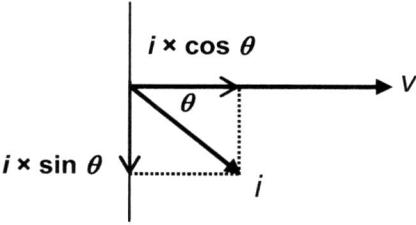

Figure 2.6 **Complex current components.** Current is shown to lag voltage by the angle θ. The component $i \times \cos\theta$ is the projection of the current phasor on the *x*-axis and thus the component of the current that is in phase with the voltage. The projection $i \times \sin\theta$ on the *y*-axis is the component of current that is in quadrature with the voltage.

Reactive power is the component associated with the cyclical storage of energy in the magnetic fields of inductors and the electrostatic fields of capacitors and then the subsequent return of the energy to the circuit. This oscillation of energy is a result of the time-variation in the voltages and the currents in an AC circuit. Unlike real power, reactive power is not associated with performing work; therefore, reactive power is not converted to any other form of power. But the reactive power requirements of complex impedance devices do contribute to the overall loading of a power system's conductors and other facilities, such as transformers. The basic unit of reactive power is volt-amperes reactive (VAR). The reactive power component is determined by

$$Q = V \times i \times \sin \theta \qquad (2.10)$$

where Q = power in VARs
V = voltage in volts
i = current in amps
θ = angle between voltage and current

For the load device in Figure 2.4, the reactive power delivered to the load is 120 volts × 8 amps × sin (36.87°) = 576 VARs.[11]

The real and reactive components of power are phasors; thus, they must be added vectorially

$$S = P + jQ = S \angle \theta \qquad (2.11a)$$

or

$$S = \sqrt{P^2 + Q^2} \qquad (2.11b)$$

Power Factor

Power factor is a mathematical relationship between real power and apparent power and is determined by

$$PF = \frac{P}{S} = \cos \theta \qquad (2.12)$$

Power factor can be determined easily if any two of the quantities of W, VARs, or VA are known. The computation of power factor produces a result of 1.0 or less.

---

[11] The reactive power can also be determined by $Q = i^2 \times X$.

Prerequisite Concepts 35

Typically, power factor is expressed as a percentage. Thus, the power factor of the load in Figure 2.4 would be 768 watts ÷ 960 volt-amperes = 0.8, or 80.0%.

When the load is a resistance, the power factor will be at unity (1.0 or 100%), since the total current would be in phase with the voltage. However, it is possible for a complex load to appear to be only resistive when inductive and capacitive reactance are also present. If the inductive and capacitive reactance values are equal ($X_L = X_C$), then the power factor will also be at unity (the circuit is said to be in resonance). If the power factor is not at unity, it must be either lagging or leading. If the reactance of the load is primarily inductive ($X_L > X_C$), the result is a *lagging power factor* (the current lags the voltage as shown in Figure 2.3). If the reactance of the load is primarily capacitive ($X_L < X_C$), the result is a *leading power factor* (the current leads the voltage as shown in Figure 2.3). The lagging and leading aspects of power factor follow from the impedance characteristics of the load, i.e., the power factor angle is the same as the impedance angle. The power factor for the load in Figure 2.4 is a lagging power factor since X_L is greater than X_C.

Electrical Losses

As stated previously, all conductors contain some amount of resistance, even if it is extremely small. Current flowing through a conductor is opposed by its resistance which converts electrical energy into thermal energy, or heat. As a result, an energy and power loss occurs. As the current flow increases, the losses increase exponentially. Electrical losses are measured in terms of watts by

$$P_L = i^2 \times R \qquad (2.13)$$

where P_L = losses in watts
i = current in amps
R = resistance in ohms

Note that Equation 2.13 is of the same form as Equation 2.9 which described the real power consumption of a load device. Electrical losses are realized in generation equipment, in the supply conductors and transformers (i.e., the transmission and distribution system), and in customer circuits and load devices.[12]

In an electrical circuit, conductor losses represent an economic cost since production energy is required to "serve" these losses. Losses in power equipment are significant since the resulting heating effects constrain the power output of

---

[12] Rotating machines are also subject to power losses that are related to mechanical factors, such as friction and windage. In addition, transformers have core losses, which are caused by the cyclical effect of the magnetic field and induced circulating currents.

generators and the power ratings of motors. On the other hand, the resistance component of a load device may be intended, by design, to convert electrical energy to thermal energy. For example, a kitchen toaster is simply a large resistor which generates heat for the purpose of browning bread. In this example, the conversion of electrical power to heat is put to practical use.

Efficiency

One way in which the operating performance of an electrical system or load component can be expressed is by comparing power output to power input. Any difference between these two parameters is caused inherently by the electrical loss characteristics of the system or component. This measure is known as *efficiency* and is determined by

$$EFF = \frac{P_{OUT}}{P_{IN}} = \frac{P_{IN} - P_L}{P_{IN}} \qquad (2.14)$$

where EFF = efficiency
 P_{OUT} = power output
 P_{IN} = power input
 P_L = losses

Efficiency has a value between 0 and 1.0, and it is typically expressed as a percentage.

A fluorescent lamp is designed with a wattage rating that indicates its rate of energy utilization during nominal operation. For example, one standard rating for fluorescent lamps is 40 watts. But not all of the 40 watts of electrical input is converted to useful light. Only about 10 watts of the total input wattage may be actually converted to light. The remaining 30 watts is lost in the form of heat which is both radiated from the bulb and dissipated by convection and conduction throughout the fixture. The electrical efficiency of the lamp then is 10 watts ÷ 40 watts = 0.25, or 25%.

2.3 FUNDAMENTAL LOAD CHARACTERISTICS

An electrical *load* is an appliance or other end-use apparatus (often referred to as utilization equipment) that requires electricity to perform any number of specific duties, such as toasting bread or heating a building or melting scrap iron. A load device converts electrical energy into another useful form of energy, which, in reference to the three examples listed here, includes thermal energy. This energy conversion process is often referred to as *energy consumption.*

Most utility customers have a variety of different electrical devices which together make up a system of loads. Groups of similar customers, e.g., domestic homes, make up a higher order system of loads, i.e., the residential class load. The

Prerequisite Concepts

customer classes combine to form the utility's total load system. A thorough understanding of the characteristics of both individual load devices and loads in aggregate is an essential aspect of electricity pricing.

Individual Electrical Loads

Load devices are designed to function under specific electrical conditions. In particular, loads are rated for use at a given voltage, which might be AC or DC. If the device operates on AC voltage, as do most electrical loads, then it is necessary to know the frequency (50 or 60 Hz for most power systems) and the phase (e.g., single-phase or poly-phase).[13] Loads must be connected to an adequate power supply and in the appropriate configuration to ensure safety and optimal performance. Residential and commercial loads typically operate at voltages levels below 600 volts. Special industrial loads may operate at voltages above 1,000 volts.

Such information can be found on the load device's *nameplate*, which is an affixed label that also lists such data as the power requirements. For example, the nameplate on a desktop copy machine lists the following information:

- 120 volts @ 60 Hz
- 9.5 amps
- 1,100 watts

From this data, other useful information can be determined. For instance, the apparent power is 120 volts × 9.5 amps = 1,140 VA, and the power factor is 1,100 watts ÷ 1,140 volt-amps = 0.965 or 96.5%.

Load can be described electrically in terms of amps, watts, or volt-amperes. Some loads, such as motors, may also be described in terms of a mechanical power rating, such as horsepower (hp). The mechanical to electrical conversion is 745.7 watts per horsepower. The full load current drawn by a single-phase AC motor can be determined by

$$i = \frac{hp \times 745.7}{V \times PF \times EFF} \qquad (2.15)$$

where i = load current in amps
hp = rated motor horsepower
V = applied voltage in volts
PF = motor power factor
EFF = motor efficiency

---

[13] Applications of single-phase and poly-phase power systems and loads are discussed in detail in Chapters 4 and 5.

When a load device is energized, it utilizes electricity at its power rating. If a device is energized for 30 minutes, the total energy it consumes in watt-hours (Wh) is equal to its wattage rating multiplied by ½ hour. Thus, an incandescent light bulb rated at 100 watts would consume 50 Wh of energy during a 30 minute period. The device is simply on or off, and it consumes energy at a constant rate while it is on.

Multi-Stage Load Devices

Some individual load devices use power in stages. For example, many copy machines have multiple modes of operation, as illustrated in Figure 2.7. When the copier is first turned on, it must go through a short period of time during which certain of its components are heated to a minimum standby state. When a copy is "requested" by pressing a switch, the fuser must be heated to its operating temperature. When ready, the camera scan commences, a blank sheet of paper is driven by a motor through the mechanism, and the copy is produced. The copier uses its maximum amount of power during the copy production stage. After a short period of time during which additional copies are not requested, the copier goes into a reduced power standby state in order to conserve energy while maintaining a readiness condition. If additional time passes and no further copies are requested, the copier returns to its minimal standby state in order to conserve the most amount of energy. While this operating cycle appears to be fairly sophisticated, the energy consumption characteristics of the machine represent a series of simple on and off states of each of the electrical components that make up the machine. Energy is consumed at a constant rate during each stage of operation.

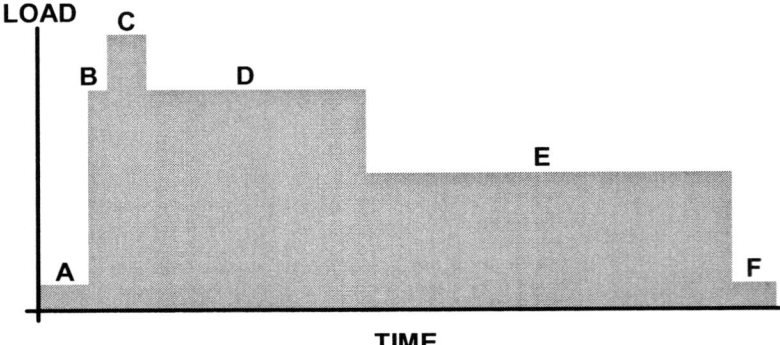

Figure 2.7 A multi-stage load device. A copy machine as an example of an individual load device which utilizes different levels of power during various stages of operation. Operation cycle: A. Power On Mode, B. Fuser Warmup, C. Copy Made, D. Hot Standby Mode, E. Energy Saver Mode, and F. Power On Mode. The load could be expressed in amps, watts, or volt-amperes.

Prerequisite Concepts 39

Load Device Operating Patterns

While some load devices may operate continuously, and other devices may operate randomly, most utilization equipment operates in some cyclical form. These usage cycles occur as a result of lifestyles and practices. People establish routines which drive their requirements for electricity. For example, an electric stove and/or a microwave oven is used for cooking food prior to serving meals. People establish a pattern of mealtimes which then dictates the frequency of usage of the cooking appliances. Not all meals require cooking and not all meals are prepared at one's home, but viewed over a sufficient period of time, a definite pattern of use will be apparent.

Some load devices cycle automatically. For example, a water heater operates periodically in order to maintain the temperature of water in the tank. A thermostat is used to control the on and off operation of the heating elements. As shown in Figure 2.8, hot water demands established by daily routines require the heating of colder water that replenishes the tank as a result of use. When the thermostat senses a drop in water temperature, the heating elements are energized until the desired temperature setting is satisfied. The wattage rating of the water heater is constant, e.g., 4,200 watts. The heater is merely on or off, but the heater is on for long periods of time when the demand for hot water is significant, as with showers or dish washing. Thus, the thermostat regulates the total amount of energy that must be consumed to meet the demand for hot water.

The operation of some devices is heavily dependent upon weather. As shown in Figure 2.9, an air conditioner cycles much like a water heater during times of the day when temperatures are relatively constant and low, as in the morning hours. The air conditioner operates periodically to maintain the internal temperature as heat slowly infiltrates the building. But as afternoon temperatures rise, the thermostat starts to require both longer and more frequent operations of the air conditioner. As with a water heater, the wattage rating of the unit is fixed. It is the amount of energy consumed during each operation that is variable.

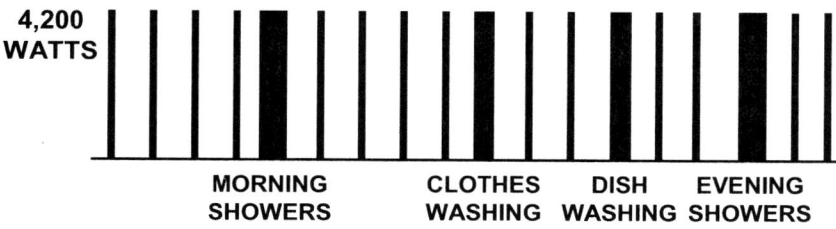

Figure 2.8 A cyclical load pattern. The cyclical pattern of a water heater is driven by both the demand for hot water at certain times to meet specific customer needs and short, periodic operations which are required in order to maintain a preset temperature level, which are necessary because of heat loss through the tank.

Figure 2.9 A weather sensitive load. An air conditioner operates to maintain a preset temperature level in a building. The air conditioner's run time is a function of the outdoor air temperature. On a hot day, the air conditioner may operate frequently even after the outside air temperature has started to decrease in the evening because of thermal buildup during the day. During the cooler hours of the day, early morning and late evening, an air conditioner operates periodically as heat infiltrates a building at a slower rate.

Systems of Electrical Loads

Customers typically have many individual end-use load devices, and the energy requirements of these devices tend to be varied in both the magnitude of the loads and their operating periods. Residential customers have a variety of appliances which serve a number of lifestyles. Beyond such basics as lighting, refrigeration, and clothes washing, residences include a large number of devices in each room of the house. Some devices, like a hair dryer, may be used on a daily basis, while other devices, such as Christmas tree lights, are used only for a limited period of time during the year. The appliances and other electrical devices collectively make up a group or system of loads.

If one were to inspect a residential dwelling and make an inventory of all electrical devices found within, the list could be rather extensive. This list would include the power requirements of each appliance and other electrical devices, whether or not each device is actually plugged in to a convenience outlet at that time. The sum of the power requirements of all such electrical devices located on the premises is called the *connected load*. The connected load represents the total amount of load within a defined system that would be using electricity if all devices were operating at their nameplate ratings at the same time. In a typical residence, the connected load can easily exceed 50,000 watts.

Prerequisite Concepts 41

Electric Load Diversity

As a result of lifestyles and practices, all connected load devices in a house, building, or factory do not operate simultaneously. This behavior among loads in a system is known as *load diversity*. With a reasonably large set of loads, it is likely that some of the devices may be operating at the same time, some perhaps in an overlapping manner, but not all of the utilization equipment is turned on at once.

Figure 2.10 illustrates how a limited group of six common business office devices and appliances could operate during a period of one hour, which for this scenario, is between 11:00 AM and noon. Susan works in a small office setting where she operates a PC, printer, and a copy machine, and she also has a desktop fan located in her workstation. The office facilities include a coffee maker and a small microwave oven that Susan uses every day. For simplicity, lighting, HVAC equipment, and other obvious office and building electrical apparatus are excluded from this example. Figure 2.10 shows the load profiles, in terms of amps, of each of the six devices.

At the beginning of the hour, Susan's computer is already running and continues to run throughout the hour. Her fan is also running, but after about 20 minutes, she feels cool and turns it off. She then realizes that she has not had a chance to drink her cup of coffee which she filled from a carafe that was freshly brewed a few minutes after 11:00 AM. Since the coffee in her cup has cooled off, Susan puts it into the microwave oven on the high setting for two minutes in order to warm it. The hot plate on the coffee maker has been operating continuously in order to keep the coffee in the carafe warm. At about a quarter to noon, Susan puts her lunch into the microwave oven and sets it to run for four minutes. She also notices that the coffee carafe is nearly empty, and so she brews a new pot so fresh coffee will be available during the noon mealtime. Throughout the hour, Susan has been working expeditiously on an important quarterly report for which she has had to print two documents and make copies of various attachments. At the end of the hour, Susan is ready to take refuge in the nearby conference room in order to enjoy her lunch during the noon work break.

By summing the ampere ratings of each of the six load devices shown in Figure 2.10, the connected load for this group is found to be 42.5 amps. This is the amount of current that would be measured if all six devices were operating simultaneously at their full power requirements. But if these loads are summed for every minute across the hour, the greatest amount of load that is observed for the group is only 25.4 amps, which occurs during minutes five and six, as illustrated in Figure 2.11. When viewing a composite load shape as in Figure 2.11, some of the end-use load distinctions become less apparent.

This example illustrates the diversity of operation between individual load devices within a defined group or system of loads. The level of diversity depends on the number and types of loads contained within the group under investigation. The period of time is also a key factor in evaluating diversity. For example, on a hot July day, the air conditioners in a neighborhood of 100 homes operate with much more diversity at 6:00 AM than they do at 6:00 PM.

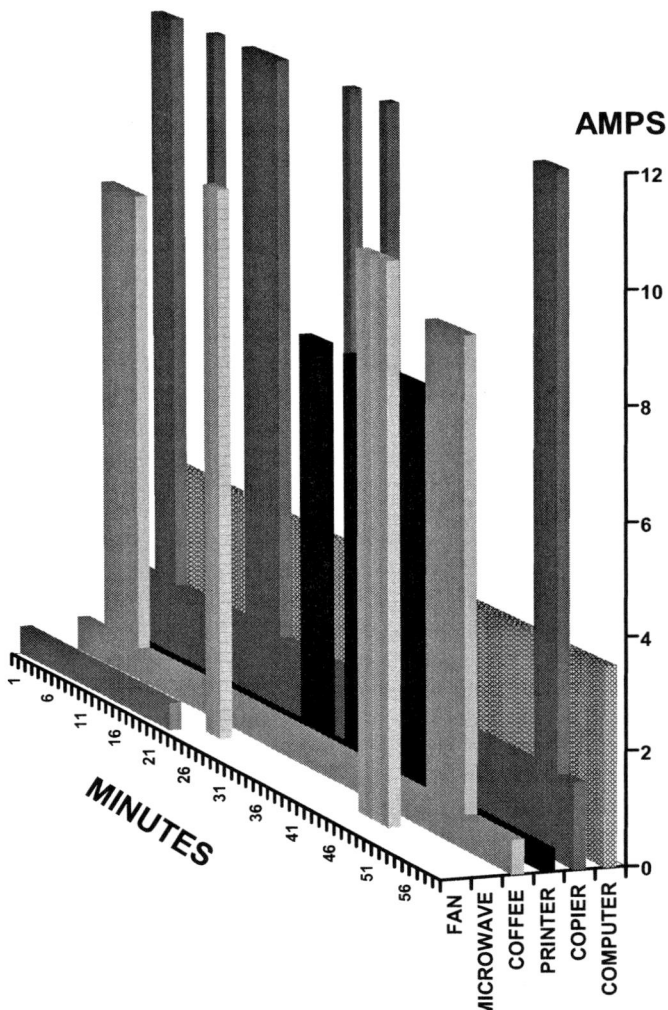

Figure 2.10 A system of load devices. A scenario for the operation of six common load devices found in a typical business office setting is represented by load shapes during a 60-minute period. The loads are illustrated by their operating current ratings and include:

1. Tabletop Fan, 0.5 amps
2. Microwave Oven, 10 amps
3. Coffee Maker, 9 amps
4. Printer, 7.5 amps
5. Copier, 12 amps
6. PC Computer, 3.5 amps

Prerequisite Concepts 43

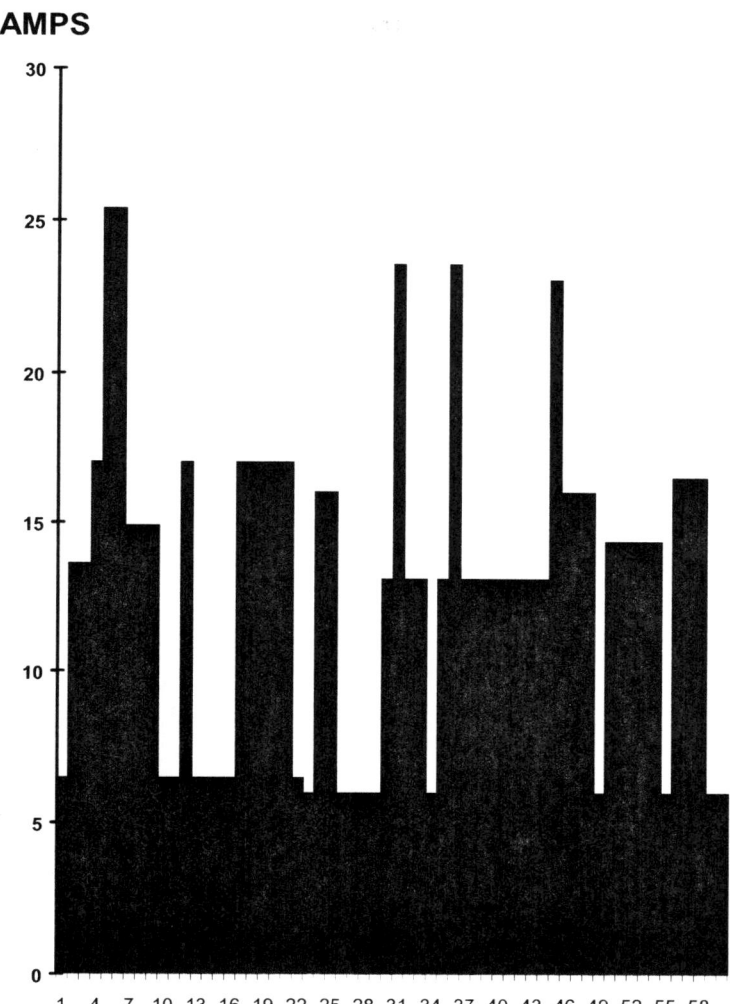

Figure 2.11 **A composite load shape.** The six individual loads represented in the example of Figure 2.10 are summed across the hour on a minute-by-minute basis in order to illustrate the aggregate load shape of the group. The resulting maximum load of the group is found to be 25.4 amps, which occurs during the fifth and sixth minutes of the hour. The connected load for the six devices is 42.5 amps; thus, a significant amount of load diversity is occurring during this hour.

Diversity Factor

The diversity of a group of load devices can be measured in terms of a load *diversity factor*, which is determined by

$$DF = \frac{\sum_{i=1}^{n} L_i}{L_{MAX}} \quad (2.16a)$$

where DF = load diversity factor
  L_i = maximum load rating of the ith load of the group composed of n loads
  L_{MAX} = maximum load of the group

The diversity factor always has a value of 1.0 or greater. If the diversity factor was exactly 1.0, each of the individual loads in the group would be operating coincidentally at their maximum ratings, i.e., with no diversity at peak load.

The diversity factor for the group of six loads illustrated in Figures 2.10 and 2.11 is 42.5 amps ÷ 25.4 amps = 1.67. In this example, the L_i loads are represented by six individual end-use load devices; therefore, the ΣL_i is the connected load of the group. The diversity factor can also be applied to higher order groups of loads. For example, a diversity factor could be determined for a utility's system level retail load which is composed of a group of customer loads represented by the residential class, the commercial class, the industrial class, etc.

Coincidence Factor

A reciprocal manner of describing the diversity of load is in terms of a *coincidence factor*, which is determined by

$$CF = \frac{1}{DF} = \frac{L_{MAX}}{\sum_{i=1}^{n} L_i} \quad (2.16b)$$

where CF = coincidence factor
  DF = diversity factor (from Equation 2.16a)

The coincidence factor always has a value of 0 to 1.0. If the coincidence factor was exactly 0, the loads in the group would be operating with total diversity, while a coincidence factor of 1.0 indicates no load diversity. The coincidence factor is often expressed as a percent; thus, the coincidence factor for the group of six loads illustrated in Figures 2.10 and 2.11 is 1 ÷ 1.67 = 0.6 or 60.0%. Coincidence of load is a key principle upon which cost-of-service study allocations of demand-related costs is based.

Prerequisite Concepts 45

Load Shape Characteristics

The load shape illustrated in Figure 2.11 was based on amperes, but it could also have been based on watts or volt-amperes. Most often load shapes for an individual customer, a customer class, and a utility's system as a whole are represented in an appropriate unit of watts because of the relationship of real power to energy consumption. Thus, load is expressed as kilowatts (10^3 watts), or kW, megawatts (10^6 watts), or MW, and even as gigawatts (10^9 watts), or GW. Likewise, energy is expressed as kilowatt-hours (kWh), megawatt-hours (MWh), and gigawatt-hours (GWh) depending on the volume of energy being represented. A load shape represents the rate at which energy is consumed over time and provides data needed to evaluate and compare the characteristics of different load types.

Energy Consumption

Estimation of the energy consumption during a specified period of time for an individual load device is rather straightforward given the knowledge of its rated kW and temporal operating cycles. But a large system of loads is much more complex because it can fluctuate significantly on an instantaneous basis as its individual load devices cycle on and off, as illustrated in Figure 2.12. The area under the load curve between two points in time represents the total kWh consumed during that period as determined by

$$E = \int_{t_0}^{t_1} p(t)dt \quad (2.17)$$

where E = energy in kWh over time duration t_0 to t_1 in hours
       $p(t)$ = instantaneous load in kW

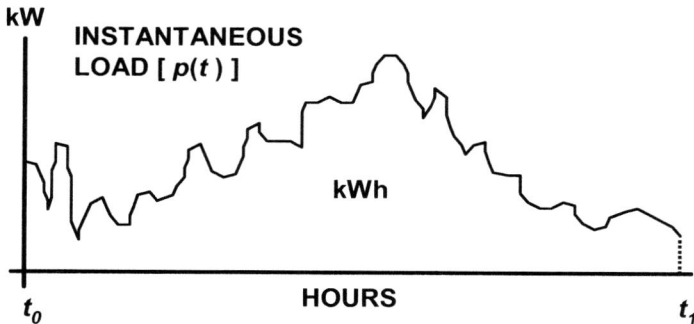

Figure 2.12 Energy consumption of a varying load shape. The area under the instantaneous load curve $p(t)$ represents the kWh of energy consumed for the period shown.

Demand

As noted previously, a system of loads often fluctuates dramatically over a relatively short period of time, and the peak load can be much higher than the load at other times. *Demand*[14] is the average load realized over a defined interval of time as determined by

$$D = \frac{1}{t_k - t_i} \int_{t_i}^{t_k} p(t)dt = \frac{E_{\Delta t}}{\Delta t} \tag{2.18}$$

where D = demand in kW
$p(t)$ = instantaneous load in kW
$t_k - t_i$ = demand interval in hours (Δt)
$E_{\Delta t}$ = energy in kWh during Δt

The period of time over which the average load, or demand, is calculated is referred to as the *demand interval*. Demands for large loads, such as customer classes or the total system load, are often computed on the basis of an hourly period. Commercial and industrial customer demands are typically measured on a 15- or 30-minute basis for billing purposes. An integrating demand meter or a recording interval meter is often used to measure such demands.

Figure 2.13 compares the hourly demand with the instantaneous load for the set of six loads previously illustrated in Figure 2.11. The maximum instantaneous load that occurs during the 60-minute interval is slightly more than 3 kW while the demand for the same period of time is found to be 1.55 kW, which is indicated by the line labeled "D." Another way to visualize demand is to move all of the energy that is above line D to the empty spaces which lie below line D. The result would be a solid block with a height of 1.55 kW and a width of one hour.

A common misconception about demand is the notion that the demand in a given interval is greatly increased when a motor is started during that period. This concern arises from the realization that a motor's starting, or inrush, current is approximately 600% of its running current. While starting a single motor may create a large instantaneous load spike, it is only momentary. The area under this motor start curve, i.e., the motor start energy utilization, is comparably small. When this incremental energy is added to $E_{\Delta t}$ in Equation 2.18, the impact on demand is minor and in many cases insignificant. On the other hand, if numerous motors were started within the interval, a measurable impact on demand might be noted.

---

[14] The phrase "instantaneous demand" is often misused when "instantaneous load" is intended. Instantaneous load is represented by a single point on a load curve. Demand, by definition, is an average which infers a time base for determination; therefore, demand cannot be instantaneous.

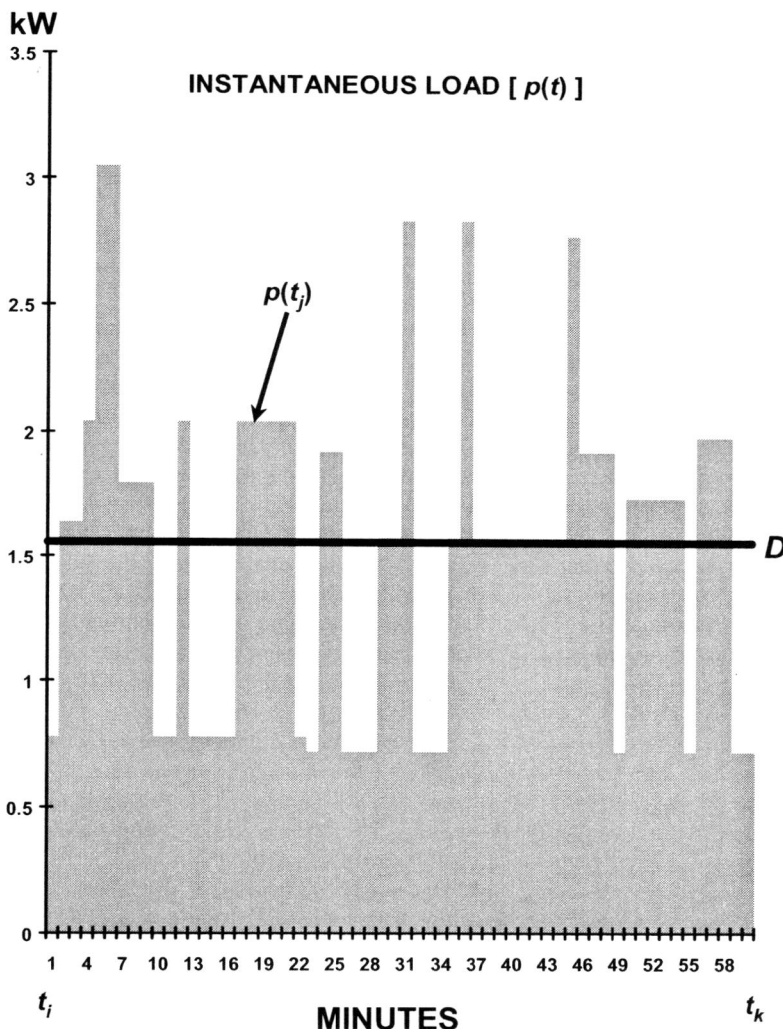

Figure 2.13 Demand vs. instantaneous load. Any point on the curve $p(t)$ represents the instantaneous load, e.g., $t = t_j$. The integration of $p(t)$ from t_i to t_k determines the area under the curve, or the energy (kWh) consumed during that interval. Demand (D) is the average load for the period over which $p(t)$ is evaluated (a 60-minute demand interval in this example). The demands for customer class loads are typically calculated on a one-hour demand interval basis, while billing demands for commercial and industrial rates are typically calculated on a 15- or 30-minute basis. For a one-hour demand interval, the magnitudes of the demand and energy are equal (in this example, D = 1.55 kW and E = 1.55 kWh).

Load Factor

Load factor is the ratio of the average load over a prescribed period of time to the peak load which occurs during that same time period as determined by

$$LF = \frac{L_{AVG}}{L_{MAX}} = \frac{\frac{E}{t_1 - t_0}}{L_{MAX}} \qquad (2.19)$$

where LF = load factor
  L_{AVG} = average load in kW
  L_{MAX} = maximum load in kW that occurs during the time period
  E = energy consumed in kWh during the time period
  $t_1 - t_0$ = time period in hours

The computation of load factor produces a result of 1.0 or less, but it is typically expressed as a percentage.

Load factor is a measure of energy utilization with respect to peak demand. When a large volume of energy is utilized with respect to the maximum load, the load factor is high. Conversely, a small volume of energy use relative to the maximum load results in a low load factor. Another perspective of load factor is that it describes the ratio of the actual amount of energy used over a period of time compared to the amount of energy that could have been used over the period if the maximum load established during the time period was sustained over the entire period of time.[15]

Figure 2.14 illustrates how two different load shapes having the same amount of energy can have much different load factors. A high load factor load will have a shape that is generally flat with minimal difference between its maximum load and its average load. Conversely, a low load factor shape will have a large difference between its peak load and its average load.

When considering load factor, the period of time over which load factor is determined is of utmost importance. To say that "load factor is equal to 65%" presents an incomplete picture. As shown in Table 2.1, the load factor associated with a maximum load can decrease considerably as the time period is lengthened. In this example, the system maximum load for the year occurred on July 9. The load factor for the peak day is fairly high at 81.5%. However, the load factor for the month of July is less than 80%, and the load factor on an annual basis is much less at just under 66%. The key to this variation is in the differing proportions of energy utilized in each of the time periods relative to the available hours, i.e., the average load varies. Thus, in order to accurately comprehend load factor, one should clearly understand the time basis upon which the calculation is made.

---

[15] As an analogy, the airline industry refers to the load factor of an airliner as the ratio of the number of seats with passengers to the total seat capacity.

Prerequisite Concepts

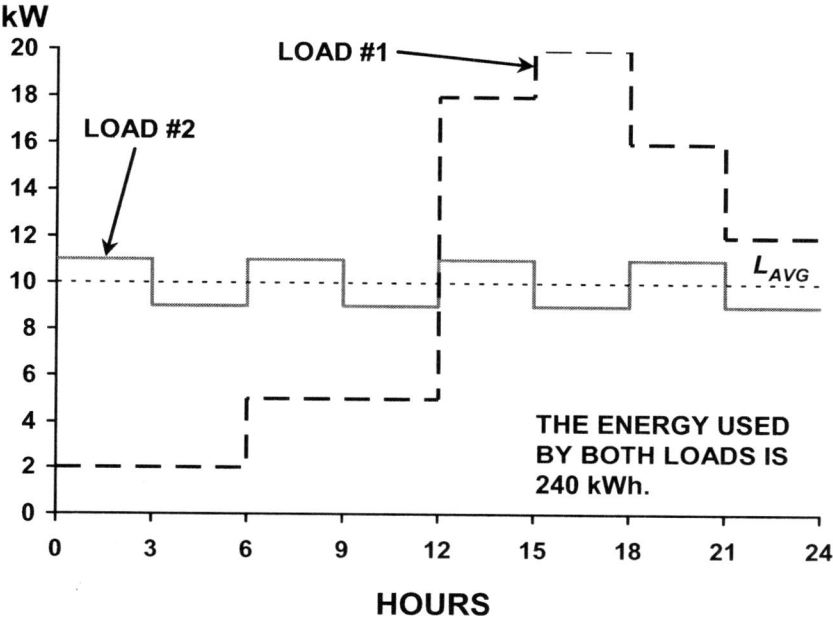

Figure 2.14 Load factor. Both loads use the same volume of energy over the 24-hour period, and they both have an average load of 10 kW. However, their daily load factors are significantly different. Load #1 has a peak load of 20 kW and thus a load factor of 50%, while load #2 has a peak load of 12 kW and thus a load factor of 83%.

Table 2.1 System Load Factor Calculations for Different Time Bases

Load Factor Basis	MWh	Hours	MW_{AVG}	%LF
Peak Day	43,394	24	1,808	81.5
Peak Month	1,274,849	720	1,771	79.8
Year	12,813,378	8,760	1,463	65.9

A maximum load of 2,219 MW occurred on July 9 during the hour ending 6:00 PM.

2.4 ELEMENTS OF UTILITY COSTS AND PRICING

Electric utilities are very capital intensive because they represent a major infrastructure of assets which include generating stations, transmission lines, substations, and distribution lines and facilities. Other general capital expenditures are required to support this infrastructure, including such items as land, office buildings and facilities, transportation equipment (e.g., line trucks and other service vehicles), and communications equipment. All of these assets are referred to as electric *plant* or *investment*.

The cost of plant has three main components: the cost of materials, the cost of labor, and overheads. The cost of materials includes the purchase price of the item plus storage, freight, and handling costs. The installation cost is labor and any associated equipment rentals. Suppose a common distribution pole costs $150, and its installation, including loading and transportation, averages about 2 hours at a base labor rate of $45 per hour. The total *installed cost* of this pole, assuming 25% overheads, would be $300, which is the amount of plant that would be capitalized on the books. The cost of the line truck needed to auger a hole and set the pole in place is capitalized separately. The installed cost of facilities constructed by the utility does not contain any profit. However, if the utility were to outsource the project, a contractor would charge the utility for its materials, labor, equipment, overheads, and a profit margin to install the facilities.

Once installed, further expenditures attributed to the facilities are categorized as *operation and maintenance (O&M) expenses*. For example, a wood pole may require straightening due to leaning or support stubbing because of deterioration of the wood at the ground line. O&M expenses are incurred in association with all functional levels of the power system. In addition, expenses arise from other operating functions, including customer accounts, customer service, sales, and general administration.

An Accounting and Financial View of Utility Costs

Under the Federal Power Act, the Federal Power Commission (FPC) adopted a "Uniform System of Accounts" which were prescribed to be used by electric utilities subject to the provisions of the Federal Power Act. The FERC, as successor to the FPC, continues to adopt this Uniform System of Accounts. This accounting scheme serves as a means for a utility to systematically document in its books and records all of its investments and expenditures.

Table 2.2 shows the fundamental organization of the Uniform System of Accounts. The 100 series of accounts represents Assets and Other Debits. The 200 series of accounts represents Liabilities and Other Credits. Taken together, the 100 and 200 series of accounts depict the utility's financial balance sheet. The 300 series of accounts detail all electric plant in service, which rolls up as a major account entry on the balance sheet under Item 1. Utility Plant. The 400 series contains income, retained earnings, and revenue accounts.

Prerequisite Concepts

Table 2.2 Organization of the FERC's Uniform System of Accounts

Electric Utility Accounts

Balance Sheet Chart of Accounts

ASSETS AND OTHER DEBITS
Accounts 100-190
1. Utility Plant
2. Other Property and Investments
3. Current and Accrued Assets
4. Deferred Debits

LIABILITIES AND OTHER CREDITS
Accounts 200-283
5. Proprietary Capital
6. Long-Term Debt
7. Other Noncurrent Liabilities
8. Current and Accrued Liabilities
9. Deferred Credits

Electric Plant Chart of Accounts

Accounts 300-399.1
1. Intangible Plant
2. Production Plant
3. Transmission Plant
4. Regional Transmission Market Operation Plant
5. Distribution Plant
6. General Plant

Income Chart of Accounts

Accounts 400-432, 434-435
1. Utility Operating Income
2. Other Income and Deductions
3. Interest Charges
4. Extraordinary Items

Retained Earnings Chart of Accounts

Accounts 433, 436-439
Retained Earnings

Operating Revenue Chart of Accounts

Accounts 440-457.2
1. Sales of Electricity
2. Other Operating Revenues

Operation and Maintenance Chart of Accounts

Accounts 500-598
1. Power Production Expenses
2. Transmission Expenses
3. Regional Market Expenses
4. Distribution Expenses

Accounts 900-935
5. Customer Accounts Expenses
6. Customer Service and Informational Expenses
7. Sales Expenses
8. Administrative and General Expenses

The 500 series of accounts represents O&M expenses associated with the production, transmission, and distribution functions. The 900 series of accounts represents O&M expenses associated with customer accounts, customer service, sales, and administrative and general (A&G).

The Uniform System of Accounts is used by electric utilities to present financial and cost data in routine reports such as the annual Form 1, monthly operating reports, and other documents. These reports serve as a principle source of information for use in preparing a cost-of-service study. A detailed list of the Uniform System of Accounts is contained in Appendix B.

Revenue Requirements

The *revenue requirement* represents the level of total income that is necessary in order for a utility to recover its costs of providing electric service and, in the case of investor-owned utilities, earn a margin of profit. The annual revenue requirement is determined by

$$RR = R + IT + EXP \qquad (2.20)$$

where RR = revenue requirement (in applicable currency)[16]
  R = return (from Equation 2.21)
  IT = income taxes
  EXP = O&M expenditures

Return is the amount of money that must be earned from the sale of electricity after accounting for all operating expenditures and applicable state and federal income taxes. This residual amount is used to pay interest on debt and dividends to stockholders. All of the earnings available for dividend payments are not necessarily distributed to the stockholders each quarter. A portion of the return amount might be held by the utility as retained earnings which can be used to ensure a specified dividend payment in future periods when earnings are lower than expected. Earnings levels may be greatly impacted by variable weather conditions due to the sensitivity of heating and air conditioning equipment which typically make up a large portion of a utility's system load.

Return is the product of a utility's net assets, or *rate base*, and its weighted cost of capital, or *rate of return*, as determined by

$$R = RB \times r = (P_{VAL} - D_{ACC}) \times r \qquad (2.21)$$

where R = return in dollars
  RB = rate base
  r = rate of return as a percent
  P_{VAL} = gross valuation of plant assets
  D_{ACC} = accumulated depreciation

A previous example assumed the installed cost of a common distribution pole is $300. This pole and its installed cost would be booked to FERC Account 364: Poles, Towers, and Fixtures in the year in which it was placed in service.

---

[16] For discussion purposes, all monetary units are expressed in U.S. dollars and cents throughout this book. Any currency could be substituted in place of this standard.

Prerequisite Concepts

Assuming this pole was installed ten years ago at a gross cost of $300 and has an expected life of 25 years, its net value today would be $180 or 60% of its original installed cost, based on straight-line depreciation (i.e., its accumulated depreciation would be [$300 ÷ 25 years] × 10 years = $120). If the rate of return was 10% on an annual basis, then the amount of return allowed for this pole today would be $180 × 0.1 = $18 per year. The addition of income taxes and associated operating expenses would complete the determination of the revenue requirement for this pole.

The revenue requirement for an individual item declines each year as a result of depreciation. The example distribution pole's value declines at a rate of $12 per year ($300 ÷ 25 years). When fully depreciated, the revenue requirement would then be just the annual O&M expenditures and taxes for as long as that particular pole remained in service. FERC Account 108: Accumulated Provision For Depreciation Of Electric Utility Plant is used to accrue the annual depreciation expense of all investments. Depreciation of utility plant in service is booked by utility function, e.g., generation (by type), transmission, distribution, general, and common.[17]

For pricing purposes, the revenue requirement is based on all of the assets together. A utility's revenue requirement varies from year to year due to dynamic changes in rate base, including depreciation, plant additions, and plant retirements. Operating costs also differ from year to year as labor rates change, fuel costs vary, etc. A number of other factors cause the revenue requirement to change. When utilities are experiencing large rates of load growth, system expansion requirements and the cost of new financing may have a significant impact on the revenue requirement. If the revenue produced from the sale of electricity under current price levels falls short of the revenue requirement, a price increase would be needed. On the other hand, if utilities experience low rates of load growth, maturing construction programs, and moderate cost escalations, eventually the prices could generate revenues consistently above the revenue requirement, and a price decrease might then be in order.

The revenue requirement is typically determined on the basis of a twelve month period of time, which is often referred to as the *test year* or *test period*. The test period does not have to be a calendar year as it may represent, for instance, "the most recent twelve months." In addition, the test period may be based on historical data, projected data, or a combination of historical and projected months. The test period methodology varies from one regulatory jurisdiction to another.

By conducting a detailed cost-of-service study based on a specified test period, the amount of return yielded by each of the customer or rate classes, relative to the overall revenue requirement, can be determined. The class returns are often found to vary significantly.

---

[17] Common plant represents facilities that are jointly used by two different utility operations. An example of common plant would be an office building which houses both gas and electric system departments.

Cost of Capital

Just like other businesses, utilities must raise capital for construction. The primary sources of capital funds are realized through the issuance of debt (long-term bonds and short-term interest bearing securities such as commercial paper), preferred stocks, and common stocks, or equity.

The utility's *capital structure* is the outstanding dollar amounts of each type of financial security, as shown in Table 2.3. The ratio represents the percent of each capital item amount to the total amount. Each type of financial security has a cost rate. The 7.76% cost rate for debt, listed in Table 2.3, is an average rate which represents a variety of bond issues having different fixed interest rates and different years when the notes are payable.[18] The 8.09% cost rate for preferred stock represents a fixed dividend that is set when the security is issued. The 12.5% rate for equity is a factor that varies based on certain conditions. In general, equity represents the type of investment that carries the most risk; therefore, the rate for equity should be higher than the interest rate on bonds. Financial experts utilize various methodologies to determine a fair rate of equity for a particular utility. However, determination of an equity rate is not a simple formulary calculation (like Ohm's Law perhaps). As a result, different cost of capital experts will derive different positions on the value of equity based on varied assumptions, perspectives, and stock value modeling methods.

Under regulated ratemaking, the Commissions have jurisdiction over approving a utility's equity cost rate to be used for setting prices for electric service. The equity cost rate approved for wholesale service by FERC typically varies from the rate which is approved for retail service by the state Commission, for a given utility that renders both categories of electric service. The authorized equity rate is set forth in a Commission order as a result of regulatory hearings in which the utility and the case intervenors present testimony on the cost of capital, as well as rate base, O&M expenditures, cost-of-service study results, and proposed rate schedule revisions, i.e., a rate case proceeding.

For the example in Table 2.3, the authorized equity rate has been established at 12.5%. The capital structure ratios are multiplied by their associated cost rates to derive the weighted cost rates. These results are then summed to determine the overall *weighted average cost of capital*, or *return on investment* (ROI), which for this example is 9.974%. This total weighted cost rate would be the value of r that would be used in Equation 2.21 in order to determine the return amount.

As mentioned previously, the cost-of-service determines the amount of return produced by each customer or rate class. By dividing a class return amount by its assigned share of rate base, the class ROI is determined. The demonstrated return on equity can also be determined on a class basis.[19]

---

[18] When possible, utilities buy back high interest rate bond issues and re-issue lower interest rate bonds in order to manage debt cost down as much as practicable.

[19] The equity cost rate is referred to as the *return on equity* (ROE).

Table 2.3 Calculation of the Weighted Cost of Capital

Capital Structure

Capital Item	Amount	Ratio
Debt	$797,768,844	45.25%
Preferred Stock	152,148,608	8.63%
Equity	813,058,963	46.12%
Total	**$1,762,976,415**	

Weighted Cost of Capital

Capital Item	Ratio	Cost Rate	Weighted Cost
Debt	45.25%	7.76%	3.511%
Preferred Stock	8.63%	8.09%	0.698%
Equity	46.12%	12.50%	3.511%
Total			**9.974%**

An Economics View of Utility Costs

Another aspect of utility costs relates to the economics perspective of quantifying costs. Considering just the production level of a utility, generating units are planned and installed in a timely fashion in order to maintain sufficient capacity to meet a growing peak demand requirement. Each utility's conditions are unique, but as an example, the average original cost of all generating facilities installed over the last 50 years for a moderate size system could be on the order of $300 to $325 per kW. Taking into account the accumulated depreciation over this time period, the average net cost of generation might then be about $175 to $200.[20] Average costs such as these are referred to as *embedded costs*. In contrast to the existing system of generation, the next unit that a utility would be required to construct is likely to be a combined cycle unit, which could be on the order of $450 per kW. The cost of the next increment, which in this case refers to new production capacity, represents the *marginal costs*.

---

[20] These numbers attempt to give a relative indication of what the current value of generation is likely to be at this time for a utility having between five and ten generating stations, including a mix of coal-fired units and combustion turbines. The magnitude of the average costs per kW is dependent on a number of factors which often vary from utility to utility (e.g., load growth rates, types of units, unit vintages, etc.). In addition, a number of utilities have accelerated the write off of assets in order to lower their potential for stranded costs in the event of restructuring.

Marginal cost is defined as the change in total cost due to the output of one additional unit as determined by

$$MC = \frac{\Delta TC}{\Delta Q} \qquad (2.22)$$

where MC = marginal cost in dollars per unit
ΔTC = change in total cost
ΔQ = unit change in output

In an electric utility, the marginal costs are typically based on the next unit of generation, transmission, or distribution capacity ($/MW or $/kW), the next unit of energy ($/MWh or $/kWh), or the next customer.

The basic economic theory behind marginal cost pricing is that it promotes the efficient allocation of resources. If the marginal cost-based price is high, consumers are more likely to respond by limiting their usage, which could help defer the need for the next increment. For example, if the existing production capacity is nearly fully utilized, additional load growth would ultimately require that a new unit of generation be added. Setting prices based on the marginal cost (e.g., $450/kW) rather than on the embedded cost (e.g., $200/kW) sends a much higher price signal for customers to respond. But if the marginal cost-based price is low, consumers are encouraged to utilize more. The marginal cost would be low in times when an abundance of generation capacity exists. Since the need for an additional increment of capacity would be far in the future, the net present value of the next generating unit could be much smaller than the embedded cost. Thus, a lower price based on the marginal cost of the next unit would encourage customers to increase their usage and thus make better use of the existing capacity.

With embedded cost pricing, rates are based on the cost of all units. For example, if a new 1,000 MW unit is added at a cost of $450/kW (or $450 million) to an existing system of 2,500 MW of generation capacity with a net average cost of $200/kW (or $500 million), the total cost of the system with the additional unit would then be $950 million, or a little more than $270/kW. Under embedded cost pricing, the existing rates would have been based on $200/kW while the new rates would be based on the $270/kW amount.

A problem that exists with the practice of marginal cost-based pricing is the resulting revenue imbalance. Most often a utility's revenue requirement is established on the basis of the original installed cost of all units in service (further recognizing that some amount of construction work in progress might be included in some jurisdictions). The marginal cost-based prices are likely to be higher or lower than the average embedded cost-based prices. Thus, compared to the authorized revenue requirement, the marginal cost-based prices would either over-collect or under-collect the necessary revenue amount. A solution to this problem is Ramsey pricing, which adjusts the price of the least elastic component in the rate structure, which typically is the customer charge, so that the revenue requirement is met. However, by making these adjustments, a bit of price distortion is introduced.

Despite the revenue requirements constraint, marginal cost data can be used to enhance price signals while actually using embedded costs for pricing. For instance, time differentiation of the embedded cost revenue requirement on a daily or hourly basis is a recognition of marginal costs. In other words, the pattern of marginal costs can be used to modulate an otherwise average rate across time and communicate much more information about the temporal variations in costs to the customer.

A Pricing View of Utility Costs

Chapter 1 defined three pricing components that were reflective of a cost causation principle, namely: customer, demand, and energy. These costs are quantified in the cost-of-service study. Customer costs are associated with a customer's connectivity to the power system. Thus, customer-related costs include (a) investment in a meter, a service drop, and a portion of the secondary and primary distribution system, (b) the O&M expenditures associated with these facilities, and (c) customer services expenditure, such as meter reading, billing, and customer accounting and assistance.

Demand costs are associated with a customer's maximum load requirements, i.e., kW or kVA. Thus, demand-related costs include the investment and expenditures associated with all facilities which provide capacity to meet the load requirements of customers throughout the distribution and transmission systems and at the production level.

Energy costs are associated with a customer's requirement for a volume of electricity, i.e., kWh. Energy costs consist primarily of the expense of fuel that is actually consumed in the process of generating electricity and certain O&M expenditures that are a function of energy output, such as operating lubricants and oils. Fuel stock, or the fuel on hand, actually represents an investment item since, simply stated, it is inventoried in a manner similar to other materials and supplies. Thus, money is invested by the utility to maintain a fuel stock.

Another way to view utility costs is by the classifications of *variable* and *fixed*. Variable costs change with output. For example, as energy output increases (decreases), the amount of fuel required for generation also increases (decreases). When a generating unit is operating at its maximum capacity, its fuel requirement is also at a maximum. When the generation output is reduced, the fuel requirement is also reduced. An analogy is gasoline usage in a limousine in which there is a direct correlation between the gallons required and the miles driven. Likewise, there is a direct correlation between the tons of coal or Mcf of natural gas required to produce a volume of megawatt-hours.

Fixed costs are investments and expenses that do not vary with output. Thus, the costs of the generation assets are fixed. As the production output varies between a plant's maximum capability and zero, the cost of the asset remains constant. The plant investment is a *sunk* cost of ownership (unless the asset were to be sold at some time). The expenses of the operations personnel also represent fixed costs since the complement of operators must be maintained as long as the plant is operational. Drawing upon the limousine analogy, the loan payments for the vehicle and the salaries of its drivers represent fixed costs which do not vary with mileage.

Pricing for Cost Recovery

Consider a pricing scheme for Allen's Limousine Service. One rate could be set for recovering the variable cost of gasoline on a "per mile" basis. Another rate could be set for recovering the fixed costs of investment and expenses on a per rental basis, where the contract period could be defined to be "per day" or "per six hour period." The amount to be charged for each contract period would be determined by dividing the total fixed costs by the expected number of rentals. The primary risk assumed by the limousine owner is in estimating a realistic number of rentals. Otherwise, this dual rate mechanism would be very efficient at tracking the costs of this service.

Alternatively, both the fixed and the variable costs could be recovered concurrently through a single "per mile" rate. This price mechanism would be more risky because the fixed cost recovery would be totally dependent on the miles driven. Predicting the total number of miles likely to be driven for all of the anticipated rentals is a more complex problem that just predicting the number of rentals alone. In addition, the use of an average rate per mile gives rise to an equity issue between the different users of the service. Long drives would over-collect fixed costs while short drives would under-collect fixed costs; thus, a subsidy between customers is created.

Now consider a methodology for pricing the sale of electricity from a single 25 MW generating unit to two directly connected customers (i.e., no requirements for transmission and distribution facilities are assumed in order to simplify the example). Similar to the limousine service example, the variable costs of fuel and the variable O&M costs could be charged on a per kWh basis. The fixed cost of the unit and the fixed O&M costs could be charged on a contract capacity basis with one customer contracting for 10 MW and the other contracting for 15 MW. As with the limousine service example, this dual rate mechanism would perform with great efficiency. The major risk to the supplier is the length of time that each customer would remain, since the rate for capacity reflects a unit cost per kW based on an investment that is considered to be very long term in nature, i.e., typically 30 to 40 years.

If the recovery of the fixed and variable costs of the electric service were based upon a single kWh rate alone, the supplier would assume a much higher level of risk. Similar to the limousine example, the customers' energy usage patterns could be different, both between each other and individually on a month-to-month basis. Accurately predicting the energy usage is key to ensuring the proper level of cost recovery. Weather sensitivity of the loads would induce an additional risk factor. If these customers have different load factors, the customer with the higher load factor will subsidize the customer with the lower load factor under the single kWh rate mechanism.

This example points to the need to recognize the three cost elements of customer (fixed), demand (fixed), and energy (variable) when designing rates. These three components of cost are combined in a variety of pricing structures which are targeted to meet various objectives. It should also be recognized that these different pricing structures carry with them a varying level of risk that is shared by both the seller and the buyer. Matching pricing structures with the characteristics of the customer markets is a key factor for adequately meeting customers' electricity requirements while recovering costs and compensating investors.

2.5 SUMMARY

This chapter has reviewed basic terminology and the elementary concepts of a number of different topics in the disciplines of utility accounting, economics, engineering, finance, and regulation. The reader should now have a broad understanding of how these seemingly unrelated topics represent the foundations of electricity pricing.

Individual customer end-use electrical devices create an aggregate demand for electricity that is characterized by unique load shapes. These load shapes are key factors for understanding the cost of electricity since they require a capital intensive utility infrastructure capable of providing reliable electrical service at all times. The power system must be planned, constructed, financed, and operated efficiently so that customers receive electricity that is economical while earning an adequate return for investors. Expenditures must be carefully accounted for since cost is the primary driver for regulated prices. A properly structured rate is crucial for recovering costs in an effective and impartial manner.

The pricing expert must master the basic concepts of this chapter in order to ensure excellence in rate design. Perhaps the singularly most important concept of all is load factor because it explains so much of the behavioral relationship between load, costs, and prices. The remainder of the book will build upon load factor along with all of the other prerequisite concepts discussed herein and illustrate their practical applications for electricity pricing.

3
Electric End Uses

3.1 INTRODUCTION

Heat is a critical need in a person's everyday life. Heat warms bath water, cooks food, and dries clothes. In winter, heat is a dire necessity when temperatures plunge to subfreezing levels. Many manufacturing operations also make use of heat in production processes, such as melting and bonding materials, boiling liquids and solutions, and curing and drying products. Heat is thermal energy, and it is put to use in many ways in order to meet different types of needs. Other energy forms are also applied in various ways to meet other kinds of needs.

Such energy applications are known as *end uses*. In order to realize an end-use benefit, an end-use technology, or energy conversion device, is required. Thus, furnaces warm air for comfort, lamps produce light for illumination, and drive systems run machines. A variety of technologies may often be utilized to produce end-use results, and the primary energy inputs are electricity and fossil fuels.[1] While electricity is the only practical choice for a large number of applications, it competes directly with other fuels to serve a number of end uses. A customer's main energy focus is the need for Btu's to operate an end-use application. The choice of technology and energy source to provide the Btu's necessary is most often a matter of basic economics.

The operating characteristics of some end-use load devices can have a significant impact on prices for electricity. For instance, the large amount of air conditioning load connected to the power system created constrained summer operat-

---

[1] Fossil fuels primarily include natural gas, propane, coal, and fuel oil. Other petroleum-based products, such as gasoline, kerosene, and diesel fuel, can be used to provide some end uses in certain applications (e.g., a gasoline powered air compressor for use in remote locations).

ing conditions throughout much of the United States during the late 1990s. Wholesale electricity prices spiked as utilities required record setting amounts of power in order to maintain service during high temperature periods.

Electricity pricing requires a fundamental understanding of customer end-use needs, applications, and characteristics. Pricing structures which target specific end uses are necessary in a competitive environment in order to gain market share. While low prices will certainly attract customers, competitive sales must also be profitable. Thus, prices must be set at a point which optimizes market share while insuring the recovery of fixed costs.

3.2 CUSTOMER END-USE NEEDS

People seek the means to make life easier, increase productivity, and enjoy more entertainment. In so doing, the quality of life is improved. Electricity has been a notable aspect of this development because electricity provides customers with the means to realize significant lifestyle and productivity benefits. Electricity by itself is merely a flow of electrons. But electricity in application creates value. The greater the value, the more people will want or need a particular end use.

In one way, electricity is like gasoline which is a different kind of energy source. No one buys gasoline as a final product. People buy gasoline to power their cars and boats, mow their yards, and run a variety of engine-driven yard machines. Likewise, people buy electricity to be cool in the summertime, to enjoy television after work, or, perhaps at work, to melt and refine scrap iron. These applications represent only a small sample of the many end uses that customers derive from electricity.

End-Use Applications of Electricity

Electricity is utilized as an energy source for accomplishing a number of practical functions. As discussed in Chapter 2, utilization equipment, or load devices, convert electricity into another useful form of energy of which there are four basic types:

- Heat energy
- Light energy
- Mechanical energy
- Chemical energy

The fact that electrical energy is converted to heat energy, for example, is simply a matter of physics. What is most significant is how this heat energy is used for a well-defined purpose, even as simply as heating water in order to brew a cup of tea. Thus, the results of these energy conversions are the realization of beneficial end-use products and services.

Electric End Uses

Heating End Uses

Heat is used by customers for many different purposes. For example, heat is used for space conditioning by simply heating the air within a building to maintain a comfort level for occupants and to protect critical building systems which would otherwise be damaged by subfreezing temperatures. Electric space heating can be achieved by means of a simple resistive heater using the i^2R effect. The point-of-use conversion of electrical energy to heat in a resistor is essentially 100% efficient with one kWh producing 3,413 Btu of heat energy.

Electric space heating can also be achieved with a heat pump which extracts heat from the outside air, from water (such as in a well), or from earth, and utilizes that heat to warm the air inside a building via a heat exchange process. To move (pump) heat in this manner requires electricity to power a compressor which circulates a refrigerant in a thermodynamic cycle. The use of a compressor is actually a mechanical energy conversion since a motor is required to pressurize the refrigerant.[2] But regardless of the type of technology employed, the end result from the customer's perspective is warm air.

Cool air is attained by running the heat pump's thermodynamic cycle in reverse in order to move heat out of the building, thus cooling the inside air. Cool or cold air is merely air which contains little thermal energy. Conventional air conditioning operates on the same principle as the heat pump's cooling cycle. The cooling capacity of an air conditioner is rated in Btu per hour, and 12,000 Btu/hr is equal to one ton of refrigeration. The Btu rating is an indication of how much heat can be moved in an hour.

The same basic heat transfer process that occurs in electric space cooling applies to refrigeration equipment where heat that infiltrates the walls of a refrigerator or freezer and which enters when the doors are opened, must be pumped out in order to maintain the thermostat's internal temperature setting. The most common end result of refrigeration devices is chilled and frozen food products and water.

The relative efficiency of heat transfer equipment, such as heat pumps and air conditioners, is designated by the *Coefficient of Performance* (COP), which is the ratio of thermal energy extraction to the work input, which is expressed in terms of the electrical energy input by

$$COP = \frac{Q_{OUT}}{E_{IN} \times 3,413} \qquad (3.1)$$

---

[2] Air source heat pumps often are supported by supplemental resistive elements which are necessary for maintaining comfort when outside temperatures dip below the heat pump's balance point (where it is difficult to extract enough heat from the air). Supplemental heat is less of a necessity with water source and ground-coupled heat pumps because of the nearly constant temperatures of these respective mediums.

where COP = Coefficient of Performance
Q_{OUT} = quantity of heat moved in Btu
E_{IN} = input energy in kWh

The COP is dimensionless, and it can exceed one (1.0) because a high efficiency piece of cooling equipment can extract more Btu of heat than it consumes in an equivalent Btu of electricity. This phenomenon does not imply that electric-driven cooling equipment is more than 100% efficient. These devices do not create heat, they only move heat, which exists naturally. Calculation of the COP for a heat pump in its heating mode should exclude the effects of any supplemental heating energies.

Another measure commonly used for rating the efficiency of cooling equipment is the *Energy Efficiency Ratio* (EER)[3] which is the thermal output divided by the electrical power input as determined by

$$EER = \frac{Q'_{OUT}}{P_{IN}} \quad (3.2)$$

where EER = Energy Efficiency Ratio
Q'_{OUT} = rate of heat moved in Btu/hr
P_{IN} = power input in watts, i.e., the rate of energy consumption

When comparing specific end-use devices, such as window air conditioners, having the same Btu capacity, the unit which has the higher EER rating will provide the same cooling effect as the lower EER rated unit, but it will use less electrical energy.

Both the design values of COP and EER represent efficiency measures that are realized by a piece of equipment operating at its full load condition. Efficiencies are less when the equipment is only partially loaded as climatic and building occupancy conditions change throughout the day and the seasons. In the case of an air source heat pump, the efficiency decreases as the outside air temperature drops, which is a result of the heat available in the cooler air being diffuse and thus more difficult to claim.

Since heating and cooling systems are not fully loaded all the time, operating efficiencies are often viewed on a seasonal basis. The *Seasonal Performance Factor* (SPF) is often used to relate a heat pump's heating output during its normal annual heating period to its total electrical power input for the same period. Similarly, the *Seasonal Energy Efficiency Ratio* (SEER) is utilized to relate air conditioning and heat pump cooling cycle operational output to the electric input during the annual hours of cooling operation.

---

[3] Like the COP, the EER is a dimensionless term; furthermore, EER = 3.4121 × COP.

Electric End Uses

Heat is also needed for food preparation, and several different kitchen appliances serve as heat sources for cooking foods in a variety of ways. Resistive heating is used for stove tops, toasters, and ovens. With these devices, the food is cooked by means of heat convection and radiation. Microwave ovens utilize electronic components to generate radio frequency waves which cause a heating effect within the food itself.

Water heating is required for a number of domestic uses such as bathing, dish washing, and clothes washing. Hot water is typically produced in volume by means of a resistive type central water heater (e.g., 40 gallon tank) and maintained over time at a temperature readiness state by means of a thermostat control. The actual energy conversion process is 100% efficient, but the efficiency of the water heater itself is less because auxiliary energy use is required to make up thermal losses through the tank (as illustrated in Figure 2.8). Both internal and external tank insulation is used to minimize these radiated losses and improve the overall operating efficiency. In addition to tank type water heating, decentralized, instantaneous water heating is used to produce small quantities of very hot water on demand. Examples of these applications are hotel guest room hot water dispensers and water temperature boosting devices contained within dishwashers.

A quick inspection of a typical home would reveal a number of other devices which are designed to convert electrical energy to heat energy for specific domestic end uses. Applications include such diverse tasks as clothes drying, water bed heating, hair curling, and driveway snow melting.

Heating energy requirements in the commercial sector include some of the same primary end-use applications used for domestic purposes, but their requirements are generally of a much greater scale. For example, the space conditioning requirements of commercial establishments can be much more, obviously, because most commercial buildings are substantially larger than the average home. A commercial building's heating, ventilating, and air conditioning (HVAC) systems also see more intense loads due to such factors as the number of occupants, lights, and equipment such as computers. Commercial establishments are very concerned about the comfort of their customers who have come to shop, dine, or partake of other services.

Hot water and steam is a major requirement of a number of commercial operations, such as hospitals, hotels, laundries, and restaurants. Code requirements for ensuring sanitary conditions may contribute to the hot water system design and operating methods, since very high temperatures are required for some commercial processes. Commercial hot water systems are similar in concept to residential applications as the conventional storage water heater is a popular choice, but large operations may call for distributed hot water heating as opposed to a single central tank. Some commercial systems utilize auxiliary storage tanks into which only heated water is placed (i.e., cold makeup water is not introduced) in order to increase the available hot water capacity.

Commercial hot water may also be derived from reclaimed heat from other heat transfer equipment such as heat pumps, refrigeration equipment, and ice making machine. Heat which is extracted for cooling purposes is otherwise wasted and can be recaptured with a heat exchange mechanism. Larger commercial customers often have the scale of simultaneous cooling and thermal energy requirements necessary to make such a heat reclaim system practical.

In the industrial sector, heat is needed for numerous manufacturing and materials processes. Some examples of industrial heating applications and methods are listed in Table 3.1.

Table 3.1 Various Industrial Electric Process Heat Applications

	End-Use Applications	Electric Heating Methods
Heating of Air	Baking, curing, drying, warming	Convection ovens
Heating of Liquids	Oils, solutions, water	Resistance element heaters: external and immersion type
	Steam production	Boilers: resistance element and electrode type
Surface Heating	Metals, plastics	Resistance element heaters: direct contact and radiation type
Pipeline Heating	Freeze protection, maintenance of product temperature or state	Electric resistance trace lines Impedance heating
Infrared Heating	Baking, curing, degreasing, drying (e.g., surface coatings), moisture removal, preheating, thawing	Quartz type reflector lamps Infrared Ovens: ceramic heater and glass panel heater types
Melting of Materials	Soft metals (e.g., lead, tin, solders), glass	Crucible heaters
	Ferrous metals	Arc furnaces Arc welders
Induction Heating*	Bonding materials, extruding, forging, heat treating, metal refining	Induction coils Induction furnaces
Dielectric Heating**	Drying (paper, textiles, wood), Embossing, sealing	Dielectric ovens

* Electrically conductive materials
** Electrically nonconductive materials

Electric End Uses

Lighting End Uses

Lighting was the first commercial application of electric power. In the late nineteenth century, Thomas Edison built his historic Pearl Street generating station in New York City. Wires were strung throughout the nearby streets, and lamps in residences and offices were connected. The era of central station electricity production and power distribution was born, and electrification began in the United States. The first load device was an incandescent lamp. Incandescent lamps are still widely used today, particularly in homes. The modern incandescent lamp consists of a simple tungsten filament mounted in a glass bulb from which all of the air has been evacuated. Incandescent lamps are rated in watts.

In response to the drive for energy conservation, more efficient compact fluorescent lights (CFLs) have been developed as an alternative to incandescent bulbs. A CFL is typically constructed as a small discharge tube mounted over a ballast and fitted with a standard screw-in base. The tube may be ring-shaped, folded, or otherwise compacted. Although larger than an incandescent bulb, the CFL is generally of a size that will fit many tabletop lamps and wall/ceiling mounted fixtures. The CFL has a longer life but costs significantly more than an incandescent bulb. Some people have difficulty adjusting to the color and brightness of the CFL lamps, particularly in home settings, because they are accustomed to or prefer the softer glow and warmer color of incandescent light. As a result, public acceptance of this substitute form of lighting has been mixed.

Conventional fluorescent lamps are widely used, especially in commercial settings. Fluorescent lighting is achieved from a low voltage discharge through a glass tube containing a low pressure atmosphere of mercury vapor. A phosphor coating on the inside of the tube converts the resulting radiation to visible light. Like incandescent bulbs, fluorescent tubes are rated in watts. A fluorescent lamp emits much more light than a comparable incandescent lamp of the same wattage rating. But the total load of a fluorescent lamp is higher than just the tube wattage because of the need for a ballast, which is an inductive device required to start the voltage discharge. The ballast is also responsible for the often low power factor characteristics of fluorescent lighting.

High intensity discharge (HID) lamps provide the most efficient source of light. Compared to fluorescent lamps, HID lamps are constructed of a small volume arc tube containing higher pressure vapors. Typical vapors consist of mercury, metal halides, and sodium.[4] These lamps are used primarily outdoors, but they do have indoor applications, such as high bay illumination of factories and warehouses. The high intensity discharge lights require a few minutes of time to warm-up before they achieve their full output.

---

[4] The Energy Policy Act of 2005 prohibits the import or manufacture of mercury vapor lamp ballasts after January 1, 2008. Low pressure sodium vapor lamps are used in limited applications because of their high light emissions. But compared to high pressure sodium vapor, the color rendition is rather poor.

Lighting efficiency is expressed in terms of *efficacy*, or lumens per watt. A *lumen* (lm) is a measure of luminous flux, and it is a unit of photometric power as compared to the watt which is a measure of electrical power. Although efficacy is not a dimensionless term, it is the ratio of two power quantities. The lumens used to calculate efficacy are typically the initial lumens, which represent the light output of a new lamp. Over time, a lamp's light output will deteriorate based on the number of burning hours that it experiences.

The lighting technologies previously discussed have different efficacy characteristics, as shown in Figure 3.1. The lumens per watt output of a lamp generally increases with wattage rating; thus, a 100-watt incandescent bulb would have a higher efficacy than a 40-watt incandescent bulb. The incandescent type bulbs have the lowest range of efficacy ratings, while high pressure sodium lamps have the highest range of efficacy ratings.

Lighting represents a major end use in all societal sectors. Inside buildings, lighting is used primarily for different tasks, safety (such as inside hallway and exit sign illumination), and different ambiences. Outside, lighting is used for illumination of roads and signs and for architectural purposes (such as to highlight building facades). All of these types of lighting devices produce visible light for illumination. Visible light lies within the electromagnetic spectrum between ultraviolet and infrared radiation. Ultraviolet is a light energy source that can be used for special applications, from curing inks to discothèque mood lighting.

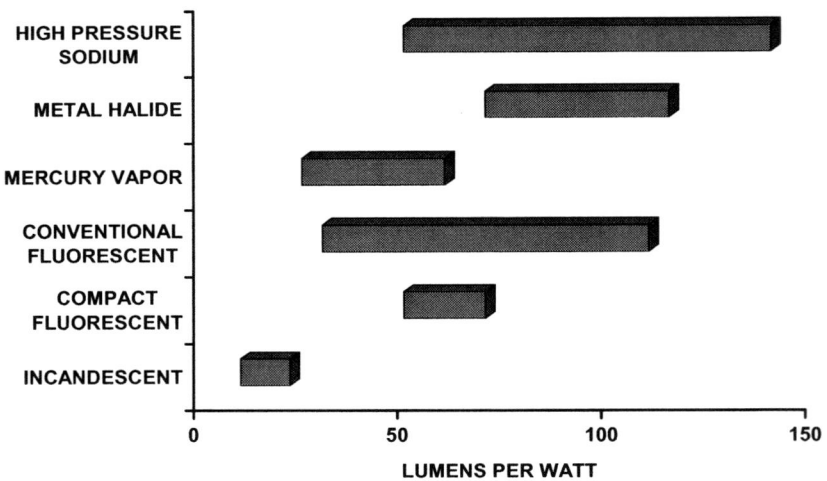

Figure 3.1 **Typical range of efficacies of various lighting technologies.** The lumens per watt of a lamp varies depending upon the wattage rating. Source: U.S. Department of Energy.

Electric End Uses

Mechanical End Uses

Through the action of magnetic fields, electricity is used to create motion and force. Mechanical motion can be linear, as in the inline action of a solenoid, or rotary, as in the spinning of a motor shaft. Magnetic fields can produce a tremendous amount of force, which can be controlled and put to use in a variety of ways.

Motors are used extensively in all customer sectors to provide a number of mechanical end uses. Electric motors drive fans, pumps, compressors, and a seemingly endless number of simple to complex machines. General-purpose motors are designed for common use while special-purpose motors are designed for nontypical drive applications, such as in unique manufacturing processes.

An electric motor consists of two basic parts: a *stator* or stationary component mounted within a frame and a *rotor* to which is connected a drive shaft. The shaft rotates because the motor produces torque as a result of the electrical to mechanical energy conversion process. Fundamentally, torque is developed when two magnetic fields attempt to align with each other. This effort is similar to two iron bar magnets, which will automatically turn in order to reorient themselves in a north pole to south pole configuration, when placed in close proximity. The stator and rotor of a motor also contain north and south poles. The magnetic fields are created about conductor windings in the stator and rotor through which electric currents flow.[5]

This magnetic field interaction is most fundamentally represented by the operation of a basic DC motor, as shown in Figure 3.2. In a DC motor, the stator has poles which are polarized in a north to south orientation, and the magnetic field between these poles is stationary.[6] The magnetic field of the rotor is created by a DC voltage source which flows current through the rotor field windings. Since the rotor is free to spin, the magnetic attraction and repulsion of the rotor and stator fields (like the two bar magnets) cause the rotor to turn. In essence, the magnetic field drags the rotor windings along with it. Since the windings are physically attached to the rotor, the rotor shaft is also dragged along. As a result, the north pole of the rotor aligns in series with the south pole of the stator while the south pole of the rotor aligns in series with the north pole of the stator.

At this point, the rotation would cease; however, the motor has a means for reversing the direction of the rotor field by 180° so that once again, the poles of the stator and rotor are found to be out of series, and rotation continues in order to align the poles. A *commutator* attached to the motor shaft is used to convert the input current, which is DC, into an alternating current flow within the rotor field windings. The rotor field, which is perpendicular to the plane of the rotor windings, reverses its north to south polarity each time the rotor current reverses its direction.

---

[5] In small DC motor, a permanent magnet is often used for the poles of the stator.

[6] Motors typically have more than one set of poles for a steadier and more powerful operation. A 2-pole machine is exemplified for the sake of simplicity.

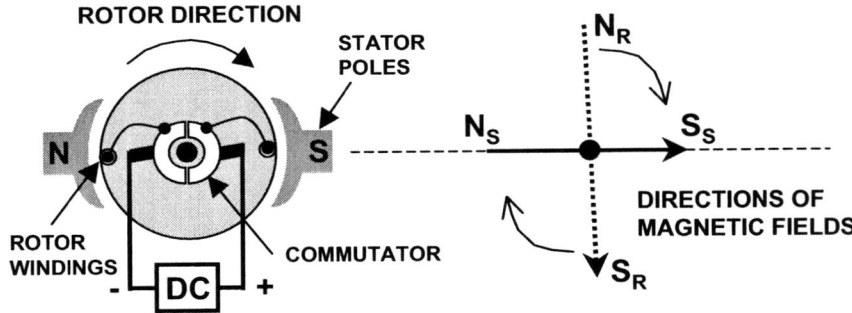

Figure 3.2 A fundamental 2-pole DC motor. The poles of the stator are formed from either a permanent magnet or windings connected to a DC voltage source. The resulting magnetic field of the stator ($N_S \rightarrow S_S$) is stationary. A DC voltage source supplies current to the rotor windings via a split-ring commutator which creates a magnetic field having a direction that is perpendicular to the plane of the rotor windings. The rotor turns as its magnetic field revolves due to the attraction of opposite poles (N_R-S_S and S_R-N_S) and the repulsion of like poles (N_S-N_R and S_S-S_R). The rotor's magnetic field maintains its direction until the commutator's air gap revolves past the DC source's contacts (carbon brushes), at which point, the polarity of the voltage applied to the rotor windings is reversed. As a result, the direction of the current flow through the rotor's windings reverses, thus causing the rotor's magnetic field direction to change (i.e., N_R and S_R exchange places). In order for the motor to function properly, the commutator and rotor windings must be positioned correctly so that the rotor field reversals occur at just the right times.

DC motors are well suited to variable speed control methods. Basically the motor's speed is proportional to the voltage applied to the rotor, given that the stator field strength is held constant. Thus, if the rotor voltage is increased by a factor of two, the speed will double. But a change in the stator field strength will also affect the speed. The speed is reduced when the stator field is strengthened and increased when the stator field is weakened. A number of methods are available for speed control depending upon the motor application requirements (e.g., motor starting, torque, etc.).

AC motors also operate on the principle of magnetic attraction and repulsion and are based on rotating magnetic fields. When an AC voltage is applied to the terminals of the stator pole windings, the resulting current flows through the windings in a sinusoidal fashion. As a result, a pulsating magnetic field is created between the north and south poles. The magnetic field reaches full intensity when the current is at its maximum value. When the winding current changes direction at every half cycle, the direction of the magnetic field between the pair of poles changes by 180° (i.e., the north and south poles exchange places). Thus, the magnetic field revolves in circular fashion. A bar magnet located between the poles would spin in order to align its fixed north and south poles with the stator's rotat-

Electric End Uses

ing magnetic field. The magnet would continue to turn in synchronism with the revolving stator field, and torque would be produced. In an AC motor, the "magnet" is the rotor. Although both magnetic fields rotate, the rotor field is stationary with respect to the stator field.

There are two types of AC motors: *synchronous* and *induction*. In a synchronous motor, an AC voltage is applied to the stator windings in order to create a rotating magnetic field. A DC voltage source is applied to the rotor windings in order to create a unidirectional magnetic field (similar to the bar magnet). The DC source is typically a DC generator, known as an *exciter*, which is usually powered by and mounted on the motor shaft.

A synchronous motor operates at constant speed, based on line voltage frequency, regardless of its mechanical load, as determined by

$$n = \frac{f \times 60}{\frac{P}{2}} \quad (3.3)$$

where n = mechanical speed in rpm
 f = line voltage frequency
 P = number of poles

Thus, the speed of a 4-pole synchronous motor supplied by an AC line frequency of 60 Hz would be 1,800 rpm.

The synchronous motor has virtually no starting torque. When the rotor is stationary, it essentially requires an auxiliary means of motoring at start up in order to bring the motor up to synchronous speed. The running torque of the motor depends on its horsepower and speed. The motor also has a maximum torque requirement, which if exceeded, will cause the motor to lose synchronism.

In an induction motor, an AC voltage source is applied to the stator only while the rotor is not connected electrically to the source. The current flow in the stator windings induces a current flow in the rotor as the lines of magnetic flux cut across the rotor windings. The induced rotor current produces a rotating magnetic field having the same number of poles as the stator. The rotor field rotates at the same speed as the stator field so torque is produced. But the speed of the rotor itself is less than the synchronous speed of the rotating magnetic fields.[7] The difference between the synchronous speed and the rotor shaft speed is referred to as *slip*, and it is usually expressed as a percent of the synchronous speed.

---

[7] If the actual speed of the rotor was the same as the rotating magnetic field, the rotor windings would not be moving through the magnetic lines of flux, and this action is required in order for induction to take place.

There are two types of induction motor rotor designs: *squirrel-cage rotors* and *wound rotors*. The squirrel-cage configuration consists of narrow conductive bars that are mounted in parallel with each other and which are situated around the axis of the rotor thereby forming a "cage." The ends of all of the bars are connected together at both ends of this assembly in order to form a complete loop through which the induced current flows. The use of conductive bars makes it possible for the rotor to handle a fairly large current. In general, squirrel-cage motors are most applicable when starting torque requirements are in the low- to medium-range. Squirrel-cage motors are widely used because they are relatively low cost, versatile, and durable.

A wound rotor motor has insulated conductor type rotor windings that are similar to the stator windings and have the same number of poles. Unlike the squirrel-cage motor, the windings of the wound rotor type are often made "accessible" by means of slip rings, mounted on the motor shaft, to which the rotor windings are connected. Carbon brushes in contact with the slip rings are connected to an external terminal block. These connections can then be used to insert a resistance into the rotor circuit in order to handle starting requirements and speed control. Wound rotor motors are well suited to applications which require a high starting torque. They also work particularly well in applications which require frequent starts. Although more functional, the wound rotor design is more expensive than the squirrel-cage rotor design.

Matching motor sizes and types to the characteristics of the loads that they are intended to drive is important for safety, economy, and energy conservation. High efficiency motors have become more popular because their higher installed cost can result in long-term operating savings. Variable or adjustable speed drives (ASDs) are also being used more, despite their relatively high cost, because of their flexibility to drive varying loads efficiently. ASDs are used to drive induction motors by varying the AC line frequency to control speed and torque for a more optimal operation of the load. ASDs basically consist of a rectifier which produces DC power, a filter for regulation and smoothing, and an inverter which produces AC power for the motor at various frequency. But while they improve the efficiency of a motor-driven process, ASDs produce harmonics (line voltage distortions) which may cause problems with the operation of other equipment.

Power factor is a major issue with respect to motors. A large amount of induction motor load gives rise to low power factor problems, particularly during times when the motors are running at less than full load. These motors require a reactive current component in order to produce the induced voltage in the rotor circuit; thus, induction motors must be supplied with VARs. Conversely, a synchronous motor can operate at a high power factor despite its loading. When the exciter supplies the exact amount of current to produce the required magnetic field, the motor operates at unity power factor. In fact, if the current is increased, the motor will become overexcited, and it will actually produce VARs which will flow out of the machine and on to the line. An unloaded synchronous motor may be used in this manner for the express purpose of improving power factor.

Chemical End Uses

The conversion of electrical energy to chemical energy has a much more narrow field of applications than the other basic forms of energy conversion. Electricity can be used to force an oxidation-reduction reaction in an electrolytic cell. A DC voltage source is applied to an anode (positive terminal) and a cathode (negative terminal) which are submersed in an electrolytic solution. The oxidation process occurs as negative ions in the solution are discharged as they give up electrons to the anode. The reduction process occurs as positive ions in the solution are discharged by acquiring electrons from the cathode. These simultaneous ion transfer processes represent a conduction of electricity between the terminals of the electrolytic cell. The chemical changes which result from applying electricity to the electrolyte is called *electrolysis*.

Electrolysis is used for decomposing various compounds. For instance, by adding a small amount of sulfuric acid to water to produce hydrogen and sulfate ions, the resulting electrolytic solution is a good conductor of electricity. By applying a voltage to the electrolyte, two molecules of water will release two molecules of hydrogen at the anode and one molecule of oxygen at the cathode. The two gasses that bubble about the terminals can be collected with a simple apparatus. A salt water solution can be processed in a similar manner to obtain chlorine gas. In addition, electrolysis can be used in specially constructed cells to extract metals, such as magnesium, potassium, and sodium, from their ores.

Electrolysis is also used for electroplating. The object to be electroplated, such as a spoon, is used as the cathode. The metal to be deposited, or plated, on the spoon, such as silver, is used as the anode. The cell would also contain an electrolytic solution of the plating metal, such as silver nitrate for this example. As electricity is applied, metallic ions are formed, the anode is used up, and the cathode (i.e., the spoon) is plated with silver. The amount of metal plated is a function of the amount of current which flows through the electrolytic cell. Faraday's Law states that 96,500 coulombs (6.023×10^{23} electrons) of electricity will reduce or oxidize one gram-equivalent of a substance, such as silver in the example above.[8]

Another electro-chemical application is battery charging. A number of rechargeable battery technologies are utilized in both small and large applications throughout all sectors. Rechargeable batteries are used extensively for a variety of transportation purposes, such as golf carts, forklifts, and airport passenger shuttles.

---

[8] In the late 1800s, Thomas Edison invented an electricity "meter" that worked on the principle of electroplating. One coulomb is the amount of electricity resulting from a current flow of one ampere for one second; therefore, a change of one gram-equivalent in the weight of one of the electrodes represents 26.8 amp-hours of current flow. By measuring the weight of either the anode or the cathode at the beginning and end of a month, the amount of electricity use during the period could be determined for billing purposes on the basis of the electrode's weight change. Innovative but crude, Edison's chemical-based meter was soon replaced by more practical electro-mechanical devices.

Small battery driven vehicles are in limited use today in business fleet operations, such as postal service delivery trucks. Battery applications are emerging in the general automotive consumer markets, particularly in the form of electric/gas propulsion hybrids. Much research and development in battery sciences is underway in an effort to make electric vehicles a practical alternative to the conventional internal combustion engine.

Electronic End Uses

Another major end-use type is electronics. Unlike the four principle end uses discussed previously, electronic devices are not designed to directly convert electrical energy into another basic energy form because they deal with electricity itself. Electronic end-use devices basically manipulate electricity by amplifying, modulating, or switching voltages and currents. For instance, a rectifier, such as in a power supply circuit, transforms an AC source into a DC output, while an inverter does just the opposite. In either case, it is still a fundamental process of electricity in and electricity out.

Electronic circuits usually interface with energy conversion devices. For example, a computer combines several of the basic end-use functions into one device. A major component of the computer is the monitor, which typically employs a liquid crystal display (LCD) technology. Screen illumination thus represents an energy conversion which produces light. Even more simply, computers have a few small glow lamps which, among other things, let the operator know that the machine is turned on and receiving power. One aspect of electronic devices is that they produce heat, but the heat in this case is losses. In order to deal with this unwanted heat and protect the machine, the computer employs mechanical energy, via a motor-driven fan, for heat dissipation. The computer also contains small motors for operation of the various disc drives which interface with electronic control circuits. A final mechanical device is a speaker that serves as an annunciator, which sounds when the user makes certain key strokes and which plays audible media.

Applications of electronic devices have been growing rapidly. More and more, computers and intelligent circuits are being integrated with ordinary appliances, equipment, and machines to improve performance and efficiency and to automate both simple and complex processes. Consumer electronics have brought a whole new wave of entertainment and convenience methods. Many devices consume low levels of power continuously to provide a ready-state mode, e.g., instant-on television sets. A number of electronic devices are battery-operated portables. But much of the electronic world depends on the outlet in the wall, and those outlets continue to see more and more electronic types of load everyday.

End-Use Energy Requirements

The primary retail customer classes -- residential, commercial, and industrial – all utilize electric heating, lighting, motors, and electronic devices to meet their spe-

Electric End Uses

cific end-use requirements. But the end-use deployment needs of each class vary. For example, lighting is typically viewed as a lesser load for a residential customer, particularly with respect to the air conditioning or water heating requirements. However, lighting represents a significant end use for the commercial class. In addition, the heat from lighting is a major element of the "cooling load" which must be air conditioned in large commercial buildings. With the industrial class, lighting is almost a trivial load compared to the end-use devices associated with heavy manufacturing processes. Thus, while lighting is necessary in all customer classes, its energy intensity relative to other end uses is widely varied.

Each customer class can be described in terms of its mix of end-use energy requirements, with the end uses represented through various electrical appliances and other equipment. While some aspects of end-use energy utilization may be common between similar types of customers, other aspects may vary widely due to the influence of such factors as weather, location, and socio-economic conditions. Thus, a greater segmentation of customers can yield a much more accurate depiction of the various energy use characteristics. A reliable understanding of the energy use profiles of the different customer types is essential for designing price structures that efficiently recover the costs of electricity while effectively targeting end uses characteristics.

Residential Sector

Residential customers represent a fairly homogeneous group because, as a whole, they own comparable load devices and generally have similar lifestyles. In a very broad sense, the main difference between residential customers is a matter of scale. Larger homes have bigger heating and cooling equipment than smaller homes. Larger customers often have more than one of a single type of appliance. The average home in the United States has about 1.2 refrigerators, which means that several homes report having two or more such devices.

Electricity utilization in the residential sector can generally be divide into two major categories: *base use* and *seasonal use*. Overall, residential base use is non-weather sensitive and is generally uniform from month to month. The types of appliances and equipment typically categorized as base use load devices are shown in Table 3.2. Water heating is sometimes excluded from base electricity usage because it is a fairly competitive end use that is often served by another type of fuel, such as natural gas.

Residential seasonal use is highly weather sensitive as it represents heating and cooling requirements. The average annual usage for central and room air conditioners and for main (primary) and secondary heat sources are shown in Table 3.2. The data provided is based on adjusted 1993 weather which was nearly characteristic of the 30 year average in terms of both heating and cooling *degree days* (DDs). DDs are a measure of how hot or cold a location is over time relative to a base temperature of 65°F. The DDs for a single day are represented by the absolute difference between the day's average temperature and 65°F. Thus, for example, a day having an average temperature of 25°F would yield 40 heating DDs.

Table 3.2 Typical Residential End-Use Devices (U.S.)

End-Use Device	Average Annual kWh Usage (Per Device)	Percent of U.S. Households with End-Use Devices
Base Use Devices:		
Refrigerators	1,155	119.77%*
Lighting	940	100.00%
Televisions	360	205.28%*
Clothes Dryers	875	56.63%
Freezers	1,204	34.58%
Ranges/Ovens	458	60.35%
Microwave Ovens	191	84.16%
Clothes Washers	99	77.12%
Dishwashers	299	45.24%
Swimming Pool Pumps	2,022	4.76%
Well Pumps	228	13.46%
Hot Tub/Spa Heaters	482	1.97%
Dehumidifiers	370	9.42%
Personal Computers	77	23.40%
Waterbed Heaters	960	15.11%
Other Home Appliances	1,364	100.00%
Water Heating	2,671	38.30%
Seasonal Use Devices:		
Air Conditioning:		
Central Air Conditioners	2,667**	42.44%
Room Air Conditioners	738**	34.27%*
Space Heating:		
Main Heat Source	4,541**	25.88%
Secondary Heat Source	400**	12.53%

* Devices which some customers often report owning more than one unit.

** Actual results have been adjusted to reflect 30 year average weather usage.
  - Actual 1993 heating DDs were 0.38% above normal.
  - Actual 1993 cooling DDs were 6.95% above normal.

Source: Energy Information Administration's 1993 Residential Energy Consumption Survey (RECS).

Electric End Uses

The energy utilization of a base use type of load device can be estimated fairly accurately by first determining the hours of expected operation and then by multiplying these hours by the device's nameplate rating in terms of kW. For example, a 1,600 watt microwave oven that is operated for 20 minutes each day on average would consume approximately 195 kWh per year. An estimation of the energy utilization of seasonal use devices is much more complex. Location is very important because of geographical differences in heating and cooling DDs. The thermal characteristics of the building structure is another key element. The orientation of the building, degree of glass usage, color of walls and roof, etc. all have a bearing on the solar gain. In addition, the operation of auxiliary equipment such as fans and pumps must be analyzed in order to determine the entire energy use of the HVAC equipment. Detailed computer models have been developed which provide reasonable estimates of building energy use based on a wide variety of input variables.

Table 3.3 represents the average end-use energy utilization of a typical U.S. residential customer. All of the end-use devices shown in Table 3.2 are included, but the typical energy usage of each device has been multiplied by the percent of households with the device in order to derive the weighted energy utilization profile of an average customer.[9] The total annual usage of 9,965 kWh per year represents the average U.S. customer based on 96.6 million residences, many of which utilize fossil fuels to operate one or more appliances. In comparison, the Energy Information Administration (EIA) reports in its 1993 Residential Energy Consumption Survey (RECS) that the average annual electrical energy usage of three specific U.S. home types are:

1. 15,639 kWh for all-electric homes;
2. 10,700 kWh for homes which have either electric space heating or electric water heating; and
3. 7,152 kWh for homes with both nonelectric space and water heating.

Location is a key factor to the amount of electricity usage that can be expected as indicated by the average household usage in different U.S. regions:

- Northeast   7,071 kWh per year
- Midwest   9,327 kWh per year
- South   13,212 kWh per year
- West   8,131 kWh per year

Climate conditions, the price and availability of competitive fuels, and other location factors have a major influence on annual electricity usage.

---

[9] Individual devices with a usage level less than 2.5% of the total are included under the category of Miscellaneous.

Table 3.3 Average Annual Residential Energy Use (U.S.)

End-Use Device	Average Annual kWh Usage (Per Household)	Percent of Total Household kWh Usage
Base Use Devices:		
Refrigerators	1,383	13.9%
Lighting	940	9.4%
Televisions	739	7.4%
Clothes Dryers	495	5.0%
Freezers	416	4.2%
Ranges/Ovens	276	2.8%
Miscellaneous	2,083	20.9%
Subtotal	6,332	63.5%
Water Heaters	1,023	10.3%
Seasonal Devices:		
Air Conditioners	1,385	13.9%
Space Heaters	1,225	12.3%
Total	**9,965**	**100.0%**

Source: Energy Information Administration's 1993 Residential Energy Consumption Survey (RECS).

Figures 3.3 through 3.5 illustrate the average annual energy usage of major residential appliances by U.S. region.[10] Figure 3.3 indicates that the energy intensity of both central and room air conditioners is pronounced in the southern areas as compared to all other regions. In fact, the annual usage in all other regions is well below the national average. Elevated humidity conditions coupled with high average temperatures contribute to the AC energy requirements in the south.

---

[10] *Northeast* includes New England (CT, ME, MA, NH, RI, and VT) and Middle Atlantic (NJ, NY, and PA). *Midwest* includes East North Central (IL, IN, MI, OH, and WI) and West North Central (IA, KS, MN, MO, NE, ND, and SD). *South* includes South Atlantic (DE, DC, FL, GA, MD, NC, SC, VA, and WV), East South Central (AL, KY, MS, and TN), and West South Central (AR, LA, OK, and TX). *West* includes Mountain (AZ, CO, ID, MT, NV, NM, UT, and WY) and Pacific (AK, CA, HI, OR, and WA).

Electric End Uses 79

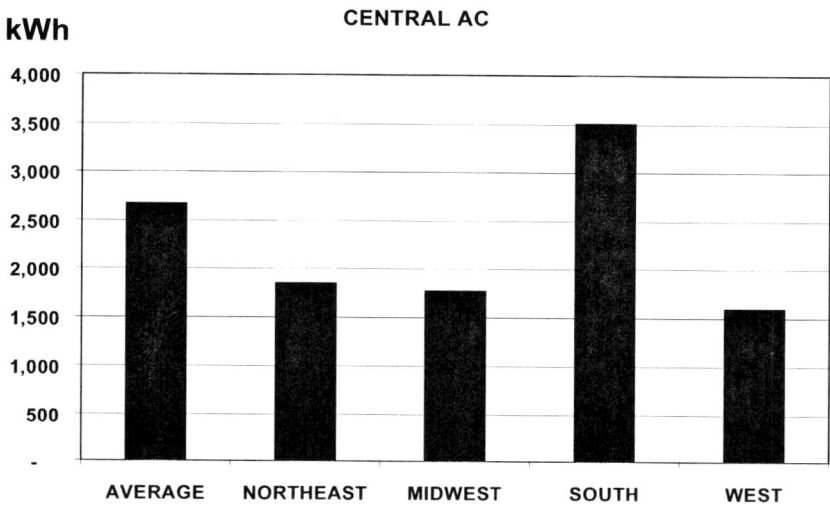

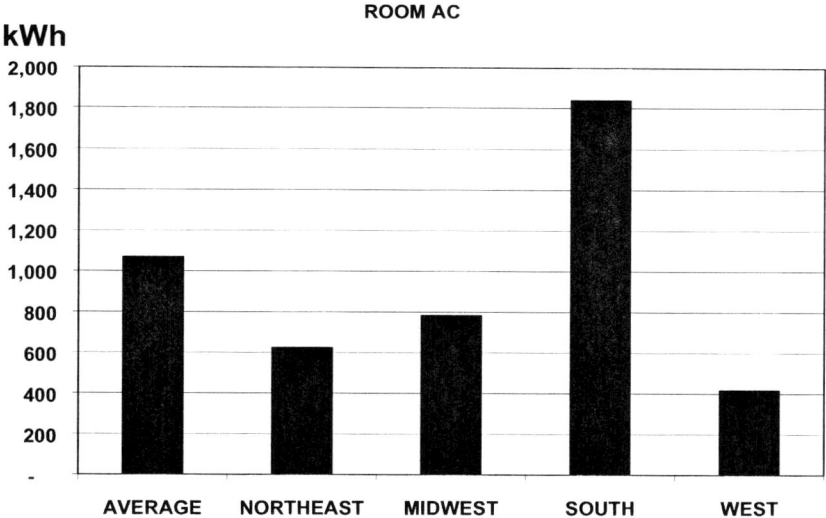

Figure 3.3 Average annual kWh use of air conditioners by U.S. region. Source: Energy Information Administration's 1993 Residential Energy Consumption Survey (RECS).

80   Chapter 3

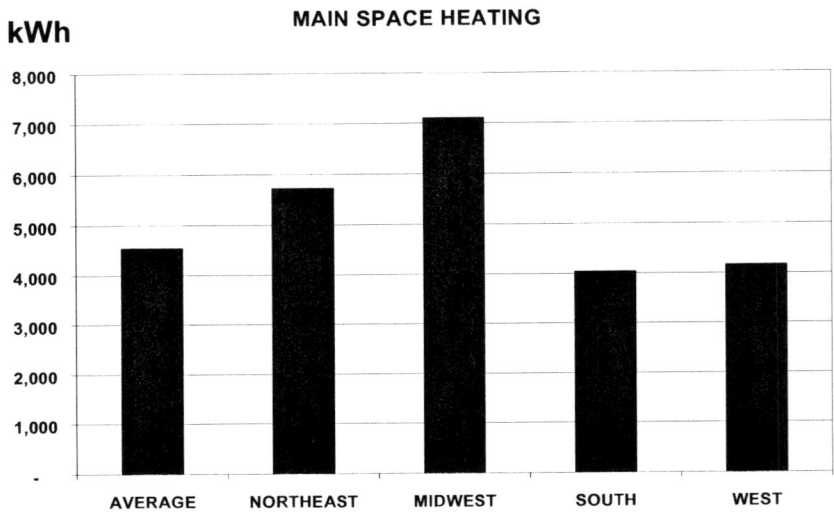

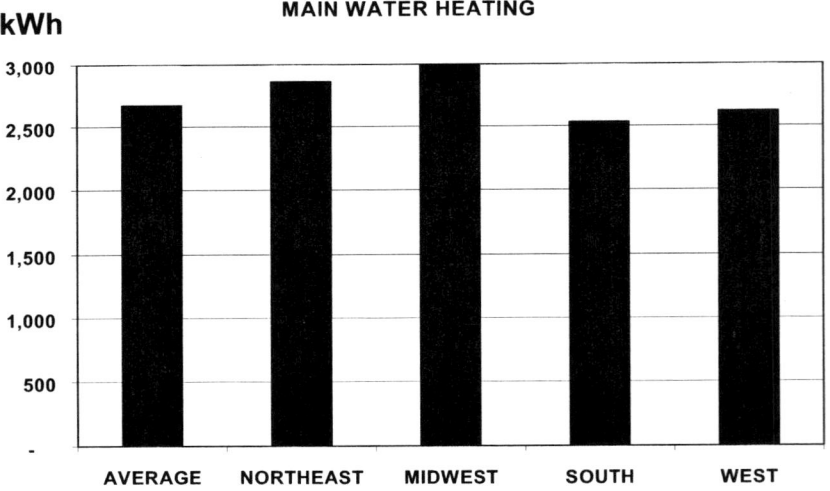

Figure 3.4 Average annual kWh use of space and water heaters by U.S. region. Source: Energy Information Administration's 1993 Residential Energy Consumption Survey (RECS).

Electric End Uses

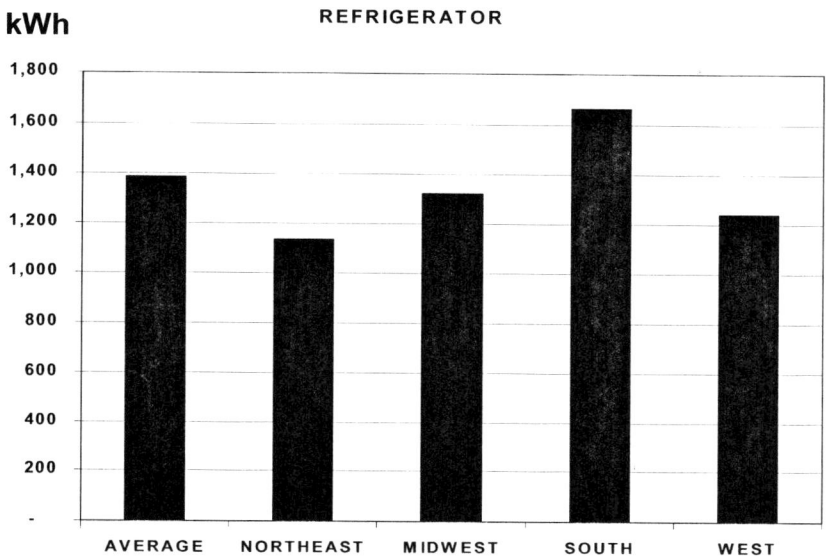

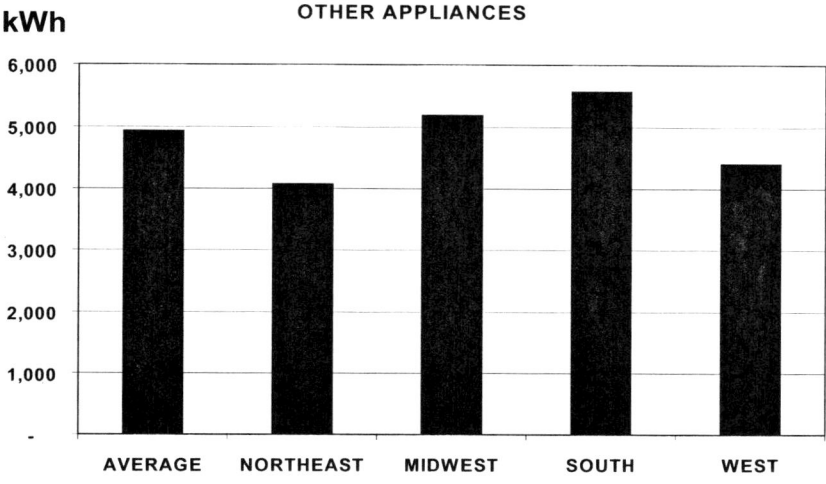

Figure 3.5 Average annual kWh use of refrigerators and other appliances by U.S. region. Source: Energy Information Administration's 1993 Residential Energy Consumption Survey (RECS).

The upper chart in Figure 3.4 shows the geographical differences in the annual energy consumption of main or primary electric space heating. The Midwest region has the greatest electric heating usage due to arctic air masses which typically plunge deep into the central portions of the United States throughout the winter season. The northeast region also experiences fairly cold winters and thus exhibits heating energy usage in excess of the national average. Heating requirements in both the south and the west are less than the national average due to more moderate winter temperatures.

The lower chart in Figure 3.4 illustrates the annual energy consumption characteristics of electric water heaters. The variation in electricity use between regions is minimal, as should be expected, since water heating is generally unaffected by outside temperatures. Energy usage in the Midwest is highest at about 12% above the national average and lowest in the south at about 5% below the national average.

Figure 3.5 shows the energy consumption requirements of base use appliances with refrigerators exhibited separately in the upper chart. The variation in use between regions is within about ± 20% compared to the national average. The usage in the south is higher than the average while all other regions are below average. However, the average number of refrigerators per household reported for the south region is 1.1 compared to the national average of 1.2. The Midwest and the west both report 1.2 refrigerators per household while the northeast reports 1.1 per home. A number of factors could cause such an apparent inconsistency for the south region. One might rationalize that refrigerators in the southern areas are accessed more often for cold drinks and ice for refreshments because of overall warmer climate conditions, thus causing a higher energy usage. Or perhaps, the inventory of existing refrigerators is older and less energy efficient than in other regions. The information necessary to determine the underlying causes of this observed trait are not readily available.

The lower chart in Figure 3.5 represents the annual energy usage of a wide variety of miscellaneous household appliances and electrical devices. The variation in usage between the regions with respect to the national average is within about ± 17%, which appears to be reasonable given the broad diversity of device types included. Since no one appliance in this group dominates the others in terms of energy usage, the exact reasons for the variation between regions cannot be quantified definitively.

Thus far, the discussion of end-use energy requirements has been based on average annual results. It is also important to understand how end-use energy requirements change across the year, particularly with respect to seasonal loads. Figure 3.6 illustrates an example of the monthly kWh consumption for various residential customer class end-use based segments. The "Standard" classification consists of base appliances and air conditioning, which is minimal but noticeable in the summer months. However, the electric space heating energy requirements are much more extreme. Such a pattern suggests a locale with high heating DDs and low cooling DDs.

Electric End Uses

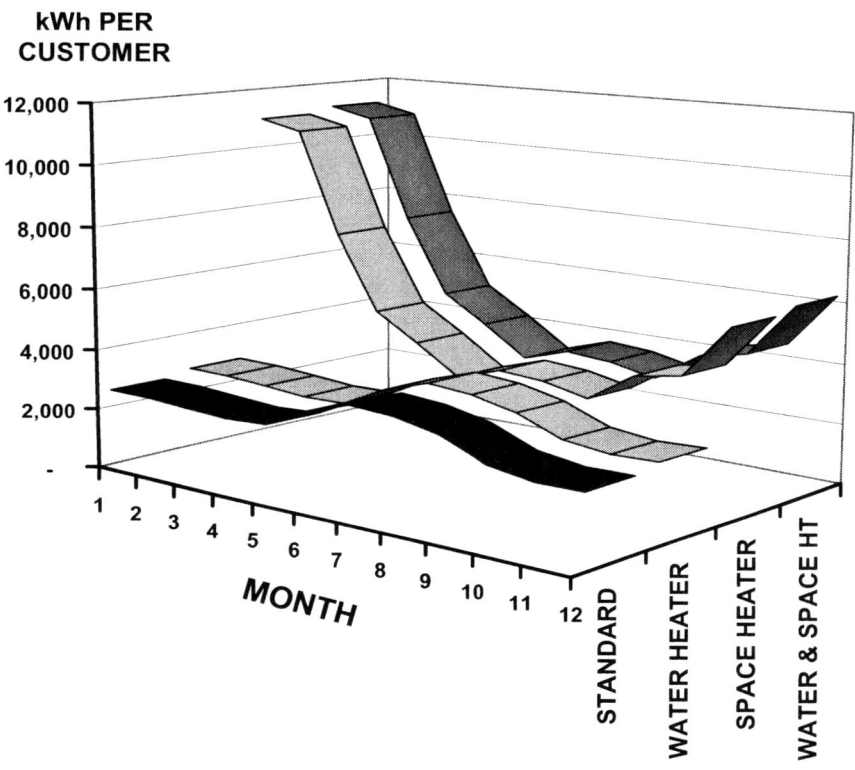

Figure 3.6 Example of monthly average kWh usage for different residential customer types. The "Standard" classification includes base appliance use and air conditioning, which is minimal because of the low amount of cooling degree days for the area. The other classifications include Standard use plus other end use(s) as indicated.

All of the energy usage statistics discussed above provide a good foundation for understanding the end-use requirements of residential customers. They are fairly well representative of energy usage under normal weather conditions. Actual weather circumstances could significantly impact a customer's monthly and annual HVAC energy usage. Specific appliance ownership and lifestyle practices also have a major influence on a particular customer's electrical energy requirements. End-use data should be well understood as it is often derived from surveys and sample measurements, which can be expected to have a statistical variance.

Commercial Sector

The end-use energy requirements of commercial customers are much more diverse than residential customers. A small commercial customer, such as Jimmy's Deep South Coffee Café located in a downtown area, might consist of only a pair of coffee brewing machines, a few lights, a window air conditioner, a CD player for enhanced ambience, and perhaps a couple of electric resistance heaters for those occasional cool days. While these types of devices are not uncommon to the basic household, their operating requirements are likely to be much different due to the nature of the business. For instance, the coffee shop may open only on weekdays in the early morning hours in order to serve the commuting working crowd and then close in the late afternoons.

In stark contrast, a large commercial operation, such as the Sears Tower in Chicago, Illinois, represents a very high density of a wide variety of electrical end-use devices, including banks of lighting systems, large-scale HVAC equipment, and a host of business machines. Large building operators typically startup the cooling system in the morning in anticipation of the heating load from lights, machines, and workers during the business hours. The HVAC system might be throttled back in the early evening, but the lights may remain on well into the night as office cleaning crews make their rounds.

In between the small and the large commercial customers lies a very diverse group of business functions, each with their own unique end-use energy requirements. Don's Grocery Warehouse requires a considerable amount of refrigeration to maintain frozen food products. Perry's Homestyle Cafeteria utilizes infrared lamps to keep fried foods hot and large quantities of hot water for steam tables and for washing cookware, dishes, and utensils. Eddie's Amusement World utilizes motor drives throughout the park to power the latest and greatest rides. Sharon's Jewelry Boutique has a large quantity of special high-definition lighting to showcase all of her custom-made adornments. Each of these business examples features a different emphasis on the major end-use device employed and its operating requirements. For instance, frozen food lockers run constantly while cafeteria cooking and serving periods follow in line with general mealtimes. Weekends are prime time for amusement park thrill seekers, but the jewelry store lights are turned off from early Saturday evening until mid-morning on Monday.

Because of the wide diversity of load devices and operations, the commercial class of customers as a whole is much less homogeneous than the residential class. The variation in scale alone, even between customers in the same line of business, accounts for different annual energy needs. Thus, when evaluating commercial end-use characteristics, it is helpful to group customers by similar business types and to relate energy requirements to a common parameter, such as floor space. The entries in Table 3.4 provide relative electricity intensities of major end uses for the average customer in various commercial customer segments. In considering these results, one should note that these figures are averages and thus do not differentiate such key aspects as weather sensitivity, end-use technology types, or percent of end-use device ownership.

Table 3.4 Commercial Sector Electricity Intensities (U.S.)

Annual kWh Per Square Foot of Building Floor Space

Customer Group	Space Heating	Space Cooling	Ventilation	Water Heating	Lighting
Education	0.5	1.4	0.5	0.2	4.7
Food Sales	N/A	3.9	1.3	0.7	9.9
Food Service	1.1	5.7	1.6	1.1	10.8
Health Care	0.4	2.7	2.1	0.3	11.5
Lodging	0.9	2.3	0.5	1.0	6.8
Mercantile and Service	0.6	1.7	0.7	0.1	6.9
Office	0.6	2.6	1.5	0.2	8.2
Public Assembly	0.8	1.8	1.0	0.3	6.4
Public Order and Safety	0.1	1.8	0.7	N/A	4.8
Religious Worship	0.4	0.6	0.3	0.1	1.5
Warehouse and Storage	0.2	0.3	0.1	0.1	3.0
Other	0.6	2.5	2.5	0.1	7.9

Customer Group	Cooking	Refrigeration	Office Equipment	Other	Total
Education	0.1	0.3	0.4	0.3	8.4
Food Sales	0.3	32.5	0.4	1.8	54.1
Food Service	1.8	9.3	0.8	4.0	36.0
Health Care	0.1	1.4	4.5	3.5	26.5
Lodging	0.1	0.7	1.1	1.7	15.2
Mercantile and Service	0.1	0.3	0.8	0.6	11.8
Office	0.0*	0.1	4.5	1.2	18.9
Public Assembly	0.1	0.5	0.7	1.0	12.7
Public Order and Safety	0.0*	0.1	1.7	2.1	11.3
Religious Worship	0.1	0.2	0.1	0.3	3.4
Warehouse and Storage	0.0*	0.5	1.4	0.8	6.4
Other	N/A	0.2	4.5	3.9	22.1

* Value rounds to zero in the units displayed.
N/A - Data withheld because less than 20 buildings reported or the relative standard error is greater than 50%.

Source: Energy Information Administration's 1995 Commercial Buildings Energy Consumption Survey (CBECS).

Industrial Sector

The industrial sector consists of factories or plants which manufacture and assemble products. Plant outputs may be finished goods, such as completed automobiles, or they may be intermediate components, such as crankshafts which fit the engines that go into the automobiles. Like the commercial sector, industrial customers are very diverse in terms of end-use needs and operations, even within the same industrial classification. An industrial operation can be very small, such as Bob's Custom Furniture Shop, or very large, such as the Eastman Kodak plant in Rochester, New York. However, there is no direct correlation between energy utilization and the area of a factory. For example, as reported in EIA's 1994 Manufacturing Energy Consumption Survey (MECS), an average fruit and vegetable canning operation has about 3 acres (43,560 ft^2 per acre) under roof while an average meat packing plant covers approximately half that size. But the meat packing plant has an average energy intensity of about 80 kWh per ft^2, which is three times the average energy use of the canning plant.

U.S. industrial and commercial operations are often categorized by product or activity types under an economic accounting system such as the Standard Industrial Classification (SIC) codes. The manufacturing sector is classified by 20 major groups as represented by

20	Food and Kindred Products
21	Tobacco Products
22	Textile Mill Products
23	Apparel and Other Finished Products Made from Fabrics
24	Lumber and Wood Products
25	Furniture and Fixtures
26	Paper and Allied Products
27	Printing, Publishing and Allied Industries
28	Chemicals and Allied Products
29	Petroleum Refining and Related Industries
30	Rubber and Miscellaneous Plastics Products
31	Leather and Leather Products
32	Stone, Clay, Glass, and Concrete Products
33	Primary Metal Industries
34	Fabricated Metal Products
35	Industrial and Commercial Machinery and Computer Equipment
36	Electronic and Other Electrical Equipment and Components
37	Transportation Equipment
38	Instruments; Photographic, Medical, and Optical Goods; Clocks
39	Miscellaneous Manufacturing Industries

In 1997, the North American Industrial Classification System (NAICS) replaced the SIC system. Business reports may include data and information based on either or both of these classification systems. A high level listing of NAICS and SIC codes is contained in Appendix C.

Industrial energy end-uses can be considered in terms of main process and facility uses. Process uses relate to the actual manufacturing operations, such as a rolling mill, while facility uses relate to building systems, such as lighting and space conditioning. As seen in Table 3.5, the majority of industrial electricity use occurs within the processes. By far, the greatest amount of electricity consumed by manufacturing industries is used to operate motors to drive machines of all types. Table 3.6 portrays the variation of end-use electricity requirements between the major manufacturing SIC groups. Variations in end-use requirements also exist between sub-classes of operations within the same industry group. For example, in the primary metals group (SIC code 33), the amount of annual kWh used in electro-chemical processes by the average primary aluminum production plant is nearly 59 times the amount of electricity consumed in the electro-chemical processes of an average primary copper smelting and refining facility.

Table 3.5 Industrial Sector End-Use Electricity Requirements (U.S.)

Percent of kWh Consumption for SIC Groups 20-39

	Percent of Main Uses	Percent of Total Use
Process Uses:		
Electric Process Heating	11.84	9.43
Process Cooling/Refrigeration	6.40	5.09
Machine Drives (Motors)	67.80	53.98
Electro-Chemical Processes	13.18	10.49
Other Process	0.79	0.63
Total Process Use	100.00	79.62
Facility Uses:		
HVAC	47.30	7.47
Lighting	40.58	6.41
Support	10.24	1.62
On-Site Transportation	0.86	0.14
Other Nonprocess	1.02	0.16
Total Facility Use	100.00	15.79
Other Uses	100.00	4.59
Total Use		100.00

Source: Energy Information Administration's 1994 Manufacturing Energy Consumption Survey (MECS).

Table 3.6 End-Use Electricity Requirements By Industry Type (U.S.)

Percent of kWh Consumption by SIC Classification

SIC CODE:	20	21	22	23	24	25	26	27	28	29
Process Uses:										
Electric Process Heating	3.99	0.83	4.40	2.97	4.84	3.69	1.91	2.66	3.29	2.27
Process Cooling/Refrigeration	22.94	2.29	6.05	1.21	0.49	1.26	1.37	4.09	6.37	5.30
Machine Drives (Motors)	50.08	55.61	57.01	40.14	71.13	53.64	80.17	45.10	61.63	80.03
Electro-Chemical Processes	0.16	0.00	0.16	N/A	0.06	0.45	0.86	0.29	17.89	0.07
Other Process	0.25	4.20	0.67	N/A	0.74	1.20	0.98	0.16	0.38	0.17
Subtotal	77.41	62.93	68.29	44.31	77.25	60.24	85.29	52.30	89.56	87.85
Facility Uses:										
HVAC	7.01	25.92	13.96	23.32	4.53	12.31	3.81	19.83	3.78	3.32
Lighting	6.58	8.73	9.69	16.76	6.44	15.55	3.74	12.60	2.89	2.74
Support	1.82	2.04	2.01	3.21	1.26	2.88	0.94	4.53	0.81	0.80
On-Site Transportation	0.23	0.00	0.15	0.13	0.08	0.21	0.06	0.26	0.03	0.03
Other Nonprocess	0.24	0.00	0.05	N/A	0.09	0.04	0.07	0.20	0.13	0.02
Subtotal	15.86	36.69	25.86	43.43	12.40	30.99	8.63	37.42	7.64	6.92
Other Uses	6.73	0.38	5.84	12.27	10.35	8.77	6.08	10.28	2.79	5.23
Total Uses	100%	100%	100%	100%	100%	100%	100%	100%	100%	100%

SIC CODE:	30	31	32	33	34	35	36	37	38	39
Process Uses:										
Electric Process Heating	16.39	3.37	23.30	23.99	12.55	9.67	16.54	8.91	6.81	11.23
Process Cooling/Refrigeration	6.98	2.40	2.33	0.83	2.19	2.79	5.76	4.12	7.84	5.00
Machine Drives (Motors)	52.10	58.89	58.97	29.35	49.30	29.17	39.64	27.95	36.91	53.98
Electro-Chemical Processes	0.22	0.00	0.23	36.13	4.51	0.34	4.05	1.54	0.85	0.84
Other Process	0.47	*	0.03	0.47	0.81	0.49	0.77	2.80	1.70	0.29
Subtotal	76.16	64.66	84.86	90.77	69.35	57.18	56.29	57.01	45.16	54.27
Facility Uses:										
HVAC	8.46	11.90	4.62	3.08	10.11	17.36	21.90	17.86	25.80	16.80
Lighting	7.67	11.06	5.43	3.52	10.98	14.32	13.31	16.26	19.25	16.07
Support	1.97	1.68	1.31	0.84	2.11	3.96	3.93	3.41	5.70	3.73
On-Site Transportation	0.24	*	0.13	0.10	0.34	0.24	0.24	0.41	0.28	0.27
Other Nonprocess	0.08	N/A	0.08	0.11	0.09	0.33	0.11	0.64	1.36	0.93
Subtotal	18.42	24.64	11.57	7.65	23.62	36.22	39.50	38.57	52.38	37.80
Other Uses	5.42	10.70	3.58	1.58	7.03	6.60	4.21	4.41	2.46	7.93
Total Uses	100%	100%	100%	100%	100%	100%	100%	100%	100%	100%

* Value rounds to zero in the units displayed.
N/A - Data withheld because the relative standard error is greater than 50%.

Source: Energy Information Administration's 1994 Manufacturing Energy Consumption Survey (MECS).

Electric End Uses

Based on EIA's 1994 Manufacturing Energy Consumption Survey (MECS), the three most electricity intensive manufacturing groups are Chemicals and Allied Products which used 22% of all annual manufacturing electricity requirements, Primary Metal Industries which consumed 17%, and Paper and Allied Products which utilized 13%. The Food and Kindred Products group used 7%, and all other groups generally used 5% or below.

3.3 END-USE VALUE

The end-use requirements of consumers relate to value. People need heat in the wintertime to remain comfortable, if not to remain alive during frigid conditions. Thus, people overwhelmingly place a very high value on having heat, at least in the winter season. Conversely, people place a high value on having the ability to be cool in the summer, particularly in hot climates.

People place a high value on being able to watch television. Table 3.2 indicates that the average household has 2+ television sets. Imagine the value of a television screen to the avid horseracing fan who, in the late afternoon on the first Saturday in May, is seated front and center in anticipation of watching the annual running of the Derby in Louisville, Kentucky. Consider further that at the end of the post parade, just prior to the race, the picture blinks off due to a malfunction of the television, or a cable TV technical difficulty, or a power outage. The exact cause for the interruption at this point is somewhat irrelevant because, in a little over two minutes, the greatest race in sports will be over. The need, or perhaps the desire, for an operable television is urgent at this point in time.

A review of Table 3.2 reveals which domestic, electricity-powered end-use devices have a high value to people as defined by the number of households that report owning them. In addition to television sets, lighting and refrigerators have high value. Microwave ovens have made preparing meals simpler and faster and thus provide a great deal of value to people with busy schedules. Not everyone owns a swimming pool, but for those who do, a circulating pump is a critical component of the filtration system. As a whole, the base use devices have a fairly high value all during the year.

However, an observation of the seasonal use devices indicates that homes have many more air conditioners than electric heaters. This condition does not imply that cooling is more important than heating (quite possibly the opposite is true since virtually every household has a heating source of one form or another). What is implied is that the heating end-use need can be met by a variety of alternative fuels, whereas the only practical way to air condition a home, at least at this time, is with electricity. Thus, electricity easily captures virtually all of the end-use value in summer while it must compete for end-use market share in the winter.

A profile of residential end-use value from an electricity perspective is suggested in Figure 3.7. The high value of base use devices exists each and every month because most of these devices are employed on a daily or weekly basis.

The value of water heating is also constant across the year, but, like space heating, it is an end use for which electricity must compete. The value of electricity as an energy source depends on the nature of the actual end uses.

Commercial and industrial end uses also have a value system. As noted previously, a considerable amount of electricity is used for industrial drives. However, electricity does not serve industrial drives exclusively as some drives are powered by steam turbines or fossil-fueled reciprocating engines. Lighting is an obvious captive market for electricity. But a number of process heating applications can be met economically by other heating methodologies. While impractical in residential applications, commercial and industrial cooling can sometimes be met by alternative thermal solutions. In the end, electricity serves both niche markets and competitive markets, depending upon the end use as well as the availability and economics of alternative technologies.

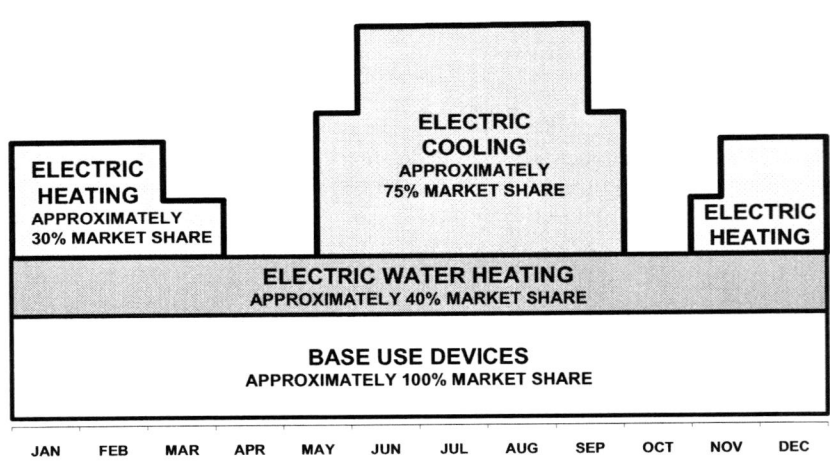

Figure 3.7 **The concept of end-use value of residential electrical devices over a year.** In this example, the *market* is defined to be all residential customers. Electricity exclusively supplies all of the energy requirements for base use devices such as refrigerators, lighting, and televisions. In addition, electricity supplies virtually all of the cooling load that is available in the market. But not every customer has a need or desire for air conditioning, thus a portion of the market is naturally left "unserved." On the other hand, nearly all of the customers desire space and water heating, but competition with fossil fuels leaves electricity with well less than a full market share for these end uses. Seasonal end-use value depends on weather, and its intensity increases with temperature extremes. Space heating and cooling needs have little value in the opposing seasons, but customers make a long-term commitment when they invest in specific technologies which will provide these end-use requirements.

A customer's need for energy is driven by discrete end-use requirements. In fulfilling these requirements, the customer must determine which technologies will be used to supply the end uses. The selection of a particular technology is often based on economics in which the life-cycle costs of one method is compared to other methods. For instance, the installed cost of a heat pump may be higher than a gas-fired furnace and an air conditioner. But if the cost of electricity to operate the heat pump in the winter was cheaper than the cost of natural gas over time, the net present value comparison would select the electric heating option.

Other factors also contribute to selecting a technology. Some people feel that an electric-driven heat pump does not afford the same level of comfort as provided by natural gas or propane. On the other hand, some people feel that gas is unsafe because of a potential for explosion or carbon monoxide build up. Thus, noncost value perceptions may weigh more heavily than the economics in the end-use technology decision making process.

The end-use value profile shown in Figure 3.7 closely resembles the typical load shape experienced by a number of utilities. End-use operations collectively forge the load patterns that power systems must be designed to serve. The sheer numbers of weather-sensitive residential and commercial customers have a major influence on the magnitude and shape of the system peak loading characteristics. In this example, air conditioning drives the annual system peak load.

3.4 CUSTOMER END-USE RESEARCH

Many utilities conduct market research on an ongoing basis in order to learn more about their customers and to determine what types of marketing programs and electric service pricing options that customers find of value. As part of this research, customer end-use load devices are often investigated. These studies typically include tracking of customer end-use trends and the development of end-use load shape characteristics.

End-Use Trends

Electric end-use load devices, such as appliances, are similar to other consumer products in terms of their life cycles. Figure 3.8 illustrates a classical time-based profile of the lifespan of new products and technologies. When a new technology is introduced to the market, it may take a while to "catch on." For example, when compact discs (CDs) first entered the consumer music marketplace, vinyl records and cassette tapes represented the prevailing music format. The prices of CD players were initially rather expensive, and the inventory of CDs was quite meager. A few years went by with only a minimal consumer acceptance of the new music format. But over time, the prices of CD players started to go down and the stock and diversity of CD selections began to increase. The popularity of CDs then began to grow at a very fast rate, and they achieved a high level of market penetration.

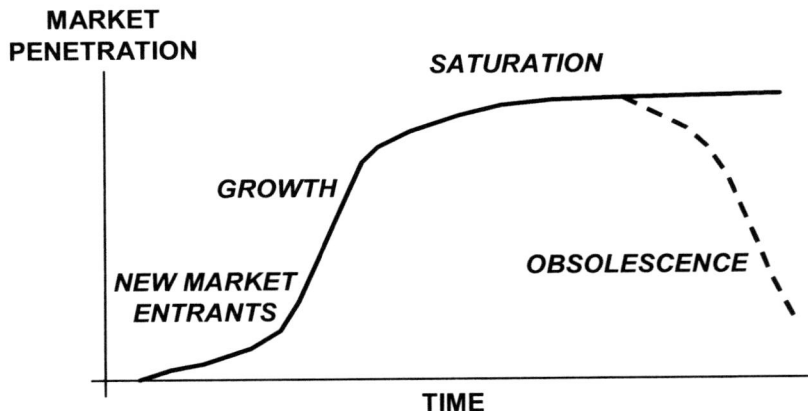

Figure 3.8 **Phases of a technology life cycle.** New technologies, such as appliances, progress through a series of life cycle phases. After entering the market, a technology that is well accepted by consumers grows in terms of market penetration until a saturation level is reached. Over time the technology may be made obsolescent by a new technology and disappear from the market. The portion of the cycle from inception to saturation follows an elongated "S" shape.

The curve in Figure 3.8 is characteristic of an elongated "S" shape from the point of market inception to market saturation. Different technologies follow different "S" shapes depending on market acceptance characteristics. Some technologies experience a high rate of growth immediately while others gain acceptance at a rather slow pace. At some point in time, a product may be rendered outmoded by an improved version or a new technology altogether. Such is the case with the CD music format, which took the market from cassette tapes and vinyl records and is now threatened itself by the wide acceptance of Mp3 players.

Electric end-use devices follow the same types of life cycle phases and the characteristic "S" shape of market penetration change as other technologies. The graphs in Figures 3.9 through 3.11 show a 25 year history of market penetrations of various residential appliances for customers of a moderate size U.S. electric utility. New market entrants represented by heat pumps, microwave ovens, and home PCs are shown in the upper graph of Figure 3.9. Heat pumps were just being introduced in the early 1970s, and after 25 years were accepted by less than 30% of the market because of competition with other technologies (including electric resistance heat). Microwave ovens, which entered the market in the mid 1970s, quickly grew to a high level of saturation by 1995. Home PCs introduced in the late 1980s had also reached about a one-third market penetration by the end of the period. Given the reduction in PC prices in the late 1990s, acceptance of home computers should be expected to continue at a rapid rate.

Electric End Uses

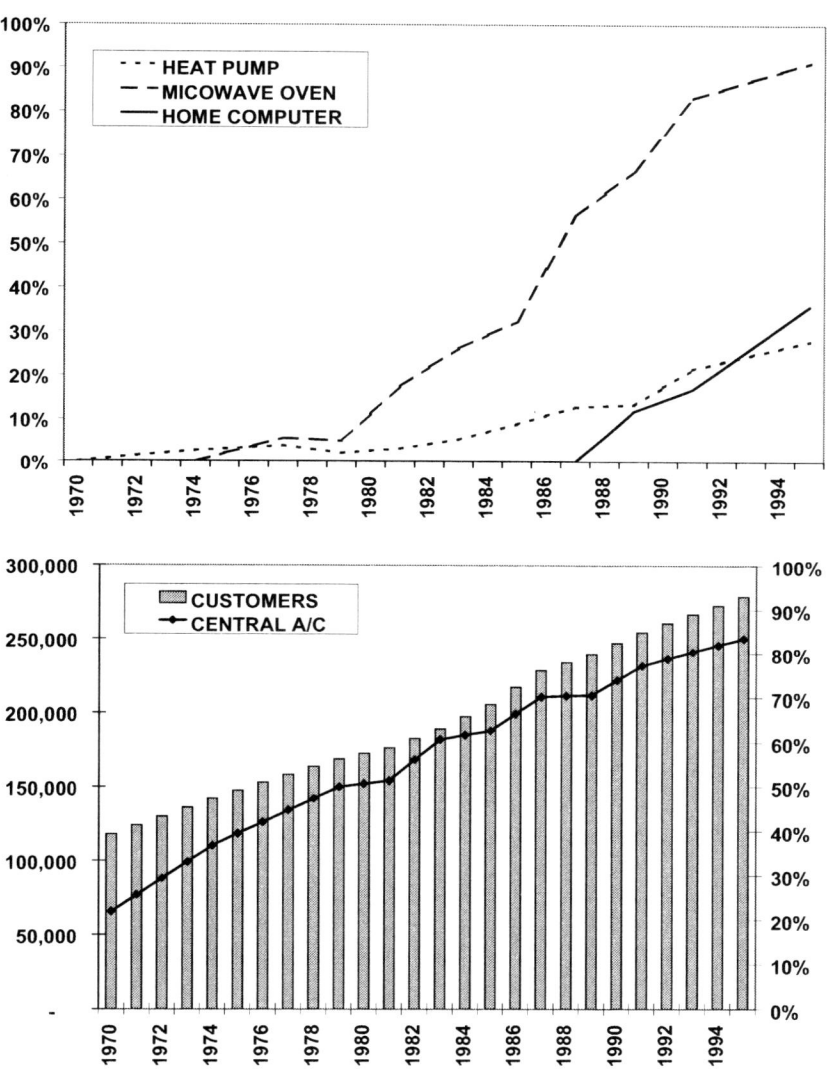

Figure 3.9 Market penetration of newer technologies. Latest residential technologies as represented by heat pumps, microwave ovens, and home personal computers (upper graph). While heat pumps indicate a relatively slow market entrance, microwave ovens and computers gained acceptance rather quickly. Growth in the percent of central air conditioning market penetration climbed quickly and in pace with the growth in numbers of customers (lower graph), which indicates that nearly every new home built had central AC.

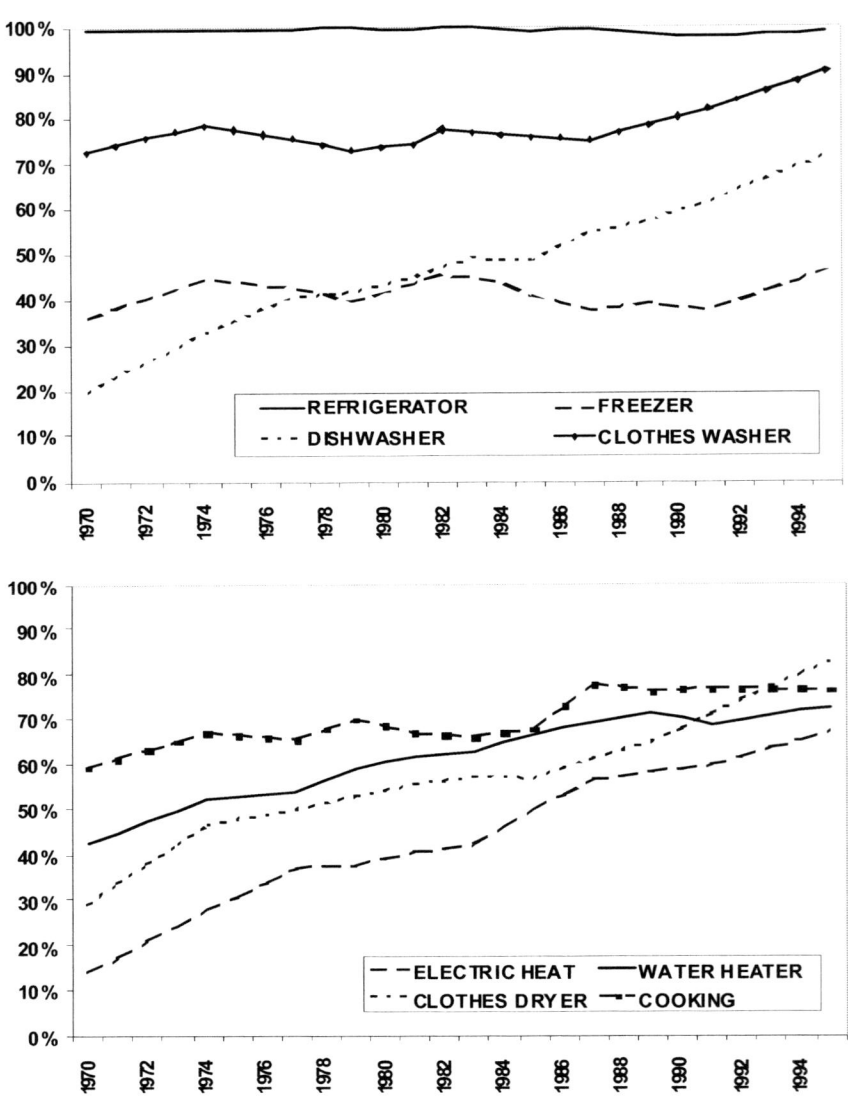

Figure 3.10 Market penetration of older technologies. Represented are electric-only end uses (upper graph) and competitive end uses, which are often served by an alternative energy source such as gas (lower graph). Some appliances have reached a point of saturation while others are still in a long phase of growth.

Electric End Uses

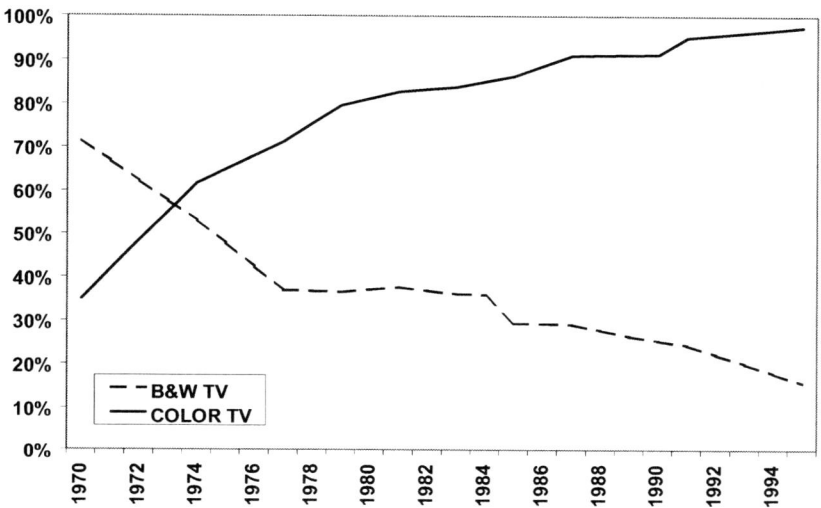

Figure 3.11 An example of technology obsolescence. Color televisions have replaced nearly all black-and-white televisions.

The lower graph of Figure 3.9 shows a nearly linear increase in residential customers over the 25-year period. Air conditioning was an established technology before the 1970s; thus, it is well into its growth phase during the 25 year period shown. From a starting point of about 20% market penetration, the increase in central air conditioners over the period follows the growth in numbers of customers at nearly the same rate, ending at just below an 85% market penetration level. Thus, a majority of the new customers each year has been installing central air conditioning systems in their homes.

Figure 3.10 shows the market penetrations of a variety of common electric household appliances, which contribute to base use of electricity, and electric heat, which includes both heat pumps and electric resistance furnaces that are used as the main source of heating. The upper graph represents electric-only types of appliances. Refrigerators, freezers, and dishwashers are essentially in a state of saturation over the 25-year period, while the market penetration of dishwashers grew from about 20% to just over 70%. The lower graph of Figure 3.9 shows the market penetration rates for electric appliances that provide end uses that can also be provided using fossil fuels as the primary energy source. Except for electric cooking, the market penetration of each competitive end use grew considerably over the period.

Figure 3.11 illustrates the obsolescence of black-and-white televisions in favor of color televisions, whose market penetration grew dramatically over the 25 year period. Obsolescence of a technology also occurs for reasons other than customer preference. For example, as older refrigerators fail and are replaced, the new refrigerators have a much greater level of energy efficiency than the units that are taken out of service. The least efficient models today are much more energy efficient than the refrigerators of times past. Thus, low energy efficient refrigerators represent a technology that is ultimately destined to become obsolete when the inventory of older refrigerators in customers' homes is finally turned over.

End-Use Load Shapes

Another way to study electrical end uses is by developing their characteristic load shapes, where load shape scenarios would be devised for each end-use technology of interest. These load shape scenarios reveal meaningful attributes of a particular end-use technology, such as the

- amount of operational load diversity that exists between similar end-use devices;
- principal seasons, days, or times of a day in which an end-use device exhibits its maximum and minimum demands for energy; and
- proportion of the whole customer load and energy use that the specific end use is responsible for.

Special metering is required in order to develop end-use load shapes. Recording meters are installed on the end-use devices in order to measure their electricity usage on a 15 minute interval basis. Since it would be impractical to locate meters on all installations of a particular end-use technology (e.g., every central air conditioner in the residential class), the energy consumption data is collected from a group of sample customers. The end-use samples must be properly designed in order to obtain meaningful and statistically accurate results.

Once the data is collected, the end-use load shape is developed by first summing the kWh usage, from all of the sample sites, in each and every interval and then dividing each of the interval totals by 4 (since there are four 15-minute intervals in one hour). The resulting profile represents the diversified interval loads (in kW) of the end-use sample customers as a whole. Further dividing the kW in each interval by the total number of customers yields the kW load shape of the *average* end-use device. The per unit load shape can be derived by dividing each interval by the magnitude of the maximum interval.

Figure 3.12 illustrates a per unit load shape of residential electric water heaters for one particular day. The example shows that the water heating demand was greatest in the evening hours. The early morning load just before 6:00 AM mostly results from the normal maintenance cycling (refer to Figure 2.8) with minimal demand for hot water. A research question would be: "How much load does electric water heating represent during the hours of the system peak load conditions?"

Electric End Uses

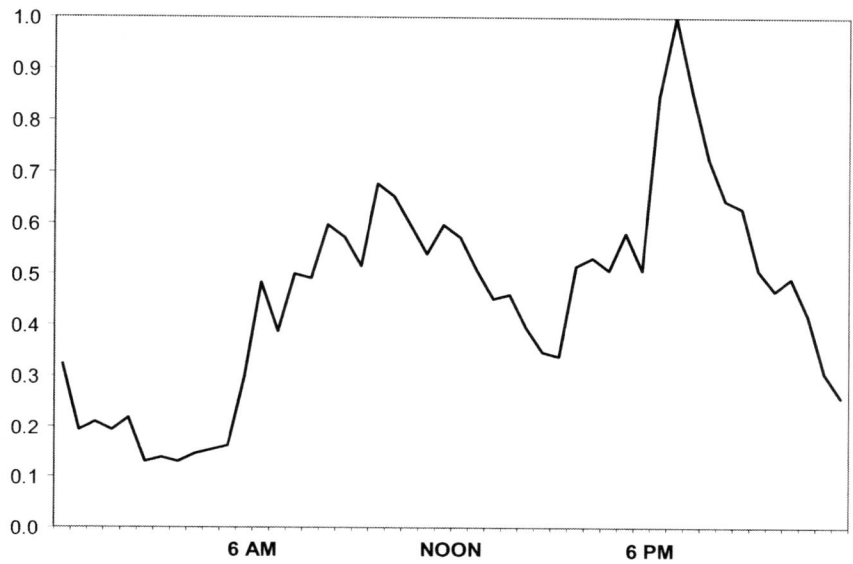

Figure 3.12 A per unit load shape of residential electric water heaters. The diversified end-use load shape based on the metered operation of electric water heaters from a sample of residential customers is plotted for a 24-hour period. Hot water demand is noted to be highest in the evening due to after dinner dishwasher operation and showers.

Figure 3.13 illustrates the effect of temperature on the load characteristics of a weather-sensitive end-use device. The per unit load shapes of residential air conditioners for two different day types shows a variation in both the timing and magnitude of the peak load and the duration of the peak period.

End-use analysis can also be conducted essentially at the whole customer level. Customers can be differentiated by nature of their major appliance or equipment ownership characteristics. For example, homes whose total energy requirements are met by electricity have different load and energy use characteristics than homes which utilize a fossil fuel for space and/or water heating. Thus, all electric vs. non-all-electric represents one way in which to view end use differences in residential customers.

The same situation applies to the commercial and industrial classes of customers as well. While specific end-use processes are sometimes metered in order to assess their unique load characteristics, customer-level end uses can also be defined by other segments, such as NAICS or SIC codes. Thus, for example, sawmills could be grouped in order to develop that industry's average load shape characteristics.

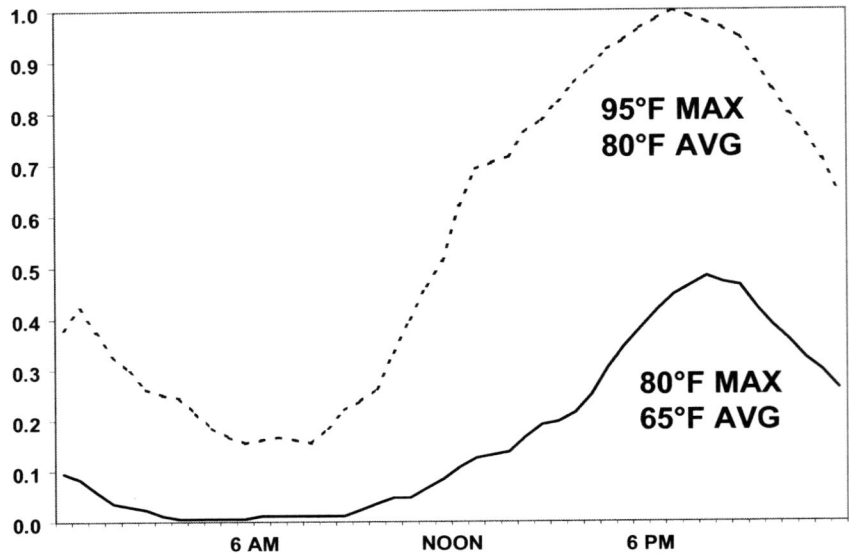

Figure 3.13 An example of air conditioning load on two days having measurably different temperature conditions. The hour in which the air conditioning load peaks on each day is different. Both the magnitude of the peak load and the duration of the peak period is greater on the warmest day.

Customer and Rate Class Load Shapes

Individual customer end-use load operations add up to define the aggregate customer and rate class load shapes. Each class load shape will exhibit unique characteristics that are caused by both the mix of end-use devices and the operating requirements of those end uses (i.e., domestic lifestyles, business processes, etc.).

With high market penetrations of residential air conditioning, the load shape of the entire residential class of customers is typically defined by the dominating effect of this end use, particularly at higher temperatures. The impact of residential AC unit operation on the class is illustrated in Figure 3.14. Note how much the residential class summer load shape resembles the higher temperature AC end-use load shape shown in Figure 3.13.

A significant presence of residential electric heat is also noted in the residential class winter peak day load shape shown in Figure 3.14. This typical electric space heating profile is a reaction to the colder morning and evening temperatures and the often milder mid-day temperatures when buildings often realize some heat gain under sunny conditions. As a result, the residential winter peak day profile typically is observed to have dual peaks, as shown in Figure 3.14.

Electric End Uses

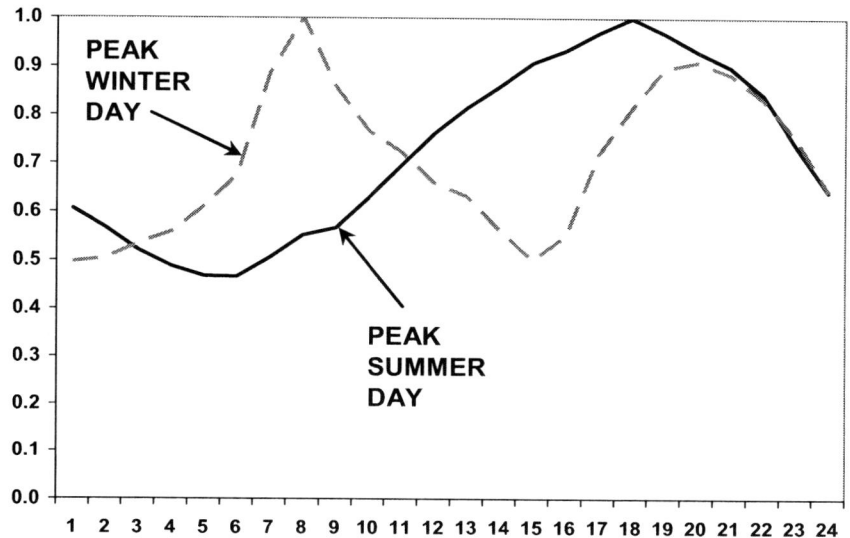

Figure 3.14 Typical per unit peak day summer and winter load shapes for the residential rate class. The impact of AC unit operation in the late afternoon is well noted in the summer load shape. The presence of electric heating is apparent in the winter load shape. Electric heating usage typically manifests itself as morning and evening peaks in the class load shape.

Depending on a utility's particular mix of customers and their end-use devices, the load characteristics of the residential class are often found to drive the system peak day load shapes, particularly since residential customers are vast in numbers. From a cost-to-serve perspective, production, transmission, and distribution substation costs, which are demand related in nature, are foremost a function of system peak loads. As a result, the load contributions of the customer or rate classes at the time(s) of the system peak(s) are used to allocate the associated costs.[11]

Load Research

Load shapes for all customer or rate classes are not readily available. While large commercial and industrial customers are often metered on an interval basis for billing purposes, customers in the residential and small- to medium-commercial

---

[11] The coincident peak demand cost allocation methodology is discussed at length in Chapter 11.

classes typically have much less sophisticated revenue meters. In order to develop the class load shapes required for demand cost allocation, sample meters must be installed that will provide reasonably accurate estimations of the class load shapes, particularly at the times of the monthly system peaks. The process of sample design, data collection, and data analysis is known as *load research*.

The four major steps of a load research effort are shown in Figure 3.15. Load research is a rather costly process since it involves meter procurement, installation, maintenance; data retrieval; processing and storage; and analysis and results evaluation. Thus, the goals of load research must be well thought out and defined. One goal of developing class load shapes for cost-of-service study purposes is to specify the resolution or the level of detail desired, i.e., customer class level only, rate schedule level only, or even sub-class or sub-rate level. Another goal is to specify the level of sampling accuracy desired, which might also be prescribed through regulatory requirements. Higher degrees of accuracy require more meters and higher program administration costs; thus, there is a tradeoff to be considered. As a general rule, at least 30 meters are needed for a sample; however, to achieve a reasonable accuracy for a class of thousands of customers, two to three hundred sample points may be required.

The sample design step determines the number of meters that are required in each customer segment in order to achieve the level of accuracy specified.[12] To be statistically valid, each sample must be based on a random selection of customer participants. In practice, issues occur that tend to bias the sample results. For example, a sample customer may not participate for a full year (which is required in order to capture the 12 monthly system peaks). During the data collection year, a sample customer may move away leaving the premises vacant, or a different classification of customer may set up operation in that location. Thus, alternative participants are selected as part of sample design to preserve the integrity of the sample as much as possible in the event of such factors.

The interval data is collected from the sample meters and processed by means of a translation program to convert meter pulses into engineering units, i.e., kW, kWh, etc. Finally, the sample data is expanded to estimate the load magnitude and shape of the associated population, e.g., the residential customer class.

A sample is designed to achieve an estimate of a population determinant, e.g., the residential mean per unit load (kW per customer) at the time of the system peak, within the bounds of a *confidence interval*. For cost of service, a commonly accepted standard is ± 10% accuracy at the 90% confidence level,[13] which can be interpreted as saying that "90% of the time, the true mean per unit value of the population will fall within a range of ±10% of the mean per unit value of the sample." A series of formulas are used to calculate the confidence interval.

---

[12] Population stratification is often used to obtain more homogeneous samples.

[13] The 90% ± 10% confidence interval standard is generally recognized by the industry as a reasonable level of accuracy in light of the costs necessary to achieve that level.

Electric End Uses

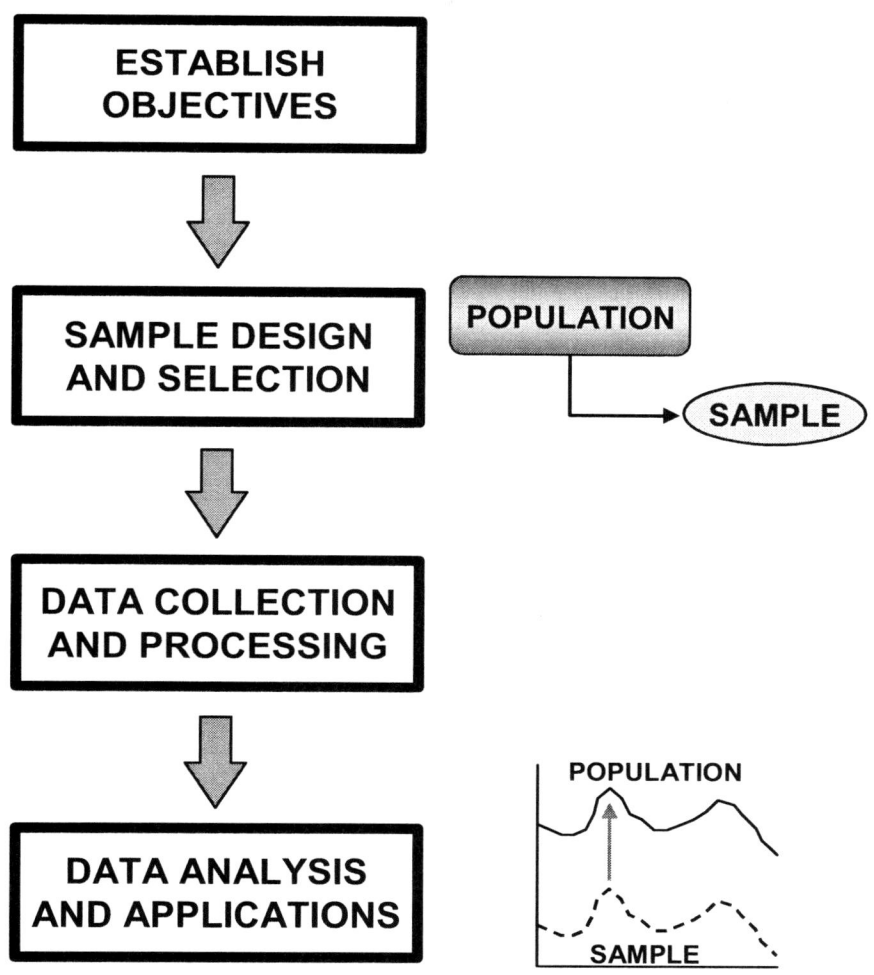

Figure 3.15 **The four major steps of the load research process.** Load research is necessary to develop critical input data for the cost-of-service study, rate design, and other key utility functions. Thus, load research program goals and objectives must be carefully scoped to ensure meeting these needs adequately. A sample is designed to estimate the desired population parameters within a specified degree of accuracy. Data is collected over the period of a year and then processed and analyzed so that the resuls can be applied.

The quantification of the confidence interval begins with the calculation of the sample mean, as determined by

$$\bar{y} = \frac{1}{n}\sum_{i=1}^{n} y_i \qquad (3.4)$$

where $\bar{y}$ = mean per unit, e.g., kW per customer
y_i = sample observation, e.g., i^{th} customer's kW load at system peak
n = number of sample observations, e.g., customers

The standard deviation of the sample, s_y, is then calculated by

$$s_y = \sqrt{\sum_{i=1}^{n} \frac{(y_i - \bar{y})^2}{n-1}} \qquad (3.5)$$

The standard error of the mean,[14] $s_{\bar{y}}$, is then calculated by

$$s_{\bar{y}} = \sqrt{\frac{s_y^2}{n}} \qquad (3.6)$$

Finally the confidence interval, CI, is determined by

$$CI = \bar{y} \pm \left(s_{\bar{y}} \times Z_{\alpha/2}\right) \qquad (3.7a)$$

For a 90% confidence interval, the tail area of a normal distribution curve is 1 − 0.9 = 0.1, i.e., α = 0.1; thus, $\alpha/2$ = 0.05 represents the area of one tail of the normal distribution curve, as shown in Figure 3.16. From a table of normal distribution tail areas, $Z_{\alpha/2}$ is found to be 1.6495. Thus, for a 90% confidence interval standard

$$CI_{90\%} = \bar{y} \pm \left(s_{\bar{y}} \times 1.6495\right) \qquad (3.7b)$$

Equation 3.7b represents the estimation of the population mean per unit within a bandwidth, and the quantity ($s_{\bar{y}} \times 1.6459$) represents accuracy. To achieve a ± 10% accuracy, the sample must be designed to have a sufficient number of survey recorders. Target accuracy is an input to the sample size determination formulas.

---

[14] If multiple samples were repeated for the same population, each sample would yield a different mean value. The distribution of these mean values would tend to form a normal distribution, i.e., the Central Limit Theorem. In essence, the standard error of the mean represents the standard deviation for the distribution of these mean values.

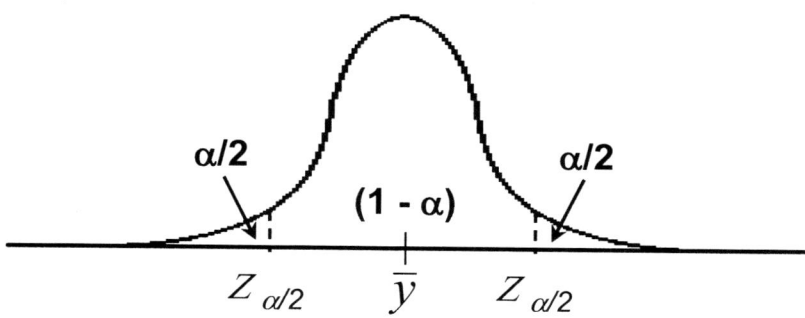

Figure 3.16 The normal distribution curve. The normal distribution curve is symmetrical; thus, its two tails are equal in area. The total tail area is α, and α is governed by the selection of a level of confidence. At 90% confidence, $\alpha/2$ is equal to 0.05, and $Z_{\alpha/2}$ is equal to 1.6465.

The mean per unit calculation must be expanded to estimate the ultimate population determinant. For example, if the average load in kW per residential customer at the time of system peak has been calculated for the sample, expansion would yield the total residential class load that is coincident with the peak. One expansion method simply multiplies the mean per unit value obtained for the sample by the total number of customers in the population. Another expansion method utilizes a ratio based on a correlated determinant, such as energy. For instance, the average load in kW is multiplied by the ratio of the population kWh to the sample kWh in order to estimate the total population kW. The sample and population kWh values must be taken from the same period of time, e.g., calendar months would be used for cost-of-service load shape development of system peak class contributions.

Aside from cost-of-service study load shape development, load research data proves invaluable for a number of rate design issues, including, but not limited to, the determination of:

- the relationship between coincidence factor and load factor;
- the relationship between demand and energy use;
- energy usage based on time-of-use rating periods; and
- special rates based on specific end uses or NAICS/SIC codes.

Load research is also useful in several other utility applications. For instance, load shape data can be used to refine standards for transformer sizing. Load research is key for designing marketing programs and then for evaluating the performance of the programs by means of pilot vs. control groups. Characteristic load shapes are also used to evaluate the profitability of economic development prospects.

3.5 SUMMARY

Customers utilize a vast number of end-use devices in order to meet a host of specific needs. Each customer class or rate group has its own unique end use driven characteristics in terms of demand and energy utilization. An understanding of what these devices are and how they are operated is one of the essential cornerstones to pricing electricity effectively.

The value of a particular end-use device to a customer varies depending on time and need. Who could easily manage without refrigeration? On the other hand, how much worth can be derived from an air conditioner in the dead of winter? End use value governs the operation of the devices which in turn establishes the customers' energy utilization requirements.

End-use devices operating on demand create the characteristic load shapes of each customer, each customer or rate class, and the system as a whole. The types and capabilities of the power system assets required to meet customers' needs reliably is highly dependent on the diversity of operation of the customer end-use devices. In other word, through load, customer end-use needs are responsible for the fixed costs that are required for a utility to provide dependable electric service. As such, a cost-of-service study relies on accurate knowledge of loads in order to fairly and equitably assess these fixed cost responsibilities to the customer or rate classes.

End-use information is gathered from customer survey data, on-site inspections, and by means of load research science. In addition, the concurrent collection of building characteristics (such as heated floor space), demographic data (such as household income levels), and customer opinions about energy usage can provide invaluable links for understanding end-use characteristics. Geographical and local economy differences contribute to the variation in customer and end-use characteristics experienced by utilities.

End-use data is valuable for pricing electricity. Pricing structures can be designed to adequately recover the utility's costs while promoting the applications of end-use technologies which create and deliver customer value.

4
Customers: Electric Service Requirements

4.1 INTRODUCTION

Once a customer has committed to specific end-use equipment, certain requirements for electric service emerge. The customer's total electricity needs represent the combined energy requirements of all end-use load devices based on their electric power ratings and operations.[1] The rate of energy use by some customers is fairly constant. For example, the load factor of a 24-hour grocery store is high because the predominant end-use loads, refrigeration, lighting, and HVAC, operate on a fairly continuous basis. Conversely, the rate of energy use is highly variable for low load factor customers. A car crushing operation, for instance, tends to use most of its energy requirements over very short periods of time. During crushing sequences, a very high rate of energy transfer is needed for the compaction drives. But in between times, the energy requirements for the auxiliary load operations are comparatively quite low.

Customers' electrical needs are very diverse; thus, the power system must be able to adequately serve a wide range of load characteristics. *Electric service* encompasses both (a) the physical arrangements necessary for safely and dependably delivering the required amount of electricity to a customer's building or operation and (b) the administrative functions that establish and support a customer's account for service. Customers must submit an application for electric service in order to activate an account. In areas where restructuring has been im-

---

[1] Many end-use devices which operate on an alternative energy source, such as natural gas or propane, also require some electricity (e.g., the blower motor in a furnace or the tumbler motor in a clothes dryer).

plemented, customers typically apply to both their local utility for *power delivery service* and to a competitive supplier for their *energy service* (and in some cases to a third party for metering services). The lead time for establishing electric service for large, new customers may be several months because of the time required to procure specific equipment, such as transformers, and to build the local service facilities. In order to plan properly for such facilities, the types of end-use devices and their electricity requirements must be well known.

4.2 CUSTOMER ELECTRICITY REQUIREMENTS

When an electric service account is established, a physical point of delivery is defined. The *delivery point* is the location where the utility's and the customer's respective wiring systems are coupled and where the transfer of electricity from the utility to the customer occurs. Ideally, the electric service billing meter is located directly at the delivery point, but it may be located elsewhere for reasons of economics or practicality. A number of technical power system parameters, from the production plants to the billing meter, must be met in order to satisfy a customer's requirements for electricity.

Energy and Capacity

Customers have the basic need for a volume of kWh over a given period of time, such as a month, in order to supply the energy requirements of their end-use load devices. Customers with weather-sensitive loads typically require a greater amount of energy during the cooling and/or heating months than during other months of the year. Other seasonal processes, such as agricultural businesses, require varying energy amounts in concert with the annual cycles of the particular operation (e.g., irrigation pumping during the growing seasons).

Customers also have the need to take or consume energy at a specific rate of use. This rate depends on the amount and size of end-use load devices that are being powered at any point in time. The rate of energy use may vary significantly from moment to moment as the customer's load devices are switched on and off. Low load factor customers exhibit a significant amount of usage variation while high load factor customers use energy at a nearly constant rate. A customer's average load over a short time interval (e.g., 15 or 30 minutes) is demand, and the highest demand established, or expected, represents a customer's maximum load requirements.

Capacity is the ability of the power system components to adequately serve a customer's maximum load requirements, and it is referenced in units of kW or, more appropriately, kVA. Reactive power can represent a significant portion of the capacity requirement of a customer. Induction motors, transformers, fluorescent lighting ballasts, and other load devices with inductive components require VARs in order to develop their magnetic fields. Uncorrected, many such load devices exhibit relatively low power factors.

Customers: Electric Service Requirements

KVA is the vector sum of kW and kVAR, and it represents the true thermal load on the power system. Thus, the capacity of a 100 kVA transformer would be fully utilized under each of these loading scenarios:

- 100 kW + $j0$ kVAR = 100 kVA, power factor = 100%
- 80 kW + $j60$ kVAR, = 100 kVA, power factor = 80%
- 60 kW + $j80$ kVAR, = 100 kVA, power factor = 60%
- 40 kW + $j91.65$ kVAR = 100 kVA, power factor = 40%
- 20 kW + 97.98 kVAR = 100 kVA, power factor = 20%
- 0 kW + $j100$ kVAR = 100 kVA, power factor = 0%

The local service facilities in particular must be sized to carry the maximum load regardless of the time of day or the season of the year during which the maximum load occurs. Service transformers are manufactured with various standard kVA capacity ratings. A transformer is selected to serve a single customer (or a few small customers) based on the expected maximum load requirements of the customer(s). The generation must also provide ample capacity to serve a customer's load; but, unlike a service transformer, a utility's system of generators is sized to meet the diversified load requirements of a large aggregate number of customers. Thus, a customer is provided generation capacity more on the basis of load requirements during the time when the system load is peaking rather than on the basis of the customer's maximum demand. A customer's capacity requirements, in terms of kW, kVAR, and kVA, are exemplified in Figure 4.1.

Voltage and Frequency

Customers require a service, or delivery, voltage that is suitable for the magnitude and mix of their end-use load devices. Most all residential and small commercial customers can be served with a basic single-phase voltage. In the United States, a 120/240 volt, three wire service operating at a frequency of 60 Hz is standard. This arrangement consists of two "hot legs" and a neutral. The voltage measured across the hot legs is 240 volts, and the voltage measured between either of the hot legs and the neutral is 120 volts. Convenience outlets in homes and buildings provide 120 volts for lamps and other plug-in devices, while special circuits provide 240 volts to major appliances such as ranges, clothes dryers, water heaters, and electric heating and air conditioning systems.

Many other countries operate 50 Hz electric systems and often power single-phase convenience outlets at 220 to 240 volts. A transformer may be required to make equipment designed for one type of system compatible for operation on the other type of system. Any difference in frequency also has an affect on the speed of motors. Since motor speed is directly proportional to the line frequency, a 50 Hz motor operating on a 60 Hz system would run 20% faster. Modern devices are more often designed to operate over wider voltage and frequency ranges.

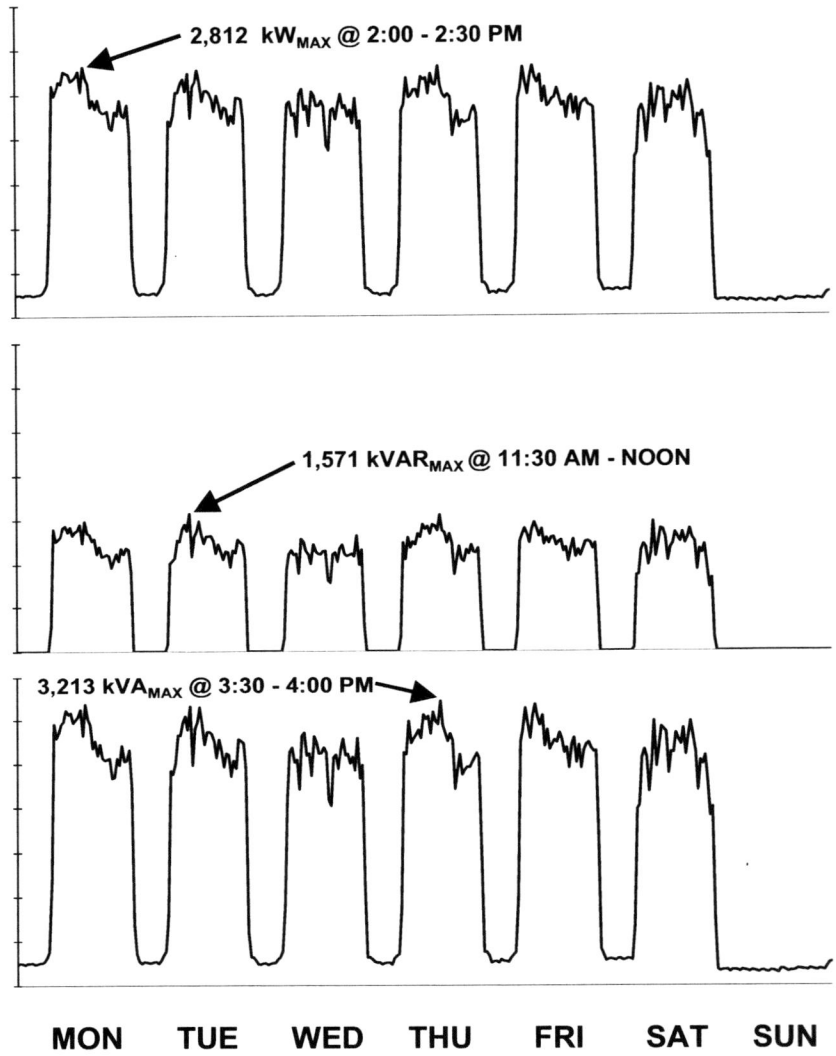

Figure 4.1 **Example of real, reactive, and apparent power loads of an industrial customer.** Customer load shapes are determined from interval meter data.. The kVAR requirement is consistently high when the kW requirement is high; however, the customer's peak power requirements are not necessarily coincident on a component basis. For the one week period shown, the kW, kVAR, and kVA maximums all occur at different times on different days.

Customers: Electric Service Requirements

Because of the types of load devices being used, larger customers typically require a poly-phase service, of which three-phase (3ϕ) voltage is most common.[2] As illustrated in Figure 4.2, three-phase voltage is essentially a combination of three single phases (1ϕ). Each phase goes through its sinusoidal cycle sequentially at a displacement of ±120° with respect to the other two phases. The three different phases are typically referenced by the letter designations A, B, and C.[3]

As a general rule of thumb, ordinary motors of five or more hp are usually of a three-phase design. Three-phase service is more efficient for large motor operations as it provides a much smoother flow of power to the load, as compared to single-phase service. Using an engine as an analogy, if single phase is equivalent to a single piston drive, three phase is like having three pistons connected to the same drive shaft with a spatial orientation of 120° apart. Smaller scale commercial establishments and very large residences sometimes have three-phase motors due to their intensive cooling requirements (thus a need for big compressors) or special devices, such as elevators. Larger commercial and most industrial customers require three-phase service primarily because of large motors for HVAC and manufacturing processes. Common three-phase motor voltages are 208 volts and 480 volts. In addition, higher density loads are more economically served by three-phase voltage systems.

Most customers that require three-phase service for power loads, such as large motors, also have a need for single-phase service for convenience outlets, lighting, and other smaller load applications. The electric service can be configured to furnish both single-phase and three-phase voltages. Thus, for example, a 480Y/277 volt, four-wire service provides 480 volt three-phase power and 277 volt single-phase power. Many commercial and industrial fluorescent lighting systems are often operated at 277 volts because some economies are gained compared to 120 volt lighting systems. Additional transformers are necessary for this type of electric service in order to provide 120/240 voltages for convenience outlets and other smaller utilization equipment. Voltages of less than 600 volts are typically classified as secondary voltages.

Bigger customers, such as large buildings and factories, often take service from the utility at a primary voltage class, such as 12,470/7,200 volts. These primary voltages are applied to three-phase feeders which are routed to load centers throughout the facility. At each load center, the primary voltage is transformed to a lower voltages which then powers the utilization equipment located in the adjacent area.

---

[2] Poly-phase voltage requirements other than three-phase service are sometimes utilized in special applications. For example, six-phase systems are used at times for supplying DC rectifiers. Two-phase applications were common at one time, but they have now virtually disappeared.

[3] Three-phase power systems are connected in two primary configurations: wye (Y) and delta (Δ). These connection methods are discussed in Chapter 5.

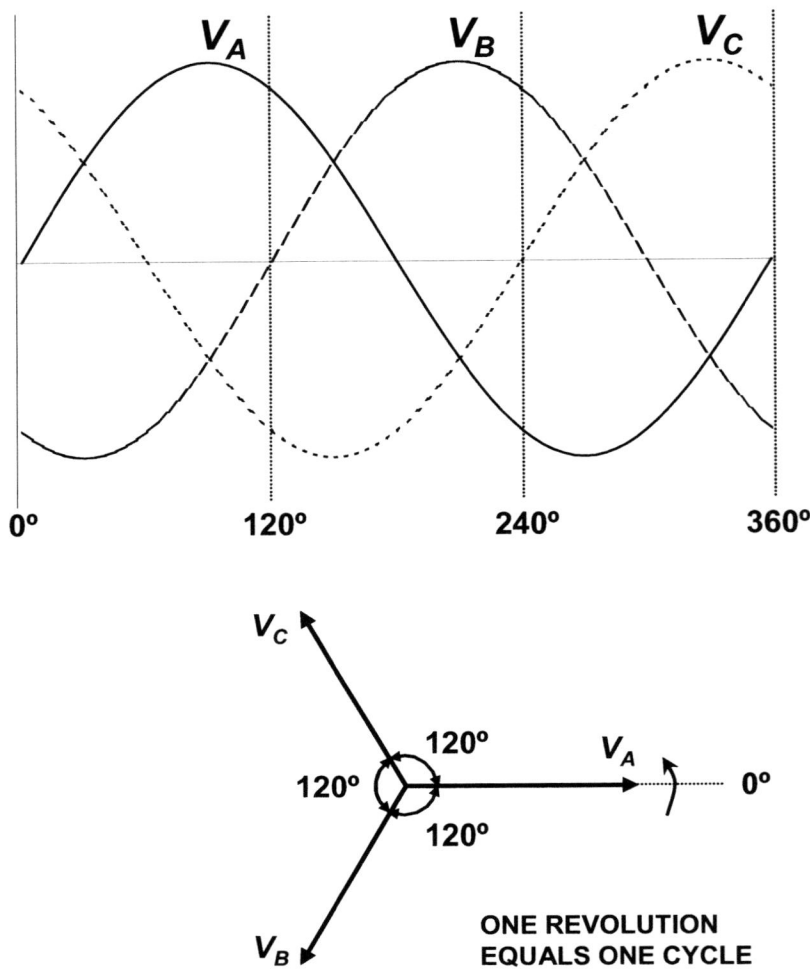

Figure 4.2 Three-phase voltage. Each voltage vector revolves through its cycle 120° ahead of the next phase in sequence.

Some special applications motors are designed to operate directly at a primary voltage class, such as 4.16 kV. Very large customers, such as heavy industrial operations, may be better yet served by a much higher transmission class of voltage, such as 115 kV. The choice of a primary or transmission voltage is usually constrained by the corresponding available voltages and capacities of the adjacent utility lines.

Service Reliability

Electricity is such an integral part of life that people generally take it for granted. With the exception of some locales where the electricity supply and demand relationship is out of balance, one does not normally wonder whether or not the lights will come on when the switch is thrown. Society expects exceptional performance of the power system. Even a temporary outage of electrical service causes a loss of productivity as business and manufacturing processes are idled, confusion and distress as traffic signals are darkened, and general dissatisfaction as customers are inconvenienced in many different ways.

In reality, service interruptions are not 100% avoidable since the power system is exposed to a number of environmental hazards, such as animal intrusions and automobile accidents. In addition, power equipment failures, which may result in an outage, can and do occur at times. To fully guarantee uninterrupted power supply at all times is neither practical nor economical. In fact, power systems incorporate protection components which, by design, de-energize portions of the system when certain electrical problems occur in order to ensure continued service to as many customers as possible as well as to maintain safe conditions. But overall, standard power systems, constructed on the basis of industry best practices, perform as expected and are generally highly reliable.

Service reliability is a measure of the continuity, or lack of continuity, of electricity supply at the delivery point of a customer. Customer outages can be extremely short as in the loss of electrical power for a few voltage cycles, a mere fraction of a second. On the other hand, outages can be extended for days in the aftermath of a catastrophic weather event.

Customers are affected differently by the frequency and duration of outages. All customers desire a high degree of service reliability, but some customers have a genuine stake in having high reliability. For example, when continuous industrial processes, such as extrusion machines or glass melting furnaces, are inadvertently shut down by a short-lived interruption of power, the customer can realize a major loss in terms of productivity and/or product, either of which can be costly. Under the same loss of power conditions, the typical result for a residential customer is a momentary shut off of the television as all the digital clock displays in the house demand in unison to be reset. The latter case is certainly an inconvenience, but not an economic impairment.

Thus, the criteria for service reliability varies among different customers. Costly power system enhancements, including dual source feeds and circuit throwover switches, can be installed in order to provide a higher than standard level of service reliability. Service improvements such as these are practically beneficial for only a limited number of applications, e.g., hospitals, and are not in demand by the majority of customers. These types of system enhancements increase the costs of providing electrical service. Thus, since only a few customers require such improvements, those customers alone should bear the responsibility for the increased costs of nonstandard or premium reliability service.

Other practical solutions to higher reliability needs lie on the customer's side of the meter. Backup generation systems are sometimes used and even required by hospitals and other critical types of operations. An Uninterruptible Power Supply (UPS) is a much more practical and economical solution than distribution system improvements for dealing with momentary outages which impact computer installations. Again, these reliability improvement measures are directed at the special requirements of a few customers. Each customer must assess reliability requirements, weigh the costs of implementation, and select the optimal mix of utility-side and/or customer-side options which meet that criteria.

Power Quality

While the basic customer end-use needs have remained somewhat unchanged, the proliferation of electronic devices connected to the power system has greatly impacted the requirements for electric service from a quality standpoint. While voltage may be generated with a high degree of purity with respect to its sinusoidal waveform and frequency, events do occur on the power system which give rise to voltage irregularities that cause operating problems for some electronic equipment. Anyone who has purchased a PC from a customer-focused vendor has been encouraged to purchase a surge suppressor to protect both the computer's power supply and telephone modem from high impulse voltage transients caused by lightning strikes.

More subtle voltage variations occur more frequently on the power system due to routine activities including the switching of capacitors and circuits. Switching often produces an oscillatory type of transient. Figure 4.3 illustrates the impact of *transients* and *surges* on the normal per unit voltage waveform.

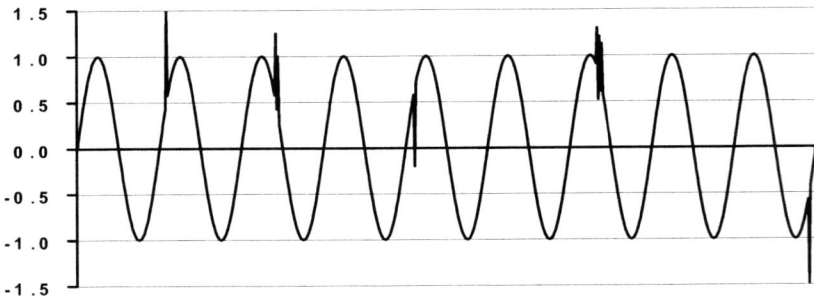

Figure 4.3 **Voltage transients and surges.** Voltage surges are caused by incidents such as lightning and circuit switching operations. The nominal voltage waveform is shown with a magnitude of 1.0 per unit. Surges and transients can create instantaneous voltages that are many times greater than normal.

Customers: Electric Service Requirements 113

Transients and surges are extremely short duration voltage excursions. High impulse voltages can destroy sensitive electrical and electronic components. Other momentary events include *sags* or *dips* in the voltage, as shown in the upper graph of Figure 4.4. Voltage sags are commonly caused by system short circuits or by switching on a large magnitude of load. When a large induction motor is energized, the starting or inrush current causes a momentary voltage sag. Such voltage irregularities can be noticed as *flicker*, a fluctuation in the output of a light connected to the same electrical source. The supply voltage may also momentarily *swell* above its nominal rating as a result of system fault current (i.e., short circuit) conditions, as shown in the lower graph of Figure 4.4. Generally, short duration voltage aberrations last less than a minute and often less than a second.

Voltage variations may also be of a long-term duration, as shown in Figure 4.5. A long duration *undervoltage* can occur at times when circuits are carrying loads beyond their nominal design parameters or when capacitance is needed on the circuit to support the voltage under substantial inductive loading conditions. Conversely, *overvoltage* conditions can occur on a distribution feeder during light loading times due to an excessive amount of capacitance located out on the circuit. One common cause of this situation results from a large customer's motors being switched off after the last production shift while the power factor correction capacitors remain engaged with the circuit. The resulting leading power factor causes the feeder voltage to rise during low load periods.

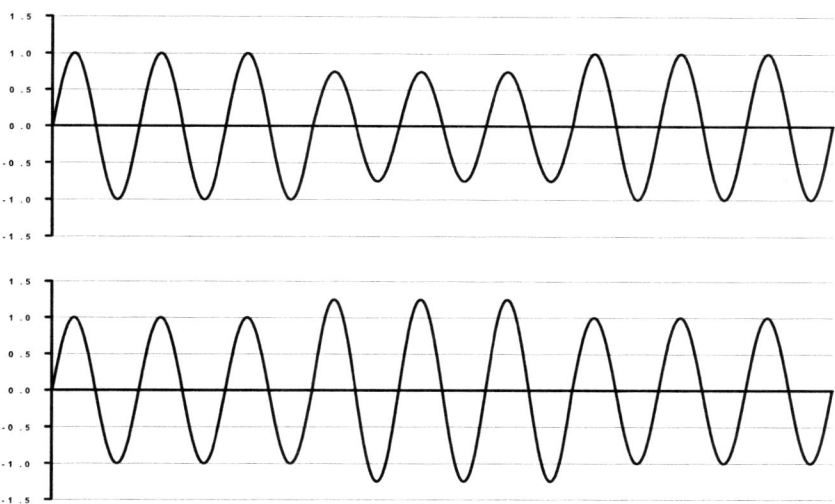

Figure 4.4 **Short-duration irregularities in voltage magnitude.** Voltage may sag (upper graph) or swell (lower graph).

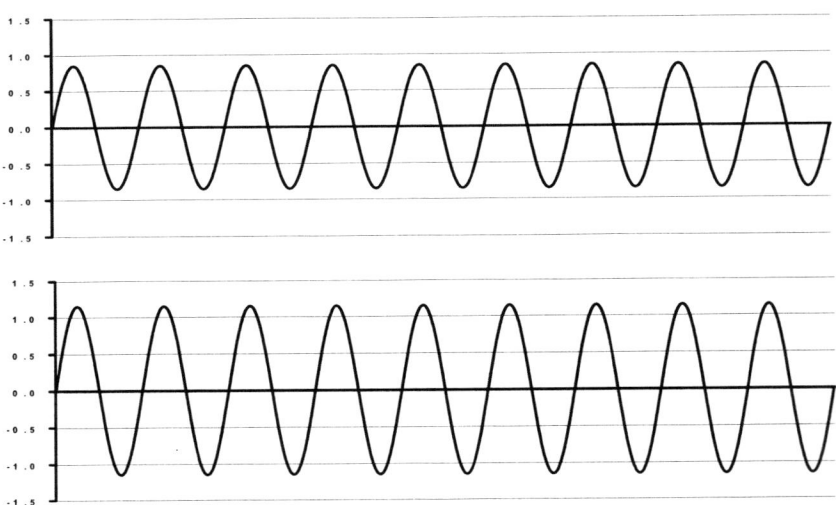

Figure 4.5 Long-duration irregularities in voltage magnitude. An undervoltage condition may occur (upper graph) or an overvoltage condition may occur (lower graph).

The continuation of long-duration voltage irregularities affects both the performance and the lifespan of utilization equipment in various ways. In general, an applied voltage that is below the nameplate rating results in less performance of an end-use load device. Thus, under lower voltages, induction motors produce less torque, and lamps yield less light output. At the same time, lower voltages cause a temperature rise in motors, but an increase in life for incandescent bulbs. The heat output of a resistance type heater goes down nearly 2% when the applied voltage is reduced by 1% (due to the i^2R effect). Under higher voltage, the above effects are generally reversed (i.e., the motor's torque increases while its operating temperature decreases; the light output increases, but lamp life is reduced; and heater output increases in proportion to the square of the voltage). Some electronic equipment include voltage regulators to ensure proper internal operating voltage for sensitive components in the event that moderate variations in the applied voltage were to occur.

The above issues are directed at the variation in voltage magnitude. In addition, variations in the voltage waveform can interfere with the operation of sensitive electronic devices which have been designed under the assumption of a pure supply voltage. Waveform distortions can also introduce electrical noise in adjacent communications circuits or create other problems, such as overheating of electrical components. Perhaps the most vital power quality issue is harmonic distortion. *Harmonics* are voltage frequencies which are integral multiples of the system frequency (i.e., 120 Hz, 180 Hz, 240 Hz, etc. based on a 60 Hz system).

Harmonic waveforms are created by the operation of *nonlinear loads* such as rectifiers, inverters, adjustable speed drives, and arc-discharge devices (e.g., fluorescent lights) and are reflected into the power system.[4] Harmonic distortion of the voltage is caused by the currents drawn by such nonlinear loads. The effects of harmonics on a sinusoidal voltage waveform is shown in Figure 4.6.

Harmonics interjected onto the power system by a customer's nonlinear load devices can create operating problems not only for sensitive internal loads but also for loads of other customers connected to the same utility distribution circuit. Because of the large number of nonlinear load devices which customers commonly utilize these days, a certain level of harmonic distortion is prevalent throughout the standard power system. Customers who generate excessive harmonic distortion levels are often required by the local utility to invest in mitigation measures in order to prevent causing problems for either the power distribution system or adjacent customers.

Most customers are not highly sensitive to the nominal level of harmonics and voltage irregularities that exist on the power system. Thus, as with the issue of reliability, each customer must assess the impacts of power quality phenomena on his particular operation. Both utility-side and customer-side power conditioning options exist for mitigating problems, and quite often the lowest cost solutions for customers are those which correct problems at the load.

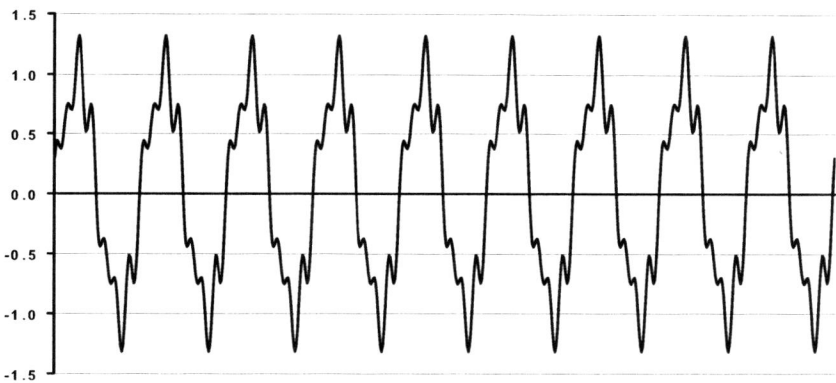

Figure 4.6 **Distortion of voltage due to harmonics.** The waveform shown is a composite of the fundamental voltage and the 3rd, 5th, and 7th harmonics.

---

[4] A nonlinear load is a device that has an impedance which varies during its operation. A linear load then is a device that has a constant impedance when operating in a steady-state condition.

Ancillary Services

Electricity service has been unbundled in many different areas. In the U.S., the FERC has required unbundled services and tariffs for wholesale transmission transactions. Network service and point-to-point service are two types of transactions that allow wholesale customers to secure capacity from a transmission provider in order to wheel power procured from an energy supplier to the load that is to be served.

Aside from the capacity component of transmission, the FERC has identified six ancillary services to be priced separately:

1. Scheduling, System Control, and Dispatch Services
2. Reactive Supply and Voltage Control Services
3. Regulation and Frequency Response Service
4. Energy Imbalance Service
5. Operating Reserve: Spinning Reserve Service
6. Operating Reserve: Supplemental Reserve Service

The FERC requires that transmission providers offer all six of these ancillary services, although customers are only required to purchase the first two services directly from the transmission provider. The customer then has the option to acquire the remaining ancillary services competitively from a third party or to self-provide the services from customer-owned resources.

Retail customers participating in competitive energy markets are also served under unbundled tariffs and thus have separate electric service contracts for energy supply (i.e., generation or production) and power delivery (i.e., transmission and distribution).[5] In addition to the primary ancillary services identified for transmission, distribution system ancillary services may also be unbundled as part of retail access. Distribution ancillary services include such functions as reactive compensation for power factor correction, voltage regulation, emergency backup, and load related losses. Some distribution ancillary services may be designed to be offered competitively.

4.3 NONPHYSICAL SERVICE REQUIREMENTS

In addition to the physical requirements necessary for powering a set of end-use load devices, customers have other needs relative to receiving electric service. These other requirements relate to electric service pricing issues, alternative billing options, and access to auxiliary customer services.

---

[5] In some cases, metering services, which are traditionally a part of the distribution function, are also unbundled and thus might be provided competitively by a third party.

Price: Choice, Options, and Terms

Many consumers relate prices for the goods and services that they purchase with a perceived value. If a product has a high value to some customers, they will pay a relatively high price. If a product has low value, these same customers will seek alternatives in order to satisfy their high value needs. Merchants and vendors differentiate their products and services in order to project a value perception in the minds of consumers.

Value-seeking customers are willing to pay higher prices for a specific product even though lower priced, lower value options might readily exist. With electricity, value is created by differentiating on the basis of such features as end-use performance, service reliability, power quality, and other amenities, such as supplementary products and services. Value-seeking electricity customers will readily pay higher prices for options in order to realize benefits which exceed standard service levels. They will also compare prices in conjunction with service attributes when selecting an electricity supplier when choice is available.

Some other customers are loyal to certain suppliers because they recognize a brand image. Better prices or better value may be found elsewhere, but they stay with the "tried and true" products that have served their needs with satisfaction in the past. Loyal electricity customers will choose to remain with their incumbent utility when industry restructuring allows customers to select their energy supplier of choice. Other customers will remain with their local service provider simply because of the inconvenience of trying to analyze and compare the service features and prices of alternative suppliers.

The value of a product to a price-seeking customer is simply low prices, and product differentiation features or brand image is virtually meaningless. In commodity markets, price shoppers search for the lowest prices available even though that particular product may be of lesser caliber than other similar products that are also available. In competitive energy markets, price shoppers will switch suppliers as soon as a lower price for electricity can be secured despite the potential for taking on a greater service risk.

Some customers are willing to receive a lesser grade of service in exchange for a lower price. For example, the direct control of residential air conditioners is sometimes offered to customers by a utility so that it can lower its system demand during peak load periods. The cycling of an air conditioner may result in slightly warmer temperatures within the building, since the unit's operation is restricted at times when the thermostat is normally calling for cooling. Load control customers typically receive a bill credit for participating in this type of program.

Even under price regulation, utility customers want rate options because customers desire to have the ability to make a choice (even if they do not always exercise their right of choice). Different, but equivalent, pricing structures can be developed for the same class of service. For example, a time-differentiated residential rate which offers prices that vary by season and by time of day can be designed as an alternative to a simple average rate that does not vary throughout the whole year. Some of those customers having the right kind of load characteristics or operational flexibility may then choose the time-of-use rate option in order to take advantage of the economic benefits that such a rate structure offers. Others who qualify for the option may appreciate having a choice but then select the less complex structure because it better meets their needs.

The term of the utility service contract is important to many customers, particularly larger commercial and industrial businesses whose electricity requirements represent a major operating cost. Such contracts for service typically have an initial term of one or more years. Price shoppers seek the shortest term possible so that the ability to quickly and easily switch suppliers is not compromised. Even in areas where utility industry restructuring is unlikely in the near term, some customers negotiated with their local utility for an agreement with a provision for contract reopening before the term is expired in the event that the choice of suppliers becomes available. Some other customers perceive value in long-term contracts in order to gain price protection or other benefits.

Billing Options

Utility invoices or bills are typically rendered on a monthly basis. Customers with monthly usage variations resulting from weather-sensitive, end-use operations or seasonal load patterns likely experience significant variations in their bill amounts. To avoid large monthly or seasonal variations, some customers prefer to pay their bills on the basis of a uniform amount each month. Many utilities offer a *levelized billing* program for residential and small commercial customers. Under levelized billing, customers "overpay" in months when usage is low, such as in the spring and fall seasons, and "underpay" in those months when usage is usually higher due to more intensive heating and cooling requirements. Over the course of a year, the overpayments and underpayments net out to nearly zero.[6]

Levelized billing offers customers the advantage of more easily budgeting for utility payments against a widely varying usage pattern. This benefit suits many residential and small commercial customers. But larger commercial and industrial customers may not realize a significant benefit under levelized billing. Large commercial and industrial customers usually have very large monthly bills; therefore, the cash flow of these customers is a major consideration. If large amounts of money were "deposited" into a levelized billing account, the time value of money would not be enjoyed. Interest that could otherwise be earned by the customer would be lost. The complexity of rate schedules for larger customers also makes levelized billing a difficult feature to manage.

Monthly utility bills are determined for each service location and mailed to the customer at a designated billing address (which is most often the service loca-

---

[6] Since usage is estimated and weather effects can not be precisely projected, a slight mismatch between the levelized bills and the actual charges will occur. This mismatch is often reconciled in the next period's calculation of the levelized payment.

tion). However, some customers have numerous service points (and thus numerous accounts) within a utility's service area, e.g., local telephone company switching stations and administrative buildings, corporate fast food restaurants, retail outlets, etc. The associated meters are typically read at different times during the month because of meter reading cycles.[7] Subsequently the monthly bills would be received by the customer over the course of several days. These customers often find it necessary to write several checks to the utility during the month, and they must collate the bills as they are received in order to maintain their accounts payable records. A billing option of value to multiple account customers such as these is *summary billing*. With summary billing, the individual bill amounts for all of the service locations are consolidated in a summary statement by the utility with a single due date. The customer can then write a single check each month in order to pay for the consolidated bills of all locations.

Another billing option is to allow large customers to choose their monthly meter reading dates so that bills reflect a specified time period of use. Other customers may request to select a "bill due date." Such customers prefer synchronizing major expenses, such as utility bills, with calendar months or with other specified dates to facilitate special corporate reporting requirements.

Some customers favor bill payments through automatic bank account drafts while others prefer paying bills via electronic fund transfers. More and more, utilities are offering the Internet as an optional medium for issuing bills and accepting electronic payments.

Customer Services

As with suppliers of other products and services, utility customers need access to a customer service function. Customer service provides a channel for customers to communicate with the utility. Utilities support customers in a number of ways in a variety of methods, including personal contact, telephone calls, e-mail and Internet applications, bill stuffers, and media advertisements. Some of the typical customer service functions offered by utilities include:

- Service initiation and termination
- Assistance with billing issues and pricing options
- Information on electrical safety and energy utilization
- Auxiliary metering information and real-time metering data
- Power services consulting and electrical maintenance plans
- Energy auditing and infrared scanning surveys
- Economic development assistance
- Service outage reporting

---

[7] Manual meter readings are taken on a route basis. About 5% of the meters are read each business day. Meter reading and cycle billing issues are discussed further in Chapter 7.

Utilities typically have maintained local offices throughout their service territories so that customers could have convenient personal access to the available services. Even in the fast growing "e-business" world of today, some customers routinely visit their utility's local offices to make their monthly payments -- often in cash. Small town customer service offices have helped to symbolize the utility as a local corporate citizen. Some utilities have elected to offer alternative payment options in order to close a number of their local customer service offices to reduce costs. Such decisions are difficult to make because it does negatively impact a segment of customers who rely on or prefer the utility's local availability.

4.4 THE CUSTOMER AS ENERGY MANAGER

Many years ago, electricity was quite expensive, but the industry was in a declining cost mode as electrification in all customer sectors required the construction of many new large-scale central production plants. The more electricity customers used, the less expensive the unit price of electricity became. Load growth was occurring, and load growth was good. People thought more about the added value of new end-use loads than the amount of the electric bill for those conveniences. But this perspective changed dramatically and rather abruptly.

The OPEC oil embargo of the early 1970s coupled with the period of high inflation that followed caused electricity prices to increase significantly. Utilities started converting oil-fired generation to other fossil fuels. Growth was still calling for construction of new central station production facilities which inevitably triggered the need for a sequence of rate increases. Even utilities that were not faced with a large construction program experienced the need to seek rate relief every two to three years as operating expenses continually increased under the pressure of inflation. Rate case activity was significantly elevated. Customers became much more aware of their electric bills.

Still today, the typical residential customer has little realization of the electricity price structure. Most residential customers have never seen a rate schedule and would likely require assistance in order to begin to understand the pricing structure and terms within the schedule even though residential rates are relatively simple. The residential customer most often responds to the total bill amount: it is either "reasonable" or "too high." A customer's total bill amount will increase when moving from a low use spring (fall) month to a high use hot summer (cold winter) month, even under a nonseasonal price structure. This increase is often perceived by many customers to be a rate hike, although in actuality the bill increase is driven by a higher consumption level and not by unit price changes.

In comparison, larger commercial and industrial customers are more informed when it comes to electricity pricing and billing. The more sophisticated of these customers truly understand their (often complex) rate schedules and utility invoices as well as how they utilize energy throughout their entire processes. These customers are particularly capable of managing their energy bills.

Customers: Electric Service Requirements

All electric utility customers are responsible for (a) purchasing energy resources which adequately meets the energy needs of their operations and (b) utilizing energy in an efficient manner in order to minimize their costs. Thus, customers may be thought of as energy managers, some more effectively so than others. The role of energy manager has been formalized as a major employment position in many large business operations. Energy intensive firms often have a staff of professionals whose duties consist of procurement and management of all utility services, particularly fuels and electricity.

Energy Procurement

The energy manager is concerned with all types of energy resources, not just electricity. Residential, commercial, and industrial customers often utilize other fuels to meet their needs. Residential customers frequently use natural gas for space and water heating and for cooking. In rural areas, where natural gas lines generally do not exist, many of these customers use propane to fuel some of their thermal end uses. Many commercial and industrial customers also use natural gas for space heating and processes. But these sectors also use propane, gasoline, fuel oil, and coal for some applications. Electricity is often competitive with alternative fuels for many end uses. The customer must evaluate meeting end-use requirements in terms of both the cost of the alternative fuel types, including electricity, and the investment and maintenance characteristics of the appliances or equipment that meets the end-use demands.

The electricity requirements of a particular customer influence purchasing characteristics, at least in part. Energy intensive industries utilize large amounts of electricity and pay significant sums of money for their monthly usage. For example, an air separation process distills common air into its fundamental gasses, including oxygen, nitrogen, and argon, typically in a liquid form. The primary raw input – air – is free. The modern air liquefaction plant is highly automated and requires a minimal amount of labor force. But a large quantity of electricity is required to compress and chill the air so that the individual liquefied gasses can be extracted. The cost of the electricity to produce these air-based products may represent well over 60% of the total cost of the gaseous products. Thus, price is a critical driver in the procurement of electricity.

The utility customer's competitive position is another key buyer characteristic relative to the purchase of electricity. For instance, the aluminum industry is energy intensive, and aluminum is extremely price competitive in world markets. Thus, the market-driven price of aluminum exerts great pressure on the price of electricity needed in order for aluminum manufacturers to retain a viable position. As a result, aluminum manufacturing often has been compelled to locate in areas where low cost hydroelectric power production can be procured.

Customer competition also may exist within the same corporation. With capital funds in limited supply, a multi-plant organization often expands or shifts operations at its plants which have demonstrated low operating costs, of which, electricity may be a significant factor. At the same time, high cost plants may lose funding or be subject to closure, events that would certainly impact the local economy including the revenues of the local utility supplier.

Some individual customers are small electricity purchasers locally (e.g., a department store) but represent large users when consolidated on a regional, national, or international scale. These "chain accounts" are often supported by a corporate energy manager function which seeks to minimize the overall cost of electricity through supplier selection, where authorized, and by ensuring service under the most economical rate schedule available at each location.

Power Management

Power management is the process of optimizing the utilization of electricity to extract efficiency while minimizing its cost. The prices of electric service may offer an impetus for controlling, where possible, the overall operation of load devices. If bill amounts are high, invoking changes in energy intensities or electricity usage patterns can prove to be economical even when capital expenditures may be necessary to achieve such operating versatility. On the other hand, if bill amounts are relatively low, the small potential for economical gains would not serve as a primary driver for customers to change their methods of energy utilization. Thus, changes in electricity usage depend upon a customer's degree of operational and economic flexibility.[8]

From an economical perspective, the degree of power management depends heavily upon the pricing structure. A basic flat energy rate, as often used to charge for residential electric service, provides only one dimension for control – total monthly kWh usage. This type of rate neither encourages nor discourages any particular time at which customers should consume energy; therefore, the only way to minimize costs is to use less electricity in total. The degree of control can be increased by pricing structures which are time differentiated. Seasonal rates offer two or three different prices per year. Time-of-day rates offer two or three different prices each day. Real-time pricing offers a different price each and every hour of the year. The inclusion of a billing demand in the rate structure also increases the dimensions for control. Typically, a lower average rate per kWh can be realized by improving load factor which can be controlled by adjusting the energy and/or demand usage relative to each other.

The control of electrical energy is accomplished by making specific end-use devices or energy consuming systems more efficient. For example, a high EER air conditioner uses less energy than a low EER unit of the same Btu rating. Increas-

---

[8] Some customer types have a great deal of operational flexibility and can function in a manner that takes maximum advantage of an electricity pricing structure (e.g., factories that can produce at any time and store their products). Some other customers may have virtually no operational flexibility. For example, restaurants must serve meals when their patrons demand and, therefore, must use the required energy at those specific times.

ing the efficiency of a building with increased insulation and minimization of air loss around doors and windows also results in the use of less energy for heating or cooling. Many such efficiency improvements will also serve to reduce the maximum demand requirements although this effect is generally modest in comparison to the energy use impact.

The control of demand is best realized by rescheduling the operation of specific end-use devices in order to reduce the rate of energy use during otherwise high load periods. This result can be achieved by either a dramatic shift in operation (e.g., running a swimming pool pump only at night once the air conditioning use has passed its peak for the day) or a slight alteration in operation (e.g., cycling ventilation fans in a large facility by switching them off for ten minutes out of every half hour during peak building energy use times). Both of these demand control strategies are based upon the same fundamental principle – increased end-use load diversity by intervention.

The magnitude of demand reduction during a particular demand interval due to load device rescheduling or cycling can be determined by

$$\Delta D = \frac{1}{\Delta t} \sum_{i=1}^{n} (L_i \times t_i) \qquad (4.1)$$

where ΔD = demand reduction in kW
Δt = demand interval in hours
L_i = kW load rating of the ith load of the group composed of n loads
t_i = time in hours the ith load is curtailed during Δt

The energy saved by curtailing load in a particular interval (or in several intervals) may or may not represent a permanent energy savings in total. For instance, if ventilation fans are cycled, the energy that would have been consumed during their periods of curtailment is truly conserved. However, the cycling of devices which are thermostatically controlled generally pay back the energy savings later because of deferred operation of the end-use load device; thus, the total electricity usage is often unchanged.

Some large industrial operations have the operational flexibility to shut down production lines or processes for a few hours at a time without incurring a major economic consequence. Such operations are well suited to an *interruptible service* contract between the customer and the utility or energy service provider (ESP). During system emergencies or high operating cost periods, the utility or ESP calls for the curtailment of customers' loads under the terms and conditions of their interruptible contracts in order to preserve the security of the power system, reduce the costs of electricity production or imports, or meet some other operating objective. While such a load reduction lowers a customer's demand at times, it does not necessarily occur at the time that the customer establishes the maximum rate of energy usage for billing purposes, i.e., the maximum demand. However, interruptible load contracts typically include a demand charge credit that is applicable even if an interruption is not called during the month; thus, the customer can take advantage of a rate feature that lowers the overall bill amount without adverse impacts.

In addition to energy and demand, reactive power requirements can be controlled by managing power factor. Large power service rate schedules often incorporate an auxiliary reactive power clause, which includes monthly charges for poor power factors. Rates having demand charges based on kVA provide a direct incentive for customer improved power factor. Power factor improvement, or correction, is most commonly achieved by installing capacitors throughout the customers electrical system to compensate the reactive requirements of induction motors. Ideally, appropriately sized capacitors should be installed at each major motor site and switched on and off as the motors are operated. Other capacitors can be installed on building or plant circuits and at the service entrance to provide power factor correction of general reactive loads requirements.[9]

Electricity Auditing, Accounting, and Monitoring

Another key role of the energy manager is to conduct certain administrative functions to ensure maximum energy value. An energy audit includes (a) a detailed inspection of electrical equipment (e.g., nameplate data) and an analysis of end-use load and operating characteristics and (b) a summary and analysis of historical electric service invoices and pricing structures. The audit provides a baseline upon which energy usage and billing improvements can be assessed. The audit also provides a means for evaluating what methods are necessary to ensure optimal use of electricity against available pricing options.

The ongoing accounting and monitoring of electricity use is critical for measuring the physical and economic effects of energy action items, e.g., power factor correction, insulation improvements, high efficiency equipment replacement, etc. Operating metrics or ratios, such as ¢/kWh, kWh/ft^2 of conditioned space, or kWh/lb. of product, can be applied to compare electricity usage characteristics from month to month or between similar facilities.

4.5 CUSTOMER-OWNED POWER SYSTEMS

In many ways, the customer's electrical wiring system resembles the utility's power distribution system. A basic electrical system schematic diagram for a medium-size commercial or industrial facility is shown in Figure 4.7. Such custom-

---

[9] Switching of capacitors based on reactive power demand avoids the problem of having too much capacitance in place relative to the real power requirements of the building or facility. As discussed previously, an excessive amount of capacitance on the circuit causes distribution system operating problems.

Customers: Electric Service Requirements

ers usually have an option to take service at either a secondary voltage level or a primary voltage level. In the latter case, the customer furnishes and maintains the transformer and associated facilities necessary to receive service at a high voltage. In essence, the customer's wiring represents an extension of the local utility's distribution system. Each end-use load device has a complete electrical path all the way back to a generation source.

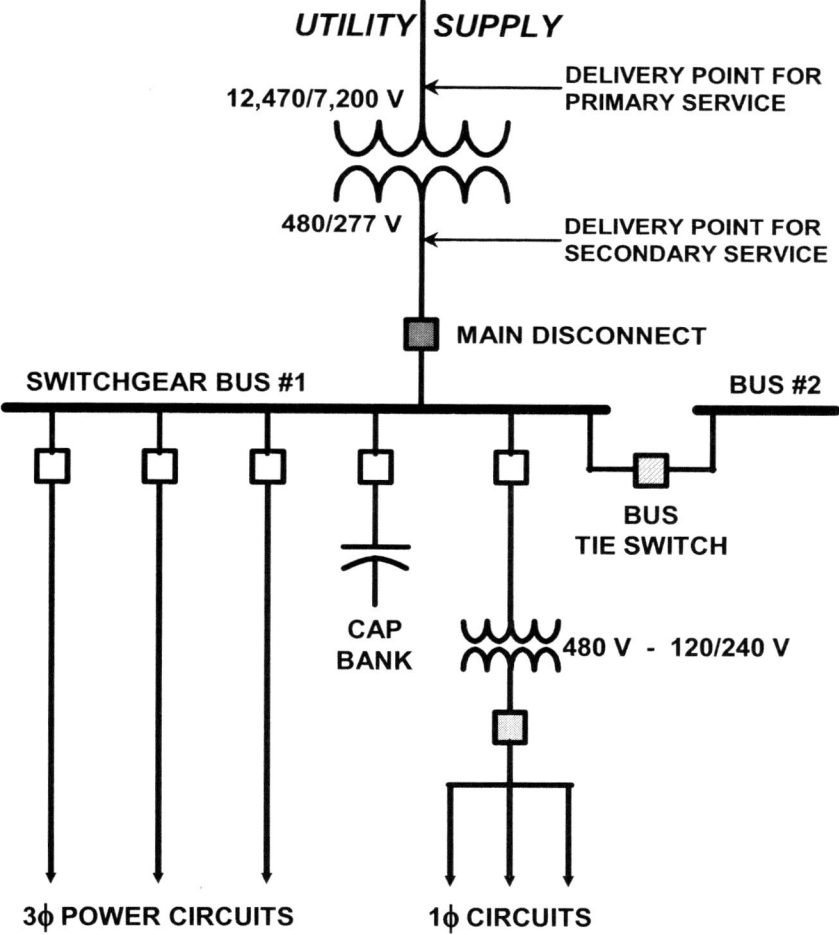

Figure 4.7 A single-line diagram of a 3φ low voltage customer-owned power system. Some customers elect to own the transformer and receive service at a primary voltage rate.

Responsibilities of an Electricity Customer

In order to receive electric service, customers must comply with all applicable national and local laws for electrical safety and fire protection. The principal electrical safety precept in the United States is the *National Electric Code* (NEC). The state Public Service Commissions also have established rules and regulations which include certain responsibilities of utility customers.

Utilities customarily publish, as a section of the electric tariff, a set of electric service rules based in part on authorized legal codes and regulations. These rules are intended to ensure electrical safety, adequacy of service for all customers, and a proper business relationship between the utility and its customers. The responsibilities of electricity customers contained within a utility's rules and policies often include the following items:

Application for Service
- Provide utility with electric service requirements (in a timely manner)
- Select a rate schedule for service (when options are applicable)
- Execute a contract for service (typically waived for small customers)
- Pay a service deposit (where applicable)
- Advance funds for line extensions (where applicable)

Service Connection
- Provide approved facilities necessary to connect systems (e.g., meter base, service entrance conductors, foundations and conduit for underground service, etc.)
- Conduct site preparation work (e.g., trenching to specifications for underground cables)
- Provide easements for utility lines (as necessary)

Beyond the Delivery Point
- Maintain all electrical facilities according to best practices
- Agree not to connect any end-use device that is detrimental to the electric system or to the service of other customers
- Install appropriate protective devices
- Abide by special rules and regulations for on-site generating equipment (where applicable)
- Allow local utility access to its meters and facilities

Use of Service
- Agree not to share or resell electric service (except under special authorized conditions)
- Maintain an adequate power factor (e.g., 90% or 95% lagging) and acceptable levels of harmonic currents (e.g., as prescribed by IEEE Standard 519)

Customers: Electric Service Requirements 127

- Refrain from intermittent or violent fluctuations in the use of service
- Notify utility supplier(s) of material changes in load conditions
- Agree not to tamper with meters or utility facilities

Payment for Service
- Pay monthly bills for electric service by due date (including any taxes, franchise fees, etc.)
- Pay for all auxiliary services (such as nonstandard facilities)

As a service provider, utilities have comparable responsibilities for rendering electric service. Both the customers and utilities must fulfill their respective terms and conditions in order to ensure the successful physical and financial transactions of electricity between the parties.

Customer-Site Power Production

In order to meet all or a portion of its electric service requirements, a customer may install and operate on-site power production facilities. There are several reasons why a customer would choose to produce electricity instead of or in addition to buying it from the electric utility system. The two main purposes for having on-site generation are for reliability and economics.

Emergency Generation and Backup Systems

As discussed previously, backup generation systems are utilized by customers having high reliability requirements, at least for a portion of their loads. As a result, the emergency generators must be sized with sufficient capacity in order to carry the intended load. These generators allow for the continued operation of critical end-use devices or the orderly shutdown of crucial processes or systems. Emergency generation is typically designed to operate only when there is a loss of main electrical supply from the utility system. Thus, emergency generation is not necessarily designed for continuous (long-term) operation.

Emergency generators produce standard single-phase and three-phase voltages which are compatible with customer distribution systems and end-use load devices. A load transfer switch is required for connecting critical loads to the generator during emergency conditions. When a customer's main source of electric power is interrupted or the supply voltage sags deeply for several cycles, the specified loads are switched from the utility source to the on-site generator, and the generator is started, either manually or automatically. As shown in Figure 4.8, the transfer switch physically disconnects the loads from the utility source so that power generated on site does not flow into the utility's system, thereby creating a safety hazard. Once the supply voltage is restored to normal condition, the load is transferred back to the main power source and the generator is turned off.

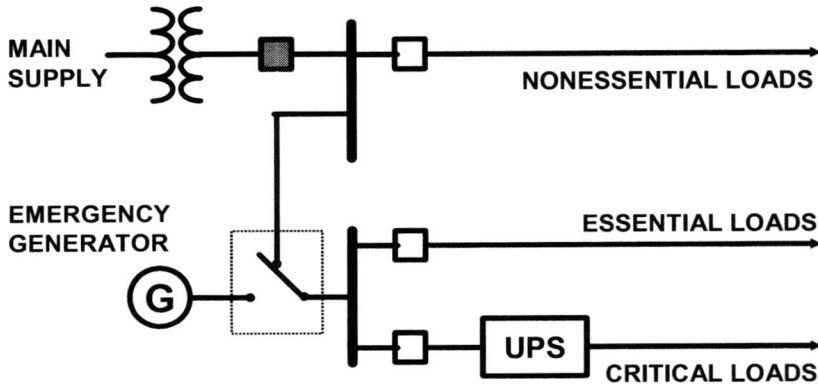

Figure 4.8 Emergency generator and Uninterruptible Power Supply (UPS) backup system. A load transfer switch isolates the customer's generation from the main supply source to prevent power from flowing out onto the utility's lines, which could create a safety issue at times when the grid loses power.

Emergency generators can be driven by internal combustion or diesel engines or by gas turbines. Protective devices, controls, and metering (volts, amps, kW, etc.) are included as part of the generator package. Voltage regulation equipment is also necessary to ensure the proper operation of sensitive load devices. Routine maintenance and operational testing of generator components is necessary to ensure performance during emergency backup conditions.

For backup of critical loads, such as computer centers, a UPS system provides a means to ride through a primary source interruption. As shown in Figure 4.8, a UPS system provide continuous service to critical loads during the interval between the loss of the main electric power supply and interim restoration from the emergency generator.[10] Thus, critical loads do not see an interruption in electrical service.

Cogeneration

In addition to emergency backup, customers may also utilize on-site generation for economic purposes. Some industrial processes have a significant requirement for heat, such as steam, which can be generated from a variety of thermal sources. Electricity can sometimes be economically co-produced along with this process heat, and then used to power end-use loads that would otherwise be served by pur-

---

[10] Other technologies such as an M-G (motor generator) set with a mechanical energy flywheel system can also be used to "ride through" short duration main power interruptions.

chased electricity. The feasibility of cogeneration is often highly dependant on the customer's thermal energy requirements.

Cogeneration of electricity and process heat is accomplished in two basics methods: *bottoming cycles* and *topping cycles*.[11] The reference to bottoming and topping refers to the general order in which each product is produced. As seen in Figure 4.9, the bottoming cycle first generates process heat from an input fuel, such as natural gas, and then uses the resulting waste heat from this process (e.g., the flue gases) for electricity production. The burning of supplemental fuel in the heat recovery boiler raises the steam temperature which increases the potential for electrical output.

In the topping cycle, electricity is produced by means of a reciprocating engine or a gas turbine, as shown in Figure 4.10. The exhaust and/or the heat from the jacket cooling water is recovered by means of a heat exchanger and used to produce steam which can then be used in a process. The electric generator must be sized sufficiently in order to ensure meeting the thermal requirements of the process.

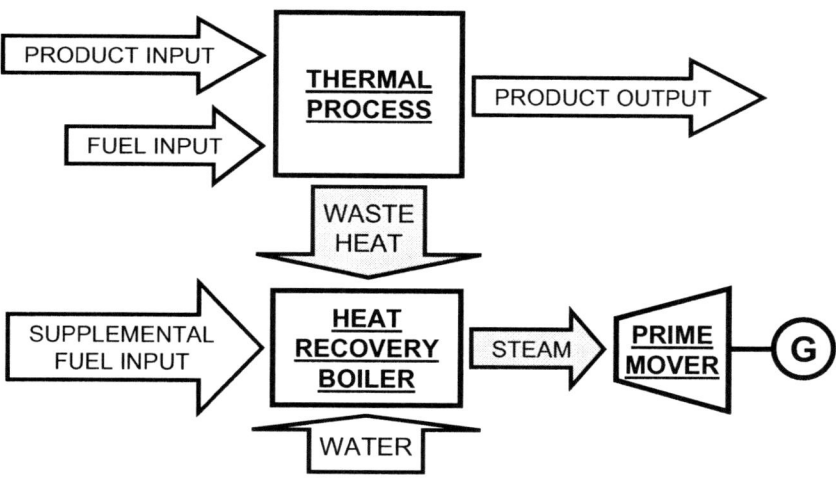

Figure 4.9 **An example of a basic cogeneration bottoming cycle.** Heat for use in a thermal process is produced first. The waste heat is then captured and used to produce steam via a heat exchanger. A steam turbine can then be used as a prime mover to drive an electric generator.

---

[11] A third methodology known as *combined cycle* is a union of the topping and bottoming cycles. With its high efficiency electrical output, the combined cycle is most often used by generating utilities and independent power producers.

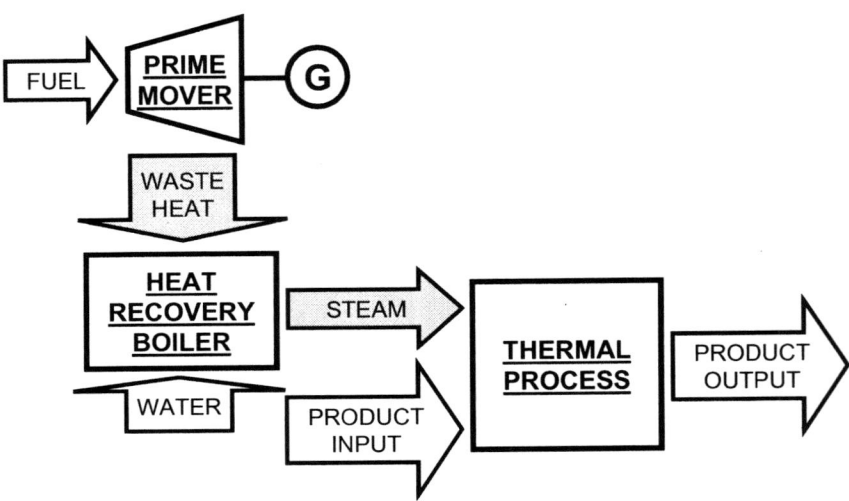

Figure 4.10 Example of a basic cogeneration topping cycle. Electricity is produced first by means of an engine or turbine in which fuel is burned. Waste heat from the prime mover is then captured by means of a waste heat recovery boiler that produces steam that can be used in a thermal process.

Unlike most emergency backup generation systems, cogeneration systems are designed typically for more of a continuous operation. Many cogeneration systems are operated while interconnected with the local utility's electrical system. *Parallel operation* of a customer-site generator yields benefits. Under parallel operation, the customer's generator would effectively be connected to an *infinite bus* (a "stiff" system created by the running generators of all of the interconnected utilities), which would provide the customer with voltage stability and frequency regulation that would be greater than if operating in isolation. The utility receives any excess energy on the grid that reduces the fuel burn at central station plants.

In order to operate in parallel with the utility's system in a safe and effective manner, the customer-site generation must produce a compatible voltage and be equipped with certain facilities and protection devices. Successful synchronization of the generator with the utility's system is achieved with instruments that ensure that the generator is in phase with the utility's system and operating at synchronous speed. The generator must be connected to the system in a manner that avoids undesirable transients and voltage aberrations and prevents damage to either the customer's or the utility's equipment. Utilities often require that a synchronization relay be installed to supervise the interconnection process. The relay would shut down the process if any problem occurred during the attempt to parallel the systems.

Customers: Electric Service Requirements

Automatic disconnection of the customer-site generation is necessary in the event that the utility's power delivery system looses its primary generation source, such as when a substation circuit breaker opens and the distribution feeder then becomes de-energized. Connected to the de-energized feeder, the customer-site generator would produce a voltage across the line and *back feed* power, which would create a safety hazard for line workers.

Qualifying Facilities

Under the U.S. PURPA legislation, the FERC developed specific criteria by which a cogenerator or a *Small Power Production* (SPP) facility[12] can secure a *Qualifying Facility* (QF) status and receive certain benefits. Since the primary intent of the PURPA law is to promote energy efficiency, a new QF must meet certain minimum energy standards in order to be qualified. The FERC reviews and certifies applications for QF status, which must be submitted to the FERC by the potential qualifying facility.

For cogeneration topping cycle operations, at least five percent of the total annual energy output of the facility must be in the form of useful thermal energy. But for facilities fired by any oil or natural gas, all of the annual electrical output and one-half of the annual useful thermal output can not be less than 42.5% of the annual energy input. Furthermore, if the annual useful thermal output is less than 15% of the total annual output, then all of the annual electrical output and one-half of the annual useful thermal output cannot be less than 45% of the annual energy input. For cogeneration bottoming cycle operations, the annual electrical energy output can not be less than 45% of any supplemental gas or oil firing applied in the heat recovery process. Fuels used in the main thermal process are not a consideration for determining qualification of the facility.

For qualifying SPP facilities, the use of natural gas, oil, or coal is limited to no more than 25% of the total annual energy input. An SPP must utilize biomass, waste, or renewable resources for more than 50% of its annual energy input.

Utility Service Options for QFs

The PURPA law requires utilities to allow QFs to interconnect and to receive certain utility services. QF customers are required to pay all of the costs necessary for making a safe interconnection for parallel operation.

A QF may use some or all of its electrical output to serve its end-use load processes. In some cases, the on-site generation may not provide the total capacity to serve the customer's entire load. Some QF operations are *load following* in nature, and thus the generation is operated only when there is a thermal requirement. In such cases, the other load of the facility would require electricity from another source. In order to satisfy the need for electricity in addition to the on-site generation output, a customer would contract with the utility for *Supplementary*

---

[12] An SPP facility produces electrical energy by utilizing biomass, waste, and/or renewable resources (e.g. solar, wind, or water) as its primary energy input.

Electric Service. Supplementary Service could be used in two ways. The customer could buy electricity to meet base load requirements and then utilize the generation to serve peak demands, a strategy referred to as *peak shaving.* If the generation must operate continuously, the customer would then purchase electricity from the utility in order to serve peak demands.

A customer might have excess generation capacity at times (or even continuously). This excess power will flow onto the utility's system through the interconnection established for parallel operation. The PURPA law requires a utility to purchase the electricity from a QF at the utility's *avoided costs* of capacity and/or energy. The avoided energy cost and payment would depend on the utility's mix of fuel and purchased power costs across the year, i.e., the marginal cost of the next unit output of energy. However, the avoided capacity cost depends greatly on the utility's capacity position at the time the contract is considered. The avoided capacity cost would be minimal for a utility that is long in generation capability and relatively high for a utility that is short in generation capability. The amount of avoided capacity cost payment to a QF would depend upon the dependability of the generator's output, and/or its ability to be dispatched like a utility resource. *As available* power is not as valuable as *firm* power.

QFs often require two additional utility services known as *standby.* In the event of an emergency forced outage of the on-site generator, the customer may require *Backup Electrical Service.* With this service, the utility would provide replacement electricity to the customer that would normally serve the load by means of the on-site generator. The price of electricity for this service is often based on the probability of failure of a typical benchmark generation facility. Standby service is also provide for scheduled outages of a customer's generator. *Maintenance Electric Service* is provide for the periodic shut down of the customer's generator for purposes of routine overhauls and repairs. Since these outages can be planned outside of utility peak times, the replacement electricity can be priced relatively low.

4.6 WHOLESALE AND COMPOUND RETAIL CUSTOMERS

Most retail customers are characterized by a primary type of operation and are broadly categorized as either residential, commercial, or industrial. Within each of these major classes, more specific definitions of customer types emerge. For example, the commercial class includes such diverse business activities as retail products and services, offices, and warehousing. The industrial class includes such prominent activities as manufacturing and mining.

A number of particular businesses are not easily categorized as one of the three basic customer classes. Thus, agricultural operations and electric transportation, for example, are often depicted as separate customer classes. Outdoor lighting is also a unique customer class which includes street and area lighting, traffic signals, and a variety of other applications such as billboard and telephone booth lighting. Public authorities are mostly like the commercial class, but they are often set out as a separate class.

Some retail customers are categorically more complex than others. For example, an airport (NAICS code 481111) is classified as a single commercial entity, but it actually represents a composite of a number of different business operations. The typical metropolitan airport consists of offices, retail shops, restaurants, hangars (warehousing), runway lighting, and a fire department. Some airports also have a fairly sophisticated electric transportation system for shuttling people between concourses and the terminal. Another example of a mixed-use customer is a military base. Large military bases include residential operations (i.e., homes and barracks), offices, retail shops (e.g., the PX), warehousing, a hospital, machine shops, and a host of other support services. A third example would be a large university which would include a mix of residential and various commercial operations. Even a large industrial facility may include some commercial operations, such as offices and a cafeteria.

These *compound retail* customers have a total requirement for electric service that represents the combined end-use needs of a diverse number of operations, all of which are located on the same, but often expansive, premise. Such customers frequently have extensive distribution systems in order to route power to a number of widespread points of use. In order to manage energy use, compound retail customers often estimate or even submeter electricity consumption at key locations throughout the operation. Energy usage and cost then can be allocated to different internal departments of the customer's operation. Some industrial customers utilize such measures to evaluate and compare the operating ratios of two or more comparable production lines.

A single retail delivery point may also serve several unaffiliated customers. For example, a high-rise building owner typically has a number of commercial tenants (a variety of offices, restaurants, and retail shops). Power is distributed from the building's main bus to each of the tenants through a common distribution system. In many locales, it is illegal for the building owner to submeter each tenant and render an invoice for electric service based on use. Such a practice would constitute a sale of electricity and, if allowed, would consider the building owner to be a "utility" subject to the jurisdiction of the local regulatory agency. As a result, the costs of utilities, along with the other occupancy costs, are included in the monthly lease of floorspace.

Conversely, a *wholesale* customer is a utility which is authorized to resell electricity it purchases to its ultimate customers. A wholesale customer, such as an electric cooperative or a municipal utility, may purchase all or a portion of its energy requirements under FERC regulated tariffs and/or from the wholesale electricity markets. Wholesale customers typically take service at a transmission or distribution voltage level and often have multiple points at which it receives service. The wholesale customer charges its ultimate customers for the cost of purchased power along with its costs of operation, including any generation, transmission, distribution, and customer services.

4.7 SUMMARY

Utility customers have specific requirements relative to their electric service. These requirements encompass both the physical attributes associated with the delivery and receipt of electricity and the various customer services which support the ongoing transactions between the utility and the customer. The prudent customer manages energy utilization and evaluates the available pricing options in light of needs and specific modes of operation in order to optimize usage and the costs of electricity.

Electric service requirements are important to electricity pricing because both the specific needs of each customer and the aggregate needs of all of the customers are primary drivers for determining a power system's structural and operational requirements and thus its costs of providing service. The amount of power taken at one voltage level has a different cost than the same amount of power taken at a higher or lower voltage level.

Most customers are successfully served by the basic power system design which is built and operated according to best practices standards of the industry. But due to the unique characteristics of certain end-use loads or processes, some customers have heightened requirements that necessitate investing in additional service enhancements to ensure adequate electrical service. The customer may incur this incremental obligation as a utility supply-side price solution (e.g., distribution system upgrades) and/or as a customer point-of-use cost solution (e.g., on-site generation).

5
Power Delivery Systems: Transmission and Distribution

5.1 INTRODUCTION

The electric power system represents the infrastructure that produces electricity and transports it to customers in order to serve the energy requirements of end-use load devices. The classical, vertically integrated utility structure was briefly discussed in Chapter 1, and the typical power system's functional organization was illustrated in Figure 1.1. In short, electricity is generated at various central station production plants, transmitted over often long distances to major load centers throughout a utility's service territory, and distributed over relatively short distances to individual customers. These functions are fundamental and common to both a fully regulated utility operation and a restructured, competitive electric industry. Electric power always flows along the chain from the points of generation to the points of end use, even when some functions of the power system are provided as a regulated service of the incumbent utility, while other functions are provided by alternative suppliers as a competitive service.

Most electric service customers are served at the secondary voltage distribution level, which, electrically speaking, is farthest removed from the energy production source. To reach a secondary level customer, electricity must travel through numerous functional segments of the power delivery system, as shown in Figure 5.1. In the course of this journey, the same power is moved at varying levels of voltages and currents. Generated voltage is stepped up for long distance transmission in order to minimize the electrical losses caused by current flow through a conductor's resistance, i.e., the i^2R effect. The voltage is progressively lowered as power is distributed locally and ultimately delivered to a customer's low voltage wiring system where end-use loads are connected.

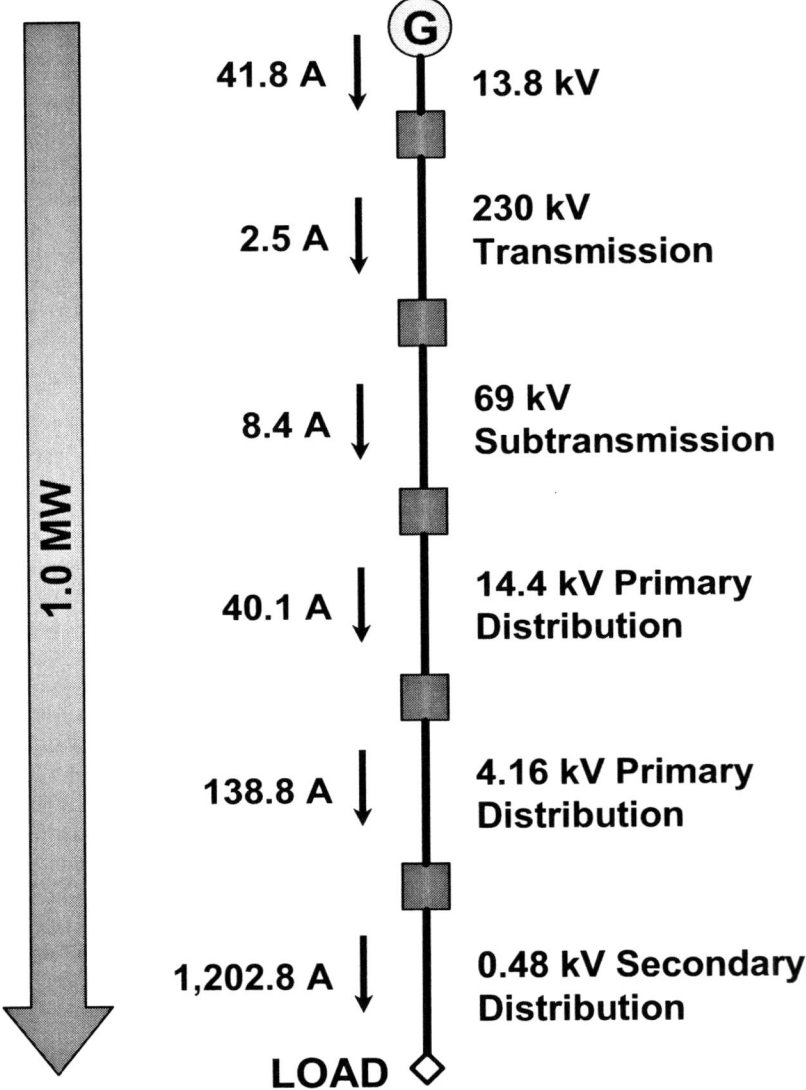

Figure 5.1 Relationship of voltages and currents throughout the power system. A megawatt of three-phase power moving from the generation through the various voltage levels of a power delivery system to serve an end-use load requirement. The load current in any line segment is indirectly proportional to the operating voltage of that segment. *For simplicity, this example ignores the electrical losses that are realized by the flow of electricity through lines and transformers, and power factor is assumed to be 100%.*

Power Delivery Systems: Transmission and Distribution

A key component utilized throughout the power system is the *transformer*. Transformers change voltage and current levels at vital locations. A *generation step-up* (GSU) transformer is used at a production plant to increase voltages. Other transformers are used to *step down* the voltage at various points. *Power transformers* are used to transfer blocks of power between high and medium voltage segments of the system, such as in substations having transmission to subtransmission or subtransmission to primary distribution voltage conversions. *Line transformers* are used to transfer power from the primary to the secondary voltage distribution systems.

Transformers have both an input set of terminals and an output set of terminals, as shown in Figure 5.2.[1] These external terminals connect to internal coils of insulated wire that are wrapped around a magnetic core, typically composed of iron. When a voltage is applied to the input terminals, the resulting current that flows through the input coil creates a magnetic field. The core material significantly intensifies this magnetic field. The magnetic field in turn "cuts across" the output coil thereby inducing a voltage, which causes a current to flow in the output side. Thus, a transformer can be viewed as a device that connects two electrical circuits by means of a coupling magnetic circuit.

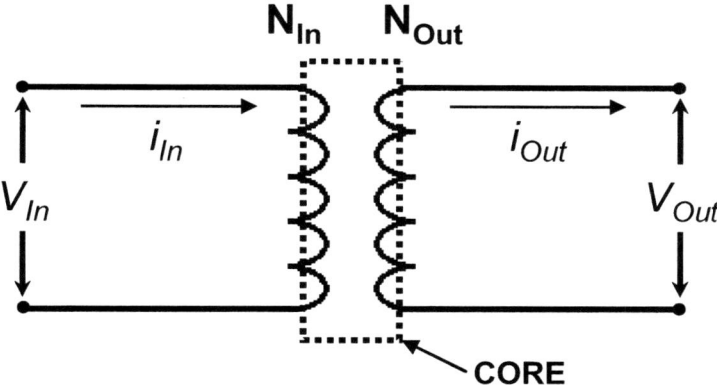

Figure 5.2 Elementary operation of a basic transformer. A transformer consists of coils of wire wrapped around a core of magnetic material, such as iron. The ratio of the input voltage to the output voltage is directly proportional to the ratio of the turns (N) of the input and output coils. Thus, a single-phase transformer having a 30:1 turns ratio, which is connected to a 7,200 voltage source, would yield a 240 output voltage.

---

[1] Transformers are often referred to as having a "primary side" and a "secondary side." This terminology can be confusing considering that transformers are used in transmission, subtransmission, primary distribution, and secondary distribution applications. Thus, the terms "input" and "output" are more descriptive.

With a perfectly efficient transformer, the volt-ampere power output would be exactly equal to the volt-ampere power input as given by

$$V_{IN} \times i_{IN} = V_{OUT} \times i_{OUT} \tag{5.1}$$

where V_{IN} = AC voltage applied to the input coil
i_{IN} = current flow in the input coil
V_{OUT} = AC voltage induced in the output coil
i_{OUT} = output coil current flow to a connected load

But because of losses transformers do not operate with perfect efficiency. Transformer losses consist of two components: *core losses* and *load losses*.

When a transformer is energized (i.e., an input voltage is applied), a component of the total current known as an *exciting current* flows in the input circuit. The exciting current is present even if no load is connected to the output side of the transformer. A portion of the transformer's exciting current, referred to as the *magnetizing current*, lags the applied AC voltage by 90° in the same manner as observed with a simple inductor. The magnetizing current creates the fluctuating magnetic field, or flux, which induces a voltage across the output coil of the transformer. The remainder of the exciting current is due to the power requirements of the core losses.[2] Core losses consist of two elements: *hysteresis losses* and *eddy-current losses*. Hysteresis losses are caused by the continuous 180° realignment of the magnetic domains within the core as the applied AC voltage causes the magnetic field to change direction. Hysteresis losses depend on the type of magnetic material used in the construction of the core. Eddy-current losses are caused by electrical currents that are induced directly in the core material itself. These losses are reduced to an acceptable level by constructing the core with insulated laminations instead of a solid material. As a result, the induced currents in each of the laminations add up to a total eddy-current that is less than the induced eddy-current of a solid core of the same size. Core losses require power to be supplied and thus are measured in units of watts.

A transformer's excitation current also creates a minimal i^2R loss due to its flow through the resistance of the transformer's conductor windings. However, the load-related portion of the total current flows in a transformer results in significant i^2R losses. These load losses are present in both the input and output coils of the transformer.[3] In comparison, the full load losses of distribution class transformers are generally on the order of 3 to 5 times the no-load losses.

---

[2] Core losses are also referred to as *iron losses* or *no-load losses*.

[3] Load losses are often referred to as *copper losses* because of the prominent use of copper wire in constructing transformer windings.

Transformer losses create heat, which must be adequately dissipated in order to prevent untimely failure of the unit. Therefore, power system transformers typically use oil-filled tanks into which the core and its windings are immersed. The oil provides a uniform insulating medium that conducts heat away from the core to the tank and beyond to the outside air, thereby preventing internal hot spots. In addition, insulated bushings are used to connect the internal windings to external terminals so that high voltages can be safely applied.

Line conductors are another principle source of i^2R losses that are significant in an electric power system. A smaller diameter conductor has a greater measure of resistance than a larger diameter conductor of the same material. Copper conductors have less resistance than aluminum conductors having the same cross-sectional area. The resistance of a conductor is a function of temperature, frequency, and, in some cases, current density. As the temperature increases, the resistance of a conductor also increases. The percent increase in a conductor's resistance between cold winter conditions and extreme summer conditions is fairly significant.

The resistance of a conductor carrying alternating current is also affected by the phenomenon known as the "skin effect" where the current flows predominantly along the outer circumference of the conductor. The skin effect is measurably increased with higher frequencies in some conductor types and sizes. The reduction in conductor cross-sectional area due to the skin effect gives rise to a higher resistance. Some conductors incorporate magnetic types of metals, such as steel. The inductive effects of current flow on these magnetic materials further impact a conductor's resistance. Thus, conductors carrying higher levels of amps per area exhibit slightly higher resistance characteristics.

Conductor resistance is commonly measured in ohms per mile. Thus, for the same load current, a longer line would have higher losses than a shorter line built with the same conductor. Greater losses are also realized as load increases. Customer loads (impedances) are connected to the power system in parallel rather than in serial fashion. For example, a 120-volt circuit feeding a characteristic household living room would have connected some lamps, a stereo system, and a television set. When plugged into the convenience outlets, these devices would all be connected in parallel. When two or more impedances are connected in parallel, the equivalent combined impedance is less than the sum of the individual impedances; therefore, the total supply current increases. End-use devices having power factors that are less than 100% also cause an increase in losses because of an increase in load current requirements.

The transmission and distribution systems are mainly composed of lines and transformers, which are essential for connecting customers to electricity production sources. The power required to serve customers includes all of the associated losses that are incurred by these facilities as power moves from the generators to the loads. Losses are offset by generating sufficiently more power than the end-use load actually requires. Figure 5.3 illustrates a model of real power flow through a utility system composed of several operating voltage levels.

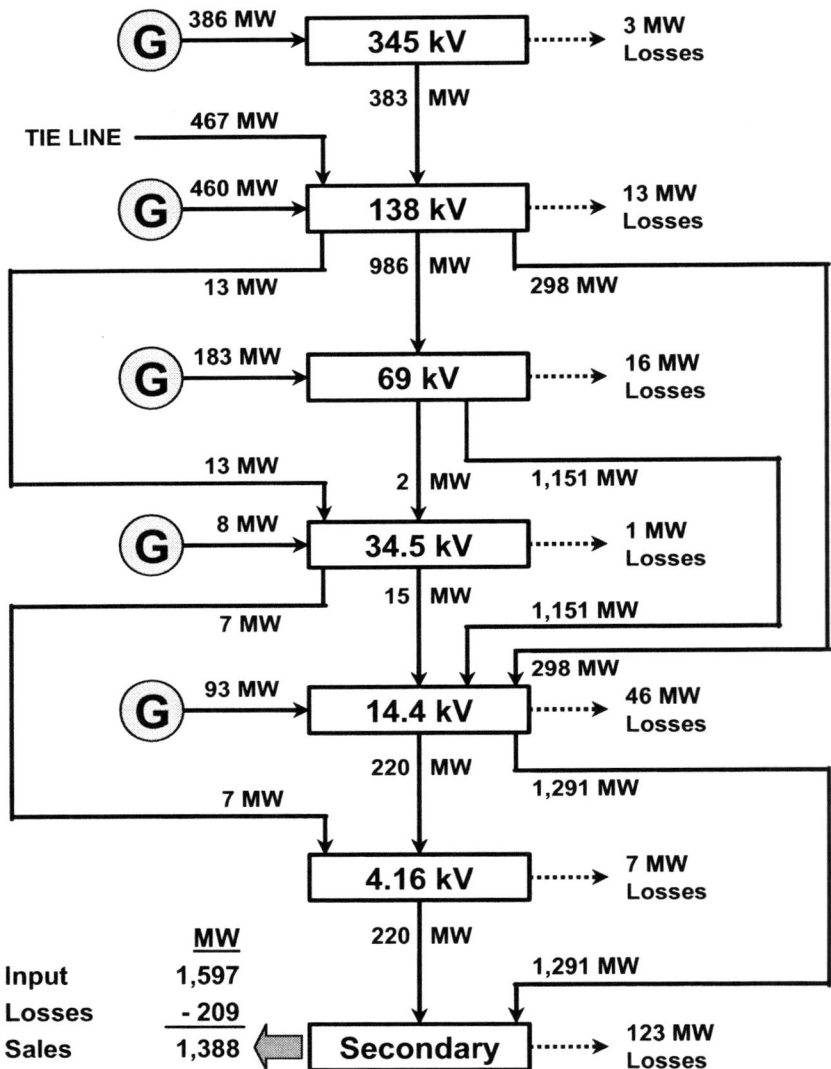

Figure 5.3 A model of real power flow through the voltage levels of a power delivery system. The power flow and losses associated with just the secondary voltage level demand sales at the time of system peak are shown. Losses are realized at each level due to lines and transformers. Due to the complexity of the system, power flows along multiple paths including bypasses around some voltage levels. A complete model of the system would include both demand and annual energy sales and associated losses at all voltages levels. Such models are developed as part of the cost-of-service study allocation process.

Power Delivery Systems: Transmission and Distribution

Several important observations can be made in reference to the example power system shown in Figure 5.3. Transmission and distribution system layouts are often complex. As a power system evolves, higher voltages are meshed with the existing voltages while lower voltages are sometimes phased out of service. Transmission voltage levels have increased significantly since the early 1900s. What was once considered to be a transmission voltage may now be classified as subtransmission. Voltage level bypasses have been created as new transformer designs have made possible direct voltage conversions from the new higher transmission levels to primary distribution levels and from higher distribution voltages to the customer's utilization voltage requirements. While the 14.4 kV and the 4.16 kV systems coexist (even overlap) in the same locations, eventually, as load is transferred to the higher distribution voltage, the 4.16 kV system will be fully retired. The elimination of this voltage level would reduce system losses.

Figure 5.3 only shows the power flowing to serve the secondary distribution customers' demands during the hour of system peak. About 13.1% of the generators' 1,597 MW output is required to "serve the losses" while delivering 1,388 MW to the secondary distribution customers. Most of the 209 MW of losses occurs in the distribution system rather than in the subtransmission or transmission systems due to the lower operating voltages and thus higher load currents. Losses have a compounding effect. The losses that occur in the lower voltage level materialize as load to the higher voltage levels. Thus, a portion of the 14.4 kV losses is caused by the power that must be supplied to the losses at the secondary level. While a majority of a utility's customers take service at low voltages, larger commercial and industrial customers often take service at a primary, subtransmission, or transmission voltage. These higher voltage customers would realize much less percent losses, and the overall system losses would be something less than 9.5%. The loss characteristics of the system depend greatly on how the system is structured and its scale as well as where customers are connected.

Generators may be connected to the system at several different voltage levels (not just at the highest voltages). This "distributed generation" positions a portion of the electricity supply in closer proximity to the load, thereby minimizing losses to some extent. The generators, as well as the power delivery system, must have adequate capacity to meet customers' demands and the associated losses. In addition to the demand losses, energy losses must also be served. While demand losses are determined at the time of the system peak load, energy losses occur in all hours of the year. Thus, fuel is consumed at the generators in order to generate electrical energy that eventually will be released from lines and transformers as heat. Models of demand and energy power flows are an integral step in the development of demand and energy-related cost-of-service study allocation factors.

The following sections address the functional structure and operation of the electric power transmission and distribution (T&D) system in more specific detail. But this review of the system proceeds in a less conventional manner by starting from a customer's point of view -- at the secondary distribution delivery point -- and then working upwards to the genesis of electricity production.

5.2 POWER DISTRIBUTION

A power distribution system is comprised of lines, equipment, and facilities which route electrical energy from a local source of supply, such as a substation, to adjacent customer delivery points. Thus, electricity is distributed conjointly to a group of customers who are all connected in common. From another perspective, a distribution system acts as a natural *aggregator* because it "gathers" together customers (and their loads) and "ties" them to a centralized point. Distribution system customers have varying electric service requirements, but they all require a comparable means of electricity transportation.

While the fundamental concept of distributing electric power to customers is relatively simple, the structure and operation of a distribution system is fairly complex. A key issue of power distribution is reliability since the typical distribution system is exposed to a wide variety of hazards. Most customer outages occur because of events within the distribution system. Planning and designing distribution systems that perform acceptably requires a considerable amount of knowledge about customers' electric service requirements, local geographical constraints, and alternative technological methods for delivering power.

Distribution System Facilities

Distribution systems are characterized by main feeders and lateral tap lines, overhead and underground construction methods, and primary and secondary operating voltages. Line construction is both an electrical and a mechanical problem. For example, conductors must be sized to adequately carry the electrical load. But, larger wires require more mechanical support in the form of stronger poles and higher tension-bearing downguys and anchors. Safety is ensured by proper line and facility clearances, such as those prescribed in the *National Electrical Safety Code* (NESC). In general, higher voltages require greater physical clearances between (a) the wires and electrical equipment and (b) the ground, trees, buildings, and other adjacent structures. Higher voltages also require greater levels of insulation to prevent electrical flashover.

While a variety of voltages are utilized in power distribution systems, there are two basic subsystems of distribution based on the levels of operating voltage: *secondary* and *primary*. Secondary distribution represents low voltage levels at which most utilization equipment is directly powered.[4] Because of voltage drop restrictions, secondary distribution lines are limited to short extensions. Primary distribution is represented by higher voltage levels, which are utilized to move a relatively large amount of power over distances of up to several miles in order to serve a great number of customers. Transformers are used at or close to customer locations in order to complete the delivery of power at a low voltage level.

---

[4] A limited number of end-use load devices operate directly at voltages in excess of 1,000 volts. Some examples include special purpose motors and arc furnaces.

Power Delivery Systems: Transmission and Distribution

A distribution customer receives service at either a secondary or a primary voltage. Customers taking service at a primary voltage must provide their own transformations to secondary utilization voltage levels at one or more locations throughout their distribution systems. Depending on the sizes and types of end-use load devices utilized, a customer requires single-phase service, three-phase service, or both. Both "wye" and "delta" configurations are frequently used for three-phase applications, as illustrated in Figure 5.4.

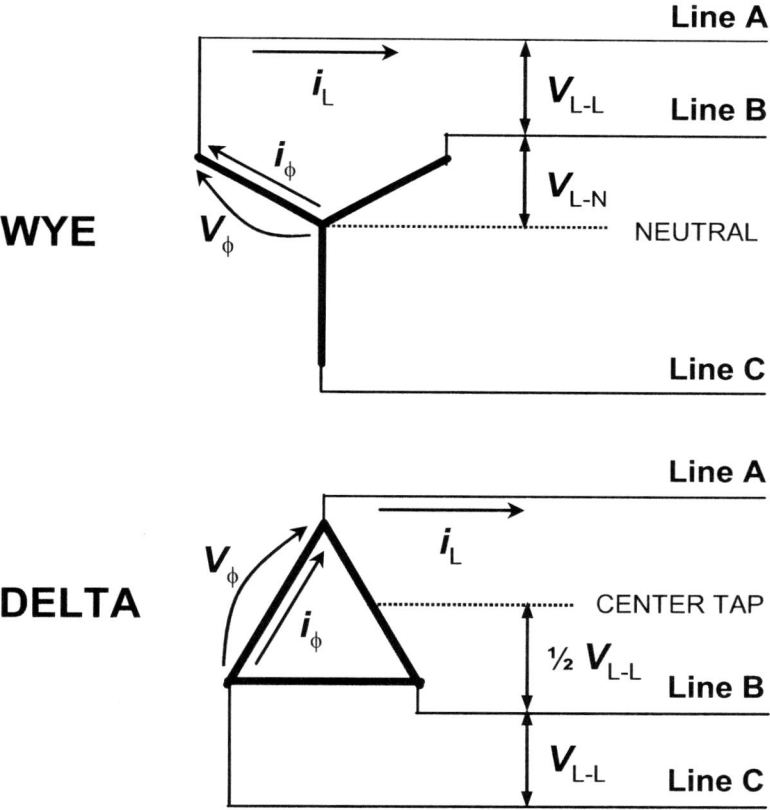

Figure 5.4 Common three-phase wiring configurations. The "wye" and "delta" shapes represent the connection of internal transformer coils to external lines designated as Lines A, B, and C. The wye connection provides both single-phase and three-phase service. The delta connection is used typically for balanced three-phase loads, but a center tap of one of the phases can be used to provide service for small single-phase loads.

Wye-Connected Systems

Wye-connected (Y) systems produce a combination of single-phase and three-phase voltages. As shown in Figure 5.4, the wye configuration consists of four conductors: three individual *line* wires (designated as A, B, and C) and a *neutral* wire. In essence, the wye configuration is the result of combining three single-phase systems with a common neutral.[5] The AC voltage of each phase is displaced by ±120° with respect to the other two phases, as previously illustrated for three-phase voltage in Figure 4.2.

The relationship of the single-phase voltage to the three-phase voltage in a wye-connected system is given by

$$V_\phi = V_{L-N} = V_{L-L} \div \sqrt{3} \qquad (5.2)$$

where V_ϕ = voltage across a transformer phase
V_{L-N} = voltage between any line wire and the neutral
V_{L-L} = voltage between any two line wires

For example, a wye-connected 480 volt, three-phase system yields a single-phase voltage of 277 volts as measured between any of the line wires and the neutral conductor. Thus, three single-phase transformers each having an output rating of 277 volts, arranged in a wye fashion and connected to a three-phase voltage source of their rated input, will output a 480/277 volt three-phase service.

In the wye-connected system, the current flowing in a line wire is the same as the current flowing through the phase, i.e., $i_L = i_\phi$. Since apparent power, S, is the product of voltage and current, the three-phase power of a balanced wye-connected system is calculated by

$$S_{3\phi} = 3 \times V_\phi \times i_\phi \qquad (5.3a)$$

or

$$S_{3\phi} = \sqrt{3} \times V_{L-L} \times i_L \qquad (5.3b)$$

The wye-connected system is normally used for loads containing a mix of significant three-phase and single-phase voltage requirements. Three-phase end-use load devices, such as large motors, are connected to all three of the line wires while single-phase loads, such as convenience outlet circuits, are connected between one of the three line wires and the neutral. Three-phase devices represent a balanced load, but single-phase devices and branch circuits must be alternatively connected to the three different line conductors of a power distribution circuit in

---

[5] With a perfectly balanced three-phase load (i.e., each phase carrying the same amount of load), the current flow in the neutral conductor would be zero.

order to ensure a reasonable balance in the loading of the three-phase system. A balanced load is important for minimizing losses and ensuring the proper operation of line equipment.

Delta-Connected Systems

The delta (Δ) configuration consists of three individual phase wires but no neutral wire.[6] Delta-connected systems are typically utilized where the load is composed of strictly balanced three-phase devices and where any associated single-phase voltage requirement is minimal. An example of such a load is a water pumping station with a large three-phase motor-driven pump and a security light.

With the delta configuration, a single-phase voltage can be obtained by center tapping the transformer coil between any two of the phases, as shown in Figure 5.4. Thus, a 240-volt delta system can also produce a 120-volt single-phase voltage. Alternatively, service can be provided by using independent delivery methods for a customer's three-phase and single-phase loads.

In a delta-connected system, the line-to-line voltage (V_{L-L}) is the same as the phase voltage (V_ϕ). The relationship of the phase current to the line current in a balanced delta-connected system is given by

$$i_L = i_\phi \times \sqrt{3} \qquad (5.4)$$

where i_L = current flowing on the line wire
  i_ϕ = current flowing through a transformer phase

As with the wye configuration, the apparent power in a balanced delta-connected system can be determined from either the line or phase voltages and currents using Equations (5.3a) and (5.3b).

A major advantage of the delta configuration is that if one of the service transformers fails, it may be possible to continue three-phase service with only the two remaining transformers.[7] If the three-phase load is minimal in comparison to the single-phase load requirement, an "open delta" configuration using only two transformers may be utilized. An open delta service arrangement is sometimes used as the initial service installation when the three-phase load is expected to grow; thus, a third transformer would be added at a future time.

---

[6] A ground wire used for safety purposes is not considered to be a neutral although neutral wires are typically grounded.

[7] Under emergency conditions when a transformer fails, the other two transformers can be reconnected as an "open delta" in which 57.7% of the total capacity of the original three transformers is available.

Secondary Distribution

Secondary distribution is a customer or utility electrical system that has operating voltages that are generally less than 1,000 volts. Typical single-phase and three-phase secondary voltage applications are shown in Table 5.1. Secondary distribution employs two-wire, three-wire, and four-wire configurations, depending on the particular service methodology. Most two-wire, single-phase services are used to connect outdoor lighting fixtures to the distribution system, although a number of residences have older wiring systems that have not been upgraded to a three-wire service.

An example of the layout of a secondary distribution system is shown in Figure 5.5. The secondary system is connected to the primary system by means of a line transformer that reduces the voltage to a utilization level, such as 120/240 volts. Service wires, often referred to as service drops or service lines, connect customers to the secondary conductors. An individual service line is necessary for each customer, but in this example, four customers share the transformer and spans of secondary conductors. In a rural setting where individual customers are separated by long distances, each customer would require a dedicated transformer to which the service line is directly connected. Thus, secondary lines may be necessary when more than one customer is being served, such as in a neighborhood where homes are located close to each other. The distance between the line transformer and a customer is limited due to voltage drop conditions. If a customer locates too far away from the primary system, a primary line extension would be necessary in order to locate the point of transformation closer to the load site to ensure an adequate delivery voltage.

Table 5.1 Common U.S. Secondary Distribution Service Voltage Applications

	Single-Phase Voltages	Three-Phase Voltages	Combined Voltages
Two-Wire Configuration:	120		
Three-Wire Configurations:	120/240	240Δ	
		480Δ	
Four-Wire Configurations:			208Y/120
			240Δ/120 *
			480Y/277

* One phase of the delta is center tapped in order to produce the equivalent of a 120/240 volt, single-phase system.

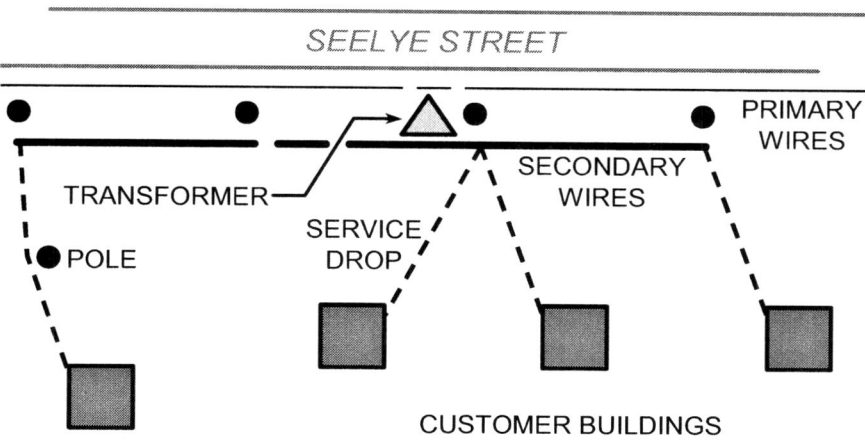

Figure 5.5 Schematic diagram of an overhead, single-phase secondary system. Four low voltage customers, located in close proximity to each other, are served from a single line transformer, which reduces the primary voltage to a secondary voltage for ultimate delivery of power to the customers. Secondary conductors along the street act as a bus from which service drops are used to connect customers.

Low voltage, secondary customers may be connected to the power system by means of an overhead or underground service methodology. An overhead service drop requires an appropriate point of connection to the customer's building, such as a service masthead extending above the roof.[8] Overhead secondary and service lines may be installed as open wires, separated by space, or as an aerial bundle in which the wires are twisted around each other ("n-plex" cables). In either case, the hot leg wires are covered with a weatherproof insulation while the neutral conductor is left bare. A *duplex cable* consists of one covered wire and one bare wire, and it is most often used to serve outdoor lighting fixtures and other 120 volt types of loads. *Triplex cable* consists of two covered wires and a bare neutral, and it is commonly used to serve three-wire, single-phase applications, such as 120/240 volt loads. *Quadruplex cable* has three covered wires with a bare neutral, and it is used for three-phase secondary and service applications.

Some underground secondary and service conductors are quite similar to overhead n-plex cables, but the underground neutral conductor is usually covered with an insulation that distinguishes it from the hot leg conductors. The conductors may be twisted together or bound in a flat configuration. Another type of wire used for underground secondary and service conductors is a concentric neutral

---

[8] At the masthead, a customer's service entrance cables connect the service drop to the meter cabinet.

cable in which one, two, or three low voltage, insulated conductors are enveloped in a protective jacket. Strands of neutral conductors are then spiral wrapped around this bundle.

An underground service line to a customer may originate from a primary underground system or from the overhead system. From either an overhead secondary line or directly from the output terminals of a pole-mounted transformer, a service line is taken underground by directing it down the side of a pole within a protective conduit. For the most part, underground cables are designed to be direct buried, although buried conduits are beneficial for such situations as crossing under pavement (as a contingency for future cable replacement). A section of conduit is also necessary at the customer's building in order to route the service cable up and into the meter cabinet. Because of trenching, conduit under roads and driveways, and the type of cable required, an underground service line is typically more expensive to install than an overhead service line.

Service conductors must be sized appropriately in order to provide the capacity necessary to serve the intended load. Each type of conductor has an *ampacity rating* which specifies its capacity in terms of a current carrying capability. For instance, a No. 2 sized, stranded copper conductor is rated for about 230 to 240 amps.[9] Instead of inventorying a full array of commercially available conductor sizes and types, utilities standardize on a limited number of wires and cables, which are selected to meet the requirements of its typical customers. Thus, even though conductors may be (in the strictest sense) somewhat oversized relative to the service requirements of some customers, the practice of standardizing facilities results in an overall economical methodology for providing electric service.

Distribution Line Transformers

Line transformers are utilized to transform primary distribution system voltages to secondary distribution system voltages in order to serve low voltage customers. Line transformers are designed for pole mounting in overhead systems or pad mounting in underground systems.[10] Large overhead transformers are often mounted on an elevated transformer platform constructed between two poles along with a primary and a secondary conductor bus system. Transformer pads are made out of concrete or as a fiberglass box which is mostly submerged in the ground.

Line transformers are fashioned in both single-phase and three-phase varieties, and they are rated in units of kVA, as shown in Table 5.2. As with conductors, utilities inventory a limited number of transformer sizes in order to minimize the overall costs of providing service.

---

[9] A conductor's actual ampacity depends on a number of factors, including the ambient air temperature.

[10] Submersible (i.e., direct buried) transformers have also been utilized in some underground distribution systems.

Table 5.2 Typical Standard kVA Capacity Ratings of Distribution Transformers

Pole-Mounted Transformers		Pad-Mounted Transformers	
1φ Units	3φ Units	1φ Units	3φ Units
1.5			
3			
5			
7.5			
	9		
10			
15	15		
25		25	
	30		
37.5		37.5	
	45		
50		50	
75	75	75	75
100		100	
	112.5		112.5
	150		150
167		167	
200			
	225		225
250		250	
	300		300
333			
500	500		500
			750
			1,000
			1,500
			2,500

Not all of the transformers listed in Table 5.2 would be recognized as standard construction items given current procurement criteria. What was considered to be a standard at a previous time may not be considered to be a standard today. While a utility may no longer install, for example, a 3 kVA overhead transformer, property records could likely show that several of these transformers are still in service at some locations throughout the system.

In overhead systems, single-phase pole-mounted transformers are often installed as a "bank" in order to form a three-phase service. Two single-phase transformers can be used to construct an open-delta service, while three single-phase transformers can be used for a full delta or a wye configuration. In the United States, transformer banking is generally a more common practice than the use of individual unit overhead type three-phase transformers.

The application of overhead transformers is limited when high capacity levels are required for service. As the capacity of a transformer increases, its weight likewise increases; thus, the physical load bearing capability of a pole or platform would be ultimately exceeded. With a high-density load, a pad-mounted transformer offers a more practical solution. As seen in Table 5.2, the kVA ratings of large three-phase pad-mounted transformers are significantly greater than the largest pole-mounted transformers.

The selection of transformer size in terms of providing the capacity required for service depends on the characteristics of the load to be served. For instance, in southern regions of the United States, air conditioning requirements typically drive the peak loads of residential and commercial customers. Properly sized for summer conditions (including all of the other end-use requirements of refrigerators, televisions, lighting, etc.), the service transformer can often handle any electric heating requirement that may also exist. In northern areas where air conditioning may not be prominent, electric heating may be the peak load driver; thus, the transformer would be sized on the basis of winter load conditions. The key point is that the transformer must be sized to meet a customer's maximum kVA demand regardless of the time of year or, for that matter, the time of the day at which the peak load occurs.

In areas of high customer density, such as in a neighborhood of residential homes, a single line transformer may serve multiple customers. The sizing of the transformer again depends on the customers' end-use requirements; however, the load diversity between the customers is also a sizing consideration. As an example, four adjacent homes having similar end-use load devices each exhibit a typical peak load of 12 kVA. But, because of slight variations in the time at which each home reaches its peak, a coincidence factor of 0.5 is realized by the group. Thus, the peak load of the group is estimated to be 12 kVA × 4 homes × 0.5 = 24 kVA, and a standard size 25 kVA transformer would be selected to serve the combined loads of the four homes even though the nondiversified load is 48 kVA.

If a single home, located in a low customer density area, having a peak load of 12 kVA required its own service transformer, it may be possible that a standard size 10 kVA transformer could be specified. Transformers are capable of carrying loads in excess of their continuous kVA ratings for short periods of time without realizing a significant loss of life due to heating. A home typically operates at a fairly low load factor; thus, the full load capacity of the service transformer is under-utilized most of the time. A short-time load excursion above the transformer's capacity rating would not result in a detrimental thermal impact. Economies are gained by carefully estimating loads and sizing transformers accordingly.

Conventional line transformers must be protected from short-circuit, or fault, conditions and from lightning strikes in order to prevent equipment damage. Fused cutouts and lightning arresters are installed on crossarm or pole brackets at each overhead transformer location.[11] An alternative type of transformer often referred to as a *Completely Self-Protected* (CSP) transformer is also used on distribution lines in some applications. CSP transformers are constructed with a tank-mounted lightning arrester and an internal primary fuse and a secondary circuit breaker. Conventional line transformers typically are not installed with a secondary-side protective device.

Primary Distribution

Primary distribution is the portion of the power delivery system that connects line transformers to distribution substations. Compared to a secondary distribution system, which is confined to a small domain with only a few customers, the primary system is vast in both the area and the numbers of customers served. A single primary distribution feeder may run for several miles and energize hundreds of line transformers. This scale is possible since primary distribution facilities are operated at lower current levels and at much higher voltages as compared to the secondary system.

Many of the three-phase distribution service voltages used for primary distribution are listed in Table 5.3. The most widely used primary line voltage in the United States is 12,470Y/7,200 volts. Many of the other voltages shown in Table 5.3 are commonly used, but not all of these voltages are considered to be standard by today's utility practice. Operating voltages are grouped into *voltage classes*. Line equipment and devices, such as lightning arresters, are rated according to the voltage classes. Thus, a 15 kV class lightning arrester would be used for distribution systems operating in the range of 11 kV to 14.4 kV.

Most of a utility's primary distribution lines consist of poles, which support aerial wires on insulators. Primary wires are often held by crossarms, but modern line construction methods specify the use of small fiberglass brackets or standoff insulators, which result in a lower profile structure of higher esthetic value. A commonly used material for poles is wood that is treated to minimize the effects of ground contact decay. Poles come in various heights, which are necessary to achieve safety clearances, and classes, which refer to the diameters of the pole. A 40-foot, Class 5 pole is commonly used for basic three-phase primary distribution line routing on generally level terrain. When heavy transformers are mounted on a pole, the diameter of the pole may need to be increased in order to provide appropriate mechanical support. A larger diameter pole is also necessary when its conductors pull off at sharp angles, thereby stressing the wood. Distribution pole class sizes typically run from 1 to 7, with Class 1 having the largest diameter. Other materials used for distribution poles include steel and concrete.

---

[11] Transformers in underground systems are also protected with overhead cutouts and arresters at the point where the underground taps off of the overhead system. Thus, a single cutout and arrester may protect several pad-mounted transformers on a loop feed. In addition, pad-mounted transformers may also have an internal primary fuse.

Table 5.3 Common Three-Phase Primary Distribution Voltage Applications

Voltage Class	Three-Wire Δ Systems	Four-Wire Y Systems
2.5 kV	2,400	
5.0 kV	4,160	4,160/2,400
	4,800	
8.66 kV	6,900	
	7,200	7,200/4,160
		8,320/4,800
15 kV	11,000	
		12,470/7,200
	13,200	13,200/7,620
	13,800	13,800/7,970
		14,400/8,310
25 kV		22,860/13,200
	23,000	
		24,940/14,400
35 kV		34,500/19,920

Some distribution poles are required to be guyed in order to compensate for mechanical stresses due to uneven conductor tensions, such as caused by acute conductor angles, changes in conductor sizes, and conductor deadends at the end of a line or at a lateral line tap point. Steel cables, connected to anchors embedded in the ground, are utilized to counteract the uneven tensions so that poles will not lean or break.

A number of aerial conductors are used in overhead primary distribution lines. While copper wire has less resistance than aluminum wire of the same cross-sectional area, it is also heavier and more costly. Therefore, aluminum conductors are used more often now for new primary line construction. However, a considerable number of copper conductors currently exists in older portions of distribution systems. Copper conductors do have an application in modern distribution construction, such as for jumper wires for connecting transformers and other line equipment to the primary lines. Copper wires are also used to ground electrical equipment and the neutral conductor by means of copper ground rods driven into the earth at the base of poles where prescribed by construction standards. A list of many of the sizes and types of aerial conductors that are manufactured for distribution applications is provided in Table 5.4.

Table 5.4 Typical Conductors for Overhead Distribution Applications

Conductor Size		Copper Conductors			Aluminum Conductors		
AWG	MCM	CW	Bare Cu	WPCu	ACSR	Bare Al	WPAl
8	16.51	X	X	X	X		
7	20.82	X	X				
6	26.24	X	X	X	X	X	X
5	33.09	X	X	X	X		
4	41.74	X	X	X	X	X	X
3	52.62	X	X	X	X		
2	66.36	X	X	X	X	X	X
1	83.69	X	X	X	X	X	X
0 (1/0)	105.6	X	X	X	X	X	X
00 (2/0)	133.1	X	X	X	X	X	X
000 (3/0)	167.8	X	X	X	X	X	X
0000 (4/0)	211.6	X	X	X	X	X	X
	250	X	X	X		X	X
	266.8				X	X	X
	300	X	X	X	X	X	X
	336.4				X	X	X
	350	X	X	X		X	X
	397.5				X	X	X
	400		X	X			
	450		X	X		X	X
	477				X	X	X
	500	X		X	X	X	
	556.5				X	X	X
	600		X			X	
	636				X	X	X
	666.6				X		
	700		X			X	
	750		X			X	
	795				X	X	X
	900		X		X	X	

CW - Copperweld
Bare Cu - Bare Copper
WPCu – Weatherproof Copper
ACSR - Aluminum Cable, Steel Reinforced
Bare Al – Bare Aluminum
WPAl - Weatherproof Aluminum

Conductors are partially identified by their physical sizes. The American Wire Gauge (AWG) is used to specify, from small to large, conductors of sizes No. 50 to No. 0000, or 4/0. Beyond 4/0, conductor sizes are specified in terms of circular mils or thousands of circular mils (MCM).[12] As noted previously, utilities do not inventory a full line of conductors and thus each establishes its own standard list of wire sizes and types.

---

[12] One mil = 0.001 inch. A 4/0 conductor, for example, has a diameter of 0.460 inches, or 460 mils. A circular mil (CM) is equal to the square of a conductor's diameter in mils. Thus, the circular mil area of a 4/0 conductor is 211,600 CM, or 211.6 MCM.

Special purpose conductors for high current carrying capacities are manufactured in sizes as large as 2,000 MCM with a diameter of about 1.6 inches. In comparison, common household wiring requires No. 14 AWG (4.11 MCM) conductors for basic service outlets and on up to generally a No. 6 AWG (26.24 MCM) set of conductors for a kitchen range. Small utility-class conductor sizes are often a single solid wire, while larger sizes are stranded, i.e., made by bundling small wires together, in order for the conductor to be flexible.

Most primary distribution overhead conductors are bare wires that are supported by pole-mounted insulators. Some conductors have a weatherproofing protective covering that is not considered to be insulation. A special purpose conductor referred to as *tree wire* is sometimes used to prevent cutting down or severely trimming special species of trees, which are often safeguarded by protective covenants in some urban areas. Tree wire is a conductor that is covered with a heavy plastic material which protects the conductor from physical abrasion and provides a minimal amount of insulation value.

In three-phase construction, the neutral conductor is usually specified to be one standard size below the line conductors. Therefore, if the main line conductors are selected to be 477 MCM ACSR (Aluminum Cable, Steel Reinforced), the neutral would typically be selected to be 336.4 MCM ACSR. The neutral conductor can be downsized if the three-phase load is reasonably balanced, since it would then carry very little current. The line wires of a feeder are designated as "A," "B," and "C." As single-phase line transformers are connected to the primary distribution system, the load should be evenly proportioned between the three different lines so that the overall three-phase loading is fairly uniform and each line carries about the same amount of current. Balanced loading helps to minimize electrical losses.

In many areas of the overhead distribution system, short sections of secondary lines are mounted on poles beneath the traversing primary lines. In these instances, the two different voltage systems can share a single neutral conductor. In addition, two primary voltage feeders carried on the same set of poles can share a common neutral.

Special direct burial cables are utilized in primary underground distribution systems, and their conductors are more often made of aluminum. Underground cables are rated by voltage class and must be insulated accordingly. Cross-linked polyethylene is a common type of insulation material used in the manufacture of primary rated cables. An insulation layer applied uniformly around the length of the conductor is covered by a shielding jacket over which a concentric neutral is often wrapped. Since the cables are normally in direct contact with the ground, which is at a potential of zero volts, a significant voltage gradient is created across the insulation. Any flaws in the quality of the cable's insulation, including a slight nick, can lead to ultimate failure as the high voltage differential at the point of the defect puts an uneven electrical stress on the insulation. Thus, underground cable assemblies often include a durable outer jacket that helps to protect the cable from the abrasive impacts that may occur during cable installation.

Splicing of an underground cable is a critical process since the integrity of the insulation must be maintained in light of the high voltage stress condition. The ends of an underground cable present an equally important problem. The bare end of a primary cable would easily flash over without the installation of an insulating terminator capable of managing the voltage differential between the energized conductor and the outer jacket, which is at ground potential. Thus, a special termination device is used at the point where an underground cable is attached to an overhead primary line. An "elbow" terminator is often used for attaching or detaching energized cables to the high voltage terminals of a pad-mount transformer under load conditions.

Compared to overhead conductors, utilities generally inventory even fewer types and sizes of underground conductors. Cable referred to as underground residential distribution (URD) cable is widely used for undergrounding distribution systems in whole neighborhoods. A typical URD system in a subdivision consists of a single-phase underground loop that begins and ends at two adjacent connection points on the same primary overhead line. Pad-mount transformers are connected to the primary loop at locations where service lines can be easily routed to nearby customers. Fused switches (cutouts) are used at the underground loop's termination points with one end open and the other end closed. In the event of a cable or transformer failure, both ends of the loop can be closed, and the loop can then be sectionalized to isolate the fault area while maintaining service to as many customers on the circuit as possible.

Power type cables are used for large three-phase underground primary distribution system applications. As with URD, an overhead feeder may be tapped with an underground extension to serve an area of commercial or industrial customers, often in a looped fashion. While most distribution feeder systems are of an overhead construction, an entire feeder system may be underground, from the substation to the customers, such as in the urban core of a large city. The high load density of a downtown area of large buildings often renders an overhead feeder system impractical. The clearances and rights of way required for overhead lines are not available. As a result, primary cables are often routed under the streets in concrete duct banks and conduits to special transformers strategically located in underground vaults. In the event that a river or another body of water must be traversed, *submarine cable* can be used with certain advantages over an overhead conductor crossing.

In a rural setting, a distribution feeder system is fairly simple. A feeder may extend for several miles with various three-phase and single-phase tap lines, or laterals, branching off at various points to reach adjacent customers and then finally terminate. This type of feeder layout is called a *radial* system. Power flows from the substation to the last customer on the line. The feeder conductors may be tapered by initially using large conductors near the substation, where the load current is greatest, and reducing their size along the length of the feeder as the load current decreases. In higher density areas, substations are typically closer together, and their feeder systems butt against each other, as show in Figure 5.6.

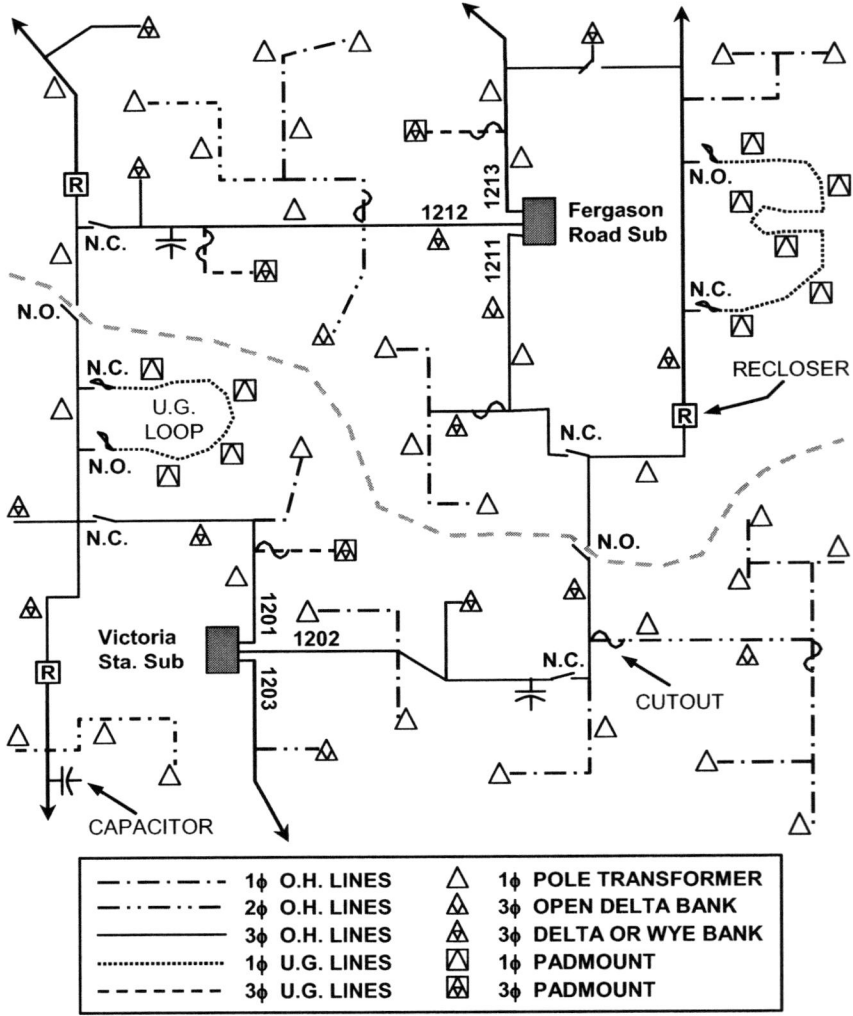

Figure 5.6 Single-line schematic diagram of a small primary distribution system. Three-phase feeders extend from substations and serve as a backbone of the distribution system. Three-, two-, and single-phase laterals branch off of the main feeders to connect adjacent customers. Primary capacitors produce VARs for reactive loads. Fused cutouts and reclosers provide coordinated system protection in the event of fault conditions. The feeders of one substation can connect to feeders of the other substation by means of line switches. Switches are normally open (N.O.) or normally closed (N.C.). The feeder systems define geographical service boundaries of the substations as indicated by the dashed line running across the middle section of the diagram.

As shown in Figure 5.6, a distribution system with multiple substations can have feeders which interconnect by means of line switches. As with a single rural type distribution line, the feeders shown in Figure 5.6 are also radial lines because they similarly extend from a substation and terminate at an open switch. Power flows from the substations toward customers located near to these open points in the lines. However, the switching capability provides some operating benefits. Load can be transferred from one substation to another by opening and closing switches that are strategically located within the feeder system.[13] Thus, service to customers can be maintained even if a substation is down because of equipment failure or routine maintenance. Load might also be transferred on a seasonal basis in order to capitalize on changes in customer load diversity that are driven by the variation in customer types and their different load characteristics. In other words, switching re-mixes the proportions of end-use devices on a feeder thus changing the feeder's loading characteristics.

Maintaining the primary distribution system voltage within an acceptable operating range is necessary to ensure a basic level of service quality. As current flows down a feeder from the substation, the impedance of the conductors causes a drop in the voltage. This drop will increase under heavier loading conditions. The voltage drop is not necessarily linear along the entire length of the feeder, as illustrated in Figure 5.7. The feeder's actual voltage profile depends on the magnitude and distribution of individual load points throughout the feeder. The feeder voltage must be managed to ensure a reasonable voltage spread at each customer's service delivery point considering that additional voltage drops will be realized in the primary laterals, line transformers, secondary lines, and services lines.

Feeder voltage can be regulated at the substation and/or out on the lines using *voltage regulators*. A voltage regulator is essentially an autotransformer which often has a step-type variable output that incrementally raises or lowers the voltage by changing the autotransformer's tap settings. Some substation power transformers have a load tap changing (LTC) functionality built into the unit that typically allows for regulating within ±10% of the rated feeder voltage.

Another method of voltage regulation is achieved through the application of shunt capacitors at key locations on the feeder system. A capacitor is a leading power factor device that counteracts the inductive reactance of the feeder's load thereby changing the circuit impedance. As a result, power factor improves, the line current is reduced, the voltage drop across the conductor's resistance decreases, and thus the line voltage rises. The reduction in line current also serves to reduce losses and to release system kVA capacity "upstream" from the point at which the capacitor is installed.

---

[13] By transferring loads from one feeder to another, the line current at the feeder interconnection switch point can be significantly high. Thus, the ability to taper the conductor sizes is necessarily limited. In addition, the overall current carrying capability (or reserve capacity) of the feeders must be increased to handle the larger amount of load during a transfer condition.

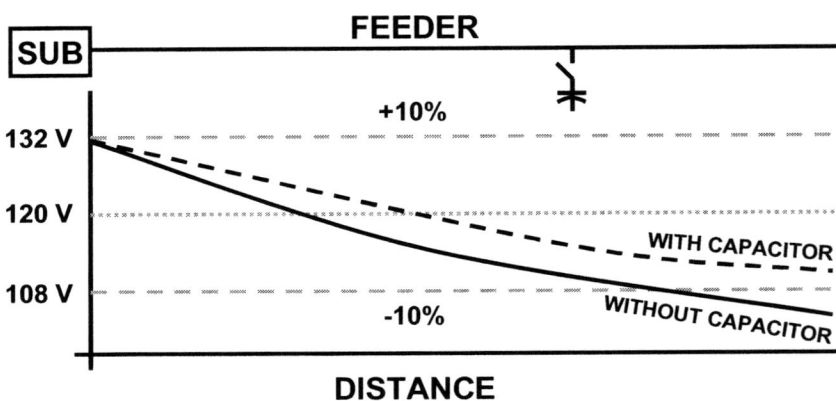

Figure 5.7 **Voltage profile of a primary distribution feeder during a peak load situation.** The profile is viewed on a 120 volt basis. The voltage profile will vary over time depending on the type of loading conditions and the dispersion of loads along the length of the feeder. Voltage regulation devices, such as capacitors, are used to improve a feeder's voltage profile. Voltage drops in the line transformer and the secondary and service lines would further affect the voltage at a customer's delivery point.

A capacitor actually provides VARs locally to the inductive loads on a feeder; therefore, less VARs need be imported through the transmission system from the generators. Capacitors installed on distribution feeders for area power factor correction and voltage control should be sized and operated to match the feeder's inductive VAR load shape. VAR load typically varies with the cycles of a day and week in somewhat of a direct proportion to the likewise varying kW requirements, as shown previously in Figure 4.1 for a single customer. With the large quantity of customers served by even a single distribution feeder, the diversity in the operation of customer inductive end-use devices creates a kVAR load profile that is characteristically similar to a diversified kW load profile, as shown in Figure 5.8.

When compensating a feeder's inductive load requirements, it is important that too much capacitance not be permanently installed. When the capacitive reactance exceeds the inductive reactance, the feeder voltage at the capacitor will rise to the point of surpassing the feeder voltage at the substation, and VARs will flow upstream towards the generators. From the substation's perspective, the feeder's kVA load could actually increase. To prevent an overcapacitance situation, capacitor banks should be installed on feeders in combinations of fixed and switched units in order to track the changing VAR requirements. Voltage and current sensing are commonly used in capacitor switching methodologies.

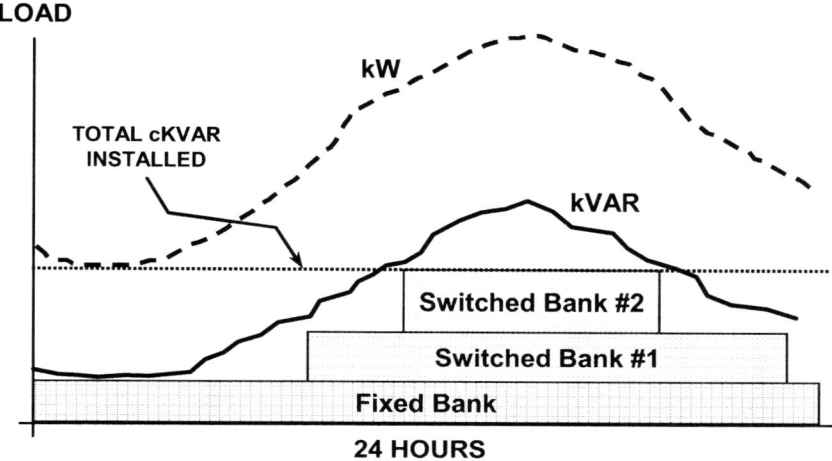

Figure 5.8 **Compensation of a distribution feeder's VAR load.** Fixed capacitors provide VAR support at all times while other capacitor installations are switched on and off in concert with the time variation in load. If capacitor banks #1 and #2 were also fixed instead of switched, the night and morning periods would experience load and voltage problems associated with having too much capacitance online.

Protection of distribution lines and facilities from overvoltage and overcurrent conditions is a critical matter. A frequent cause of potentially damaging voltage transients is lightning, which can initiate a high voltage wave that can travel a considerable distance along a distribution line. Lightning arresters are routinely installed for protection of equipment, such as line transformers. Lightning arresters are also installed at strategic locations along a feeder. A lightning arrester diverts a voltage surge to ground potential thereby dissipating the destructive level of energy produced by the lightning strike.

Overcurrent conditions are caused by short circuits, or faults, when a line conductor comes into direct or indirect contact with adjacent lines or the ground (which includes the neutral conductor). A line-to-ground fault on a single-phase lateral is easiest to visualize: the line conductor falls across and contacts the neutral. At the fault location, the line voltage essentially drops to zero, thus all of the loads connected to the line become short-circuited (the applied voltage is virtually zero). The full voltage of the source (i.e., the substation bus) is now impressed across the very small impedance presented by the line conductor segment between the bus and the fault location. As a result, the line current increases substantially and rushes to the fault. In addition, motors operating on the line at the time of a fault begin to slow down under their zero applied voltage condition, and they stop motoring and start generating due to their rotational inertia. Temporarily the motors then act as a source that contributes to the fault current. The cumulative impedance of the line between the substation and the fault location (Ω/mile × miles) provides some limitation to the fault current. Thus, the *available fault current* diminishes somewhat between the substation feeder bus and the end of the feeder. In other words, a short circuit located 10 miles from the substation will produce less fault current than the same short circuit condition located only ¼ mile from the substation.

Knowing the available fault currents at key locations in a feeder system is crucial for designing an overcurrent protection scheme. A feeder and its laterals will have several protection devices in place, as shown in Figure 5.9. The substation feeder breaker, reclosers, sectionalizers, and fused cutouts are used to provide various levels of protection. The difference in available fault current throughout a feeder facilitates the coordination of such multiple protection devices. Beyond the basic task of protecting the lines and facilities from overcurrent conditions, the main objective of the protection scheme is to maintain service to as many customers as possible in the event the circuit is faulted.

Reclosers are automatic line-type circuit breakers, which attempt to save fuses, since a blown fuse requires manual replacement. Reclosers typically operate through a sequence of one or two fast (e.g., 2 cycles) and two delayed (e.g., 20 cycles) trips to give the fault a chance to clear. If the fault has cleared upon reclosing, service is maintained to the whole circuit and the recloser resets. If the fault persists, the reclosing action allows the appropriate fuse to isolate the faulted section of the line. If the fault does not clear, the recloser ultimately locks out.

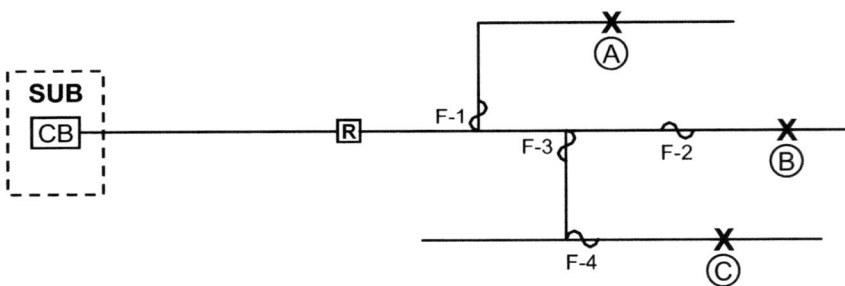

Figure 5.9 **A circuit protection scheme.** The fuses are coordinated with the recloser. In the event of a fault at A, B, or C, the recloser would operate through its programmed on-off cycles in an effort to allow the fault to clear before a fuse blows. If the fault is temporary, the recloser operation will preserve the fuse. If the fault is permanent, the fuse will blow before the recloser locks out (F-4 is sized to clear a permanent fault at C before F-3 blows). The faulted section is thus isolated from the remainder of the circuit.

Distribution Substations

A substation is a facility in which voltage is transformed from one level to another in order to move a block of power between two systems (e.g., between transmission and distribution). A substation is classified as a *distribution substation* if the output of its transformer(s) is a primary distribution voltage. A distribution substation's input could be a transmission, subtransmission, or even a higher distribution voltage. A distribution substation is constructed to serve either a single large customer directly or numerous small- to medium-sized customers by means of a distribution feeder system. In addition to voltage transformation, a distribution substation often has some transmission and/or primary distribution switching capability either within the facility or directly adjacent to it.

A small substation with a single transformer and one or two distribution feeders requires a minimal amount of land area, and such examples are sometimes situated directly within the urban area that it is serving. The feeder system is generally confined to the local area because of the relatively high density of customers and loads. For esthetic reasons, the urban substation may have a low profile, a decorative fence, and underground conductor entries and exits. A substation can even be contained within a structure that is disguised as a house or building that is characteristic of the area. Single transformer substations are also found in rural settings where the customer and load density is much lower than in urban settings. In contrast, rural feeders typically extend out for several miles in order to connect distant customers.

Larger distribution substations, with two or more transformers, are more commonly found in urban and suburban settings and in areas of large commercial and industrial operations. In very large substations, some or all of the *feeder getaways* may need to be placed underground in order to prevent physical congestion of the numerous distribution lines and facilities needed to route the distribution system outward from the substation. An example of a typical electrical layout for a distribution substation is shown in Figure 5.10.

In addition to the power transformers, a substation contains a large number of other electrical facilities. High-side and low-side bus structures made of large copper wire, solid bars, or tubing are used for interfacing the transformers and other equipment with the incoming transmission and the outgoing feeder conductors. Circuit interruption devices, such as circuit breakers and load break switches, are needed for operational switching, isolating equipment for maintenance, and protection of facilities. Substations may also contain voltage regulation equipment such as regulators, capacitors, synchronous condensers, or even small generation units, which provide VARs for local voltage support. Common line transformers, connected to a distribution voltage bus, furnish low voltage station service for use in substation maintenance, lighting of the substation yard, battery charging, and operation of auxiliary equipment. Storage batteries are used to provide DC voltage for substation control equipment. Substations also include relays for sensing voltage and current conditions and tripping of circuit breakers under specified abnormal operational conditions.

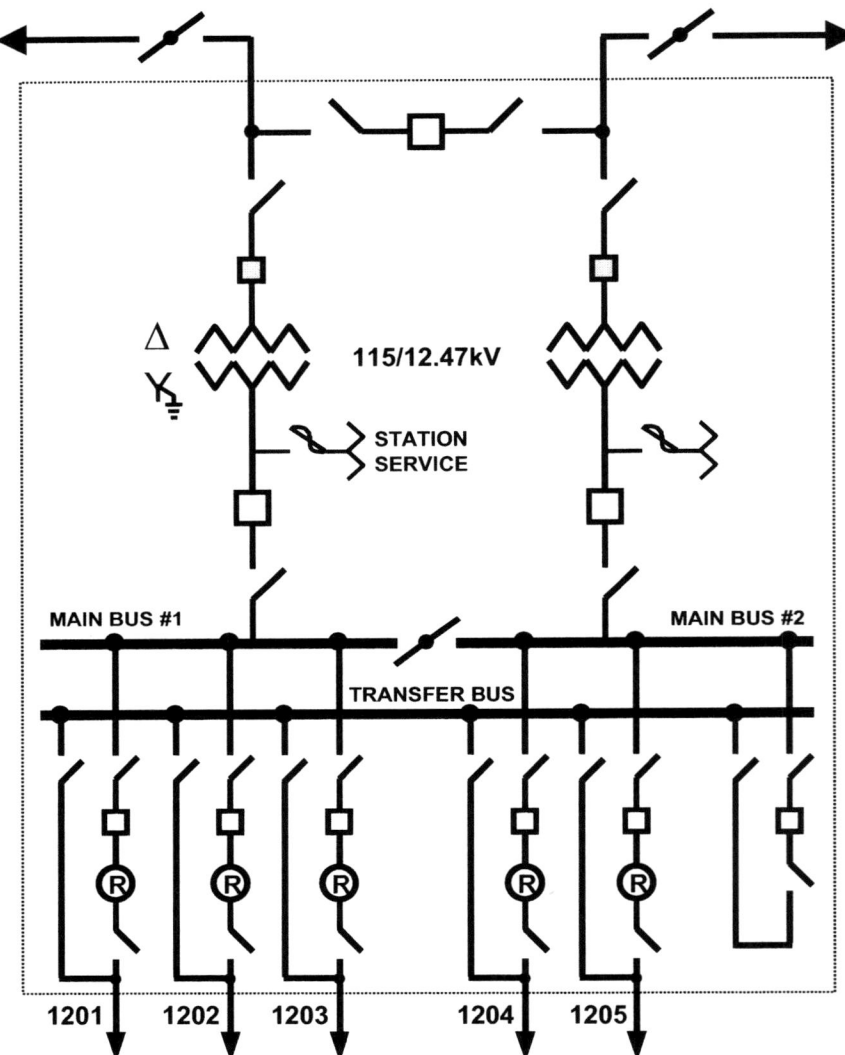

Figure 5.10 Single-line schematic diagram of a typical distribution substation. Transmission is looped through the high side of the substation providing a feed from two different directions. Dual transformers provide both greater capacity and backup capability in the event that one of the units was to fail. A transfer bus scheme provides greater operating flexibility. Circuit breakers provide protection for substation equipment. Line transformers are used in the substation to provide low voltage station service for lighting of the yard, charging storage batteries, and other operating purposes.

A number of other facilities and structures are necessary for the substation. Controls, storage batteries, and communications equipment are housed within a small building located within the confines of the substation yard. For security and public safety, substations are enclosed by a fence. The yard is covered with a thick base of gravel for swift drainage of rainwater. A copper grounding grid is buried beneath the gravel base, and all substation facilities are connected to it. Overhead static wires are used for lightning protection of the substation.

Metering is often used for monitoring of substation and feeder loading conditions. It is important to know the characteristics of loads placed on substation facilities relative to their current carrying or capacity ratings. The i^2R effect of electricity flow creates harmful heat that must be dissipated. Like most line type transformers, substation power transformer cores are immersed in an insulating oil that provides for the transfer of heat to the surrounding air. The heat dissipation is enhanced by the passive flow of oil through hollow cooling fins or external radiators, which create a greater surface area between the oil and the air. Oil-immersed, air cooled transformers of this type are designated as "OA." Blast fans can be attached directly to the radiators for increasing the airflow thus enhancing the cooling effect. The designation for transformers with forced air-cooling is "OA/FA." In addition, pumps can be installed to actively circulate the oil through the transformer and its external radiators. Transformers with both forced oil and forced air-cooling are designated as "OA/FOA."

Heat serves as a limiting factor for a transformer's load carrying capability. Power transformers are designed with a basic kVA capacity rating which, under the proper conditions, can be exceeded for short periods of time without a reduction in life expectancy that would be caused by a continual elevated thermal loading. The use of forced air and forced oil cooling extends a transformer's basic kVA capacity rating for peak load conditions. Thus, power transformers can have various capacity ratings depending on the auxiliary cooling methods applied. For example, a distribution substation power transformer might be rated as

- OA capacity rating 24,000 kVA
- OA/FA capacity rating 24,000/32,000 kVA
- OA/FA/FOA capacity rating 24,000/32,000/40,000 kVA

Distribution feeders operating at a higher voltage class provide more capacity than lower voltage classes since current levels are less for the same amount of load, and thus line losses are less of a constraint. A 12,470 volt line can generally serve two to three times as much load as a 4,160 volt line. Thus, given a region of uniform load density (kVA per mi.^{2}), a higher distribution voltage feeder would be able to serve a larger geographical area than feeders of lower distribution voltages. As a result, the operating voltage of the distribution feeder system has a major bearing on the area that a particular substation will serve, as illustrated in Figure 5.11. A substation's feeder system layout defines its service boundaries.

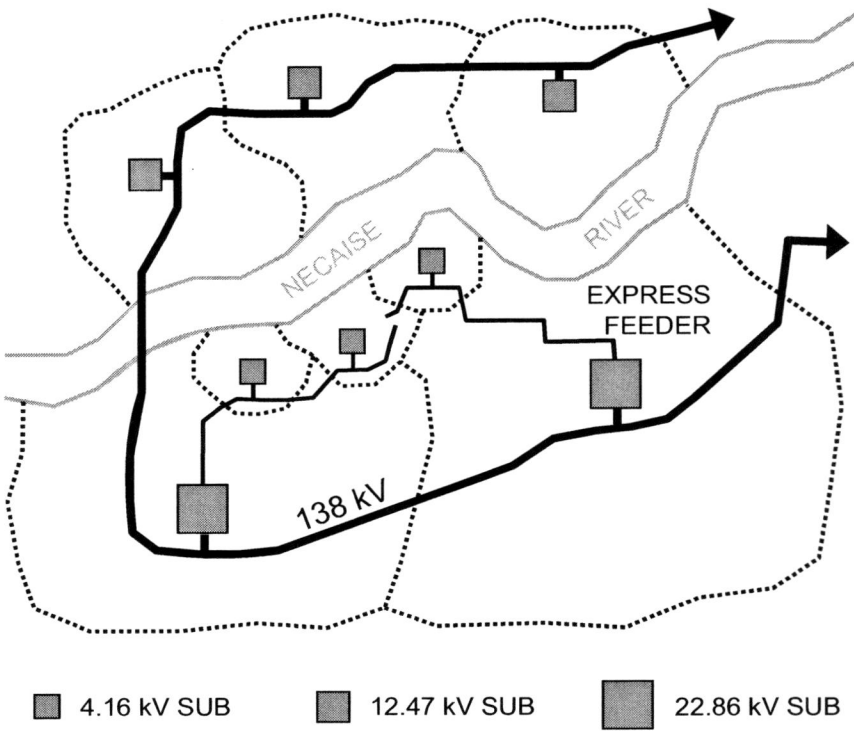

Figure 5.11 **Substation service areas.** Substation boundaries are defined in part by local geographical conditions. In addition, the voltage selected for distribution helps to fix the limits of a substation's service area since higher voltages can carry greater amounts of load. Losses also provide a spatial limiting factor when lower operating voltages, and thus higher load currents, are utilized. Notice the use of higher voltage distribution (acting as a subtransmission function) to serve lower voltage distribution substations.

A system of distribution substations will be optimal if the substations are located at the load centers. Matching total substation capacity to total peak load requirements is relatively simple. But several substations are necessary to serve a utility's distribution loads because customers are dispersed, load density is not uniform, and customer and load growth variations occur even within a single metropolitan area. As a result, a number of load centers emerge across a utility's service area. If a substation is not located at the load center, distribution feeders will have to be extended further and thus higher losses will be incurred. The spatial and temporal aspects of customer and load growth further complicate the substation siting/expansion process.

Spatial Attributes of Distribution Systems

Distribution systems can vary dramatically in terms of their structures and configurations, the types of customers and end-use loads served, and the vintages of facilities in service. A great deal of these variations depend on the intrinsic differences that occur from one location to the next. When viewed on an area-by-area basis, these spatial variations give rise to different cost structures throughout a utility's distribution systems.

A distribution system must reach out to connect customers; thus, the course of a distribution feeder system is driven to a large extent by where customers choose to establish homes, businesses, factories, and other structures (e.g., radio towers or bill boards). In urban areas, streets and roads usually tend to follow somewhat of a uniform grid shape, and the distribution lines usually track close to this structured pattern. However, rural area distribution lines tend to sprawl and appear more random in form because customers are located further apart and with less uniformity than customers in urban areas. But local geographical features also have a great influence in routing distribution lines. A number of obstacles must often be negotiated while attempting to minimize the overall length of distribution lines. Instead of being able to construct a distribution feeder in a relatively straight line, sometimes the route must be diverted because of large bodies of water, rough or steep terrain, or restricted land use areas, such as wildlife preserves, wetlands, parks, and cemeteries. Easements for directly crossing certain tracts of land may be impractical in cases such as airport expanses, restricted governmental and military installations, and other large-scale customer facilities.

Spatial differences in the types and dispersion of customers and their end-use devices have a bearing on the characteristics of load served by a distribution line. While a feeder would unlikely serve just one type of customer exclusively, some feeders may serve a predominance of residential, commercial, or industrial types of customers. Consider a hypothetical substation having one residential feeder with a group peak load of 1,000 kW and one commercial feeder also with a group peak load of 1,000 kW. Because of load diversity between these two customers classes (e.g., commercial operations typically peaks a few hours earlier than residential), the substation may realize a diversified peak load of 1,500 kW. Operating differences among the residential customers on one circuit and similar differences among the commercial customers on the other feeder cause the only feeder load diversity. But if the two feeders each had an equal mix of the residential and commercial customers, then the diversified peak load of each feeder would become 750 kW, while the total peak load now seen by the substation would be the same total diversified load of 1,500 kW.

Because of differences in customer density, the number of customers per circuit mile is greater in urban and suburban areas than in rural areas. Lot sizes are smaller in cities, so houses are much closer together, and apartments are more common in metropolitan areas. In high-density areas, several customers often share a single line transformer while rural customers usually require their own individual transformers. In high-density areas, loads are concentrated; thus, feeders are relatively short and their conductors are large in size (e.g., 477 to 795 MCM). Rural circuits are comparatively much longer and their conductors are usually smaller in size since the amount of load carried is typically less than for high density area feeders. Smaller conductors and longer circuits result in higher losses per customer in the rural areas.

The different customer class types have distinct location preferences. Urban core commercial and residential customers are typically high-rise office buildings and apartments. Suburban commercial is characterized by low-rise buildings and strip shops, while residential is mostly single-family homes and two- to three-story apartment buildings. Also in specific suburban areas, commercial office parks and shopping malls are found in locations having ready access to major roads and highways so workers and shoppers can easily commute. In general, a commercial entity locates in an area where it can best serve its market. Both single industrial customers and industrial parks also need access to transportation means including not only roads and highways but also railroads and waterways. These location needs drivers fundamentally explain why metropolitan areas are structured in a particular manner. Cities generally have similar attributes and types of customer classes, but their different dispersions require unique distribution system layouts in order to serve the resulting spatially varying loads.

The densities of end-use load devices also vary geographically. A subdivision of large homes, on the order of 2,500 or more square feet, would require larger space conditioning equipment than smaller homes (furthermore, large homes generally have more end-use load devices than smaller homes). In areas where competitively priced natural gas is available, electricity is less likely to be used for meeting all of the space and water heating requirements. Low customer density will not economically support a natural gas delivery system in sparse rural areas, so electricity is more apt to be used for heating applications (the likely source of competition would come from propane, which is delivered by trucks and stored on-site). On the other hand, high-density apartments are often electrically heated as a result of builders avoiding the extra cost of installing gas service lines.

Variations in local climate conditions serve as an underlying factor in the spatial differences in some end-use load densities. For example, at one extreme, a utility system could have southern coastal service areas where prolonged high temperatures and humidity levels compel virtually all customers to install and frequently operate air conditioning equipment. At the other extreme, the same utility could have inland mountainous areas with a much more temperate summer climate. The installation and use of air conditioning equipment in these areas would be comparatively much lower than on the coast.

An example of the spatial variations in electric load is shown in Figure 5.12. All of the effects of the dispersion of customers and their end-use devices combine to create a unique map of load density for a major metropolitan area. Siting substations in close proximity to the load centers and designing feeder systems to minimize losses would optimize the distribution system serving such a load.

Power Delivery Systems: Transmission and Distribution

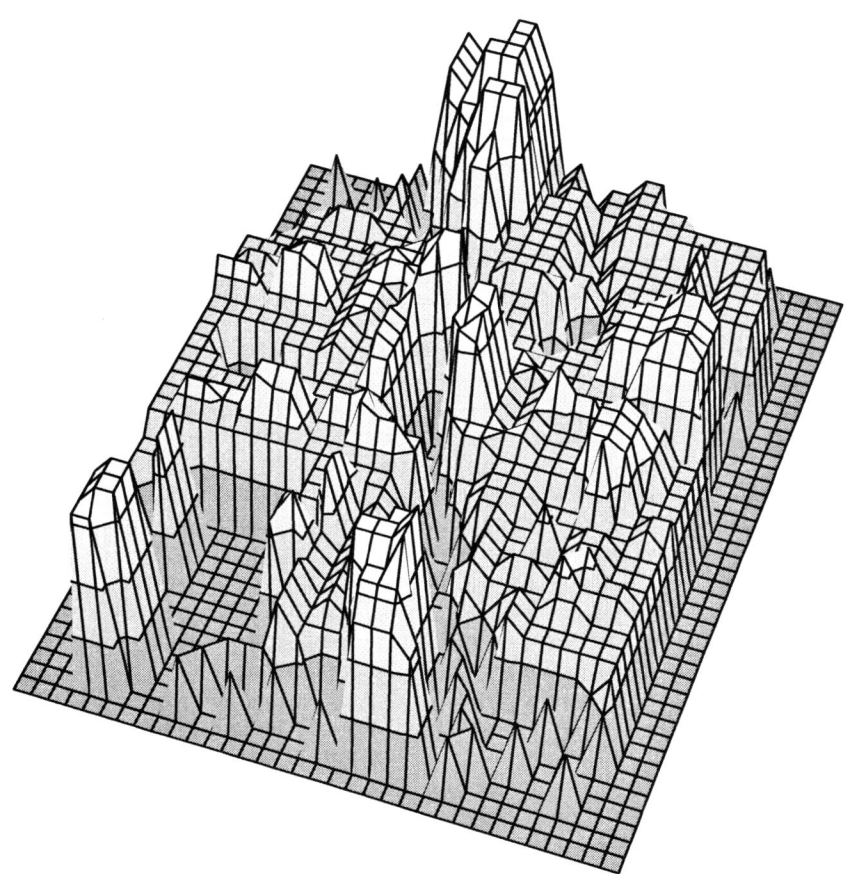

Figure 5.12 **A 3-D plot of distribution load.** Load map of a dense metropolitan area in which the height represents kVA per square mile. A downtown core area of large buildings located on each side of a river is situated in the center of the chart. The large peak at the top of the graph is an area of heavy industrial operations also located on the river. The "hole" in the lower left corner of the map is an airport surrounded by commercial and industrial facilities. Other "holes" representing large parks can be noted. The remainder of the chart is a mix of high to low density residential and commercial. This view is for one point in time; thus, the magnitude and pattern of the loads would change dramatically during daily, weekly, and seasonal load cycles.

Load growth in terms of new customers occurs as buildings and facilities are constructed on previously vacant land. As a town or city expands, the heart of the city fills up thereby requiring new growth to occur around the periphery. Load growth at the distribution system level proceeds in a manner similar to customer ownership of end-use load devices – in an S-shape fashion. An open area, such as on the outskirts of a city begins to attract new customers, and distribution lines must be extended into the area to serve the new load. As load continues to grow in the area, a new substation may eventually be required, and the feeder system would be expanded, enhanced, and reconfigured to optimize the manner in which customers are served. Over time, the growth in the area will eventually use up the available land; thus, that area will reach a point of load saturation. Additional growth will then start in a new area further out from the city.

Cities tend to grow concentrically, except for obvious geographical constraints, such as mountain ranges or oceans; however, growth may be more prominent in a particular direction because of some inherent attractiveness of that area. For example, Atlanta, Georgia as a whole experienced considerable development in the 1980s and 1990s, but the north side realized a predominance of that growth even though land was generally available in all directions.

Growth is also not likely to be uniform around a utility's service territory. One city or town may realize a high rate of growth as another similar size city or town may become stagnant or even experience a decreasing load under a weak economy. The requirements for the electric power system are driven by such factors. A significant amount of new distribution resources is needed in the growing area while the existing infrastructure in a depressed area essentially may become underutilized. The cost to downsize distribution capacity is mostly prohibitive. While a substation power transformer may sometimes be moved and reused in a different location because of less load, large feeder conductors could not be economically replaced with smaller conductors.

With all of the spatial variations that occur in load type and growth, a utility's distribution system is represented by a spectrum of vintage equipment and facilities. Recent high growth areas have wholly new capital installations while older areas have facilities that may be for all practical purposes fully depreciated, but which have useful remaining life because of ongoing maintenance efforts. As a result, the cost of distribution systems varies from one location to the next.

Distribution System Performance

Power delivery systems are subjected to a number of factors that cause aberrations in distribution reliability and power quality performance. Many of these factors are beyond the control of responsible system design and operation. For instance, lightning strikes on distribution lines create temporary overvoltage conditions. Also, as discussed previously, system protection equipment operates during fault conditions in an effort to preserve service to as many customers as possible while safeguarding line facilities. But these types of incidents can sometimes create temporary outages for some customers. A number of common events periodically cause outages within the distribution system, as shown in Table 5.5.

Table 5.5 Distribution Outage Examples – Customer Outages during a Year

Cause of Outage	Lee Street Sub Circuit 1404	Blake Street Sub Circuit 1202	Myrtle Street Sub Circuit 1303
Weather:			
- Lightning	126	1,317	3,608
- Wind/Rain	24	133	5
- Snow/Ice	-	-	-
Nature:			
- Trees	67	22	160
- Animals	46	-	480
Public:			
- Vehicles	-	-	636
- Acts of Others	2	29	38
Utility:			
- Scheduled	-	29	2
- Unscheduled	2	-	-
Equipment Failure:			
- Deterioration	345	-	386
- Contamination	-	-	-
- Loose connection	-	-	-
Other:			
- Overload	-	-	-
- Unknown	2,971	47	26
Outage Total	3,583	1,577	5,381
Total Customers	2,774	917	1,321

As observed in Table 5.5, the outage characteristics of distribution feeders are not uniform. The causes of distribution outages differ from one feeder to another because of any number of spatial varieties. For instance, some lines may be routed mostly in the open while some other lines may pass through vast wooded areas. Some lines are located only a couple of feet from heavy vehicular traffic while other lines may follow the edge of a road right of way that is several yards from the edge of the pavement. Lightning strikes on distribution facilities occur randomly in nature. Throughout the entire distribution system, the different causes of outages can also be expected to vary over time, as indicated by the three-year outage statistics for the distribution system shown in Figure 5.13.

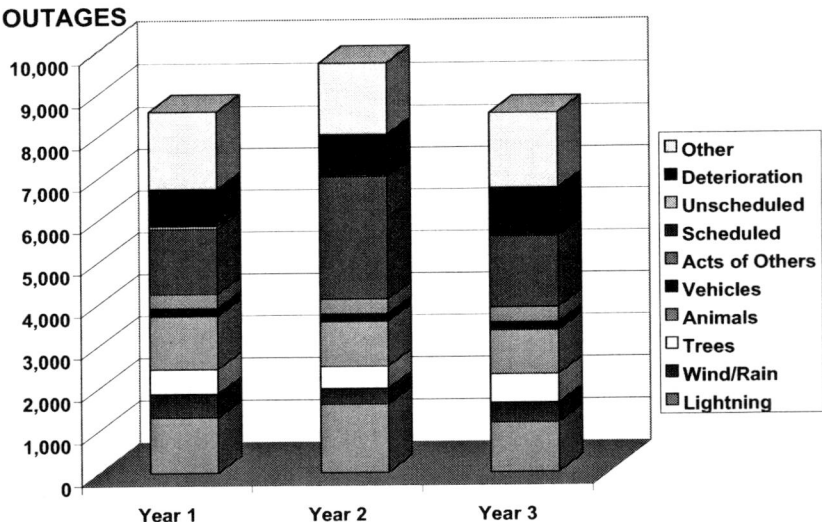

Figure 5.13 **Three-year outage statistics for a large distribution system.** The total number of outages differs from year to year, as do the individual causes of outages. Weather is a major cause of distribution outages. Outages due to trees often vary with tree trimming cycles. Annual system expansion and maintenance requirements affect scheduled outages.

Utilities differ slightly with their definitions of an "outage."[14] An interruption in service may be *momentary* or *sustained*, and the dividing line among these categories is typically established between one and five minutes. Momentary interruptions are resolved through system automation means (e.g., most recloser operations) or by manual switching that can be effected at an instant. Sustained interruptions usually require the dispatch of a line operator or a crew to conduct field switching and/or line repairs.

The level of detail at which sustained interruptions are recorded also varies among utilities. The simplest measurement of outages occurs at the feeder circuit breaker located at the substation. The most sophisticated system of measure logs outages at the customer level. The measurement of feeder level momentary interruptions or transient events is not widely practiced because of the requirement for special data recording equipment.

---

[14] In contrast to general utility standards, some customers may consider a temporary event such as a sag in voltage over a few cycles to be an outage due to its effects on voltage-sensitive equipment. Transient events are generally viewed as power quality issues.

Power Delivery Systems: Transmission and Distribution

Reliability Indices

For benchmarking and comparison purposes, the utility industry has adopted a series of indices that provide different measures of distribution system reliability-related performance. Evaluation is typically based on an annual period. The more common of these metrics for sustained outages include:

System Average Interruption Frequency Index – SAIFI

$$SAIFI = \frac{\sum_{i=1}^{n} C_{INT_i}}{C_{TOT}} \qquad (5.5a)$$

System Average Interruption Duration Index – SAIDI

$$SAIDI = \frac{\sum_{i=1}^{n} (C_{INT_i} \times T_{INT_i})}{C_{TOT}} \qquad (5.5b)$$

Customer Average Interruption Duration Index – CAIDI

$$CAIDI = \frac{\sum_{i=1}^{n} (C_{INT_i} \times T_{INT_i})}{\sum_{i=1}^{n} C_{INT_i}} = \frac{SAIDI}{SAIFI} \qquad (5.5c)$$

Customer Average Interruption Frequency Index – CAIFI

$$CAIFI = \frac{\sum_{i=1}^{n} C_{INT_i}}{N_{INT_i}} \qquad (5.5d)$$

where
    n = number of outages during the evaluation period
    C_{INTi} = number of customers interrupted during the ith outage
    C_{TOT} = total number of customers served
    T_{INT} = interruption duration in hours or minutes during the ith outage
    N_{INTi} = total number of customer interrupted during the evaluation period (affected customers are only counted once)

A momentary outage has a duration of less than one minute. Momentary interruption indices are also evaluated by some utilities. For example, SAIDI is recast as MAIFI (the ratio of the total number of customer momentary interruptions to the total number of customers served).

A problem of inconsistency often exists when comparing utilities because of the variation in the level of detail recorded. In addition, the use of these indices provide minimal assistance to a customer who is considering investing in reliability improvement projects because they are based on systemwide averages. For example, a utility's SAIDI of 75 annual outage minutes per customer would be exceptional compared to the U.S. national average. However, because of spatial performance variations, some of the customers could realize an interruption time of less than 10 minutes per year while other customers could be interrupted for 10 hours or more per year. Each customer's service requirements are unique, so a large number of annual outage minutes may or may not present a major problem depending on how they occur, e.g., all at once or 100 seconds per day every day. A reliability improvement solution may be more effectively found with such intelligence, but such detail is not usually readily available.

A key driver to reliability is exposure. The more that distribution facilities are subjected to hazards such as traffic and weather, the less reliably they can be expected to perform. Exposure is also a function of distance from the source. Customers situated near the end of a distribution circuit can expect to be interrupted to a greater extent than customers located on the same circuit but just outside of the substation.

General distribution system reliability can be improved and sustained by a number of capital and maintenance projects, including

- Installing additional substations with shorter feeder systems
- Using primary and secondary distribution network configurations
- Placing distribution lines and facilities underground
- Automating the distribution system
- Installing selective distributed resources
- Improving system protection and sectionalizing capabilities
- Routinely inspecting lines and replacing damaged and failing facilities
- Maintaining tree trimming cycles

In addition, a number of reliability enhancement methods can be customized for meeting specific customer needs. Such solutions include nonstandard distribution system enhancements, customer-side point-of-use enhancements, or a combination of both. Hospitals and other consequential operations, for example, have emergency generators for their critical loads, but they also often have the local utility provide a dual primary feed (preferably from two different substations) with a transfer switch that operates automatically when power is lost from the principal source.

5.3 POWER TRANSMISSION

Transmission is structurally similar to distribution in that it is a wires system mainly composed of conductors supported by wood and metal poles and towers and sometimes cables placed underground in large urban areas. As with distribution, transmission's fundamental mission is to transport power. Transmission interconnects central station production facilities with large load centers from which power is distributed to individual customers.

However, the role of the transmission system is much broader than simply moving power from one point to another in order to serve load. Transmission is used to interconnect a utility's power system with other power systems to increase reliability, gain economies through the interchange of power, and realize other technical benefits associated with a large-scale synchronized operation.

The issue of delineating between transmission and distribution is noted in the regulatory arena as well. Generally, the level of the operating voltage is used by utilities to classify power delivery facilities as transmission and distribution. However, in its Order 888, the FERC did not specify transmission and distribution functions by explicit voltage levels. Instead, the Commission proposed the following seven indicators of local distribution to be evaluated on a case-by-case basis:

1. Local distribution facilities are normally in close proximity to retail customers;
2. Local distribution facilities are primarily radial in character;
3. Power flows into local distribution systems; it rarely, if ever, flows out;
4. When power enters a local distribution system, it is not reconsigned or transported on to some other market;
5. Power entering a local distribution system is consumed in a comparatively restricted geographical area;
6. Meters are based at the transmission/local distribution interface to measure flows into the local distribution system; and
7. Local distribution systems will be of reduced voltage (FERC Stats. & Regs. ¶ 32,514 at 33,145).

As shown in Figure 5.14, lines and buses form a basic transmission network or grid. Generators and loads are connected to the network at various buses. Within this grid, both real and reactive power can move in different directions depending on the system's particular loading and generating conditions at the time in question. If a generator or line is outaged, the power flows redistribute throughout the grid based on the electrical properties of the transmission lines. The grid must be designed for such contingencies in order to prevent subsequent line overloads. Both AC and DC methods are utilized to form the transmission network, although AC is by far the most prominent application in this era.

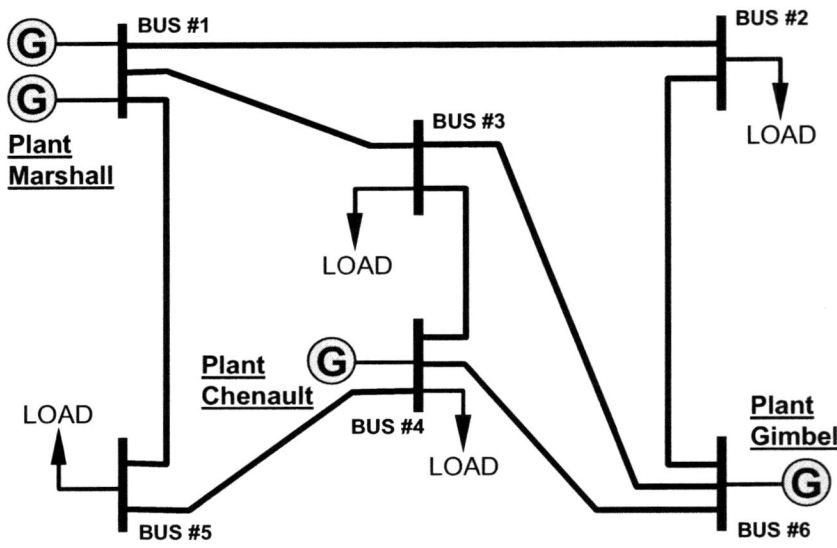

Figure 5.14 Transmission network example. The transmission grid is comprised of lines which connect buses within the system. Power production facilities and load centers are connected to the network of buses.

AC Transmission

Transmission is subdivided into two major functions: *bulk transmission* and *subtransmission*. Bulk transmission forms the backbone of the power system, and it is capable of transporting huge blocks of power both within and through the network. Overall, bulk transmission lines tend to extend over relatively long distances. The bulk transmission system feeds power to subtransmission substations and large distribution substations often in a "loop through" manner. Subtransmission on the other hand is somewhat like a hybrid of bulk transmission and distribution. It moves blocks of power across the network while distributing power at smaller load centers. Compared to bulk transmission, subtransmission tends to extend over shorter distances and may be configured more in a radial fashion than in a network configuration.[15]

Transmission lines are typically connected in a delta configuration. Transmission has no neutral conductor, but overhead static wires are used for lightning

---

[15] Some high voltage line segments (transmission "taps") operate as distribution in that they route power radially from the network to a distribution substation or a specific customer.

shielding of aerial lines. Typical three-phase voltages used for bulk transmission and subtransmission system applications are shown in Table 5.6. Large utility systems that cover vast areas containing heavy load centers typically operate higher transmission voltages than smaller sized utilities. In large systems, 115 kV may be considered to be a subtransmission voltage, while 46 kV and 69 kV may be regarded as bulk transmission voltages in small power systems.

ACSR conductors are ordinarily used for aerial transmission lines because the steel core provides the high tensile strength needed for very long spans between supporting structures. The distance between some transmission structures may be several hundred yards. The sag of these conductors in high temperature summer conditions can be several feet. As with distribution lines, the safety clearance from transmission voltages is a paramount issue.

For lines up to 230 kV, a single conductor is typically used for each phase line. Above 230 kV, multiple wires separated by spacers that form a conductor bundle are used for each phase line. Generally, two conductors per phase are used in 345 kV to 500 kV construction, and four conductors per phase are used for 765 kV. The paralleling of conductors in this manner provides more line capacity while reducing the reactance of the lines. Aerial conductors are fastened to single pole structures by either stand-off or suspension insulators connected to brackets or mast arms. Suspension insulators are used to attach conductors to wooden H-frames and steel or aluminum lattice towers. An appropriate amount of insulation is required for the applied voltage. Sometimes a transmission line is constructed with an insulation level higher than required for its initial operating voltage in anticipation of increasing the applied voltage at a future date. Some transmission support structures may carry more than one circuit, and distribution circuits are often underbuilt on transmission structures, particularly in urban areas.

Table 5.6 Common U.S. Three-Phase Transmission Voltage Applications

Voltage Class	Medium Voltage	High Voltage	Extra-High Voltage
Subtransmission	34,500		
	46,000		
	69,000		
Bulk Transmission		115,000	
		138,000	
		161,000	
		230,000	
			345,000
			500,000
			765,000

As demonstrated previously in Figure 5.1, if the operating voltage of a transmission line were increased, the result would be a decrease in current for a given amount of load. Thus, higher voltages would increase a transmission line's load bearing capability since its conductors are thermally limited by the amount of current that they can carry without annealing. Increasing the number of conductors per phase line also contributes to a higher capacity since more current can be carried safely. The increase in line capacity yielded by higher operating voltage applications is significant. A 765 kV line, for example, has a thermal rating in excess of 5,000 MVA as compared to a single circuit 230 kV line, which has a thermal rating of less than 500 MVA.[16]

The reliability of the transmission system is generally greater than in distribution systems. The network aspect of transmission, compared to the mostly radial aspect of distribution, provides greater flexibility for system operation during overloads and disturbances. Transmission facilities have less exposure to trees and vehicular traffic. The distance between conductors and between conductors and grounded structures are relatively great in transmission lines; thus, incidents due to animals and birds are much less likely as compared to distribution. As a result, the transmission system experiences many fewer outages than does the distribution system.

As discussed previously, distribution systems aggregate customer loads to a common source, such as a substation. The transmission system carries load aggregation to a much higher order. Whereas a single distribution substation and its feeder system may serve hundreds to thousands of customers, a transmission system serves tens to hundreds of thousand of customers as it aggregates multiple distribution systems. Very large individual customer facilities are often served at a transmission delivery point either directly or with a dedicated substation in a manner similar to a common distribution substation.

Voltage and the Transport of Complex Power

Both real and reactive power flow in the transmission system. However, due to the network aspect of the transmission system, the same line may transport power in different directions at different times. The specific flow of real and reactive power is a consequence of certain voltage conditions at transmission nodes or buses. These relationships are most easily understood by a simple example of a "short" transmission line.[17]

---

[16] Generally, higher voltage lines also require more right-of-way than do lower voltage lines because of safety clearances. The transmission line construction method (e.g., poles vs. towers) also has a bearing on the right-of-way requirements.

[17] For analysis purposes, transmission lines are classified as short, medium, or long. With a short transmission line, i.e., generally less than 50 miles, the effect of the conductors' capacitive reactance is minimal and thus can be ignored without sacrificing accuracy. The problem is therefore greatly simplified.

A two-bus system connected by a single transmission line is shown in Figure 5.15 (a). Bus 1 is a source bus, and Bus 2 is a load bus; thus, current must flow from Bus 1 to Bus 2 in order to deliver power to the load. V_1 and V_2 represent the voltages at the buses. The current flow through the conductor's impedance causes a voltage drop across both the resistance and the inductive reactance components. These voltage drops are represented by V_R and V_X.

With transmission lines, the magnitude of the inductive reactance is generally much more than the magnitude of the resistance. The resistance and reactance of a transmission line is distributed in nature, i.e., Ω per mile. However, with a short transmission line, the resistance and inductive reactance can be treated accurately as a lumped impedance between the two buses ($Z = R + jX$). As a result, the voltages at the two buses can be related simply by

$$V_1 = V_2 + IZ = V_2 + V_R + V_X \tag{5.6}$$

where $V_R = I \times R$ and $V_X = I \times X$. Thus, all of the voltages have been identified, and can be summed to solve Equation 5.6 when treated appropriately as vectors. Either V_1 as the *sending-end voltage* or V_2 as the *receiving-end voltage* can be selected to serve as the reference voltage, which is plotted along the x-axis (i.e., at an angle of 0°). V_2 has been chosen as the reference voltage for all examples shown in Figure 5.15 and will be held constant throughout.

Consider first the case of a unity power factor load connected to Bus 2 with a real power requirement of P; thus, the apparent power S is equal to P. Since the reactive requirement of the load, Q, is zero, the load at Bus 2 appears as only a resistance. The load current then must be in phase with the bus voltage V_2, as shown in Figure 5.15 (b). Likewise, the voltage drop across the resistance of the line, albeit small, is directly in phase with the current. The voltage drop across the inductive reactance of the line leads the current by 90°, as V_X must be in quadrature with V_R. Note that the magnitude of V_1 is slightly greater than V_2 and that V_1 has a leading angle of δ degrees with respect to V_2 (the convention for vector rotation being counterclockwise).

Next consider a second case where the load is represented by the same real power requirement as before but at a lagging power factor. The addition of a reactive power requirement, $+jQ$, at Bus 2 increases the apparent power load, thereby causing an increase in the current flow. As shown in Figure 5.15 (c), both the magnitude and the angle of the current have changed as the current now lags V_2 by the power factor angle θ. The increase in the current further causes an increase in the magnitudes of both of the line voltage drops, V_R and V_X. In addition, V_R now lags the reference bus voltage by θ as it remains in phase with the current, while V_X, in quadrature with V_R, is now found to be at an angle of $90° - \theta$ with respect to the reference bus voltage. The magnitude of the resulting sending-end voltage at Bus 1 has increased over the unity power factor case, and V_1 maintains a leading angle compared to V_2 (δ has decreased slightly compared to Case 1).

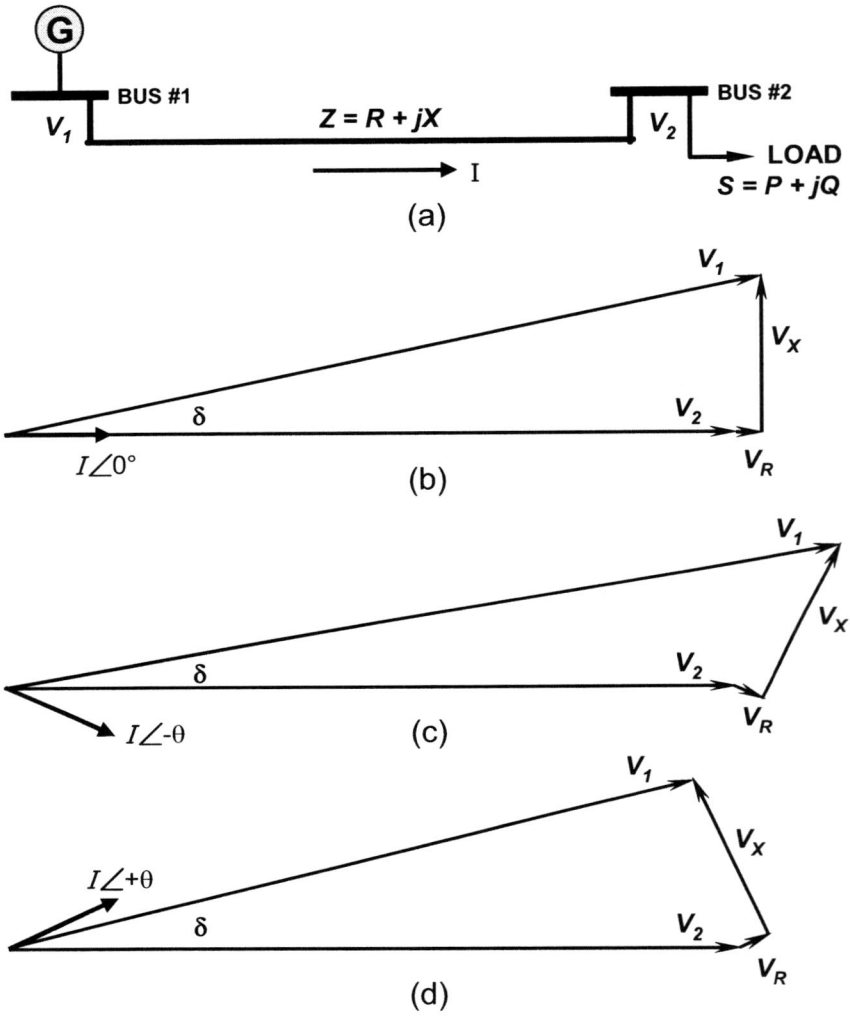

Figure 5.15 Phasor diagram of a short transmission line. Vector diagrams of voltages for a short transmission line connecting a source bus to a load bus. In Case 1, diagram (b), the load is operating at unity power factor. In Case 2, diagram (c), the load is operating at a lagging power factor. In Case 3, diagram (d), the load is operating at a leading power factor. In all three cases, the sending end voltage, V_1, has a leading angle, δ, on the receiving end voltage, V_2; thus, real power, P, flows from the source bus to the load bus. Reactive power, Q, flows from the bus with higher voltage to the bus with lower voltage. Therefore, in Case 3, reactive power flows from the load bus to the source bus and must be absorbed by the generator.

The final case merely changes the load's reactive power requirement to a leading power factor situation ($-jQ$). Both the apparent power and the current increase as before; however, the current is now found to lead the reference bus voltage by the angle θ. As expected, V_R now leads V_2 by θ as it remains in phase with the current, while V_X, in quadrature with V_R, is now found to be at an angle of 90° + θ with respect to the reference bus voltage. The Bus 1 voltage, V_1, once again maintains a leading angle compared to V_2 (δ has increased slightly compared to Case 1); however, the magnitude of V_1 is now less than V_2.

On a transmission line connecting two buses, *real power flows from the bus with the leading most voltage angle to the bus having the lagging most voltage angle*. In the three cases presented in Figure 5.15, the angle of the voltage V_1 exceeds the angle of the voltage V_2 by δ. Thus, under the conditions presented in all three of the cases, P flows only from Bus 1 to Bus 2. In a complex system of multiple loads and sources connected to buses within the network, the flow of real power between points is managed by controlling bus voltage angles.

The flow of reactive power (VARs) is also governed by differences in bus voltages but in a manner different than with real power flow. On a transmission line connecting two buses, *VARs flow from the bus having the higher voltage magnitude to the bus having the lower voltage magnitude*. In the cases presented in Figure 5.15, the magnitudes of the two bus voltages are shown to change dramatically in relation to the reactive requirements of the load connected to Bus 2. Looking first at Case 2, the load has a lagging power factor requirement, and the magnitude of V_1 exceeds the magnitude of V_2. As a consequence, VARs flow from Bus 1 to Bus 2. Conversely in Case 3, where the load has a leading power factor requirement, VARs flow from Bus 2 to Bus 1 as the magnitude of V_2 exceeds the magnitude of V_1.

Another way to view the relationship between voltage and reactive power is to realize that VARs provide crucial voltage support because an injection of reactive power raises the voltage at that point. In the case of the lagging power factor load, the generator at Bus 1 is required to produce reactive power to supply the inductive load requirements at Bus 2. The VAR injection causes the voltage at Bus 1 to increase to a level that is higher than Bus 2, and consequently, reactive power flows from the generator bus to the load bus. In the case of the leading power factor load, VARs are injected into the system at Bus 2 due to the highly capacitive aspects of the load (e.g., from an overabundance of online capacitors). The voltage at Bus 2 consequently rises above the voltage at Bus 1, and reactive power now flows from Bus 2 to Bus 1. In this case, the generator is required to absorb VARs. In either case, VARs are required to maintain voltage magnitudes. Without the injection of sufficient VARs, the system would experience a voltage collapse.

In Case 1, the magnitude of V_1 slightly exceeds the magnitude of V_2 even though the load at Bus 2 is at unity power factor and thus requires no reactive compensation to be supplied from the generating source at Bus 1. Nor does the load at Bus 2 produce reactive power that must be absorbed by the generator. As with Cases 2 and 3, this bus voltage differential results primarily from the inductive reactance of the transmission line, i.e., jX_L (the effect of the line's resistance component is virtually negligible and the capacitive reactance, $-jX_C$, is negligible for short lines). The inductive reactance of the line itself requires and thus absorbs VARs in a manner similar to the inductive reactance of a transformer coil or an inductive end-use load device, such as an induction motor. Consequently, the generator must produce VARs in order to satisfy the reactive power requirement of the line and thus maintain the reference voltage at Bus 2.

With medium to long transmission lines, the shunt capacitive reactance of the lines can not be ignored. Under light loading conditions, the line capacitance becomes dominant, and it yields reactive power that must be absorbed by generators and other inductive system components. Under heavy loading conditions, the inductive reactance of the lines becomes dominant, and it absorbs reactive power in the manner described for short transmission lines. However, under the certain condition known as *surge impedance loading* (where the line is carrying a specific amount of real power representing a portion of its loading capability), the effects of a line's distributed capacitive and inductive reactances balance each other out. Thus, at the surge impedance loading point, the line operates at unity power factor with the voltage and current in phase along its entire length. Under this condition, the voltage at either end of the line has virtually the same magnitude considering that the voltage drop across the line resistance is effectively zero.[18]

In transporting power on a transmission line, both real and reactive losses are realized. Since the reactance of a line is much greater than its resistance, the reactive losses (i^2X) greatly exceed the real power losses (i^2R) at high or low loading levels (i.e., above or below the surge impedance loading point). Providing VARs to meet a load's reactive power requirements and/or to support voltages over long distances is impractical because of the elevated level of transmission losses that would be incurred. In addition, the transport of reactive power utilizes a portion of the line's capacity that could be used for real power transmission. As a result, the transport of reactive power on heavily loaded lines is highly constrained. It is more effective to produce VARs locally where they are needed as opposed to trying to import VARs from distant sources. An optimal power system has a balanced mix of reactive power production capability from both generation resources and distributed static compensation devices, such as capacitors on transmission and distribution lines.

Power Flows in a Network System

As described previously, the transmission system is structured in a network or grid configuration. The flow of power within the grid depends to a great extent upon

---

[18] At the surge impedance loading point, a phase angle does exist between the sending end voltage and the receiving end voltage due to the time difference required to propagate the voltage wave across the length of the line.

the voltage characteristics at the various buses located throughout the system. A complicating factor is that there are often parallel paths over which power may flow from one point in the system to another. The path with the lowest per unit impedance will transmit more power than the corresponding paths in the same parallel circuit.[19] Thus, for two lines in parallel, the power flow will divide in inverse proportion to the per unit line impedances as given by

$$P_1 = \frac{Z_2}{Z_1 + Z_2} \times P_{IN} \quad \text{and} \quad P_2 = \frac{Z_1}{Z_1 + Z_2} \times P_{IN} \qquad (5.7)$$

where P_1 = power flow in Line 1 in MW
        P_2 = power flow in Line 2 in MW
        P_{IN} = total input power in MW
        Z_1 = per unit impedance of Line 1
        Z_2 = per unit impedance of Line 2

Parallel paths need not be of the same voltage, as shown in Figure 5.16. For example, Line 1 and both buses may be operating at a nominal voltage of 138 kV, while Line 2 may be operating at 230 kV. Thus, a transformer is necessary at each end of Line 2 in order to connect it to each bus. The transformer impedances must be included in determining the Line 2 per unit impedance.

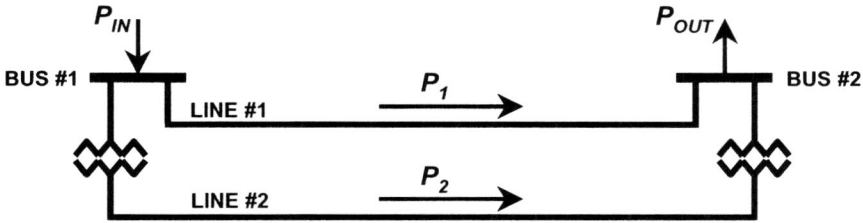

Figure 5.16 Parallel paths between two transmission system buses. The input power, P_{IN}, divides between Lines 1 and 2 based on their respective impedances (including the impedances of any transformers as shown in Line 2). Assuming a lossless line (i.e., $R = 0$), $P_{IN} = P_1 + P_2 = P_{OUT}$.

---

[19] The *per unit* methodology is used in the analysis of power systems. Discrete quantities of power, voltage, current, and impedance are converted to per unit quantities by dividing by selected base factors. One need only choose an MVA_{BASE} (e.g., 1,000 MVA) and a kV_{BASE} (e.g., 230 kV 3ϕ) even though multiple voltages may be involved. Given these two base amounts, an I_{BASE} and a Z_{BASE} is computed (e.g., 251 amps and 529 ohms). Thus, an actual line impedance of 31.5 ohms would be represented as 0.06 per unit.

As load is increased on a transmission line, it will ultimately reach its thermal limit. Each transmission line is assigned normal and emergency condition loading ratings. A relay protection scheme is used to prevent damage to lines due to overcurrents, and a line will be tripped out if it tries to transport too much power. In parallel paths, one line may reach its maximum loading limit while capacity remains available on the other line(s). Consider in Figure 5.16 that Line 2 is operating exactly at its limit, due to its low impedance, while Line 1 is loaded to 85% of its rating, given a power input to Bus 1 of 100 MW. If the power input were increased slightly to 105 MW, Line 2 would attempt to carry its proportionate share of the incremental load, because of the principle of power division, and thereby reach an overload condition. Meanwhile, Line 1 picks up its proportionate share of the incremental load and still has capacity available. Since Line 2 presents a limiting factor, the transmission corridor made up of these two paths is incapable of handling the load increase, even though additional capacity is available in Line 1. If Line 2 tripped, Line 1 would attempt to transport all of the input power, and then it too would reach an overload condition causing it to trip out.

This simple example attempts to indicate the complexity of transporting power through a large transmission network. As system load increases, certain lines or corridors may become congested and further load increases threaten the security of the system. Under high load conditions, particularly during peak summer air conditioning use, an overtaxed transmission system could lose critical lines and ultimately suffer cascading outages. The increase in power transmission due to competitive wholesale energy markets has heightened this concern and has created the need for additional transmission facilities so that power can be transported reliably.

Since parallel lines of unequal impedances will not naturally carry the same magnitude of load, they could be made to transport the same amount of power by changing the impedances to exactly match each other. Using capacitors and inductors for series compensation of the line can accomplish this task. Placing a capacitor in series with the line increases the flow of power by reducing the reactance. Conversely, placing an inductor in series with the line increases the reactance and thus decreases the flow of power. Of course adding impedance to a line is less desirable because of the additional voltage drop and losses that would be incurred. Nevertheless, by changing the line impedances in this manner, power flow in transmission lines can be controlled. Since capacitors and inductors have fixed ratings, banks of these devices must be switched in and out of the circuit in stages in order to provide the proper degree of compensation.

To transport equal amounts of power to a common receiving end bus having a specified voltage requirement (reference voltage), while leaving the unequal line impedances alone, would mean that each line would call for a different voltage at the common sending end bus. These two required voltages would neither be in phase nor have the same magnitude. The line with the greater impedance would require a greater sending end voltage. To resolve this sending end voltage disparity, a voltage equal to the vector difference between the Line 1 and Line 2 required sending end voltages must be added. Simply raising the lower required voltage to the magnitude of the higher required voltage will not work because they are out of phase. A *phase-shifting transformer* can be used to introduce a voltage, which provides the necessary phase angle compensation. The phase shifting transformer has special tap-changing coils that inject a ± quadrature voltage into the circuit, which thus serves to increase or decrease the voltage phase angle as required.

Flexible AC Transmission Systems

The electrical equipment used for voltage regulation and phase angle control that has been discussed thus far is operated by means of mechanical switching devices. By its nature, mechanical switching takes a relatively long time to execute. In addition, repeated switching operations cause wear and tear on mechanical contactors. Solid state switching devices solve both shortcomings of mechanical apparatus by providing greater dynamic response with no moving parts. The application of solid state control devices along with conventional electrical components, such as inductors and capacitors, adds a whole new dimension to the operation of AC power systems. *Flexible AC Transmission Systems* (FACTS) make use of these technologies in a variety of ways in order to extend the capabilities of the power transmission network. For example, with FACTS control devices, the system can be operated closer to its thermal limitations without compromising reliability. Thus, the power transfer capability of a transmission network may be increased without otherwise having to upgrade line conductors.

FACTS control devices provide a number of power system operating functions, including:

- Real and reactive power flow control
- Voltage and current control
- VAR compensation
- Transient and dynamic stability
- Fault current limitation
- Oscillation damping

Several FACTS controller types have been developed. These devices are installed in shunt, series, or combined shunt-series configurations. For instance, the *Static Synchronous Compensator* (STATCOM) is a shunt-connected VAR compensation apparatus, which provides for greater power transmission capability by means of controllable inductive or capacitive current independent of voltage. The *Static Synchronous Series Compensator* (SSSC) increases the voltage across the line impedance by injecting a controllable voltage in quadrature with and independent of the current. The *Unified Power Flow Controller* (UPFC) combines the STATCOM and the SSSC devices. The UPFC therefore provides a wide range of control to transmission line impedances, voltage levels, and bus voltage phase angles, and thus provides extensive control over real and reactive power flows.

Two major events are helping to promote the development and implementation of FACTS control methods as transmission systems are being called upon both to carry a greater volume of power and to move power in many different directions. Much of the transmission grid was built to serve an integrated utility structure and not originally designed for the types of transactions brought about by the transition to widespread competition in the generation markets. In addition, environmental and other social concerns provide constraints on system expansion and the routing of new transmission lines. FACTS has the potential to yield better system performance under such constraints.

DC Transmission

The commercialization of DC power transmission began in Sweden in 1954. Since then, several high voltage DC (HVDC) projects have been implemented around the world. While somewhat limited in application, certain advances in HVDC technologies would allow for many more practical installations. As in the case of FACTS, the evolution of competitive energy markets and environmental issues serves to promote the potential of HVDC.

DC has not kept pace with AC for general-purpose transmission applications because cost has been a limiting factor. To have an economic advantage over AC, HVDC lines must extend over very long distances.[20] The Pacific Coast Intertie traverses nearly 850 miles between northern Oregon and southern California. While DC lines themselves are actually less expensive than AC lines on a per mile basis, the terminal point converter stations, which integrate a DC line with the AC systems, add a considerable cost.

In special applications, HVDC has been found to be more practical than AC transmission. For instance, long high voltage cables, such as the submarine cables used in some parts of Europe, require so much charging current because of their high level of capacitance that their available capacity for transporting power would be diminished significantly before load is even connected. Although limited opportunities exist, short DC lines can be used to interconnect two asynchronous power systems (e.g., 50 Hz to 60 Hz) or two systems having incompatible frequency control. "Back-to-back" DC systems, where the AC to DC and the DC to AC conversions are accomplished in the same location without a transmission line, are used for interconnecting very large-scale AC systems. For example, operating the entire U.S. and Canadian transmission systems as a single synchronous power grid is not technically feasible. As a result, several east-west and north-south back-to-back DC ties are being used for AC system interconnection.

HVDC offers several advantages over AC transmission. In their simplest form, DC transmission lines can be operated with only one line conductor (or con-

---

[20] The breakeven distance for underground lines is much less than for overhead lines. However, a long distance underground installation of either AC or DC lines is more costly than the comparable overhead line construction.

Power Delivery Systems: Transmission and Distribution

ductor bundle), whereas AC transmission requires three conductors (or conductor bundles) for normal three-phase systems. A *monopolar* DC line has a single insulated conductor "pole" with an earth return. The voltage of the line is measured with reference to the ground. A *bipolar* DC line uses two conductor poles and voltage is again measured with respect to ground with one conductor being at a positive voltage and the other conductor at a negative voltage. At this time, such lines are operating generally in the range of ±500 to ±600 volts, but DC voltages in excess of ±1,000 are expected to be possible in the future. In the event of an outage of one of the conductor poles, the remaining conductor pole may be capable of operating in an emergency monopolar fashion using earth return. A third possible arrangement is a *homopolar* DC line that has two or more conductor poles operating at the same positive or negative polarity with an earth return. DC lines do not have to transport reactive power, and they generally have lower losses than AC lines. In addition, DC lines do not have the voltage stability problems associated with AC line operations.

HVDC also has disadvantages compared to AC transmission. The converter stations required at each end of a DC line are much more expensive than the typical substations that terminate AC lines. Voltage level transformations must be made on the AC side. The process of converting between AC and DC creates a considerable amount of harmonics, and the AC systems at each end of the line must provide VAR support in order to furnish energy for these harmonics. The reactive power compensation requirement can be more than half of the real power flow being transmitted. In addition, breaking the DC circuit presents a different problem since the current flow is continuous in one direction and does not cycle periodically through zero as in an AC circuit.

Converter stations operate in two different modes. At the sending end of the line, AC is *rectified* to produce DC. At the receiving end of the line, DC is *inverted* to produce AC. Both the sending end and receiving end converter station have essentially the same layout; thus, power can be transmitted in either direction on the DC transmission line. A schematic diagram of the major components of a converter station is shown in Figure 5.17. Mercury arc rectifiers (or valves) represented the earliest methodology for converting voltage; however, more recent technology makes use of solid-state elements known as thyristors for the conversion of voltage between AC and DC. A limiting factor in using solid-state devices for power system applications has been the high voltages and currents that must be tolerated without breaking down.

The basic rectification process is essentially just like the functioning of a power supply in an electronic end-use load device. When a single-phase AC voltage is applied to a single rectifier device, such as a diode, its operation only passes current during the positive portion of the alternating voltage cycle. It blocks the current during the negative portion of the alternating cycle of the input voltage. Thus, the rectifier's pulse output is unidirectional (and thus DC) but not constant. However, this half-wave rectification is highly rippled due to its harmonic content (i.e., an AC component rides on top of the DC voltage).

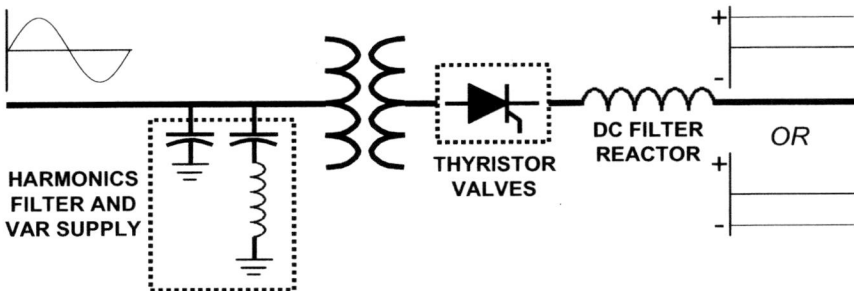

Figure 5.17 A schematic diagram of an HVDC converter station. The station connects an AC line to either a positive or negative DC line. The station acts as a rectifier when converting AC to DC and as an inverter when converting DC to AC. Two converter stations in tandem, with opposite DC polarities, would be used to operate a bipolar DC line.

The DC output of a power supply can be improved considerably by adding a second diode to the circuit. When connected in the proper arrangement, full-wave rectification of both halves of the AC cycle input occurs. Both the positive half of the cycle and the negative half of the cycle of the AC input voltage appear as a positive voltage in the output. The result is an improved DC output having a higher frequency AC ripple component as compared to the half-wave rectifier. To provide a constant or near-constant DC voltage profile, the output must be filtered to remove the harmonics.

In HVDC converter stations, rectifiers consist of a "bridge" network of thyristors. Numerous thyristors are connected in series in order to create a "valve" that is capable of handling the stress of high voltages. A thyristor is similar to a diode in that it has an anode and a cathode, but it also has a "gate." The thyristor conducts current under a forward voltage bias (i.e., when the applied voltage of the anode is positive with respect to the cathode) like the diode but only when it is fired by a gate signal. The gate provides an aspect of control. Modern HVDC converters are referred to as "6-pulse" and "12-pulse" rectifiers, which are capable of outputting six or twelve DC current pulses for each AC input cycle. The use of three-phase AC voltage for the input ensures a DC output with minimal harmonics. A reactor-type filter is then used for smoothing the DC voltage.

The converter station can also be operated in a reverse fashion to invert a DC input to an AC output with the same thyristor valve circuits used for rectification. With a DC input, the output of a thyristor valve is a square wave. With commutating circuit components, a simple thyristor bridge is made to produce an alternating square wave. The use of 6-pulse and 12-pulse designs yields a more trapezoidal-shaped alternating waveform that is the result of summing the square wave outputs of the various thyristor valves (which are fired in time sequence stages). This output consists of the fundamental frequency (e.g., 60 Hz) plus harmonics, the content of which depends on the number of valves utilized in the converter station. AC filters are used to absorb the harmonic currents. The resulting AC output current leads the voltage, and thus the converter requires compensating reactive power.

HVDC is a proven methodology that can be expected to improve and become more prominent in transmission applications as cost-reducing technological advances are made. Globalization in the electric power business will further promote HVDC as a workable solution. In Europe, for instance, a number of additional HVDC projects between neighboring countries are under construction or in the planning stages.

Interconnected Transmission Systems

Adjacent utilities have interconnected their transmission systems in order to gain technical benefits.[21] For example, the conjunctive operation of a large-scale transmission network allows all interconnected utilities to maintain system frequency in unison. The loss of a generating unit or block of load within the large system would have much less impact on the frequency than the loss of that same unit or load in a smaller or isolated system.

A number of economic benefits are also realized through large interconnected utility systems. Historically, power interchange capabilities have allowed utilities some flexibility in planning and scheduling the construction and startup of new generation facilities. As a utility started commercial operation of new generation to serve its load growth, it would initially have temporary excess capacity available for sale, for a while, since the new capacity provided a margin for anticipated future loads. An adjacent utility could enter into an agreement to purchase this power and thus have a dependable resource that would allow it to complete its construction program in a less than exhortative manner. At another point in time, the seller-buyer situation of the two utilities would likely be in the reverse orientation. From a broad perspective, coordination of utility construction schedules in this way helped to decelerate the overall requirements for new generation to the benefit of both the utilities and their customers. In some cases, two utilities have also been able to capitalize on their seasonal load diversities by exchanging power at different times of the year each using their existing fleet of generating units. Some utilities have joined together in formal *power pool* agreements that allow for coordinated planning, operations, and scheduling of outages for generator unit maintenance, all of which is enabled through their transmission interconnections.

Short-term operating advantages are also gained by transmission interconnections when a utility, itself having adequate capacity, can forgo a unit startup or suspend its own online generation for even a few hours when less expensive en-

---

[21] A single utility may have interconnections with numerous other utilities and have multiple connection points with one or more of those utilities.

ergy is available on the grid and can be imported from other sources. Such spot market arrangements again provide an economic benefit to both the utilities and their customers. The power supply for such transactions may originate from a third party's generator with one or more intermediary utility transmission systems used as an avenue for transporting the power to the receiving utility. Such transactions are referred to as transmission *wheeling* arrangements.

Another fundamental reason for having transmission interconnections is for emergency backup conditions. For example, a sudden loss of specific generating resources at the time of a system peak could put a utility in a negative capacity position with respect to its load. With such a loss, the affected load would immediately begin to draw power across the transmission tie-lines from the neighboring utilities as their generators would detect an increase in demand and attempt to compensate. Without transmission interconnections, the afflicted utility could have wide-scale system stability problems and major customer outages. Utilities have typically entered into formal agreements with their neighboring utilities to exchange emergency backup power (if available) for such contingencies and thus ensure a higher degree of system reliability. With large interconnected systems, the ability to compensate for the loss of a major generator is easier than with smaller scale systems.

A voluntary organization known as the North American Electric Reliability Council (NERC) was formed in 1968 in the aftermath of a large-scale blackout which occurred on November 9, 1965 in Ontario, Canada and parts of the Northeastern U.S. NERC was created to promote the reliability of electricity supply throughout North America.[22] As of 2009, NERC consists of eight Regional Councils, as shown in Figure 5.18. NERC's membership consists of utilities, RTOs, and a number of other industry companies and organizations. NERC is organized with a Board of Trustees and several technical committees, which develop planning, and operating standards and policies on behalf of the membership.

The interconnection of transmission systems does introduce some technical operating issues and problems. For example, when each interconnected utility's generation exactly matches its load, the entire system operates in balance and no power flows across the tie-lines. However, if the generation and load get slightly out of balance somewhere in the system, some *inadvertent energy* will flow across the tie-lines. While not as consequential as the loss of a major generating unit in the system, as described above, inadvertent energy flows do take up capacity and increase losses in the transmission systems of the interconnected utilities. Inadvertent energy is accounted for by tie-line metering.

---

[22] In 2007, FERC granted legal authority to NERC to enforce reliability standards with United States, bulk power system owners, operators, and users. FERC made compliance with those reliability standards mandatory and enforceable. NERC has similar authority in Ontario, and New Brunswick and is further seeking similar authority in Mexico and the other providences of Canada, which ultimately would provide a consistent reliability standard for all of North America.

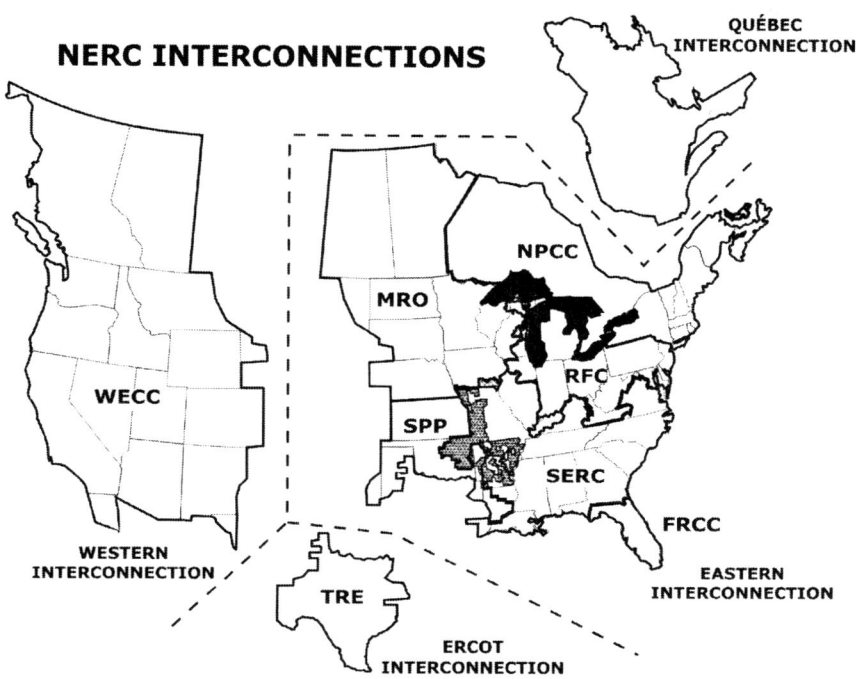

Figure 5.18 A chart of the North American Electric Reliability Council. NERC regions include: **FRCC** - Florida Reliability Coordinating Council; **MRO** – Midwest Reliability Organization; **NPCC** - Northeast Power Coordinating Council; **RFC** – Reliability-*First* Corporation; **SERC** - SERC Reliability Council; **SPP** - Southwest Power Pool, RE; **TRE** – Texas Regional Entity; and **WECC** - Western Electricity Coordinating Council (includes a portion of Baja California Norte, Mexico). Source: North American Electric Reliability Council.[23]

Inadvertent flows also occur in the normal course of scheduled power interchanges between two interconnected utilities. To transfer power, the sending utility increases its level of generation while the receiving utility lowers its generation by an equivalent amount thus causing a specific amount of power to flow between the two utilities. However, the interconnected transmission grid is complex, and

---

[23] This U.S. and Canadian chart of the NERC regions from the North American Electric Reliability Corporation web site is the property of the North American Electric Reliability Corporation and is available at www.nerc.com. This content may not be reproduced in whole or any part without the prior express written permission of the North American Electric Reliability Corporation.

several parallel paths often exist between the power source and the load. The power flow divides and follows the parallel paths, frequently through other utility systems that are located at great distances away from the end points of the transaction. Such behavior is referred to as *loop flows* because of the way that the power flows circumnavigate the most direct connection between the sender and the receiver. Without a FACTS type of device for controlling the flow of power, the *contract path* of transmission designated for the scheduled interchange may carry only a very small portion of the actual power transfer.

The effort to make energy supply markets competitive has significantly increased the number of wholesale transactions taking place on the transmission system. In such an environment, the transmission system becomes a common carrier that can be used by any number of sellers and purchasers of electricity. In England, all of the transmission systems have been consolidated under a single entity, government-owned National Grid system. In the United States, a number of *Regional Transmission Organizations* (RTOs) have been established. With an RTO, transmission owners may enter into a variety of agreements that serve to provide comparable open-access transmission and ancillary services on a large-scale basis.[24]

Historically, transmission systems have been planned, built, and operated to meet the needs of a vertically integrated utility structure. Interconnection of transmission systems provided the economic and technical benefits discussed above. The individual transmission systems were not built with the added purpose of supporting a myriad of wholesale transactions; thus, it is not surprising that without significant modifications that the interconnected transmission system would be constrained at times. Congestion management, for instance, is a common necessity because the system has yet to be optimally configured to function in a manner that is required in today's competitive wholesale power markets.

5.4 SUMMARY

The power delivery system, which consists of a transmission and a distribution function, provides the means to transport electricity economically from generators to customers. While most customers are connected to the distribution system at secondary voltage levels, some customers receive service directly at a primary distribution or transmission voltage. Transformers are used to reduce voltages from higher to lower levels as necessary in order to connect customers in accordance with their particular electric service requirements. The movement of elec-

---

[24] An RTO may take several forms. An Independent System Operator (ISO) consists of a nonprofit organization that only plans and operates a transmission grid on behalf of member transmission owners. A Transmission Company (TRANSCO) is an autonomous for-profit organization that owns, plans, and operates its transmission grid. Other RTO variations have elements of ISOs and TRANCOs.

Power Delivery Systems: Transmission and Distribution

tricity through the power delivery system causes both energy and demand related losses to be incurred. Thus, the capacities of lines and equipment must be sized to accommodate both the load and its attendant electrical losses.

The attributes of distribution feeders often vary from one location to another as a result of such factors as the mix of end-use load devices, the growth rates of new customers and per capita energy usage, and the local geographical conditions. In addition, distribution reliability is affected by numerous variations in local environmental factors. All such spatial influences together create unique and distinct characteristics throughout the distribution system.

Unlike distribution, transmission systems are designed to move large blocks of power to major load centers, such as portions of large cities, and to provide interconnections with other transmissions systems for power interchange and system reliability improvement. Most transmission systems are operated with AC voltages; however, DC transmission applications are found to be feasible in a number of special situations. Since electricity follows the path of least resistance, transmission systems must be controlled in order to optimize power flows throughout the interconnected network. The formation of Regional Transmission Organizations has served as a means for promoting a competitive generation market.

A comprehensive understanding of the structure and operation of T&D systems is an essential requisite of electricity pricing. Power delivery systems are highly complex, and cost-of-service study processes deal with numerous T&D system principles at great length in order to develop engineering-based cost allocation methodologies which in turn support the development of equitable pricing structures for electric service.

6
Electric Power Production

6.1 INTRODUCTION

The power generation function is responsible for the production of electricity, which includes producing both real and reactive power. As shown previously in Figure 5.3, generators can be connected at various voltage levels within the transmission and distribution systems. Generation step-up (GSU) transformers are utilized to match a generator's output voltage (e.g., 13.8 kV) to the applicable voltage of the local transmission system.[1] *Central station* power plants consisting of one or more large generating units may be located in close proximity to major load centers, as within a city, or at great distances away. The siting of central station power plants depends on major factors such as land availability, fuel transportation methodologies, and environmental impacts. *Distributed generation* (DG) utilizes much smaller units that typically are located directly at or in close proximity to a customer's facilities.

A variety of technologies are used for generating electricity in both small and large scales. However, with some exceptions, the production of electricity is accomplished by means of a rotary powered synchronous generator that is driven by a *prime mover*. A generator is an electric machine that is quite similar to a synchronous motor that operates in a reverse manner. With a generator, torque must be applied to the rotor shaft to cause it to spin. As electrical load is applied to the generator, the prime mover must supply a compensating level of mechanical power in order to satisfy the electrical load requirement.

---

[1] A GSU substation is fundamentally similar to a distribution substation, which transforms voltage between the transmission system and the primary distribution system. The current FERC Uniform System of Accounts categorizes the GSU transformers as a part of transmission plant. Functionally, the GSU transformers should reside with generation.

A simple schematic diagram of a typical synchronous generator is shown in Figure 6.1. An *exciter* provides DC current to the *field* windings located on the machine's rotor thereby creating a magnetic field. The resulting rotating magnetic field induces sinusoidal electrical currents in the *armature* windings of the stator. Three-phase generation requires at least three sets of stationary poles in the armature that are physically spaced at 120° with respect to each other. Thus, the generated voltage would have the three-phase waveform characteristics as illustrated previously in Figure 4.2. The number of poles and the speed of revolution govern the frequency of the generated voltage. For example, a three-phase machine having two poles per phase and operating at 3,600 revolutions per minute would produce a frequency of 60 Hertz (reference Equation 3.3). A generator's internal impedance, which is mostly due to the reactance of its windings, causes a voltage drop as current flows out of the machine. Thus, the voltage at the generator's terminals is slightly less than the actual generated voltage.

The production of real power is accomplished by the mechanical power input to the rotor. Thus, throttling of the prime mover controls the real power output. The production of reactive power is accomplished by the level of applied excitation. A synchronous generator can be made to produce VARs (an overexcited state) or to absorb VARs (an underexcited state) by adjusting the strength of the field current. A synchronous generator is designed to operate at a specific power factor, such as 85% lagging. Thus, at the limit of its prime mover capability, such a generator could produce reactive power at a rate of 0.62 MVAR per MW of output. A generator is typically sized with an MVA capacity in excess of the real power capability of its prime mover in order to provide reactive capacity for the control of system voltage.

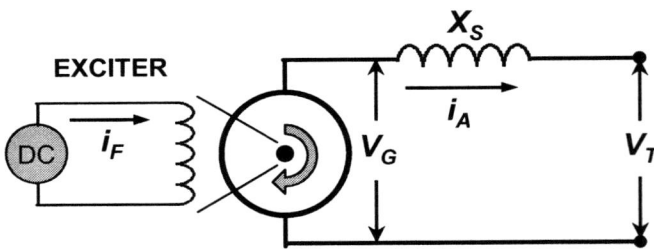

Figure 6.1 **A basic one-line schematic diagram of a synchronous generator.** A DC exciter supplies current (i_F) to the field coils located on the generator's rotor (rotor in axial view). The resulting magnetic field, rotated by a prime mover, induces an AC current flow (i_A) in the armature coils. X_S represents the generator's synchronous reactance (the resistance of the armature windings is typically very small compared to the reactance). V_G is the generated voltage, while V_T represents the voltage, which appears at the terminals of the machine.

A generator's ability to produce both real and reactive power is limited by thermal heating effects caused by the i^2R current flows in the field and armature conductors. Figure 6.2 exemplifies the capacity output or loading capability of a synchronous generator designed to operate at an 85% lagging power factor. Below 85% power factor, i.e., at higher overexcitation states, the generator's output is limited by the heating of the field windings (Point A to Point B). Above an 85% lagging power factor, through unity, and into a slight leading power factor condition (about 98% in this example), the output is limited by the heating of the armature windings (Point B to Point C). For higher underexcitation states (Point C to Point D), power output is constrained by the machine's steady-state stability limit, which is primarily a function of the generator's synchronous reactance (X_S) and the equivalent reactance of the connected system (X_E). In addition, heating of the iron components of the machine caused by eddy current induction provides constraints to the generator's output in this area. As shown in Figure 6.2, the range of operation for underexcitation is much less than for overexcitation.

Proper removal of the generator's heat is important for obtaining a high degree of unit capacity. Large generating units typically use hydrogen gas under pressure for the direct cooling of the machine's components. Water and oil are also used sometimes as a media for cooling purposes, occasionally in conjunction with hydrogen cooling. The hydrogen atmosphere is maintained at a very high level of purity with concentrations in the upper 90 percent range, which are too high to support combustion. The hydrogen is pressurized to improve its heat transfer capability. Typical hydrogen operating pressures include 15, 30, 45, and 60 psig (gauge pressure) or even higher. Increasing the hydrogen pressure is a means for increasing a generator's capacity, and generators typically carry multiple ratings based on the operating pressure of the hydrogen. Thus, a generator's power production capability is represented by a family of operating curves of the type shown in Figure 6.2.

6.2 POWER PRODUCTION TECHNOLOGIES

The production of electricity is accomplished by means of an energy conversion process using a fuel or other basic form of energy. With most production technologies, the input energy resource is ultimately transformed into mechanical energy, which turns a generator's prime mover. The fundamental energy resources used to develop the mechanical input energy for powering a generator include pressurized steam, combusted gases, moving water, and wind currents. Steam power and gas power are produced mostly by means of depletable fossil fuels while water power and wind power are both considered to be renewable energy resources. Expendable nuclear fuel is another resource used to provide a source of heat for use in steam power production. Solar radiation energy is a renewable resource that can be used in a few specialized power production applications, including the direct conversion of light energy to electricity.

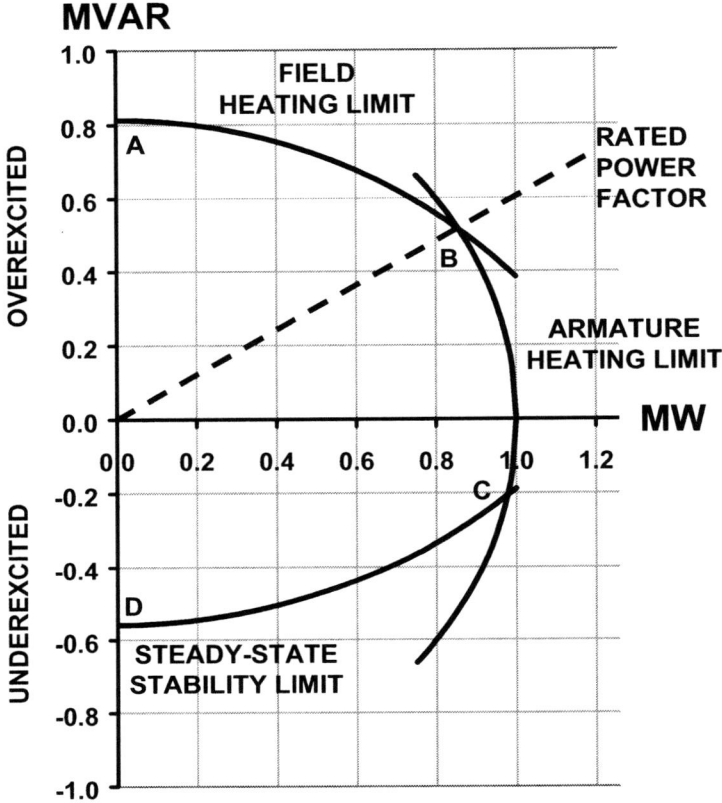

Figure 6.2 **Synchronous generator loading or capability curves.** In an overexcited state, the generator produces VARs (like a capacitor) that flow out and into the connected system. In an underexcited state, the generator absorbs VARs (like an inductor) from the system. The scale is shown in per unit; thus, the output in terms of actual MW and MVAR is found by multiplying the per unit values by the machine's MVA rating. The graph represents the intersection of three circles, which together define the capacity limits of machine operation due to various thermal effects. The field and armature limitations are based on the safe maximum levels of current-carrying capability of their windings. In the underexcited state, the output of the machine is limited due to steady-state stability constraints imposed by the power system to which the generator is connected. An additional limitation results from heating in the iron components of the armature due to induced eddy currents. The output capacity of a machine is also a function of its cooling system. The real and reactive capacity characteristics shown above assume a hydrogen-cooled generator operating at a hydrogen pressure of 60 psig. For lower operating pressures, such as 30 or 45 psig, the machine's capacity would have to be derated.

Steam-Powered Electricity Generation

A steam-driven prime mover consists of three major components: a steam production device, a turbine, and a condenser. These devices are essential to the operation of the Rankine cycle necessary to produce mechanical power to be applied to the generator. With steam production, the chemical energy of a fuel source is converted to heat energy, which elevates water (i.e., the working fluid) to a vapor state at a high pressure and temperature. This steam is then fed into a high-pressure steam turbine where it undergoes the expansion portion of the thermodynamic cycle. The steam expands to a state of low pressure and low temperature, which exerts a force against the stainless steel or alloy blades of the turbine thereby causing rotary motion of the rotor shaft. The applied steam temperatures and pressures are constrained by the metallurgical limits of the turbine. In large turbine operations, the exhaust steam from the high-pressure turbine is often passed through a *reheater* and then through an intermediate turbine and/or low-pressure turbine all of which are attached to a common rotor. The waste steam exhausted from the turbine(s) is then passed through a *condenser* that reduces the steam back to a liquid state to complete the thermodynamic cycle. The principle energy inputs for steam-driven prime movers are fossil and nuclear fuels.

Fossil-Fueled Power Plants

Combustible fossil fuels used for electric power production consist primarily of coal, natural gas, and petroleum. A variety of coal types are used (bituminous, lignite, etc.). Steam production using fossil fuels takes place in a boiler. The fuel is injected into the boiler where it is combusted with air. In the case of coal, pulverizing mills are used to grind the coal into an extremely fine powder. The coal may be augmented with gas or oil under some conditions in order to stabilize the ignition. Water is passed through a system of boiler tubes where it is heated under pressure and converted to steam. Boiler efficiency is improved through the use of a *superheater*, which is a system of tubes located at the top of the boiler that uses hot combustion gases to further heat the steam. Additional boiler efficiency is gained by the use of heat that is extracted from the flue gases by an *economizer*, which transfers heat to the boiler feedwater, and by an *air preheater*, which transfers heat to the combustion air input stream.

Transportation of fuel to a power production plant site is a significant factor for electricity generation. Coal is most economically transported by barge and rail. A *unit train* delivers 100 railcars or more of coal at a time with each railcar hauling up to about 60 metric tons of coal. A truck delivery of coal to large central station plants is usually only practical for a *mine-mouth plant*. Coal slurry pipelines have also been used in a few applications. Coal is stockpiled in the open on the grounds or in silos. Fugitive dust from outside coal piles must be controlled. Oil is also delivered by barge or railcar and is stored on-site in large storage tanks. Natural gas is delivered by pipeline and used as required although a storage methodology may be used as well.

The combustion of fossil fuels, particularly coal, results in stack emissions that must be controlled to minimize environmental pollutants, including particulates, sulfur dioxide (SO_2), and oxides of nitrogen (NO_X). Particulates are removed by means of mechanical or electrostatic *precipitators*. The use of a cyclonic burning pattern can help to reduce the amount of fly ash produced by coal firing. Disposal is always an issue, but fly ash can be reclaimed and utilized in the production of concrete products.

SO_2 emissions are directly related to the sulfur content of the fuel, and they can be minimized either during the fuel firing process or in the flue gas stage preceding the stack. *Flue gas desulfurization* (FGD) is accomplished in a *scrubber* by passing the stack gases over or through an alkaline sorbent that reacts with the SO_2 to form a solid compound (and sometimes, with additional investment, a marketable by-product such as sulfur or sulfuric acid). Limestone ($CaCO_3$) or lime (CaO), which is calcinized limestone, in a slurry are the agents used for the scrubbing process. Alternatively, limestone can be introduced into the boiler where the lime is carried into the scrubber by the flue gas. While this method of FGD reduces the cost of calcination, the chance of boiler fouling or other problems may be at issue. The output of the scrubber process is a calcium sulfite ($CaSO_3$) and calcium sulfate ($CaSO_4$) sludge to which fly ash can be added to help stabilize the waste for handling. Scrubber sludge is surface impounded at the plant site, shipped to an appropriate waste disposal landfill, or used for recycling purposes, such as in the production of gypsum wallboard.

The major factors contributing to the creation of plant NO_X emissions is the nitrogen content of the fuel and the characteristics of the combustion process, including the amount of excess air and the firing temperature, which are directly related to the boiler's design and operating characteristics. A common technique for controlling nitrogen oxide emissions is a modification of the combustion process using low-NO_X retrofit burners. Several power plants are utilizing a method known as *selective catalytic reduction* (SCR), which is a post-combustion treatment of the flue gases. With SCR, ammonia (NH_3) is injected into the flue gas stream using honeycomb or plate-type catalytic beds thereby creating a reaction with the NO_X gases that results in the release of free nitrogen (N_2) and water (H_2O) in vapor form. Both temperatures and ammonia injection amounts must be carefully controlled in the SCR reactor to ensure effective operation and avoidance of creating undesirable substances, such as ammonium sulfate (($NH_4)_2SO_4$) and ammonium bisulfate (NH_4HSO_4), particularly when burning high sulfur coal. Although possible, a combined process for removing NO_X and SO_2 together is both complex and more expensive at this time than the application of separate FGD and SCR methods.

Nuclear-Fueled Power Plants

Fissile fuel, such as U^{235} (a heavy isotope refined from uranium ore), is commonly used as an energy source for nuclear-powered steam production. In a reactor core, these heavy atoms are split into two dissimilar atoms (e.g., cesium, iodine, krypton, strontium, xenon, etc.) along with the release of one or more neutrons and heat. The freed neutrons cause other atoms to split in a controlled chain reaction that effectively provides enough thermal energy to produce steam sufficiently to drive a large turbine. Once loaded into the reactor, the nuclear fuel lasts for quite some time and partial refueling of the reactor core is typically performed on an annual basis.

Two different systems of nuclear steam production are used for commercial electric power generation: *boiling water reactors* (BWRs) and *pressurized water reactors* (PWRs). Both the BWR and PWR systems utilize a steel pressure vessel that contains a core of fuel bundles consisting of the fissionable material. Control rods made of neutron-absorbing cadmium are used to control the intensity of the reaction by insertion and extraction from the reactor core. As the control rods are removed, a point will be reached when the reactor goes critical, i.e., a fission chain reaction is sustained. Reinsertion of the control rods slows the reaction. Both the BWR and PWR systems utilize ordinary water (as opposed to heavy water, D_2O) as the working fluid and thus are referred to as *light water reactors* (LWRs). Heat from the fission reaction is transferred to the water as it flows through the reactor core.

In a BWR system, boiling water and steam caused by the reactor's heat, flow out of the top of the reactor and into a steam separator which dries the steam. The resulting high temperature steam is then transported out of the pressure vessel by a steam line directly to the turbine and then through a condensing stage and finally back into the pressure vessel in a totally enclosed-loop system. Thus, the steam is common to both the reactor and the turbine. In the event of a leak and subsequent loss of water from this loop system, an emergency core cooling system would engage to prevent a potentially hazardous reactor core meltdown situation.

In a PWR system, a closed-loop piping system exits the pressure vessel, passes through a *steam generator*, and then reenters the pressure vessel to complete a primary loop. Water is constantly circulated through this system. The high temperature water caused by the fission reaction is kept in a liquid state by means of high pressure in the loop. The steam generator is essentially a heat exchanger in which heat is transferred from the high temperature water in the primary loop to water flowing in a secondary loop in order to create steam. This steam passes through the turbine and condenser and then the condensate flows back into the steam generator to complete the secondary loop. Thus, in the PWR system, the steam is isolated from the reactor core.

Although used world wide, among the biggest concerns with nuclear power plants are public safety and the potential exposure to radioactive materials. Major incidents, such as those that have occurred at the Three Mile Island and Chernobyl nuclear power stations, have resulted in extra safety requirements that have significantly increased the expenditures of utilizing the nuclear option. Initially promoted as "too cheap to meter," the costs of new nuclear generation reached the point where further installations came to a standstill. However, the nuclear option is once again being considered as an effective energy alternative.

The capital cost of nuclear power production exceeds the cost of other conventional generation alternatives.[2] Nevertheless, nuclear plants that are operational today generally produce electricity at a relatively low variable cost and thus hold a competitive position in many energy markets.

Geothermal Power Plants

Steam from geothermal resources can also be used for driving a turbine in order to generate electric power. The first commercial production of electricity using geothermal energy was accomplished in Italy in 1904. Various forms of geothermal energy resources exist, but the hydrothermal variety is the primary resource used for electric power production today. Subterranean reservoirs of steam and hot aqueous fluids are created as ground water circulates through the porous spaces of rock that is heated by magma. The steam and hot water are brought to the surface under its own pressure by means of an extraction well. The topside production methodology then depends on the particular characteristics of the hydrothermal resource.

Some limited sources, such as the geyser areas in California, produce "dry steam" that can be piped right out of the ground and directly into a low-pressure turbine. In more common cases, water at temperatures in excess of about 200°C rises to the surface where most of it can be flashed to steam as its pressure drops in a *flash tank*. The steam is separated from the residual water (brine), which is returned to the underground thermal rock zone through an injection well to help sustain the hot water resource. A third method is used for water resources that are above the boiling point but less than the temperature required for a flash steam operation. A *binary cycle* operation extracts the hot water through a heat exchanger and immediately injects it back into the thermal rock zone to complete a closed primary loop. A secondary closed loop having a low boiling-point working fluid, such as isobutane, also passes through the heat exchanger. The vaporized fluid is then used to power a turbine in the secondary loop system. In each of the three processes, the excess heat of the steam or vaporized fluid, after the turbine stage, is ejected to the atmosphere using a conventional cooling mechanism. Condenser water is typically injected into the ground.

Solar-Thermal Power Plants

Concentrated solar power can be used effectively to produce steam for powering a small-scale turbine generator. Using mirrors, direct sunlight can be collected and focused to create high temperature heat. One method for concentrating the sun's radiant energy is with a long parabolic-shaped collector, or *linear trough*. A *receiver pipe* runs axially along the focus point of the trough's parabola where the

---

[2] The capital and operating costs of coal plants are also increasing due to environmental requirements for the reduction of stack emissions. With changes in technologies and regulations, the relative competitive positions of energy alternatives are subject to change.

reflected energy is highly concentrated. The pipe contains a heat transfer fluid that is ultimately circulated through a heat exchanger to produce steam. Numerous collector troughs are arranged in rows and columns in a north-south orientation thereby forming a solar field that provides for optimal sunlight collection. The collector troughs are rotated about the receiver pipe as the sun travels from east to west to ensure maximum output during the daylight hours. Fossil-fired fuels and associated equipment may be used to supplement the solar energy collection process, thus forming a hybrid plant whose production is not limited solely to daytime operation.

Another method for producing steam with solar power is by using a vast array of large flat mirrors, or *heliostats*, that together focus the sun's energy at one central heat transfer point atop a tower-type of structure. The use of molten salt (a mixture of sodium and potassium nitrates) as the working media has been found to be both an efficient means for the heat transfer process and an effective method to provide thermal storage due to its heat retention properties. Thus, the electrical output capability of this methodology can be extended somewhat beyond the normal daylight period. However, hybridization with a fossil-fired alternative is a prudent strategy in the early stages of this emerging technology in order to mitigate some of its financial risks.

Solid Waste-Fueled Power Plants

Solid waste products, typically burned in conjunction with fossil fuels, provide both a source of useful thermal energy and a potential alternative to conventional landfill disposal. Useable solid waste products include such items as municipal refuse, sewage sludge, agricultural and forest industry waste, and other organic resources. The most practical means of transportation of solid wastes to the point of use is by large hauling capacity trucks.

Preparation of the solid waste materials prior to combustion involves a number of required steps. The materials must be dried and shredded to an appropriate size, which typically requires a primary and a secondary shredding process. Inorganic materials, including aluminum, ferrous metals, and glass, are then sorted out by means of magnets and air streams that suspend the burnable products while allowing glass and other heavier inorganic materials to fall out and be collected for recycling purposes. Air pollution can be a significant issue with the combustion of some waste products as they typically emit more SO_2, NO_X, and heavy metal toxins (e.g. lead and mercury) as compared to a coal-fired plant which has been fitted with state-of-the-art abatement facilities.

In order to utilize the treated waste products in a conventional steam production process, the coal, gas and/or oil-fired boiler must be modified by the installation of solid waste ports for injecting the waste fuel as a supplement to the standard fuel. The fuel mixture is adjusted so that the boiler ignition is sustainable. Given the favorable Btu content of the processed solid wastes, their use as a supplemental fuel source can yield a worthwhile reduction in the use of conventional fossil fuels for steam-powered generation.

Steam Cycle Condenser Operations

Regardless of the fuel or the methodology used to initially produce the working steam, the low-pressure, low-temperature steam exhausted from the turbine must reject its remnant heat and thus be condensed to a liquid state in order to complete the thermodynamic cycle. Several cooling methods are used to reduce the spent steam to a condensate. A *once-through system* takes water from an adjacent large source such as a river, lake, or ocean, circulates it through the condenser where it absorbs the rejected heat from the steam, and then returns the warmed water to the source. The water source must be large enough so that the rejected heat does not create an unacceptable rise in its temperature. *Cooling ponds* are often used when a large water source is not available. The pond is a reservoir that serves as the water source and sink for condenser operation, and its size depends on the production capacity of the plant. Since water will naturally evaporate from the pond, makeup water must be available from another source.

Both "wet" and "dry" *cooling towers* are also used for condensing steam. Large natural draft wet towers cool condenser water by direct contact with air that drafts naturally as a result of a hyperbolic-shaped concrete chimney structure. A mechanical draft wet tower uses fans to force or induce a draft across water droplets which cascade down through a series of decks composed of wood or other materials. Both types of wet towers cool on the principle of evaporation and thus both types require a source of makeup water. A dry cooling tower uses a closed system where the condenser water is circulated through a set of finned cooling coils, much like a radiator, over which air is blown by a fan. No water is lost from this system, but large surface areas are required to provide a sufficient level of heat transfer.

Gas-Powered Electricity Generation

Natural gas is a major energy resource used for electric power generation. Natural gas is composed primarily of methane (CH_4) but, unless cryogenically removed, also often consists of butane, ethane, propane, and other hydrocarbons (C_XH_X). A number of impurities, such as gaseous sulfur compounds, are typically removed as a phase of the natural gas production process. As a result, the combustion of natural gas generally produces a much cleaner emission as compared to coal. As described previously, natural gas is a fossil fuel that is commonly utilized in steam turbine operations, but gas is also burned in gas-fired turbine applications. Natural gas is transported to the plant site by means of pipelines capable of providing high volume delivery to the generator. Gas turbine generation can be brought online much faster than steam-powered generation.

Gas-Fired Turbines

An open-cycle combustion turbine (CT) used for electricity production is virtually the same as the engines utilized for jet aircraft propulsion. A gas-fired turbine operates on the Brayton thermodynamic cycle, and it has three major components:

a centrifugal or axial *compressor*, a combustion chamber or *combustor*, and the *turbine blade assembly*. The turbine's working fluid is actually air, which is compressed and delivered to the combustor under high pressure where it is mixed with the ignited gas. The gas combustion process significantly increases the air's temperature. The high-temperature, high-pressure air and combusted gases next expand through the turbine blades thus causing rotary motion of the drive shaft to which an electric generator's rotor is attached. Some of the mechanical power delivered to the shaft is required to drive the compressor, which is axially mounted to the shaft.

Cooling of turbine exhaust is not required as the spent gases can be vented directly to the atmosphere, as is the case in the simple-cycle mode. However, by incorporating a *regenerator*, a heat exchanger located between the compressor stage and the combustor stage, heat from the exhaust can be recouped and used to preheat the compressed air stream to improve the cycle thermal efficiency. Exhaust heat can also be recovered in a *heat recovery steam generator* (HRSG) to produce steam for use in an industrial heating process and/or for driving a steam turbine, the latter being an example of cogeneration. *Combined cycle* (CC) operations generate electricity using both a CT-type gas turbine and a steam turbine in this manner and thus have a combined cycle efficiency than is much greater than either the simple or the regenerative cycles.

When ambient temperatures are high, the density of air decreases. Thus, the mass airflow into the CT's compressor stage is less under hot season conditions than during other cooler periods. As a result, the performance of the CT is constrained somewhat by elevated temperature conditions. By cooling the air at the compressor's intake, to increase its density, the capacity output of a CT can be boosted. Turbine inlet cooling can be accomplished by either evaporative cooling methods, such as water sprays, or by refrigeration applications. The type of refrigeration system used depends in part on the expected time requirements for which cooling is needed. For continuous cooling requirements, such as during extended peak summer conditions, a chiller system is often used. If the inlet cooling is only required for a few hours of the day, a cool storage system that chills and stores a heat transfer media or makes ice during off-peak hours may offer a more practical solution.

Another gas resource used for generation is landfill gas, which is produced by the anaerobic decomposition of organic waste products and contains a moderately high level of methane. The amount of landfill methane production depends on such factors as the type and content of waste products, the age and size of the landfill site, etc. Viable landfill sites are generally capable of supporting the production of from less than one to a few megawatts of power on a fairly continuous basis, depending on their size. The methane gas is extracted from the ground from various wells and collected and delivered to the generator by means of a piping system, which may or may not be buried. The generators are located at the landfill site since the cost of running pipelines a great distance would be uneconomical. Since the load requirements of a landfill operation are generally quite low, more power is produced than can be utilized at the site. Consequently, landfill generation units are usually interconnected with the transmission grid or local distribution system so that the excess power can be marketed.

Locally sited, gas-powered generators, which are typically sized to serve a single customer or a small group of customers, represent a major component of *distributed generation* (DG), or in a broader perspective, *distributed resources* (DR). Advances in *microturbine* technology have reduced the footprint of these units thus making them more practical for smaller customer applications. Microturbines have a capacity range of a few kW to a few hundred kW, but multiple units can be configured in the same installation to collectively produce a much higher capacity. High-pressure natural gas is often used as a fuel source for microturbines, although they can be operated on other gaseous and liquid fuel sources. Low-pressure gas sources, such as from landfill methane, require the use of auxiliary compressors. Microturbines typically are air-cooled, have one moving part, and operate at very high rotational speeds while producing standard secondary level voltages (e.g., 480Y/277 volts).

The exceptional rpm characteristics of the microturbine operation result in a high frequency voltage output that must be converted to a standard frequency of 50 or 60 Hz. A back-to-back AC-DC/DC-AC converter can be used for this purpose as well as for providing a DC point to which a battery can be attached for independent starting (i.e., black-start capability). Otherwise, interconnected microturbines can use the incoming line as a power source for motoring startup. For safety purposes, microturbines operating in parallel with the distribution system must be able to sense the loss of line power and immediately disconnect in order to prevent backfeeding power onto the distribution lines.

Like larger gas-burning generation units, microturbines have very clean exhaust emissions. In addition, some microturbines can be augmented with an exhaust heat recovery system in order to produce hot water for "micro-cogeneration" applications. The noise level of package unit microturbines is generally acceptable for customer site installations and can be reduced further with ancillary noise abatement controls.

Gas-Powered Fuel Cells

Fuel cells represent another genre of DR that is well suited to customer site installations. Although hydrogen-oxygen fuel cells have long since been a proven technology for powering spacecraft, the implementation of fuel cells for commercial power production is still at a prototype stage. Fuel cell research is being conducted, not only for electric power applications but also for transportation. Ultimately, on-site fuel cells or an integrated fuel cell/microturbine package could become a major source of electricity, at least for larger customers.

A fuel cell does not utilize a prime mover since it produces electricity directly from a high temperature oxidation-reduction reaction in which the fuel is not burned as in a combustion turbine. Fuel cells combine ionized hydrogen and oxygen to produce electricity while emitting heat and water vapor as by-products.

The fuel can be provided from a variety of hydrogen-rich resources, such as natural gas, propane, or landfill methane, while the oxygen is simply taken out of the air. A *reformer* is used to extract hydrogen gas from hydrocarbon compounds. With processing, a variety of liquid fuels (e.g. gasoline) could also be used as a source of hydrogen. Various fuel cell technologies are in demonstration or under development, including: phosphoric acid, molten carbonate, solid oxide, and proton exchange membrane.

A fuel cell is similar to a battery in that it has an anode and a cathode; however, it produces electricity as needed and does not store energy. Fuel cells can be combined in series and/or parallel configurations to achieve any desired voltage, current, and power rating. Since fuel cells produce a DC voltage, an inverter and power conditioner is required to provide a compatible AC output. Fuel cells produce a constant supply of power and cannot be operated to follow load swings in the manner possible with conventional prime mover generation methods; thus, auxiliary batteries may be required in certain cases.

A variety of commonly available gas resources used for electric power generation has been discussed. Like other nonrenewable fuels, these gas reserves are limited. Coal reserves are in abundance, and coal gasification is an area of research that could ultimately prove successful in enhancing the implementation of gas-powered electricity production.

Water-Powered Electricity Generation

Water-powered generation represents a sizeable nonthermal, renewable energy resource. The force of moving water has been harnessed for electric power production in various ways for both central station and small-scale generation applications. Water-powered generation facilities have a relatively high installation cost but a very low running cost since no fuel is required in their operations.

Conventional Hydroelectric Power Plants

Conventional hydroelectric dams provide a means for channeling the flow of water from an upper basin (lake, reservoir, or river) through a *penstock* that leads down to a hydraulic turbine, or water wheel, which in turn powers a generator. The discharge water is carried away to a lower basin by the *tailrace*. A dam may not be used exclusively for power production. Dams also impound water for flood control, irrigation, municipal water supply, and recreation. Meeting these needs may at times affect the ability to produce power. *Run-of-the-river plants* are only capable of minimal water storage; thus, their generating potential is subject to river flow conditions. Seasonal snow and rainfall amounts also has a bearing on hydroelectric power production.

When the amount of water in the upper pool exceeds the capability of the generating facilities, the excess amount may be released through a spillway to prevent the eroding of the dam due to water that would otherwise flow over the crest. Conversely, some dams are designed with the capability to withstand overflow conditions. Small dams are typically built of compacted earth and clay while large dams are constructed of reinforced concrete. The power output potential of a hydraulic turbine is a function of the water flow rate and head (i.e., the vertical elevation over which the water falls); therefore, turbines are located at the base of the dam in order to maximize their generating potential.

There are two basic types of hydraulic turbines: *reaction* and *impulse*. In the Francis-type of reaction turbine, water from the penstock enters into and floods a spiral water case from which it exerts pressure against the curved buckets of a completely enshrouded water wheel impeller or *runner*. The Kaplan-type of reaction turbine uses propeller blades, which are either fixed or adjustable, in an unshrouded water passage. Kaplan units are typically used in applications of low hydraulic heads while Francis units are used for medium to high hydraulic heads. Impulse turbines utilize nozzles to transform the static head into high-velocity water jets that impinge on bowl-shaped buckets located around the periphery of the runner. An adjustable needle in the nozzle controls the water jet action. Impulse turbines, such as a Pelton wheel, are used in very high head applications (i.e., typically greater than 1,000 feet). Varying the flow of water through the runner controls turbine speed. Hydraulic turbines turn at speeds generally of 600 rpm or less, which is considerably slower than steam or combustion turbines. Also, while steam and combustion turbine-driven generators are typically configured in a horizontal fashion, hydraulic turbines are quite often arranged with the shaft in an upright position with the generator mounted above in a *powerhouse*.

Pumped-Storage Hydroelectric Power Plants

Another application of hydroelectric power generation is *pumped storage hydro*, which is a closed-cycle method of storing potential energy in the form of elevated water. Reversible hydraulic turbines of the Francis type, which can be motored by their generators from power taken off the grid, are used to pump water from a low level water source, such as a river, to a high level retention pool, such as a reservoir constructed at the top of a mountain. The water in the upper reservoir is later released back through the turbines to generate electricity.

Unlike conventional hydroelectric generation that is highly efficient, pumped storage hydro operations are much less efficient as about 3 kWh of pumping energy is required in order to realize 2 kWh of generated energy. However, pumped storage hydro is economical when off-peak, low-cost generation resources (e.g., nighttime nuclear power) are used for the pumping process. The pumped storage hydro water is released during on-peak periods to displace higher cost generation, such as a combustion turbine.

Tidal Power Plants

Another method of employing water power to generate electricity is the use of two-way turbines that are turned by oceanic tidal forces. Power is generated as the tide comes in, and water is impounded using a damming system, such as a barrage, thus creating a hydraulic head. At high tide, the potential to further generate ceases as the water level on both sides of the dam is nearly the same. After the tide recedes to a lower level, the captured water can be released back to the sea through the turbines to again produce electric power. The generating potential varies with the natural rhythm of the tides, and thus the maximum output of a tidal power plant will often occur during off-peak periods when the energy has a low value on the grid. Tidal power generation is in commercial use in a few select locations, including France and Canada. Although tidal power technologies may someday be advanced, its potential as a major resource for electricity generation is currently limited because of site requirements, high construction costs, and environmental concerns.

Environmental issues are of concern in many hydroelectric applications because of the vast amount of lands that may be flooded and the wildlife habitats that are directly affected. However, a major environmental benefit of hydroelectric power is that it requires no fuel to be burned in the production of electricity, and it may facilitate nearby irrigation.

Wind-Powered Electricity Generation

A small amount of the solar radiation that reaches the earth's atmosphere warms the air masses and is converted into wind energy that exists at all altitudes. In general, wind varies dramatically in terms of velocity and direction. However, some locations experience much more consistent wind patterns on average than others. In a few of these sites, prevailing winds close to the ground are being utilized to power aeroturbines for the production of electricity. The power output potential of a windmill is directly proportional to the cube of the wind's speed. Given frequent variations in wind speed, the annual amount of energy production is perhaps a better measure of a windmill's performance than its maximum power output capability.

Applications of wind-powered electricity production range from stand-alone distributed windmills to large power grid-connected windmill farms. Large generating farms require a considerable amount of land as the windmills must be spaced out to meet rotor diameter clearance requirements and to minimize wind interference between units due to diminished wind-power density in the windmill's wake. Generator capacities usually range from 100 kW to 1 MW.

A minimum amount of wind velocity, referred to as the cut-in speed, is required for the windmill to reach a turning speed at which it can begin to produce electrical power. At a high enough wind velocity, the windmill will reach a cut-out, or furling, speed at which the turbine requires shutting down to prevent mechanical damage. The optimal extraction of the wind energy depends to a great extent on the geometry of the aeroturbine's airfoil blades, and the blades may have either a fixed pitch or variable pitch, which allows the blades to be feathered. Similar to an airplane wing, the movement of the blade is based on aerodynamic lift and drag forces.

A windmill's major components are the blades and rotor, a gearbox, the electrical generator, and a mechanism for providing directional orientation (yaw control) of the rotor blades with respect to the wind direction. The more common windmills in use are horizontal axis machines having two or three metal blades attached to a hub. Each blade has an axial twist, which provides compensation for the variation in rotational velocity along its length. Stall-controlled windmill blades may utilize automatic deployment of blade tip brakes to stop the machine under excessive rotor velocities. With large turbines, the blades may have a rotor diameter on the order of 150 feet; therefore a large tower is required for adequate support of the windmill's rotor blade assembly and nacelle which contains the gearbox and generator equipment.

Although not in widespread service, a variety of vertical axis windmills have also been designed and tested. Such machines utilize S-shaped and other upright airfoil structures. The main advantage of the vertical axis type of structures is that they are well suited to wind from any direction without requiring reorientation control of the blade mechanism. In addition, maintenance of the generator and gearbox is simpler than for a horizontal axis windmill as the equipment is located at ground level.

Unlike most electric power production technologies, aeroturbines most often drive an induction-type of AC generator, which is exactly like an induction motor having some form of prime mover. When the applied mechanical power causes the machine's speed to exceed the synchronous speed of the electric power system to which it is connected, it behaves as a generator and produces power. The electric system governs the machine's frequency. The induction machine does not require a DC exciter for its field like a synchronous machine since its rotor is excited by the AC flow in its stator. In other words, the induction machine uses the connected power system as its source of excitation. The generator's VAR requirements then must be provided from the system. The use of induction generators are generally limited to lower power output applications thus making them ideal for many windmill projects. In addition to induction generators, windmills may also power synchronous and DC generators.

Solar-Powered Photovoltaic Generation

Solar radiation represents a significant inexhaustible energy resource. To a minor extent, the heat and wind produced on earth by the thermal properties of solar energy are being harnessed with current technologies to generate electric power, as previously discussed. In addition, solar energy can be converted directly into electricity by means of *photovoltaics* (PV). Photovoltaic energy conversion is accomplished by means of solar cell arrays, which are made of individual, p-n junction semiconductors that are typically composed of silicon. During fabrication, a silicon crystal is doped, or infused, with impurities such as phosphorous to create an n-type layer and boron to create a p-type layer. The result is a light-sensitive photodiode that has a high electric field across its p-n junction. When the photodiode is illuminated, "electron-hole pairs" are created as the photon energy causes electron excitation. These charge carriers are separated by the p-n junction's electric field with the electrons being attracted to the n-type layer and the holes being attracted to the p-type layer. This charge separation creates a potential difference or voltage of about one-half of a volt depending on the amount of illumination available (insolation). If connected to a load, current would flow and the photodiode would deliver power.

By combining individual photodiodes in a series and parallel fashion to form an array, practical utilization levels of voltage and current can be derived. Optical concentrators, such as Fresnel lenses, can also be utilized to focus a greater amount of light on the solar cells thereby increasing their level of performance. The output of a solar cell array is DC, which must be inverted to produce AC for most electric power supply applications. Furthermore, inverting to AC is imperative for grid interconnection of PV operations. Since the output of the solar cell arrays is at low voltage, a step-up transformation to a primary voltage level is also required for interconnection with the distribution system.

A central station or grid-support PV facility requires a considerable amount of land for the installation of a field of solar cell arrays or modules. Each module is encapsulated in a transparent covering that provides protection from the elements while allowing the transmission of light through to its photodiodes. The PV modules may be installed with a fixed tilt as a best compromise for daily and seasonal variations in the sun's path or with a mechanized single-axis or dual-axis tracking system that optimizes the reception of the solar radiation at all times. The PV modules may be swiveled to a stowing configuration at nighttime and to prevent damage by minimizing the profile of the solar cell modules under high wind conditions.

PV can also be utilized as a distributed or stand-alone type of power supply application. In remote areas, where grid connection is impractical, a small PV installation may be designed to serve an individual delivery point or a specific end-use load device, such as a DC pump or fan required during daylight hours. For such applications as roadside telephone call boxes, highway sign and warning lighting, and distant monitoring stations, PV is often used to charge batteries, which ensure power availability at all times. Larger PV installations can also be used to meet a portion of a building's energy requirements in conjunction with purchased electricity or other on-site generation and battery storage facilities. Stand-alone PV systems are often roof- or wall-mounted and thus require no additional land for their deployment.

PV provides electricity with minimal environmental impacts as it requires no fuels and has no emissions or other waste products that must be managed. Like fuel cells, PV is a well-proven technology in space program applications. However, the potential of PV as a major electricity resource is limited at this time as only a few locations have a feasible level of insolation and because the current level of cost associated with PV is considerably greater than for conventional power generation methods. Like other emerging energy technologies, solar power research efforts could ultimately yield more economical applications of PV as an effective, commercial electricity resource.

Other Power Generation Technologies

Various other technologies, fuels, and energy resources are used to generate electricity. For instance, a widely used prime mover power production methodology is accomplished by means of internal combustion. Reciprocating engines burn a variety of petroleum products, such as diesel fuel or gasoline, but they can also be engineered to burn gases, such as natural gas and landfill gas. Reciprocating engines do not have the power capability of turbines and thus are limited to smaller, more distributed generation applications (including providing black start capacity for central power stations).

While fossil fuels represent a major source of energy for thermal processes, many other combustible materials are used to produce heat for steam generation. Commercial and industrial operations often yield wastes and byproducts that are incinerated as a means of disposal. Example waste materials include such items as tires, railroad ties, old utility poles, digester gas, waste alcohols, wood wastes, and various other biomass substances. Biomass and wood wastes are burned directly or first gasified. Black liquor, a byproduct of wood pulp processing, has been used by forest products businesses for a long time to generate electricity for both on-site utilization and export to the electric grid.

Many propositions for additional energy sources for power production have been offered. The liquid metal fast breeder reactor, for example, would appear to be ideal since it yields more fuel than it consumes. A breeder reactor fueled with U^{238}, produces Pu^{239}, a fissile product. However, plutonium is a highly toxic substance, and large quantities of the isotope would have to be handled in an extremely safeguarded manner. All nuclear fuels remain highly radioactive for many years; thus, a long-term, safe method of storing used fuel assemblies is required. Because of the lack of a permanent strategy for transporting and disposing of spent nuclear fuel, it must be stored at the generation plants. With such major issues as this, the nuclear power generation option has been limited.

Additional use of the energy in the oceans is being researched. Waves have tremendous energy that could be harnessed by various floating mechanisms whose bobbing motions could be transferred into rotary power for electricity generation. The thermal gradient between the surface water and the ocean depths provides a means for heating and cooling a working fluid that can be used to power a turbine in a closed-loop operation. Such ocean technologies are only in their infancy at this time.

Energy storage is a key issue relative to electric power generation. Pumped storage hydroelectric generation was discussed previously as a means to expend electricity to store water during low cost periods that could then be later utilized to generate electricity during high cost periods. An analogous methodology involves the use of electricity for the storage of compressed air in large underground caverns, which can be extracted and expanded through a turbine-generator at a later time. Electrical energy can also be stored as mechanical energy in the form of a high-speed flywheel or stored as thermal energy in a working fluid. Electricity itself can be stored in chemical form by charging batteries. The successful development of advanced battery technologies could eventually have a major impact on the operation of power generation systems.

Mix of Generation Resources

A wide variety of generation resources are used to produce electricity to meet the load requirements of customers. Table 6.1 shows the current characteristics of U.S. generation capability, based on nameplate capacity, in terms of both the energy resources and the prime mover technologies utilized. Table 6.2 shows the existing inventory of U.S. generation for both utilities and nonutilities (NUGs) in terms of the number of operational units and their installed capacity ratings.

Table 6.1 U.S. Generation Nameplate Capacity Characteristics as of 2007

Percent of Capacity Based on Energy Resources		Percent of Capacity Based on Prime Movers	
Combustible Resources		Steam Turbines	53.763%
• Natural Gas	41.312%	Combined Cycle	20.329%
• Coal	30.892%		
• Petroleum	5.736%	Gas Turbines	14.527%
• Wood Products	0.690%		
• Other Biomass	0.444%	Hydraulic Turbines	9.009%
• Other Gases	0.245%		
		Wind Turbines	1.526%
Noncombustible Resources		Internal Combustion	0.793%
• Nuclear	9.723%		
• Water	9.009%	Binary Cycle	0.028%
• Wind	1.526%		
• Geothermal	0.297%	Other	0.022%
• Solar Radiation	0.046%		
		Photovoltaics*	0.003%
Other Resources	0.080%	* Direct energy conversion (no rotary prime mover involved)	

Source: Energy Information Administration, Form EIA-860, "Annual Electric Generator Report."

Table 6.2 U.S. Fleet of Generation as of 2007

	No. of Units			Nameplate MW		
	Utility	NUG	Total	Utility	NUG	Total
Steam Turbines						
Nuclear	104	--	104	105,764	--	105,764
Fossil Fuels						
• Coal	1,192	274	1,466	331,513	3,896	335,409
• Natural Gas	523	139	662	93,040	1,335	94,374
• Other Gases	--	43	43	-	666	666
• Petroleum	141	47	188	31,250	704	31,954
Solid/Liquid Wastes						
• Wood	80	100	180	1,883	1,265	3,148
• Biomass	11	13	24	358	140	498
• Refuse	76	23	99	2,341	382	2,723
Gaseous Wastes						
• Landfill Gas	13	--	13	305	--	305
• Biomass Gases	--	1	1	--	18	18
Industrial Wastes						
• Black Liquor	--	166	166	--	4,363	4,363
• Waste Heat	5	41	46	152	1,245	1,397
Geothermal	161	--	161	2,929	--	2,929
Solar Thermal	11	--	11	465	--	465
Other	1	38	39	37	779	816
TOTAL STEAM	**2,318**	**885**	**3,203**	**570,037**	**14,791**	**584,828**
Gas Turbines						
Fossil Fuels						
• Natural Gas	1,838	335	2,173	126,960	6,356	133,316
• Other Gases	--	20	20	--	627	627
• Petroleum	644	9	653	23,738	139	23,877
Gaseous Wastes						
• Landfill Gas	48	1	49	161	5	166
• Biomass Gases	5	2	7	22	11	33
TOTAL GAS	**2,535**	**367**	**2,902**	**150,881**	**7,138**	**158,019**
Internal Combustion (IC)						
Fossil Fuels						
• Natural Gas	687	244	931	1,930	455	2,385
• Petroleum	2,545	324	2,869	4,794	491	5,285
Solid/Liquid Wastes						
• Biomass	--	4	4	--	7	7
Gaseous Wastes						
• Landfill Gas	985	24	1,009	841	33	874
• Biomass Gases	19	51	70	27	52	80
TOTAL IC	**4,240**	**643**	**4,883**	**7,600**	**1,031**	**8,630**

Table 6.2 U.S. Fleet of Generation as of 2007 – Cont.

	No. of Units			Nameplate MW		
	Utility	NUG	Total	Utility	NUG	Total
Combined Cycle (CC)						
Fossil Fuels						
▪ Coal	4	--	4	631	--	631
▪ Natural Gas	1,475	177	1,652	208,981	9,755	218,735
▪ Other Gases	5	3	8	334	36	369
▪ Petroleum	30	--	30	1,234	--	1,234
Gaseous Wastes						
▪ Landfill Gas	14	--	14	63	--	63
▪ Biomass Gases	3	4	7	23	75	97
Industrial Wastes						
▪ Waste Heat	1	--	1	10	--	10
TOTAL CC	1,532	184	1,716	211,275	9,865	221,140
Binary Cycle						
TOTAL GEOTHERMAL	63	--	63	304	--	304
Hydroelectric						
Conventional	3,742	250	3,992	76,929	715	77,644
Pumped Storage	151	--	151	20,355	--	20,355
TOTAL HYDRO	3,893	250	4,143	97,284	715	97,999
Wind Electric						
TOTAL WIND	389	--	389	16,596	--	16,596
Solar Electric						
TOTAL PV	26	1	27	36	1	37
Other Electric						
▪ Compressed Air	1	--	1	110	--	110
▪ Other Gen.	5	10	15	37	91	128
TOTAL OTHER	6	10	16	147	91	238
TOTAL RESOURCES:	15,002	2,340	17,342	1,054,159	33,632	1,087,791
Percent of Totals	86.5%	13.5%		96.9%	3.1%	

Notes: Utility is defined as entities classified as NAICS Code 22.
All numbers do not crosscheck due to rounding.

Source: Energy Information Administration, Form EIA-860, "Annual Electric Generator Report."

In Table 6.1, it can be noted that nearly 80% of the input energy resources for U.S. generation (based on nameplate capacity) are from combustible fuels. Some of these resources are considered to be renewable resources, such as biomass wastes, industrial byproducts, and landfill gas. The remaining 20% of the input energy resources are noncombustible fuels of which a little more than one-half of those are from renewable resources (water, geothermal, wind, and solar). Nearly 90% of the prime mover nameplate capacities are based on thermal types of generation processes with the remaining 10% representing water, wind, and photovoltaic technologies.

In the U.S. at this point in time, the most common choice for new generation is natural gas-fired turbines and combined cycle units. The current technological advantages of gas generation are low emissions, relatively low capital requirements, and quick permitting and construction cycles. Of the renewable energy resources, wind and landfill gas generation have been increasing the most, although together they represent only a very small portion of the installed capacity.

In Table 6.2, it can be observed that, at this time in the U.S., NUGs, identified as entities classified as all NAICS codes other than 22 (Utilities), own about 13.5% of the generating units, comprising about 3.1% of the nameplate capacity. Recent industry restructuring activities have caused some utilities to divest some or all of their generation in order to mitigate the issue of market power. Some of this generation has been purchased by other utility subsidiaries from other regions.

Individual utilities plan, build, and operate a mix of generation resources to serve their native loads in the most economical fashion. This system of generation resources is often supplemented with power purchases from other parties through its interconnections. Purchased power transactions may be long-term agreements (years), spot market purchases (hourly), or somewhere in between. Even a distribution-only utility must ensure the proper type of generation mix through its purchases in order to optimize its power supply against varying load conditions.

6.3 CLASSIFICATIONS OF GENERATION

Generation is categorized in accordance to how it is operated to serve load. Some generators are run continuously, except during periods of maintenance, while other generators are run intermittently. Such operations are driven by the mission to serve the load at the lowest cost. The types of generation required in order to meet this objective can be illustrated by viewing the characteristics of the load.

The upper graph in Figure 6.3 exemplifies a system load shape for a calendar year as a plot of all 8,760 hourly loads. The load swings through daily and seasonal cycles and is observed to be generally higher during the summer. The highest load for the year occurred on August 17th at 5:00 PM. Consequentially, the lowest load for the year occurred on October 8th at 5:00 AM. When the hourly loads are rank-ordered from highest to lowest and plotted accordingly, the result is a *load duration curve* as shown in the lower graph in Figure 6.3.

Electric Power Production

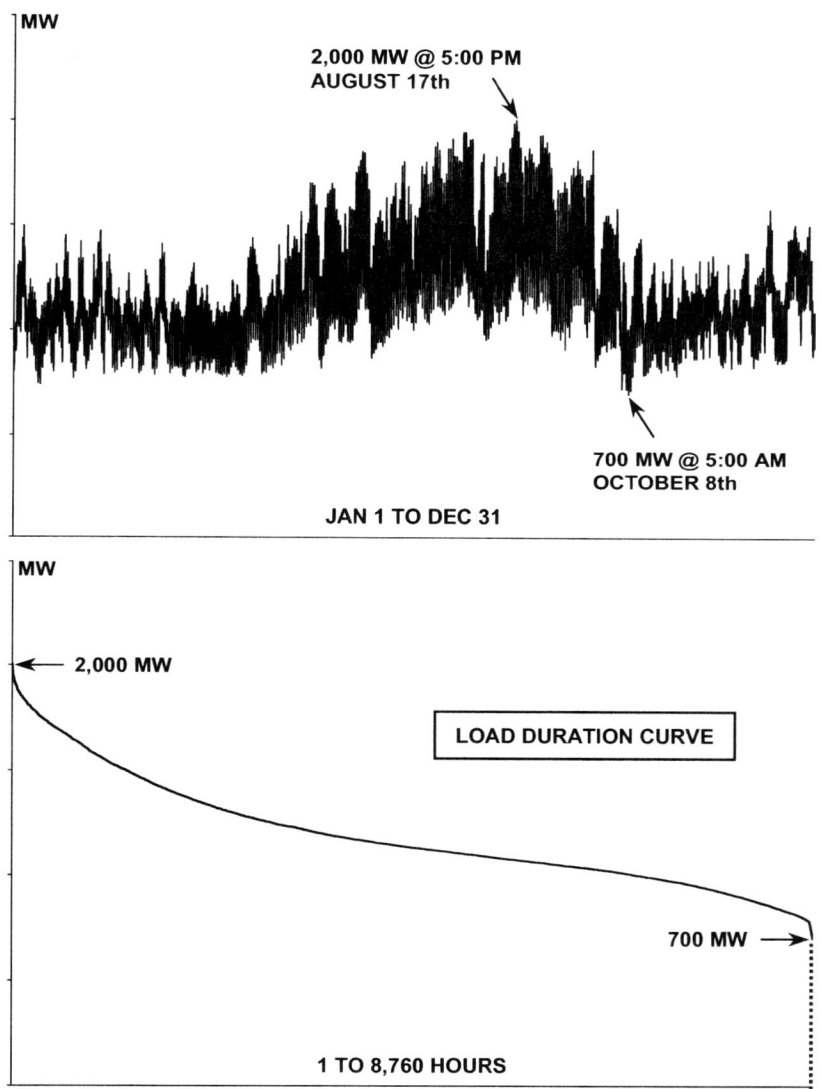

Figure 6.3 An annual load duration curve for a system of loads. A load duration curve is plotted by rank ordering each hourly load (upper graph) from highest to lowest. In this example, the load is always greater than 699 MW, and the peak hour of the year is 2,000 MW. The characteristic shape of the load duration curve is used to help define the categories of generation require to serve the load economically.

The load duration curve illustrates certain characteristics that are inherent in the aggregate electrical loads of a system. As observed in the example in Figure 6.3, the load in any hour is never less than 699 MW; thus, at least 700 MW of generation capacity must be provided continuously during the year. This certainty begins to describe an amount of the system load that is referred to as the *base load*, which consequently is served by *base load generation*. Base load is not wholly defined by the absolute minimum load of the year. Even generators, identified as base load generation, must be scheduled for a maintenance outage during the year. As a result, base load is (imprecisely) represented by the lower part of the load duration curve, as depicted in the upper graph of Figure 6.4.

The left-hand side of the load duration curve (i.e., near to the *y*-axis) contains the hours of the year in which loads are at extreme peak levels. In Figure 6.4, the load duration curve decreases somewhat exponentially for the first few hours before transitioning into more of a gradual and near linear-like decline. The load duration curve for systems having a high market penetration of weather-sensitive, end-use load devices often depict just a few hours of relatively high loads as a result of temperature extremes in some years. In such cases, the load duration curve initially declines rather dramatically before flattening out. Regardless, the initial hours of the load duration curve (imprecisely) define the need for generation that is only required to run for short periods of time, such as during weekday afternoons in the summer when temperatures are at their highest. Such intermittent generation is referred to as *peak load* or *peaking generation*.

The remaining area of the load duration curve that lies between the peak load area and the base load area is referred to as *intermediate load*, which is served by *intermediate load* or *cycling generation*. The generation units used to meet this part of the load requirement must be capable of frequent ramping up and down in order to follow the typical daily load excursions.

The load duration curve gives an indication of what type of generation mix would be best suited to serve a particular system of loads in the most economical fashion. *Generation screening curves* are used to plan new generation sources as the magnitude of the load grows and the characteristics of the load changes. Example screening curves for just three types of generation technologies are shown in the lower graph of Figure 6.4. The screening curves are plotted in terms of the fixed and variable costs per MW for each type of unit as a function of *annual capacity factor*, which is determined by

$$GCF = \frac{E_{GEN}}{P_{GEN} \times N} \times 100 \qquad (6.1)$$

where GCF = annual generation capacity factor as a percent
E_{GEN} = annual energy production in MWh
P_{GEN} = maximum rated capability in MW
N = number of hours in the year

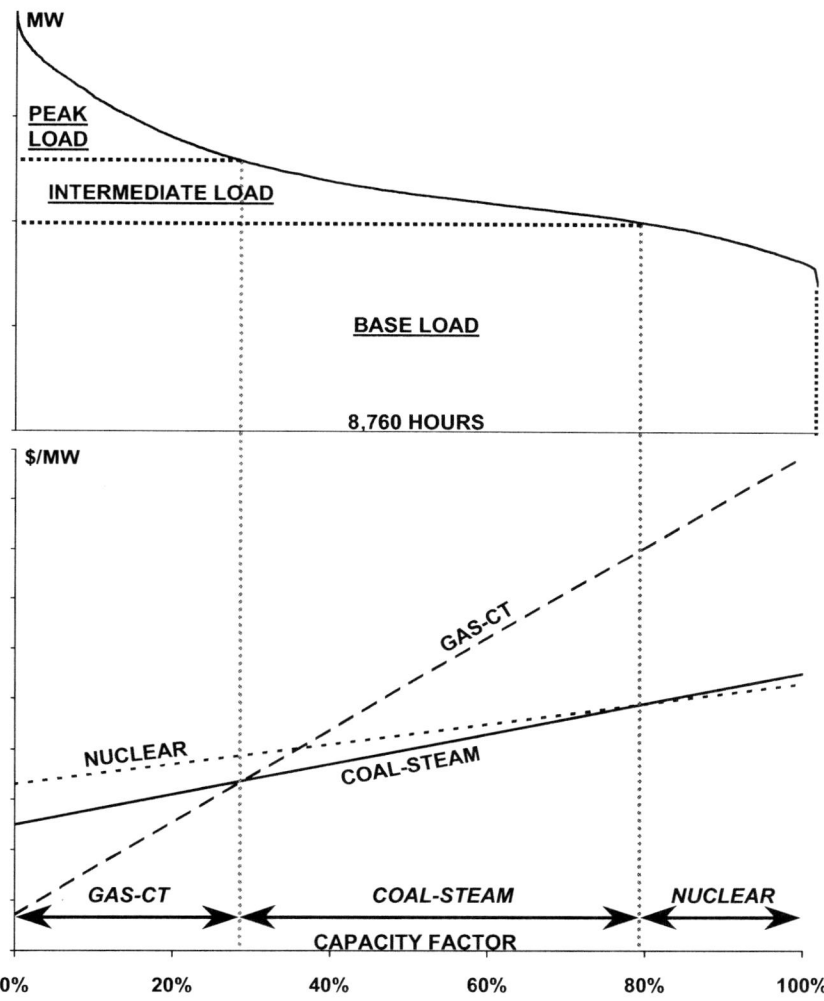

Figure 6.4 Generation screening curves. The curves in the lower graph illustrate the fixed and variable cost characteristics of three different generation technologies. Nuclear generation has a high fixed cost of capacity (installation or capital cost) but a low variable cost (fuel and variable O&M). In sharp contrast, a natural gas-fired combustion turbine has the opposite characteristics, while coal-fired steam generation lies in-between. The screening curves help to determine the generation technologies that best serve load considering the nature of the load duration curve. Several different technologies may be evaluated in order to plan the optimal generation mix.

The capacity factor for a generating unit is the ratio of its actual production to its theoretical production assuming a continuous output throughout the year at its maximum rating. The capacity factor can also be viewed as the percent of the year that the unit operates at its equivalent maximum output. In reality, generating units do not normally operate at a 100% capacity factor. All generating units require scheduled annual maintenance outage periods, and they are also highly likely to experience unscheduled or forced outages at various times. Nuclear units require periodic outages for refueling. While base generation operates the most amount of time, cycling and peaking generation operate a lesser amount of time in order to follow the load.

The screening curves in Figure 6.4 show how each example unit's total cost, on a dollar per MW basis, increases with respect to its cumulative hours of operation. These cost relationships have crossover points with each other. Thus, a gas-fired CT represents the lowest cost resource up to a little less than about a 30% capacity factor. Then a coal-fired steam unit becomes the lowest cost resource up to about an 80% capacity factor at which point a nuclear unit continues as the lowest cost option. In this example of limited resources, nuclear units would serve the base load, coal-fired steam units would serve the intermediate load, and gas-fired CTs would serve the peak loads. At this time, nuclear and gas-fired CTs are technologies that more or less represent the extremes of the cost spectrum of the more large-scale generation alternatives. By adding other resources to the screening curve analysis, such as a CC unit or a renewable energy resource, additional crossover points would be created. By assessing the characteristics of the load duration curve, generation technologies can be planned in order to produce an optimal mix of generation resources.

Generation Reserves

Because of unscheduled generation outage events, a utility must have available an amount of production capacity that exceeds its peak loads in order to ensure that an adequate supply of electricity is maintained at all times. This excess amount of generation is referred to as *reserve capacity*, and the percent of the total capacity that lies above the peak load requirement represents the *reserve margin*, as determined by

$$GRM = \left[\frac{C_{INST}}{L_{PEAK}} - 1\right] \times 100 \qquad (6.2)$$

where GRM = generation reserve margin as a percent
C_{INST} = installed generation capacity in MW
L_{PEAK} = system peak load in MW

Electric Power Production

Reserve capacity may be available from the utility's own generation fleet, or it may be secured through its interconnections under agreements from other providers. As in the situation cited previously, where high peak loads sometime occur for only a very few hours of the year, it would likely be more economical to buy capacity as needed rather than to build new capacity. Such purchased capacity could either be in the form of generation supply or as a contract with customers who are willing to interrupt their operations to reduce load on the system in a matter of minutes, i.e., interruptible rates.

An amount of reserve capacity is maintained online at all times to accommodate abrupt increases in system load or to compensate for the sudden loss of generation. With the expectation of a unit contingency, *operating* or *spinning reserve* is typically maintained as a percent of the load or at an amount equal to the capacity rating of the largest generator on the system or some other risk-based criterion. Rather than operating a single source of backup, spinning reserve is ordinarily spread over several units, which are running to meet the load requirements, to improve the response rate for picking up the incremental load.

A fairly quick response can be achieved from CT units whose turbines are not spinning at the particular time when called upon. Such units can be brought up to synchronous speed and connected to the system grid very quickly. *Hot reserves* are provided by thermal generation whose turbines are not spinning, but whose boilers are operating in a ready mode. However the response rate for such generation may be several minutes to hours. *Cold reserves* are thermal units that are available to run, but which are off-line. The start up of such idle units could take several days and thus are really considered to be a practical source of reserves only in an extended operating period sense.

6.4 GENERATION OPERATION AND CONTROL

Generation expansion planning focuses on constructing and procuring an economical mix of resources from a long-term perspective. However, the function of generation operations focuses primarily on short-term scheduling of units and controlling the available (online) resources on a momentary basis. As system load normally increases or decrease through its daily cycles, the decision must be made as to which unit(s) will be ordered to compensate for such incremental changes. *Economic dispatch* is a process of assigning load to each of the generation resources in real time in order to minimize the total cost of operating the system. To dispatch generation in this manner, each unit's specific running requirements, including the cost of fuel and variable O&M, must be determined.

Thermal generating units have a *heat rate* characteristic, which is the amount of Btu of fuel that must be burned to produce a kWh (or MWh) of output. Thus, heat rate is a measure of the efficiency of electricity generation. A heat rate of 9,000 Btu/kWh yields a higher generation efficiency than a 10,000 Btu/kWh heat rate. Heat rates vary between different generation technologies and between specific units within the same technology, e.g., one coal-fired steam unit is more (or less) efficient than another. The heat rate of a specific unit also varies somewhat as a function of its level of power production (MW). The heat rate is highest at the unit's minimum output rating because of the amount of fuel required to stabilize the burn and spin the turbine. As the power output of a unit increases, its heat rate declines somewhat but not in a precise linear relationship, and it then increases to some extent as the output approaches the unit's maximum rating.[3]

By multiplying the heat rate characteristic of a unit (Btu/MWh) by its output (MW) for various operating levels, an *input-output curve* representing the Btu per hour of fuel required to produce each level of MW can be determined. By knowing the cost of fuel ($ per Btu), the input-output curve can then be expressed as the cost per hour as a function of output to operate the unit. The incremental cost required to increase a generator's output by one additional unit of capacity is determined by

$$\lambda = \frac{dF}{dP} \tag{6.3}$$

where λ = incremental cost of a unit in dollars per MWh
F = fuel cost in dollars per hour
P = net power output in MW

Power output is stated as a net amount since a portion of the gross output of the generator is required to supply power to plant auxiliary equipment.[4] Aside from marginal fuel cost, other variable cost elements related to the operation of the unit may be included as part of the incremental cost for unit dispatch consideration. Typical additional costs include variable operation and maintenance (VO&M), SO_2 emission allowances, and fuel handling.

Since heat rates and fuel costs vary from one unit to another, the incremental cost curves of the units will likewise vary. Economic dispatch operates on the fundamental principle that the total cost of production is minimized if each generating unit is operated at the same incremental cost.[5] Being constrained to the same λ determines how much load each unit should carry relative to the total output requirement. Figure 6.5 illustrates an example of optimized loading for a simple system of three generating units.

---

[3] Heat rate curves can be determined from unit test data points to which a polynomial function is fit.

[4] Auxiliary power requirements have increased significantly, particularly as equipment and systems have been added in order for production plants to meet environmental regulations.

[5] It is well demonstrated that loading the generating units simply on the basis of fuel efficiency does not ensure the lowest cost of production, as two units having the exact same efficiency characteristics could have different fuel costs.

Electric Power Production

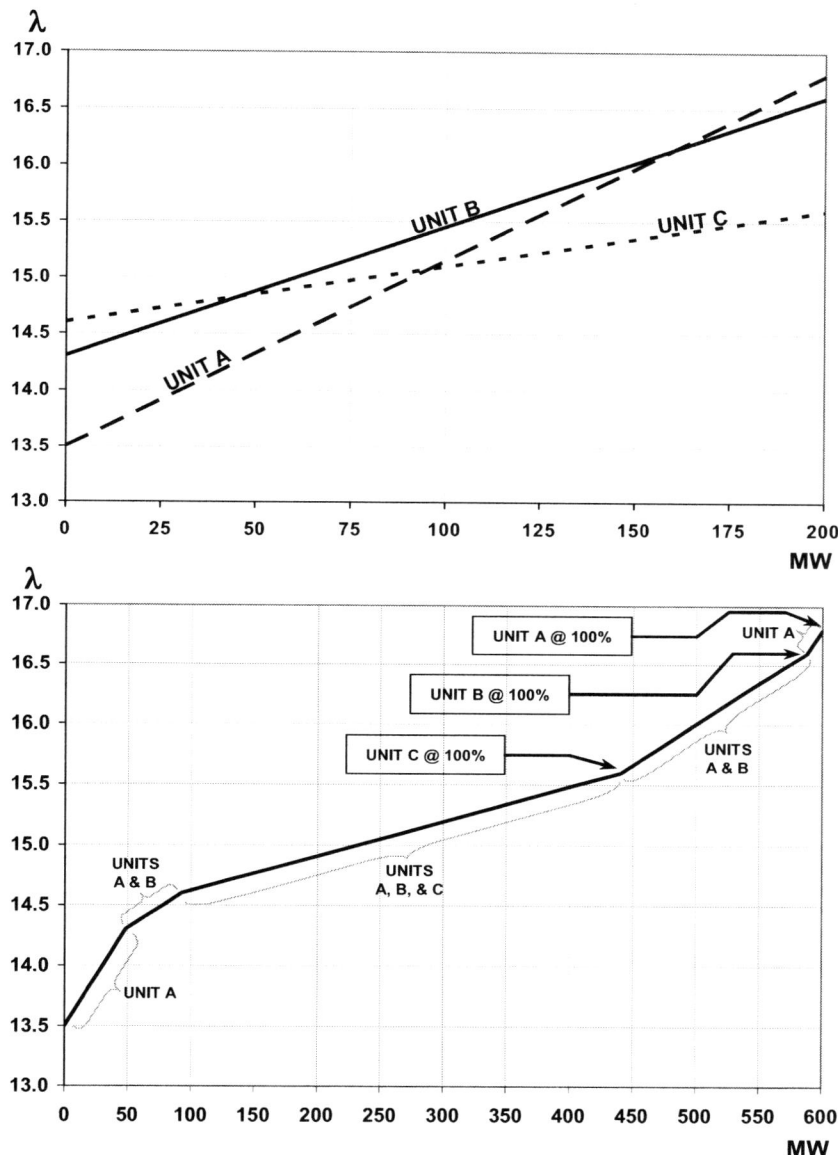

Figure 6.5 Economic dispatch example for three 200 MW generating units. Unit A alone picks up the first 49 MW of load. Unit C becomes fully loaded at 440 MW and Unit B becomes fully loaded at 588 MW. Then Unit A alone picks up the last 12 MW.

The incremental cost curves of three 200 MW generating units are shown in the upper graph of Figure 6.5. Unit A's cost at its minimum output is lowest at $13.5/MWh, so it is dispatched first to serve load. As shown in the lower graph of Figure 5.6, Unit A alone picks up load up to 49 MW, at which point, Unit A's incremental cost is equal to Unit B's incremental cost at its minimum output, i.e., $14.3/MWh. Above 49 MW, Unit's A and B begin to share in serving the next increments of load. Compared to Unit A, Unit B will pick up proportionately more of the next increment of load because the slope of Unit B's incremental cost curve is less than the slope of Unit A's incremental cost curve. For example, for a 10 MW load increase from 49 MW to 59 MW, Unit A would pick up 4.1 MW while Unit B would pick up 5.9 MW. Note how the slope of the cost curve decreases above 49 MW due to the weighting of Unit A and Unit B.

Units A and B continue to share the load increase up to 93 MW, at which point the incremental cost for Units A and B are now equal to Unit C's incremental cost at its minimum output, i.e., $14.6/MWh. Above 93 MW, all three units share proportionately in serving the next load increments with Unit C picking up the larger proportion, and again the slope of the cost curve decreases. Once the load rises to 440 MW, Unit C has reached its maximum capacity of 200 MW while Unit B is at 113 MW and Unit A is at 127 MW. Once the load reaches 588 MW, Unit B is at its maximum capacity of 200 MW. Unit A reaches its maximum capacity of 200 MW as it serves the remaining 12 MW of the total 600 MW. Dispatched in this manner, the costs are minimized at all levels of output.

The discussion of generation dispatch thus far is germane to units within the same production facility. Production plants are located at different points around the system and are integrated by means of the bulk transmission network (refer to Figure 5.14). Power flow from a lower cost unit at a given plant site to a particular load bus may incur more transmission line losses than a higher cost unit located at another plant site. Thus, the generating unit dispatch algorithm must take into the increased cost per MWh associated with these losses (often referred to as a "transmission penalty factor"). The result of the optimization is a *merit order* by which units are loaded. At times, a unit may be required to run out of merit order for the express purpose of providing system voltage support. In addition, spot market and economy energy transactions by means of the system interconnections are factored into the dispatch process as such opportunities become available.

While economic dispatch represents the means for loading units to meet load changes in real time, *unit commitment* is a process for scheduling the mix of units to be available to serve load that is forecasted for the short term. Operating costs can be saved by not having units running that are not needed for dispatch; start-up costs alone can be significant. Unit commitment considers such factors as availability (some units may be off-line for maintenance in the unit commitment planning horizon), minimum unit run time and minimum unit down time, spinning reserve requirements, power system security, and the cost of purchased power that may be available. The unit commitment problem is most effectively solved through the application of dynamic-programming algorithms.

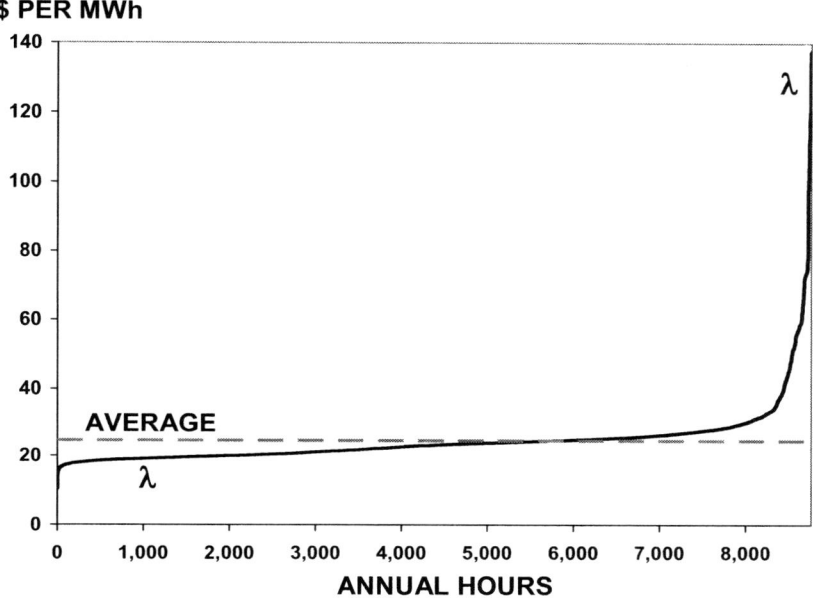

Figure 6.6 Hourly incremental fuel costs vs. the average cost of fuel. Hourly lambdas for a whole year calculated from a production costing analysis and plotted in rank order from the lowest cost hour to the highest cost hour. The average cost of fuel for the same year is also shown.

Production cost modeling is a planning function that serves a number of purposes, one of which is to establish an annual fuel budget. Production cost modeling is a probabilistic analysis which simulates the operation of a system of generating units to serve an expected load shape for one or more future years. A probabilistic approach is needed since the projection of hourly loads and generating unit unavailability (due to forced outages) has a degree of uncertainty. Production cost modeling incorporates the principles of both unit commitment and economic dispatch. The projected cost of fuel on an hourly basis, i.e., system λ, is determined in the analysis. An example of projected hourly lambdas, rank ordered from the lowest cost hour of the year to the highest cost hour of the year, is shown in Figure 6.6. The average hourly cost of fuel across the year is also shown in Figure 6.6.

System λ can also be viewed as a utility's avoided cost of energy. Projected hourly lambda values are often combined into a seasonal or seasonal time-of-use pricing structure for use in applications with a qualifying facility (QF) that is interconnected with the utility's system. At those times when the customer-owned generator is producing more power than is required for the customer's load, energy will flow into the utility's system. The avoided energy costs are paid by the utility to the QF for energy delivered to the grid.

The cost of fuel is recovered through monthly electric service billing, typically as a separate and identifiable charge on a ¢ per kWh basis.[6] Commission-approved fuel rates may be applied on a monthly, quarterly, semiannual, or annual basis, although in some jurisdictions, established fuel rates are allowed to continue until such time as a fuel cost over/under-recovery condition warrants a revision of the rate. Regardless of the particular period utilized, the fuel rate is determined by dividing the projected fuel cost by the projected kWh billing units of the customers for the projected period of application. However, the fuel cost for the projected period is first adjusted to reflect the over/underrecovery of costs from the prior period that is caused as a result of actual costs and kWh usage not exactly matching their associated projections, i.e., a true-up mechanism. For basic rate applications, the fuel cost recovery rates are applied as average values and not as hourly amounts.

On the other hand, *real-time pricing* (RTP) electric service rates for commercial and industrial customers are based on hourly lambdas, which are actually projected in near real time on a one-hour ahead basis or on a day ahead basis. Each day, the day ahead RTP option determines 24 kWh-based rates, which are applicable to the hourly energy usage on the following day and which are provided to the customer before the close of business on the previous day. In addition, hourly lambdas can be averaged over seasonal and/or time-of-day rating periods, as noted above, and then used for structuring and pricing time-differentiated rates for all classes of service.

Power Pools and Exchanges

A *power pool* is an operating arrangement between affiliated (holding company operating units) and/or nonaffiliated utility companies under which the participating utilities' generation resources are pooled together and commonly dispatched as a single, larger system. In so doing, a superior operating performance is often achieved as compared to the best optimizations of the pool member generation systems operating independently. With pooling agreements, the member utilities retain ownership and the associated responsibility for their specific generating units and facilities.

The composite load of the larger system will have somewhat of a unique characteristic to the extent that load diversity exists between the individual utility systems (i.e., the load duration curve of the pool system will be defined by the individual utilities' coincident hourly loads). This pool system load characteristic along with the merit reordering of all of the participating utility generation resources will likely have an operating impact on a number of the generating units in

---

[6] Some of the cost of fuel may be embedded in the rate schedule's kWh charges.

the pool. Some units may run for longer periods of time while other units may run less time. Furthermore, some units that were originally operated to serve base load may be moved up in the stack to function more typically as cycling units. Some cycling units may move up the stack to a peaking category while some peaking units may now serve only in a generation reserve mode by running only under emergency conditions. In other words, the operation of the generating units is re-optimized to serve the composite pool load as economically as possible. System λ then is defined by the merit order of the units participating in the pool.

Because of the changes to the load and to the operating characteristics of individual generating resources, a pool member utility (owning ample capacity to meet its peak load conditions) may at times have more native load to serve than owned capacity in service. Units that would normally be running, absent a power pool, could be off-line in favor of more economical units elsewhere in the system. Thus, to meet its capacity requirement, the utility would buy from the pool at such times to compensate for the discrepancy. Conversely, utility members that are in an overgeneration condition, because of having extra capacity in service to meet the merit requirements of the pool, would sell to the pool. Depending on load conditions and the resources available in the pool, a particular utility may be a buyer at times and a seller at other times.

A *power exchange* (PX) is an independent generation resource coordinator that conducts daily auctions in order to acquire a sufficient supply of energy in order to serve expected hourly loads of participating customers. In other words, the PX provides a market in which buyers and sellers of electricity can operate together. PXs typically accept bids from potential utility and nonutility generators on a "day-ahead" and a "day-of" basis. Bids may also be received from customers wishing to buy energy at a stated price level.

With the basic PX operation, the participating generators submit a schedule of their available capacity and a bid price for the energy (in \$/MWh). The PX qualifies the bids and rank-orders the resources by bid amounts in order to accumulate a sufficient amount of resources to meet the demand at the lowest cost. Since high bids would not be competitive against low cost producers, the exchange functions so as to minimize the cost of energy for customers.

6.5 SUMMARY

Power production is the system function in which electricity is generated by means of an energy conversion process. Except for solar photovoltaic production, generators are connected to prime movers driven by a variety of power sources, including steam, combusted gases, moving water, and wind. Most of the existing production capacity is provided from traditional fossil fuel sources. Thus far, renewable energy sources have made only a minor contribution to meeting customers' overall demand for electricity due to limited scale technologies and relatively high costs of capacity installation. Nevertheless, a portfolio of diverse generation technologies is required to meet demand in an economical manner and to address environmental regulations. Overall, power production is a very capital intensive portion of the industry.

The variable costs associated with the operation of fuel-based production are also significant, and can account for as much as one-half or more of a typical bill for electric service. An optimization approach to scheduling and operating a utility's fleet of generating resources results in substantial cost savings for customers. Even more savings may be realized through utility participation in power pool and power exchange operations.

System operators must ensure that adequate unit capacity is readily available to meet the real time variations is customers' demands. The ability to provide reliable electric service at the customer's meter begins with a viable source of electricity at the generators.

Electric service rates are designed to recover the fixed costs of production assets by means of capacity-related charges. The variable cost of production energy, which is primarily the cost of fuel, is recovered through volumetric (kWh consumption) charges. The temporal variations in fuel cost, which are driven by dispatching a fleet of generating resources to meet system daily and seasonal load excursions, serves as a basis for the design of various time-differentiated pricing structures. Such structures provide customers with price signals that help them manage their load and energy usage.

7
Revenue Metering and Billing

7.1 INTRODUCTION

The fundamental measures of electricity usage for electric service billing purposes are *energy* and *demand*. Energy usage relates to the overall consumption of electrical energy, typically over the period of a month, while demand relates to the rate of use of electrical energy, commonly determined on a subhourly basis. Although the energy and demand characteristics of an electric service could be estimated given certain knowledge of key end-use load devices along with their operating characteristics, metering equipment is utilized to explicitly measure customers' electricity usage.[1] Metering accurately captures the variations in energy and demand due to weather-sensitive loads and utilization changes. Thus, by using suitable metering equipment, equity between customers can be maintained when applying pricing structures to compute electric service invoices.

Meters are designed to measure the physical and temporal attributes of electric service usage, as shown in Table 7.1. A variety of meters having different levels of functionality are employed depending upon the applicable form of the pricing schedule. Large commercial and industrial customers are often billed under sophisticated rate schedules which require the measures of kWh, monthly maximum demand in terms of kW or kVA, and power factor. The energy and/or demand might also be billed out on the basis of time-differentiated rating periods. In contrast, residential and small commercial customers are often billed under rate schedules which require only the measure of monthly kWh usage; thus, only a very basic meter is necessary when rendering the electric service.

---

[1] Some electric services are not economical to meter and, owing to their simplicity and predictability, can be practically measured by means of engineering estimation. A common example is outdoor lighting applications.

Table 7.1 Attributes of Electric Service Metering

Electrical Attributes

Electrical Energy	Electric Demand
kWh	kW
kVARh	kVAR
kVAh	kVA

Time-of-Use Attributes

Annually
Monthly – Summer, Winter, and Spring/Fall Seasons
Daily – On-Peak, Off-Peak, and Intermediate-Peak Periods
Hourly or Subhourly

Equipment used to measure electric service for billing purposes is referred to as *revenue metering*. Ideally, revenue meters are installed directly at the electric service delivery point regardless of the level of the operating voltage at which the customer takes service. In so doing, an accurate measure of the customer's energy and demand usage can be made. A meter in and of itself is a low voltage and low current device (i.e., ≤ 480 volts and $\leq$ about 400 amps). *Self-contained meters* can be installed directly in the service line to the customer. Metering at high voltages and/or currents requires auxiliary equipment known as *instrument transformers*, which provide to the meter safe levels of voltages and currents that are proportional to the actual line conditions. In some situations, economics or various technical reasons dictate that the meter should be installed at a voltage level that is either higher or lower than the delivery voltage. In such cases, a methodology to compensate for losses is required in order to accurately determine the customer's energy and demand usage characteristics relative to the rate.

Aside from billing purposes, metering is used extensively to support a number of operations throughout the utility system. Meters are installed at key positions within the generation, transmission, and distribution functions. The data collected from these sites are invaluable for planning and operating the systems effectively and for accounting for critical factors, such as system losses. This information, along with metered customer load research data, provides inputs to the analysis of costs and the design of rates.

Metering technology has advanced significantly from the late 1800s when the first measuring equipment was being developed. Thomas Edison introduced a meter that measured current usage by means of electroplating. Arthur Wright utilized the thermal expansion of air against an acid medium to measure demand in a

Revenue Metering and Billing

graduated glass tube (in some respects, similar to a thermometer). By today's standards, even the more recent and widely used applications of magnetic tape recording meters are considered to be old technology. The expanding use of electronics has brought a significant amount of functionality completely "under the glass." As advances are made in communications, automatic meter reading may yet emerge as the standard for routine data collection.

7.2 METERING OF ELECTRICAL ENERGY

The most basic instrument for measuring electrical energy use is the *watt-hour meter*. A watt-hour meter functions in a manner similar to an electric motor. As shown in Figure 7.1, the watt-hour meter contains a voltage coil and two current coils which are wrapped around poles formed within an iron stator. A thin aluminum disk positioned within the air gap between the pole faces is connected to a shaft (similar to a rotor), which in turn is connected to a register via a gear train. The rotation of the disk is based on the principle of electromagnetic induction.

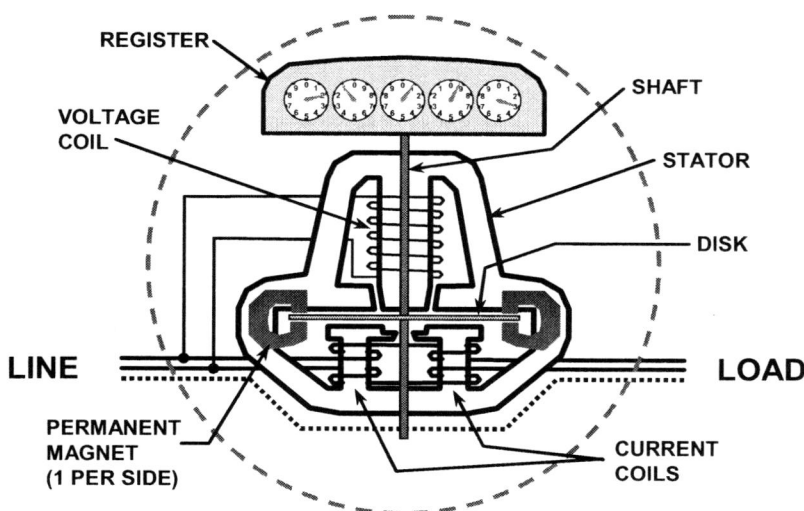

Figure 7.1 Diagram of the major components of a common induction watt-hour meter. An aluminum disk rotates in the air gap between the voltage coil pole face and the two current coil pole faces at a speed that is in direct proportion to the rate of energy usage by the load. The disk in turn rotates a shaft which drives a register that records kWh use by means of a calibrated gearing mechanism.

When power flows through the meter, an AC magnetic flux is created in each of the stator's poles. These fluctuating stator pole fields cut across the aluminum disk (an electrical conductor) thereby inducing eddy currents in the disk, which in turn produce magnetic fields. The interaction of the magnetic fields of the stator and the induced magnetic fields of the disk creates a torque, which causes the disk, and thus its shaft, to rotate in a counterclockwise direction when the meter is viewed from above. The average torque is proportional to the average real power (i.e., the product of the line voltage, the load current, and the cosine of the angle of displacement between them). In a manner similar to an induction motor in which real power produces torque to turn a rotor, the real power flowing through the meter produces a torque on the disk which can be connected to a register to measure energy usage in terms of kWh.

However, the torque on the disk tends to cause a constant acceleration of the disk, which if not limited would produce a very high speed of disk rotation that would make accurate registration difficult. If the speed of the disk was proportional to the driving torque, then the speed would be proportional to the power flow. To make the disk speed proportional to the power flow and thus register the energy usage appropriately, a pair of permanent magnets are used. The interaction of the flux associated with the eddy currents and the fields of the permanent magnets create a retarding or breaking torque that is counter to the driving torque and proportional to the disk speed. The driving torque is thus balanced so that the disk will rotate at a constant speed given a constant level of power flow and will increase or decrease in direct proportion to the power. The total number of revolutions of the disk is then proportional to the total amount of watt-hours of energy that have flowed through the meter.

Under a no-load condition, the current coils obviously produce no flux. At the same time, the voltage coil continues to be energized by the line and thus produces an AC magnetic field which results in a small but measurable amount of torque on the disk. To prevent an inaccurate "trickle" registration, the disk contains a pair of anti-creep holes that cease the rotation under no-load conditions. The holes are located in such a manner on the disk as to provide a sufficient resistance to the eddy currents created by the voltage coil.

Electromechanical meters are typically *detented* or ratcheted so that the disk will rotate only in one direction in order to prevent a reverse registration in the event that power were to flow backwards through the meter. Without detents in place, the single register meter would record only the net flow of power over a period of time. Thus, the register could show virtually no change in registration even though large quantities of energy could have been transported in the opposing directions through the meter during the measurement period. In applications of two-way power flow through the same delivery point, i.e., an interconnection with a cogenerator or small power producer, the accurate measure of energy usage would be achieved through the installation of either a set of back-to-back meters (one meter for each direction) or a bidirectional electronic meter that is programmed to measure separate power flows.

Revenue Metering and Billing

Watt-hour meters have a number of adjustment points that allow for the proper calibration of the meter so that it will function accurately under all normal operating conditions. The registration mechanism must also be designed to accurately record the energy consumed by the load. The register translates the disk revolutions, by means of a gear mechanism, to a readable kWh amount. One revolution of the disk is equal to a specific amount of watt-hours, which is noted by the disk constant, K_h. A K_h rating of 7.2 is prevalent for residential electromechanical watt-hour meters, while the K_h rating for the electronic versions is 1.

The common clock-type meter register, as shown in Figure 7.2, is driven by a set of interconnected gears which are motored by the shaft that is connected to the disk. The gears of each dial are meshed with the gears of the adjacent dials such that the 1's place dial drives the 10's place dial which in turn drives the 100's place dial and so forth. Watt-hour meters typically have four or five dials. In an alternating fashion, the dials have either a clockwise or a counterclockwise notation. The pointers move in either the clockwise or counterclockwise direction because of the way the sets of gears are configured. A more complex gearing structure could allow for all pointers to turn in the same direction, but such an enhancement would increase the meter's cost and introduce additional friction that would require further compensation. In addition to the clock-type of register, the cyclometer-type of register is sometimes used. The cyclometer register is also driven by a gearing mechanism, but it utilizes rotating drums which display discrete readout numbers much like an automobile's mechanical odometer.

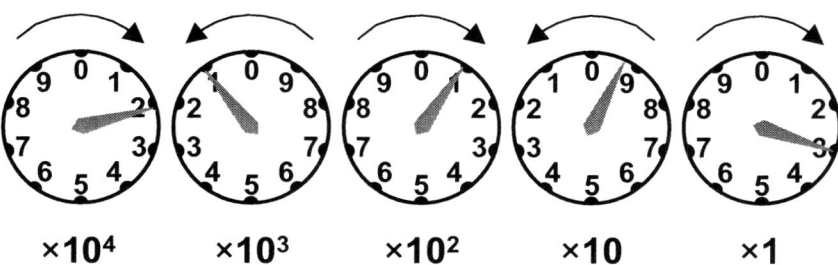

Figure 7.2 A clock-type meter register for displaying accumulated kWh usage. The gear train is so structured that it causes the pointers on adjacent dials to turn in the opposite directions. As the pointer on each dial makes a complete revolution, the pointer on the dial to its immediate left moves one-tenth of a revolution (i.e., from one number to the next higher number). The dials are read from right to left taking the number which each pointer is just past, given the indicated direction of rotation. Except for the far right-hand dial, when a pointer appears to be directly on top of a number, the reading of the dial to its immediate right should be consulted to ensure the proper reading. For instance, the middle dial appears to have reached 1, but since the dial to its right has not quite completed a full revolution, the reading of the middle dial would be accurately taken as 0. Thus, the correct total reading for this example is 21,093 kWh.

Although a watt-hour meter is designed to only register kWh usage, it can be employed to provide a spot estimate of demand (i.e., the average load) for a selected period of time. The kW demand of the load can be easily determined using a watt-hour meter by manually counting the revolutions of the meter's disk that occur during a chosen time interval. The estimated kW demand is determined by

$$P_{AVG} = \frac{R \times K_h \times [PT \times CT] \times 3.6}{\Delta t} \quad (7.1)$$

where P_{AVG} = average load or demand in kW
R = number of disk revolutions counted
K_h = meter constant in watt-hours per revolution
$[PT \times CT]$ = instrument transformer ratios ($PT \times CT$ = 1.0 for self-contained meters)
Δt = period in seconds over which the disk revolutions are counted

Fundamental Watt-Hour Meter Billing Applications

Watt-hour meters have been used extensively for revenue metering of small, single-phase customers, which are predominantly residential and small commercial in nature. Although it would be highly desirable to also measure the maximum loads of these small customers, the cost of a demand meter (including a measure of energy) comes at a premium when compared to the cost of a simple watt-hour meter.[2] Thus, the billing of demand-related charges for these small sized customers are accomplished through the design and administration of nondemand rate structures that require a measure of kWh only.

Figure 7.3 illustrates the schematic diagram of meter connections for two-wire and three-wire low voltage, single-phase services. Two types of meter structures are shown: a bottom connected A-Base type and an S-Base socket type. The A-Base meter has a bottom mounted terminal where the incoming and outgoing wires are connected. As a result, the meter is hard wired into the line making it somewhat difficult to remove for servicing.

The socket meter is widely used for self-contained metering applications. The back of the meter has a set of lug connectors which plug into jaw type terminals that are pre-wired inside the meter base, which makes meter removal easy and safe. Some socket bases have jaws which close when the meter is out, which allows service to be maintained to the load. Socket bases also allow for the installa-

---

[2] Demand metering used for billing purposes also incorporates the function of a watt-hour meter in a single device.

Revenue Metering and Billing

tion of special equipment, such as a service entrance surge suppressor, which can be inserted between the meter and the meter socket.

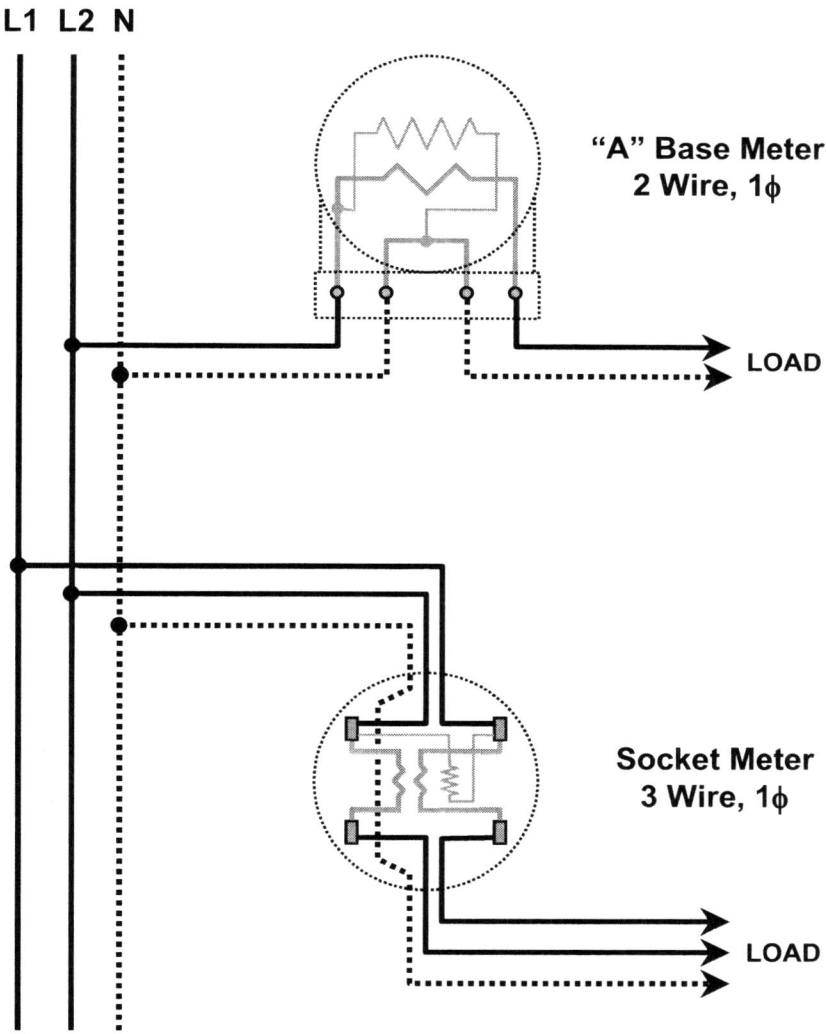

Figure 7.3 Diagram of low voltage, single-phase watt-hour meter installations. Shown are a two-wire (A-Base) meter example and a three-wire socket (S-Base) meter example. Such meters are used for nondemand rate applications.

Watt-hour meters are also used in three-phase service applications. Three-phase watt-hour meters are constructed with two or three stators which share the disk and register as an integral unit. To measure the energy use on a three-wire, three-phase delta service, for example, requires a two-stator meter. A four-wire, three-phase delta or wye service requires a three-stator meter, although it is possible to utilize a two stator meter in special situations.

Metering of Reactive Energy

Since some rate schedules incorporate a provision for power factor, a measure of reactive power consumption must be obtained from suitable metering facilities. Due to the added cost of metering the reactive component, such rate schedules are designed for large commercial and industrial classes applications. The basic watt-hour meter that is utilized to measure kWh can be readily augmented and applied to measure reactive energy (kVARh), as shown in Figure 7.4.

In order for a watt-hour meter to measure kVARh, the voltage applied to its voltage coil must be in quadrature (i.e., lagging by $\varphi = -90°$.) with the line voltage (V_L) and have the same absolute value. Since (inductive) VARS lag watts by 90°, as given by $P = V_L \times i \times \cos \theta$ and $Q = V_L \times i \times \sin \theta$, then the voltage component $V_L \times \sin \theta$ also lags the voltage component $V_L \times \cos \theta$ by 90°. Thus, by means of phase shifting, the voltage applied across the kVARh meter's voltage coil, $V\varphi$, is made to be in quadrature with V_L.

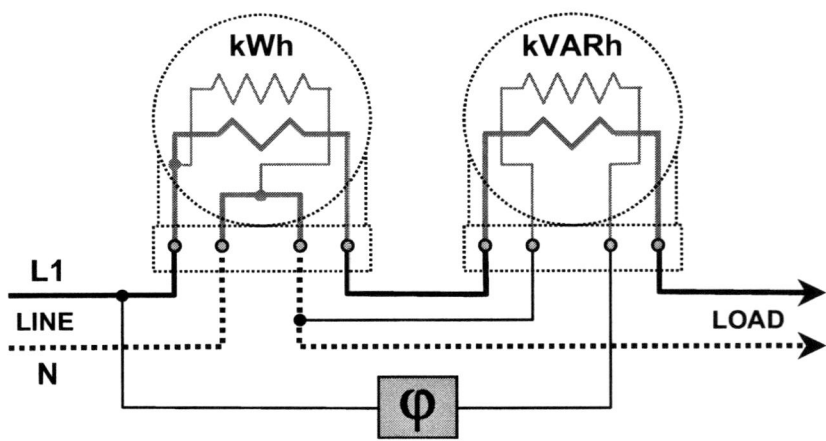

Figure 7.4 Application of watt-hour meters to measure both kWh and kVARh. A phase-shifting device, indicated by the block marked with φ, is required to provide a voltage to the meter measuring reactive energy that is in quadrature with the line voltage.

Revenue Metering and Billing

Two basic types of voltage phase-shifting devices are used for reactive metering installations. For typical single-phase installations, a unit having a resistor and capacitor connected in parallel is placed in line with the reactive meter's voltage coil. Three-phase installations utilize autotransformers, often referred to as a *phase-shifting transformer*, in order to lag the voltage by the required 90°. With a two-phase system, no specific phase-shifting device is needed owing to the connatural relationship of the voltages. By cross phasing in which the polarity of one of the voltage coils is reversed and the voltage of each phase is made to interact with the current from the opposite phase, the 90° shift can be realized. Cross phasing can also be used in three-phase applications.

The meter registrations of kWh and kVARh can be easily used to determine the average power factor over a period of time as given by

$$PF_{AVG} = \cos\left[\tan^{-1}\left(\frac{kVARh_{\Delta t}}{kWh_{\Delta t}}\right)\right] \quad (7.2)$$

where PF_{AVG} = average power factor
$kWh_{\Delta t}$ = kWh registered during time period Δt
$kVARh_{\Delta t}$ = kVARh registered during time period Δt

As with the kWh meter, the reactive meter may be detented in order to prevent reverse registration, which in the case of reactive metering would occur under leading power factor conditions.

Time-Differentiated Metering of Energy

Some rate schedules contain pricing structures in which the price per kWh varies by different time periods. In the most general sense, seasonal energy pricing establishes different kWh rates based on periods indicated as *summer* and *non-summer*, *winter* and *nonwinter*, and sometimes *shoulder* (Fall and Spring). The standard watt-hour meter can be used for these applications because the seasonal price changes coincide with monthly billing periods (e.g., summer may be considered as June through September while non-summer is October through May). But for price structures that vary during a monthly billing cycle, meters that are capable of simultaneously maintaining multiple registrations of energy use are needed.

Time-of-Use Metering

While "time of use" can be broadly used to include seasonal pricing or even sub-hourly pricing, it most often refers to pricing that is based on the time of day. Time-of-use pricing periods are considered to be fundamentally *on-peak hours* and *off-peak hours*; however, *intermediate* or *shoulder hours* are sometimes defined to be a discrete pricing period. The characteristics of the system demand are used to define these time periods. On-peak hours are those in which the demand is high, typically during the daytime hours occurring on weekdays. Off-peak hours occur when demand is low, such as during nighttime hours and often over all hours of weekend days. The intermediate hours represent a transition between on-peak and off-peak time periods.

Time-of-use meters are designed to accumulate not only the total kWh but also the on-peak and off-peak kWh consumptions using separate registers. Thus, the meter must be able to perform a time clock function in order to engage and disengage the on-peak and off-peak kWh registers at the appropriate times. In addition, the time-of-use rating periods may change on a seasonal basis. For example, a summer day on-peak period might be from Noon to 7:00 PM while the winter day on-peak period might be from 6:00 AM to 10:00 AM and then also from 6:00 PM to 10:00 PM. Thus, the meter must also be capable of switching time periods based on billing months. Accommodation of daylight savings time shifts can be handled either by programming the meter accordingly or by the data processing function once the data is retrieved.

Interval Metering

While time-of-use metering records energy use by time categories, interval metering records energy use in discrete time sequence intervals. Interval meters are typically programmable so that the duration of the interval can be chosen from among a variety of standard settings, including 1-, 5-, 15-, 30-, or 60-minute intervals. For electric service billing purposes, 15- and 30-minute periods are the most commonly used intervals for recording energy use data.

In actuality, the interval meter records electronic pulses that can be interpreted as kWh use. As energy is used, a *pulse initiator* generates serial data pulses which are recorded in a storage medium, such as in electronic mass memory. In a similar manner, reactive energy pulses can be recorded simultaneously in a separate memory or channel. Meters having additional recording channels could be used to measure other interval parameters, such as the local temperature, given the installation of a suitable transducer.

An additional pulse is recorded at the end of each interval in order to define the time base. By using the power system line frequency, a time-keeping device counts the number of zero crossings of the voltage. With two crossings per cycle, 120 crossings is equal to one second for a 60 Hz system. Thus, a 15-minute interval contains 54,000 cycles or 108,000 zero crossings of the voltage. When the correct amount of zero crossings are counted, a time pulse is generated and recorded. A backup battery and an oscillator are used to maintain accurate time-keeping in the event that line voltage is temporarily lost. Time keeping in this manner is of exceptional accuracy.

The pulse data recorded in the meter's storage system must be periodically acquired and then processed by means of a *translation* system. The translator converts the raw energy pulses back to kWh (kVARh) use by interval.

Revenue Metering and Billing

While the interval meter fundamentally records energy use, by nature of its time period measurements, it also yields a profile of interval demands, as shown in Figure 7.5. For example, a meter programmed for 30-minute intervals will produce 1,460 kWh readings in a typical month of 730 hours. By dividing the kWh reading in each of the intervals by 0.5 hours, the average load or demand in kW for each of the intervals is computed. Such high resolution metered load data can be used for demand rate billing determinations, real-time pricing (RTP) applications, and supporting customer class load shape development programs.

7.3 METERING OF DEMAND

As noted above, interval recorders can be used to determine a customer's demands for billing purposes. While interval data is highly desirable, its associated metering cost is relatively high for small to medium-size customers who are typically served under a maximum demand type of rate structure. Thus, interval metering is generally limited to large customers and for special rate applications.

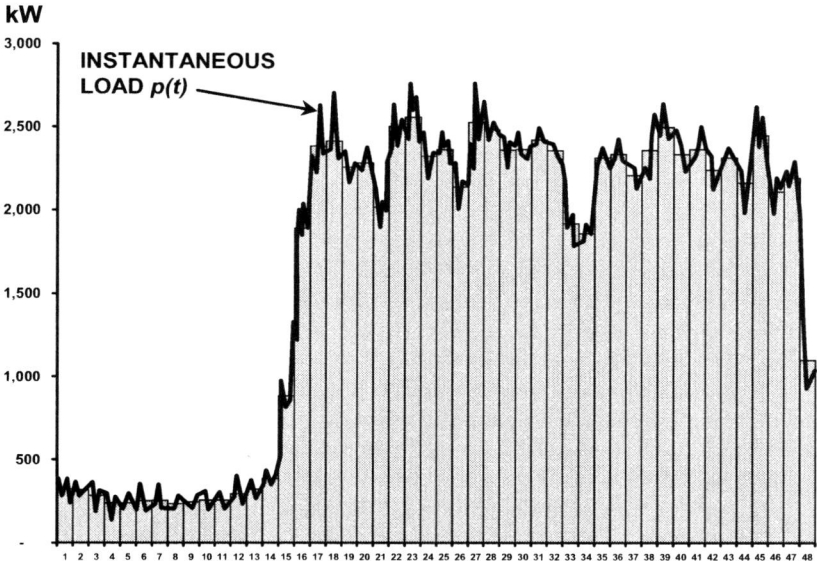

Figure 7.5 A load shape based on interval meter data. A 24-hour plot of an industrial customer's operation depicted by both an instantaneous load curve, $p(t)$, and an interval load profile. The interval profile was developed from 30-minute interval metered data.

Fundamental Demand Meter Billing Applications

A widely used economical method of measuring smaller commercial and industrial customers' demands for electric service billing is the application of *integrating demand meters*. The integrating demand meter may function by using either a mechanical or an electronic mechanism. The integrating demand meter's register is driven by means of a watt-hour metering device; thus, for billing applications, the energy and demand measurement and display functions are typically combined in a single metering unit.

The demand meter integrates the customer's instantaneous load over short periods of time known as *demand intervals*, as defined in Chapter 2. In essence, the demand register accumulates the kWh under the instantaneous load curve during each demand interval and divides each interval's sum by the width of a single interval in terms of hours (i.e., 0.25 hours for a 15-minute demand interval or 0.5 hours for a 30-minute demand interval) in order to compute each interval's average load or demand (reference Figure 2.13). At the end of each demand interval, the demand meter reading is reset to zero in preparation for measuring the next interval's demand. This methodology produces a series of chronological (or synchronous) demand measurements based on uniform time intervals. As previously discussed, the demand measurements are similar to those which are processed by interval metering, but unlike an interval meter's recorded data, the integrating demand meter generally only retains the greatest of the demand readings during the billing period.

A mechanical demand meter has a gearing mechanism which drives a registration device in direct proportion to the watt-hours consumed during each demand interval. The basic demand meter is of the *indicating type* in which the maximum demand attained during the billing period is indicated on a register that is read manually. Clock type registers, similar to those illustrated in Figure 7.2, and cyclometers are sometimes used for displaying demand.

Another commonly used demand register is shown in Figure 7.6. This type of register uses two pointers and a graduated scale for indicating the demand. The pointer-pusher hand is an active indicator in that it displays the demand of the current interval. It moves from zero at the start of the interval and reaches a point on the scale at the end of the interval that is commensurate with the average load measured over the interval. At the end of each interval, the pointer-pusher hand automatically resets to zero. Conversely, the pointer hand is a passive indicator that floats and thus moves up the scale by means of the pointer-pusher action.

At the beginning of the billing period (which is also the end of the previous billing period), both indicator hands are set to zero by the meter reader. As the pointer-pusher hand moves during the first interval, the pointer hand is pushed along with it. At the end of the interval, when the pointer-pusher hand resets to zero, the pointer hand retains its location on the scale. During subsequent intervals, the pointer hand will ratchet up the scale only if the pointer-pusher hand measures a demand that exceeds a previous reading. In this manner, the passive pointer hand records the highest demand measured during the billing period, and its reading on the scale is recorded for billing purposes. After recording the meter's maximum demand reading, the meter reader sets both hands back to zero as discussed above.

Rolling Demand Metering

The basic integrating demand meter determines average loads for a series of consecutive intervals across the billing period. For billing purposes, the objective is to determine the maximum interval demand occurring during that period, consistent with the actual peak load. However, since the demand intervals are fixed in time series sequence, i.e., in end-to-end order, it is quite possible for the meter to not adequately capture the peak since the consecutive demand intervals are not necessarily synchronized with changes in the load shape. If the actual load were to increase significantly near the end of a particular demand interval and then decrease significantly after the start of the next demand interval, the demand determined for either of the adjacent intervals could be measurably less than if that peak occurred in time so as to be contained within the bounds of a single interval. This peak-splitting problem can be resolved by the use of electronic registers, which are capable of determining average loads on a *rolling demand* or *sliding window* basis. This method requires data to be collected with subintervals that represent subdivisions of the stated billing demand interval.

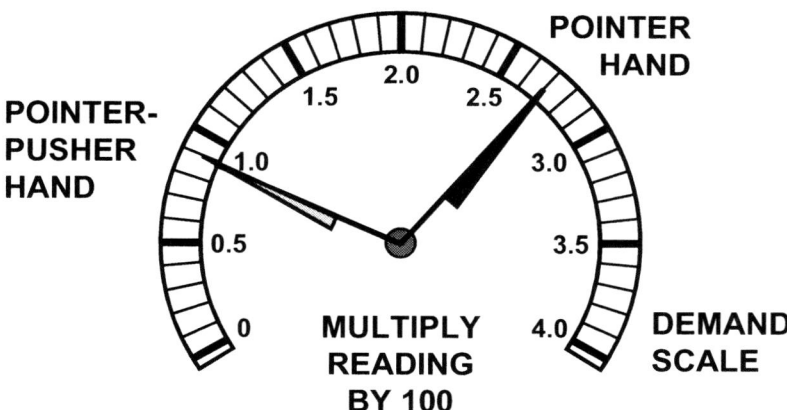

Figure 7.6 Pointer-type register for an indicating demand meter. The pointer-pusher hand represents the current interval's demand reading while the pointer hand represents the highest interval demand reading that has occurred since the meter was reset to zero at the end of the prior month billing period.

Figure 7.7 compares the rolling demand method (lower graph) to the synchronous interval method (upper graph). The gray background in both graphs represents the plot that would result if the load was metered on a 5-minute subinterval basis over the duration of one hour. A 15-minute maximum demand is stated for billing purposes; thus, three 5-minute subintervals must be averaged together in order to compute a 15-minute demand. For example, the first three 5-minute subintervals in the upper graph are averaged together to determine the 15-minute demand for the first of the four consecutive demand intervals. The first demand interval then is about 2,200 kW. The maximum demand determined in the upper graph using the four consecutive 15-minute intervals for the hour is found to be with the third demand interval at 2,489 kW.

In the lower graph, the first 15-minute interval demand is determined in the exact same manner as for the first 15-minute interval in the upper graph; thus, the two resulting demands are the same (i.e., about 2,200 kW). Then this 15-minute "window" slides to the right by dropping the first 5-minute subinterval data and adding the fourth 5-minute subinterval data. The average of the second, third, and fourth subintervals results in a 15-minute demand that is slightly less than determined for the first 15-minute period. The window continues to slide to the right based on 5-minute increments so that at the end of the hour, ten 15-minute overlapping demand interval periods have been produced.[3] The maximum demand determined in the lower graph using the rolling demands for the hour is found to be with the eighth demand interval at 2,545 kW, which is 56 kW higher than the maximum demand determined using consecutive demand intervals. The rolling demand method has a higher resolution and thus more closely follows changes in the load than the consecutive demand interval method.

Electronic meters can be programmed to compute the rolling demands in the field throughout the billing period. Alternatively, rolling demands can also be determined in house at the end of the billing period if the subinterval data is recorded. The translation system is used to compute the sliding window demands and then select the maximum demand for billing purposes.

Time-Differentiated Metering of Demand

Like energy, demand can be metered on a time-of-use basis, including seasonal and time-of-day rating periods. Time-of-day demand meters record the maximum demands which occur during the on-peak time period and during the off-peak time period of each billing period. The higher of the on-peak demand or the off-peak demand would be equivalent to the maximum demand that would be measured by a standard demand meter. Even with simple on-peak and off-peak measures, a variety of time-differentiated demand rate structures can be fashioned.

---

[3] For this example, the hour chosen was the first for the billing period; thus, the first 15-minute demand could not be determined until the first three subintervals were completed. The next hour would actually span twelve 15-minute demand measures.

Revenue Metering and Billing 241

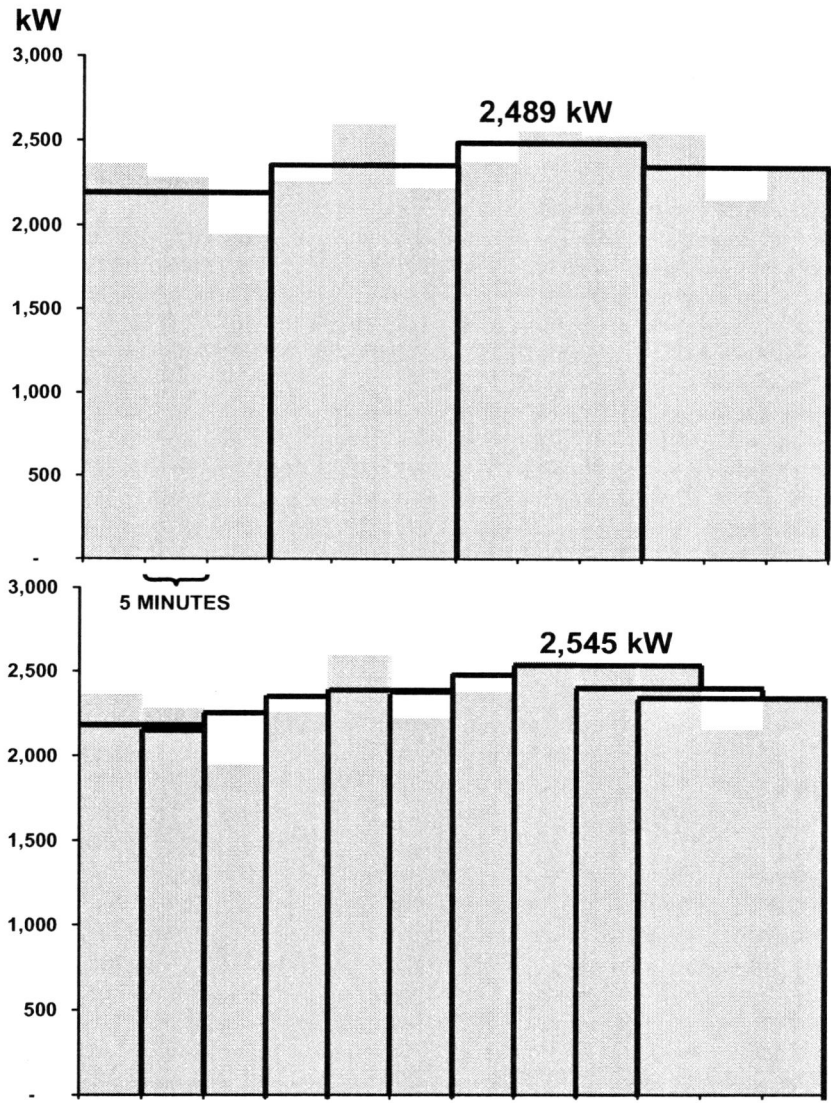

Figure 7.7 Comparison of synchronous interval and rolling demand measurements. Determination of the 15-minute maximum demand based on consecutive interval measurements (upper graph) and rolling demand measurements (lower graph). The solid background depicts the actual load as measured on a 5-minute subinterval basis. The rolling demand method yields a more precise measure of the maximum load condition.

Metering of Reactive Demand

Reactive demand can be determined in a manner similar to the measurement of reactive energy using integrating demand meters in place of watt-hour meters for the metering arrangement shown in Figure 7.4. The consequence of this measurement would be a maximum reactive demand in kVAR that would not necessarily coincide in time with the maximum kW demand (reference the time difference of the peak kW and kVAR readings for the loads shown in Figure 4.1). The use of these noncoincident kW and kVAR readings to calculate other load properties would result in estimations (e.g., a *hybrid kVA* and a *hybrid power factor*).

Reactive demand and power factor can also be determined on an interval basis. Figure 7.8 illustrates the variation in power factor that can occur over time as the reactive demand requirement changes with the operation of end-use load devices, such as induction motors, in an industrial customer's daily operation. The power factor is at unity during the time when the plant is not operating but drops to a minimum of about 87% during the 30 minute interval in which the kVAR requirement is highest relative to the coincident kW requirement.

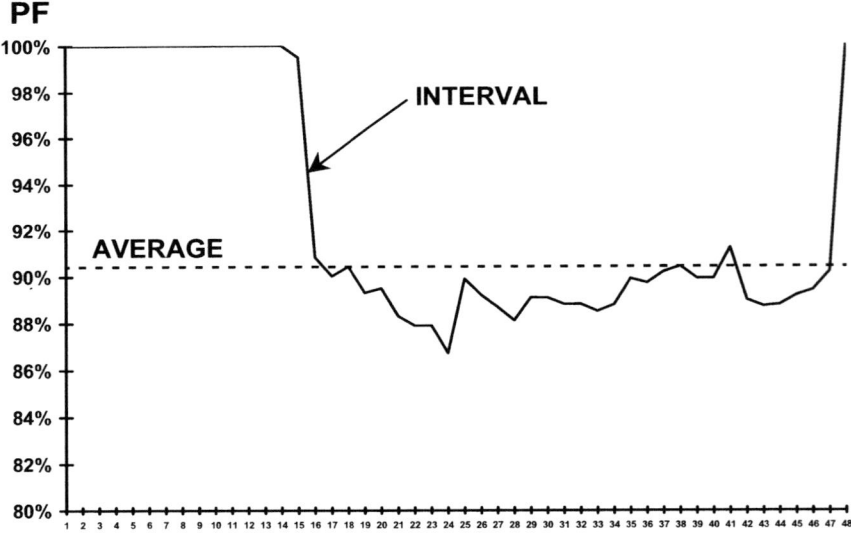

Figure 7.8 Interval power factor vs. average power factor. Power factor across the period of a day is calculated on a 30 minute interval basis, using interval kW and kVAR readings, for an industrial customer having a two shift operation. In the early hours of the day, the reactive load requirement is zero; thus, power factor is at 100% during that time. The average power factor for the period shown is determined from the kWh and kVARh accumulated during the period.

Q-Metering

A disadvantage of the 90° voltage phase shifting methodology used for metering reactive power, as discussed previously, is that only lagging power factor conditions can be measured using a single meter. The reactive meter would typically be detented in order to prevent reverse registration due to those times when the customer's load is predominantly capacitive and thus has a leading power factor, such as might occur during daily or weekend operational shutdown periods. Without such prevention, the load could appear to be at or near a unity average power factor even though the load is actually in a leading power factor condition.

Both lagging and leading power factor loads are of concern to the utility because the kVA capacity requirement increases as power factor moves away from unity in either direction. Most customer loads tend to function predominantly with a lagging power factor. However, large commercial and industrial customers are more likely to impact the power system during times of leading power factor operation. Such occurrences are common when capacitors used for compensating reactive end-use loads are not switched off during shutdown periods, as previously discussed.

By displacing the voltage by an angle greater than 0° but less than 90°, it is possible to measure both leading and lagging power factors, within a limited range, using a single meter referred to as a *Q-hour meter*.[4] A 60° displacement of the voltage obtained by means of cross-phasing is commonly used for Q-metering, and it provides for an accurate measurement for lagging power factor conditions and beyond unity to an 86.7% leading power factor. The power factor characteristics of the load must be carefully considered when selecting the Q-meter's voltage displacement angle for customer billing applications. Otherwise, if the power factor of the load were to exceed the meter's leading power factor limit, the meter reading of Q would be incorrect.

The Q-meter provides readings in units of Q (and Qh); thus, by knowing the corresponding amount of watts, the VARs can be determined by

$$\text{VARs} = \frac{Q}{\sin\varphi} - \frac{\text{Watts}}{\tan\varphi} \qquad (7.3a)$$

where φ = the voltage angle displacement from a reference of 0° used to measure Q

---

[4] The "Q" in Q-metering should not be confused with the unit *Q*, which is used to denote reactive power. The torque of a watt-hour meter is (a) proportional to watts at a voltage displacement of 0° and (b) proportional to VARs at a voltage displacement of 90°. At any other angle of voltage displacement between 0° and 90°, the watt-hour meter's torque is proportional to a unit defined as "Q" (as in reference to **Q**uantity).

Based on a 60° displacement of the voltage (i.e., $\varphi = 60°$), Equation (7.3a) simplifies to

$$\text{VARs} = \frac{2Q - \text{Watts}}{\sqrt{3}} \tag{7.3b}$$

The disposition of the power factor can be determined from the relationship between the watts and Q measurements:

- $[(2 \times Q) - \text{Watts}] > 0$ ⇒ Lagging Power Factor
- $[(2 \times Q) - \text{Watts}] = 0$ ⇒ Unity Power Factor
- $[(2 \times Q) - \text{Watts}] < 0$ ⇒ Leading Power Factor

Volt-Ampere Metering

The technique of voltage displacement used for measuring kVAR and kQ can also be used to measure kVA. By displacing the voltage without changing its magnitude, so as to bring it in line with the angle of the load current, a watt-hour demand meter will yield readings of kVA and kVAh. This angle corresponds to the power factor angle of the load; thus, if the power factor remains fairly constant, a reasonably accurate reading of kVA and kVAh will be attained. However, if the power factor is expected to vary significantly, other methods of determining kVA and kVAh are necessary.

The simultaneous measurement of kWh and kVARh in which these quantities are summed vectorially by a special meter mechanism will produce kVA and kVAh readings directly at the meter's register. Such readings can also be produced on an interval basis from separate recordings of kWh and kVARh pulses which are interpreted and vectorially summed using a translator system.

7.4 SPECIAL METERING CONSIDERATIONS

The methodology utilized for rendering electric service becomes more complex with larger-size commercial and industrial customers. The delivery voltage may be at a primary or a transmission service level. The point of metering may necessarily be at a location different than the point of delivery. Multiple delivery points may be require to provide adequate capacity to serve the load. As a result, more sophisticated metering schemes and special metering equipment and facilities are required to ensure accurate energy and demand measurements. A very high degree of metering precision is of particular importance for large-use customers since only a very slight error in measurement can have a significant impact on the electric service billing.

Metering with Instrument Transformers

Electric meters are low voltage and low current measuring devices. The design necessary for a self-contained meter to safely handle the electrical stresses and heating due to high voltages or currents would be both impractical and uneconomical. Nevertheless, it is important to site the revenue meter at the delivery point if reasonably possible even if the point of delivery is high at a high voltage. But even at a secondary voltage, the load current could exceed the coil ratings of a self-contained meter. Metering of services at high voltages and/or currents is effectively accomplished through the use of basic watt-hour and demand meters in conjunction with *instrument transformers*.

There are two types of instrument transformers: *current transformers* (CTs) and *potential* or *voltage transformers* (PTs or VTs). Like any transformer, these devices operate on the basis of electromagnetic induction. The input of an instrument transformer is connected at the point on a circuit where the measurement is desired. The output of the instrument transformer is connected to a transformer-rated meter. The PTs and CTs respectively transform high voltages and high currents into low voltages and low currents suitable to the meter.

For metering of low voltage, high current services (generally in excess of 200 amps), CTs are required.[5] A simple example of a single-phase, secondary voltage CT installation is shown in Figure 7.9 (meter located in the lower part of the figure). For low voltage applications, window-type (or "donut") CTs are typically used. These CTs consist of a coil of wire with an air-core through which a service conductor is passed; the service conductor actually acts as the input (primary) coil for the CT. The load current flowing in the service conductor induces a low ampere current in the CT output (secondary) coil. CTs are typically designed for a maximum rating of 5 amps and are manufactured with a variety of current reduction ratios in order to satisfy a broad operating range requirement. For example, a 200:5 CT would output 5 amps if 200 amps were flowing through the service conductor (other standard ratios include 400:5, 600:5, 800:5, etc.). The CT's output current varies proportionately to the load current according to the current reduction ratio and has the same phase relationship.

Small gauge wire can be used to connect the CT's output terminals to the terminals of the meter's internal current coils. A separate CT is required for each line of a 120/240 volt service, as shown in Figure 7.9, and on each line of a three-phase service.[6] When two service lines are installed in parallel to meet higher capacity requirements, a single set of CTs with larger windows can be used for the conductor pairs as long as the same conductor polarities are maintained.

---

[5] PTs may also be required on 480-volt systems for safety reasons even though the meter is insulated for that level of voltage.

[6] A single CT could be used for a 120/240-volt service if the two line conductors are passed through the CT in opposite directions (due to the difference in polarity). This application changes the current reduction ratio (e.g., from 200:5 to 200:10).

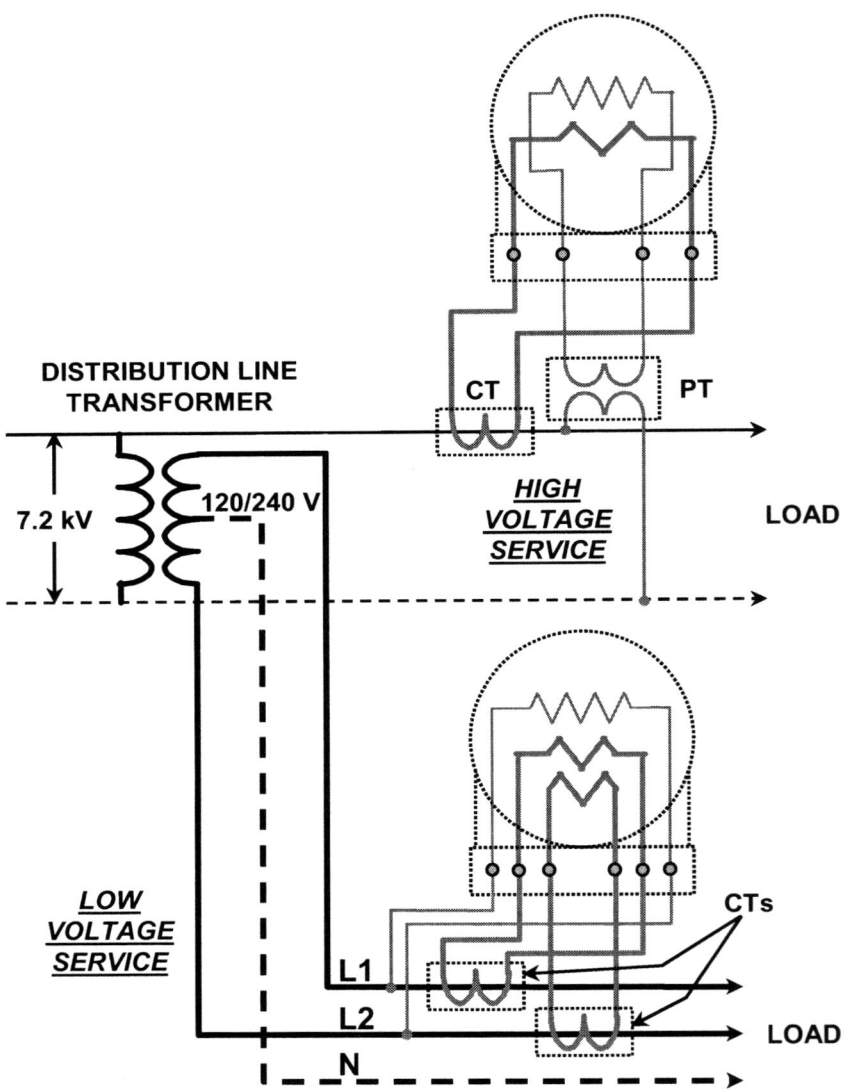

Figure 7.9 Basic applications of instrument transformers. The low voltage service in the lower portion of the figure requires CTs due to current levels that exceed the rating of the meter. Metering of the primary service in the upper portion of the figure requires both a CT and a PT. Instrument transformers provide the meter with low magnitude currents and voltages that are proportional to the actual line currents and voltages. Single-phase metering examples are shown for clarity although three-phase applications are more common.

Revenue Metering and Billing

High voltage metering requires both CTs and PTs. CTs used in high voltage applications are typically of the "bar type" design which has a copper bus bar that is permanently mounted within the window of the output coil. The bar-type CT is installed in-line with the high voltage circuit conductors by means of the bus bar's terminal type ends. Like window-type CTs, bar-type CTs are designed in several sizes to accommodate the metering of various load ranges and are also designed with a 5-amp maximum output current as a standard.

PTs are needed in high voltage metering in order to produce low voltages that are proportional to the circuit voltage while maintaining the proper phase relationship. A PT is similar in concept to a distribution transformer having input and output coils about an iron core, but it is of a much lower capacity rating. PTs are typically designed to produce an output of 120 volts and are manufactured with various winding ratios in order to accommodate metering of different high voltage applications. For example, a PT with a 60:1 ratio can be used to meter a 7,200 volt, single-phase line where the input side of the PT would be connected between the line and the neutral conductors. As with the CT meter connection, small gauge wire can be used to connect the PT's output terminals to the terminals of the meter's internal voltage coil. A simple example of a single-phase, primary voltage PT and CT installation is shown in Figure 7.9 (meter located in the upper part of the figure).

When metering with instrument transformers, the uncompensated meter will register less kWh use than is actually consumed; thus, the reading must be adjusted by the PT and/or CT ratios. For example, if a 200:5 CT were used on a low voltage circuit, the meter's register reading would be multiplied by 40. If a 200:5 CT and a 60:1 PT were used on a high voltage circuit, the meter's register reading would be multiplied by [40 × 60] = 2,400. Self-contained meters have PT and CT ratios of 1.0.

Loss Compensation Metering

Electric service rates are based on energy and demand delivered to the customer at a prescribed voltage level. Ideally, the revenue meter should be located at the delivery point where the customer's electrical system receives power from the utility's system, as discussed previously. However, in some cases, economics, ease of access, or physical constraints require a departure from this criterion, and the meter must then be placed elsewhere on the circuit.

As exemplified in Figure 7.10, a customer receives service at a high voltage, perhaps at a transmission level, provides a transformation to a lower voltage, and then distributes power to the load via its own distribution system. Because of the relatively high cost to meter the service at the exact point of delivery, the utility may elect to install the revenue meter on the low voltage side of the transformer. In so doing, the meter would not register any of the power usage associated with the customer's transformer. Specifically, the transformer uses power in the form of both load and no-load losses.

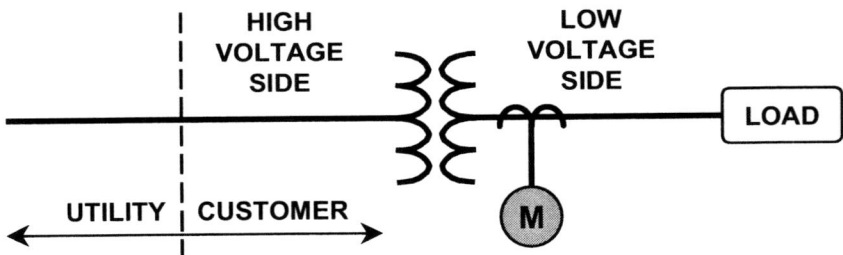

Figure 7.10 Metering point vs. delivery point. An example where the revenue meter is installed at a location different than the delivery point. The meter does not see the load of the transformer, which is part of the customer's distribution system. Thus, the meter readings must be compensated for losses.

Transformer core or no-load losses are constant while i^2R load losses are proportional to the current flow through the transformer's windings. If the meter is displaced a long distance from the delivery point, load-related losses associated with conductors can also become significant. Buses may produce another source of load losses that can be of consequence.

Note also from Figure 7.10 that if the delivery point was at low voltage and the meter was at high voltage, the meter would register the transformer load which is already anticipated in a low voltage rate for electric service. Thus, the meter's basic registration would not be a true representation of the customer's actual usage of energy and demand.

Under the types of circumstances described above, uncorrected meter registrations must be compensated by adding or subtracting losses so the energy and demand readings would be the same as if the meter was located at the delivery point. Loss compensation requires specific watt and VAR loss information from data which is provided from the transformer manufacturer and derived from tests on the transformer. A standard mechanical meter can be made to read the unmeasured losses by means of an external loss compensator which is calibrated to provide the meter with the losses necessary to correct the meter's registration. Special electronic loss compensation meters incorporate algorithms which can be programmed with specific loss information. Alternatively, the loss adjustments can be implemented as a design feature of the rate schedule where uncompensated meter readings are increased or decreased by means of average percentage adjustments for energy and demand during the billing process.

Metering Totalization

An electric service rate is typically based on providing a single customer with a single delivery point through a single meter. One exception to this rule is when a three-wire, three-phase power service provided to a customer from a delta connected transformer bank is accompanied by a single-phase service from a separate transformer for lighting and other small single-phase end-use devices. The power consumption from these two service lines can be mechanically combined, or totaled, by means of a multistator meter. This *totalization* automatically accounts for the diversity of load existing between the two delivery points that would not be recognized through the operation of two separate integrating demand meters.

Another application where metering totalization is necessary is when a customer's load exceeds the largest standard capacity available through either an individual three-phase transformer or a bank of three single-phase transformers. Thus, two side-by-side transformer installations are required to adequately serve the load. While primary metering of the feeder supplying the transformers would allow for a single meter, a physical constraint or other overriding aspect might dictate that the metering be established on the secondary side of the transformers (where totalization is achieved by paralleling the outputs of low voltage CTs).

An additional example of using metering totalization is shown in Figure 7.11. A single industrial customer located on a large contiguous tract of land has a load requirement of a magnitude that could justify transmission service to a dedicated substation serving only that customer. However, several situations could prevent this approach: transmission is too expensive to extend to the site, routing of transmission is impeded, or land at the site for the dedicated substation is inadequate or unavailable. Without a direct transmission source, the customer must be served from the local primary distribution system supplied by area distribution substations that supply service to other adjacent customers. Again, the load is significantly large and requires the combined capacity of three distribution circuits, two feeders from one substation and one feeder from another substation. Thus, the customer receives primary voltage service at three separate delivery points which is then totalized to a single meter.

Metering totalization is often accomplished by accumulating, to a single revenue meter in real time, pulses which represent energy use and which are generated by each of the delivery point meters. Thus, small gauge wires for pulse transmission must be routed from each of the delivery point meters to the totalizing meter. Pulse initiators within the meters are used to generate the pulses for transmittal and totalization. Pulse initiation can be implemented by simply "counting" revolutions of the meters disk, such as with an optical device, and then creating contact closures of a low voltage AC or DC circuit to produce electrical impulses. Either a three-wire (KYZ) circuit, known as a Form C, consisting of a single-pole, double-throw switch or a two-wire (KY or KZ) circuit, known as a Form A, consisting of a single-pole, single-throw switch is used for pulse generation. The meter pulse represents a quantity of energy, and the value of a pulse is typically expressed in terms of watt-hours. Totalizing relays are often required when several pulse sources must be totalized. These relays serve as intermediate totalizers which sum pulses from two or more pulse initiators and then transmit the combined pulses to the location of final pulse totalization.

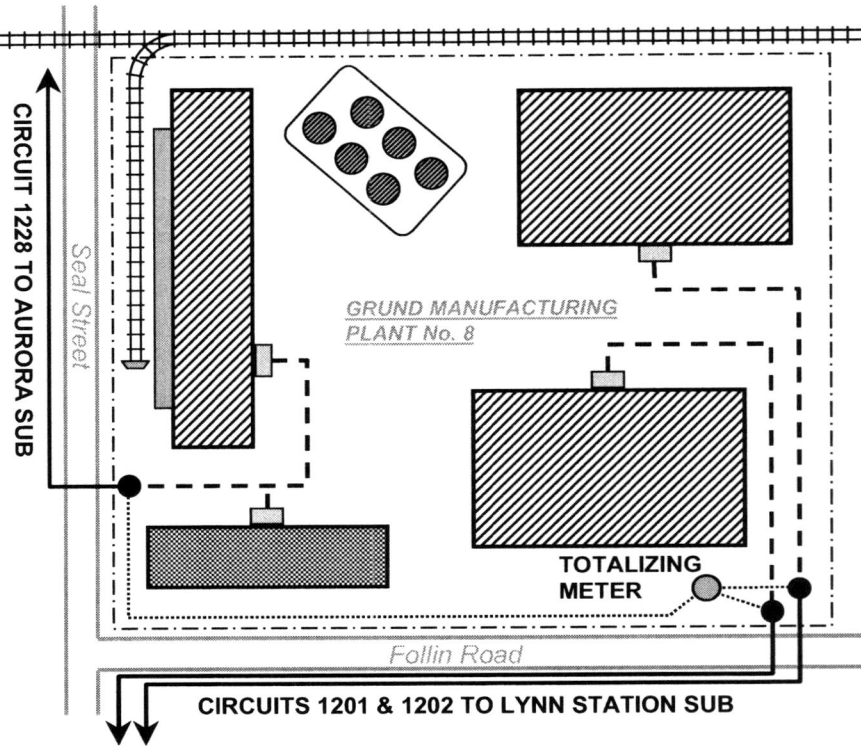

Figure 7.11 Example of meter totalization. A single customer's large electric load requirement is served at a primary voltage by three feeders from two different distribution substations located in the vicinity. The three delivery points are individually metered with pulse initiator metering. Since the load is located on a single premise owned by one customer, pulses from the three delivery points are transmitted over small wires for conjunctive registration at a totalizing meter for billing under the applicable rate schedule or special contract.

Totalization can also be accomplished after retrieving the metered data, as at the end of the billing period, using the pulse translation system. To ensure full load diversity, the end-of-interval pulses recorded at the individual meters should be synchronized to the exact same time.

Metering totalization and conjunctional billing are generally not intended to be used for combining the meter registrations of multiple delivery points, which may be provided by the utility to serve separate loads on a single premise, when a single delivery point is otherwise feasible. In other words, the customer has the responsibility of wiring out the facility. If the utility does provide multiple delivery points in lieu of the customer expanding the distribution system, each delivery point should be separately metered and individually billed under the applicable rate schedule. In addition, totalization in the context of this discussion should not be confused with the concept of *account aggregation* in which the energy use of a single customer, having loads on multiple, noncontiguous premises and separately metered, is combined. Account aggregation is most often used for billing the energy supply portion of electric service where customers have a choice of energy suppliers. Conjunctional billing of the delivery portion of electric service to multiple premises should not be done except under a rate that is specifically designed to accommodate aggregation of power delivery services.

Nonrevenue Applications of Metering

Meters are required for a number of power system applications that are not revenue related but which have a bearing on setting electricity prices. For instance, the basic cost-of-service study used for supporting profitability assessment and rate design requires customer or rate class load shapes which are applied in the cost allocation process. To further allocate costs to usage levels within a customer or rate class for rate structuring purposes requires the development of special load relationships, such as average customer peak kW demand as a function of kWh and coincidence factor vs. load factor. In order to develop such load information, interval data metering is required.

Interval recorder revenue metering is typically installed for large commercial and industrial customers; thus, the data collected can support both the billing process and load analysis applications. However, it is not cost effective to use interval meters as a standard for all customers. The load data required for cost of service studies and other load analyses for small and medium size customers must then be developed from load research. Load research is based on a statistical sampling methodology that yields load shape results within a prescribed range of accuracy, which in part is determined by the number of survey meters contained in the sample. A typical load research sample for a large, fairly homogeneous customer class, such as residential, would require on the order of two- to three-hundred load survey recorders to achieve an accuracy criterion on the order of $\pm 10\%$ at a 90% confidence level. Quality load data is essential for producing credible cost-of-service allocations and rate designs.

A utility company must utilize electricity to operate the business. Like customers, identifiable company-use power consumption can be metered at such locations as office buildings, warehouses and shops, and station service at generating plants and transmission and distribution substations. Meters are also installed on generators in order to measure their energy output. In a very general sense, by knowing the MWh output of the generators and the associated MWh of sales at the metered delivery points of customers, the difference then represents the energy losses throughout the transmission and distribution systems. These losses consist of technical losses (e.g., i^2R), unidentifiable company use (e.g., energy used to operate motor-driven transmission and distribution switches and equipment), and theft of service (often referred to as *current diversion*). While electrical losses exist throughout transformers and lines and can not be measured directly, meters placed strategically within the power system can help to quantify these losses by general voltage levels. Determining losses by voltage or functional levels is another crucial element of the cost-of-service study process.

Meters are also located at the transmission tie-line interfaces in order to measure power interchanges and inadvertent energy flows into and out of interconnected systems. Special metering is also needed at key locations within the power system to support Supervisory Control and Data Acquisition (SCADA) functions. Transducers measure electrical quantities (volts, amps, watts, and VARs), and remote terminal units (RTUs) then transmit the data by telemetry to the supervisory master unit for processing and control.

7.5 METER DATA ACQUISITION, PROCESSING, AND BILLING

Meters are installed for the purpose of measuring and registering/recording key electrical attributes for routine use in both electric service billing of customer accounts and supporting a variety of other utility operations. Metering is an extremely data intensive function. Metered data must be collected regularly from numerous sites which are spread over a large geographical area. The raw metering data must then be validated, properly formatted, and transferred to the various users of metered information for further processing and application.

Meter Reading

The procedure of acquiring information from meters is referred to as *meter reading*. Meter reading must occur on a periodic and timely basis so that bills can be rendered to customers on a monthly frequency, although not necessarily on a calendar month period. Meter data can be acquired in a variety of ways as listed in Table 7.2.

The most widely practiced method of reading a customer's meter by local utilities is for a meter reading worker to physically visit the meter, take a visual reading, and make an on-site database entry of the register readings using a handheld electronic instrument. An electronic recording meter requires a data download to a handheld instrument or a PC using either an optical or wire connection port. With older types of recorders, a memory component must be retrieved and replaced with a "fresh" component for recording the next month's data.

Site visit meter readings are accomplished in cycles. The meter reader work force reads nearly five percent of the meters each business day. Accounting for week ends and holidays, all of the meters can be read over a typical 21-cycle sequence within a month (meter reader staffing requirements vary depending on the size of the utility in terms of number of customers and geographical area). Meter reading routes are designed to follow the 21-cycle sequence.

Table 7.2 Meter Data Acquisition Methods

Customer Site Visits

- Handheld Electronic Terminal
    - Key Pad Data Entry
    - Optical Port Data Download
- Portable PC Terminal (Optical or Wire Connection Port)
- Solid-State Memory Module Retrieval*
- Magnetic Tape Cartridge Retrieval*
- Handwritten Book Entry*

Remote Data Interrogation

- Mobile (Drive-By or Walk-By) Meter Reading
- Power Line Carrier
- Direct Telephone Dial Up (Outbound or Inbound Calling)
    - Conventional and Fiber Optics Land Lines
    - Cellular Communication
- Radio Communication
- Internet
- Satellite Relay

Other Data Retrieval Methods

- Customer Read and Reported (Mail-in Post Card)
- Pre-Pay Metering (Download to Electricity Purchase Card)

* Older technologies that may be in use in some locations.

Metered data can also be acquired by both short-range and long-range remote interrogation of the meter's registration using a variety of communications methods. In lieu of walking up to each meter's specific location, data from radio-equipped meters can be harvested when polled with special instruments while either walking or driving a vehicle along the front street or back alley accesses to the customers' premises. Power line carrier (PLC) technologies using the distribution system lines can also be used to transmit metered data from several adjacent customers over short distances to a local collection point or node. The data can then be retransmitted by other long-distance communication methods to a central location for processing.

For both short-range and long-range communications, conventional copper, fiber optics, and cellular phone lines with modem connections are used to interrogate meters and download data from memory. Systems can be configured either for a central computer to call the meters or for the meters to call in to the central computer at specific times. Either a dedicated line or the customer's normal phone line can be used for the communications channel. When using the customer's phone line, special conditions must be considered. For example, when the computer calls for data, the connection to the meter should be completed without ringing the customer's phone. If in process, the data download should cease whenever the customer picks up the phone and continue without loss of data once the customer hangs up the phone. Such conditions can be met with special equipment and coordination with the customer's local phone system. Alternative methods of transmitting data from customers' meters to the central computer is provided by the Internet and satellite communication systems.

Remote interrogation of meters is also known as *Automatic Meter Reading* (AMR). Initially, AMR installations were generally of small scale and served niche applications, such as real-time pricing and interruptible contracts where the costs of RTU-type metering facilities are justified and recovered directly from the subscribing customers through the design of the rate schedule. Applications of *Advanced Metering Infrastructure* (AMI) technologies, which includes AMR, are now moving from the pilot study phase to broad implementation. Beyond the obvious savings realized by eliminating the manual reading of meters, AMI facilitates the implementation of new rate and energy management options for customers and T&D system automation that improves reliability and control.

A minimal-cost meter data retrieval method is for customers to read their own meters and report monthly usage for billing purposes by means of a mail-in postcard provided by the utility or energy service provider. In particular, this method has been used in rural, low customer density settings where the cost of utility meter reading, on a per customer basis, is exceptionally high. However, a utility meter reader visits each customer's site occasionally in order to verify that the meter is functioning properly and that readings are being reported correctly.

Another method for retrieving metered data via customers is through a special application referred to as *pre-pay metering*. Pre-pay metering allows customers to purchase a quantity of electricity in advance of use from a kiosk terminal unit located at a utility customer service office or at other convenient locations, such as 24-hour retail stores. The customer inserts a prepay data card (similar in size and design to a credit card) into a card reader along with money into a cash receptor. The card is credited with the amount of the deposit. In addition, the card reader downloads prior energy use data stored on the card by the customer's pre-pay metering installation. Back at the customer site, the card is inserted into a reader on a small terminal unit which is connected to the meter by means of wires or PLC. This action enables the meter (i.e., the service) to remain connected until the pre-paid funds are depleted. At the same time, the meter downloads to the card the energy usage information gathered since the previous card reading.

Metered Data Analysis

Once metered data is retrieved, it must be processed to ensure its validity and to properly format it for various applications. Simple kWh and indicating demand meter readings can be verified for reasonableness within a database or the billing system by programming evaluation criteria for comparison. Such criteria may look at trends in prior usage levels and set a flag on a customer's account when anomalies occur (e.g., actual usage deviates from the use in the corresponding month of the previous year by specified set points while taking weather differences into account). Such irregularities may indicate a misread of the meter or a metering hardware problem that warrants a field investigation and correction.[7]

Interval meter data requires a much more detailed evaluation process. Instead of simple energy and demand register readings, interval data is gathered as recorded pulses which have a specific Wh, VARh, VAh, or Qh per pulse value. Pulses are interpreted as engineering units by means of a computerized translation and data storage system, as shown in Figure 7.12. Validation of interval data is accomplished with a variety of criteria. For instance, a comparison is made between the actual number of intervals counted and the number of intervals expected to be found. The interval recorder readings of maximum demand and energy are compared to a check meter or a redundant metering installation. Other validation inquiries include comparing current energy and demand use to historical data (as described previously), identifying any abnormally large increases or decreases in usage from one interval to the next, and recognizing any apparent loss of data.[8] If all validation criteria are met within an acceptable tolerance, the data can be processed for applications.

When the interval data fails validation, it must be edited before it can be utilized in application. Editing may be relatively simple or substantially complex depending on the severity of the data defect. An abnormally high or low value, including the total loss of data, in just a single interval or in a few consecutive intervals can be reasonably corrected by interpolation using the adjacent intervals of reliable data. However, a significant loss of interval data requires that a load profile be synthesized by scaling historical data which correlates to the period in question or using load profile estimation algorithm.

As noted previously, the translation system can be used for totalization of individually metered delivery points for conjunctive billing. Load profiles for load research studies are also developed by totalizing the interval data of individual customers within a specified sample.

---

[7] A faulty electromechanical meter most often slows down thus registering less than the actual usage. In order for a meter to speed up its registration, the permanent magnets must be affected. Such an occurrence is rather uncommon.

[8] Total or partial interval data can be lost or corrupted due to events such as metering equipment faults (e.g., failure of an instrument transformer or memory component), data overflows, or incorrect meter constants, multipliers, configurations, etc.

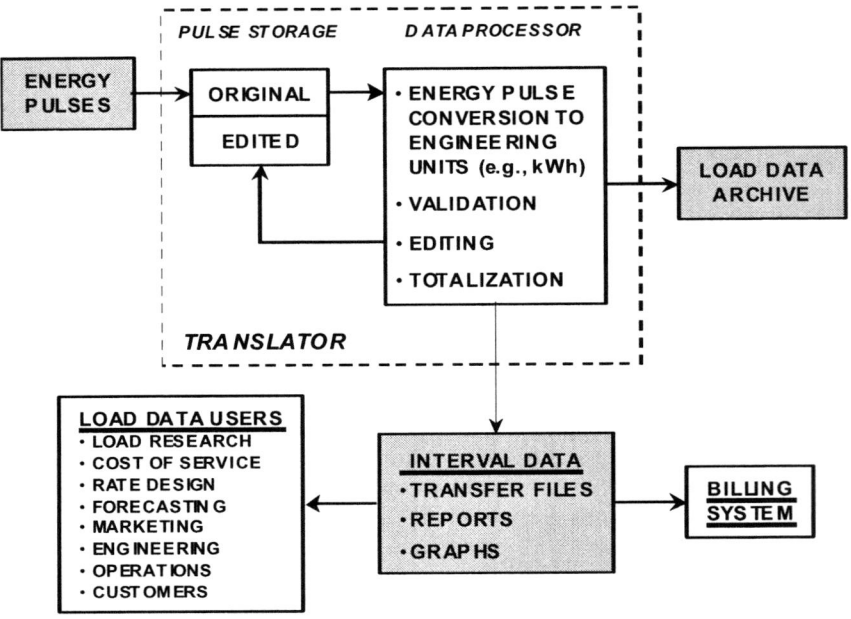

Figure 7.12 Meter pulse data translation system. By means of a translator, energy pulses acquired from meters are converted to engineering units, such as kWh, kVARh, kW, kVA, and power factor; validated and edited to correct for corrupt or missing data; and totalized when conjunctive billing is applicable. The processed data is then transported to the billing system and made available to various load data users.

Once interval data has been validated, edited, and totalized where appropriate, it can then be formatted for applications by a number of load data users. Quality load and energy data is critical to the monthly billing function. Data may be transferred to the billing system where it may be used directly in interval form or summed to time-of-use periods or monthly totals of energy and maximum demand as required for specific rate schedule and special contract applications. Special reports and graphs can be generated for use in support of cost-of-service studies, rate design analyses, and a number of other utility functions.

Electric Service Billing

An invoice or bill for electric service is typically rendered to a customer on a monthly basis within just a few days of reading the meter. Since manual meter readings occur in cycles, as previously discussed, the associated bills are also determined in cycles, as shown in Figure 7.13.

Revenue Metering and Billing

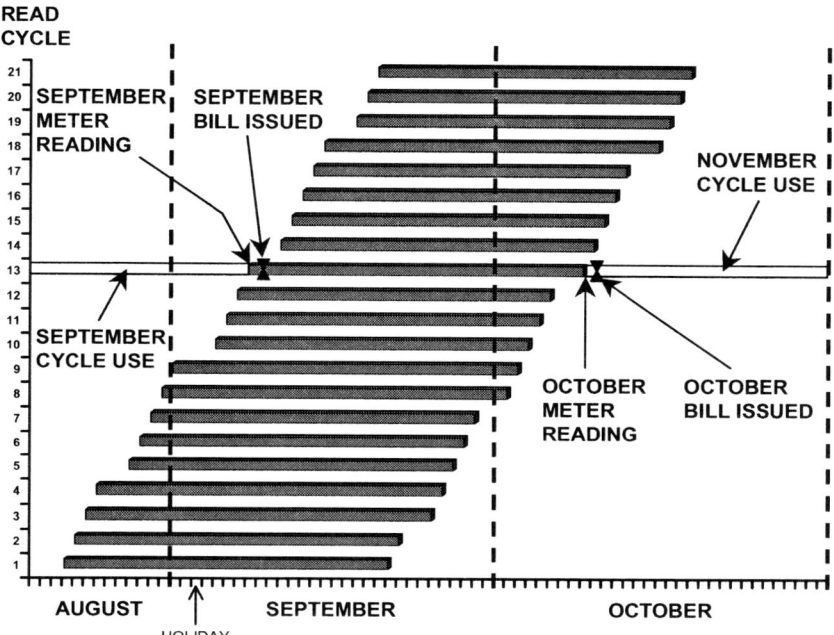

Figure 7.13 **An example of cycle billing for the billing month of October.** The horizontal bars show the days of consumption between the September billing month cycle meter reading dates and the October billing month cycle meter reading dates. The October bills are also issued in cycles as each meter reading cycle is processed. Although two-thirds of the meters are read in the calendar month of October, most of the actual energy consumption for the October billing month occurs in the calendar month of September while some of the consumption actually occurs in the calendar month of August.

As noted in Figure 7.13, the energy consumption, or sales, of each cycle occurs over a period of around 30 days (± a day or so depending on weekends and holidays). However, the energy consumed in all of the cycles, starting at the beginning of Cycle No. 1 and ending with the completion of Cycle No. 21 spans a duration of approximately 60 days. As a result, the sales and subsequent bills for Cycles 1 through 21 relate to a *billing month* rather than a calendar month.[9] Thus, each calendar month contains sales which are billed and sales which are *unbilled*. Adjustment methods are used to estimate what the actual month sales would be since sales and revenues are accounted for and booked on a calendar basis.

---

[9] Meters of the largest customers and AMR meters are often read on a calendar month basis. Full scale AMR implementation would eliminate meter reading cycles.

The adjustment of sales from a cycle basis to a calendar month basis is critical for aligning costs with energy consumption. For example, the energy produced by generators is measured on a calendar month basis by means of interval metering. Overall, the customers' energy consumption is typically less in October than in September as air conditioning requirements at the end of summer begin to wane with declining average temperatures. As a result, the production of energy to serve the load is less in October than in September. But, the metered sales for the billing month of October occurred mostly during the hotter month of September when energy requirements were high. Thus, without adjustment, the booked sales for the October billing could exceed the actual amount of energy produced.

Estimated and Prorated Bills

In the event that a meter can not be read at the end of a billing period as scheduled, due to temporary inaccessibility (e.g., such as from a severe ice storm), an estimated bill is usually rendered. An estimated bill is the result of applying specific rate schedule prices to an estimate of demand and/or energy usage. Estimated bills are based typically on either the previous month's usage characteristics (such as the directly proceeding month or the corresponding month of the prior year) or on some other reasonable basis for estimation of consumption. Since an estimate of usage for the end of one billing period also serves as the previous usage reading for the subsequent billing period, two sequential estimated bills result for a single estimate of usage. Usually, the meter will be read as scheduled at the end of the next billing period following the estimation.

Once the meter is next read, the amount of total consumption for the extended period spanning the two readings will be exactly correct; however, the total revenue of the multiple billings may not be exact depending on the nature of the rate applied. If the rate structure is a single, uniform price per kWh during the extended period, the total billed revenue for the extended period would be accurate, although the estimated individual monthly billings are likely to be inexact (the monthly billing errors cancel out over the period). On the other hand, if the rate has a structure where either incremental price changes occur on a volumetric basis or the price(s) vary due to an associated transition to a different season (e.g., the price structure changes between the last summer month and the first fall month), the total revenue of the bills will be in error somewhat owing to the estimate of usage. In this case, the errors do not necessarily cancel out.

Except for the effects of weekends and holidays, meters are typically expected to be read approximately every 30 days, since rates are designed on the basis of an average month. But service may not be in place during the entire billing cycle. For example, a new service may be initiated or an existing service may be terminated at sometime during the middle portion of the billing cycle. Sometimes, reading of the meter may not be possible until several days beyond the basic 30-day billing period. Thus, when the actual meter reading for the end of the billing period occurs outside of a reasonable range, such as less than 25 days or more than 35 days, the monthly bill should be computed on a pro rata basis.

Revenue Metering and Billing

The concept of *bill proration* is to first estimate what the bill would be if computed for usage based on a complete billing cycle duration and then scale the estimated bill according to the actual days of use. A simple and straightforward method of bill proration is demonstrated in the following two examples.

Example 1: Proration of a Nondemand Rate Structure

Rate: $15.00 per Month
First 650 kWh @ 5.0 ¢/kWh
Over 650 kWh @ 3.0 ¢/kWh

Usage: Energy = 1,008 kWh
Usage Period = 36 Days
Bill Cycle = 31 Days

Estimated energy usage for 31 days = 1,008 kWh × (31 ÷ 36) = 868 kWh

Bill for 31 days:
$15.00
650 kWh × $0.05 = 32.50
218 kWh × $0.03 = 6.54
868 kWh $54.04

Bill for 36 days: $54.04 × (36 ÷ 31) = $62.76

Example 2: Proration of a Demand Rate Structure

Rate: $6.30/kW
All kWh @ 2.0 ¢/kWh

Usage: Energy = 7,500 kWh
Demand = 1,000 kW
Usage Period = 10 Days
Bill Cycle = 30 Days

Estimated energy usage for 30 days = 7,500 kWh × (30 ÷ 10) = 22,500 kWh

The maximum demand measured during the 10-day period is assumed to be the maximum demand that would have occurred during a whole billing month and is used for the bill calculation (unless provisions of the specific rate schedule require consideration of other billing demand criteria).

Bill for 30 days:
1,000 kW × $6.30 = $6,300.00
22,500 kWh × $0.02 = 450.00
$7,350.00

Bill for 10 days: $7,350.00 × (10 ÷ 30) = $2,250.00

As a final note, if a bill is neither *estimated* nor *prorated*, then it is deemed to be *normal*. Classifying bills in this manner is of particular significance relative to the rate design and analysis processes.

Billing Statements

The monthly billing statement provided to the customer for payment should contain certain types of information. Regulatory agencies that govern electric service often promulgate rules that specify bill formats and/or contents. At a minimum, the bill statement should show: the account number; meter number(s); previous and current meter reading dates; the register readings from those dates; any register multipliers necessary to compute the metered usage; the total actual usage amounts; the designation of the applicable rate schedule; the gross and/or net amount(s) of the bill; the date the bill is due; and any other essential information upon which the bill is based.

Certain billing components that are assessed "outside" of the principal tariff charges for electric service, such as specific taxes, franchise fees, and extraordinary surcharges and credits, might also be itemized on the billing statement. The actual details of the rate applications against the billing determinants are not typically included on residential and small commercial billing statements because such details generally create confusion due to a lack of consumer understanding of electricity tariffs. Such inexpert customers are typically more responsive to the total bill amount than to the individual price components. On the other hand, billing statements for medium to large commercial and industrial customers are customarily quite detailed and include all steps of the rate applications. Usually these more dexterous customers are keenly interested in tracking their usage characteristics and comprehending rates and bill calculations because they are often seeking ways to minimize their costs of purchased utilities.

An example of a fairly complex bill calculation for a medium-size general service customer is shown in Figure 7.14. This billing statement is well organized, and its presentation of the calculation steps closely emulates the layout of the pricing components contained within the associated rate schedule(refer to Figure 8.16), thus making the bill easy to examine.

Billing statements may include basic statistical information to assist customers in putting the bill amount in perspective without having to fully understand how rates work. For example, a bill may display the kWh use on a per day basis for not only the current month but also for the previous month and the corresponding month from the previous year. The customer can then recognize the consistency or the deviation in use characteristics and thus have a fundamental basis for rationalizing why usage might vary. Was load added since last year? Did the replacement of the old air conditioning unit with a more efficient unit measurably reduce the energy requirements for cooling? Was this month warmer or cooler than previous months? Weather information, such as monthly degree days correlated with energy use, can provide an often compelling statistic that can be shown in a simple graph on the bill statement for weather-sensitive customers such as residential and small commercial. Larger commercial and industrial bills can be accompanied with simple but effective graphs of billing demand history, load factor and power factor characteristics, etc. All such information can help customers to better understand their electric service and its costs.

Revenue Metering and Billing

Account No.: 1228-8635875				Service from JUN 7 to JUL 9		
Rate Code: C-410						
Meter No.	Type	Present Reading	Previous Reading	Difference	Constant	Metered Usage
R25-900522	kW-DEM	3.4	--	--	80	272.0
R25-900522	kWh	4577	2854	1723	80	137,840
X10-530611	RKVAh	0773	0482	291	80	23,280

Average Power Factor = 98.6%   Maximum Demand Previous 11 Months = 308 kW

BILLING UNITS:

Billing Demand:	Metered Maximum kW Demand		272.0 kW
	Plus 1.0% for Metering at Secondary		2.7 kW
			274.7 kW
	Power Factor Adjustment:		
	274.7 kW × [90% Base ÷ 98.6% Actual] =		**250.7 kW**
Billing Energy:	Metered kWh Energy		137,840 kWh
	Plus 2.5% for Metering at Secondary		3,446 kWh
			141,286 kWh

BILL CALCULATION:

Customer Charge:	For Three-Phase Secondary Service			$ 80.00
Demand Charges:	First	50.0 kW @ $4.59/kW		$ 229.50
	Next	200.7 kW @ $4.37/kW		877.06
	Secondary Service Charge:	250.7 kW @ $0.45/kW		112.82
				$1,219.38
Energy Charges:	First	25,000 kWh @ 3.11¢/kWh		$ 777.50
	Next	100,000 kWh @ 2.61¢/kWh		2,610.00
	Next	16,286 kWh @ 1.90¢/kWh		309.43
	Load Factor Credit:			
	[141,286 kWh - (250.7 kW × 400 kWh/kW)] @ 0.3¢/kWh			(123.02)
				$3,573.91
			BASE BILL	**$4,873.29**

Adjustment Clauses:

Fuel & Purchased Power Costs:	141,286 kWh @ 1.91¢/kWh	$2,698.56
Environmental Costs:	1.6% of [Base Bill + Fuel Adjustment]	121.15
Demand-Side Management Costs:	141,286 kWh @ 0.06¢/kWh	84.77
		$2,904.48
	TOTAL BILL	**$7,777.77**

Figure 7.14 An example monthly billing statement. A bill calculation for a medium-size general service customer showing metered information and application of the rate schedule prices and billing provisions. The bill is based on rates shown in Figure 8.16.

7.6 SUMMARY

Metering is a critical aspect of providing electric service since metered data is directly utilized for billing customers for their usage. While small-size customers can be billed effectively under a rate schedule that requires only a measure of kWh, large size customers are commonly served under complex schedules that consider not only energy usage but also maximum kW or kVA demand and power factor. A variety of meters have been designed to measure energy and demand for both real and reactive power flows. Interval recording meters provide detailed load profiles of customer energy usage. Such metering is applied both for rendering service under certain types of rate schedules and for gathering load research information that is essential for analyzing the costs of providing electric service and designing effective rate structures.

Billing meters are intended to be located at the physical point of delivery where a customer takes service from the utility since rate structures are inherently based on delivering power at a specific voltage. Instrument transformers permit the safe and accurate metering of high voltages and currents which exist under certain service arrangements. When real conditions call for placement of the meter at a location other than at the physical delivery point, special metering facilities can be used to compensate for the disparity in losses that exists between the two positions. Very large customers may require two or more delivery points due to the size of the load; thus, totalization of the metering of the multiple feeds is used to acquire a single measurement of the combined usage as though all of the power flowed through a single delivery point for billing purposes.

Metered data can be retrieved in a number of fashions through both onsite and remote acquisition methods. Before use in billing and other application, metered data must be processed to ensure its integrity.

It should be noted that in places where industry restructuring has been implemented in order to give retail customers a choice of electricity suppliers, revenue cycle services (i.e., metering and billing) and some customer services may also be provided as individual competitive services that are subject to customer choice. To offer such services competitively, the local utility's rates should be unbundled and the services individually priced so that customers have a basis for comparing alternative service providers. Since metering and billing data is utilized in a host of other power system applications, multiple parties require access to the same information. The issues of data formatting requirements, data transference between different parties, and data ownership and security must be resolved to ensure that all service provider needs are met in an effective manner.

Metering technology is advancing at an exceptional pace. While older technologies are still in use in some locations, ultimately *smart meters*, along with AMI, will likely transform the industry in terms of the types of data that can be effectively measured and retrieved and the pricing options that can be offered to customers to help them manage their usage and their bills.

8

The Rate Schedule: Pricing Methods and Risk

8.1 INTRODUCTION

Power systems are structured and operated to meet the electric service requirements of customers in a practical, economical, and reliable fashion. In so doing, fixed and variable costs are incurred by the regulated utility or, in restructured markets, by various competitive service providers. Since power systems are generally both capital and fuel intensive, these costs must be recovered in an adequate manner to minimize risk and ensure the viability of the electric service provider(s). The rate schedule is the fundamental mechanism for billing customers in order to recover the costs of providing electric service.

The costs of providing electric service are categorized as customer-related, demand-related, and energy-related components. The customer component is the fixed costs associated with providing access to electric service and is not directly related to the amount of electricity required. The demand component is the fixed costs associated with providing capacity to meet peak energy requirements. The energy component is the variable costs associated with providing electrical energy or kWh. Pricing is based on these three fundamental components and thus the principle of cost causation is inherent to rate design.

Figure 8.1 illustrates how the customer, demand, and energy charges of a bundled service rate relate to the various functions of providing electric service. Quite a number of discrete power system functions are bundled together under the customer and demand rate components. Some power system functions represent multiple cost-causation elements. For example, a line transformer provides a customer with both the access to an energy supply (customer component) and the capacity required to receive the energy (demand component).

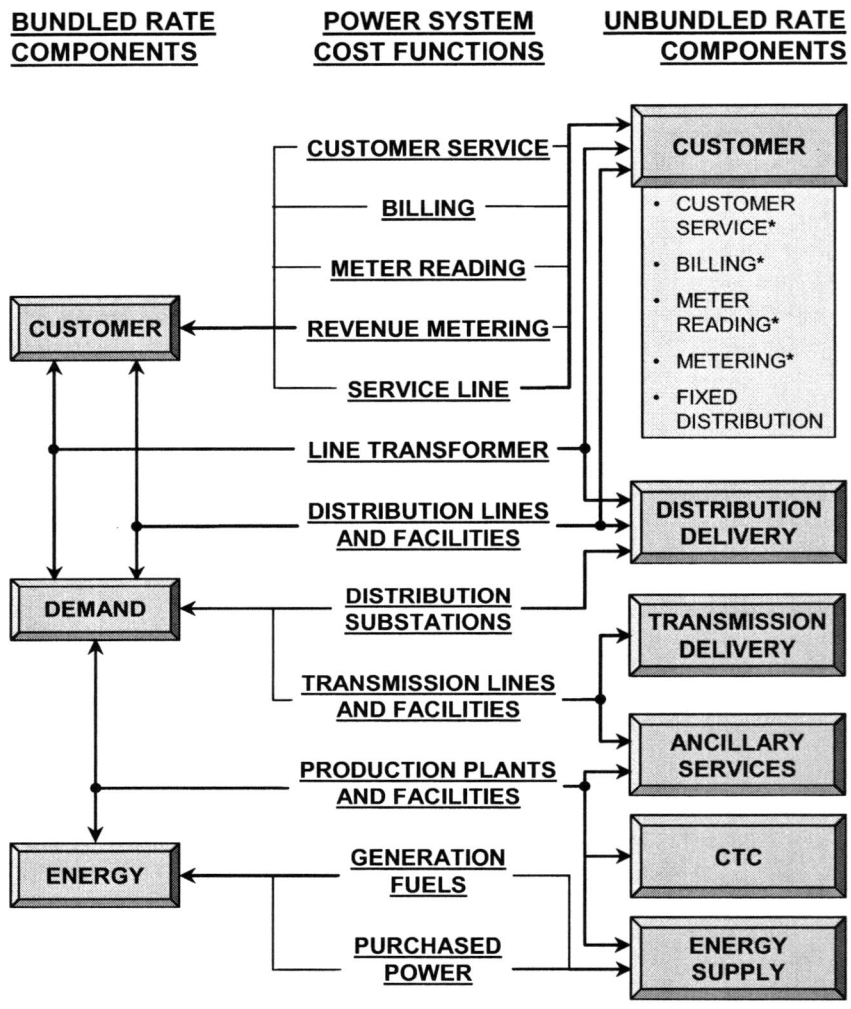

Figure 8.1 Typical pricing components for bundled and unbundled service rates. Rates are typically bundled when all electric service functions are subject to regulatory jurisdiction and unbundled in markets where customers are allowed to choose between energy service providers. In some competitive electricity markets, certain customer service and revenue cycle services (as denoted by an asterisk) may also be provided by competitive suppliers. Thus, they too would be priced separately by the incumbent utility responsible for providing regulated power delivery services. The Competitive Transition Charge (CTC) is a temporary fee imposed at the onset of competition.

Figure 8.1 also illustrates how the components of an unbundled rate relate to the power system functions. The labels of the unbundled rate components tend to be more descriptive of the actual services as compared to the cost component labels of customer, demand, and energy used with bundled rates. The restructuring of electricity markets to offer customers the choice of energy service providers (ESPs) requires the separation of prices between competitive and regulated utility services. Customers might also have the choice of suppliers for some of the attendant revenue cycle services (e.g., metering) and certain customer services; thus, these services may be individually priced in the incumbent utility's rate schedule so customers can see the "price to beat." These prices for competitive service offerings become shopping credits in the event that the customer selects an alternative service provider while the incumbent utility continues to provide the power delivery service under regulated rates.

Although a pricing structure may be unbundled along the framework of the power system, as shown in Figure 8.1, it should be noted that the functional rate components still pertain to the basic customer, demand, and energy components of cost. Thus, for example, the power delivery charges are related to a demand cost causation and are often designed with rate structures and other pricing/billing features that resemble the demand charges used under bundled rates.

Unbundled structures are not commonly utilized when all functional utility services are subject to jurisdictional regulation; however, some aspects of price unbundling often exist under full regulation. For instance, generation fuel is an energy-related variable cost that is often charged wholly outside of the base rate structure.[1] Depending on the development of market structures for Regional Transmission Organizations (RTOs), the unbundling of rates (or perhaps the partial unbundling of transmission from rates) may be required eventually, even in the absence of retail competition.

Tariff Organization

The tariff is a document that governs the sale of electricity. Similar to a contract for the sale of a major product or service, the tariff contains:

- Technical specifications for electric service
- Standard pricing structures and options
- Terms and conditions for service

Electric service provided to a larger commercial or industrial customer is typically rendered under a formal contract that is a binding agreement between the customer and the utility (and any other service provider as applicable).

---

[1] Fuel may also be partially unbundled in the sense that a base amount of fuel may be embedded in the energy charges of a rate while a separate adjustment factor would be used to periodically compensate for changes in the cost of fuel.

The nucleus of the tariff consists of a series of rate schedules designed for the provision of electric service to a variety of distinct customer classifications. To ensure equitable pricing to customers, several rate schedules are required for the typical utility since different customers receive power at different voltage levels, utilize energy with varying intensities at different times, and have diverse metering requirements. A cost-of-service evaluation which considers investment and expense requirements, customer usage characteristics, and system losses clearly indicates that cost differentials exist between the residential, commercial, industrial, and outdoor lighting sectors. In addition, the tariff may contain a number of rate riders which modify one or more of the provisions of the standard rate schedules in order to provide for additional service requirements and offerings (e.g., interruptible service).

The tariff also includes a set of terms and conditions or rules and regulations governing electric service. Such criteria are applicable in a general nature to all classifications of electric service. For example, a utility's rules would address such issues as meter reading, billing, and payment protocols, restrictions in a customer's use of the service, and other general customer and utility obligations and liabilities regarding electric service.

In addition to the fundamental documents identified above, the tariff may include a variety of other information relative to the provision of electric service, such as:

- Map of the service territory
- List of cities, towns, and communities served
- Definitions of rate, contract, and electrical terms
- Standard contract forms
- Bill format examples
- Rate application charts and tables
- Marketing policies and procedures

Rate Schedule Anatomy

The rate schedule establishes specific rules, requirements, and pricing for a designated class of electric service. A rate schedule consists of three major sections:

1. Qualifying Service Provisions – the basic stipulations for application of the rate schedule to a customer (e.g., contract term);

2. Central Rate Structure – the fundamental billing mechanism configured from customer, demand, and energy charges; and

3. Attendant Billing Provisions – clauses which address any special conditions of the service (e.g., power factor adjustments) and/or clauses which incorporate billing components developed outside of the rate schedule and common to multiple rate schedules (e.g., fuel cost adjustments).

The Rate Schedule: Pricing Methods and Risk

The framework of a rate schedule for a large commercial, industrial, or general service class typically includes the following types of provisions:

- Availability of Service
- Applicability of Service
- Character of Service
- Monthly Rate for Service
- Determination of Billing Demand
- Delivery Voltage Adjustment Clause
- Metering Adjustment Clause
- Power Factor Adjustment Clause
- Energy Cost Adjustment Clause
- Special Adjustment Clauses
- Tax Clauses
- Minimum Bill Clause
- Term of Payment Clause
- Term of Contract

Rate schedules for residential and small general service customers incorporate less technical billing provisions.

8.2 QUALIFYING SERVICE PROVISIONS

A rate schedule is designed for electric service provided to a particular type of customer group. The customers within such a group should have rather specific, or at least reasonably similar, attributes that distinguish them from other customer groups. These distinctions may be related to such factors as cost-of-service differences, customer load and energy use characteristics, or special requirements of the service.

Qualifying service provisions are used to ensure that only those customers who meet certain criteria are provided service under a particular rate schedule. In so doing, the qualified customers will form a homogeneous group to which the rate will function in an effective and equitable fashion as intended.

Availability

The *Availability Clause* indicates the locations in which a rate schedule applies. In general, most rate schedules apply uniformly throughout a utility's service territory given that adjacent power lines of adequate capacity currently exist. In some cases, a utility's rates for the same classes of service may be differentiated geographically because of significant differences in the cost to serve (e.g., as with an underground distribution network system) or because the service territory spans multiple regulatory jurisdictions. An experimental rate schedule might be made available only in a specific geographic area in order to limit the scope of a pilot program so that a controlled evaluation can be made prior to a full-scale program implementation.

Instead of geographical restrictions, the total number of customers or the quantity of their combined loads subscribing to an experimental or special rate schedule may be used to limit the availability of the rate. The limitation could be for a maximum number of customers or amount of load after which the schedule would be closed to new customers. Conversely, a limitation might be imposed for a minimum quantity of customers or load required under which the schedule would not be implemented. A minimum limitation would help to ensure cost recovery where the service provided requires a significant capital investment (e.g., an optional renewable energy resource) or extraordinary administrative expenses (e.g., an RTP program). The availability of experimental or special rates is often limited by a *date certain* beyond which customers must choose between continuing service under the experimental rate's successor schedule (if the program is extended) or switching to another appropriate standard rate offering.

Sometimes a long-standing rate schedule may be targeted for withdrawal and thus is *frozen* or closed to new customers. The customers may be grandfathered for continued service under the frozen schedule but eventually migrated to other appropriate rates as the existing contract terms expire or as other favorable conditions arise (such as changes that occur in a customer's load characteristics that favor a different type of rate structure). The Availability Clause would be used to indicate that a schedule is frozen and unavailable to new customers after a given date.

Applicability

The *Applicability Clause* stipulates the types of customers that can be served under the rate schedule. The clause may denote applicability in terms of broad customer classes (i.e., residential, commercial, industrial, or public authorities). For example, industrial service may be distinguished from commercial service by a requirement that building lighting load be incidental to the total load (e.g., 10% or less). The rate schedule may apply only to specific end-use load devices such as water, space, or process heating, to specific business operations such as schools, mining, or irrigation, or to specific industry classifications as denoted by industry codes (e.g., NAIC or SIC). The Applicability Clause would also indicate any "all electric" qualifications for the rate schedule by stating that all energy requirements of the customer must be met entirely through the electric service.

The Applicability Clause also stipulates any restrictions based on the customer's load characteristics. For example, a rate may be applicable only to those customers having a metered maximum demand of at least 25 kW but less than 500 kW. Qualifying electric heating loads may be required to achieve a minimum stated demand (e.g., 250 kW). High load factor rate schedules may designate that the customer must maintain a minimum load factor requirement (e.g., 80% per month or 75% on an annual basis).

Maximum energy usage and/or demand restrictions may be imposed for small commercial services (e.g., usage may not exceed 5 kW or 1,500 kW per month). Residential customers whose expected maximum loads exceed a specified threshold (e.g., 75 kW) may be required to be served under a small commercial or general service rate schedule. A house used for both domestic and small business purposes may be required to take the entire service under a small commercial or general service rate schedule unless the customer separates the wiring so that individual delivery points can be established and metered for billing under the appropriate rate schedules.

Character of Service

The *Character of Service Clause* indicates the types of voltages that are provided by the utility for electric service. While some special rate schedules may offer a DC voltage for limited applications, most electric service applications are based on AC voltages The Character of Service Clause describes the service available in terms of the:

- Frequency (i.e., 50 or 60 Hertz)
- Phase (i.e., single or three phase)
- Standard voltages and configurations

The standard voltages may be specifically identified on the rate schedule itself, or they may be defined in either the rules section of the tariff or in the utility's electric service specifications documents. The available voltages follow from the utility's transmission and distribution construction standards and inventory practices, and they generally meet the requirements for most electric service applications. Customers are ordinarily required to provide all of the facilities necessary to receive service according to applicable electrical safety codes. Customers requiring a nonstandard voltage for serving unique end-use load devices would be required to provide their own transformations unless the utility was willing to provide for such exceptions under a special service agreement.

Rate schedules for residential and small business services are often limited to single-phase voltages while customers with three-phase voltage requirements are served typically under a general service rate schedule. The Character of Service Clause may specify restrictions regarding motor applications. For example, single-phase service may only be provided for motors of 5 Hp or less while motors greater than 5 Hp may be required to have starting compensation to minimize disturbances to other adjacent customers. Commercial and industrial customers having large loads that begin to exceed the service capacity of a primary distribution feeder may be required to take service at a transmission voltage. The Character of Service Clause may also indicate whether or not the electric service is regulated (in terms of voltage). Delivery point and metering restrictions may also be specified (e.g., one point of service through one meter).

Term of Contract

The *Term of Contract Clause* indicates the minimum or initial period of time to which the customer must commit to taking service under the rate schedule. The term of contract generally increases with the magnitude of the load. For example, rates designed for small-business customers generally require a one-year contract. Rates for medium-size commercial and industrial customers may require a three-year contract, while rates for large commercial and industrial customers may require a five-year contract. This incline in the initial term commensurate with the size of the load is based on the increased requirements for facilities which the utility must install to adequately serve the customer. Small loads generally require only a minimal transformer installation and a service line. The termination of the service after a few months of operation would have a definite albeit relatively minor financial consequence. Conversely, large customers often require extensive investment in local facilities in order to establish service. In addition, under an obligation-to-serve criteria, the utility's generation expansion plans can be impacted measurably by single large load additions. The initial contract term provides some assurance of cost recovery in the event that the customer decides to curtail or cease operations, thereby stranding the utility's investment. Given the small size and somewhat transient nature of residential customers, contracts are not generally required for domestic service.[2]

Upon completion of the initial term, the contract may be extended in various ways. The initial term may self-renew if not terminated by either the customer or the utility. The contract may continue on a month-to-month or a year-to-year basis until terminated. The contract may continue indefinitely until terminated. Typical notice periods for termination are a month, three months, six months, or a year. A rate schedule for interruptible loads may require a termination notice period of several years so that the utility has adequate time to construct new generation or secure other resources that will be needed to replace the interruptible capacity after the load has converted to firm service. In another situation, an interruptible contract may be offered for a term of one or two years only and without the implication for renewal as new resources are expected to be available subsequent to that contract period.

An existing customer may add load or modify its operations in such a way as to find that benefits could be realized by service under a different rate schedule with a longer term of contract. If the customer otherwise qualifies for the new schedule and no additional facilities are required for the customer to continue service, credit for time accumulated under the original rate schedule may be applied under the new schedule. Sometimes pricing and rate structure changes may affect the relationship between different rate schedules in a manner that compels certain existing customers to switch from their current rate schedule to another rate

---

[2] Residential customers served under a special rate option may be subject to a contract term, particularly if additional investments are required in order to provide the service.

The Rate Schedule: Pricing Methods and Risk

schedule under which they would otherwise qualify for service. In such cases, these customers may be given the option to be released from their existing contract and migrate to another rate schedule if they accept the specific terms and conditions of their new schedule.

In areas where industry restructuring for competition is considered to be a contingency, a *meet or release* clause may be utilized in contracts that otherwise necessitate a relatively long term. Once competition has started, this provision would allow a customer to choose another energy service provider and be released from the existing electric service contract if the customer has received a bona fide lower price offer that the incumbent utility chooses not to match.[3] The meet or release provision may also be applied symmetrically by also allowing the incumbent utility to terminate an existing customer's energy supply contract if it has received from another customer located outside of its service territory a bona fide offer to buy energy at a higher price. Under this arrangement, the existing customer would then choose whether or not to match the offer price and thus continue or terminate service with the incumbent utility.

8.3 RATE STRUCTURE FORMS

The rate structure is the primary billing mechanism of the rate schedule. Its purpose is to recover a significant portion of the costs of providing electric service to customers within a specific class in a fair and equitable manner.[4] As previously discussed, the fundamental cost-of-service components are defined as: customer, demand, and energy. Together these three components represent revenue requirements associated with providing the electric service based on different cost-causation drivers. To help ensure the fairness and equity criterion, it follows that the monthly billing elements of a rate structure should be patterned after the three basic cost components. The most straightforward rate structure design that meets this criteria is a *three-part rate* which was introduced around 1900 by Mr. Henry L. Doherty. The Doherty rate consists of discrete Customer, Energy, and Demand Charges and is often referred to as a *CED rate*.

While the direct and pure design of the Doherty rate structure has great appeal and is used in many electricity pricing applications, other factors make it necessary to develop a variety of other rate structure forms. For instance, the Doherty rate design operates on a monthly measurement of a customer's maximum load and thus requires a demand meter. Since it is not economically feasible to install

---

[3] With competition, it is also possible that the law might abrogate the existing contracts so that customers could be allowed to choose an alternative supplier without waiting out the term of the current contract.

[4] All or a portion of the fuel costs are often recovered outside of the basic billing mechanism. Some taxes and other special cost elements are also typically collected through external clauses.

demand meters on all classes of service, particularly residential and small business, it becomes necessary to recover the associated demand costs along with the energy costs on the basis of monthly kWh usage. Thus, a *two-part rate* consisting of a Customer Charge and combined Energy/Demand (kWh) Charges is commonly used for small customers.

With large commercial and industrial customers, energy and demand costs may also be collected together through kWh-based charges that function on the *hours use of the monthly maximum demand* (discussed below in Section 8.3), along with a separate Customer Charge. The result is a two-part rate with multiple kWh price steps.[5] In addition, a two-part rate structure may be developed by consolidating the customer cost component with the demand cost component while retaining a separate Energy Charge structure.

Each of the fundamental rate structure components may not be designed strictly for the sole recovery of their associated cost components. To meet the needs of the regulatory environment, residential and small business Customer Charges are often set below the amount justified through a cost-of-service analysis. Under such circumstances, a portion of the customer costs must then be recovered on the basis of kWh in order to realize the full revenue requirements for the residential class. Such a design creates an intra-rate subsidy as customers having a higher volume of kWh usage bear some of the fixed cost responsibility associated with providing electric service to the lower use customers.

Even in the case where separate Demand and Energy Charges are utilized in a rate design, some of the demand costs are typically recovered through kWh-based charges. This practice is known as "tilting the rate," and its use helps to balance the risks between the customer and the service provider that are associated with applying a single rate structure to a group of customers having a wide variation in load factor. Higher load factor customers are best served under a relatively high Demand Charge—low Energy Charge rate structure (as might be derived directly from a class cost-of-service study) while lower load factor customers are best served under a relatively low Demand Charge—high Energy Charge rate structure (by departing from the direct cost-of-service relationship between the two cost components).[6] By taking load factor into account when segmenting commercial and industrial customers into homogeneous rate groups, rate tilt can be applied effectively by adjusting the Demand Charge—Energy Charge relationship separately in consideration of each group's load characteristics. Alternatively, a single large group of customers having a wide variety of load factors can be served effectively by means of an individual, although more complex, hours use of demand rate structure. Either approach will yield the lowest practical average price per kWh (commensurate with the level of load factor) while ensuring for adequate cost recovery.

---

[5] A maximum demand measurement is required for the *hours use* rate application.

[6] Rate tilt theory and application is discussed further in Chapter 12.

The Rate Schedule: Pricing Methods and Risk

The Demand and/or Energy Charges of fundamental rate structures may be levelized across the year (i.e., an average or uniform rate), or they may be time differentiated by season and/or time of day. With seasonal differentiation, a rate structure and its demand and/or energy prices for a specific class of service may vary considerably between the summer, winter, and spring/fall periods of the year. In some cases, a time-of-day rate structure may be utilized (mandated) as the standard offer or default pricing mechanism for electric service, particularly for larger customers. In other cases, a time-of-day structure may be offered as an option to a more uniform rate structure.

A rate structure's Customer, Demand, and Energy Charges are established on the basis of a designated level of delivery voltage. When different service voltage levels are provided for under the same rate schedule, additional charges or credits are necessary to maintain equity between customers.

A variety of rate structure forms are used in practice, each having the fundamental objective of recovering the associated customer, demand, and energy costs in an equitable method. The following discussion will address the most common as well as some of the alternative structuring methods utilized for each of the three basic billing elements.

Customer Charge Structures

The *Customer Charge* is intended to recover a number of fixed costs that are not directly dependent on either the monthly volumes of energy use or the capacity required by the customer, although the customer-related costs of larger customers may indeed be greater per delivery point than for smaller customers. Specifically, customer-related costs include the investment and expenses in metering equipment and service lines along with a minimal portion of the secondary distribution lines, line transformers, and the primary feeder system of poles, conduit, and conductors. Certain line facilities, such as reclosers, are also deemed to be related to the customer component of costs. The Customer Charge costs also include the expenses associated with meter reading, billing, and customer accounting, sales, and service. The Customer Charge is often designated by other descriptive nomenclature, including *Base Charge, Service Charge, Facilities Charge,* and *Access Charge*.

The Customer Charge is most commonly expressed as a fixed fee per month (e.g., $18.25 per month). However, the Customer Charge may also be represented as a fixed fee per day (e.g., [$18.25/month × 12 months/year] ÷ 365 days/year = 60¢/day). With the daily fee approach, the actual Customer Charge amount billed will vary slightly from month to month depending on the number of days in the billing cycle. In addition, a daily Customer Charge essentially prorates itself when meter readings or estimates are made outside of the range of days associated with a normal billing cycle.

Multiple Customer Charges may be identified on a single rate schedule to account for the variations in customer-related costs incurred when rendering the same class of electric service to customers having differing characteristics. For example, customers on the same commercial/industrial rate schedule may take service at various delivery voltage levels. Compared to a secondary single-phase service, a secondary three-phase service requires additional costs relative to the meter, the service line, and the transformation facilities. While a primary or transmission level service requires no voltage transformation facilities to be installed by the utility to deliver power, instrument transformers necessary for metering high voltages add additional cost. As a result, the Customer Charge may be differentiated on the basis of delivery voltage and designated in terms of general voltage classifications (i.e., single-phase and three-phase secondary, single-phase and three-phase primary, subtransmission, and transmission) or specific voltages or voltage ranges (e.g., 2.4 kV to 12.47 kV or >1,000 volts).

Customer-related costs also tend to be higher for rate groups consisting of large load customers. More sophisticated metering facilities, such as reactive power measurement and interval recording, are generally standard. The maintenance requirements for such metering equipment is rather intricate. For monthly billing purposes, a specialized meter technician may be required to download information stored in the meter to a laptop computer, or a remote interrogation system may be used to acquire the metered data. Information processing is much more involved with complex metered data than with simple billing cycle meter readings. In some cases, large customers served under more complicated rates may be manually billed or require a certain amount of nonautomated support to complete the billing process. In addition, the customer service costs associated with large commercial and industrial customers are typically higher than for smaller customers due to the practice of assigning specific Account Executives to key customer accounts and providing technical marketing programs.

When a rate schedule is used for service to a group where the magnitude of the loads of individual customers may differ substantially, the Customer Charge can be structured to recognize the associated variations in customer costs. One method for differentiating the Customer Charge is to base it on the customer's monthly demand using discrete charges that increase with designated load ranges (e.g., $750/month for demands less than 1,000 kW, $1,500/month for demands between 1,000 and 5,000 kW, and $3,000/month for demands over 5,000 kW). Alternatively, the Customer Charge could be based on a monthly fixed amount plus a per unit charge applied to the demand (e.g., $500 plus 50¢/kW). This formulary approach can also be combined with designated load ranges to moderate the sharp step differentials of the first method and the pure linear effect of the second method (e.g., $750/month for demands less than 1,000 kW, $300 plus 45¢/kW for monthly demands between 1,000 and 5,000 kW, and $1,050 plus 30¢/kW for monthly demands over 5,000 kW).[7]

---

[7] It should be noted again that the customer-related costs of service do not vary based on customers' monthly demands, but they can vary based on the relative size of customers, as discussed above. The demand-based methods for recovering customer revenues merely provide a means for differentiating prices according to basic cost causation principles.

The Rate Schedule: Pricing Methods and Risk 275

With small-size customers, the Customer Charge may be differentiated on the basis of the actual type of metering equipment. Some customers served under the same small business rate schedule may have an indicating demand meter while others have just a simple watt-hour meter, or some special circumstance customers could be literally unmetered (monthly kWh usage must be consistent and easily estimated). Space or water heating usage can be priced as a separate delivery provided to an electric service customer through an additional (companion) meter and included within a single rate schedule for service. The Customer Charge would then be differentiated on the basis of the customers' end-use requirements (i.e., with and without space and/or water heating service). The differential would reflect the difference in the cost of having either one meter or two meters.[8]

The Customer Charge may also be differentiated on the basis of the electric service methodology. Overhead service lines are generally more economical than underground service lines, particularly for residential and small business loads. Utilities often require a *contribution in aid of construction* (CIAC) payment from the customer (or builder) in order to provide underground service when overhead service represents the standard, and the CIAC requirement would apply to both new construction and service conversions. The CIAC payment would reimburse the utility for the difference in capital costs between the two service methods, and the customer would then pay for electric service under the basic rate schedule applicable to overhead service customers. Alternatively, the Customer Charge can be differentiated by having a higher rate for underground service customers on a perpetual basis. For a faster recovery of the incremental capital, the Customer Charge rate for underground service can be increased further still and include a term (e.g., 96 months) after which the customer would pay the basic Customer Charge. However, since none of these approaches address the issue of the ongoing incremental costs of maintenance for the underground service lines, further consideration for differentiating the Customer Charge between overhead and underground services is warranted.

The Customer Charge can be differentiated to recognize the differences in costs that occur due to customer density. The number of customers per mile of distribution line is typically much greater in an urban area as compared to a rural setting. In urban and suburban neighborhoods with generally small-size lots, a single line transformer may be used to serve two, three, or more customers. In outlying areas where homes and buildings are often widely spaced apart, a transformer must be provided nearly exclusively for each customer. More poles are necessary to route primary voltage conductors over longer distances to reach all customers in the rural areas. In addition, the per customer cost of manual meter reading in widespread, low density areas would be considerably more than in urban areas due to the travel time to reach the meters. As a result, the Customer Charge may be specified separately for urban areas and for rural areas.

---

[8] The differentiation in Customer Charges can be represented as two or more discrete charges or as a single charge for the basic service plus adders for the special conditions.

Customer Charges based on customer density might also be applied in the case of multiple-tenant buildings, such as with apartments. The Customer Charge would be differentiated and specified on the number of dwelling units within the building. For example, the per customer Customer Charge would be less for customers residing in buildings of three or more units than for customers residing in single-family or duplex homes.

Some locations may experience a fair number of seasonal usage customers (summer occupants or winter occupants) whose dwellings remain connected for service throughout the entire year, but which utilize an insignificant amount of electricity, if any, in the off-season. To accommodate such cases, the Customer Charge can be differentiated between seasonal and year-round customers. The seasonal Customer Charge would be applied only during a specified period of months, and it would be stated as a higher monthly rate than the year-round Customer Charge.

A variety of ways to differentiate and represent a Customer Charge have been discussed thus far. Some of these methods can be combined within a single rate schedule in order to recognize multiple cost causation differences within the particular class of service. For example, Customer Charges differentiated by voltage type may be further distinguished by overhead and underground, urban and rural, or seasonal and year round. Theoretically, a rather complex matrix of Customer Charges could be produced to set prices more in line with costs given the various service features. However, such a detailed approach should be considered carefully in light of understandability and the ease of administration.

As discussed previously, customer-related costs are fixed costs incurred on an ongoing basis regardless of the amount of electricity utilized by the customer, i.e., these monthly costs are the same in a billing period in which a customer consumes zero kWh as they are in another billing period in which the customer consumes 10,000 kWh. However, for rate design purposes, the Customer Charge for a residential or small business customer class may be structured to include a small amount of energy use. For example, the Customer Charge may be stated as being $20.00 per month, including the first 25 kWh or less (other charges would apply to usage exceeding 25 kWh). Customers having a consistently low monthly energy utilization would realize the same base bill amount each billing period under this design. This type of structure is referred to as an *initial charge rate*, and it is in relatively limited use today. The initial charge rate, which originated in the early 1900s, was most likely introduced as a more acceptable means of ensuring the recovery of customer-related costs at a time when having a separate charge for "the readiness to serve" was politically unappealing, particularly given the low usage characteristics of the customers of that era.

Demand Charge Structures

The Demand Charge is intended to recover the investment and the fixed operation and maintenance costs associated with nonfuel production, transmission, and the portion of distribution costs not included in the Customer Charge (i.e., the capacity-related part). In a bundled rate, the recovery of these functionally-related costs are typically consolidated into a single per unit charge or set of charges in which it is not possible to specifically identify what amount of the Demand Charge relates to each of the functional components. On the other hand, an unbundled rate contains multiple Demand Charges which deliberately correlate with the functional components of the power system, and which are designed to recover their associated costs, as previously illustrated in Figure 8.1.

Demand Charges are applicable to the *billing demand* which must be determined in accordance with the provisions of the specific rate schedule. The metering criterion for the maximum demand is commonly a 15-minute or 30-minute interval measurement, although a 60-minute demand may be utilized in some special rate applications. The Demand Charge can be designed on the basis of either kW or kVA. Demand Charges based on kVA units are customarily reserved for large commercial and industrial customer rate classes since the associated metering is significantly more expensive as compared to kW metering. There are two fundamental ways to structure a Demand Charge: as an *explicit demand* charge and as an *hours use of demand* system of charges.

Explicit Demand Charges

The explicit demand method of structuring a Demand Charge results in a simple price per kW (kVA) that is applicable directly to all kW (kVA) of billing demand, e.g., $6.11 per kW for all kW of billing demand. In the late 1800s, Dr. John Hopkinson introduced a rate structure consisting of separate, explicit Demand and Energy Charges as an improvement over charging for all of the electric service on the basis of kWh usage only. The explicit demand structure is often referred to as the Hopkinson Demand rate. A variation of this structure is the Block Hopkinson rate in which the per unit Demand Charge decreases over a specified level of use, e.g., $7.09 per kVA for the first 250 kVA of billing demand plus $5.22 per kVA for all over 250 kVA of billing demand. One objective for "blocking" the Demand Charge is ensuring better recovery of any customer-related costs that are intended to be collected on the basis of the billing demand and not through a discrete Customer Charge. With either variety of the Hopkinson rate, the Energy Charges might also be established in a block format.

The Demand Charge may also be bifurcated for application to different measures of demand, even within a bundled rate structure. For example, one Demand Charge could be applicable to the maximum demand of the customer which occurs during the billing period, while a second Demand Charge could be applicable to the customer's demand at the hour of the utility's system peak load during the billing period. This structure provides price signals that offer a rudimentary demarcation between the production and transmission costs, which are more closely related to the system peak load, and the distribution costs which are more related to the customer's peak load. Bifurcated Demand Charges are also use in time-differentiated rate structures.

Time-Differentiated Demand Charge Structures

A rate structure may be time-differentiated by season of the year and/or by the time of day. With seasonal time differentiation, the monthly Demand Charge is simply set at a higher level during the months that are included in the season designated as being peak and at a lower level during the months included in the other seasons. Depending on the geographical location and the customer mix of weather-sensitive, end-use loads, the peak season can be distinctly either summer or winter. However, the months included in the peak season do not necessarily conform with the calendar definition of the seasons. Identification of the peak season months should be made by investigating several years of system load and local degree day history. For example, a seasonal transition month, such as October, is ordinarily considered to be a fall month in the northern hemisphere. But, over a period of several years, October might more typically exhibit the behavior of a summer month from the perspective of the system load. Once the months of the peak season have been identified, the remaining months of the year are most often designated collectively as being off peak. Figure 8.2 shows the monthly maximum system loads for a predominantly summer peaking utility.

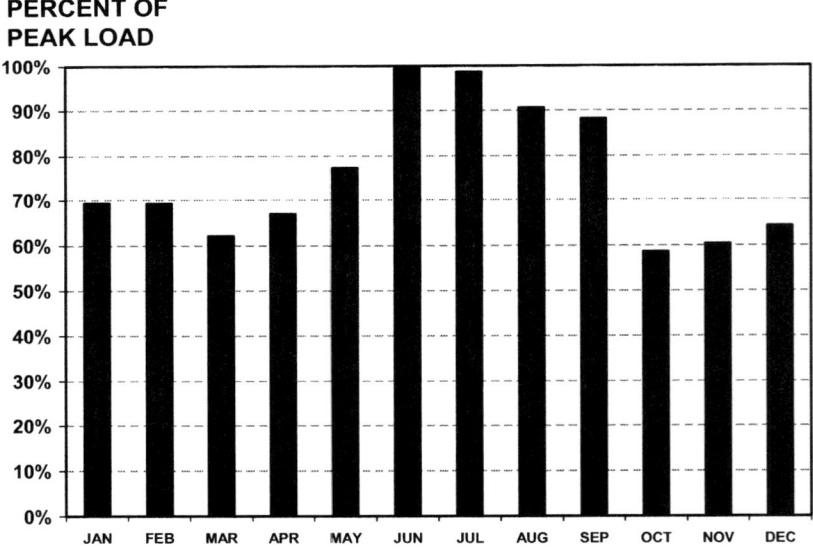

Figure 8.2 Monthly maximum system loads expressed as a percent of the highest hourly load of the year. In this example, the June through September period would be designated as the peak season for time-differentiated pricing.

The Rate Schedule: Pricing Methods and Risk

With seasonal pricing, a predominantly summer peaking utility would establish a relatively high priced Demand Charge in the summer months to reflect the high cost of capacity requirements driven by elevated air conditioning load. During the remaining months, the Demand Charge would be established at a lower level to promote efficient electric heating use in recognition of ample capacity available at that time. Oftentimes the off-peak period is designated as "nonsummer" or "nonwinter" and includes a number of the spring and fall months.

With time-of-day rate structures, on-peak and off-peak periods must be determined on a daily basis. The hours of the day considered to be on peak and off peak may vary significantly between seasons as the utilization patterns of weather-sensitive load devices shape the system load differently in the summer, winter, and spring/fall periods. Figure 8.3 illustrates a classical example of a weekly system load shape during the air conditioning season. The load is consistently cyclical from day to day, including Saturday and Sunday even though the magnitude of the load decreases substantially during the weekend.

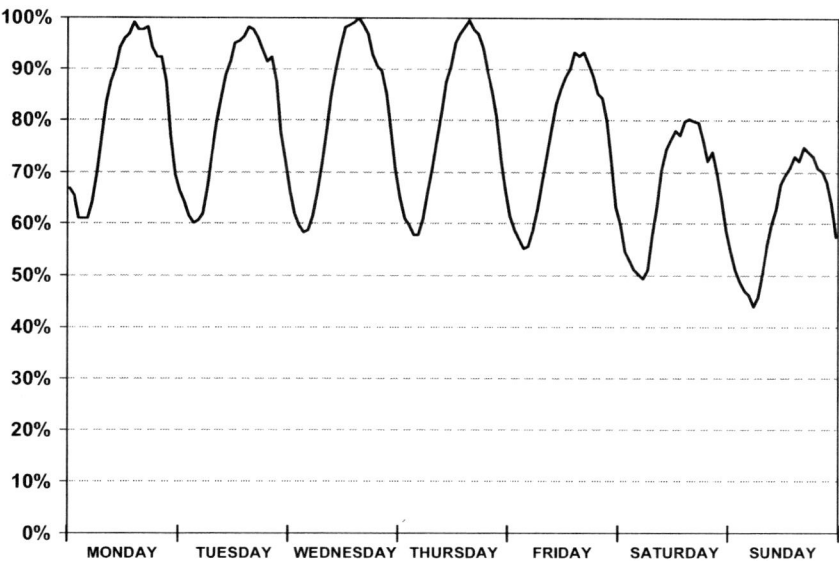

Figure 8.3 A system load for a typical week with each hour's load expressed as a percent of the peak hour occurring during the week. Monday through Friday contains uniform periods consisting of both on-peak and off-peak hours. All hours of Saturday and Sunday are typically classified as being off peak.

An analysis of the example load shape in Figure 8.3 would reveal, that during Monday through Friday, the system load routinely exceeds a specified threshold (e.g., 80% of the system peak load) during approximately the same hours each day. Thus, the probability of a system peak occurring in one of these hours is very high. Using an evaluation criterion such as this, a definition of the daily on-peak period can be made, and thus by default the off-peak period would likewise be defined. In this example, the load exceeds 80% of the peak hour nearly consistently between 9:00 AM and 10:00 PM. Typically Saturday and Sunday system loads are considerably lower in magnitude and are considered to be off peak in all hours. In addition, several legal holidays are found to be off peak as manufacturing and many businesses tend to curtail operations in recognition of key holidays.

During the heating season, it is common to observe a morning peak period and a separate evening peak period in the same day. Since daily outdoor temperatures tend to be colder at those times, the associated heating demand would be higher. In the middle part of the day, some of a building's thermal requirements are met by such factors as solar gain, increased occupancy levels, and equipment and machinery usage. Thus, the heating system may operate less frequently during the midday period, and a decrease in system load is likely to occur between the morning and late afternoon. As a result, the daily on-peak hours would be specified over two noncontiguous periods (e.g., 6:00 AM to 10:00 AM and 6:00 PM to 10:00 PM). If the market penetration of electric heating is low, then the entire heating season might be considered as off peak.

In order to bill for demand explicitly under a time-of-day rate, the revenue meter must be capable of registering specific types of load data, including the maximum on-peak period demand, the maximum off-peak period demand, and the monthly maximum demand (i.e., the same measure as for a non-time-differentiated rate). These demand measurements are illustrated in Figure 8.4 for a large customer. If the rate is also differentiated by season, the meter must be capable of switching between rating periods at the appropriate times of the year. This extra functionality causes a time-of-day demand meter to be rather more expensive than a basic indicating demand meter.

A time-of-day demand rate includes a bifurcated Demand Charge structure that applies to different measures of monthly demand, as identified in Figure 8.4. The configuration of the two Demand Charges for a time-of-day rate can be designed in a variety of ways, including:

- An On-peak Demand Charge with a monthly Maximum Demand Charge (The monthly maximum demand can occur during the on-peak period, in which case, both charges would apply to the same demand.);
- An On-peak Demand Charge with (or without) an Off-peak Demand Charge; and
- An On-peak Demand Charge with an Excess Demand Charge applicable to the amount of load that the off-peak demand exceeds the on-peak demand.

The Rate Schedule: Pricing Methods and Risk   281

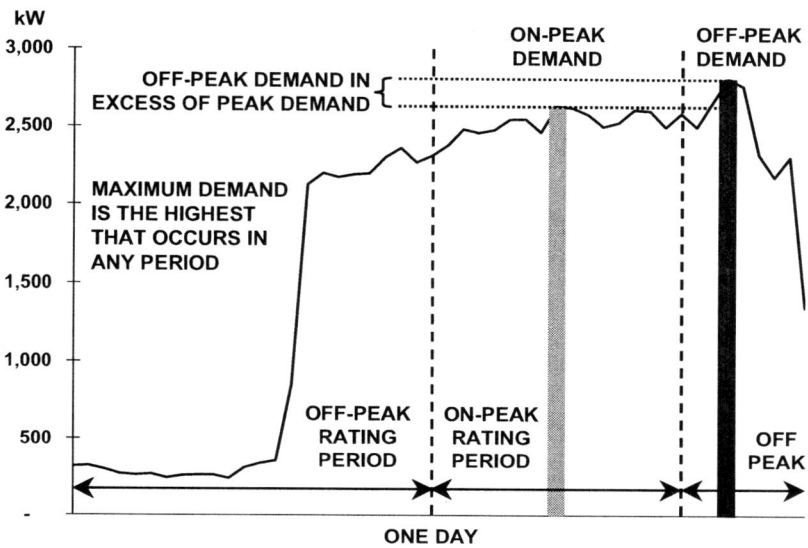

Figure 8.4 **Various monthly demand measurements for a customer served under a time-of-day based demand rate structure.** The meter must be capable of registering the maximum 15- or 30-minute maximum demand that occurs within the on-peak period. A measure of the maximum off-peak period demand and/or the monthly maximum demand is also needed for billing applications under one of the commonly used time-of-day rate structure formats.

As an alternative approach, a non time-differentiated demand rate schedule can be modified and used to bill on a time-of-day basis if a time-of-day meter is installed. Under this scheme, the billing demand would be selected as the greater of the maximum on-peak demand or a specified percentage (e.g., 50%) of the maximum off-peak demand, and the single Demand Charge of the rate schedule would then be applied. This method would allow for off-peak usage that is much greater than the on-peak usage without the customer incurring additional charges. At a 50% multiplier, the off-peak demand could be as much as twice the on-peak demand before the charges would increase further.

Hours Use of Demand Charges

During the same era that Dr. Hopkinson was promoting the explicit demand and energy component rate structure, Mr. Arthur Wright introduced an alternative demand rate design that functions on the basis of the customer's monthly load factor. To be more precise, the Wright rate form is structured around a demand and energy usage relationship known as the *hours use of demand*.

Recall that load factor is defined as the ratio of the average load over a prescribed period of time to the peak load which occurs during that same time period, as previously denoted in Equation (2.19). With rate and billing applications, load factor is based on the monthly energy and maximum demand usage characteristics of the customer as determined by

$$LF = \frac{\frac{E}{t}}{D_{MAX}} \qquad (8.1)$$

where LF = monthly load factor
E = kWh used during the month
t = hours in the month (average is 730 hours)
D_{MAX} = maximum monthly demand in kW (or kVA)

If the load factor is multiplied by the hours in the month (t), the result is known as the *hours use of demand* or

$$HUD = \frac{E}{D_{MAX}} \qquad (8.2)$$

where HUD = hours use of demand in hours
E = kWh used during the month
D_{MAX} = maximum monthly demand in kW (or kVA)

A visualization of the hours use of demand (HUD) relationship is shown in Figure 8.5. The upper graph shows a customer's load shape plotted over an average month of 730 hours. The maximum demand occurring during the month is 100 kW, and the energy is 54,000 kWh. The dotted line drawn horizontally across the graph at 100 kW references the usage that would have occurred if the maximum demand had been sustained across all hours of the month. The middle graph shows kWh utilized near the end of the month being shifted to the left-hand side of the graph in such a manner as to make up the difference between the maximum demand that could have occurred during each hour of the month (i.e., 100 kW) and the actual load that did occur. The shifting of kWh from the right to the left continues until an exact rectangle shape is attained. The area or kWh of the rectangle is equal to the area or kWh under the original load curve. As shown in the lower graph, the base of the triangle is 540 hours; thus, the hours use of the 100 kW demand is 540 hours. Five hundred and forty hours represents 73.97% of the total hours of the month (i.e., 730 hours), and this percentage is also equal to the monthly load factor.

The Rate Schedule: Pricing Methods and Risk

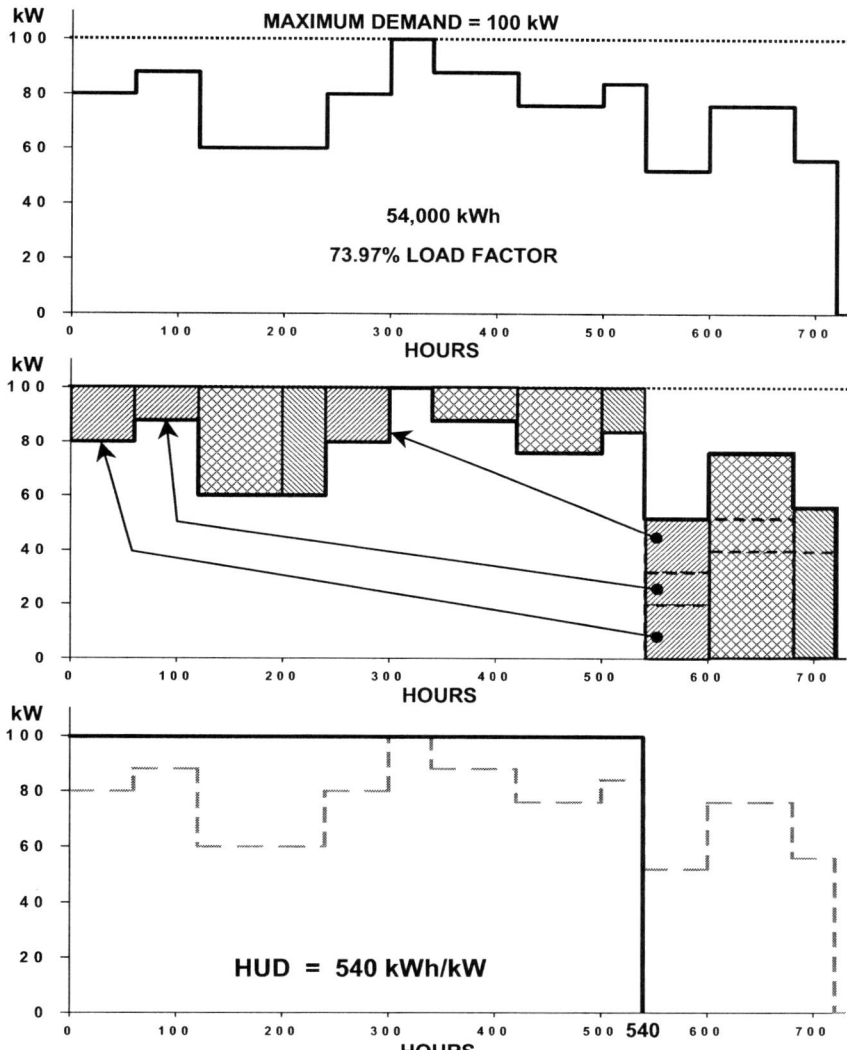

Figure 8.5 A visualization of Hours Use of Demand. The load shape shown in the upper graph is recast as a rectangle having a height equal to the monthly maximum demand of 100 kW and a base equal to 540 hours. Thus, the area of the rectangle is equal to the area under the original load curve, which is 54,000 kWh. The hours use of demand is the amount of time in hours that would transpire if the actual amount of energy used had been consumed at a continuous rate equal to the maximum demand from the very start of the month.

The Wright rate structure recovers demand-related costs on the basis of the customer's monthly HUD characteristics using Demand Charges that are expressed in terms of kWh units. The structure consists of two or more hours use of demand steps identified by kWh/kW (or kWh/kVA) breakpoints. An example of a three-step Wright rate structure is given by:

First 200 kWh per kW:
    All (associated) kWh @ 5.0¢/kWh
Next 250 kWh per kW:
    All (associated) kWh @ 3.0¢/kWh
Over 450 kWh per kW:
    All (associated) kWh @ 1.0¢/kWh

Although the Demand Charges are priced strictly on the basis of kWh units, a measure of maximum demand is required to determine how much of the kWh usage falls within each of the HUD rate steps. If a customer's HUD in a given month was less than or equal to 200 kWh/kW, the entire amount of kWh would be billed at 5.0¢/kWh. If instead the customer's HUD exceeded 450 kWh/kW, a portion of the total kWh would be allocated into each step of the rate. Thus, for a bill having 54,000 kWh and 100 kW, 20,000 kWh (200 kWh/kW × 100 kW) would be billed at 5.0¢/kWh, 25,000 kWh (250 kWh/kW × 100 kW) would be billed at 3.0¢/kWh, and 9,000 kWh (the remaining kWh) would be billed at 1.0¢/kWh. The bill amount would be $1,840.00, and the average rate realized would be 3.407¢/kWh, which is shown as Case A in Table 8.1. As the HUD is decreased or increased by changing either the total kWh with respect to the maximum kW (as in Cases B and C) or the maximum kW with respect to the total kWh (as in Cases D and E), the average rate can be observed to decrease or increase accordingly. The distribution of the kWh between the steps of the example rate structure for each of the five cases in Table 8.1 is shown in Figure 8.6.

Table 8.1 Examples of the Effect of Hours Use of Demand on the Average Rate

Case	kWh	Maximum Demand (kW)	HUD (kWh/kW)	LF	Bill	¢/kWh
A.	54,000	100	540	74.0%	$1,840.00	3.407
B.	43,200	100	432	59.2%	$1,696.00	3.926
C.	64,800	100	648	88.8%	$1,948.00	3.006
D.	54,000	120	450	61.6%	$2,100.00	3.889
E.	54,000	80	675	92.5%	$1,580.00	2.926

The Rate Schedule: Pricing Methods and Risk

	200 kWh/kW		450 kWh/kW
	5.0 ¢/kWh	3.0 ¢/kWh	1.0 ¢/kWh

A.
100 kW
54,000 kWh

20,000 kWh	25,000 kWh	9,000 kWh
37%	46%	17%

B.
100 kW
43,200 kWh

20,000 kWh	23,200 kWh
46%	54%

C.
100 kW
64,800 kWh

20,000 kWh	25,000 kWh	19,800 kWh
31%	39%	30%

D.
120 kW
54,000 kWh

24,000 kWh	30,000 kWh
44%	56%

E.
80 kW
54,000 kWh

16,000 kWh	20,000 kWh	18,000 kWh
30%	37%	33%

Figure 8.6 **Distribution of kWh by hours use of demand rate steps.** Illustrated are the five example load characteristics cases identified in Table 8.1. A change in the hours use of demand due to varying either the energy or demand results in a reallocation of the total kWh between the price steps of the Wright rate structure. The distribution percentages can also be viewed as weighting factors which when applied to their associated prices cause the average rate realized to increase or decrease as the hours use of demand increases or decreases.

Figure 8.6 illustrates how the average rate realized is affected by the distribution of the kWh between the steps of a Wright rate structure. For example, compare Case B to Case A. The maximum demand remains the same, but the total amount of energy used goes down thereby reducing the HUD. As a result, there is no longer any kWh billed in the 1.0¢/kWh step, and the proportions of the total kWh billed in the first two (higher priced) steps increase significantly. Even though the bill amount decreases from Case A to Case B due to a lesser amount of kWh, the average rate realized goes up because of the decline in the hours use. If, on the other hand, the demand remained the same while the energy increased, as in Case C, the bill amount would increase, but the average realization would go down.

In general, customers will experience some level of change from month to month in their load factors and thus their HUD characteristics will change due to variations in end-use load operations. Compared to Hopkinson's explicit Demand Charge rate, the more complex structure of Wright's hours use of demand rate results overall in more temperate incremental changes in the average rate realized as these monthly variations occur. This moderation is due to the multiple *implicit* Demand Charges embedded within the steps of the hours use rate. In reference to the three-step rate example from above, the first hours use step behaves just like an Energy Charge-only rate of 5.0¢/kWh. However, the second step of the rate behaves in the same fashion as a Hopkinson rate with a 3.0¢/kWh Energy Charge and a discrete $4.00/kW Demand Charge. Likewise, the last step of the rate behaves in the same fashion as a Hopkinson rate having a 1.0¢/kWh Energy Charge and a $13.00/kW Demand Charge. Thus, the example Wright rate is in essence a composite of three separate rate designs, each of which are applicable to a specific HUD range. The ability for the customer to naturally transition between these separate "subrate structures" within the Wright rate provides for the abated impact on the average rate realized due to monthly load factor changes. This phenomenon is further discussed and illustrated in Chapter 9.

The Wright rate structure may be designed on either a kW or a kVA demand basis. Even though the kVA may be of a different magnitude than the kW due to the power factor being other than unity, the application of kWh per kVA for an hours use of demand rate structure is mechanically the same. However, a new dimension is added in that customers having a poor power factor, despite a relatively high load factor as computed on the basis of the kW, will have a greater proportion of their total kWh allocated to the first step(s) of the rate where the unit prices are higher. In other words, the customer's kWh/kVA HUD would be less than the comparable kWh/kW HUD.

The discussion of Wright rate structures thus far has focused primarily on the Demand Charge component. Like a Hopkinson rate, the Wright rate structure would also need to provide for the recovery of energy-related costs. Since the cost of energy is defined on the basis of the volume of kWh used, each hours use step of the rate can be increased by a uniform rate per kWh to ensure recovery of the energy component.

With regard to time differentiation, the hours use of demand rate structure is often seasonally differentiated, in part or in whole, for the same reasons previously discussed. The hours use structure is not typically utilized within the design of a

The Rate Schedule: Pricing Methods and Risk 287

time-of-day demand rate application. However, it can be observed that the hours use rate can provide benefits similar to those attained with a time-of-day rate. Consider a customer whose maximum demand normally occurs during the system peak period. If the customer increases kWh consumption without a proportional increase in the maximum demand (e.g., by adding a second or third shift of operation), the incremental kWh would be applied to the lower-priced rate steps due to the increase in the customer's HUD. The average rate realized would decrease under the hours use rate in a manner similar to the decrease that would be realized under a time-of-day rate. Furthermore, a part or all of the additional kWh consumption could occur in the off-peak period. In this case, the hours use rate would emulate the basic operation of a time-of-day rate, and both the customer and the utility would realize benefits.

On the other hand, if a customer served under the hours use rate normally established the monthly maximum demand when the system was in the off-peak period, the addition of kWh consumption might actually occur during the on-peak period. The economic benefit to the customer would be the same as the previous customer who added consumption during the off-peak period because the hours use rate response is related directly to volumetric changes without a direct relationship to time. In this case, the hours use rate functioned counter to the time-of-day rate, and the customer would benefit at the expense of the utility.

Even considering load diversity between customers, most customers tend to set their maximum demands sometime within the system on-peak period (especially considering that the peak period may run from before noon to late in the evening). Thus, some time-related benefits are possible from an hours use of demand rate structure, even though it is not a perfect substitute for a well designed time-of-day rate.

Energy Charge Structures

The Energy Charge is intended to recover the variable costs associated with electric power production. Energy-related costs include generation fuels (coal, gas, petroleum products, and uranium); certain materials related to energy output, such as lime utilized for flue gas desulfurization; the energy-related portion of purchased power; and variable operation and maintenance expenses. Fuel is typically the most significant cost element of the energy-related component, and purchased power may also be a major operating cost. These costs are generally considered to be expenses that are passed directly through to consumers and upon which the utility does not earn a profit margin.

Practices vary regarding whether or not any of the fuel and purchased power costs are included directly in the base rate Energy Charges of the rate schedules. In some cases, a specified fixed amount of fuel and purchased power costs may be embedded in the Energy Charges, and an auxiliary rate schedule or rider is utilized as a means of billing for actual energy cost variations that occur. Many utilities have completely unbundled fuel and purchased power from the base rate Energy

Charge and recover all of these costs through the auxiliary schedule. Since fuel and purchased power costs are relatively volatile, periodic price revisions are generally required in order to prevent an unsatisfactory situation of overrecovery or underrecovery of revenue. Separating the pricing of these elements from the base billing mechanism facilitates the implementation of fuel and purchased power rate revisions without requiring changes to the base rate structure itself (which is typically more stable in terms of requiring price level revisions).

As previously noted, with a number of rate applications, some or all of the demand and/or customer components of cost are often recovered through Energy Charges based on kWh. The effective recovery of these costs is thus dependent on customers utilizing the assumed volume of energy upon which the Energy Charges are designed. Climatic conditions can be a particularly significant factor in the revenue performance of Energy Charges due to weather-sensitive end-use load devices. Inclusion of any demand and/or customer costs with the basic energy-related cost can be accomplished through a simple *flat rate* or *straight-line rate* design in which all kWh usage is billed at a uniform price per unit on an annual or seasonal basis or through more complex Energy Charge structures, which are based on a variety of different objectives.

Block Rate Energy Charge Structures
With a *block rate* design, the Energy Charge varies in conjunction with designated blocks of monthly energy consumption. There are two basic forms of the block rate: the *declining block* and the *inverted block* structures. Example of these two block rate structures are given by:

Declining Block Rate Structure		Inverted Block Rate Structure	
First 150 kWh	@ 5.6¢/kWh	First 300 kWh	@ 3.0¢/kWh
Next 350 kWh	@ 4.8¢/kWh	Next 450 kWh	@ 4.5¢/kWh
Next 500 kWh	@ 4.1¢/kWh	Next 750 kWh	@ 6.5¢/kWh
Excess kWh	@ 3.7¢/kWh	Excess kWh	@ 7.5¢/kWh

The declining block rate was originally developed as a means to promote electric applications. Electricity was first utilized for lighting in buildings, and thus rates were designed on the basis of area (e.g., floor space) or even on the number of rooms that were illuminated. As electrification of homes progressed and metering became common, the declining block rate provided a means for incrementally targeting various domestic electric end uses with discounted prices. Initially, the kWh blocks were very small in size and several in number since the goal was to entice customers to add such new devices as electric refrigeration, cooking, water heating, and other basic appliances to their homes. The *quick-break declining block* rate utilized a first step of only about 20 kWh so that the initial decline to lower priced kWh took place rapidly.

The Rate Schedule: Pricing Methods and Risk

Today, the declining block rate structure is still used to promote electric load additions. Competition from fossil fuels exists for a number of domestic end uses. Declining block rates are often used to encourage the installation of electric heating technologies, particularly by summer peaking utilities that have extra capacity available in the winter period. However, the contemporary block rate design tends to be more simplistic with fewer and larger kWh blocks compared to its early predecessor. As discussed in Chapter 3, the average consumption of electricity for basic, nonweather sensitive end uses, based on the U.S. residential sector as a whole, is generally in the range of 500 to 600 kWh per month. Basic use appliances, such as refrigerators, have long since reached their market saturations and no longer represent a major marketing opportunity for load growth. Thus, the first step of a block rate is often set at a kWh level which captures the expected average base use energy for the specific class of customers. The last step, or *tail block*, is typically set to recover the variable cost plus a minimal portion of fixed costs.

The inverted block rate structure is used to provide customers with a price signal that escalates with increasing levels of energy usage. This type of structure can be priced so as to deter excessive consumption and thus encourage customers to conserve energy. The inverted rate is commonly used for pricing when costs are high and capacity may be more constrained than at other times, particularly during the peak season. As structured, the inverted block rate places a preponderance of the cost recovery responsibility on the higher use customers that are served under the rate schedule.[9]

The inverted block and declining block structures may be combined within the same rate schedule in order to enhance seasonal price differentiation. The prices for the first step in each of the two structures may be set equal to each other in recognition that the base use energy is generally constant year round. The next steps are then inverted or declined depending on the season and generally target the weather-sensitive end uses. In addition, a flat rate design might also be incorporated for application in the spring and fall (shoulder) months as a transition between the inverted peak season rates and the declining off-peak season rates.

A number of block rate structure variations are utilized. For instance, a *stopper* mechanism can be use with a block rate to limit the average rate billed to either a minimum with a declining block structure or a maximum with an inverted block structure. An example of a stopper for a declining block rate is given by:

First 800 kWh	@	6.0¢/kWh
Next 1,200 kWh	@	4.0¢/kWh
Excess kWh	@	2.5¢/kWh

Except that the average charge for total kWh shall not be less than 3.5¢/kWh.

---

[9] The inverted block rate has also been used as a *lifeline rate* mechanism in which high use customers subsidize low use customers. The premise of such an application is that low use customers generally have a minimal income level.

With this structure, the average rate (i.e., the bill amount divided by total kWh usage) for any level of use from 1 to 800 kWh would be 6.0¢/kWh. From 801 to 1,200 kWh, the average rate would decline from 6.0¢/kWh to 4.8¢/kWh due to the lower price of the second step of the structure. Beyond 1,200 kWh, the average rate would continue to decline due to the yet lower price of the rate's tail step. Unimpeded, the average realization at a usage of 5,000 kWh would be 3.42¢/kWh. However, the stopper would be invoked once the average rate of the bill dropped below 3.5¢/kWh, which occurs at 4,600 kWh. As a result, all bills exceeding 4,600 kWh would be priced at 3.5¢/kWh for all kWh usage. This same basic approach can also be used to establish a maximum average rate for an inverted block structure. Rate stoppers can be applied year round or on a seasonal basis, or they can be seasonally differentiated.

A variety of other block rate modifications are used to directly target electric water heating and/or space heating usage.[10] The associated end-use devices are typically required to be permanently installed and to meet certain specifications in order to qualify for lower prices. For example, approved water heaters may be required to be of a minimum size (e.g., no less than 30 gallons) and to have upper and lower immersion resistance elements which are each thermostatically controlled and interlocked to prevent simultaneous operation. The lower element may be required to be located so as to heat the entire tank while the upper element must then be located to heat not more than the upper one-third of the tank. The rating of a single element may be limited in size (e.g., not to exceed 4,500 watts). Customers meeting such requirements would receive a reduced rate for a portion of their kWh consumption. A further lowering of the rate might be provided for customers that install a storage type water heater that is subject to control by means of a timing device that prevents operation during peak periods. Block rate modifications for promoting electric end-use loads are most often implemented using a declining block rate structure as the standard design. Such rate modifications often begin at a kWh level that is subsequent to the monthly energy usage that corresponds to the typical operation of the most basic appliances (e.g., lights, etc.).

Block Discounts

Block discounts provide for reduced energy prices in certain kWh blocks while maintaining the standard sizes of those blocks.[11] One or more of the blocks within the standard rate structure may be discounted for those customers meeting qualifying end-use load device requirements. There are two primary methods of discounting a block rate: the *full block discount* and the *partial block discount*.

---

[10] Alternatively, separately metered services can be used to provide energy directly and exclusively to water heating or space heating devices under a companion rate schedule or auxiliary billing provision.

[11] The revenue "loss" due to discounting a particular rate block is compensated through the pricing of the other standard unit rates within the structure such that the overall revenue requirement for the particular class of service is recoverable.

The Rate Schedule: Pricing Methods and Risk

For qualifying customers, the *full block discount* rate design merely modifies the price of an existing kWh block within the standard rate structure. An example of a full block discount is given by:

Standard Block Rate Structure

First 150 kWh	@	5.6¢/kWh
Next 350 kWh	@	4.8¢/kWh
Next 500 kWh	@	**4.1¢/kWh**
Excess kWh	@	3.7¢/kWh

[Boldface indicates blocks and/or price adjustments.]

With an approved water heater, the next 500 kWh block shall be billed at 3.5¢/kWh. [Note that the prices are now inverted between the 3rd and 4th blocks of the rate.]

With an approved space heater, all monthly kWh in excess of 1,000 kWh during the months designated as winter shall be billed at 2.5¢/kWh. [Note that this discount creates a seasonal rate structure applicable to customers having qualified space heating.]

With both an approved water heater and space heater, the next 500 kWh block shall be billed at 3.5¢/kWh and all monthly kWh in excess of 1,000 kWh during the months designated as winter shall be billed at 2.5¢/kWh.

The *partial block discount* rate design provides a lower price to only a portion of the monthly kWh in a particular block or blocks while the remaining portion of the kWh in those same blocks would be billed at their associated standard block prices. In reference to the standard block rate structure shown above, two examples of a partial block discount are given by:

With an approved water heater, 25% of the monthly kWh usage in each of the 3rd and 4th rate blocks shall be billed at 3.0¢/kWh.

With an approved water heater, 50% of the monthly kWh usage in excess of 400 kWh shall be billed at 3.2¢/kWh. [Note that only the last 100 kWh of the "Next 350 kWh" block receives the discounted price.]

Block Inserts

Block inserts actually alter the structure of the standard rate in order to provide lower unit prices for designated end-use load devices. A block insertion expands the steps of the rate structure and, in most cases, spreads the differential prices of the rate across a greater quantity of monthly kWh consumption. A variety of block insert rate structure designs can be used to promote electric water heating or space heating end uses.

The basic *block insert* places a new price block between two existing blocks of the standard rate without disturbing their sizes and prices. For example, a standard block rate might be modified for water heating by the following provision:

With an approved water heater, a 400 kWh block shall be inserted in the rate after the "Next 350 kWh @ 4.8¢/kWh" block, and all monthly kWh usage in that 400 kWh block shall be billed at 3.5¢/kWh.

The effective impacts on the standard rate structure due to the basic block insert are shown by the following rate comparison:

Standard Block Rate Structure		Modified Block Rate Structure	
First 150 kWh	@ 5.6¢/kWh	First 150 kWh	@ 5.6¢/kWh
Next 350 kWh	@ 4.8¢/kWh	Next 350 kWh	@ 4.8¢/kWh
Next 500 kWh	@ 4.1¢/kWh	**Next 400 kWh**	**@ 3.5¢/kWh**
Over 1,000 kWh	@ 3.7¢/kWh	Next 500 kWh	@ 4.1¢/kWh
		Over 1,400 kWh	@ 3.7¢/kWh

As with the water heating block discount discussed previously, the additional block is inserted in the standard rate structure at a monthly energy usage level that is commensurate with year-round electric water heating requirements.

The *tail block insert* creates an additional price block at the very end of the existing blocks of the standard rate structure. The tail block insert is directed at higher than average energy consumption levels, and thus this option is used typically for promoting weather-sensitive end-use applications. For example, the standard block rate might be modified for electric space heating by the following provision:

With an approved space heater, all monthly kWh in excess of 1,500 kWh during the months designated as winter shall be billed at 2.5¢/kWh.

The effective impacts on the standard rate structure due to the tail block insert are shown by the following rate comparison:

Standard Block Rate Structure		Modified Block Rate Structure	
First 150 kWh	@ 5.6¢/kWh	First 150 kWh	@ 5.6¢/kWh
Next 350 kWh	@ 4.8¢/kWh	Next 350 kWh	@ 4.8¢/kWh
Next 500 kWh	@ 4.1¢/kWh	Next 500 kWh	@ 4.1¢/kWh
Over 1,000 kWh	@ 3.7¢/kWh	**Next 500 kWh**	@ 3.7¢/kWh
		Over 1,500 kWh	**@ 2.5¢/kWh**

Note that in the modified block rate structure, the excess kWh block of the standard rate (i.e., the "Over 1,000 kWh @ 3.7¢/kWh") is now limited to only 500 kWh of monthly usage. Utilized in this manner, the tail block insert creates a seasonal rate application for customers having qualified end-use devices since the standard rate structure would be in effect during the nonwinter period.

The Rate Schedule: Pricing Methods and Risk 293

Another block insert option which affects the higher consumption levels of a rate structure is the *penultimate block insert* which interjects an additional price block immediately preceding the tail block of the standard rate. For example, the standard block rate might be modified for electric space heating by the following provision:

With an approved space heater, all monthly kWh usage between 1,000 kWh and 2,500 kWh during the months designated as winter shall be billed at 2.5¢/kWh. All kWh in excess of 2,500 kWh during the winter months shall be billed at 3.7¢/kWh.

The effective impacts on the standard rate structure due to the penultimate block insert are shown by the following rate comparison:

Standard Block Rate Structure			Modified Block Rate Structure		
First 150 kWh	@	5.6¢/kWh	First 150 kWh	@	5.6¢/kWh
Next 350 kWh	@	4.8¢/kWh	Next 350 kWh	@	4.8¢/kWh
Next 500 kWh	@	4.1¢/kWh	Next 500 kWh	@	4.1¢/kWh
Over 1,000 kWh	@	3.7¢/kWh	**Next 1,500 kWh**	@	**2.5¢/kWh**
			Over 2,500 kWh	@	**3.7¢/kWh**

Like the tail block insert, the penultimate block insert can be used to create a seasonal rate application in order to promote weather-sensitive end-use loads. However, since the price reduction resides in the penultimate rate block, the price then becomes inverted upon reaching the kWh usage in the tail block, as shown in the modified block structure above. An advantage of this methodology is that the rate structure not only promotes the end-use load but also telegraphs a higher price signal to the customer in the event that excessive utilization of the end-use device occurs (as during extreme weather conditions when the cost of energy supply tends to increase).

The basic *split block insert* option divides an existing kWh block within the standard rate structure into two or more new price blocks. In effect, the split block insert serves to discount the standard price of the existing kWh block but with added pricing flexibility due to the infusion of multiple prices. Thus, within the same overall range of kWh usage as the standard rate block, the split block insert allows for enhanced pricing transitions. For example, the standard block rate might be modified for electric water heating by the following provision:

With an approved water heater, the "Next 500 kWh @ 4.1¢/kWh" block shall be billed at 3.4¢/kWh for the monthly kWh usage between 500 kWh and 750 kWh and at 3.6¢/kWh for the monthly kWh usage between 750 kWh and 1,000 kWh.

The effective impact on the standard rate structure due to the split block insert is shown by the following rate comparison:

Standard Block Rate Structure			Modified Block Rate Structure		
First 150 kWh	@	5.6¢/kWh	First 150 kWh	@	5.6¢/kWh
Next 350 kWh	@	4.8¢/kWh	Next 350 kWh	@	4.8¢/kWh
Next 500 kWh	@	4.1¢/kWh	**Next 250 kWh**	@	**3.4¢/kWh**
Over 1,000 kWh	@	3.7¢/kWh	**Next 250 kWh**	@	**3.6¢/kWh**
			Over 1,000 kWh	@	3.7¢/kWh

In this example, the "Next 500 kWh @ 4.1¢/kWh" block of the standard rate structure has been wholly replaced with two equally sized blocks that span the original 500 kWh of consumption range. In addition, the new blocks have been set at different discounted price levels. Since two new prices have been inserted into the structure, it is therefore possible to invoke a relatively deep discount in the first half of the original 500 kWh range while providing a quicker return (in the latter half of the range) to the tail block price level. Recall that the basic block discount design would provide a uniform price decrease across the entire 500 kWh consumption range.

The split block insert methodology can also be used to resize an existing price block within the standard rate structure while inserting an ensuing price block that targets the range of monthly kWh usage intended for promoting an end-use load. A change in the number of kWh in the original block allows for inserting the new block at a level of use consistent with the usage characteristics of the end-use device. For instance, the standard block rate might be modified for electric water heating by the following provision:

With an approved water heater, all monthly kWh usage between 400 kWh and 800 kWh shall be billed at 3.5¢/kWh.

The effective impact on the standard rate structure due to this alternative split block insert is shown by the following rate comparison:

Standard Block Rate Structure			Modified Block Rate Structure		
First 150 kWh	@	5.6¢/kWh	First 150 kWh	@	5.6¢/kWh
Next 350 kWh	@	4.8¢/kWh	**Next 250 kWh**	@	4.8¢/kWh
Next 500 kWh	@	4.1¢/kWh	**Next 400 kWh**	@	**3.5¢/kWh**
Over 1,000 kWh	@	3.7¢/kWh	Next 500 kWh	@	4.1¢/kWh
			Over 1,300 kWh	@	3.7¢/kWh

Note that the "Next 350 kWh @ 4.8¢/kWh" block of the standard rate structure essentially has been split into two pieces. The first part of the original block has been resized to 250 kWh at the original price, and thus the latter 100 kWh of the block has been expanded and its price revised by inserting the "Next 400 kWh @ 3.5¢/kWh" price block.

The Rate Schedule: Pricing Methods and Risk

Floating Tail Block

Another type of rate structure used for promoting an electric end-use device is the *floating tail block*. A standard block rate structure is modified by designating that, for qualifying customers, a discounted price is applicable to a last fixed block of monthly kWh consumption. Typically, a minimum amount of energy usage (i.e., the monthly base use amount) must be billed under the initial block(s) of the standard rate structure before the discounted price applies. An example of a floating tail block structure for electric water heating applications is given by:

Standard Block Rate Structure

First 150 kWh	@	5.6¢/kWh
Next 350 kWh	@	4.8¢/kWh
Next 500 kWh	@	4.1¢/kWh
Over 1,000 kWh	@	3.7¢/kWh

Except that with an approved water heater, the last 300 kWh in excess of the first 500 kWh of monthly usage shall be billed at 3.5¢/kWh.

With this example, the first 500 kWh of monthly usage is always billed in the first two rate blocks. When the total monthly usage is in excess of 500 kWh but no more than 800 kWh, the discount price would apply to the last 300 kWh or less. Whenever the monthly consumption exceeds 800 kWh, the last 300 kWh of such usage becomes a floating tail block that slides further out in the rate structure thereby allowing the latter prices of the standard rate structure to apply in a normal fashion. Since the last 300 kWh block targets water heating, higher monthly usage due to weather-sensitive end-use loads would be priced by the last two blocks of the standard rate structure.

The size of the last energy block might also be varied as a function of the magnitude of the qualified end-use device to more accurately reflect its expected level of energy consumption. For example, the water heater provision might be stated in the following manner:

Except that with an approved water heater, the last [x] kWh in excess of the first 500 kWh of monthly usage shall be billed at 3.5¢/kWh, where

Minimum Tank Capacity	Last kWh Block
80 Gallon	300 kWh
100 Gallon	400 kWh
120 Gallon	500 kWh

Percentage-of-Use Energy Charge Structures

A different method for differentiating prices within an energy charge rate structure is accomplished through *percentage-of-use price blocks*. Unlike conventional fixed kWh block rate structures, the kWh dimensions of the percentage-of-use blocks vary in linear proportion to the total energy consumption, as exemplified by a three block rate structure in Figure 8.7. Typically a kWh threshold based on a minimum use requirement is included in the first and second blocks of the rate structure in order to maintain a higher average rate realization when the total monthly consumption is low. A basic three-block percentage-of-use rate structure could be defined by:

First 30% of the total kWh, but not less than 300 kWh	@	6.5¢/kWh
Next 50% of the total kWh, but not less than 500 kWh	@	4.3¢/kWh
Last 20% of the total kWh	@	2.1¢/kWh

As seen in Figure 8.7, the total energy usage must exceed the first block's 300 kWh minimum before any of the energy is billed in the second block. Furthermore, the total energy usage must exceed 800 kWh before any of the energy is billed in the last block. At a level of 1,000 kWh and higher, the energy is distributed into the three blocks exactly according to the stated block percentages.

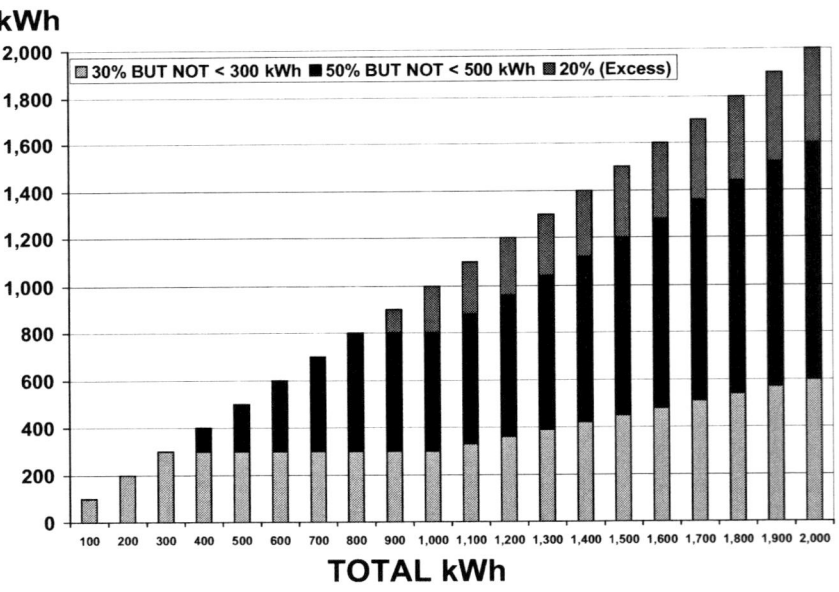

Figure 8.7 **Distribution of kWh within a three block percentage-of-use energy rate structure.** While the price blocks each have a fixed percentage of the total kWh usage, the volume of kWh contained in each block increases as the total kWh usage increases.

The Rate Schedule: Pricing Methods and Risk

One advantage of the percentage-of-use rate structure over the fixed block rate is that it recognizes the difference in magnitude of the kWh usage that would be associated with basic energy use when applied to a group of customers having a wide range of small to large loads. For instance, an average apartment or small house would have some lighting, a refrigerator, a television, and a few miscellaneous household appliances. In contrast, an average large house having more occupants would have more rooms and thus would likely have a greater lighting use and perhaps two refrigerators, a freezer, two or more televisions, etc. Thus, the monthly energy consumption of the nonweather-sensitive end-use loads of the average large house would be greater than for the average small house.

A percentage-of-use feature can also be incorporated in a standard declining block rate structure as a block insert in order to promote an end-use device, such as electric water heating, as demonstrated by:

Standard Block Rate Structure			Modified Block Rate Structure		
First 150 kWh	@	5.6¢/kWh	First 150 kWh	@	5.6¢/kWh
Next 350 kWh	@	4.8¢/kWh	Next 350 kWh	@	4.8¢/kWh
Next 500 kWh	@	4.1¢/kWh	**Next 300 kWh or up to**		
Over 1,000 kWh	@	3.7¢/kWh	**25% of the total kWh,**		
			whichever is less	@	**3.5¢/kWh**
			Next 500 kWh	@	4.1¢/kWh
			Over 1,300 kWh	@	3.7¢/kWh

The percentage requirement of the block insert results in a gradual incline in the amount of energy to which the discounted price applies, up to the maximum block size of 300 kWh. Thus, the rate structure is based on the premise that water heater energy usage would vary as a function of the total kWh consumption.

Time-Differentiated Energy Charge Structures

As with demand charge structures, the energy charge structure can be time differentiated in various ways. Either a flat or a block rate structure can be varied by season. A basic watt-hour meter is sufficient for administering seasonal billing. On the other hand, time-of-use (TOU) or *time-of-day* (TOD) energy rate structures require more sophisticated metering in order to capture energy usage by rating periods. A meter which registers kWh consumption by just the defined on-peak period and in total is adequate for billing a basic time-of-day rate since the off-peak usage is simply the difference between the total and the on-peak registrations.

TOD energy rate structures are time-differentiated versions of the flat rate structure and can be designed with either two periods or three periods. When the normal system daily load excursions between low and high load conditions occur rather rapidly, a two period on-peak, off-peak TOD structure is generally sufficient. However, when the system load shape typically is more level with longer transitions between the low- and high-load conditions, a three-period TOD structure which incorporates an intermediate-peak or shoulder-peak price period might be more fitting. An example of a two-period and a three-period TOD rate structure is given by:

Two-Period TOD Rate Structure		Three-Period TOD Rate Structure	
On-Peak kWh	@ 10.0¢/kWh	On-Peak kWh	@ 12.5¢/kWh
Off-Peak kWh	@ 2.5¢/kWh	Shoulder-Peak kWh	@ 5.3¢/kWh
		Off-Peak kWh	@ 2.0¢/kWh

TOD energy rate structures may be further time differentiated by season generally by having a higher on-peak price in the on-peak season as compared to the off-peak season. Depending on the nature of the system load, the off-peak season could even be defined as having only a shoulder-peak and/or an off-peak pricing structure. When coupled with time-differentiated demand charges, the energy charge could be structured with either time-variant prices or a flat time-invariant price. When a seasonally differentiated TOD rate is utilized, the meter must be programmable so that it can recognize any change in the on-peak and off-peak rating periods when moving between seasons.

The ultimate time differentiation of an energy charge is accomplished through a variable pricing method, such as real-time pricing (RTP) or spot market pricing. Another variable pricing structure is *critical peak pricing* (CPP), which is an energy-only TOD rate with an elevated kWh price adder that is applied at such times as when the system experiences critical loading/high cost conditions.

Rate Structure Combinations

Rate structures are designed to recover the basic fixed and variable cost elements related to the customer, demand, and energy components of electric service. Five fundamental rate structure types have been discussed previously, including:

- Flat Energy Rate
- Fixed Block Energy Rate
- Percentage-of-Use Energy Rate
- Wright Demand Rate
- Hopkinson Demand Rate

Excluding the Hopkinson demand rate, the first four rate structures have prices which are based on units of kWh consumption (although a billing demand is also required to apply the Wright rate structure for billing). As such, these four structures can be utilized in a self-contained fashion for the full cost recovery of the customer, demand, and energy components, except that all or a portion of the customer-related costs are typically recovered through the addition of a discrete Customer Charge. The basic Hopkinson rate, on the other hand, is an explicit

The Rate Schedule: Pricing Methods and Risk   299

Demand Charge structure which further requires an associated Energy Charge structure for recovery of variable costs. As noted previously, some of the demand cost recovery may be accomplished within the Energy Charges by applying a degree of demand rate tilt in the design of the kWh prices. The accompanying Energy Charge structure in the Hopkinson demand rate could take the form of a flat, block, or percentage-of-use rate design (although the percentage-of-use Energy Charge structure is generally less common with Hopkinson rate designs).

Alternatively, an explicit demand charge is often combined with an HUD structure in order to form a *Wright-Hopkinson* rate design, as exemplified by:

Wright-Hopkinson Rate Structure		Equivalent Wright Rate Structure	
All kW of Demand @ $4.32/kW			
First 200 kWh per kW:		First 200 kWh per kW:	
All associated kWh	@ 2.9¢/kWh	All associated kWh	@ 4.4¢/kWh
Next 250 kWh per kW:		Next 250 kWh per kW:	
All associated kWh	@ 1.8¢/kWh	All associated kWh	@ 2.3¢/kWh
Over 450 kWh per kW:		Over 450 kWh per kW:	
All associated kWh	@ 0.7¢/kWh	All associated kWh	@ 1.5¢/kWh

When compared to an equivalent Wright rate design, the Wright-Hopkinson structure's energy prices within its HUD blocks are noted to be lower since a portion of the otherwise embedded demand costs are recovered through the explicit Demand Charge. The comparative rates above are designed to recover the same amount of annual revenue based on a specific set of billing determinants used for their design.

Compared to the pure Wright rate design, the Wright-Hopkinson structure favors, in terms of the average rate realized, customers that have higher monthly load factors because of its explicit Demand Charge component. Again, this consequence is a function of the degree of demand rate tilt that is in effect. As the explicit Demand Charge is increased, the magnitude of and the differential between the HUD energy prices diminish, and the Wright-Hopkinson design would continue to approach a pure Hopkinson structure. In cases where a Wright-Hopkinson rate design is applied to customers having mostly mid-range to high load factors, the price per kWh of the first hours use step could actually be set at zero by properly elevating the Demand Charge. In other words, the monthly Energy Charge for all kWh falling within the first hours use step of the rate would be recovered exclusively on the basis of the kW (or kVA) billing demand.

A Wright rate structure can also include kWh blocks within one or more of its HUD steps in order to provide a finer delineation of prices across the spectrum of energy and demand usage characteristics. These blocks can be based on fixed and/or percentage-of-use structures. Consider the following general service rate structure example:

Nondemand Metered General Service

First 30% of total kWh, but not less than 300 kWh
    First 150 kWh   @   6.00¢/kWh
    Over 150 kWh   @   5.05¢/kWh

Next 50% of total kWh, but not less than 500 kWh
    First 1,000 kWh   @   4.05¢/kWh
    Next 2,000 kWh   @   2.55¢/kWh
    Over 3,000 kWh   @   1.85¢/kWh *

Next 20% of total kWh
    First 25,000 kWh   @   1.85¢/kWh *
    Over 25,000 kWh   @   1.65¢/kWh

* The kWh billed at 1.85¢/kWh shall not exceed 25,000 kWh.

Demand Metered General Service

First 60 kWh per kW, but not less than 300 kWh
    First 150 kWh   @   6.00¢/kWh
    Over 150 kWh   @   5.05¢/kWh

Next 100 kWh per kW, but not less than 500 kWh
    First 1,000 kWh   @   4.05¢/kWh
    Next 2,000 kWh   @   2.55¢/kWh
    Over 3,000 kWh   @   1.85¢/kWh *

Next 40 kWh per kW, but not less than 300 kWh
    First 25,000 kWh   @   1.85¢/kWh *
    Over 25,000 kWh   @   1.65¢/kWh

* The kWh billed at 1.85¢/kWh shall not exceed 25,000 kWh.

Over 200 kWh per kW and 1,000 kWh - Summer
    First 8,000 kWh   @   1.55¢/kWh
    Over 8,000 kWh   @   1.50¢/kWh

Over 200 kWh per kW and 1,000 kWh – Nonsummer
    First 8,000 kWh   @   1.55¢/kWh
    Over 8,000 kWh   @   1.45¢/kWh

    The complexity observed in this example rate structure exists because of the multiple objectives that were considered in its design. This type of rate is applicable to customer classes consisting of a large number of customers having a wide range of demand and energy characteristics. Smaller size customers are billed under the upper portion of the rate structure which is based strictly on monthly kWh consumption and requires only a simple watt-hour meter. This part of the rate structure is composed of a combination of fixed kWh and percentage-of-use price blocks in which all of the incremental prices decline. Larger-size customers are billed under the lower portion of the rate structure which requires a monthly maximum demand measurement. This part of the rate structure is a four-step Wright rate structure having multiple fixed kWh blocks in each step. Furthermore, the last HUD step is seasonally differentiated with summer being the obvious peak season.

While this example represents somewhat of an extreme rate design, rate structures for commercial and industrial customer classes are often found to be fairly complex. When placed in large groups, these customers exhibit a considerable amount of diversity in terms of size and usage characteristics. Given the detailed pricing configuration of the above example general service rate, it could be effectively applied to as much as 90% or more of a utility's commercial and industrial customer base. Thus, the use of a complex rate structure has the advantage of minimizing the total number of rate schedules needed for billing electric service. To apply a much simpler rate structure to such a large and diverse group would cause significant cost recovery inequities between customers within the group. As a result, the practice of using simplistic rate structures necessitates that a greater number of rate schedules be established for smaller groups of customers and whose individual rate structure designs are "tuned" to the more homogeneous characteristics of those smaller customer groups.

Another form of a combined rate structure, which can be applied effectively to a customer class having fairly diverse usage characteristics but with a much less complicated configuration than the general service rate structure discussed above, is known as a *block extender* or *stretcher block*. An example of a basic block extender is given by:

First 500 kWh*	@	5.1¢/kWh
Next 1,000 kWh	@	4.2¢/kWh
Next 3,500 kWh	@	3.1¢/kWh
Excess kWh	@	1.2¢/kWh

* *Whenever the billing demand exceeds 15 kW, the first energy block shall be increased by an amount equal to 50 kWh for each kW in excess of 15 kW.*

In essence, the block extender operates as a simple declining block rate structure when the monthly load requirement is below a specified level of billing demand and as a Wright demand rate structure when the load is above that prescribed demand. As seen in the example above, when the demand is greater than 15 kW, the first fixed block becomes a variable block, and the number of kWh priced at 5.1¢/kWh becomes a function of the kW billing demand. If a customer's monthly maximum load requirement moved above or below the threshold demand, the billing application would automatically migrate between the energy-only and the HUD portions of the rate structure. Customers whose maximum load requirements are expected at all times to be less than the threshold demand could be served under this rate structure with a simple watt-hour meter.

Modification of the basic fixed block rate structure to attain a block extender configuration can be accomplished in alternative ways. The first block can be extended (as previously shown), or a block subsequent to the first block could be extended. Multiple blocks could also be extended and at different kWh per kW increments. The kWh per kW extender could also be varied with increasing demand, for example:

Whenever the billing demand exceeds 15 kW, the second energy block shall be increased by an amount equal to (a) 75 kWh per kW for the first 85 kW in excess of 15 kW, (b) 50 kWh per kW for the next 150 kW in excess of 100 kW, and (c) 35 kWh per kW in excess of 250 kW.

A number of combined rate structures have been developed and utilized in practice by electric utilities, the more common of which have been discussed. When properly interpreted, many rate applications are found to be similar to each other in function, but which may also utilize alternative approaches to achieve the same end. The "art of rate design" allows for variations in bill calculation methodologies, rate structure terminology, and the overall presentation of the rate structures within the electric tariff.

Rate Structure Risk

Rate structures have an inherent risk from the standpoint of both the customer and the service provider. A customer's risk can be viewed in terms of the average price per kWh realized and the sensitivity of that price to change as usage characteristics vary. The service provider's risk relates to the ability to recover the costs of service adequately, particularly when customers' actual usage characteristics deviate from the volumes upon which the rate structure was originally designed. Aside from normally expected usage cycles, actual billing units are impacted primarily by atypical weather and economic conditions.

The level of risk varies significantly depending on a rate structure's methodology for recovering fixed and variable costs, including its degree of time variability. A matrix of different rate structure designs is shown in Figure 8.8. Each of these alternative structures have been designed to recover the same amount of annual revenue given a common set of billing determinants. A progression of basic rate structures is shown from left to right along with their time varying derivatives (where applicable) from top to bottom. From a customer's perspective, the annual flat energy rate presents the least amount of risk while the Hopkinson demand rate with hourly varying energy prices affords the greatest amount of risk.

The Rate Schedule: Pricing Methods and Risk

↑ INCREASING RISK TO CUSTOMER

	TYPE OF RATE STRUCTURE				
	FLAT ENERGY RATE	BLOCK ENERGY RATE	WRIGHT DEMAND RATE	WRIGHT-HOPKINSON RATE	HOPKINSON DEMAND RATE
TIME-INVARIANT STRUCTURES					
ANNUAL				$2.50/kW	$4.69/kW
	ALL kWh 6.26¢/kWh	1ˢᵀ 500 kWh 7.62¢/kWh	1ˢᵀ 250 kWh/kW 6.36¢/kWh	1ˢᵗ 250 kW h/kW 3.90¢/kWh	ALL kWh 1.75¢/kWh
		OVER 500 kWh 6.09¢/kWh	OVER 250 kWh/kW 1.99¢/kWh	OVER 250 kWh/kW 2.03¢/kWh	
TIME-VARIANT STRUCTURES					
SEASONAL	------- SUMMER -------				
				$2.75/kW	$5.03/kW
	ALL kWh 6.86¢/kWh	1ˢᵀ 500 kWh 6.27¢/kWh	1ˢᵀ 250 kWh/kW 6.56¢/kWh	1ˢᵗ 250 kW h/kW 3.90¢/kWh	ALL kWh 1.95¢/kWh
		OVER 500 kWh 7.01¢/kWh	OVER 250 kWh/kW 3.53¢/kWh	OVER 250 kWh/kW 2.20¢/kWh	
	------- NONSUMMER -------				
				$2.25/kW	$4.24/kW
	ALL kWh 5.46¢/kWh	1ˢᵀ 500 kWh 6.27¢/kWh	1ˢᵀ 250 kWh/kW 5.96¢/kWh	1ˢᵗ 250 kW h/kW 3.90¢/kWh	ALL kWh 1.61¢/kWh
		OVER 500 kWh 5.13¢/kWh	OVER 250 kWh/kW 2.75¢/kWh	OVER 250 kWh/kW 1.72¢/kWh	
TIME OF DAY:					
ANNUAL				$2.89/kW_MAX $3.01/k W_ON-PEAK	
	ON-PEAK kWh 10.20¢/kWh			ON-PEAK kWh 0.75¢/kWh	
	OFF-PEAK kWh 3.35¢/kWh			OFF-PEAK kWh 0.67¢/kWh	
SEASONAL	------- SUMMER -------				
				$2.89/kW_MAX $3.77/k W_ON-PEAK	
	ON-PEAK kWh 11.44¢/kWh			ON-PEAK kWh 0.78¢/kWh	
	OFF-PEAK kWh 3.35¢/kWh			OFF-PEAK kWh 0.67¢/kWh	
	------- NONSUMMER -------				
				$2.89/kW_MAX $2.22/k W_ON-PEAK	
	ON-PEAK kWh 8.30¢/kWh			ON-PEAK kWh 0.71¢/kWh	
	OFF-PEAK kWh 3.35¢/kWh			OFF-PEAK kWh 0.67¢/kWh	
HOURLY				$3.00/kW	
	ALL kWh VARIES			ALL kWh VARIES	

Figure 8.8 **A rate structure risk matrix.** A customer's risk in terms of price volatility increases with respect to the rate structure type and its degree of time variability.

One method for demonstrating the concept of rate structure risk is by evaluating the amount that the average rate realized can vary as customer usage characteristics change from month to month. The average rate graphs shown in Figure 8.9 were derived from the five fundamental, time-invariant rate structures shown across the top row in Figure 8.8, based on a monthly billing demand of 20 kW. An assessment of the change in the average rate realized for each of these rate designs across a range of hours use of the billing demand allows for rank ordering the comparable rate designs in terms of the risk of price variability. Obviously, no price variability exists with the flat energy rate of 6.25¢/kWh; the slope of the curve is zero. Thus with the flat rate design, the energy usage and the HUD with respect to the 20 kW of billing demand could change dramatically from one month to the next, and the average rate per kWh would remain completely unchanged.

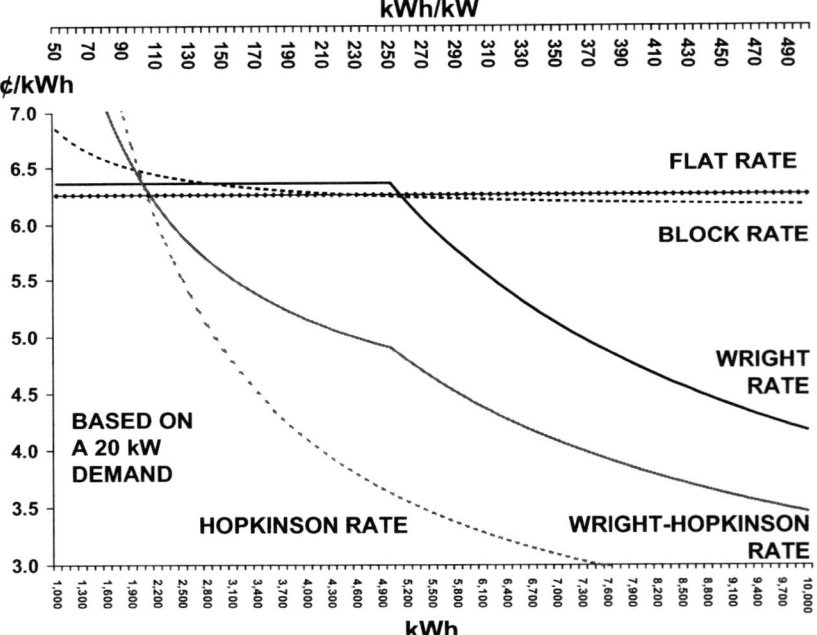

Figure 8.9 Rate structure risk comparisons. A graph of the average rate realized for each of the time-invariant rate structures portrayed in Figure 8.8 based on a billing demand of 20 kW. The average rates are plotted over the range of 1,000 to 10,000 kWh (50 to 500 HUD). The rate structures can be rank ordered for risk in terms of the price variability which occurs over the designated range of use. The flat energy rate structure exhibits zero price variability, while the Hopkinson demand rate reveals the most price variability.

The Rate Schedule: Pricing Methods and Risk 305

The average rate per kWh of the block energy rate design varies only a minimal amount across the range of use except when the kWh consumption is at the lower end of the range. The slope of the curve in the lower use range is a bit more negative compared to the higher use range. The nonlinearity of the block rate structure's average realization curve is caused by the fact that the rate is blocked. Within the first block, which is 7.62¢/kWh for the first 500 kWh, the average rate realized does not vary; it has the same characteristic as the flat energy rate. (Note that this block of use does not fall within the range of use plotted in the graph of Figure 8.9.) However, once the monthly energy usage exceeds the first block, that block becomes a fixed charge in the amount of $38.10. The average rate anywhere in the range shown in Figure 8.9 (i.e., 1,000 to 10,000 kWh) is determined by dividing the $38.10 by the kWh usage of interest and then adding the tail block rate of 6.09¢/kWh. Thus within the lower usage range, the average rate realized will be greater than it is within the higher usage range, as evidenced by the plot of the curve. Therefore, the fixed charge aspect of the block rate introduces price variability across the range of use shown.[12]

The comparable Wright demand rate behaves in a manner similar to the block energy rate but with considerably more price variability over the range of use shown. The first HUD step is priced at 6.36¢/kWh for 250 kWh per kW, which for a 20 kW billing demand is equivalent to a fixed block of 5,000 kWh and thus a fixed charge of $318.00 (which is an order of magnitude greater than the block energy rate's equivalent fixed charge). Usage in excess of 250 kWh per kW is priced significantly lower at 1.99¢/kWh. When the average rate realized is computed for consumption in the upper end of the range, it is found to decline rapidly as the effect of the $318.00 fixed charge is quickly diminished on a per kWh basis. The slope of the average rate curve becomes substantially negative. The price variability over the range shown is much greater than it is for the block energy rate. With service under this Wright demand rate structure, customers normally operating in excess of 250 kWh per kW whose energy use varied by just a few hundred kWh from one month to the next would realize a notable change in the average rate per kWh.

The Wright-Hopkinson rate design extracts a portion of the demand-related costs from the kWh charges of its comparable Wright rate structure to form an explicit Demand Charge rate component of $2.50/kW. Application of this Demand Charge to a billing demand of 20 kW yields a fixed charge of $50.00. If the 20 kW demand was based on a 15 minute demand interval, the associated kWh for the month could theoretically be as low as 5 kWh (5 kWh ÷ 0.25 hr = 20 kW). At 5 kWh, the average realization of just the fixed charge demand component would be $10.00 per kWh. Combined with the much lower energy price of

---

[12] Price variability would be introduced (even in a flat energy rate structure) merely by including a Customer Charge, which is by nature a fixed charge. Customer Charges have been excluded from this example in order to demonstrate the price variability attributed to the monthly consumption characteristics.

3.90¢/kWh in the first HUD block, the overall average rate realized drops very rapidly (the slope is highly negative in the low use range). Over the entire range of use shown, the price variability of the Wright-Hopkinson rate design is found to be much greater than for its comparable Wright rate structure. In the lower end of the range of use, a monthly deviation of a few hundred kWh would result in a significant change in the average rate per kWh.

The final rate design considered is the comparable Hopkinson demand rate structure with a $4.69/kW Demand Charge and a flat 1.75¢/kWh Energy Charge. The fixed charge due to applying the Demand Charge to the 20 kW of billing demand is $93.80, which at a theoretical monthly consumption of 5 kWh would yield an average realization of $18.76 per kWh. As seen in Figure 8.9, the Hopkinson rate design has the greatest price variability over the range of use shown of any of the five comparable rate structures. It has a negative slope in the lower end of the usage range that approaches a vertical asymptote.

This investigation of the characteristics of the average rate curves developed for the five different example rate structures illustrates how price variability is greatly impacted as a result of the design methodology utilized to recover fixed and variable costs. Creating a fixed charge component through either a discrete Demand Charge, an HUD price step, or a kWh price block causes the average rate per kWh to vary as usage changes. This variation depends on the magnitude of the fixed charge component relative to the variable charge component(s), i.e., the kWh prices. Once again, the intrinsic price variability of a rate structure is a function of the applied demand rate tilt used in its design.

Another way to view rate structure risk is to consider the billing consequences that occur when different rate designs are applied to two customers having significantly different load factors. Figure 8.10 illustrates the load profiles for two customers for the day in which each customer establishes the maximum demand for the month (note that each customer's maximum demand could occur on different calendar days). Both customers have a maximum demand of 20 kW, but Customer #1's monthly energy usage is 7,500 kWh while Customer #2's monthly energy usage is only 1,500 kWh. Customer #1 has an HUD of 75 hours (or 10.3% load factor), and Customer #2 has an HUD of 375 hours (or 51.4% load factor). Table 8.2 shows the bill amounts and the average realization for each customer's billing under the time-invariant versions of the Flat energy rate, the Wright demand rate, and the Hopkinson demand rate which were presented previously in Figure 8.8.

Consider first the effect of these rate structures on Customer #1, a relatively high load factor customer. The Flat rate design results in a high average realization of 6.26¢/kWh since the demand component of costs are fully tilted into the kWh Energy Charge. The average rate is reduced to 4.90¢/kWh with the Wright rate structure because of the fixed charge component that is created by the 250 kWh per kW HUD step, i.e., the implicit demand charge. The average rate is further reduced to 3.00¢/kWh under the Hopkinson demand rate structure which has a relatively high explicit Demand Charge and a relatively low Energy Charge.

The Rate Schedule: Pricing Methods and Risk

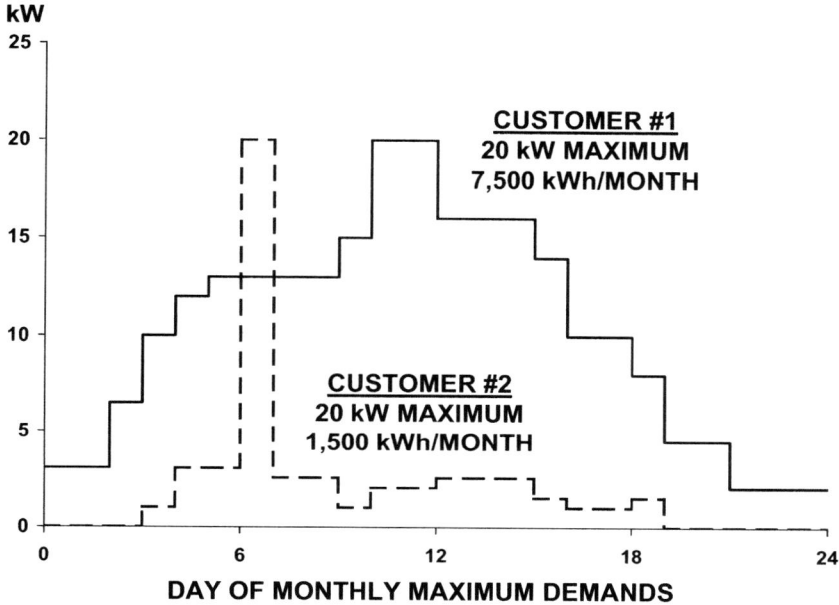

Figure 8.10 Load profiles for the day in which two different customers set their maximum demands. Both customers have the same maximum demand; however, Customer #1 has a much higher monthly load factor than Customer #2.

Table 8.2 Billing of Example Customers under Various Rate Structure Designs

	Customer #1		Customer #2	
	Bill Amount	Average Rate per kWh	Bill Amount	Average Rate per kWh
Flat	$469.50	6.26¢	$93.90	6.26¢
Wright	$367.75	4.90¢	$95.40	6.36¢
Hopkinson	$225.05	3.00¢	$120.05	8.00¢

The obvious preference of Customer #1 would be the Hopkinson rate structure. Higher load factor customers have a large volume of kWh with respect to the maximum demand. The average realization under this rate structure is lowest because the revenue associated with the explicit Demand Charge is spread over a large volume of kWh, and this per unit amount is then consolidated with an already low kWh charge to determine the overall average rate. A *high load factor rate* design utilizes the structuring principle of a high Demand Charge and a low Energy Charge to recognize the relationship between demand and energy for customers having high monthly load factors. The separate Demand and Energy Charges minimize the shifting of revenue responsibility, between customers with differing load factors, that would occur under other structures that incorporate a higher degree of demand rate tilt.[13] A high load factor customer served under a Flat rate structure would have an incentive to seek other alternatives in lieu of service under that rate, perhaps including self-generation.

Consider now the case of Customer #2. Under the Hopkinson rate design, Customer #2 would be responsible for the same amount of demand-related revenue as Customer #1, since both have the same maximum demand of 20 kW. But Customer #2 has considerably less monthly energy usage. As a result, Customer #2's overall average realization would be 8.00¢/kWh. Given such a low load factor and low monthly energy usage, the Flat rate design, at 6.26¢/kWh, presents the lowest average realization to Customer #2. Under the Hopkinson rate design, Customer #2 would have an incentive to peak shave with generation or other demand control measures in order to reduce the magnitude of the average rate realized. The result would be a severe reduction in the explicit demand-related revenue, and it is this revenue component which yields the bulk of return, i.e., profit, to the service provider.

These examples illustrate the economic interaction between a customer's load factor and the configuration of the demand and energy charges within a rate structure. A simple one-size-fits-all approach to rate structuring would seriously elevate the level of risk to both the customers and the service provider. The prudent tariff practice is to group commercial and industrial customers having similar load factors into classes of service under a series of different rate structures which are designed in correspondence with the characteristics of the classes. The Flat and Block energy rate structures fit best to low load factor customer classes, while the Hopkinson demand rate (considering the degree of rate tilt) fits best to medium and high load factor customer classes. Given the typical utility's commercial and industrial customer base, generally three or more Hopkinson demand rate classes would be required to adequately serve the medium and high load factor customers.

---

[13] A 100% load factor customer would only realize the lowest average rate per kWh possible if the Demand Charge included all fixed costs of electric service while the Energy Charge included only the variable costs of electric service. If any amount of demand rate tilt is incorporated in the kWh prices, customers having load factors less than the class average would benefit at the expense of those customers with load factors greater than the class average. High load factor rate classes often include customers with monthly load factors ranging between about 75% to 100%, and thus the associated Hopkinson rate structure requires some degree of demand rate tilt to compensate for this range.

The Rate Schedule: Pricing Methods and Risk

Depending on the number of HUD steps used, the Wright and Wright-Hopkinson demand rate designs can be effectively applied across a customer class having a much wider variation in load factor characteristics. The Wright design behaves as both an Energy Charge only rate for customers with low load factors and as a partially tilted demand rate for customers having a load factor (and thus an hours use of demand) that exceeds the HUD threshold of the rate structure (e.g., 250 kWh per kW). Note from the example in Table 8.2, that compared to the Flat rate design, the Wright rate structure substantially reduced the average rate per kWh for the higher load factor customer (#1) while only minimally increasing the average rate for the low load factor customer (#2). By using multiple HUD steps, the range of application to customers having diverse load factor characteristics can be expanded. In essence, the multi-step Wright rate structure can effectively fulfill the function otherwise required from separate Hopkinson rate designs for two or more customer classes. A mixture of all of the basic rate structure types is commonly used within the electric tariff.[14]

This discussion has focused on the inherent risk of the basic time-invariant rate structures. This risk relates to both the service provider and the customer, including the risk which exists between customers having varying load characteristics that are served under a given rate structure. Rate structure risk can also be viewed in a temporal sense, and time-differentiated rate structures can be implemented to mitigate such aspects of risk. In addition, elements of risk exist within other aspects of electricity pricing and even just by providing electric service itself, and this risk can be compensated through the design and implementation of attendant billing provisions.

8.4 ATTENDANT BILLING PROVISIONS

The basic rate structure provides a baseline for electric service billing in that it generally captures most of the costs of providing service through its configuration of Customer, Demand, and Energy Charges. Thus, the rate structure represents a key component of a rate schedule for a given class of electric service. Other billing provisions are often necessary to actually implement billing under the basic rate structure. For example, the billing demand to be applied in the base bill calculation of a demand metered rate structure must be determined before application of the unit Demand Charge(s). As part of this determination, considerations other than just the current month's meter reading are typically taken into account in order to ensure adequate revenue recovery as well as equity between customers. In

---

[14] A single but substantially intricate Wright demand rate structure approach has been utilized successfully by some utilities to render service to a vast portion of their commercial and industrial customer bases. However fully such structures address the objectives of electricity pricing, they tend to limit the flexibility that is otherwise available with separate rate schedules for separate classes.

addition, other billing provisions are necessary to account for cost variations in the electric service methodology since the basic rate structure is prescribed for power delivery and metering at a designated voltage class. Different customers served under the same rate schedule may take service at two or more available voltages, i.e., secondary, primary, subtransmission, or transmission voltages. Still other provisions are necessary for capturing other aspects of the costs of service which are intended to be recovered outside of the basic rate structure, including fuel, purchased power, various taxes, and special programs. In addition, a minimum bill methodology is necessary to ensure some level of contribution toward fixed cost recovery for times when a customer's actual usage is less than was anticipated when the facilities required for normal operation were installed.

These types of billing provisions are identified in the rate schedules for the various classes of electric service. In many cases, these provisions are fully described within the language of the rate schedule itself. In other cases, these provisions denote separate price schedules or rate riders that apply in conjunction with the basic schedule for electric service.

Billing Demand Determination

The primary consideration of demand to be used in the monthly rate application is the current month's meter reading of the 15- or 30-minute kW (or kVA) requirements from either an indicating demand or an interval type meter.[15] The most common selection criteria is the single maximum demand occurring during the monthly billing period. A lesser used method is based on the average of multiple demands occurring within the billing period (e.g., average of the three highest 15-minute demands occurring during the month). When bi-monthly meter reading is used, the demand for bi-monthly billing might be taken as a percent (e.g., 95%) of the maximum demand occurring during the two-month period.

Fully or partially unbundled rate structures and TOU demand rates contain bifurcated Demand Charges which require two or three different measures of monthly demand. For example, the customer's maximum monthly demand is used typically for application of a Distribution Demand Charge, while the customer's demand at the time in which the system peak occurs during that month might be used for application of Transmission and Generation Demand Charges. Interval metering would be needed to identify these different demands.

Since indicating demand meters for small size customers (e.g., up to about 15 kW) often have a maximum demand scale which registers to one decimal place, the determination of billing demand from meter readings should indicate a rounding protocol for billing purposes. Thus, the meter reading could be adjusted to "the nearest tenth of a kW," "the nearest half kW," "the nearest half kW which does not exceed the actual kW," or "to the nearest whole kW."

---

[15] A five-minute interval demand might be considered in the case of highly fluctuating loads, such as electric furnaces and welders.

The Rate Schedule: Pricing Methods and Risk

In some cases, the monthly demand might be estimated in lieu of direct metering. For low-use commercial customers (e.g., less than 1,000 to 2,000 kWh per month), the metered energy consumption can be divided by a representative HUD factor, such as 100 or 175 kWh per kW, to derive a demand for billing purposes. The estimate of demand for larger commercial customers might be accomplished on a customer-by-customer basis. For example, a *counted demand* approach is an engineering calculation based on the customer's connected end-use load devices. Coincidence factors are applied to lighting and power loads to establish an overall diversified maximum demand. Motor horsepower would have to be converted to an equivalent kW load taking into account efficiency and power factor. For promotional purposes, a competitive load, such as electric cooking, might be excluded from the determination of the monthly billing demand. The counted demand approach is somewhat tedious in terms of administration, and its use is less common today as basic kW demand meters can be applied to much smaller loads more economically than in the past.

When kVA-based Demand Charges are utilized for billing, the maximum kVA demand may be determined by directly metering kVA, by analysis of real and reactive usage on an interval basis, or by calculating a maximum hybrid kVA demand, as given by

$$kVA_{CALC} = \sqrt{kW_{MAX}^2 + kVAR_{MAX}^2} \qquad (8.3a)$$

where kVA_{CALC} = calculated maximum kVA demand
kW_{MAX} = maximum kW demand during the billing period
$kVAR_{MAX}$ = maximum kVAR demand during the billing period

The real and reactive maximum demands are obtained from indicating demand meter readings and thus are not always coincident with each other every billing period. This calculated kVA would then be equal to or slightly greater than the true kVA maximum that would be determined from directly metering kVA.

A hybrid kVA could also be calculated on the basis of the average power factor of the billing period, which is determined from the metered kWh and kVARh energy readings, as given by

$$kVA_{CALC} = \frac{kW_{MAX}}{PF_{AVG}} \qquad (8.3b)$$

where kVA_{CALC} = calculated maximum kVA demand
kW_{MAX} = maximum kW demand during the billing period
PF_{AVG} = average power factor during the billing period (as determined by Equation 7.2)

When based on average power factor, the hybrid kVA can be expected to be somewhat less than the true kVA maximum demand.

Power factor is an electric service issue, and compensation for low power factors is typically included in larger commercial and industrial rate schedules through either kVA-based Demand Charges, as discussed above, or as an attendant billing provision. Alternatively, power factor can be utilized to adjust the monthly metered maximum demand before the application of kW-based Demand Charges. If kVA metering is installed, the monthly kW billing requirement could be based on the product of the maximum metered kVA and a *base power factor* (e.g., 85 % or 90%). Thus, when the customer's actual power factor is less than the designated base power factor, the resulting kW demand for billing would be greater than the true maximum kW demand, and compensation for low power factor would be realized through the increased Demand Charges.

As another demand adjustment method, the ratio of the designated base power factor to the customer's actual monthly power factor can be applied to the maximum metered kW demand as given by

$$kW_{BILLING} = kW_{METERED} \times \frac{PF_{BASE}}{PF_{ACTUAL}} \tag{8.4}$$

where $kW_{BILLING}$ = adjusted maximum kW demand
  $kW_{METERED}$ = maximum metered kW demand during the billing period
  PF_{BASE} = base power factor specified in the rate schedule
  PF_{ACTUAL} = actual lagging power factor during the billing period

The actual power factor could be specified as either the average during the billing period or that which occurs at the time of the customer's maximum demand. When the actual is less than the base power factor, $kW_{BILLING}$ is greater than $kW_{METERED}$; however, when the actual exceeds the base power factor, $kW_{BILLING}$ is calculated to be less than $kW_{METERED}$. In such circumstances, some rate schedules allow a reduction in the billing demand while others rate schedules do not.

Demand Ratchets

Each of the billing demand determination methods discussed above are based on the load characteristics of the current billing period. Historical demands may also be considered when determining the billing demand for the current month. A *demand ratchet* processes a customer's metered maximum demands of the prior eleven months by applying a specified percentage to those demands in all or a portion of those months and then selects the highest resulting calculated demand as the current month's billing demand–if it exceeds the current month's maximum metered demand. An example comparing various demand ratchet percentages over the course of a year is shown in Table 8.3.

Table 8.3 Examples of Billing Demands Based on Various Demand Ratchets

Month	Actual kW				
Previous Year (#1)					
Jan	1,463				
Feb	1,540				
Mar	1,506				
Apr	1,613				
May	1,640				
Jun	1,592				
Jul	1,563				
Aug	1,506				
Sep	1,546				
Oct	1,675		**Billing Demands for the Current Year**		
Nov*	1,800				
Dec	1,502		**Demand Ratchet Percentages**		
		100%	90%	75%	50%
Current Year (#2)					
Jan	1,550	1,800 Nov #1	1,620 Nov #1	1,550 Actual	1,550 Actual
Feb	1,575	1,800 Nov #1	1,620 Nov #1	1,575 Actual	1,575 Actual
Mar*	1,650	1,800 Nov #1	1,650 Actual	1,650 Actual	1,650 Actual
Apr	1,360	1,800 Nov #1	1,620 Nov #1	1,360 Actual	1,360 Actual
May	1,200	1,800 Nov #1	1,620 Nov #1	1,350 Nov #1	1,200 Actual
Jun	1,280	1,800 Nov #1	1,620 Nov #1	1,350 Nov #1	1,280 Actual
Jul	1,225	1,800 Nov #1	1,620 Nov #1	1,350 Nov #1	1,225 Actual
Aug	1,450	1,800 Nov #1	1,620 Nov #1	1,450 Actual	1,450 Actual
Sep	1,300	1,800 Nov #1	1,620 Nov #1	1,350 Nov #1	1,300 Actual
Oct	1,242	1,800 Nov #1	1,620 Nov #1	1,350 Nov #1	1,242 Actual
Nov	1,390	1,650 Mar #2	1,485 Mar #2	1,390 Actual	1,390 Actual
Dec	1,508	1,650 Mar #2	1,508 Actual	1,508 Actual	1,508 Actual
Total	**16,370**	**21,300 kW**	**19,223 kW**	**17,223 kW**	**16,730 kW**

*Months in which the actual demand triggers a ratchet.

Table 8.3 shows the impact on a customer's billing demands for the Current Year (#2) for four demand ratchet scenarios. The ratchet is invoked when the current month's maximum demand is found to be less than a specified percentage of a prior month's maximum demand. The ratchet operates on a moving window of twelve months of demands; thus, as a new month is added, the twelfth month back is dropped from consideration. The four scenarios are plotted in Figure 8.11.

314   Chapter 8

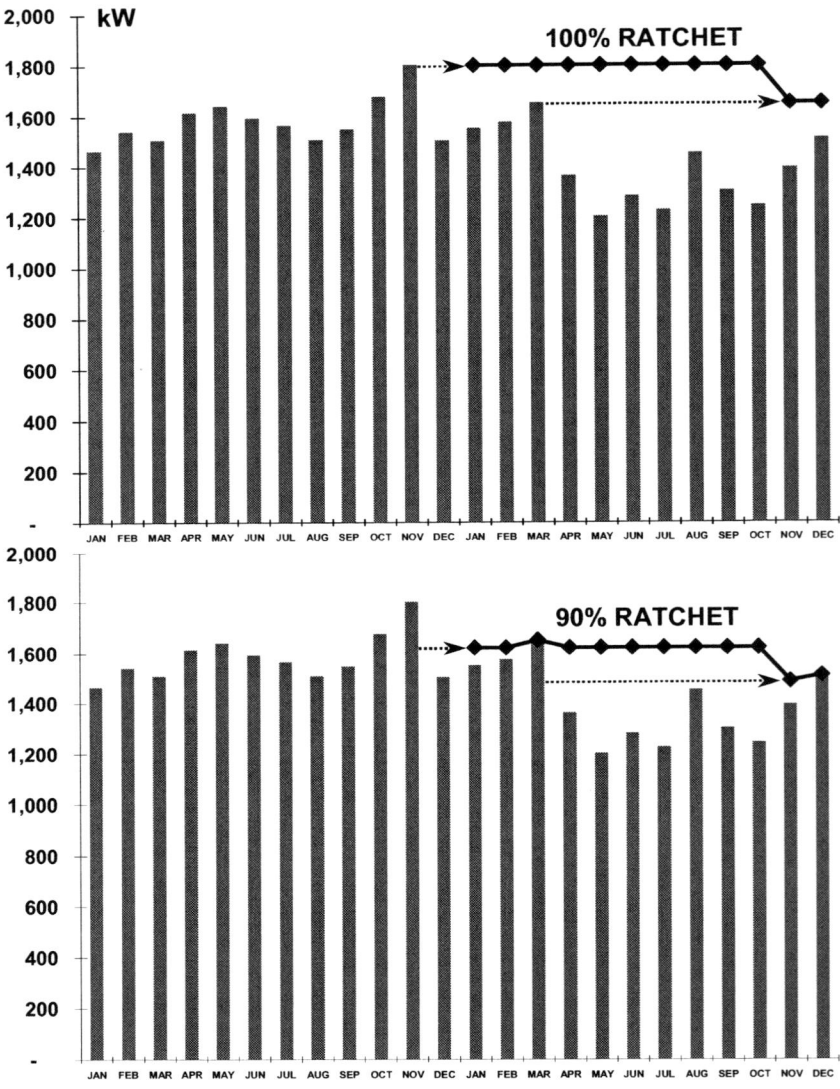

Figure 8.11 A comparison of four demand ratchets. Different adjustments are applied to determine the billing demands for the twelve months for the second year. The bars represent the actual maximum demands, while the diamonds represent the ratcheted billing demands. The ratchet is first triggered by the maximum demand in November of the previous year, and its consideration for billing demand determination terminates a year later.

The Rate Schedule: Pricing Methods and Risk

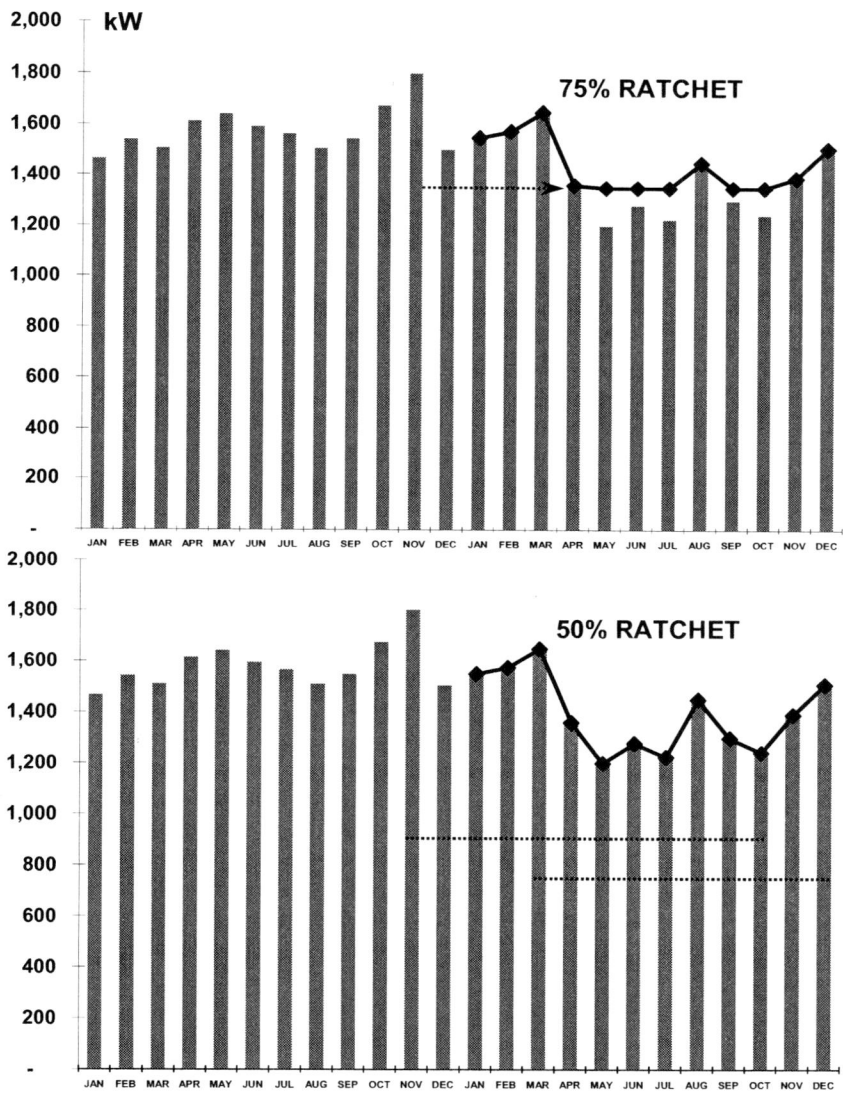

Figure 8.11—Cont. The basic objective of the ratchet is to levelize the billing demand across the year. The leveling effect of the demand ratchet diminishes as the percentage applied is lowered. Demand ratchets operate more frequently with customers that have acute seasonal loads or wide variations in their monthly maximum load requirements.

A demand ratchet provides a leveling effect to a customer's monthly billing demands, and its impact is more pronounced when a customer has wide variations in monthly demand requirements, such as with seasonal load cycles. Consider the installation of a 1,000 kVA transformation and service to a customer that is sized on the basis of the customer's expected maximum load requirements. If the customer's demand each month is reasonably close to the rating of the facilities, the Demand Charge revenues will provide an adequate recovery of the associated costs. However, if the monthly demands vary considerably from the installed capacity, the cost recovery of the facilities could be extended by a considerable number of years. In essence, a significant amount of the installed capacity could be "standing by" a considerable amount of the time. Such underutilization of the local facilities strands a portion of the related investment. With a demand ratchet operation, the billing demand becomes greater than the metered maximum demand; thus, there is an increase in billing units and the applicable Demand Charges become higher. As a result, the monthly billing revenues become more stable and provide a better assurance of recovering fixed costs over a reasonable period of time. In other words, the ratchet helps to mitigate the risk of stranded investment.

The choice of the demand ratchet percentage and how it is applied depends on the characteristics of each type of customer class and the rate structures utilized. Large, high load factor customers are typically subject to a high, if not 100%, ratchet since a considerable amount of facilities are reserved for such customers. In addition to the local distribution facilities, a substantial commitment in transmission and generation capacity is required given that high load factor loads are highly coincident with the system's peak demand conditions. The demand ratchet is typically less for smaller customers with lower load factors.

In a bundled rate, a single ratchet must be designed as a compromise between the local facilities and the transmission and generation facilities. The local facilities portion of the electric service calls for a high percentage ratchet, as discussed above, since it relates to the customer's maximum demand which can occur at anytime. However, overall the maximum demands of the lower load factor customers are more diverse, compared to high load factor customers, and the amount of transmission and generation capacity that must be reserved is a function of that diversity. Since the ratchet applies to a customer's (not necessarily coincident) maximum demand, diversity would be recognized through the use of a lower ratchet percentage. With bifurcated Demand Charges, multiple demand ratchet features can be designed independently for each of the various unbundled demand-related components.

The ratchet can also be designed to promote off-peak season usage. For example, with a summer peaking system, the ratchet percentage applied to the previous summer months could be set higher than the percentage that is applied to the previous nonsummer months. As a result, a customer could increase winter demands a certain extent, through electric heating applications for instance, without the concern for triggering a ratchet operation.

The Rate Schedule: Pricing Methods and Risk 317

Customers who experience a ratchet operation often view the ratchet feature as being punitive or as a *take-or-pay* provision. Another perspective may regard the ratchet as encouraging a customer to use power unnecessarily since its operation appears to charge the customer for power that was not actually used. However, such is not the case for the capacity requirement and cost recovery reasons discussed above.

Furthermore, the use of a ratchet promotes equity between customers within the same rate class. Without a demand ratchet, customers having consistent monthly demands subsidize other customers having widely fluctuating demands. Note in Table 8.3 how the total billing demand is greater than the actual demand under the various ratchet scenarios. When a Demand Charge is designed, the demand-related revenue requirements are divided by the sum of the billing demands. The resulting unit rate would be higher under an application which does not include a ratchet feature since the billing demands would be equal to the actual demands. With a ratchet, the unit rate for demand becomes less since the billing demands are greater than the actual demands, and customers with consistent monthly demands realize the benefit. Thus, the demand ratchet mitigates some of the risk that exists between customers having different operational patterns. In addition, a customer that is caught by a ratchet would be released automatically from its effects after a year provided that the monthly maximum demands become more consistent.

Other Billing Demand Considerations

Service agreements with larger commercial and industrial customers commonly include a kW or kVA *contract demand* consideration when determining the monthly billing demand. The contract demand might be applied at its full value or at a percentage specified within the rate schedule. Similar to a demand ratchet, the contract demand provides some assurance of fixed cost recovery in the event that the monthly actual demands do not at some point actualize as originally expected when the service was initiated. Thus, a customer's contract demand should be set in general accordance with the capacity of the facilities that are sized and installed to serve the customer's maximum demand requirements. The contract demand should be revised as conditions change, such as major customer load additions or permanent load reductions of an unusual nature.

The contract demand works in correspondence with the contract term. Suppose a customer being served under a five-year contract decided to significantly curtail operations or close the business altogether after just two years. Actual energy consumption could be reduced to only maintenance levels or cease completely. The inclusion of a contract demand would provide a basis for billing a reasonable amount of monthly Demand Charges for the remainder of the contract term after any demand ratchet has ceased to yield any further value.

A final consideration for determining the monthly billing demand would be a designated kW or kVA *rate minimum*. Under no conditions would the monthly billing demand ever be set for less than the minimum amount specified within the rate schedule. Again, this minimum is intended to provide some basic assurance of fixed cost recovery. Rate minimums vary with the classes of customers. Small general service rate schedules typically have a rate minimum of five kW. Medium general service rate schedules have a rate minimum on the order of 20 to 25 kW, while large general service rate schedules have a rate minimum on the order of 500 to 1,000 kW. Large, high load factor rates commonly have a rate minimum of as much as 5,000 to 10,000 kW.

Determination of the Monthly Billing Demand

Several if not all of the features discussed above for determining the billing demand are typically combined to form a selection criteria for identifying the monthly billing demand that will be applied in the operation of the rate structure. An example rate schedule provision for the determination of the monthly billing demand is given by the following

The billing demand each month shall be the integrated 15-minute period of maximum kW load requirements as measured by suitable instruments during each billing period of approximately 30 days, provided however that such billing load demand shall not be less than the largest of

1. *90% of the measured maximum kW load requirements established during the billing months of June through September falling within the eleven months preceding the current billing period;*

2. *60% of the measured maximum kW load requirements established during the billing months of October through May falling within the eleven months preceding the current billing period;*

3. *50% of the total contracted kW demand; or*

4. *25 kW for secondary service or 50 kW for primary service.*

Voltage Level Adjustments

Rate structures are designed under the baseline assumption that electric power is delivered to a customer and metered for billing purposes at a single designated voltage class, such as secondary, primary, subtransmission, or transmission service. Thus, if the rate structure is designated for service at secondary, the associated Customer, Demand, and Energy Charges would adequately recover the basic costs of providing service at the secondary voltage level.

In practice, commercial and industrial customers are often allowed to take service at different voltage levels under the same rate schedule whose fundamental rate structure charges have been designated for one specific voltage class. In addition, the revenue metering may be located at a voltage different from the cus-

The Rate Schedule: Pricing Methods and Risk

tomer's specific delivery voltage as a practical matter. As a result, the rate schedules must include billing provisions which provide adequate compensation in recognition of such conditions. Without proper price differentiation, some customers would be subsidized at the expense of other customers.

Delivery Voltage Adjustments

Figure 8.12 illustrates the possible voltage classes at which different customers could take electric service under a common rate schedule. A distinct Demand Charge is necessary for each voltage level in order to reflect the capacity-related costs of service for each delivery level. The differentials between voltage levels are caused primarily by the cost of providing the voltage transformations.

Alternatively, the rate structure could be designed with a single Demand Charge that represents the cost of delivering power to a designated voltage level, such as primary voltage, as indicated in Figure 8.12. Customers served at other voltage levels would then be subject to an additional per unit Demand Charge (or credit) based on the monthly billing demand to reflect the costs incurred (or saved) by the power delivery provider in supplying (or not supplying) the transformation capacity necessary for service to be rendered at the different levels.[16]

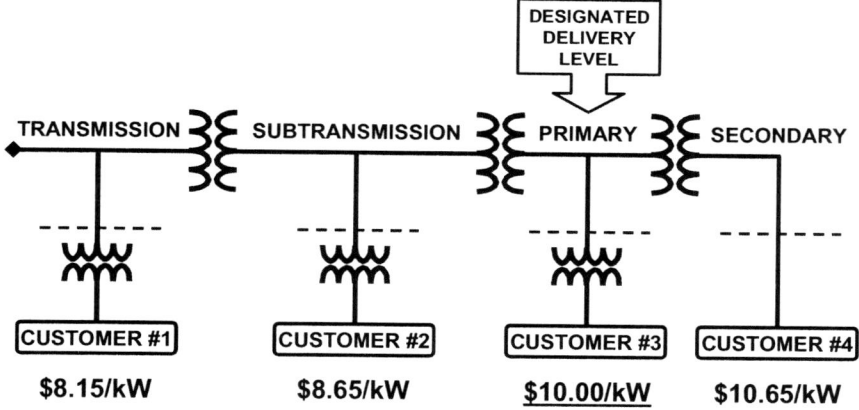

Figure 8.12 Customers taking service at different voltage levels under a common rate schedule. The Demand Charge is differentiated to reflect the costs of service by level.

---

[16] The term "voltage discount," which is often used when service is taken at a voltage higher than the voltage level designated by the rate structure, is a misnomer. The power delivery service provider is not making a price concession, as the term implies, but merely reflecting the cost-of-service savings from not having to provide a transformation.

Customers taking service at voltages higher than secondary still require voltage level transformations in order to distribute power to end-use load devices at the proper utilization voltages. Thus, the higher voltage customers internalize the capital and maintenance costs of the required transformations. Sometimes a customer may find it advantageous to take service at a higher voltage level in order to receive a lower monthly Demand Charge but at the same time prefer to lease the transformation facilities rather than to take up the responsibilities of ownership.

Separate voltage-specific rate schedules could also be utilized as a means for differentiating the prices for electric service according to voltage level. Customers naturally would then be assigned to an applicable rate schedule on the basis of their specific service voltage requirements. This market segmentation approach results in more homogeneous groups of customers having more common power delivery requirements and thus more similar cost-of-service characteristics.

Metering Voltage Adjustments

Ideally, the revenue meter should be located directly at a customer's service delivery point where it can accurately measure the customer's demand and energy requirements. However, practical or economic circumstances sometimes dictate that the meter must be installed at a voltage level that is different from the voltage of the delivery point. In such cases, compensation of electrical losses in the transformation facilities (including busses and conductors) is necessary. Compensation can be accounted for through either loss compensation metering (a hardware solution) or a rate schedule adjustment (a billing solution).

As shown in Figure 8.13, Customer A owns the service transformer and thus takes power at a high voltage level, such as primary, subtransmission, or transmission. The customer's monthly billing demand would be applied to the appropriate Demand Charges for high voltage electric service. When located at the high voltage delivery point, the meter would correctly see the transformation facilities included in the customer's total demand and energy requirements. However, if the meter is located on the low voltage side of the transformer, the meter would not measure the transformer's demand and energy losses. To compensate, the metered demand and energy units must be increased in order to attain the proper billing units.

In contrast, Customer B takes electric service at a low voltage, such as secondary, as shown in Figure 8.13. The utility provides the necessary service transformer, and the rate schedule charges for low voltage service inherently account for the transformation. When placed on the low voltage side of the transformer, the meter correctly measures the demand and energy requirements of the customer's total load. However, if the meter is located on the high voltage side of the transformer, the measured energy and demand usage would exceed the true amount for the customer's load (in essence a double counting of the transformer's losses would effectively be taking place). As a result, the metered units must be decreased in order to attain the customer's proper billing units.

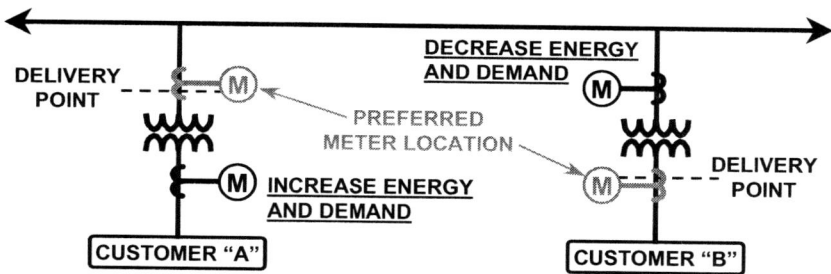

Figure 8.13 Billing unit adjustments due to meter location. When the meter is located at a voltage level other than the service delivery point, transformer losses must be accounted for by means of either loss compensation metering or rate schedule billing demand and energy adjustments.

Billing adjustments to compensate for variations in meter location are often made by applying loss correction factors to the metered demand and energy units. These factors are typically represented in the rate schedule as fixed percentages. For example, if the rate structure is designated for primary service, the metered units might be increased by about 1 to 3% for secondary delivery and decreased by about 1% for transmission delivery. The correction factors represent the average losses associated with transformation of voltage between levels.

Power Factor Adjustments

Demand Charges based on kVA billing units bundle the pricing of both real and reactive power requirements together. At unity power factor, kVA is equal to kW, and the resulting kVA-based Demand Charge revenue would be at a minimum, given the associated kW demand. With either lagging or leading power factor conditions, the kVA would become greater relative to the kW demand, thereby increasing the Demand Charge revenue. In this manner the kVA Demand Charge provides some compensation for inferior power factor load conditions.[17]

Power factor was discussed previously in regard to the determination of the monthly billing demand. Power factor can be utilized to calculate a hybrid kVA or to otherwise adjust the metered kW requirement to produce the maximum demand to be used for billing purposes. The application of power factor to adjust the kW demand represents another method for billing real and reactive power requirements within a bundled price structure.

---

[17] The use of kVA Demand Charges provides a compelling incentive for customers to invest in power factor correction equipment. However, it is important that the corrective reactive capacity be sized properly and controlled through switching devices, as necessary, to prevent leading power factor conditions that could increase the kVA billing demand.

A partial unbundling of the Demand Charge, in recognition of the customers' reactive power requirements, can be accomplished by charging for the increment of either the metered kVA demand or the metered kVAR demand that exceeds the associated base kVA or kVAR demand amount corresponding to a specified base power factor. The base power factor utilized in these types of billing provisions is typically in the range of 80% to 95% lagging. By setting the base power factor below 100%, customers are given an economic incentive to improve their power factors with less concern for creating an overcorrection response which could cause leading power factor problems within the power system.

The *excess kVA* approach is shown in the upper portion of Figure 8.14. The amount of kVA that is subject to an Excess kVA Demand Charge is determined by

$$kVA_{EXCESS} = kVA_{ACTUAL} - kVA_{BASE} \quad (8.5a)$$

and

$$kVA_{BASE} = \frac{kW_{ACTUAL}}{PF_{BASE}} \quad (8.5b)$$

where kVA_{EXCESS} = amount of the actual kVA that exceeds the base kVA
 kVA_{ACTUAL} = maximum kVA demand during the billing period
 kW_{ACTUAL} = maximum kW demand during the billing period
 PF_{BASE} = base power factor specified in the rate schedule

As shown in Figure 8.14, the excess kVA amount can be visualized by rotating the actual kVA vector so that it is in line with the base kVA vector and then noting the difference in magnitudes. An excess kVA amount exists only if the magnitude of the actual kVA is greater than the magnitude of the base kVA. The base kVA can change from month to month since it is a function of the metered maximum kW for each billing period.

The *excess kVAR* approach is shown in the lower portion of Figure 8.14. The amount of kVAR that is subject to an Excess kVAR Demand Charge is determined by

$$kVAR_{EXCESS} = kVAR_{ACTUAL} - kVAR_{BASE} \quad (8.6a)$$

and

$$kVAR_{BASE} = kW_{ACTUAL} \times \tan\left[\cos^{-1}(PF_{BASE})\right] \quad (8.6b)$$

where $kVAR_{EXCESS}$ = amount of the actual kVAR that exceeds the base kVAR
 $kVAR_{ACTUAL}$ = maximum kVAR demand during the billing period
 kW_{ACTUAL} = maximum kW demand during the billing period
 PF_{BASE} = base power factor specified in the rate schedule

The Rate Schedule: Pricing Methods and Risk

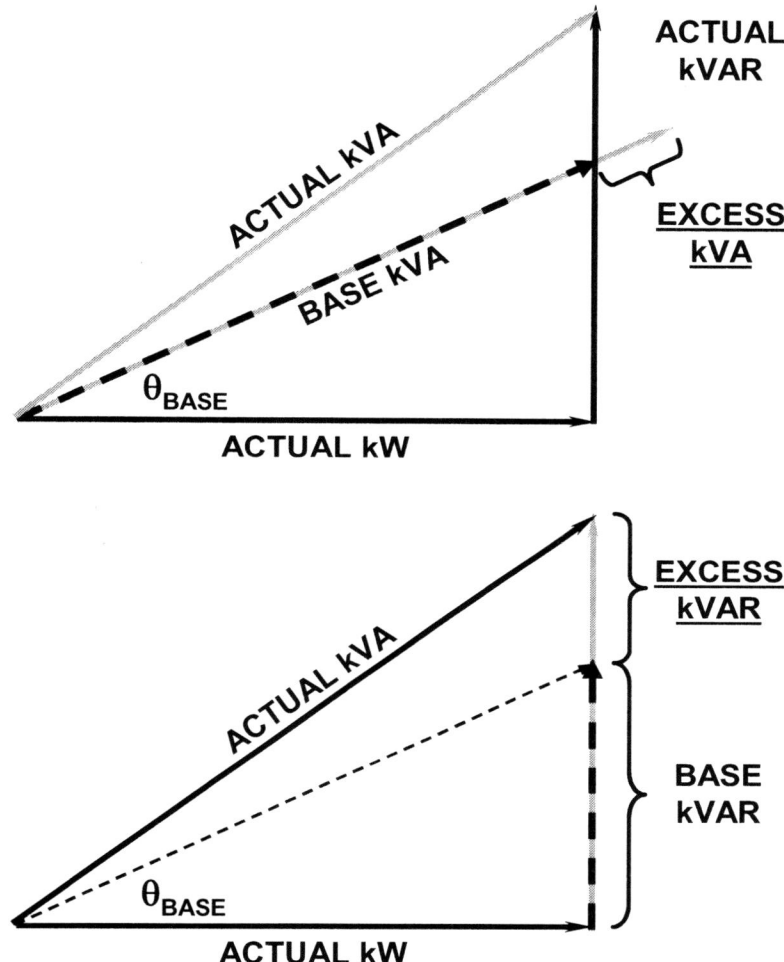

Figure 8.14 Excess kVA and Excess kVAR methods for power factor billing. Both of these methods incorporate a base power factor, as specified in the rate schedule, whose angle is indicated as θ_{BASE}. These methods also function on monthly actual (metered) units of kW, kVAR, and kVA.

An excess kVAR amount exists only if the magnitude of the actual kVAR is greater than the magnitude of the base kVAR. The base kVAR can change from month to month since it is a function of the metered maximum kW for each billing period. A significant advantage of the excess kVAR method is that it provides a direct indication of the amount of corrective capacity required to improve the power factor, since capacitors are rated in units of kVAR.

To maintain simplicity in the tariff language, the excess kVAR billing provision is often written to indicate that the Excess kVAR Demand Charge is applicable to the amount of the monthly maximum metered kVAR demand that is in excess of "X" percent of the monthly maximum metered kW demand. "X" percent represents the tangent of the base power factor angle and has the implied units of kVAR per kW. Thus, the product of "X" percent and the maximum kW is equal to the base kVAR. For example, 48.43% (0.4843) would be the tangent of an angle of 25.84°. The cosine of 25.84° is 0.90 which represents a power factor of 90% (i.e., the base power factor).

A wide variety of other methods for charging for poor power factor conditions are utilized in practice. For instance. A simple, single Reactive Demand Charge may be applied to the monthly maximum reactive demand as metered and without consideration of a base and an excess kVAR component. The charge for reactive power consumption could be based on the monthly volume of the metered reactive energy (kVARh). Still other unique formulary approaches, including variations of the basic methods described here, can be found in practice.

The Power Factor Price Signal

A key objective of charging for reactive power consumption is to encourage customers to increase their power factors and operate near unity. When the load power factor is increased, the power system realizes improved local voltage conditions, released kVA capacity in lines and transformation facilities, and lower system electrical losses. These technical benefits translate directly into savings in capital and operating expenses which lowers the cost of providing service. Responsive customers would realize a reduction in monthly electric service billings.

Metering in support of reactive power billing is quite expensive. Thus, residential and small commercial rate schedules only consider the costs of compensating the load power factor as an inherent component of the basic rate structure charges, even within unbundled rate structures.[18] High voltage capacitors at substations and on feeder systems are used to provide area correction of such customers' load power factors. On the other hand, reactive power metering is less of a cost issue with larger commercial and industrial customers. Inclusion of a power factor feature in the medium and larger power service rate schedules directly recognizes higher costs which are associated with poor power factors and further ensures some level of equity among the customers within these classes.

---

[18] However, with unbundled transmission service, a separate and specific ancillary service schedule for voltage support, with charges specified on a per kWh basis, may also apply to small customers. This charge is directed toward the costs of providing adequate VARs throughout the power system to maintain the voltages necessary to transport real power.

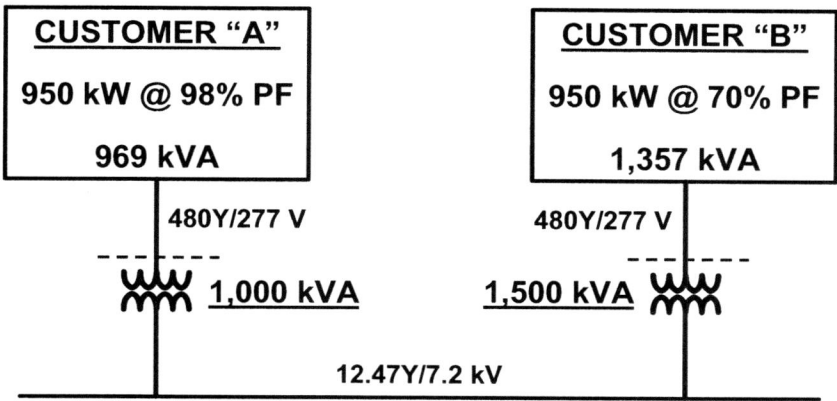

Figure 8.15 Service capacity requirements as a function of power factor. Two large customers having the same kW demand but dissimilar power factors require different standard size transformer capacities to serve the loads adequately. Since the cost to serve the poor power factor load is greater, Customer A would subsidize Customer B if an equitable power factor provision is not incorporated in the rate schedule.

Figure 8.15 illustrates the major local cost impacts due to a customer's inferior power factor characteristics. Customer A and Customer B are alike in every manner except for power factor. Consequently, their kVA capacity requirements are unequal. Since line transformers, buses, and conductors are sized to serve a load in terms of kVA capacity, Customer B requires a higher standard size transformer than Customer A. Thus the cost to serve Customer B is greater than the cost to serve Customer A. Rate structures based on either kVA Demand Charges or kW Demand Charges with an attendant power factor billing provision provide a means of recognizing the specific differences in reactive power requirements among customers served under the same electric service rate schedule.

The economics of power factor correction by customers depends on the amount of savings that can be attained through monthly electric service billings and on the costs of installing and maintaining the required corrective capacity. A driving factor is the voltage level at which power factor correction facilities can be installed by the customer. The per unit cost of secondary voltage capacitors, in terms of kVAR capacity, is approximately four times as high as for primary voltage capacitors. Thus, primary service customers have a distinct economic advantage for correcting power factor. Unfortunately, most commercial and industrial customers take service at a secondary voltage level where it is often more difficult to realize an acceptable payback for power factor improvements. Many of these customers choose instead to pay a higher monthly electric bill while investing their capital resources in other projects which offer higher returns.

Minimum Bill Provision

The minimum bill provision specifies the lowest possible charges that would be rendered under any condition of electric service use. Minimum bills are most commonly structured for monthly billing periods. However, when the load is highly seasonal in nature, such as with irrigation and crop processing operations, the minimum bill may be specified on an annual basis.

The primary objective of the minimum bill provision is for the recovery of fixed costs; thus, it is structured on the basis of Customer and Demand Charges. For standard, nondemand metered rates, the minimum bill is usually specified to be the same as the Customer Charge specified in the rate structure. If the rate structure contains no explicit Customer Charge, a fixed amount that would be equivalent to a Customer Charge which includes some kWh usage (similar to an initial charge structure) is typically specified in the minimum bill provision. Alternatively, the minimum bill provision may specify the applicable minimum charge in terms of a minimum amount of kWh per month, even if no consumption was measured.[19]

With demand metered rates, the minimum bill is usually specified as the total of the Customer and Demand Charges, where the Demand Charge is based on the monthly billing demand determination provision. Thus, with no monthly consumption, the minimum bill would be based on a ratcheted demand or a rate or contract minimum. With the Wright rate structure of implicit Demand Charges, the minimum bill provision typically specifies an explicit per unit charge per kW or kVA in addition to the Customer Charge. With this explicit Demand Charge structure, the minimum demand provision is designed to override the normal rate structure charges under very low load factor conditions (in essence it functions as a variable rate stopper).

High load factor rate schedules often include a minimum bill provision which requires billing at the level of kWh usage that corresponds to the monthly billing demand and the minimum load factor requirement. Consider a high load factor rate having a Customer Charge of $750.00, a Demand Charge of $10.00/kW, an Energy Charge of 1.0¢/kWh, and a monthly load factor requirement of 75%. The associated hours use of demand, based on an average month of 730 hours, then is 547.5 kWh/kW. The monthly bill could then be stated as the Customer Charge plus $15.48/kW ($10.00/kW + [547.5 kWh/kW × $0.01/kWh]). More precisely, the variation in the number of days in each billing period could be recognized through a monthly formulary approach for determining the minimum kWh associated with the load factor requirement, as given by

$$kWh_{BILLING} = X \times N \times kW_{BILLING} \qquad (8.7)$$

---

[19] When fuel and purchased power is excluded from the Energy Charges of the rate structure, the vast majority of the prices per kWh represents fixed cost recovery.

where $kWh_{BILLING}$ = monthly kWh corresponding to the load factor minimum
X = number of hours per day corresponding to the minimum load factor requirement (e.g., $X = 18$ for a 75% LF)
N = actual number of days in the billing period
$kW_{BILLING}$ = monthly billing demand (could also be based on kVA)

A minimum bill provision can be designed such that when it is invoked, it merely supersedes only the charges otherwise applied just by the monthly rate structure components. For instance, if the minimum bill provision overrides the rate structure, it could still be subject to voltage credits or charges based on the service voltage level, just as with the basic rate structure. In addition, the minimum bill provision does not typically override the operation of any other billing provisions that may be applicable to the rate schedule.

Special Billing Provisions

A variety of other billing provisions may be embodied along with the basic charges of the electric service rate schedule. A number of special billing provisions are often applicable to several, if not all, of the other rate schedules within the electric tariff. As a result, a special billing provision often corresponds to a separate tariff document referred to as a *contract* or *rate schedule rider*. The use of rate schedule riders provides some administrative efficiency. The rider helps to minimize the language volume of the rate schedules by placing the specifications of the common billing provision in a single auxiliary document. As a separate document, the rider can be amended as necessary without having to revise all of the rate schedules in order to incorporate a particular modification.

Rate Schedule Riders

Some rate schedule riders function to recoup expenditures that are essential to providing basic electric service, such as fuel and purchased power cost recovery. A rider can also be used to reserve funds for future specific purpose expenditures associated with the electric power system, e.g., catastrophic weather event recovery or nuclear decommissioning costs. The costs of system benefits programs, such as research and development and demand-side management efforts, may also be funded through monthly electric service billing. Likewise, customers can be compensated for savings realized by efficiency improvements in the business, such as from utility mergers, in the form of monthly billing credits.

Some rate schedule riders are implemented in order to recover the costs of regulatory or governmental mandated public policy programs that are intended to provide social benefits. Examples of such programs include consumer information, conservation and environmental initiatives, and universal service. Since rate schedule riders can address such a wide range of issues, it is not uncommon to observe the application of several riders at the same time.

Some rate schedule rider fees, charges, or credits may be specified as a fixed monthly amount per bill, while others may be applied as a percentage of the subtotal electric service billing amount. However, the most common pricing structure utilized in rate schedule rider design is one in which the unit charges or credits are expressed on a per kWh basis. The fees, charges, or credits and may also be differentiated by the applicable rate schedules or classes of service or by some other service distinction (e.g., delivery voltage level).

Energy Supply Cost Recovery Rider
The total cost of fuel used in the production of electricity varies much more frequently than most other costs of providing service. This variation in total cost is caused by fluctuations in the market price of the fuels (e.g., per ton of coal or per MBtu of natural gas) and by changes in the mix of generating resources utilized for production. Some stability in fuel procurement costs can be gained by buying forward in the energy markets as a hedge against price changes. However, the total cost of the fuel consumed is also a function of the load being served and the efficiencies of the production plants.

A fleet of generating units is operated to serve the load on an economic dispatch or merit order basis. The operating cost of each unit (e.g., $ per MWh of output) depends on both its particular input fuel and its energy conversion efficiency. As the load changes, the order of dispatch is always from the lowest cost unit to the highest cost unit; thus, the lowest total cost of production is realized at all times. From one month to the next, the system hourly loads will be different. Thus, the mix of dispatched generation will be different, and the specific generating units that are available for service may also be different. Some units could be out of service for scheduled maintenance, or a forced outage may occur. As a result, the total cost of fuel on a per kWh basis varies each month. In addition, any opportunity to augment the economic dispatch with lower cost purchased power would further impact the total cost of energy supply. On the other hand, the need to purchase higher cost power to serve the load would also affect the total cost.

With such frequent variations in the cost of energy supply, the selection of a fixed rate per kWh for the recovery of fuel and purchased power costs would eventually result in either an overrecovery or an underrecovery situation each month. The accumulated over/underrecovery of the costs of energy supply over the course of a year could be significant. In practice, fixed rates for energy supply are established, but they are adjusted periodically to correct for variations in cost recovery. An Energy Supply Cost Recovery rider is commonly used for setting the fixed rate for fuel and purchased power for all the classes of electric service. A portion of the cost of fuel is sometimes embedded as a fixed component of the Energy Charges of the rate schedules. In such cases, the periodic adjustments are made only to the incremental fuel and purchased power rate.[20]

---

[20] It is customary to adjust the embedded fuel cost amount when base rate structure changes are made so as to (initially) zero out the incremental fuel cost recovery rate.

The Rate Schedule: Pricing Methods and Risk

Changes in the pricing of the fuel and purchased power billing factor is determined through a formulary mechanism specified within the rate schedule rider. The fixed or average factor can also be differentiated by class of service, delivery voltage level, and/or time of use in order to recover revenue from customers on more of a cost-causation basis. Depending on jurisdictional regulations, the energy supply factor(s) might be procedurally revised on an annual, semiannual, quarterly, or even a monthly basis. Otherwise, the energy supply factor(s) might be implemented on an indefinite basis until such time as an over/underrecovery condition would meet an established criteria that would necessitate an adjustment.

Environmental Cost Recovery Rider

Environmental issues in the electric power industry have been elevated as a result of heightened public interest along with the enactment of more intensive governmental compliance standards. Monitoring of power plant emissions has increased. Additional improvements, such as the abatement of SO_2 and NO_X emissions at thermal power plants, are requiring significant investments and an increase in operating expenses. Large scale projects may require several years of construction. Financing of environmental projects may be supported through the issue of pollution control bonds.

Prudent expenditures incurred by power producers in order to comply with environmental regulations represent a legitimate component of the total cost of the production of electricity. These costs increase the overall revenue requirements that must be recovered through sales of electricity to customers. In lieu of making frequent changes in base rate schedules as environmental projects proceed, a separate Environmental Cost Recovery Rider can be developed and administered in a manner similar to an Energy Supply Cost Recovery Rider. However, unlike the costs of fuel and purchased power, environmental expenditures include both capital and operating and maintenance costs. The pricing mechanism must therefore include a means of recovering capital-related costs, including return on investment and depreciation.

Demand-Side Management Cost Recovery Rider

Demand-side management applications continue to be of significant interest within the utility industry, including areas where competitive retail energy markets have been implemented. A Demand-Side Management Cost Recovery rider can be utilized as a billing mechanism for recovering revenues which are then used to fund energy conservation and load management programs intended to provide social and system benefits.

Selective demand-side management programs may be directed at specific classes of customers which in turn provide the funding for those programs. For example, a home weatherization upgrade package or an appliance direct load control program may be offered exclusively to residential customers. The annual costs to support those programs might then be fully assessed to the residential class, and thus the associated monthly cost recovery factor would be made applicable only to the residential rate schedules. Depending on the types of programs offered, each customer class may have a different cost recovery factor. Large commercial and industrial customers may be excluded from the applicability of a rate schedule rider since these customers typically plan, implement, and fund demand-side management measures using their own resources. The operational complexity of these types of customers requires that demand-side management applications be customized on a case-by-case basis.

Other Special Purpose Rate Schedule Riders
The rate schedule rider approach can also be used to provide a nonstandard electric service option that is available at the customer's discretion. For example, the rider may modify the prices, provisions, or terms of the standard commercial and industrial rate schedules in order to provide the customer with seasonal service, an interruptible service option, or an economic development incentive.

Rate schedule riders can be utilized to implement any number of regulatory ordered adjustments. Such adjustments could also be applied over a limited period of time to meet specific but temporary requirements, such as the application of a Competitive Transition Charge (CTC).

Taxes and Tax Adjustments
The electric service billing process is commonly used as a vehicle for collecting a wide variety of specialized state, provincial, and local taxes. Examples of such taxes include:

- State Public Utility Tax
- State and Local Sales Tax
- Gross Receipts Tax
- Excise Tax
- Ad Valorem Tax
- Regulatory Tax
- Municipal Tax
- School Tax
- Franchise Fee

These taxes may be unbundled from the rate structure charges or partially embedded such that an external incremental charge is necessary to fully collect the taxes. The utility collects the customers' monthly bill payments and then remits the applicable tax amounts to the various taxing authorities.

At the time in which rates are designed, such as upon resolution of a rate case proceeding, some taxes are typically fully embedded in the rate schedule charges (e.g., state and corporate income tax). If subsequent changes in these tax rates occur, a billing adjustment would be necessary. However, in lieu of an immediate rate redesign, the new tax level can be properly accounted for by means of an external tax adjustment rider which increases or decreases the bill accordingly.

The Rate Schedule: Pricing Methods and Risk 331

Term of Payment

The monthly bill for electric service specifies a date upon which payment is due to be received by the service provider. For example, the due date might be 15 or 17 days after the billing date indicated on the monthly invoice. The rate schedule may specify the terms of payment of monthly bills as either a *prompt payment provision* or a *late* or *deferred payment provision*.

A prompt payment provision offers a credit amount or percent reduction from the stated charges of the rate schedule if payment is made within the allotted time. Otherwise the normal charges of the rate schedule apply. The bill might state both a net amount for early payment or a gross amount for payment after the due date. A late payment provision typically assesses an additional charge for each month in which the account is in arrears. The charge can be either a fixed fee or a percentage applicable to the unpaid balance. A deferred payment provision operates just like a late payment assessment but without the negative connotation. Instead, it offers the customer some payment flexibility for a fee.

8.5 SUMMARY

The rate schedule serves as the key component of the contract for electric service. The rate schedule presents the pricing, terms, and conditions for a designated class of service, which may be general or rather specific.

A rate schedule consists of three major sections: Qualifying Service Provisions, the Central Rate Structure, and Attendant Billing Provisions. The Qualifying Service Provisions indicate the types of customers to which the rate schedule is applicable. The Central Rate Structure is the base billing mechanism of the schedule, and it is designed to recover the fixed and variable costs of providing electric service in the form of Customer, Demand, and Energy Charges. The rate structure may consist of bundled or unbundled prices. When unbundled, the prices are structured according to the major operating functions of the power system, i.e., power production (energy supply) and transmission and distribution (power delivery). The Attendant Billing Provisions provide a means to price special aspects of electric service which are not included in the Central Rate Structure. All of the components of the rate schedule are designed to balance both the risks between the service provider and the customers and the risks between customers having different characteristics but served under the same rate schedule.

An example of a bundled rate schedule for general service is provided in Figure 8.16.[21] This type of rate schedule would be applicable to medium- to large-size commercial and industrial customers. This example summarizes the key elements of the electric service rate schedule and illustrates a typical presentation format for a rate schedule placed on file with a regulatory commission.

---

[21] The rate schedule shown in Figure 8.16 is used in the bill calculation example illustrated in Figure 7.14.

REACTIVE POWER COMPANY

RATE PPS - SCHEDULE FOR PRIMARY POWER SERVICE
Original Sheet No. 8.1

Availability: Service under this rate schedule is available on a uniform basis throughout the service territory of the Company.

Applicability: Service is applicable for general service applications where the Customer contracts for no less than 25 kW of capacity.

Character of Service: Service shall be for alternating current having a nominal frequency of 60 Hertz, unregulated, at a standard single-phase or three-phase primary voltage between 2.4 kV and 14.4 kV. Service at a standard secondary voltage will be provided under the provisions set forth within this rate schedule.

Monthly Rate for Primary Service:

Customer Charge: $225.00 for three-phase primary service
$150.00 for single-phase primary service
$80.00 for three-phase secondary service
$30.00 for single-phase secondary service

Demand Charges:

$4.59/kW for the first 50 kW of the Billing Demand
$4.37/kW for all kW in excess of 50 kW of the Billing Demand

Energy Charges:

3.11¢/kWh for the first 25,000 kWh
2.61¢/kWh for the next 100,000 kWh
1.90¢/kWh for all kWh in excess of 125,000 kWh

Load Factor Credit:

In any month in which the monthly billing energy exceeds 400 hours use of that month's billing demand, the number of kWh in excess of 400 times the billing demand shall receive a credit of 0.3¢/kWh.

Determination of Billing Demand and Energy: Energy, demand, and power factor will be measured by suitable integrating meters. The metered maximum demand shall be the average kW required during a 15-minute interval of greatest use during the month. The service will normally be metered at a primary voltage level; however, at its sole option, the Company may elect to meter the service on the load side of any transformation located on the Customer's premises. When service is measured at a secondary voltage, transformation losses will be compensated by increasing the metered maximum kW demand by 1% and increasing the metered kWh energy by 2½%. The metered maximum demand shall also be adjusted to the equivalent of an average lagging power factor of 90% prior to the application of the demand charges. The Billing Demand shall be the highest of (1) the current month's metered maximum kW demand, adjusted for power factor and any applicable transformation losses; (2) 75% of the metered maximum kW demand, adjusted for power factor and any applicable transformation losses, established during the preceding 11 months; or (3) 75% of the kW contract capacity. In no case shall the Billing Demand be less than 25 kW.

Figure 8.16 Example of a general service rate schedule for a bundled service rate.

REACTIVE POWER COMPANY

RATE PPS - SCHEDULE FOR PRIMARY POWER SERVICE
Second Revised Sheet No. 8.2

Minimum Bill: In consideration of the readiness of the Company to furnish service under this rate schedule, no monthly bill will be rendered for less than the Customer Charge plus the Demand Charge.

Delivery Voltage Adjustment: When the Company provides a transformation in order to provide service to the Customer at a voltage lower than the prevailing primary distribution voltage, an additional charge of $0.45/kW of Billing Demand shall apply.

Energy Cost Adjustment: The monthly amount of charges for electric service computed above shall be increased in accordance with the provisions of the Company's current schedule for Fuel and Purchased Power Cost Recovery.

Environmental Cost Adjustment: The monthly amount of charges for electric service computed above shall be increased in accordance with the provisions of the Company's current schedule for Environmental Cost Recovery.

Demand-Side Management Cost Adjustment: The monthly amount of charges for electric service computed above shall be increased in accordance with the provisions of the Company's current schedule for Demand-Side Management Cost Recovery.

Tax Clause: To the total of all of the above charges for electric service under this rate schedule, there shall be added applicable existing state and municipal sales taxes, any new or additional taxes, and increases or decreases in the rates of existing taxes imposed by any governmental authority upon electric service after the effective date of this rate schedule.

Order of Billing: The charges for electric service under this rate schedule shall be applied in the following sequence: Monthly Rate For Service, Minimum Monthly Bill, Delivery Voltage Adjustment, Energy Cost Adjustment, Environmental Cost Adjustment, Demand-Side Management Cost Adjustment, plus any applicable taxes.

Term of Payment: Bills rendered under this rate schedule are due upon receipt. If it becomes necessary to bill the Customer for an amount in arrears, the Company may apply a late payment charge of 1½% per month on the unpaid balance for each billing period until the amount in arrears shall have been paid.

Term of Contract: The initial term of service under this rate schedule shall be for three years and continuing thereafter from year to year until terminated by six month's written notice by either party to the other.

Service under this rate schedule is subject to the Company's Rules and Regulations Governing Electric Service.

Figure 8.16—Cont. General service rate schedule example.

9

The Mathematics of Rates

9.1 INTRODUCTION

The calculation of a bill for electric service is a series of basic mathematical operations. In general, the various monthly demand and/or energy billing determinants are each multiplied by the respective unit prices of the rate structure (along with any other applicable billing mechanisms) in order to compute an amount or charge due for each applicable rate component. The resulting charges are then summed to determine the total bill amount for the billing period of concern. Thus, rate structures, along with their attendant billing provisions and associated rate riders and clauses, function as elements of mathematical expressions.

The unit prices of a rate structure represent numerical coefficients of an equation while the billing determinants, to which these prices are applied, serve as variables of the equation. Consider a basic Doherty demand rate structure exemplified by:

> Customer Charge: $150.00 per Month
> Demand Charge: $6.18 per kW
> Energy Charge: 2.23¢ per kWh

The associated bill amount for any given monthly quantities of kW and kWh is determined by the algebraic processes indicated by:

$$\text{Bill Amount} = \$150.00 + (\$6.18 \times \text{kW}) + (\$0.0223 \times \text{kWh})$$

which is in the form of a fundamental equation given by:

$$z = Ax + By + C$$

Multi-step rate structures are represented by a system of equations. The number of equations required to fully describe a rate structure mathematically is governed by the number of blocks or steps contained within the structure. Consider the declining block energy structure exemplified by:

    Customer Charge: $14.00 per Month

Energy/Demand Charges:
First 150 kWh	@	5.6¢/kWh
Next 350 kWh	@	4.8¢/kWh
Next 500 kWh	@	4.1¢/kWh
Excess kWh	@	3.7¢/kWh

The bill amount for any monthly consumption from 0 to 150 kWh can be determined by:

Bill Amount = $14.00 + ($0.056 × kWh)

which is a linear equation in the form of $y = Ax + B$. However, when consumption exceeds 150 kWh in a month, the first block of the rate becomes a constant (i.e., it becomes a fixed charge in the amount of 150 kWh × $0.056 = $8.40). A simplified equation for the second rate block can also be developed in the form of $z = Ay + B$ by multiplying the entire amount of monthly kWh consumption by 4.8¢/kWh and by fixing a charge for the first block of the rate by the differential price between the first- and second-rate blocks. Thus, the bill amount for any monthly consumption from 150 to 500 kWh can be determined by:

Bill Amount = $14.00 + ($0.048 × kWh) + [($0.056 - $0.048) × 150]

which simplifies to:

Bill Amount = $15.20 + ($0.048 × kWh)

Simplified bill equations for the remaining blocks of the example rate structure are determined by following the same process to fix an incremental charge for each subsequent price differential. Thus, the bill amount for any monthly consumption from 500 to 1,000 kWh is determined by:

Bill Amount = $18.70 + ($0.041 × kWh)

and the bill amount for any monthly consumption of 1,000 kWh and up is determined by:

Bill Amount = $22.70 + ($0.037 × kWh)

The Mathematics of Rates

A complex rate structure, such as a multi-step Wright-Hopkinson design can be resolved as a set of elementary, two variable equations. Even time-of-use rate structures having multiple demand and energy charges can be expressed in a simplified form, albeit with more than two input variables. The basic rate structure formulas can be modified to incorporate the attendant billing provisions, such as delivery voltage adjustments, power factor clauses, and applicable rate schedule rider mechanisms. With some modifications, additional input variables are introduced; however, the resulting formulas remain as simplified resolutions of an otherwise complex billing mechanism.

Rate equations are utilized in a number of practical applications. As demonstrated above, rates can be resolved into simplified formulas for quickly calculating a monthly bill amount. With rate equations, two or more different rates can be easily compared across a broad range of usage characteristics in order to determine which of the specified rates produces the lowest bill under any given amount of kW and/or kWh consumption. Rate equations can also be used to produce graphs and charts which facilitate rate design and tariff administration. The visual depiction of rates and billing mechanisms offers great insight to how they work and interact with each other.

9.2 GRAPHICAL REPRESENTATION OF RATES

A rate can be graphed in various ways to illustrate its specific characteristics over a range of demand and/or energy usage. Three illustration methods that are particularly useful for rate design and administration include:

- Bill Charts
- Average Rate Charts
- Rate Structure Charts

Bill Charts

A *bill chart* is a Cartesian coordinate graph of the bill amounts (y-axis) which result from the application of a rate as a function of kWh usage (x-axis). The bill amount can be either the Base Bill (i.e., the amount associated with the prices and provisions of the rate schedule alone) or the Total Bill, which includes all attendant billing provisions (such as special clauses, applicable taxes, etc.). Bill charts are normally plotted using an arithmetic scale for each axis, and these graphs can be used to plot both nondemand and demand rates.

An example of a bill chart for a basic non-demand, declining block rate structure is shown in Figure 9.1. The y-axis intercept is equal to the Customer Charge, which is the minimum bill for zero kWh use. The bill amounts for various levels of energy usage are represented by contiguous straight-line segments. Each block of the rate structure yields a different line segment.

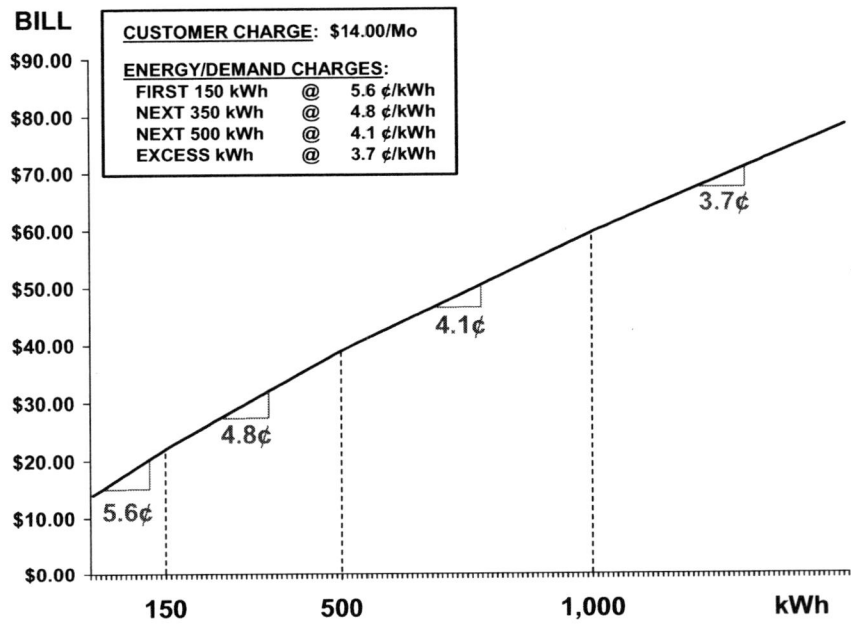

Figure 9.1 A bill chart for a basic nondemand, declining block rate structure. The bill amount is a function of one variable, kWh usage. The *y*-axis intercept is equal to the Customer Charge. The slope of the curve decreases in each of the successive rate blocks as the per unit rate declines through the structure.

The slope of the line segment in any particular kWh block is equal to the price per kWh for that block. As energy usage increases, the slope of the bill curve decreases with each subsequent block, given the full declining block structure. Conversely, an inverted block rate structure would result in an increasing slope with each subsequent block.

The use of the kWh variable usually works quite well with nondemand rate structures since the normal amount of energy usage is generally low. For example, the monthly usage of residential customers typically occurs in a range of a few hundred to a few thousand kWh, and such a scale can be depicted easily arithmetically. However, the monthly energy usage of large industrial customers can range into the millions of kWh. Rate structure detail can likely be diminished if the scale is intended to show such a volume. Thus, when graphing demand rates, it is often best to use the hours use of demand variable for the *x*-axis in order to represent energy usage. An example of a demand rate structure bill chart is shown in Figure 9.2.

The Mathematics of Rates

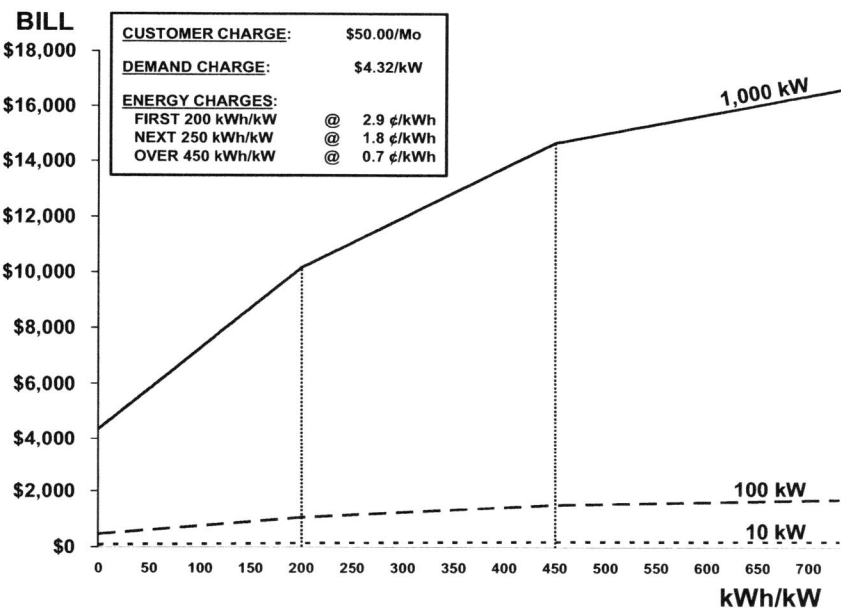

Figure 9.2 A bill chart for a Wright-Hopkinson rate structure. Bills for three different billing demands are plotted. The use of the hours use of demand for the *x*-axis allows for the illustration of bill amounts over the full range of possible monthly energy usage. A separate bill curve is required for each designated amount of billing demand.

When considering a bill chart for a demand rate structure, both kW (or kVA) demand and energy represent variables of monthly usage. Since the graph shown in Figure 9.2 is two dimensional, it is necessary to select a value representing a monthly billing demand (thereby eliminating one of the variables) in order to calculate and plot its associated bill curve. Thus, the demand rate bill chart yields a family of curves based on billing demand, as noted by the three selected demands exemplified in Figure 9.2.

A similar situation exists when plotting time-differentiated rate structure bill charts. Even with a basic nondemand, time-of-day rate, a family of curves would be necessary to represent all of the possible variations in the on-peak (or off-peak) energy usage in relation to the total monthly kWh consumption. A demand time-of-day rate would be considerably more complex, since there could be two measures of demand (e.g., on-peak demand and maximum demand) as well as two or more energy components (on-peak, shoulder, and off-peak kWh). An extensive series of bill charts would be necessary to illustrate the range of potential usage characteristics.

Average Rate Charts

The average rate realized for a given level of usage is determined by dividing the calculated bill amount by its associated total kWh. For example, a bill amount of $33.44 for 380 kWh results in an average realization of 8.8¢/kWh. An *average rate chart* is a graph of the average rate realized (*y*-axis) as a function of kWh usage (*x*-axis). An average rate chart for the previously exemplified declining block rate structure is shown in Figure 9.3.

Recall that the example declining block rate can be described by four simplified billing equations given by:

Bill Amount = $14.00 + ($0.056 × kWh) for usage between 0 and 150 kWh

Bill Amount = $15.20 + ($0.048 × kWh) for usage between 150 and 500 kWh

Bill Amount = $18.70 + ($0.041 × kWh) for usage between 500 and 1,000 kWh

Bill Amount = $22.70 + ($0.037 × kWh) for usage in excess of 1,000 kWh

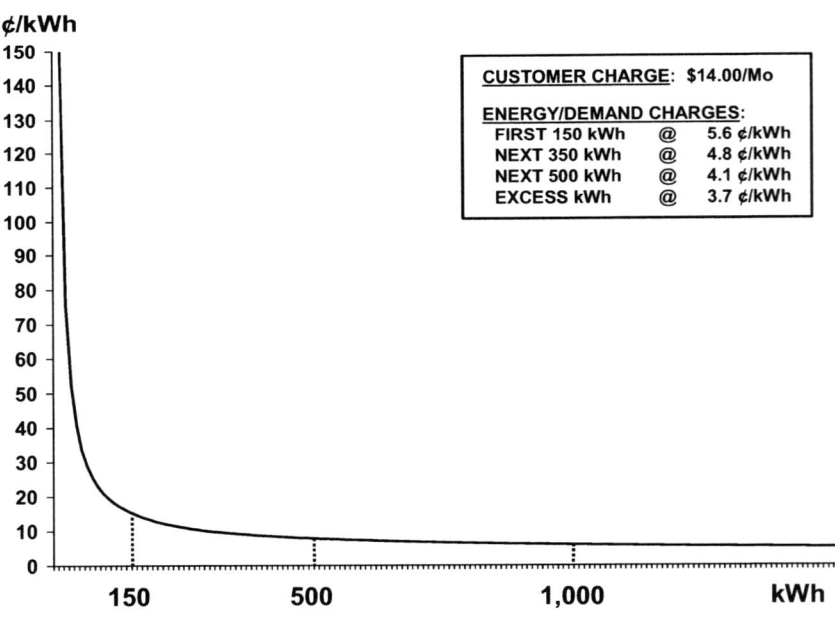

Figure 9.3 An average rate chart for a declining block rate structure. The curve is composed of segments of four hyperbolas (one for each rate block) which intersect at the block interfaces.

The Mathematics of Rates

Each of these simplified billing formulas is a linear equation in the form of $y = Ax + B$, which is redefined in terms of the rate-related nomenclature given by

$$B = r_E e + C \tag{9.1a}$$

where B = bill amount
r_E = a block specific unit price per kWh
e = total kWh usage
C = a block specific constant

By dividing both sides of Equation 9.1 by kWh, the average realization is given by

$$r_{AVG} = \frac{B}{e} = r_E + \frac{C}{e} \tag{9.1b}$$

where r_{AVG} = the average rate realized at a volume of e kWh

Equation 9.2 is in the form of an equilateral hyperbola which lies in the $+x$ and $+y$ sector of the graph. The average rate curve shown in Figure 9.3 is actually composed of segments of four hyperbolas, one for each block of the rate structure.

Recall that the bill amount for a simple Doherty rate structure can be expressed as an equation in the form of $z = Ax + By + C$. Equation 9.1 can be modified to include a Demand Charge component as given by

$$B = r_E e + r_D d + C \tag{9.2a}$$

where r_D = a specific unit price per kW (or kVA)
d = total demand in kW (or kVA)

Likewise, Equation 9.2 can be modified for the Demand Charge as given by

$$r_{AVG} = \frac{B}{e} = r_E + \frac{r_D d + C}{e} \tag{9.2b}$$

Once a billing demand has been chosen, the Demand Charge ($r_D d$) becomes a constant ($r_D d + C = C'$); thus, Equation 9.4 is also in the form of a hyperbola.

When simplified bill equations are determined for a Wright demand rate structure, each hours use step will yield a unique billing formula in the form of Equations 9.3 and 9.4. Thus, the average rate chart for the previously exemplified Wright-Hopkinson rate will be composed of three hyperbola segments for each designated billing demand, as shown in Figure 9.4.

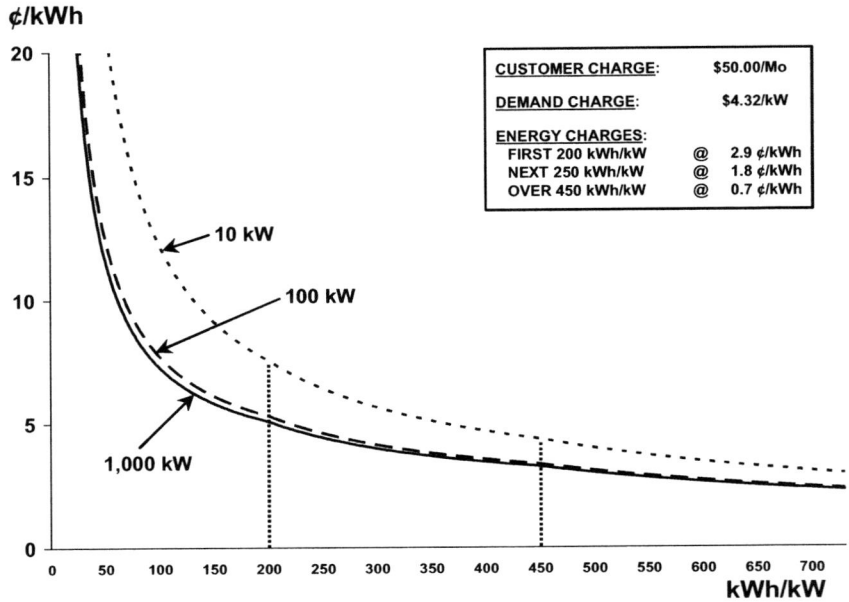

Figure 9.4 An average rate chart for a Wright-Hopkinson demand rate structure. Shown are plots for three different billing demands. Each curve is composed of segments of three hyperbolas (one for each hours use of demand step) which intersect at the HUD interfaces.

In both Figures 9.3 and 9.4, it can be observed that the average rate per kWh is significantly higher at low energy usage levels as compared to just moderately more amounts of kWh consumption. The dominating effect of the fixed charges (i.e., Customer and Demand Charges) divided by kWh is quickly diminished as energy consumption increases. At very high usage amounts, the fixed charges contribute minimally to the magnitude of the average rate.

The average realization for a bill calculated for a given set of billing determinants is a useful metric for understanding the characteristic impacts of volume or load factor changes. The average rate chart is particularly helpful when comparing the characteristics of two different rates, such as present and proposed rate structure versions, or when comparing a single rate at different demand levels. Note how the unitizing effect of the average rate in Figure 9.4 yields a much improved view of the demand rate's characteristics, as compared to Figure 9.2.

Although average rate curves can easily be drawn with the aid of a computer, the precision when interpreting readings is somewhat limited without consulting the data in tabular form. This interpretation difficulty exists because the

The Mathematics of Rates

point-to-point change in the slope of the curves tends to defy the resolution of the arithmetic scale.

The Hyperbolic Scale

Another version of the average rate chart can be constructed by replacing the arithmetic scale along the *x*-axis with a *hyperbolic scale*. An average rate curve is a hyperbola, which, due to its constantly varying slope, requires numerous data points for accurate plotting. The use of the hyperbolic scale transforms the curve of a hyperbola into a straight line. Furthermore, the complete average rate can usually be plotted with just one initial bill calculation. A hyperbolic graph of the average rate for the previously discussed declining block rate structure example is shown in Figure 9.5, while the Wright-Hopkinson demand rate example is shown in Figure 9.6.

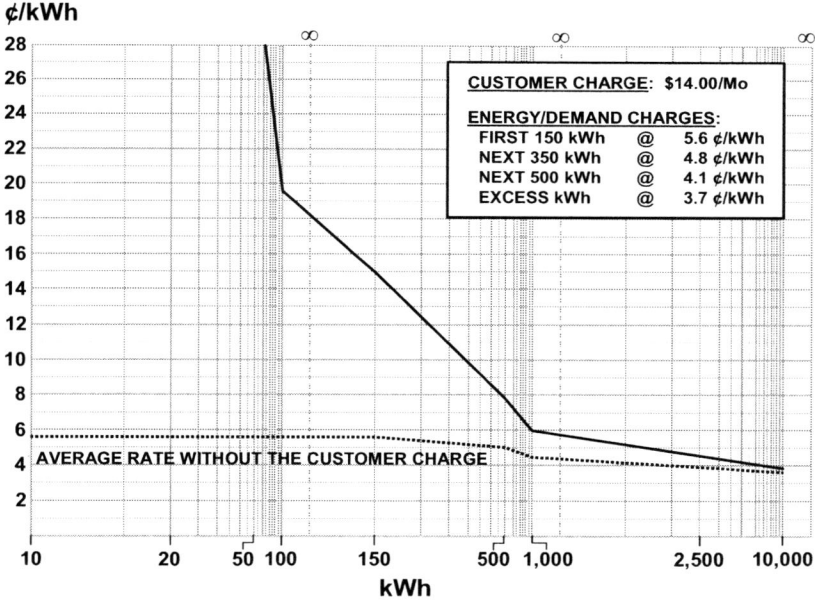

Figure 9.5 An average rate chart for a declining block rate structure using a hyperbolic scale for kWh usage. The average rate along the *y*-axis is maintained as an arithmetic scale. The *x*-axis hyperbolic scale transforms the curves of an arithmetic scale into straight lines, thereby improving the interpretation of the average realization. The rate is plotted both with and without the Customer Charge in order to illustrate the impact on the average rate due to the fixed charge component of the rate structure.

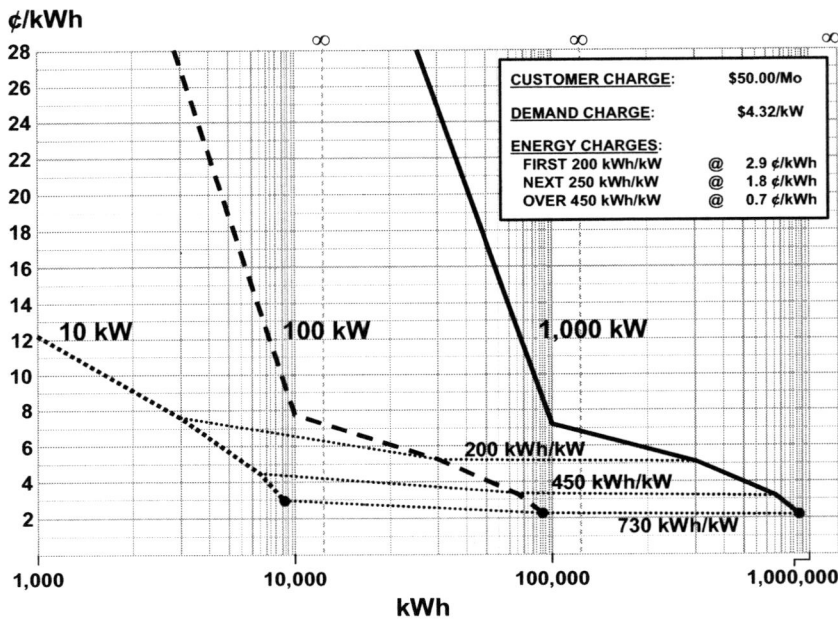

Figure 9.6a An average rate chart for a Wright-Hopkinson demand rate structure as a function of kWh. A three-cycle hyperbolic scale for kWh usage along the *x*-axis is used for the chart. The reference lines indicate lines of constant hours use for the HUD steps of the rate and for the maximum use of 730 kWh/kW (or 100% load factor).

The average rate charts in both Figures 9.5 and 9.6a utilize a three-cycle hyperbolic scale to show a wide range of usage.[1] Fundamentally, each cycle runs from 1 to 1, and the scale can be set with increasing orders of magnitude through the cycles (e.g., 1 to 10 to 100 to 1,000; 10 to 100 to 1,000 to 10,000; etc.) depending on the range of use of interest. Note that the scale does not begin at zero since the average rate for zero usage is infinite and can not be plotted. One or two cycle hyperbolic scales are also used to provide more detail for a more limited range of use. In addition, a cycle can be truncated. For demand rate structures, the *x*-axis can also be set as a hyperbolic scale representing the hours use of demand, as shown for the Wright-Hopkinson demand rate example plotted in Figure 9.6b.

---

[1] The hyperbolic graph appears to be similar to a semi-logarithmic graph. The inclusion of reference lines which represent points at infinity (∞) are not part of a logarithmic scale. Plotting an average rate accurately using semi-log scales would require several data points.

The Mathematics of Rates

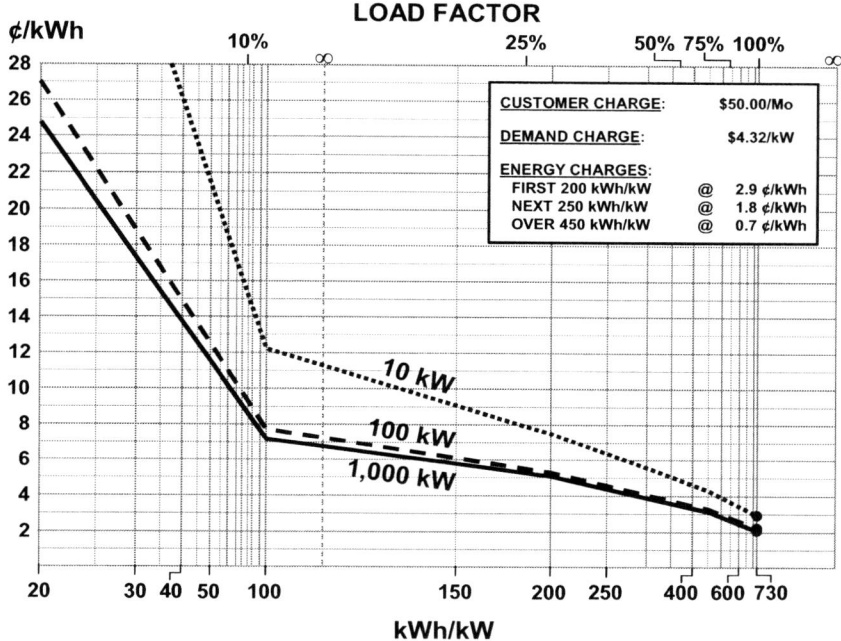

Figure 9.6b An average rate chart for a Wright-Hopkinson demand rate structure as a function of kWh per kW. A two-cycle hyperbolic scale for hours use of demand along the *x*-axis is used for the chart. Note that the first cycle is truncated below 20 hours use to allow for greater detail for higher load factors.

As discussed previously, the average rate is in the form of a hyperbola. The average rate then for a multi-block rate structure is a series of hyperbola segments, with one hyperbola resulting for each block of the rate. Figure 9.7 illustrates the two hyperbolas and their applicable segments corresponding to the average realization for a simple two block, nondemand rate. The hyperbolas continue to infinity in both the $+x$ and the $+y$ directions. Considering the mathematical form of the average rate developed in Equation 9.1b, it can be observed that if the *x*-axis energy usage was allowed to approach infinity (for either hyperbola), the average rate would approach the price per kWh of each block respectively, as determined by

$$\lim_{e \to \infty} r_E + \frac{C}{e} = r_E \qquad (9.3)$$

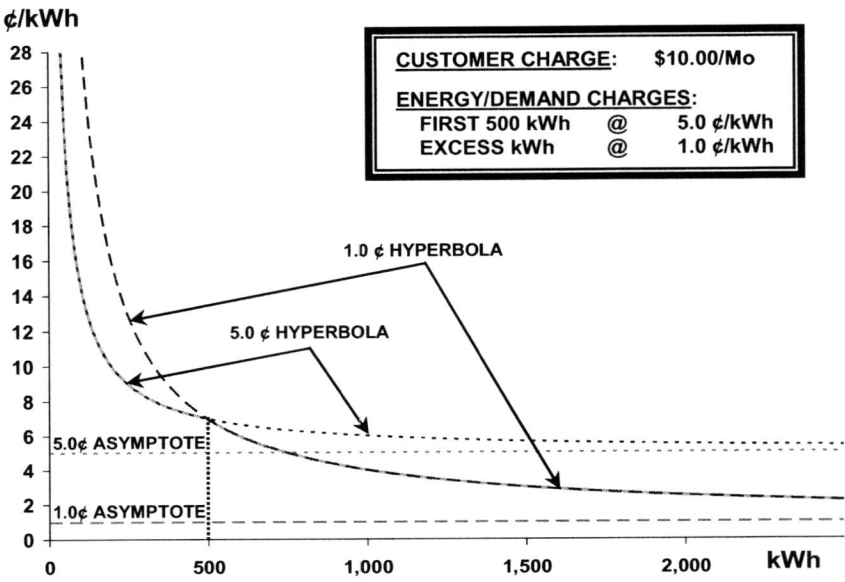

Figure 9.7 An average rate chart with asymptotes for a non-demand, two block rate structure. Shown are arithmetic scale plots of the hyperbolas which correspond to each rate block. The hyperbolas cross at the interface between the two blocks. The block prices represent the asymptotes of the hyperbolas along the *x*-axis.

The prices of the rate blocks themselves thus identify the asymptotes for the hyperbolas which run parallel to the *x*-axis. With a declining block rate structure, the asymptotes cascade down in unit price as each new block is reached (the converse would be true for an inverted block rate). With the particular rate structure illustrated in Figure 9.7, the average rate would approach but would never quite reach the tail block price of 1.0¢/kWh over the practical range of customer energy consumption.

The determination that the average realization would approach the price of each block given that each block's hyperbola approaches its associated asymptote at infinity is a key element to the simplicity of plotting an average rate on a hyperbolic graph. The hyperbolic scale contains an infinity (∞) reference point at which the rate block prices (asymptotes) can be sited without the need for any numerical calculation.

The relationship of the hyperbolic scale to the arithmetic scale is shown in Figure 9.8. The hyperbolic scale is also referred to as the *reciprocal scale* or the *inverse scale* which is obviously noted by the reciprocal calculation illustrated in Figure 9.8.

The Mathematics of Rates

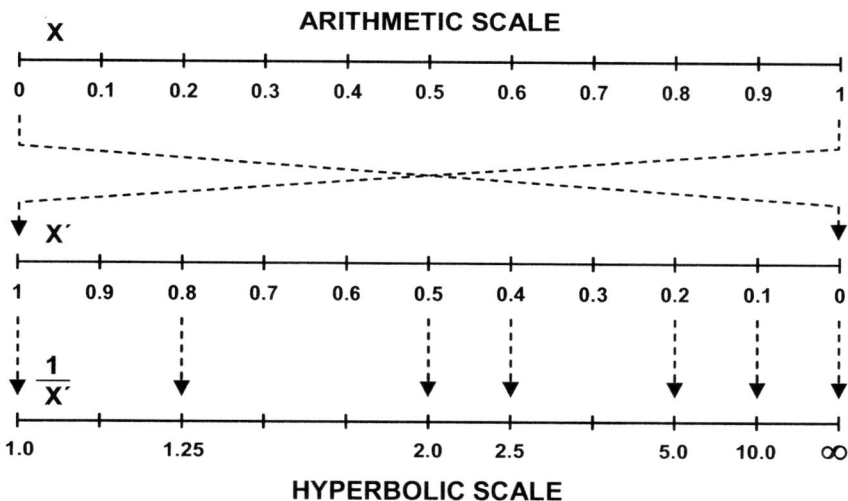

Figure 9.8 The relationship of the hyperbolic scale to the arithmetic scale. The hyperbolic scale is also referred to as the reciprocal or inverse scale due to the 1/X relationship. Note that the arithmetic scale is first plotted in reverse order so that the hyperbolic scale will increase from left to right.

The hyperbolic axis can also be scaled with reference to the arithmetic axis as determined by

$$x_H = \frac{10}{10 - x_A} \tag{9.4}$$

where x_A = a point on the arithmetic scale
        x_H = the corresponding point on the hyperbolic scale

Thus, 0 on the arithmetic scale would become 1.0 on the hyperbolic scale, 2.0 on the arithmetic scale would become 1.25 on the hyperbolic scale, and so forth.

The average rate curve shown on an arithmetic scale in Figure 9.7 for the two-block nondemand rate has been redrawn on a three-cycle, semi-hyperbolic graph in Figure 9.9. Note that there is just one block interface since there are only two price blocks in this example rate structure, and the price change occurs in the center cycle given that the x-axis starts at 10 kWh. In addition, the average rate curve traverses all three of the cycles on the graph since the (arithmetic) y-axis scale was set at a maximum value that is high enough for these price levels.

348   Chapter 9

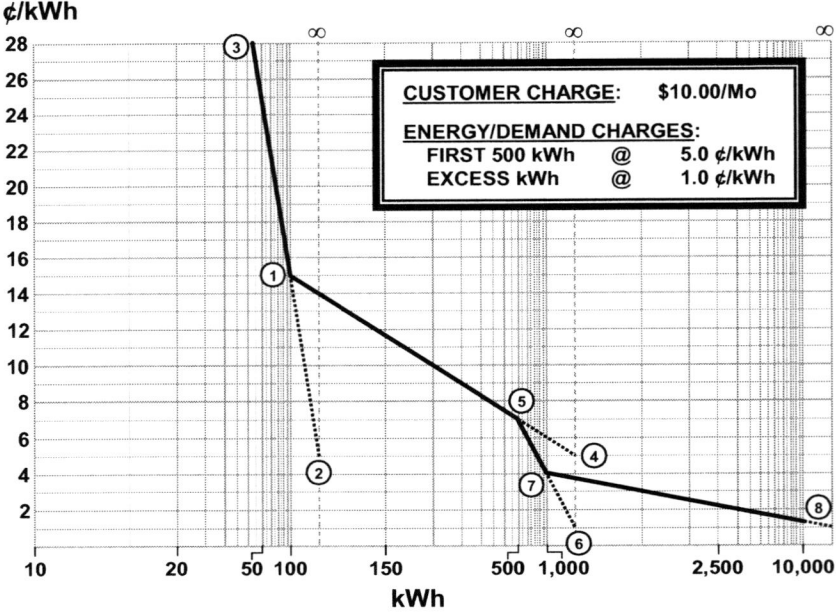

Figure 9.9 Semi-hyperbolic graph plotting coordinates for a two block nondemand rate structure. Shown are line segment end points for the example block rate plot, which represent the steps necessary to complete the graph.

To begin plotting the block rate shown in Figure 9.9, a bill amount and its associated average realization is calculated for a selected point in the first cycle of the graph. A useful starting point is at the lower of (a) the break between the first and second cycles (100 kWh in this example) or (b) the first and second block interface (i.e., 500 kWh, which occurs in the second cycle). Thus, a bill for 100 kWh would be $10.00 + ($0.05/kWh × 100 kWh) = $15.00, and the average realization would be 15.0¢/kWh, which is then sited at ①.[2] Next the first block price of 5.0¢ is sited on the first cycle's infinity line at ②. These two points define a straight line for the first cycle which is then drawn from ③ to ①. Point ② is only used for targeting the limit of the average rate of the first block at infinity.

---

[2] When plotting a block rate with a monthly Customer Charge on a first cycle scale of 10 to 100 kWh in units of dollars and cents, and when the first block of the rate is equal to or greater than 100 kWh, then the average rate at 100 kWh in cents per kWh is simply the whole dollar amount of the Customer Charge plus the first block energy price expressed in cents (e.g., 10.00 + 5.0 = 15.0¢/kWh).

The Mathematics of Rates

From the 100 kWh line at ①, the average rate curve proceeds into the second cycle of the graph. The block price per kWh does not change until 500 kWh of usage is reached. Thus, the first block price of 5.0¢/kWh is again set as the target average rate at infinity, but this point is sited now at ④ on the second cycle's infinity line. Points ① and ④ now define a straight line which traverses the second cycle. However, the line is drawn only to ⑤, which is the point at which the price changes to 1.0¢/kWh for usage in excess of 500 kWh. Still being within the second cycle, the average rate of 1.0¢/kWh at infinity is now retargeted at ⑥ on the second cycle' infinity line. Points ⑤ and ⑥ define a straight line; however, the average rate curve is only drawn to ⑦ on the 1,000 kWh line, which is the end of the second cycle.

The tail block price of 1.0¢/kWh is again sited at infinity but on the third cycle's infinity line at ⑧. Points ⑦ and ⑧ define a straight line that traverses the third cycle. Since no further price changes occur, the average rate curve is drawn from ⑦ to 10,000 kWh, the end of the third cycle. The average rate curve is now complete.

Demand rates plotted on semi-hyperbolic graphs having an hours use of demand (kWh/kW or kWh/kVA) scale follow a drawing procedure that is similar to nondemand rates plotted on a kWh scale. Once a billing demand value is selected for the chart, the Demand Charge becomes a fixed component like the Customer Charge. These charges are only necessary for the initial bill calculation, and subsequently, the energy price(s) guide the curve plotting process. However, when a demand rate includes a block energy rate structure, either with or without additional HUD steps, the hours use of demand must be determined at the end of each rate block (i.e., the point of price change) in order to draw the average rate curve. The hours use of demand at the end of the block is a function of the billing demand. For example, the Hopkinson rate with a block energy structure:

Demand Charge: $4.50/kW
Energy Charges:
First 10,000 kWh @ 3.0¢/kWh
Next 40,000 kWh @ 2.0¢/kWh
Over 50,000 kWh @ 1.0¢/kWh

can be restated as an equivalent Wright-Hopkinson rate structure for a given demand by dividing the block breakpoints by demand (e.g., First 10,000 kWh ÷ 100 kW = 100 kWh/kW or Over 50,000 kWh ÷ 250 kW = 200 kWh/kW).

Rate @ 100 kW
Demand Charge: $4.50/kW
Energy Charges:
First 100 kWh/kW @ 3.0¢/kWh
Next 400 kWh/kW @ 2.0¢/kWh
Over 500 kWh/kW @ 1.0¢/kWh

Rate @ 250 kW
Demand Charge: $4.50/kW
Energy Charges:
First 40 kWh/kW @ 3.0¢/kWh
Next 160 kWh/kW @ 2.0¢/kWh
Over 200 kWh/kW @ 1.0¢/kWh

Rate Structure Charts

A *rate structure chart* is used primarily to graph the configuration of Wright demand rates. The chart functions like a map by showing the "areas of usage" where each of the energy prices apply. The rate structure chart utilizes arithmetic scales for both its x and y axes, where the *x*-axis is defined as kW (or kVA) demand, and the *y*-axis is defined as kWh energy. Thus, a line indicated on the graph would have a slope based on the units of kWh per kW (or kVA), which is also a measure of the HUD. The HUD would be constant at every point along its length for any line that passes directly through the origin, e.g., $y = 50x$ (the HUD at $x = 10$ kW is 50 kWh, and the HUD at $x = 25$ kW is also 50 kWh/kW). However, the HUD would vary from point to point along its length for any line having a nonzero *y*-intercept, e.g., $y = 50x + 1,000$ (at $x = 10$ kW, $y = 1,500$ kWh, and the HUD at those coordinates would be 150 kWh/kW; but, at $x = 25$ kW, $y = 2,250$ kWh, and the HUD at these coordinates would be 90 kWh/kW).

Consider the multiple price rate structure, which is plotted as a rate structure chart in Figure 9.10. Only an Energy Charge structure is shown in this example. Any Customer or explicit Demand Charges that might be applicable in addition to the Wright structure's HUD price steps and blocks are not rate elements that are plotted, since they would typically apply at each and every coordinate on the chart. However, all applicable charges of the rate structure should be noted on the chart for completeness.

The chart in Figure 9.10 includes an HUD reference line which passes through the origin at a slope of 730 kWh per kW as a representation of the maximum possible usage (i.e., 100% load factor) for the average month. The "First 200 kWh per kW" and the "Next 250 kWh per kW" HUD lines of the rate would also pass through the origin except for the minimum demand requirement of 20 kW.[3] When the actual metered demand is less than 20 kW (assuming no other ratchet or contract demand requirement is applicable), the demand used for billing purposes would necessarily be 20 kW. The breakpoint between the first two HUD price steps of 5.0¢/kWh and 3.0¢/kWh is 200 kWh per kW, and at 20 kW a usage level of 4,000 kWh must then be exceeded before the 3.0¢/kWh price becomes effective in billing. Since the 20 kW minimum requirement is enforced for actual demands of 0 to 19 kW, 4,000 kWh must also be achieved in conjunction with any actual demand in this range in order to reach the billing HUD threshold of 200 kWh per kW. Thus, the effect of the minimum demand requirement is to bend the 200 kWh per kW line upward to where it runs parallel with the *x*-axis (at 4,000 kWh) between the rate minimum of 20 kW and its intersection with the 730 kWh per kW line. The same type of interaction between the rate minimum and the 450 kWh per kW line occurs at the 9,000 kWh usage level.

---

[3] Note that the two HUD steps are additive. In other words, to plot the area for the next 250 kWh per kW above the first 200 kWh per kWh, the reference HUD line must be drawn at 450 kWh per kW.

The Mathematics of Rates 351

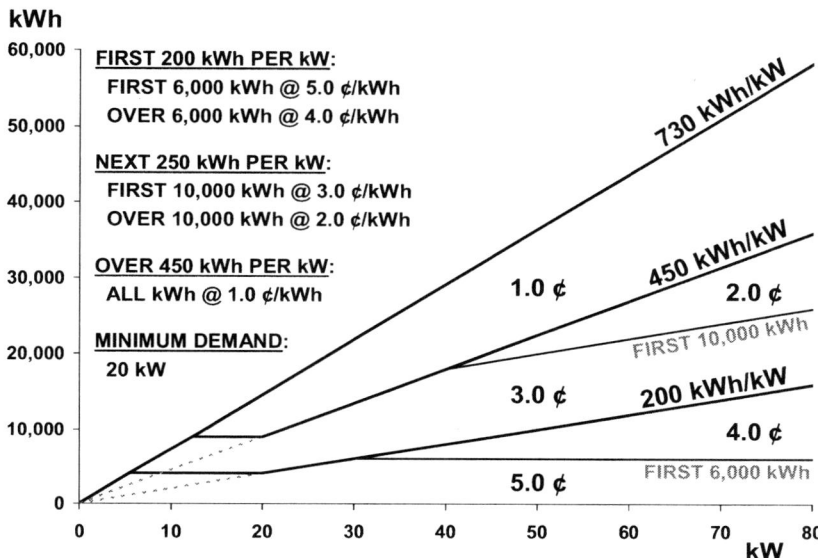

Figure 9.10 A rate structure chart for a multi-step, multi-block Wright demand rate structure.

Once the HUD lines are plotted, the other rate blocks can be graphed. The first 6,000 kWh at 5.0¢/kWh usage area within the first HUD step is represented by a line that runs parallel to the x-axis at a constant level of 6,000 kWh. The area between the first 6,000 kWh line and the first 200 kWh per kW HUD line represents the usage priced at 4.0¢/kWh.[4] Note the intersection of the 6,000 kWh line with the 200 kWh per kW line at a demand of 30 kW. Price blocks contained within succeeding HUD steps are plotted as lines that run parallel to the line of the preceding HUD step. Thus, the first 10,000 kWh of the next 250 kWh per kW step of the rate is a line that exceeds by 10,000 kWh and is parallel to the first 200 kWh per kW line, as shown in Figure 9.10. Note the intersection of the 10,000 kWh line with the 450 kWh per kW line at a demand of 40 kW.

The graph in Figure 9.10 clearly illustrates the types of billing determinant conditions under which the prices of the example Wright rate structure are applicable. It illustrates how the kWh usage is apportioned among the different prices within the rate structure. In particular, note that the 4.0¢/kWh and the 2.0¢/kWh charges would not apply to any demands at and below 30 kW regardless of the

---

[4] If the first HUD step had contained any additional price blocks, they would be plotted (additively) as lines of constant kWh usage parallel to and above the first 6,000 kWh line.

level of hours use. However, all of the prices of the rate would apply for any usage in excess of both 40 kW and 450 kWh per kW.

The rate structure chart is an aid for evaluating the billing impacts of load factor changes due to changes in a customer's operations or additions of new end-use loads. By depicting the price characteristics of the Wright rate structure in relation to demand and energy usage, it also facilitates the development of simplified billing formulas, which are used for making quick bill calculations and conducting comparisons of different rate structures.

9.3 FORMULARY REPRESENTATION OF RATES

As discussed previously, a series of simplified bill equations ($z = Ax + By + C$) can be developed for rates which have multiple unit prices, such as with block energy and/or HUD rate structures. The rate structure chart discussed above is used as a guide for determining the areas of demand and energy usage where each bill formula is applicable. The example rate plotted in Figure 9.10 will be used below for describing a methodology for formulizing a complex rate structure.

Rate Formulization Methodology

The first step for formulizing a rate is to graph its structure with enough detail so as to clearly identify the areas of demand and energy usage corresponding to each price of the rate structure (as exemplified in Figure 9.10). The lines representing various hours use levels and price blocks serve as some of the boundaries of the areas in which different billing formulas will apply. The minimum demand requirement (i.e., 20 kW in this example) also creates a boundary, and it is plotted as a vertical reference line that runs between the x-axis and the 730 kWh per kW line, as shown in Figure 9.11. In moving from left to right on the chart, a similar vertical boundary is created at each demand point where a new kWh block price becomes applicable. Specifically, the 4.0¢/kWh price only applies beyond 30 kW and the 2.0¢/kWh price only applies beyond 40 kW, whereas the 1.0¢, 3.0¢, and 5.0¢ kWh prices are applicable at all levels of demand.

The combination of the HUD lines, the kWh block lines, and the vertical demand reference lines complete the definition of the boundaries for all of the billing formula areas. Note that a billing formula area is represented by a polygon having either three or four sides.[5] Each of the formula areas for the rate shown in Figure 9.11 have been sequentially identified with a different letter of the alphabet for reference purposes. Each of the formula areas are labeled with an upper case letter except for the three areas that are situated below the minimum demand requirement. The significance of the lower-case designations will be discussed later.

---

[5] The right most formula areas of the graph are all shown to be open on the far right side; however, the boundary for closure of these polygons is assumed to exist at infinity.

The Mathematics of Rates

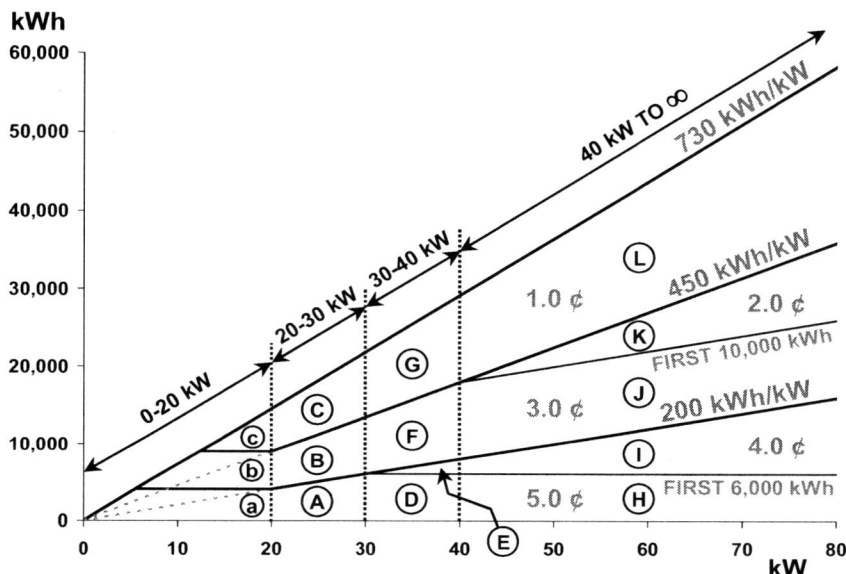

Figure 9.11 Formula areas of a rate structure chart for a multi-step, multi-block Wright demand rate. Specific areas have been identified in preparation for formulization.

In starting the formulization process, some observations can be made with reference to the graph in Figure 9.11. Note first that the selection of a formula, for the purpose of calculating a bill, is dictated by the area in which the total monthly kWh usage terminates relative to the demand. Thus, for example, a bill for 35 kW and 20,000 kWh would be calculated using the area G formula. Next, notice that the bottom row of areas (a, A, D, and H) have a common energy price of 5.0¢/kWh, and it applies exclusively to all kWh usage within any of these areas (regardless of the level of demand). The 20 kW, 30 kW, and 40 kW boundaries were established as a result of the minimum demand, the 4.0¢/kWh block, and the 2.0¢/kWh block respectively, and these limits have no bearing on the bill calculation when the total kWh usage does not exceed the upper boundaries of the bottom row of areas. As a result, the simplified billing formulas are found to be the same for all four of the bottom row areas. The mutual billing formula for areas a, A, D, and H is then simply equal to $0.05\,e$, where e represents the total monthly kWh usage variable. Further observation reveals that areas E and I also share a mutual billing formula as do areas F and J.

For presentation purposes, the billing formulas can be organized in a matrix format. Table 9.1 serves as a template for identifying the sequence of equations with respect to the associated energy and demand boundaries of each area.

Table 9.1 An Equation Matrix Template for a Wright Demand Rate Structure

	0 – 20 kW	20 – 30 kW
First 200 kWh/kW		
First 6,000 kW	a = A	A
Over 6,000 kWh	-	-
Next 250 kWh/kW		
First 10,000 kWh	b = B @ 20 kW	B
Over 10,000 kWh	-	-
Over 450 kWh/kW		
All kWh	c = C @ 20 kW	C
	30 – 40 kW	**Over 40 kW**
First 200 kWh/kW		
First 6,000 kW	D = A	H = A
Over 6,000 kWh	E	I = E
Next 250 kWh/kW		
First 10,000 kWh	F	J = F
Over 10,000 kWh	-	K
Over 450 kWh/kW		
All kWh	G	L

With the exception of the bottom row areas, all of the formulas for the areas designated with an upper case letter will contain both an energy variable, represented by e as noted above, and an explicit demand variable, represented by d. On the other hand, areas b and c were denoted with lower case letters to recognize that they are located below the minimum demand requirement of the rate. The formulas for areas b and c will contain no explicit demand variable because of the 20 kW minimum requirement, i.e., the element of demand will be represented as a fixed charge component in the formula. These fixed demand components in formulas b and c are derived by further simplifying the formulas for areas B and C) for a demand equal to 20 kW, as noted in Table 9.1.

Considering the commonality of some of the area formulas, the example rate structure can be formulized with just eight unique equations for the areas identified as: A, B, C, E, F, G, K, and L. The simple, one-kWh price solution of the bottom row area A was discussed above. Solutions of the remaining areas require an evaluation of how energy is billed among multiple kWh prices. A systematic approach for developing a formula is to first select an arbitrary point (d,e) located within the boundaries of the area of interest and then project a line straight down to the x-axis, as demonstrated for area L in Figure 9.12. This vertical line illustrates how portions of the total energy usage, e_L, would be billed through the block and HUD step prices of the rate structure. All five of the rate structure's kWh prices apply to the total energy usage terminating within area L.

The Mathematics of Rates

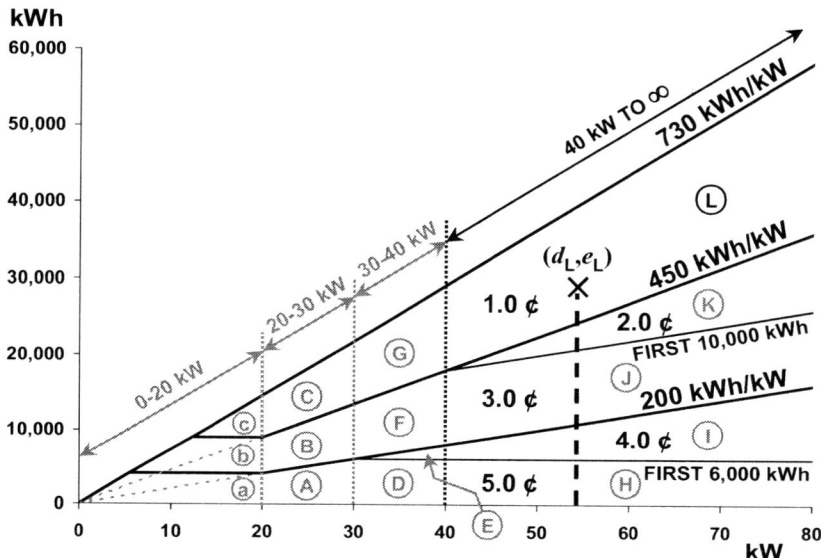

Figure 9.12 **Formulization of a specific rate area.** Selection of an arbitrary point within area L of the rate structure chart illustrates how each of the rate structure's kWh prices apply to the total energy usage, e_L.

Formulization of billing equations for areas which exceed the First 200 kWh per kW line will begin to reveal the implicit demand charges and the inherent rate tilt characteristics of the Wright rate structure. When selecting an arbitrary point (d,e) for an area located above the 200 kWh per kW line, the energy corresponding to any point on that HUD line would be equal to $(200 \times d)$, which is in units of kWh. Thus, for area B, the portion of the total energy, e_B, that would be billed at 5.0¢/kWh is then equal to $(200 \times d_B)$, while the amount of the total energy that would then be billed at 3.0¢/kWh is then equal to $[e_B - (200 \times d_B)]$ kWh. The billing formula for energy and demand usage terminating within area B is thus:

$$0.05(200\ d_B) + 0.03(e_B - 200\ d_B)$$

which simplifies to

$$0.03\ e_B + 4.00\ d_B$$

Under this formulization, all of the energy usage is billed at 3.0¢/kWh, and an explicit demand charge of $4.00 per kW emerges.

The billing formula for energy and demand usage terminating within area C is determined by three of the kWh charges:

$$0.05(200\ d_C) + 0.03(450\ d_C - 200\ d_C) + 0.01(e_C - 450\ d_C)$$

which simplifies to

$$0.01\ e_C + 13.00\ d_C$$

The inherent rate tilt characteristics of the Wright rate structure become obvious from the formulization of areas A, B, and C as together they extend over the complete range of hours use or load factor. In summary thus far,

	HUD Range	LF Range	Formula
A:	0 to 200 kWh per kW	0 to 27.4%	$0.05\ e_A + [0.00\ d_A]$
B:	200 to 450 kWh per kW	27.4% to 61.6%	$0.03\ e_B + 4.00\ d_B$
C:	Over 450 kWh per kW	61.6% to 100%	$0.01\ e_C + 13.00\ d_C$

As the hours use or load factor increases, the energy prices are noted to decline as the demand prices rise. Area A is represented by a fully tilted rate since only the energy charge is fully responsible for both the energy and demand components, while Area C's formula is a highly untilted structure. Just from these three areas, the simplified billing formulas of the Wright rate structure can be viewed as being equivalent to a system of Hopkinson type demand rates having varying relationships between their demand and energy prices, all as a function of the hours use of demand or load factor.

By further simplifying the equations for areas B and C for d = 20 kW, the minimum demand requirement, the results respectively represent the billing formulas for areas b and c:

$$0.03 e_b + 80.00 \quad \text{and} \quad 0.01 e_c + 260.00$$

Formulization of the remaining areas proceeds in the same manner as for areas B and C, except that two fixed kWh blocks are encountered (i.e., the 5.0¢/kWh price is constrained to 6,000 kWh above 30 kW and the 3.0¢/kWh price is constrained to 10,000 kWh above 40 kW). Area L encompasses all five of the energy prices, and its simplified billing formula is determined by

$$0.05(6{,}000) + 0.04(200\ d_L - 6{,}000) + 0.03(10{,}000)$$
$$+ 0.02(450\ d_L - 200\ d_L - 10{,}000) + 0.01(e_L - 450\ d_L)$$

which simplifies to

$$0.01\ e_L + 8.50\ d_L + 160.00$$

The Mathematics of Rates

Note that the 6,000 kWh and 10,000 kWh blocks yield a simple fixed charge of $160.00. In addition, the introduction of the 2.0¢/kWh and 4.0¢/kWh prices in the formulization of area L, as compared to having only three prices in area C, cause the effective Demand Charge to go down from $13.00 per kW to $8.50 per kW.[6]

The simplified billing formulas are continuous in that the formulas for two adjoining areas will calculate the same bill amount at the areas' interface, i.e., the common boundary. By calculating bills at just a few key points in the graph, the formulizations can be easily verified. For example, at (d,e) = 30 kW, 6,000 kWh, the formulas for areas A, B, D, E, and F should all yield a bill amount of $300.00; and at (d,e) = 40 kW, 18,000 kWh, the formulas for areas F, G, J, K, and L should all yield a bill amount of $680.00.

Once the formulas are developed for all areas, they are summarized in a tabular form for ease of application. Each of the simplified billing formulas listed in Table 9.2 correspond to an area identified in the equation matrix template of Table 9.1.

Table 9.2 Simplified Billing Formulas for a Wright Demand Rate Structure

	0 – 20 kW	20 – 30 kW
First 200 kWh/kW		
First 6,000 kWh	$0.05e$	$0.05e$
Over 6,000 kWh	-	-
Next 250 kWh/kW		
First 10,000 kWh	$0.03e + 80.00$	$0.03e + 4.00d$
Over 10,000 kWh	-	-
Over 450 kWh/kW		
All kWh	$0.01e + 260.00$	$0.01e + 13.00d$
	30 – 40 kW	**Over 40 kW**
First 200 kWh/kW		
First 6,000 kWh	$0.05e$	$0.05e$
Over 6,000 kWh	$0.04e + 60.00$	$0.04e + 60.00$
Next 250 kWh/kW		
First 10,000 kWh	$0.03e + 2.00d + 60.00$	$0.03e + 2.00d + 60.00$
Over 10,000 kWh	-	$0.02e + 4.00d + 160.00$
Over 450 kWh/kW		
All kWh	$0.01e + 11.00d + 60.00$	$0.01e + 8.50d + 160.00$

---

[6] The inclusion of kWh blocks within the Wright rate structure is sometimes utilized as a proxy for a delivery voltage credit, as opposed to a separate and explicit credit per kW or kVA, under the premise that customers with higher demands are served at higher voltages.

Simplifying Complex Billing Structures

The previous discussion focused on the elementary formulization of a multiple HUD step and kWh block Wright demand rate structure. Other possible features of such a rate were not taken into consideration, such as a Customer Charge, an explicit Demand Charge, and other rate schedule provisions. When other billing elements are involved, they can be either incorporated into the original formulization or, more simply, added in afterwards by modifying the basic billing formulas in a progressive manner.

Consider now a more comprehensive rate structure along with a variety of attendant billing provisions, as given by

Monthly Rate for Secondary Service

Customer Charge: $25.00 per Month

Demand Charges:
First 30 kW   @ $2.75 per kW
Over 30 kW   @ $2.25 per kW

Energy Charges:

First 200 kWh per kW
  First 6,000 kWh   @   5.0¢ per kWh
  Over 6,000 kWh   @   4.0¢ per kWh
Next 250 kWh per kW
  First 10,000 kWh   @   3.0¢ per kWh
  Over 10,000 kWh   @   2.0¢ per kWh
Over 450 kWh per kW
  All kWh   @   1.0¢ per kWh

Metering Voltage Adjustment:
When metering of the service occurs at a primary voltage level, for billing purposes the metered demand in kW will be reduced by 1.0%, and the metered energy in kWh will be reduced by 1.5%.

Delivery Voltage Adjustment:
When customer takes service at a primary voltage, a monthly credit of $1.45 per kW of billing demand shall apply.

Power Factor Adjustment:
In any month when the metered maximum kVAR demand exceeds an amount equal to 48.43% of the metered maximum kW demand, such excess kVAR shall be billed at $1.00 per kVAR.

The Mathematics of Rates

Other Adjustment Factors:
All kWh are subject to an additional charge of 1.975¢ per kWh.
[Note: Treated as a net figure in this example, a number of individual kWh-related charges and credits could apply, such as fuel and purchased power, DSM, environmental, social benefits, etc.]

Special Surcharge:
An amount equal to 0.4% shall be applied to the subtotal of the rate and all applicable adjustment factors in determination of the total bill.

Some or all of the other pricing features of the above example rate schedule can be incorporated with the basic simplified billing equations in order to determine a base bill or total bill amount. For instance, the elementary formula for area L could be modified in the following manner for metering and delivery at a primary service level and where the monthly power factor is below the base requirement of 90%.

Elementary Formula: $0.01\ e_L + 8.50\ d_L + 160.00$

Add Customer and Demand Charges:

$$0.01\ e_L + 8.50\ d_L + [2.25\ (d_L - 30) + (2.75 \times 30)] + 160.00 + [25.00]$$

$$0.01\ e_L + 10.75\ d_L + 200.00$$

Add Power Factor Adjustment Charge:
[Where d_{kVAR} is equal to the metered maximum kVAR demand]

$$0.01\ e_L + 10.75\ d_L + [1.00\ (d_{kVAR} - 0.4843\ d_L)] + 200.00$$

$$0.01\ e_L + 10.266\ d_L + 1.00\ d_{kVAR} + 200.00$$

Add Primary Delivery Voltage Credit:

$$0.01\ e_L + 10.266\ d_L + [-1.25\ d_L] + 1.00\ d_{kVAR} + 200.00$$

$$0.01\ e_L + 9.016\ d_L + 1.00\ d_{kVAR} + 200.00$$

Adjust for Primary Metering:

$$0.01\ [0.985\ e_L] + 9.016\ [0.99\ d_L] + 1.00\ [0.99\ d_{kVAR}] + 200.00$$

Base Bill: $\mathbf{0.00985\ e_L + 8.92584\ d_L + 0.99\ d_{kVAR} + 200.00}$

Add Other Adjustment Factors:

$$0.00985\ e_L + [0.01975\ (0.985\ e_L)] + 8.92584\ d_L + 0.99\ d_{kVAR} + 200.00$$

$$0.02930375\ e_L + 8.92584\ d_L + 0.99\ d_{kVAR} + 200.00$$

Add Special Surcharge:

$$1.004 \times [0.02930375\ e_L + 8.92584\ d_L + 0.99\ d_{kVAR} + 200.00]$$

Total Bill: $\quad \mathbf{0.029421\ e_L + 8.96154\ d_L + 0.994\ d_{kVAR} + 200.80}$

The results of the advanced formulization and an actual bill calculation are compared for monthly billing determinants of: 50.0 kW, 37.5 kVAR, and 30,000 kWh.

		Amount
Customer Charge:		$25.00

Demand Charges:

Metered Maximum kW Demand	50.0	
Less 1.0% for Metering at Primary	0.5	
Billing Demand	49.5	
First 30 kW @ $2.75 per kW		82.50
Next 19.5 kW @ $2.25 per kW		43.88

Primary Service Credit:

49.5 kW @ $(1.25) per kW		(61.88)

Power Factor Adjustment:

Base kVAR = 49.5 kW × 0.4843 kVAR per kW = 24.0

Metered Maximum kVAR Demand	37.5	
Less 1.0% for Metering at Primary	0.4	
	37.4	
Less Base kVAR	4.0	
Excess kVAR	13.1	
13.1 Excess kVAR @ $1.00 per kVAR		13.10

The Mathematics of Rates

		Amount
Energy Charges:		

Metered kWh Energy	30,000	
Less 1.5% for Metering at Primary	450	
	29,550	
First 200 kWh per kW		
First 6,000 kWh @ 5.0¢ per kWh		$300.00
Next 3,900 kWh @ 4.0¢ per kWh		156.00
Next 250 kWh per kW		
First 10,000 kWh @ 3.0¢ per kWh		300.00
Next 2,375 kWh @ 2.0¢ per kWh		47.50
Over 450 kWh per kW		
7,275 kWh @ 1.0¢ per kWh		72.75
BASE BILL		**$978.85**

Base Bill by Formula:
$0.00985\,(30{,}000) + 8.92584\,(50.0) + 0.99\,(37.5) + 200.00 \;=\;$ $978.92

                                                      Difference   +0.07

Net Adjustment Clauses:

29,550 kWh @ 1.975¢ per kWh		583.61
Subtotal		$1,562.46
0.4% Special Surcharge:		6.25
TOTAL BILL		**$1,568.71**

Total Bill by Formula:
$0.029421\,(30{,}000) + 8.96154\,(50.0) + 0.994\,(37.5) + 200.80 \;=\;$ $1,568.78

                                                      Difference   +0.07

A small difference exists between the line item bill calculation and the formulary bill calculation due to rounding. In this example, the difference is only seven cents (a percent difference of less than 0.01%) at both the Base Bill and Total Bill levels. Such an accuracy is more than adequate to allow for the application of simplified billing formulas in estimating monthly bill amounts and for conducting other rate analyses that are dependent on formulary methods, such as in the development of differential charts.

Geometric Interpretation of Billing Formulas

Rate structure charts are plotted in the xy-plane, where x relates to demand and y relates to energy. This two-dimensional chart shows the areas (or polygons) in which different unit prices are applicable. The chart in Figure 9.13 shows the four polygons that result from the plot of an example two-step Wright-Hopkinson rate structure. The corners of the polygons are labeled for reference with the numbers 1 through 8, where 3, 6, and 8 represent intermediate points since the right-hand boundaries of the graph are assumed to exist at infinity.

The simplified billing formulas of a rate are in the form of the mathematical expression for a plane in space: $z = Ax + By + C$. Solution of a billing equation for a given (x,y) produces a bill amount which can be plotted along the z-axis. A plot of the bills for all points in the areas defined in Figure 9.13 yields the three-dimensional view illustrated in Figure 9.14. The projection of the bill amounts above the xy-plane appears as a "roof" composed of four different panels that are connected at the lines associated with the rate's 210 kWh per kW step and 20 kW minimum demand requirement. These four plane segments have different slopes, and their shapes resemble the areas of the two-dimensional plot in the xy-plane.

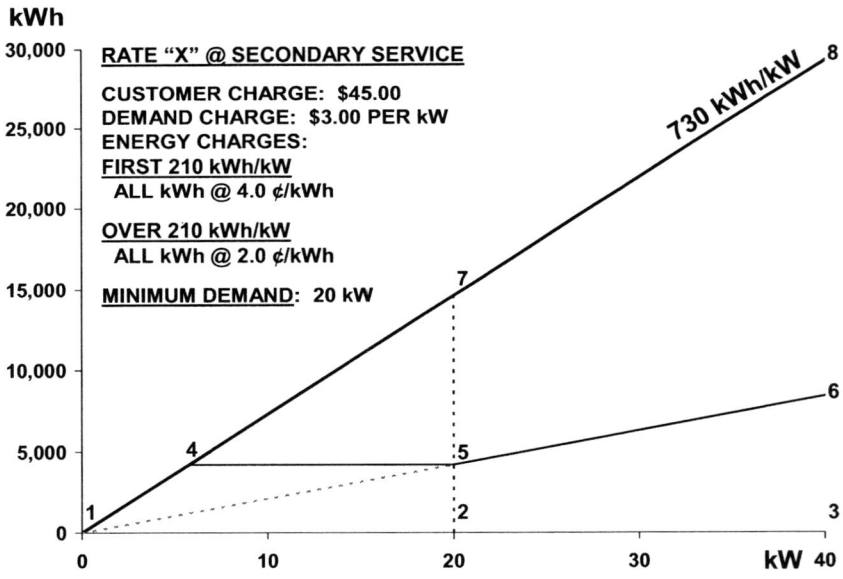

Figure 9.13 **A rate structure chart of a Wright-Hopkinson form of rate.** The labels 1 through 8 identify the corners of the different price areas or polygons of the rate (points 3, 6, and 8 are intermediate points since the actual corners are located at infinity).

The Mathematics of Rates

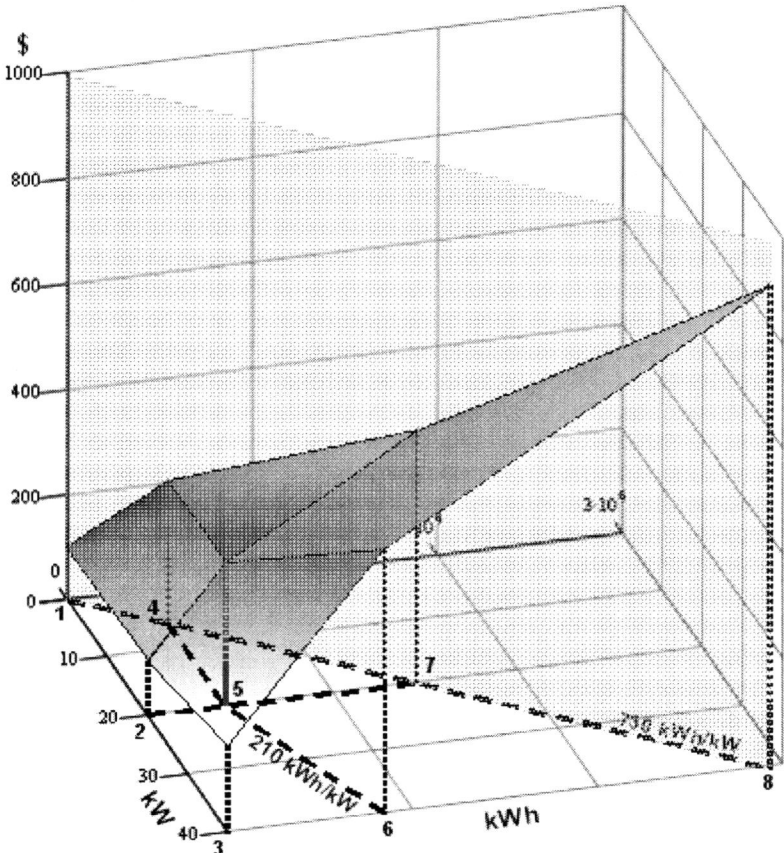

Figure 9.14 A 3-D plot of bills. The plot is produced by solving for a rate's simplified billing equations (which are in the form of $z = Ax + By + C$) for all possible demand and energy usage coordinates. The plot on the xy-plane is a rate structure chart.

The three-dimensional bill chart provides insight as to how rapidly a bill amount can change with minor changes in energy and/or demand usage. While the multi-directional slope of a plane segment is constant, the slope transformation at the plane segment interfaces can be quite significant. The visualization of the three-dimensional view of a rate structure is also helpful for understanding how two or more rates may interact with each other when more than one rate is applicable to a customer.

9.4 RATE COMPARISON METHODS

Rate comparisons are conducted for a variety of purposes. For example, with rate design, the present and proposed versions of the same rate can be compared in order to evaluate the change in the bill amount or average realization at different levels of demand and/or energy usage. Two or more different rates within a single tariff, which are all applicable to a qualified customer, can be compared in order to determine which one of the rates would produce a lower bill amount. Such an analysis would be helpful to customers for selecting between alternative rate schedules. Similarly, comparable rates from two or more separate tariffs can be compared to make a competitive assessment of different service suppliers.

Rates can be compared in both tabular and graphical fashion. bill charts and Average rate charts, as previously described, are excellent tools for comparing two or more rates; although, several charts may be necessary for making a full evaluation of the differences between demand-based rate structures. Rate structure charts are used for the production of differential charts, which provide comparisons of two or more rates across the entire range of possible energy and demand usage. Developing the simplified billing formulas for the rates is a prerequisite step for establishing both rate comparison charts and tables.

The following two rates will be used as examples for the subsequent discussion on rate comparisons:

Rate Schedule "X"	**Rate Schedule "Y"**
Monthly Rate for Secondary Service	Monthly Rate for Primary Service
Customer Charge: $45.00	Customer Charge: $150.00
Demand Charge: $3.00/kW	Demand Charge: $3.10/kVA
Energy Charges:	Energy Charges:
First 210 kWh/kW @ 4.0¢/kWh	First 360 kWh/kVA @ 2.0¢/kWh
Over 210 kWh/kW @ 2.0¢/kWh	Over 360 kWh/kVA @ 1.0¢/kWh
Minimum Demand: 20 kW	Minimum Demand: 50 kVA
Primary Service Credit: 60¢/kW	Secondary Service Charge: 50¢/kVA

In order to conduct an equitable comparison, it is imperative that both rates are expressed on an equivalent basis. Therefore, both rates must be stated at the same voltage level, and the charges must be defined on either a kW or a kVA basis. Either rate could be converted for analysis. For this example, Rate Y is selected to be transformed to a secondary service rate by adding the Secondary Service Charge of 50¢/per kVA to the $3.10 per kVA Demand Charge for a total of $3.60 per kVA. This voltage adjusted Demand Charge is then converted to a kVA Demand Charge by dividing by an assumed power factor, 90% for this example; thus,

The Mathematics of Rates

$3.60 per kVA ÷ 0.9 kW/kVA = $4.00 per kW. Other components of the rate that are based on kVA must also be converted to a kW basis. The HUD step of 360 kWh per kVA is likewise divided by 0.9 with a result of 400 kWh per kW, while the Minimum Demand requirement of 50 kVA is multiplied by the 90% power factor assumption to compute an adjusted rate minimum of 45 kW. The equivalent rates are summarized below (with modifications noted in boldface).

Rate Schedule "X"
Monthly Rate for Secondary Service

Customer Charge: $45.00
Demand Charge: $3.00/kW
Energy Charges:
  First 210 kWh/kW @ 4.0¢/kWh
  Over 210 kWh/kW @ 2.0¢/kWh

Minimum Demand: 20 kW

Rate Schedule "Y"
Monthly Rate for **Secondary** Service
[Based on a power factor of 90%]
Customer Charge: $150.00
Demand Charge: **$4.00/kW**
Energy Charges:
  First **400 kWh/kW** @ 2.0¢/kWh
  Over **400 kWh/kW** @ 1.0¢/kWh

Minimum Demand: **45 kW**

Bill Comparison Tables

Tables can be used to compare bill amounts for two or more rates for discrete demand and energy billing determinants. For example, monthly bills for a maximum demand of 40 kW and various levels of kWh usage are listed in Table 9.3 for Rate X and Rate Y. This table also shows the difference amount for each pair of bills as well as the percent difference between the bill amount for Rate X and the bill amount for Rate Y. In this format, it can be noted easily which of the rates produces the lowest bill for any given level of kWh usage.

Table 9.4 presents a bill comparison for Rates X and Y in an alternative format. Monthly bills for both of the rates are paired together in a matrix whose dimensions are defined by a range of demands and hours use levels.

Table 9.3 Bills for Two Rates at 40 kW and Various kWh Usage Levels

| kW | kWh | $ Monthly Bill Amount | | Difference | |
		Rate X	Rate Y	Amount	Percent
40	4,000	325.00	410.00	85.00	26.2
40	8,000	485.00	490.00	5.00	1.0
40	12,000	573.00	570.00	(3.00)	(0.5)
40	16,000	653.00	650.00	(3.00)	(0.5)
40	20,000	733.00	710.00	(23.00)	(3.1)
40	24,000	813.00	750.00	(63.00)	(7.7)
40	29,200	917.00	802.00	(115.00)	(12.5)

Table 9.4 Bills for Two Rates at Various kW Demands and Hours Use Levels

kW	Rate	kWh per kW						
		100	200	300	400	500	600	730
25	X	220.00	320.00	375.00	425.00	475.00	525.00	590.00
25	Y	380.00	430.00	480.00	530.00	580.00	630.00	692.50
30	X	255.00	375.00	441.00	501.00	561.00	621.00	699.00
30	Y	390.00	450.00	510.00	570.00	630.00	690.00	729.00
35	X	290.00	430.00	507.00	577.00	647.00	717.00	808.00
35	Y	400.00	470.00	540.00	610.00	680.00	720.00	765.50
40	X	325.00	485.00	573.00	653.00	733.00	813.00	917.00
40	Y	410.00	490.00	570.00	650.00	710.00	750.00	802.00
45	X	360.00	540.00	639.00	729.00	819.00	909.00	1,026.00
45	Y	420.00	510.00	600.00	690.00	735.00	780.00	838.50
50	X	395.00	595.00	705.00	805.00	905.00	1,005.00	1,135.00
50	Y	450.00	550.00	650.00	750.00	800.00	850.00	915.00
55	X	430.00	650.00	771.00	881.00	991.00	1,101.00	1,244.00
55	Y	480.00	590.00	700.00	810.00	865.00	920.00	991.50
60	X	465.00	705.00	837.00	957.00	1,077.00	1,197.00	1,353.00
60	Y	510.00	630.00	750.00	870.00	930.00	990.00	1,068.00
65	X	500.00	760.00	903.00	1,033.00	1,163.00	1,293.00	1,462.00
65	Y	540.00	670.00	800.00	930.00	995.00	1,060.00	1,144.50

In Table 9.4, the lower bill in each of the bill pairs has been highlighted, with the shaded bills representing where Rate X is lower and the unshaded bills representing where Rate Y is lower. Given the two-dimensional nature of this table (a field of demands vs. HUD levels), it is easy to observe that there is a distinct demarcation between two areas that indicate which of the two rates produces the lower bill amounts. For example, at 40 kW and 200 kWh per kW, Rate X is lower; however, at 40 kW and 300 kWh per kW, Rate Y is lower. Thus, a point must exist between 200 kWh per kW and 300 kWh per kW (for a demand of 40 kW) at which both rates produce the same bill amount. Furthermore, a line of equality that progresses through the rows and columns between the shaded and unshaded areas also exists. The exact line of equality between two (or even more) rates can be determined by further analysis of the rates' simplified billing formulas and plotted as a differential chart.

Differential Charts

A simplified billing formula for an area of a rate structure, having boundaries in the demand and energy field (xy-plane), represents a segment of a plane in space defined by $z = Ax + By + C$. When the bill amounts for a given rate are plotted along the z-axis, the bill plane segment(s) can be viewed, and the number of plane segments revealed depends on the number of bounded areas of the rate. A plot of the bill amounts for the four bounded areas of example Rate X produced the three-dimensional graph shown previously in Figure 9.14.

If the bill amounts of the example Rate Y were plotted in similar fashion to Rate X, the resulting graph would also reveal four bill plane segments since the basic structures of the two rates are similar. However, the plane segments of Rate Y would differ somewhat from the plane segments of Rate X in terms of their dimensions and orientations in space. If the bill plane segments for both Rate X and Rate Y were plotted together on the same graph, there would be coordinates in space at which the plane segments would intersect, i.e., at the xyz-points where the bill amounts of Rate X are equal to the bill amounts of Rate Y. These points of equality would be expected given the observation of the bill comparisons shown previously in Table 9.4.

The intersection of two planes is a straight line. Thus, the intersection of two bill plane segments located above the demand and energy field yields a straight line segment in space. With Rate X and Rate Y, four line segments result from the various crossings of their bill plane segments, as illustrated in Figure 9.15. Although each line segment has a different length and orientation, together the four line segments generate a continuous path in space along which the bill amounts of both rates are equal. As seen in Figure 9.15, the line of equality in space is projected downward to a line of equality on the demand and energy field. The two-dimensional view of the line of equality in the xy-plane is known as a *differential chart*.

The line of equality segments for the different bounded areas of two rates can be determined by setting the billing formulas of the rates equal to each other; thus,

Rate #1 Bill Amount = Rate #2 Bill Amount

is determined by

$$A_1 e + B_1 d + C_1 = A_2 e + B_2 d + C_2 \tag{9.5a}$$

Note that by setting the formulary bill amounts equal to each other, the z variable (i.e., the bill amount) is eliminated from the equation. Simplification of the equality results in a straight line in the xy-plane that is in the form of $y = Ax + B$, as given by

$$e = \frac{(B_1 - B_2) \times d + (C_2 - C_1)}{(A_1 - A_2)} \tag{9.5b}$$

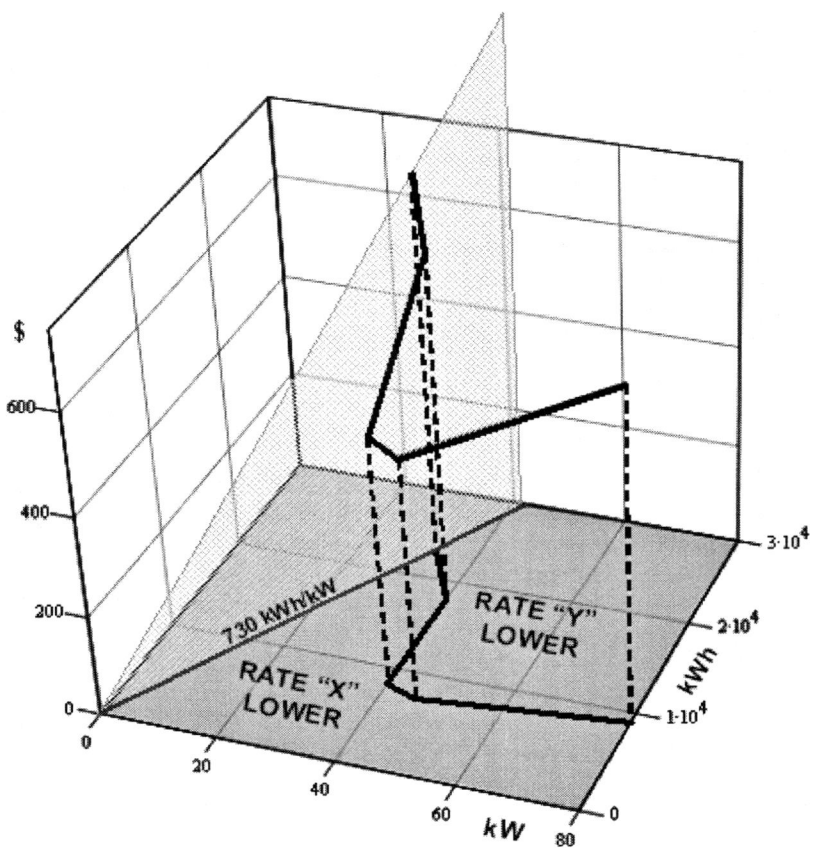

Figure 9.15 A differential chart showing the line of equality between two rates. The bill amounts produced by both rates are the same on every point along the line of equality. The bill amounts are lower for Rate X than for Rate Y on one side of the line of equality, while the bill amounts for Rate Y are lower than for Rate X on the other side of the line of equality.

Note from Equations 9.5a and 9.5b that if the demand coefficients (B_1 and B_2) were the same, the solution would be that $e = (C_2 - C_1) \div (A_1 - A_2)$ which is equal to a constant, or a straight line parallel to the x-axis. On the other hand, if the energy coefficients (A_1 and A_2) were the same, the solution would be that $d = (C_2 - C_1) \div (B_1 - B_2)$ which is also equal to a constant, or a straight line parallel to the y-axis.

The Mathematics of Rates

Although a line extends to infinity in both directions, only that portion of the line of equality that lies within boundaries defined in the xy-plane (i.e., the demand and energy field) is of valid concern. Furthermore, a solution line must pass through the bounded area in which the two bill equations are applicable in order for it to be a legitimate result.

Table 9.5 shows the simplified billing formulas for rates X and Y. Each rate has four formulas since each rate has one HUD step and a minimum demand requirement. These two boundaries create four billing formula areas within each rate. However, these specific boundaries are not the same for the two rates (i.e., 210 kWh per kW vs. 400 kWh per kW and 20 kW vs. 45 kW), so the resulting four areas of each of the rates would not align with each other if plotted to scale on the same graph. The formulas in Table 9.5 correlate to the areas of the rate structure charts that are illustrated in Figure 9.16.

In Figure 9.17, the boundaries of Rate X and Rate Y are plotted together to create the framework for graphing the differential chart. The four areas of Rate X and the four areas of Rate Y produce eight zones on the differential chart.[7] The combinations of billing formulas for the two rates which are applicable within each of the zones is also shown in Figure 9.17.

Table 9.5 Simplified Billing Formulas for Example Rates X and Y

Rate X at Secondary

	0 – 20 kW	Over 20 kW
First 210 kWh/kW		
All kWh	a_X: 0.04e + 105.00	A_X: 0.04e + 3.00d + 45.00
Over 210 kWh/kW		
All kWh	b_X: 0.02e + 189.00	B_X: 0.02e + 7.20d + 45.00

Rate Y at Secondary (90% Power Factor Assumed)

	0 – 45 kW	Over 45 kW
First 400 kWh/kW		
All kWh	a_Y: 0.02e + 330.00	A_Y: 0.02e + 4.00d + 150.00
Over 400 kWh/kW		
All kWh	b_Y: 0.01e + 510.00	B_Y: 0.01e + 8.00d + 150.00

---

[7] The number of Differential Chart zones is not necessarily equal to the sum of the billing formulas of the two rates being compared, which is eight in this example. If the minimum demand requirement of Rate X was increased to 25 kW or more, a ninth zone would emerge in the left-hand corner of the existing zone labeled as "V" even though the total number of formulas would not change.

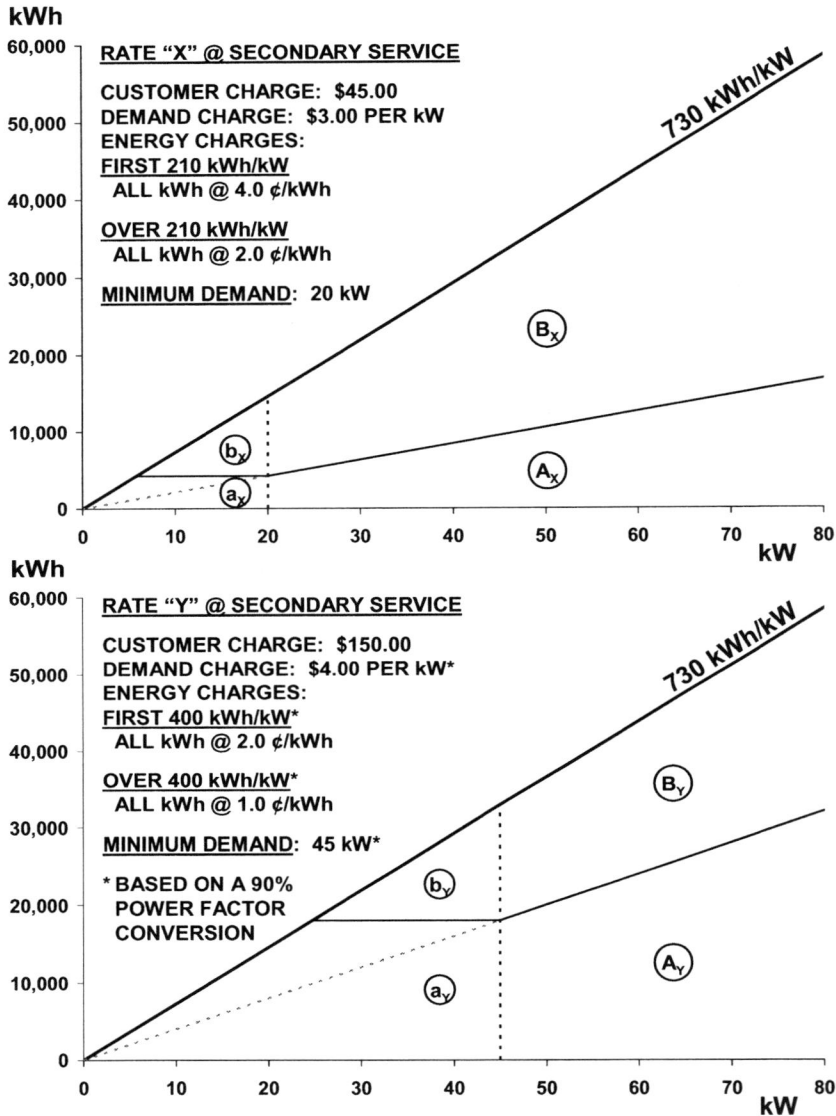

Figure 9.16 Rate structure charts prepared for a rate comparison. The formula areas for equivalent versions of Rates X and Y have been identified on each chart. The scales of both graphs are identical so that one chart can be overlaid directly with the other chart. Note how both formula areas a_X and b_X of Rate X are encompassed entirely within the one formula area a_Y of Rate Y.

The Mathematics of Rates

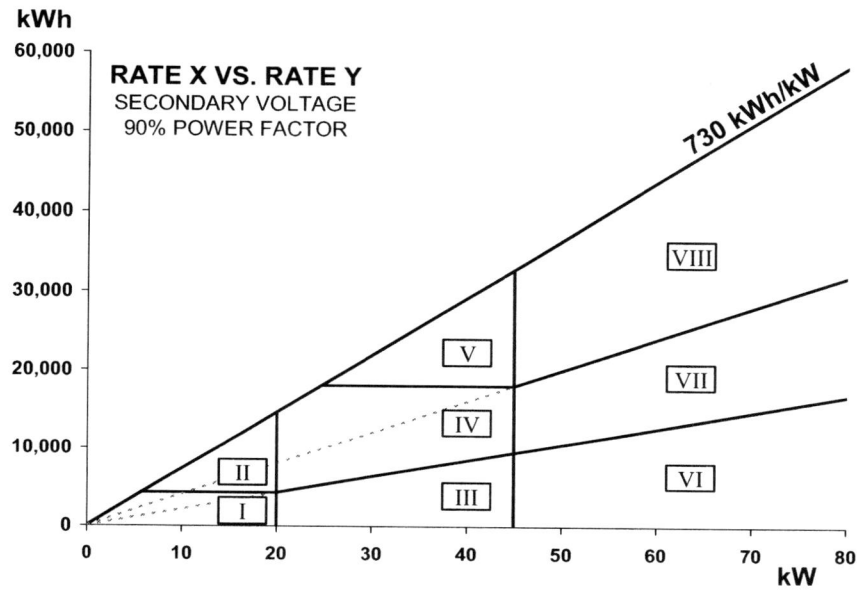

Formulas	I	II	III	IV	V	VI	VII	VIII
Rate X	a_X	b_X	A_X	B_X	B_X	A_X	B_X	B_X
Rate Y	a_Y	a_Y	a_Y	a_Y	b_Y	A_Y	A_Y	B_Y

Figure 9.17 The framework for development of a differential chart. The rate structures of rates X and Y are graphed together in order to determine zonal boundaries within which the various simplified billing formulas of the rates pertain. In this example, a total of eight combinations of formulas are applicable across the total demand and energy field.

To determine the exact route of the line of equality between Rate X and Rate Y through the demand and energy field, the pairs of formulas in each zone are equated in the manner described previously by Equations 9.7 and 9.8. The result of each equation pair must be validated by ensuring that the solution line actually traverses its associated zone by crossing two zonal boundaries (or in less common cases, follows directly in line with a zone boundary). The solution line in an adjacent zone must also intersect the other zone's solution line at the common boundary. The line of equality is therefore actually comprised of zonal line segments, each having end points at two of its particular zone's boundaries. The line of equality is completed when both ends of the continuous string of line segments reaches a terminal point (at the *x*-axis, the 730 kWh per kW line, or infinity) or when the line segments form an area of closure within the confines of the demand and energy field.

In virtually all cases, a line of equality will not pass through each and every zone, so not all formula pairs need to be equated in order to complete the differential chart. Unless the rates being analyzed consist of rather complex or unusual structures, the line of equality (if one exists) is almost certain to begin at a point along the 730 kWh per kW line and terminate at infinity. The most likely point of contact with the 730 kWh per kW line is predicted to be in that zone whose upper boundary is the 730 kWh per kW line and whose right-hand boundary is the higher of the two rates' minimum demand requirements.[8] In this example, Rate Y has the higher minimum demand requirement of 45 kW; thus, Zone V is an ideal place to initiate the search for the line of equality.

When equating the two billing formulas of a zone to find a solution line, the formula with the higher energy coefficient is placed on the left-hand side since *e* is the designation of the *y*-variable; thus, for Zone V,

$$B_X = b_Y$$

which is

$$0.02\,e + 7.20\,d + 45.00 = 0.01\,e + 510.00$$

and

$$0.01\,e = -7.20\,d + 465.00$$

which simplifies to

$$\mathbf{\mathit{e} = -720.0\,\mathit{d} + 46{,}500}$$

Zone V has three boundaries: the 730 kWh per kW line, the 45 kW minimum demand requirement line, and 18,000 kWh (a line which intersects the 45 kW line and the 730 kWh per kW line at 24.658 kW). Any two of these three boundaries could be used to validate the solution line; therefore, if *e* is set equal to 730 *d*, the coordinates of the intersection of the solution line and the 730 kWh per kW line are found to be 32.069 kW and 23,410.3 kWh. Since the solution line intersects the 730 kWh per kW line at 32.069 kW, which is between the demand boundaries of Zone V (i.e., 24.658 kW and 45 kW), the solution line does indeed pass through Zone V (and at a negative slope). Subsequently, if *e* is set equal to 18,000, *d* is found to be equal to 39.583 kW.[9]

---

[8] The use of bill comparisons such as shown in Table 9.4 can also help to narrow the search for a suitable analysis starting point, particularly with complicated rate structures.

[9] If the other boundary, *d* = 45, had been selected instead of 18,000 kWh, *e* would be found to be equal to 14,100 kWh. These (*d*,*e*) coordinates could also be used to validate the solution line even though 14,100 kWh is actually below the lower boundary of Zone V.

The Mathematics of Rates 373

The line of equality's first solution component, denoted as the α segment, has end-point coordinates of 32.069 kW and 23,410.3 kWh at the 730 kWh per kW line and 39.583 kW and 18,000 kWh at the 18,000 kWh line. The solution line and its α segment are plotted in the upper left-hand graph of Figure 9.18.

Since the α segment terminates at the 18,000 kWh interface between Zones V and IV, the line of equality must then extend into Zone IV. While the Rate X formula B_X also applies in Zone IV, the applicable Rate Y formula has changed from b_Y to a_Y. Therefore, for Zone IV,

$$B_X = a_Y$$

which is

$$0.02\,e + 7.20\,d + 45.00 = 0.02\,e + 330.00$$

and, since both of the coefficients of e are equal,

$$7.20\,d = 285.00$$

which simplifies to

$$\mathbf{d = 39.583}$$

The solution for Zone IV is a vertical line which traverses the zone at 39.583 kW. As shown in the lower left-hand graph of Figure 9.18, the Zone IV solution line intersects the Zone V solution directly at the interface of the two zones. This intersection must occur since the line of equality must be continuous. The line of equality's second solution component, denoted as the β segment, has end-point coordinates of 39.583 kW and 18,000 kWh at the 18,000 kWh line and 39.583 kW and 8,312.5 kWh at the 210 kWh per kW line, which is the interface between Zone IV and Zone III. Thus, the line of equality must next extend into Zone III.

Since Zone III lies between the 210 kWh per kW line and the x-axis, the Rate X formula changes from B_X to A_X, while the Rate Y formula a_Y continues to apply within Zone III; thus,

$$A_X = a_Y$$

which is

$$0.04\,e + 3.00\,d + 45.00 = 0.02\,e + 330.00$$

and

$$0.02\,e = -3.00\,d + 285.00$$

which simplifies to

$$\mathbf{e = -150.0\,d + 14{,}250}$$

If e is set equal to 210 d, the coordinates of the intersection of the solution line and the 210 kWh per kW line are found to be 39.583 kW and 8,312.5 kWh, which validates the solution line at the zone interface. Since the solution line has a negative slope and a positive y-intercept, it can only intersect either the x-axis boundary (i.e., the 0 kWh line) or the 45 kW boundary.

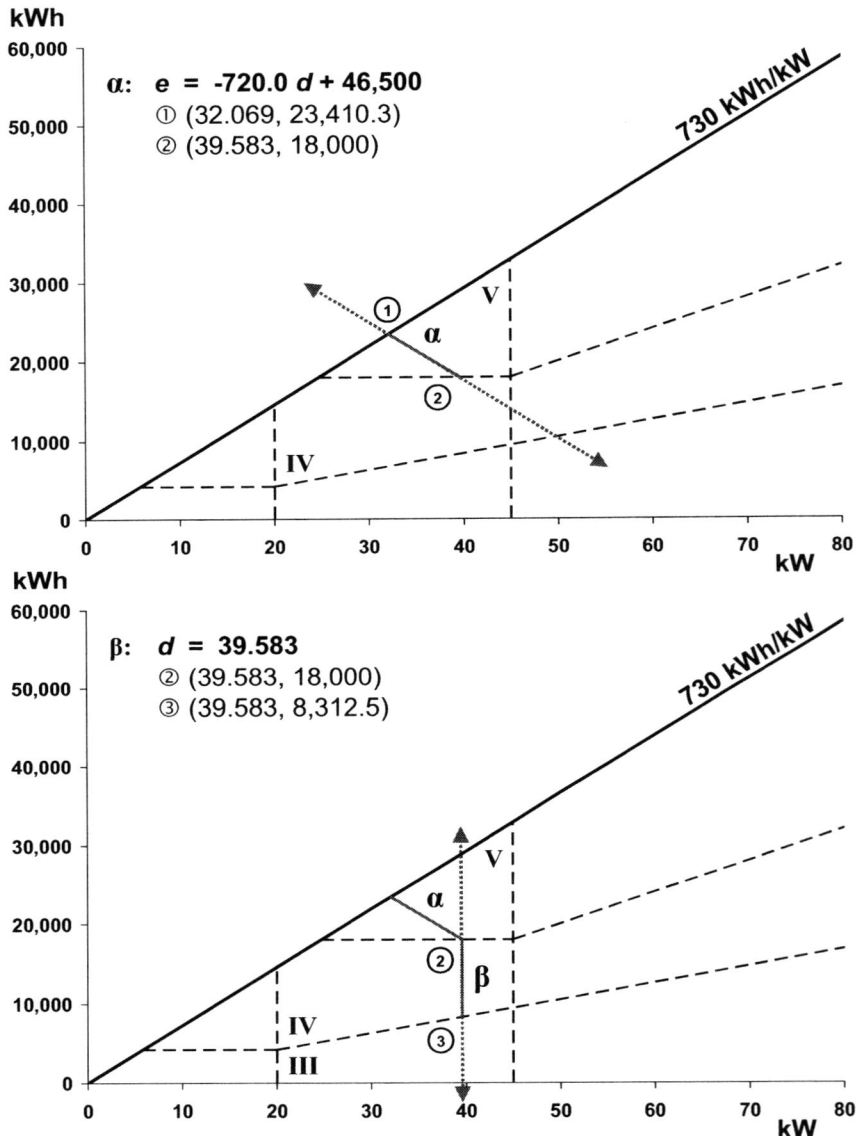

Figure 9.18 Construction of a differential chart. Rate X and Rate Y are compared n an equivalent basis. Upper chart: the line of equality begins at the 730 kWh per kW line and intersects with the 18,000 kWh line. Lower chart: the line of equality continues from the 18,000 kWh line and intersects with the 210 kWh per kW line.

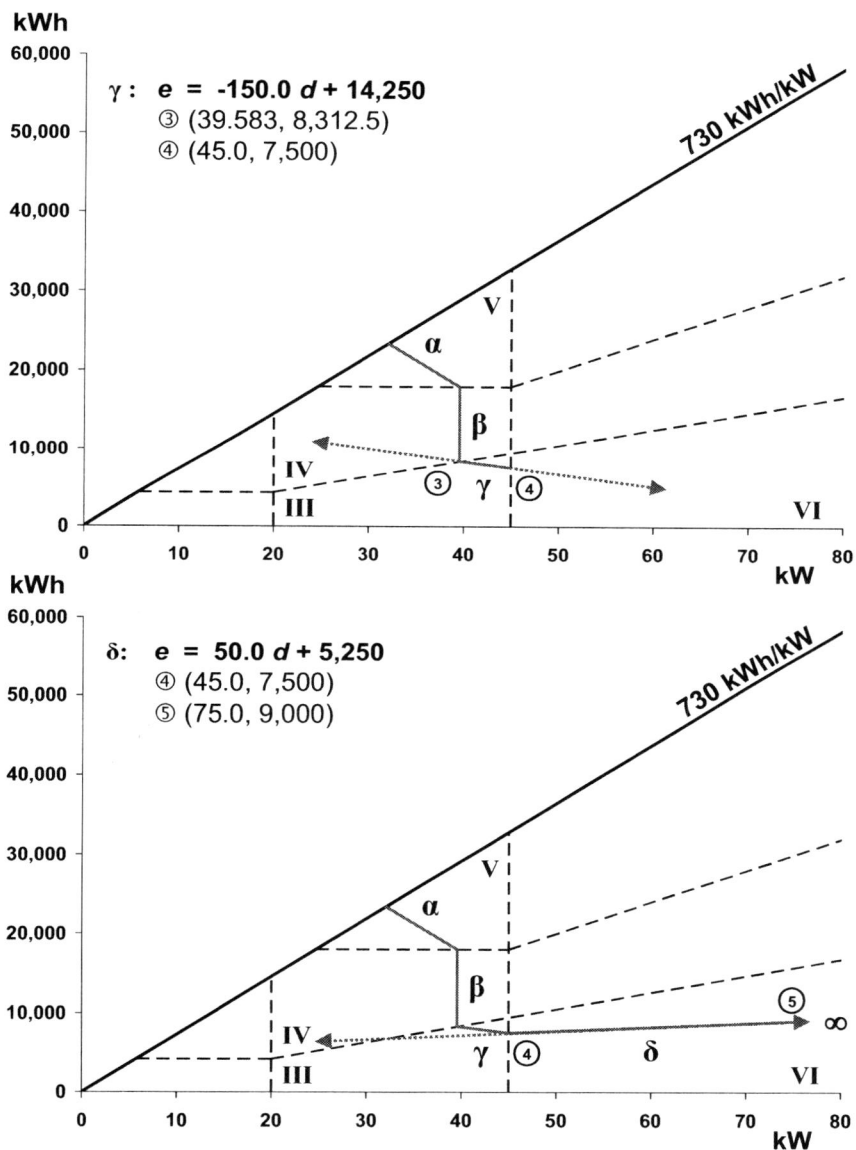

Figure 9.18—Cont. Upper chart: the line of equality continues from the 210 kWh per kW line and intersects with the 45 kW line. Lower chart: the line of equality extends from the 45 kW line to infinity. Rate X and Rate Y are compared at secondary voltage service and at a 90% power factor.

If d is set equal to 45, e is found to be equal to 7,500. Therefore, the line of equality's third solution component, denoted as the γ segment, has end-point coordinates of 39.583 kW and 8,312.5 kWh at the 210 kWh per kW line and 45.0 kW and 7,500 kWh at the 45 kW line, which is the interface between Zone III and Zone VI. As shown in the upper right-hand graph of Figure 9.18, the line of equality must extend next into Zone VI.

Zone VI also lies between the 210 kWh per kW line and the x-axis, so the applicability of the Rate X formula A_X remains unchanged. However, the Rate Y formula a_Y changes to A_Y since the zone lies beyond the minimum demand requirement. Thus, for Zone VI,

$$A_X = A_Y$$

which is

$$0.04\,e + 3.00\,d + 45.00 = 4.00\,d + 150.00$$

and

$$0.02e = 1.00\,d + 105.00$$

which simplifies to

$$e = 50.0\,d + 5{,}250$$

Validation of the solution line is confirmed by setting d equal to 45 and finding that e is equal to 7,500 at the 45 kW line. From this point, the solution line proceeds with a slight positive slope toward infinity between the 210 kWh per kW line and the x-axis, as shown in the lower right-hand graph of Figure 9.18. Furthermore, since the y-intercept of the solution line is positive, the line is found to be divergent with both the 210 kWh per kW line and the x-axis, and the solution for the line of equality is complete.[10] Since the line of equality approaches infinity as an end point, any level of demand within the scale of the graph can be chosen for a plotting reference. Thus, if d is set equal to 75 kW, e will be found to be equal to 9,000 kWh.

The completed differential chart for Rate X and Rate Y for secondary voltage service and an assumed power factor of 90% is shown in Figure 9.19. The area to the left and below the line of equality represents that portion of the demand and energy field where the bill amounts of Rate X are less than the bill amounts of Rate Y; whereas, the area to the right and above represents that portion of the demand and energy field where the bill amounts of Rate Y are less than the bill amounts of Rate X. A spectrum of HUD lines is shown for reference purposes.

---

[10] If the Zone VI solution line had a positive slope and a negative y-intercept, it would eventually intersect with the 210 kWh per kW line. The line of equality would thus extend into Zone VII (and possibly into Zone VIII) before terminating at infinity. While this condition occurs somewhat frequently with these types of rate structures, any such re-crossing of an HUD line often occurs at a level of demand that is well in excess of the practical application of the rates.

The Mathematics of Rates

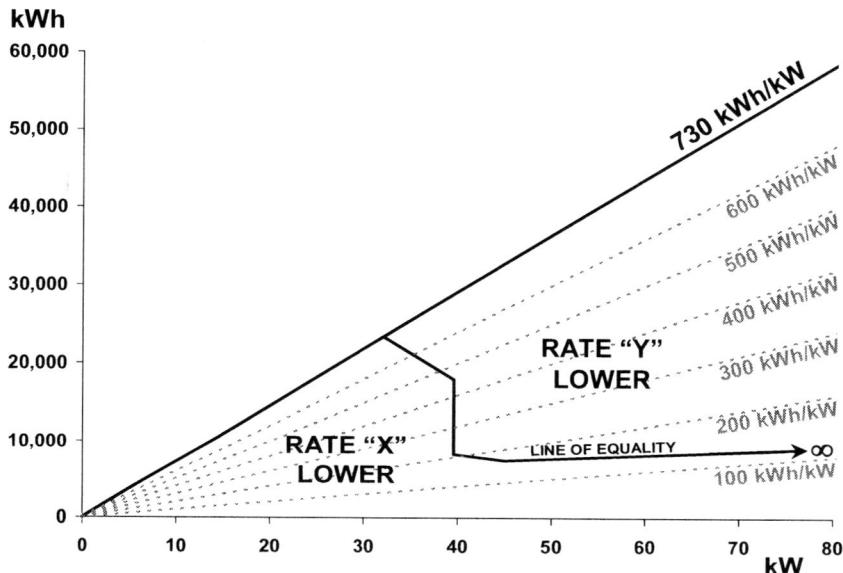

Figure 9.19 A completed differential chart for two rates. Rate X and Rate Y are compared at a secondary voltage service and at a power factor of 90%. Note the correlation of the differential chart with the bill comparison shown in Table 9.4.

The differential chart indicates the kinds of customers each rate is targeting based on demand and energy usage requirements. Rate X is generally intended for customers with relatively lower demands and load factors, while Rate Y is generally intended for customers with relatively higher demands and load factors. The rate design process takes into account the planned relationship between two (or more) rates. Both the types of rate structures and their associated unit price levels represent key criteria for positioning of the line of equality. In some cases, even minor modifications of one or both of the rates can effect a significant and perhaps undesirable shift of the line of equality.

The differential chart shown in Figure 9.19 is two dimensional since the equating of the bill formulas of the two rates eliminated the z variable from the equation. The line of equality illustrates all of the demand and energy coordinates at which both of the rates yield the same bill amounts, but the graph does not actually quantify the bill amounts since they vary from point to point along the line, as well as across the entire field of usage. However, by selecting a level of demand and plotting a bill chart (or an average rate chart) as a function of the hours use of demand, a more detailed view of the relationship between the bills of the two rates is revealed, as shown in Figure 9.20.

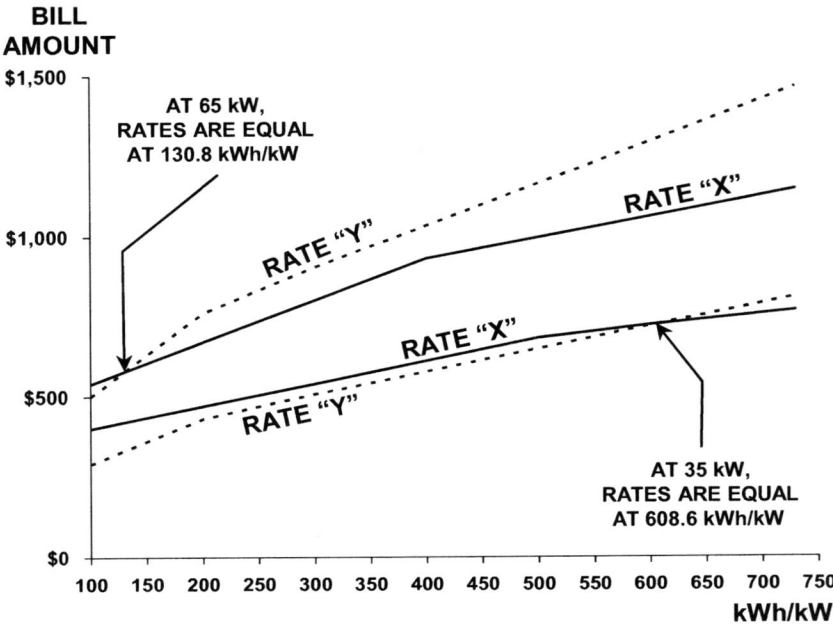

Figure 9.20 Cross-sectional views of a differential chart. The chart provides a graphical view of the bill comparison shown in Table 9.4 for two levels of demand. The bill amounts are plotted as a function of the hours use of demand. The crossing locations represent points on the line of equality between Rate X and Rate Y. The graph also gives an indication of the differential between the bill amounts of the two rates at various levels of use.

By plotting two rates together, the bill chart (or average rate chart) functions as a cross-section of the differential chart. Figure 9.20 illustrates the bill amounts for both Rate X and Rate Y associated with a demand of 35 kW and a demand of 65 kW. Note the correlation of the crossing points of each pair of bill plots with the differential chart shown in Figure 9.19.

The graph in Figure 9.20 also provides an indication of the magnitude and variation in the differential between the bill amounts of the two rates across nearly the entire range of possible energy usage. At 35 kW, there is minimal difference in the bill amounts of Rate X and Rate Y except at low load factors. On the other hand, at 65 kW, the bill amount differential is significant across virtually the full range of use. Figure 9.21 goes a step further by illustrating a topological view of the differential chart in which percent differences between the bill amounts of the two rates are shown as contour lines in the demand and energy field.

The Mathematics of Rates

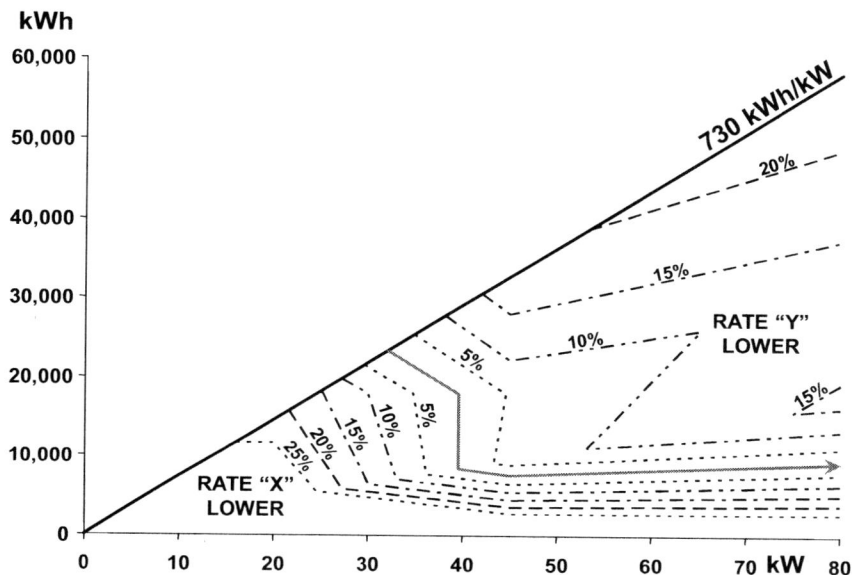

Figure 9.21 A differential chart in topological view. Illustrated is the Rate X vs. Rate Y differential chart showing contour lines which represent various percent differences between the bill amounts of the two rates. For example, the 10% contour line on the right side of the chart represents those demand and energy coordinates at which the bill amounts of Rate Y are 10% less than the corresponding bill amounts of Rate X.

The topological version of the differential chart is constructed by equating the simplified billing formulas of one rate to a percentage of the simplified billing formulas of the other rate. For example, to determine the demand and energy coordinates at which the bill amounts of Rate Y are 10% less than the bill amounts of Rate X, the Rate Y formulas would be set equal to 90% of the corresponding Rate X formulas. Thus, for the original analysis starting point of Zone V,

$$(1.0 - 0.1) \times B_X = b_Y$$

$$(0.9) \times [0.02\, e + 7.20\, d + 45.00] = 0.01\, e + 510.00$$

or

$$0.018\, e + 6.48\, d + 40.50 = 0.01\, e + 510.00$$

which simplifies to

$$e = -810.0\, d + 58{,}687.5$$

Like the original differential chart plot of the α segment, which was illustrated in Figure 9.18, this alternative α solution line intersects the 730 kWh per kW line within the boundaries of Zone V but at a slightly higher demand (38.109 kW vs. 32.069 kW). But unlike the original solution, the alternative α segment then intersects the 45 kW minimum demand requirement line before it intersects the 18,000 kWh line. As a result, the line of equality does not continue into Zone IV as before, but it proceeds instead from Zone V to Zone VIII. Further analysis shows that the line of equality next intersects with the 400 kWh per kW line where it continues downward into Zone VII. The line of equality then intersects with the 210 kWh per kW line where it continues into Zone VI and then on to infinity.

The comparable line of equality which shows the demand and energy coordinates at which the bill amounts of Rate X are 10% less than the bill amounts of Rate Y is found in the same manner. Thus, the α segment is determined from the equality: $B_X = 0.9 \times b_Y$.

Other Differential Chart Considerations

All of the comparisons between Rate X and Rate Y have thus far been based on an assumed power factor of 90%. A power factor assumption is required whenever a kVA-based rate structure is compared to a kW-based rate structure to ensure an equitable analysis. The power factor most representative of the customer group is typically used in the analysis. However, the actual power factors of individual customers often differ somewhat from the group's average value, so it is important to understand the impacts of power factor variations in the analysis.

The inherent reactive power charge of the kVA demand rate increases significantly as the actual power factor decreases. For example, consider a Demand Charge of $4.00 per kVA and a metered demand of 100 kVA at 100% power factor. The amount billed for demand would be $400.00 or $4.00 per kW at a 100% power factor, since the real power component is 100 kW while the associated reactive power requirement is zero. Now suppose that the 100 kW demand remains constant while the power factor drops to 80%. The metered demand then becomes 125 kVA as the associated reactive power requirement is now 75 kVAR. The amount billed for demand would now become $500.00, i.e., an increase of $100.00 only because the power factor decreased. In a sense, the effective charges would be $400.00 for 100 kW and $100.00 for 75 kVAR.

Assume that an alternative kW-based rate structure is also applicable, and that this alternative rate either lacks an attendant power factor provision or includes only a minimal charge for power factor. The effective reactive power charge within the kVA-based rate structure greatly increases the total bill amounts at lower power factors. Thus, when compared to the kW-based rate, the kVA-based rate offers the lower bill amounts across a decreasing area of the demand and energy field as the power factor declines. Figure 9.22 illustrates the impact on the differential chart's line of equality for Rate X and Rate Y over a range of assumed power factors. As the power factor decreases, the kW-based Rate X produces the lower bill amounts at higher demands and higher load factors.

The Mathematics of Rates

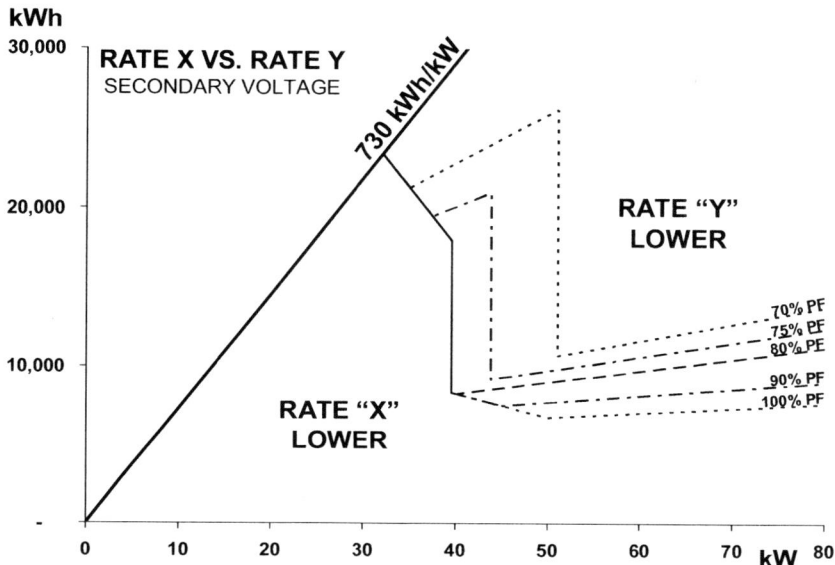

Figure 9.22 **Impact of power factor on the line of equality.** Shown is a comparison of Rate X, a kW-based rate structure, and Rate Y, a kVA-based rate structure. As power factor declines, the area of the demand and energy field, in which the kVA-based rate produces lower bill amounts, diminishes significantly.

When highly intricate rate structures are compared in a differential chart analysis, the resulting line of equality may deviate significantly from the typical "✓" or "L" shaped solution that essentially starts at the 730 kWh line, intersects with an HUD line and/or the larger minimum demand requirement line, and finally ends at infinity. A number of exceptions have been observed in practice. For example, when a complex minimum bill provision is considered (i.e., where a rate's attendant minimum bill provision overrides the rate's basic billing formula within a section of the demand and energy field), a line of equality may start at the 730 kWh per kW line and end at the x-axis. As another example, the line of equality may start at the 730 kWh per kW, wrap around in sort of a "U" shaped fashion, and then end back at the 730 kWh per kW line.

The line of equality might not intersect with either the 730 kWh per kW line or the x-axis. Rather, it may form a line of closure creating an area for one rate that is completely surrounded by the other rate, or it may appear as a "U" shape tilted toward the right-hand side of the chart with both of its end points located at infinity. Finally, there may be multiple lines of equality where one rate is lower than the other rate in two discontinuous areas of the chart.

Application of the Differential Chart

The differential chart illustrates those areas of the demand and energy field where one applicable rate produces lower bills than another applicable rate. As such, the chart provides a quick way in which to ascertain which of the rates would be the more economical consideration given a customer's expected demand and energy requirements. However, since the rate structures used in the development of the chart are most often based on monthly charges, the differential chart in and of itself represents just a single billing period view of the rate comparison.

A plot of an individual customer's actual monthly billing determinants for a year is shown in Figure 9.23. Some variation in monthly demand and energy requirements can be observed. If all of the billing determinant coordinates were located on either side of the line of equality, the rate selection which would produce the overall lower annual billing amount would be certain. However, the selection is indeterminate since one half of the bills are located on each side of the line of equality. Table 9.7 reveals that the selection, which actually produces the lowest annual billing amount, would be realized under Rate Schedule Y even though the Rate Y bills are greater than the Rate X bills in six months out of the year.

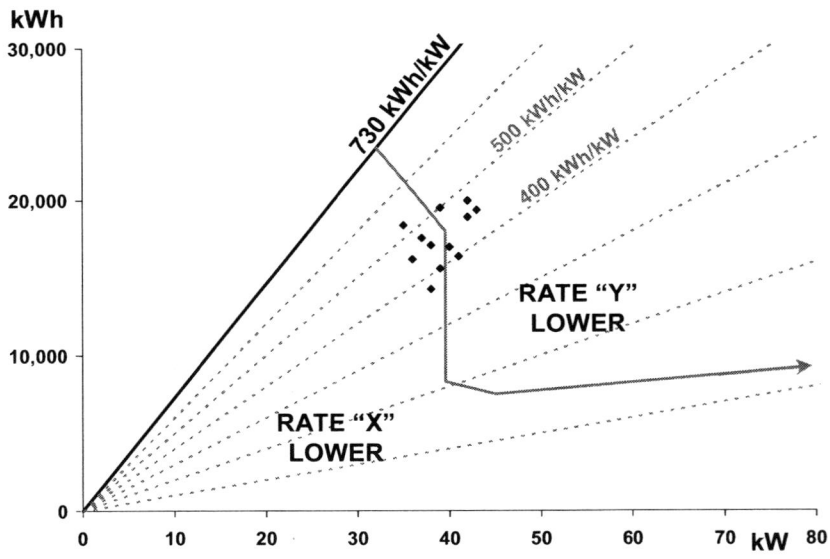

Figure 9.23 A differential chart on which a customer's twelve monthly billing determinant coordinates have been plotted. The overall lowest billing amount must be determined by comparing the sum of the twelve monthly bills under each rate.

Table 9.6 Comparison of Twelve Monthly Bills under Both Rate X and Rate Y

	kW	kWh	$ Monthly Bill Amount		Difference	
			Rate X	Rate Y	Amount	Percent
JAN	36	16,200	628.20	654.00	$25.80	4.11
FEB	37	17,575	662.90	681.50	18.60	2.81
MAR	38	17,100	660.60	672.00	11.40	1.73
APR	38	14,250	603.60	615.00	11.40	1.89
MAY	40	17,000	673.00	670.00	(3.00)	(0.45)
JUN	42	18,900	725.40	699.00	(26.40)	(3.64)
JUL	43	19,350	741.60	703.50	(38.10)	(5.14)
AUG	42	19,950	746.40	709.50	(36.90)	(4.94)
SEP	41	16,400	668.20	658.00	(10.20)	(1.53)
OCT	39	15,600	637.80	642.00	4.20	0.66
NOV	39	19,500	715.80	705.00	(10.80)	(1.51)
DEC	35	18,375	664.50	697.50	33.00	4.97
TOT	470	210,200	$8,128.00	$8,107.00	$(21.00)	(0.26)

Given that the annual billing is virtually indifferent under either rate in this example, the selection of a rate schedule under which the customer should contract for electric service must depend on other criteria. For instance, anticipated alterations in usage from end-use load additions or changes in operations could have a bearing on future billing amounts.[11] Differences in contract terms between the two rates could also be a consideration.

When a seasonally-differentiated rate structure is considered for a differential chart analysis, a separate line of equality is required for each season. However, the conventional differential chart formulary analysis becomes somewhat more intricate for a comparative analysis of more complex time-differentiated rate structures.

Time-Differentiated Rate Comparisons

The formulary comparison of rates involving time-differentiated structures is complicated by the condition of contending with more than the standard demand and energy billing determinants. A simple nondemand TOU rate has two or three different rating periods (e.g., on-peak, intermediate-peak, and off-peak usage). A TOU demand rate is often structured with two demand charges as well as multiple energy usage variables. Consider an analysis comparing a TOU demand rate to a non-TOU demands rate.

---

[11] While not particularly significant in this example given the location of the billing determinants on the Differential Chart (reference Figure 9.23), planned improvements in power factor could also be one of the considerations in the selection of a rate schedule.

Rate Schedule "M"	Rate Schedule "GT"
Monthly Rate for Secondary Service	Monthly Rate for Secondary Service

Rate Schedule "M" — Monthly Rate for Secondary Service

Customer Charge: $50.00
Demand Charge: $4.69/kW

Energy Charge: 1.75¢/kWh

Minimum Demand: 25 kW

Rate Schedule "GT" — Monthly Rate for Secondary Service

Customer Charge: $70.00
Demand Charges:
   On-Peak Demand: $2.89/kW
   Maximum Demand: $3.01/kW
Energy Charges:
   On-Peak Energy @ 0.75¢/kWh
   Off-Peak Energy @ 0.67¢/kWh

Minimum Demand: 25 kW
Minimum On-Peak Demand: 0 kW

As with the comparison of two non-time differentiated rate structures, the variables e (total kWh usage) and d (maximum demand) are used in the rate formulizations; however, three additional variables are required: d_{ON} (maximum on-peak demand), e_{ON} (on-peak kWh usage), and e_{OFF} (off-peak kWh usage). The simplified billing formulas for the two areas of the rates that are greater than the minimum demand requirements of 25 kW (i.e., $A_M = A_{GT}$) are equated, as given by

$$0.0175\, e + 4.69\, d + 50.00 = 0.0075\, e_{ON} + 0.0067\, e_{OFF} + 3.01\, d + 2.89\, d_{ON} + 70.00$$

which is

$$0.0175\, e - (0.0075\, e_{ON} + 0.0067\, e_{OFF}) = -1.68\, d + 2.89\, d_{ON} + 20.00$$

Since $e = e_{ON} + e_{OFF}$; then $e_{OFF} = e - e_{ON}$

and

$$0.0108\, e - 0.0008\, e_{ON} = -1.68\, d + 2.89\, d_{ON} + 20.00$$

At this level of simplification, one out of the five original variables has been eliminated. By making an assumption of a relationship between the two demand variables and the two energy variables, the equation can be further reduced to two variables. Numerous combinations of assumed relationships could be devised. However, one such scenario could be based on the average characteristics of the TOU rate, which for this example is assumed to be that the on-peak demand is 80% of the maximum demand and the on-peak energy usage is 40% of the total energy usage.

$$0.0108\, e - 0.0008\, (0.4 \times e) = -1.68\, d + 2.89\, (0.8 \times d) + 20.00$$

which simplifies to

The Mathematics of Rates

$$e = 60.305\,d + 1{,}908$$

and which can be plotted on a conventional differential chart. Other scenarios can be developed to illustrate the impact on the line of equality due to different rating period usage characteristics.

A different form of a differential chart can be developed for comparing non-demand rates. Consider an analysis of a non-TOU and a TOU pair of applicable rate structures.

Rate Schedule "R"
Monthly Rate for Secondary Service

Customer Charge: $14.00
Energy Charge: 6.26¢/kWh

Rate Schedule "RT"
Monthly Rate for Secondary Service

Customer Charge: $17.50
Energy Charges:
  On-Peak Energy @ 10.20¢/kWh
  Off-Peak Energy @ 3.35¢/kWh

Each rate has just a single billing formula which together equates to

$$0.0626\,e + 14.00 = 0.1020\,e_{ON} + 0.0335\,e_{OFF} + 17.50$$

Substituting $e - e_{ON}$ for e_{OFF} yields

$$0.0291\,e = 0.0685\,e_{ON} + 3.50$$

and by rearranging

$$e_{ON} = 0.4248\,e - 51.09$$

Dividing both sides of the equation by e results in a kWh usage ratio given by

$$\frac{e_{ON}}{e} = 0.4248 - \frac{51.09}{e}$$

When the ratio of on-peak kWh usage to total kWh usage ($e_{ON} : e$) is plotted as a function of e, a line of equality between the TOU and the non-TOU rate is established, as shown in Figure 9.24. The line of equality is in the form of an inverted hyperbola whose asymptotes are the y-axis and a line representing a constant value of the ratio, as given by

$$\lim_{e \to \infty} 0.4248 - \frac{51.09}{e} = 0.4248$$

The line of equality between the TOU and the non-TOU rate thus approaches an on-peak to total energy usage ratio of 42.48%.

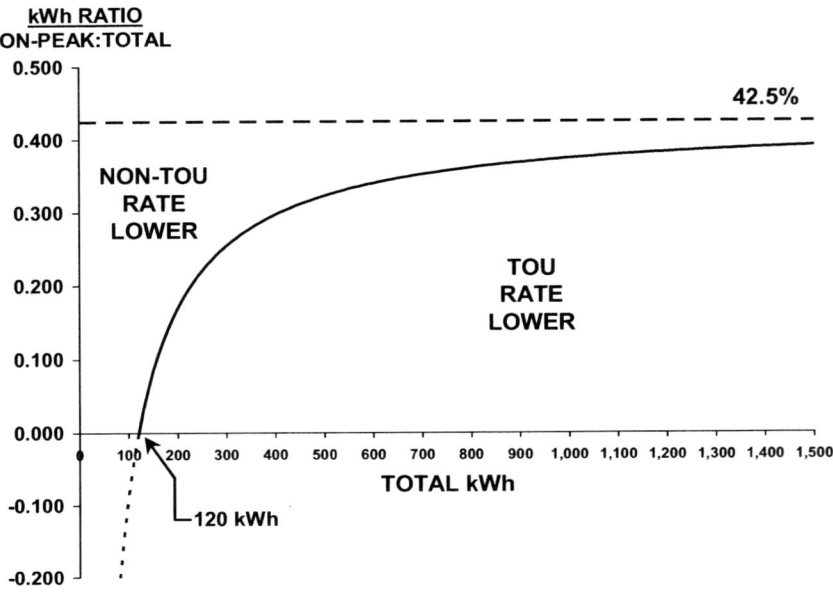

Figure 9.24 A TOU differential chart comparing a time-differentiated, nondemand rate to a straight-line energy rate.

As observed in Figure 9.24, with total usage less than 120 kWh, bill amounts of the TOU rate would be higher than for the non-TOU rate even if the entire energy consumption occurred in the off-peak rating period. At least 120 kWh of usage is required in order to overcome the differential in the fixed charge components of the rates (i.e., the Customer Charges).[12]

Objective Rate Comparisons

A rate analysis may be generic in scope, or it may be designed around a specific situation, such as in response to a current or prospective customer's solicitation. In selecting rates for the analysis, it is important that the problem is structured to assure a truly comparable solution. The provisions under which the particular rates of concern operate may differ to some extent; thus, all of the rate schedule features must be fully understood. Particular attention should be heeded when comparing rates between different utilities or service providers. Fundamentally,

---

[12] Compared to a non-TOU rate, the Customer Charge of a comparable TOU rate of the same class of service typically includes incremental costs associated with metering.

the analysis should correspond closely to the actual application and billing aspects of the rates to ensure an equitable comparison.

The Availability of Service, Applicability of Service, and Character of Service provisions should be tested to determine if a designated rate is a justified match for the problem situation. A particular rate schedule may be restricted to limited types of customers (e.g., by SIC or NAIC codes) or have a specific voltage requirement. The tariff definition of delivery voltage can vary between utilities. For example, 44 kV and 69 kV are considered to be subtransmission voltages by some utilities but transmission voltages by other utilities.

The metering methodology can have a bearing on a particular demand measurement. A 15-minute interval measure can yield a higher demand than a 30-minute interval measure. At a 100% load factor the readings would be the same; however, at lower load factors, the two maximum demand measurements of the same load shape could differ by as much as two percent or more. In addition, billing determinant adjustments may be required as a consequence of the actual voltage level at which the measurements are taken.

Power factor may be determined in different ways for billing purposes. When determined from monthly maximum kW and kVAR demands, the resulting power factor is higher than when determined as a monthly average. As a result, a calculated hybrid kVA demand would be different under the two methods.

An analysis of base rates (i.e., the application of just the basic rate structures) may be adequate for producing a functional differential chart when comparing two or more rates within the same tariff, but only if all of the associated billing provisions apply uniformly. However, when an adjustment factor, such as fuel cost recovery, is differentiated on the basis of the rate schedules themselves, it may be necessary to adjust the formulization of the rates to include those distinct factors. In comparing rates of different utilities or service providers, it is a foregone conclusion that all applicable adjustment factors and taxes be included in the analysis in order to ensure an equitable rate comparison.

9.5 SUMMARY

The calculation of a bill for electric service is a mathematical process, and as such it can be modeled as an equation derived from the rate structure and its component unit prices. Even complex rate schedules can be reduced to a system of simplified billing formulas. A three-variable billing formula for a fundamental demand rate represents a plane in space, which is characterized as a surface of bill amounts that is positioned over the field of demand and energy usage. The simplified billing formula condenses further to a two-dimensional form with basic nondemand rate structures. Simplified billing formulas are used for quickly estimating bill amounts and in advanced analyses of rate characteristics.

The mathematical characteristics of rates can be illustrated in a variety of charts and tables. The rate structure chart functions as a map for multiple price component rates by showing the different areas across the demand and energy field where each energy price applies. Bill charts and average rate charts are graphs of a rate's revenue potential expressed as a function of demand and/or energy usage. These charts are used to contrast the behavior aspects of a rate at varying points in the demand and energy field or to compare the characteristics of one rate against one or more other rates across a common range of usage.

Two or more rates can also be evaluated together through the development of a differential chart, which reveals a line of equality between comparable rates within the demand and energy field of usage. The chart defines the points of usage at which one rate produces bills that are lower than those produced by the comparable rate(s).

Understanding the mathematics of rates provides considerable insight to how electric rates function and how they interact with each other. Rate analysis is a key function which supports both the practical design of rate schedules and the administration of the electric tariff.

10
Billing Data Statistics and Applications

10.1 INTRODUCTION

The ongoing process of invoicing customers for electric service results in an accretion of billing data within a customer information system (CIS). The recording of monthly billing determinants generates a time-series database of individual customer demand and energy usage requirements. Aside from this routinely collected data, each CIS account record must also contain any additional tariff-related information that is required to execute the billing process, such as a customer's delivery voltage level. Customer account records are enhanced when they include supplementary customer-specific information, such as industry classification codes or end-use related revenue codes. A considerable number of practical statistical inferences can be harvested from a well-maintained CIS database.

The compilation and analysis of customer characteristics is a key resource for developing electricity prices. In order to minimize the risks associated with applying a common rate structure to a large base of mixed customer types, electric service should be differentiated by distinct groups which ensure a reasonable amount of homogeneity of customer characteristics. When a wide diversity of customer traits exists, several rates schedules may be necessary to price the electric services adequately. Rate schedules thus represent one principal way in which to differentiate or segment the customer base.

The universe consisting of all of the electric service customers can be segmented into many categorizations on the basis of a number of customer attributes, as exemplified in Figure 10.1. Such attributes reveal information relative to customer requirements for electric service, load and usage characteristics, and cost-of-service implications.

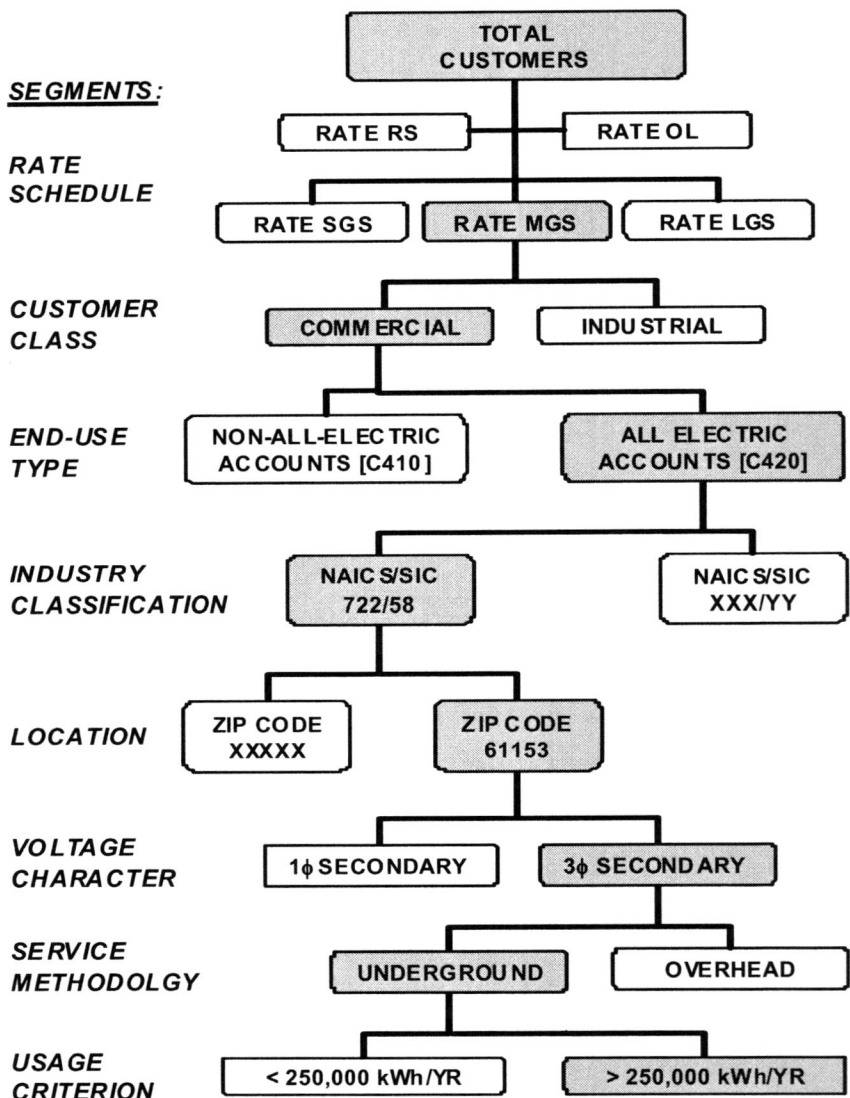

Figure 10.1 Attributes of customer account records. The universe of electric service customers can be segmented on the basis of a number of attributes, which are commonly entered as a part of each CIS account record.

Billing Data Statistics and Applications

Knowledge of billing data and other customer attributes helps guide the design of rate structures and attendant billing provisions. For instance, industrial customers tend to have higher load factors than do commercial customers. All-electric customers tend to use more annual energy than comparable non-all electric customers. Compared to longer standing customers, newer customers are more likely to be served by underground distribution systems. A group of customers having particularly unique characteristics may warrant having its own specific rate class (e.g., customers in excess of 10,000 kW demand with monthly load factors of 90% or more). In addition, various industry types may exhibit unique working patterns or have operational flexibility that is suitable for time-differentiated pricing or interruptible service options (e.g., some pumping and manufacturing operations). The ability to segment customers into homogeneous populations facilitates the administration of alternative electric rates as well as the marketing of service options.

Historical billing data and associated customer information is used not only for the support of rate design but also in the process of forecasting future revenues. The analysis of projected revenues and costs is a key input for managing the profitability of the utility or service provider business.

10.2 CHARACTERISTICS OF BILLING DATA

Time series billing data can be rearranged, viewed, and analyzed in a variety of fashions. A *frequency distribution* is one way in which to organize data systematically. A frequency distribution for a data set is developed by compiling the number of observations, such as bills, according to user defined intervals, such as kWh (per bill). For example, an analysis of July billing data for a given rate class might show that 1,961 bills had usage falling within the interval of 1 to 10 kWh, 1,953 bills had usage falling within the interval of 11 to 20 kWh, and so forth. A *relative frequency distribution* can be developed by dividing the observations of each interval by the total number of observations. Thus, if the total number of July bills for the given rate class was 24,500, the relative frequency for the 1 to 10 kWh interval would be 0.080 (or 8.0%), and so forth.

A plot of a frequency distribution is known as a *histogram*. Histograms provide a visual clue as to the usage characteristics of a particular population of interest, such as a customer or rate class. Measures of *central tendency* (i.e., an average value quantity) as well as the variability or dispersion of the data provide numerical revelations about a population's usage characteristics.

Characteristics of Residential Billing Data

The residential customer class is typically characterized by two primary end-use subclasses: electric heating customers and nonelectric heating customers. An example of an average winter month load shape for both of these customers types is illustrated in Figure 10.2.

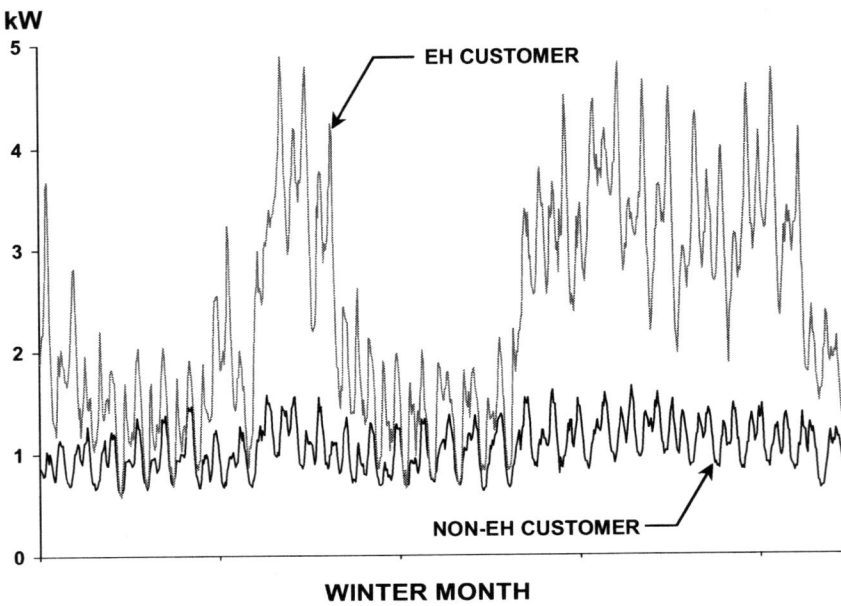

Figure 10.2 A plot of an average winter month load shape two segments of the residential customer class. Both residential nonelectric heating customers (NON-EH) and residential customers that utilize an electric end-use device as the primary source for space heating (EH) are shown. The periods of colder weather can be noted easily by the distinct increases in the electric heating customer demands, particularly during the latter half of the month.

The typical winter month demand and energy use requirements of the electric heating customer are much more intensive than the nonelectric heating customer, as evidenced by the load shape comparison in Figure 10.2. In this example, the monthly kWh usage of the electric heating customer appears to be on the order of twice that of the nonelectric heating customer (weather is a key factor to this observed relationship).[1]

Relative frequency distributions for the example winter month bills for the electric heating and nonelectric heating residential subclasses are show in Figure 10.3. Both histograms exhibit a *positive skew* (i.e., an asymmetrical distribution with a longer right-hand tail), which is typical for residential class energy usage.

---

[1] Assuming that the market penetration of air conditioning is practically the same for both the EH and the NON-EH Customers, the average summer load shapes of the two residential subclasses would be virtually identical.

Billing Data Statistics and Applications

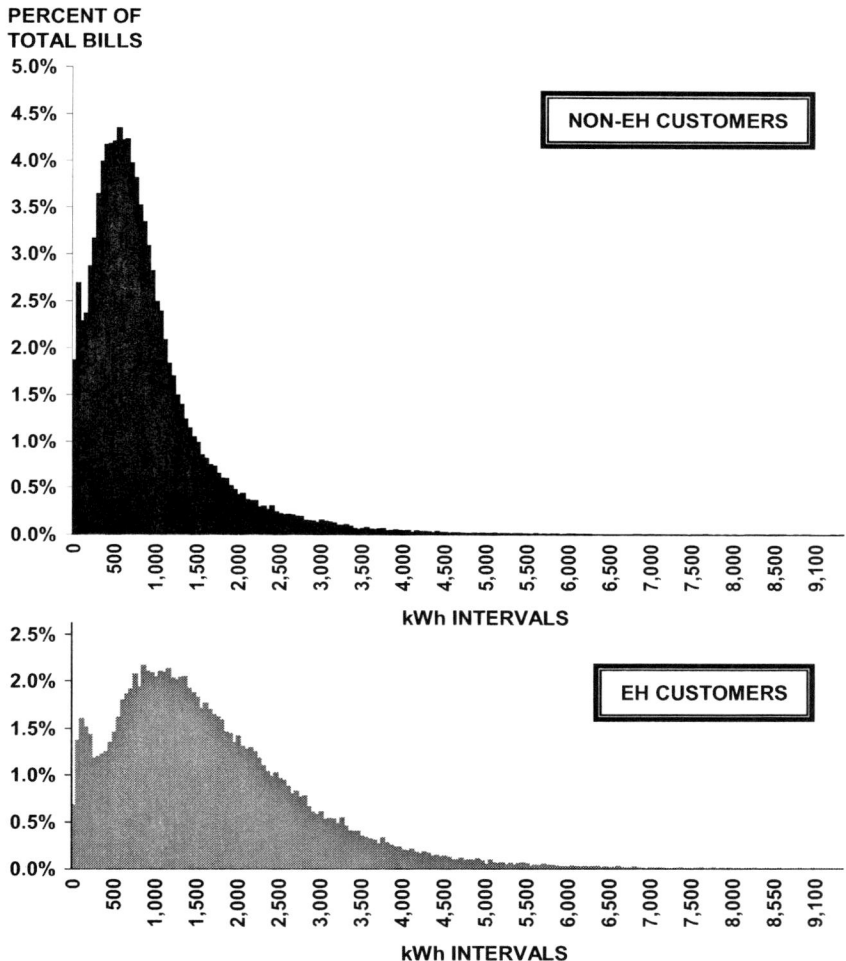

Figure 10.3 **Histograms of the relative frequency of residential bills.** The histograms are plotted by intervals of 50 kWh (per bill) for a winter month. The distribution of bills for nonelectric heating customers is shown in the upper graph. The distribution of bills for customers having electric end-use devices as the primary source for space heating is shown in the lower graph.

Although both histograms in Figure 10.3 fundamentally have a similar shape, the relative distribution of the electric heating customer bills is broader than the nonelectric heating customer bills due to the higher volume of kWh use of the average electric heating customer, as discussed previously. In other words, the electric heating service bills are spread out somewhat more uniformly across the kWh intervals and the proportion of higher usage bills is much more prominent. It can also be observed that there are a visible number of bills having zero kWh usage in both of the distributions.

The central tendency, or average value, of a frequency distribution can be quantified by three different measures: the *mode*, the *median*, and the *mean*.[2] The mode for grouped data is simply the frequency of that interval which contains the most observations. For the bill distributions shown in Figure 10.3, the mode for the nonelectric heating customer bills is the 500 to 550 kWh (per bill) interval (with 4.35% of the total nonelectric heating service bills), while the mode for the electric heating customer bills is the 800 to 850 kWh interval (with 2.16% of the total electric heating service bills). A point estimate of the mode for grouped data can be determined by

$$Mode = LL_{MoI} + w_{MoI} \times \left[\frac{F_{MoI} - F_{PI}}{(F_{MoI} - F_{PI}) + (F_{MoI} - F_{SI})} \right] \quad (10.1)$$

where LL_{MoI} = lower limit of modal interval, in kWh
 w_{MoI} = width of the modal interval, in kWh
 F_{MoI} = frequency of the modal interval, in bills
 F_{PI} = frequency of the interval preceding the modal interval, in bills
 F_{SI} = frequency of the interval succeeding the modal interval, in bills

Thus, for the distributions shown in Figure 10.3, the estimated mode for the nonelectric heating customer bills, by Equation 10.1, is found to be 526.27 kWh (per bill), and the estimated mode for the electric heating customer bills is found to be 838.92 kWh (per bill).

The median is the midpoint of all of the observations, i.e., the kWh level at which the distribution of bills is divided exactly into half. For the nonelectric heating customers shown in the upper graph of Figure 10.3, 48.2% of the bills are accounted for at the 600 to 650 kWh interval while 52.2% of the bills are accounted for at the 650 to 700 kWh interval. Thus, the 50th percentile for the example distribution occurs, by simple average, at approximately the 675 kWh per bill level. For the electric heating customers, the 50th percentile of the bill distribution occurs within the 1,350 to 1,400 kWh interval (48.3% at the 1,300 to 1,350 kWh interval and 50.2% at the 1,350 to 1,400 kWh interval) or, by simple average,

---

[2] The mode, median, and mean are found to be equal in the case of a normal distribution; thus, these measures of central tendency are not expected to be equal in residential bill distributions because of positive skew.

Billing Data Statistics and Applications

at approximately the 1,375 kWh per bill level. A point estimate of the median for grouped data can be determined by

$$Median = LL_{Mel} + w_{Mel} \times \left[\frac{\frac{N+1}{2} - \sum_{i=1}^{P} F_i}{F_{Mel}} \right] \qquad (10.2)$$

where LL_{Mel} = lower limit of median interval, in kWh
w_{Mel} = width of the median interval, in kWh
N = total number of bills in the population
F_i = frequency of the ith interval, in bills
P = the interval preceding the median interval
F_{Mel} = frequency of the median interval, in bills

Thus, for the distributions shown in Figure 10.3, the estimated median for the nonelectric heating customer bills, by Equation 10.2, is found to be 672.50 kWh (per bill), and the estimated mode for the electric heating customer bills is found to be 1,394.79 kWh (per bill).

The median can also be observed through a plot of the *cumulative relative bill frequency distribution*, as shown in the upper graph in Figure 10.4.[3] The dashed lines denote the kWh levels at the 50th percentile of each distribution. The lower graph in Figure 10.4 shows the cumulative relative frequency distribution of kWh sales (as opposed to bills) by kWh interval. The kWh sales of the electric heating customers accumulate across intervals at a much lower rate compared to the nonelectric heating customers because their winter monthly energy requirements are so much higher.

While nonelectric heating customers have some degree of weather sensitivity (because of fossil fuel furnace air circulating fans and portable resistance heater usage), the distribution of winter kWh sales across kWh intervals can be expected to be fairly stable throughout the heating season. On the other hand, the distribution of kWh sales of the electric heating customers can be expected to be more volatile because of variations in monthly heating degree days. The curve would move to the left under milder conditions and to the right under colder conditions.

---

[3] A cumulative bill frequency distribution is developed by summing the bills of all prior kWh intervals at each kWh interval. The cumulative relative bill frequency distribution is determined by then dividing the accumulated bills at each kWh interval by the total number of bills.

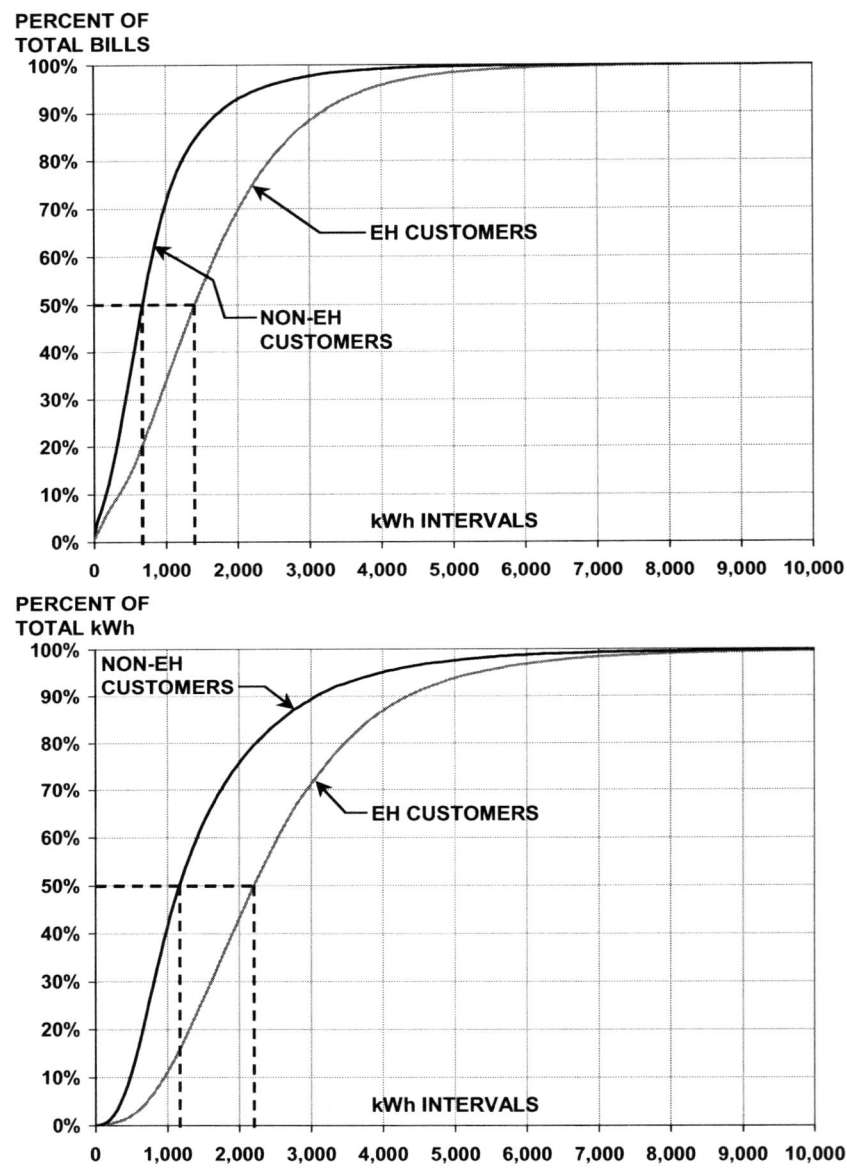

Figure 10.4 Cumulative relative frequency distributions for the residential class. The distribution of residential bills is shown in the upper graph and distribution of residential kWh sales is shown in the lower graph. The intervals are based on kWh (per bill).

Billing Data Statistics and Applications

Perhaps the most commonly used measure of central tendency, or average, in billing data analysis is the mean. The mean kWh per bill for a population of data is determined by

$$\mu = \frac{\sum_{i=1}^{N} X_i}{N} \quad (10.3)$$

where μ = population mean kWh per bill
X_i = kWh of the ith bill
N = total number of bills in the population

The mean kWh usage per bill for the example nonelectric heating customers is found to be 855 kWh, while the mean kWh usage for the electric heating customers is found to be 1,628 kWh.[4]

One advantage of using the mean rather than the mode or the median is that it can be used to calculate the dispersion of the bill frequency distribution in terms of *variance* and *standard deviation*. The population variance is determined by

$$\sigma^2 = \frac{\sum_{i=1}^{N} (X_i - \mu)^2}{N} \quad (10.4a)$$

and the standard deviation of the population is determined by

$$\sigma = \sqrt{\sigma^2} \quad (10.4b)$$

where σ^2 = population variance in (kWh per bill)^2
σ = population standard deviation in kWh per bill
X_i = kWh of the ith bill (i.e., observation)
μ = population mean kWh per bill
N = total number of bills in the population

The variance represents the average or mean value of the square of the deviations, i.e., the departures exhibited by each of the individual observations (bill kWh)

---

[4] Note that with both example distributions, the mean lies to the right of the median, which in turn lies to the right of the mode. This sequence of the measures of central tendency is characteristic of residential bill distributions due to their positive skew nature.

with respect to the population mean kWh per bill. The standard deviation is a linearization of the variance and, in this example, has units of kWh per bill. The standard deviation for the nonelectric heating customers is found to be 770.2 kWh per bill, while the standard deviation for the electric heating customers is found to be 1,197.9 kWh per bill.

The previous discussion has focused on population data, i.e., the population of residential electric heating customers and the population of residential nonelectric heating customers. Most analysis of billing data is based on total populations. However, samples might be used at times for special studies of billing data. A sample is used to provide an estimate of the mean and variance of a population. While the calculation of the sample mean is just like the calculation of the population mean, the calculation of variance of the sample is slightly different. Thus, the mean, variance, and standard deviation of a sample are determined by

Mean:
$$\bar{x} = \frac{\sum_{i=1}^{n} x_i}{n} \qquad (10.5a)$$

Variance:
$$s^2 = \frac{\sum_{i=1}^{n}(x_i - \bar{x})^2}{n-1} \qquad (10.5b)$$

Standard Deviation:
$$s = \sqrt{s^2} \qquad (10.5c)$$

where $\bar{x}$ = sample mean kWh per bill
x_i = kWh of the ith bill of the sample
n = total number of bills in the sample
s^2 = sample variance in (kWh per bill)^2
s = sample standard deviation in kWh per bill

Compare Equation (10.5b) with Equation (10.4a). The form of the equation for sample variance differs from the equation for population variance in that $n - 1$ is used instead of just n for the denominator (i.e., N is used as the denominator for population variance). When sample size n is small relative to population size N, the use of $n - 1$ makes s^2 an unbiased estimator for σ^2.

Characteristics of Commercial and Industrial Billing Data

Compared to the residential class of customers, commercial and industrial customers exhibit much greater variations in their energy and demand usage characteris-

Billing Data Statistics and Applications

tics. The composition, size, and operation of commercial and industrial end-use load devices is much different than for residential. Furthermore, such end-use variations also exist between the commercial class and the industrial class. For instance, electric space heating and lighting are present in both classes, but they are more predominant with commercial customers. Likewise, motor applications are common in both classes; however, a much greater concentration of motor-driven devices is found throughout industrial settings. The industrial class also has a lot of unique devices because of manufacturing and fabrication requirements (e.g., special process heating devices).

Two examples of an average summer month customer load shape, based on both the total commercial class and the total industrial class, are shown in Figure 10.5. A rhythmic daily operating cycle can be observed in both of the average customer load shapes; however, the industrial load shape in the lower graph indicates that operations are routinely more moderate on Saturdays and Sundays.[5] Although the commercial class customer load averages only about 8 to 9 kW across the month (as compared to about 900 to 950 kW for the industrial class), its daily excursions are proportionately much more pronounced than with the industrial class load. As a result, the commercial class hours use of demand (HUD) is observed to be on the order of 500 kWh per kW (69% load factor) compared to the industrial class HUD of about 625 kWh per kW (87% load factor).

As with residential bills, commercial and industrial bill distributions can be exhibited as histograms. Figure 10.6 illustrates histograms of the relative frequency of bills by kWh per kW, or HUD, intervals for two minor segments of the example commercial and industrial customer classes. The use of HUD intervals rather than kWh intervals helps to show effectively the distribution of the commercial and industrial bills across the whole range of demand and energy usage (kWh intervals can be used but offer less plotting resolution). The commercial class segment bill frequency in the upper graph is observed to have a bit of positive skew, similar to, but not as pronounced as, the residential bill distributions that were discussed previously. The average HUD of the commercial class segment is 342.5 kWh per kW (47% load factor), while the average HUD of the commercial class segment is 457.5 kWh per kW (63% load factor). The industrial class segment bill frequency in the lower graph approaches more of a normal distribution, but it appears to be somewhat negatively skewed.[6] These skew characteristics follow from the difference in average load factors of the commercial and industrial class customers.

---

[5] It must be realized that the curves in Figure 10.5 represent the average hourly loads of all customers within each of the classes. Thus, each load shape corresponds to the weighted result of five, six, and seven day weekly operations (as well as one, two, and three shift daily operations).

[6] Positive skew was defined previously for an asymmetrical distribution having a longer right-hand tail. Negative skew is exhibited by an asymmetrical distribution having a longer left-hand tail.

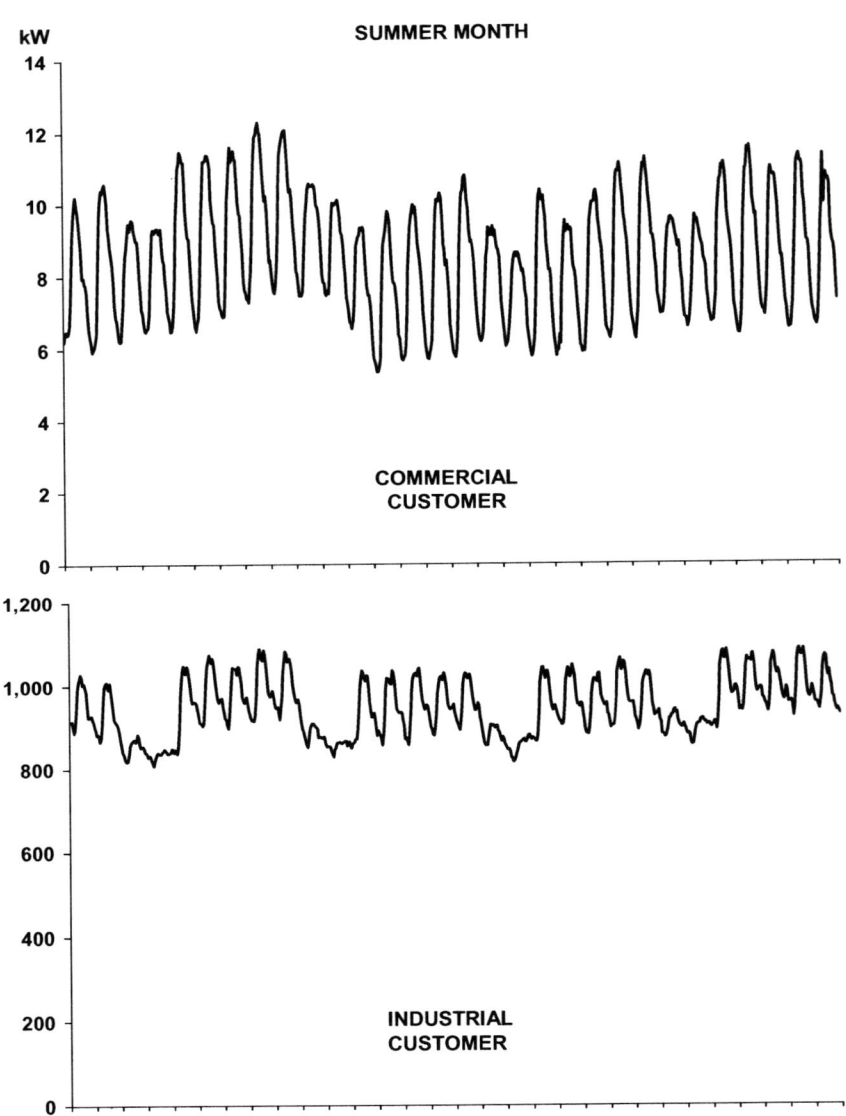

Figure 10.5 Typical monthly load shapes for the commercial and industrial classes. Average summer month load shapes for example commercial (upper graph) and industrial (lower graph) customer classes. Both load shapes exhibit distinctive daily operating cycles. Overall, commercial customer loads tend to have lower load factors compared to industrial customer loads.

Billing Data Statistics and Applications

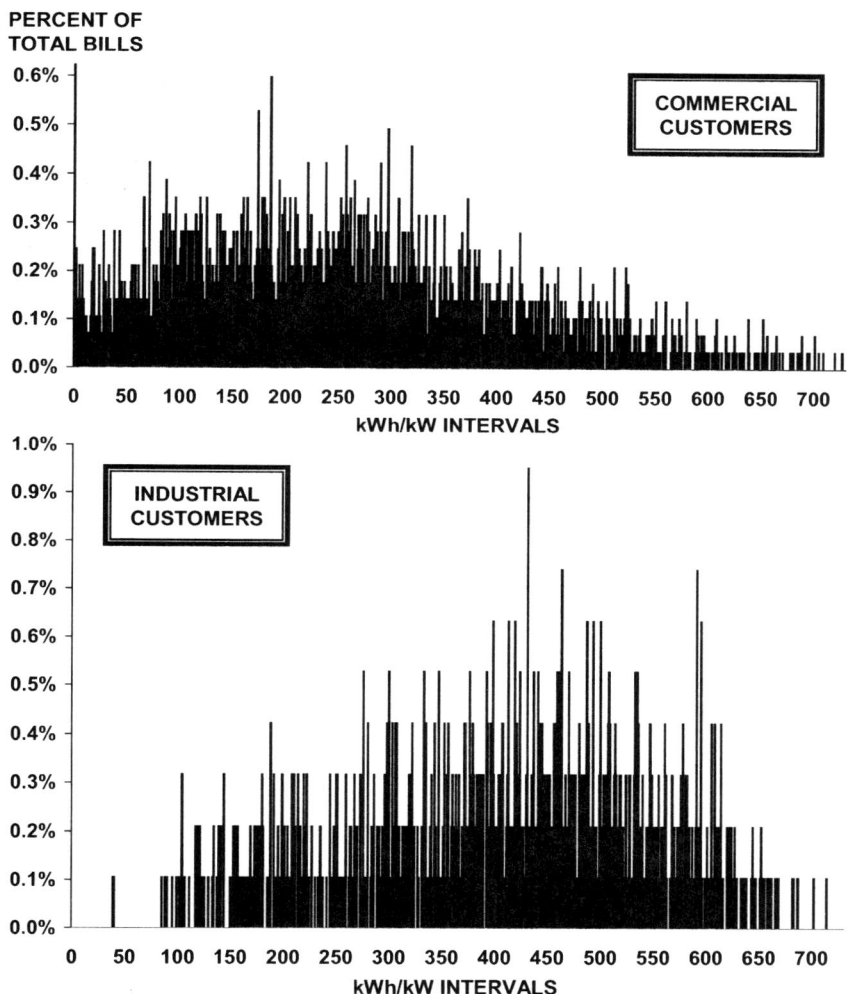

Figure 10.6 **Relative frequency histograms for the commercial and industrial classes.** Histograms of the relative frequency of commercial bills (upper graph) and industrial bills (lower graph) by HUD intervals for a typical summer month. The plots are less smooth than comparable residential plots since the commercial and industrial classes have many fewer numbers of customers. The commercial bill distribution exhibits a fairly positive skew (similar to residential), while the industrial bill distribution exhibits somewhat of a negative skew. The average load factor of the commercial class is notably less than that of the industrial class.

Delivery Voltage Characteristics

Customer demand and energy usage characteristics vary by service voltage. As shown by the scatter diagram in Figure 10.7, small to medium size three-phase customers tend to have a wider distribution of demands and load factors compared to single-phase customers. Secondary single-phase customers have mostly low demands since nearly all motor applications greater than 5 to 10 hp typically require three-phase service. The single-phase customer group shown in Figure 10.7 has an average demand of 36 kW and an average HUD of 189 kWh per kW, compared to the three-phase customer group, which has an average demand of 204 kW and an average HUD of 263kWh per kW.

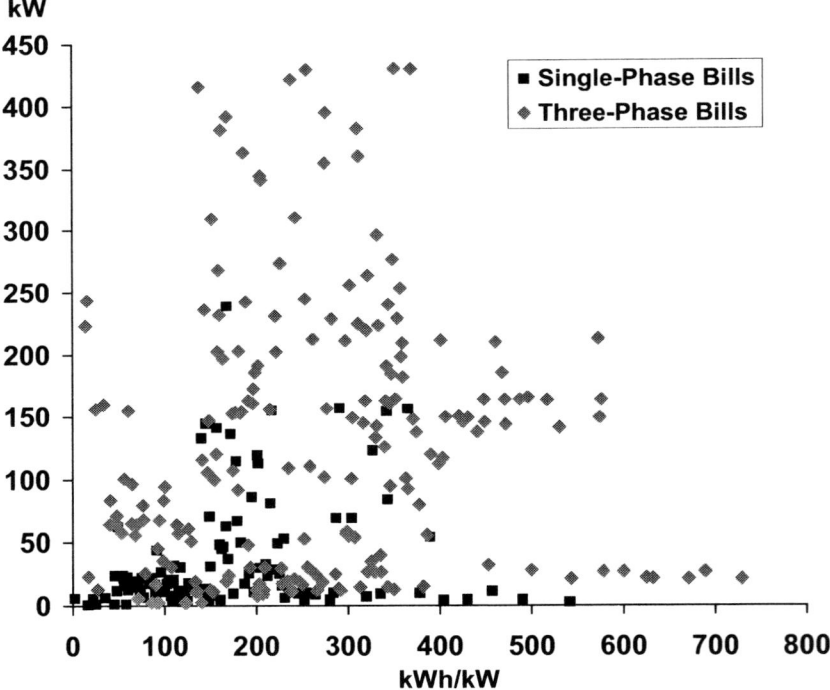

Figure 10.7 **A scatter plot of monthly demands as a function of the monthly hours use of demand.** Shown are plots of kW vs. kWh per kW for a segment of small- to medium-size commercial and industrial customers. Typically, the dispersion of demand and energy usage characteristics of such three-phase customers is found to be rather widespread. Compared to three-phase customers, single-phase customers generally have lower demands and operate at lower monthly load factors.

Billing Data Statistics and Applications

Figure 10.8 illustrates the trend in the energy, demand, and HUD characteristics for an example set of commercial and industrial customers that have been segmented by delivery voltage levels. The average monthly energy and demand (and thus the resulting HUD) requirements each increase with respect to the delivery voltage level. Note however in this example that the increase in the HUD from three-phase primary to transmission voltage is only minimal even though both the energy and demand usage amounts increase significantly (the percentage increase in the average energy usage is only slightly more than the percentage increase in the average demand). Since the number of transmission customers served by a typical utility is usually quite low compared to the thousands of primary and secondary customers also taking service, the usage characteristics of the transmission group is much more unique (i.e., characterized by very specific customer end-use load devices and operating requirements). In other words, a high demand transmission customer may, by nature of its business, operate at a fairly low load factor, and its impact could influence the average characteristics of the transmission customer segment. Thus, with groups of comparable commercial and industrial customers from other utilities, the transmission customers' HUD characteristics might be noted to increase much more.

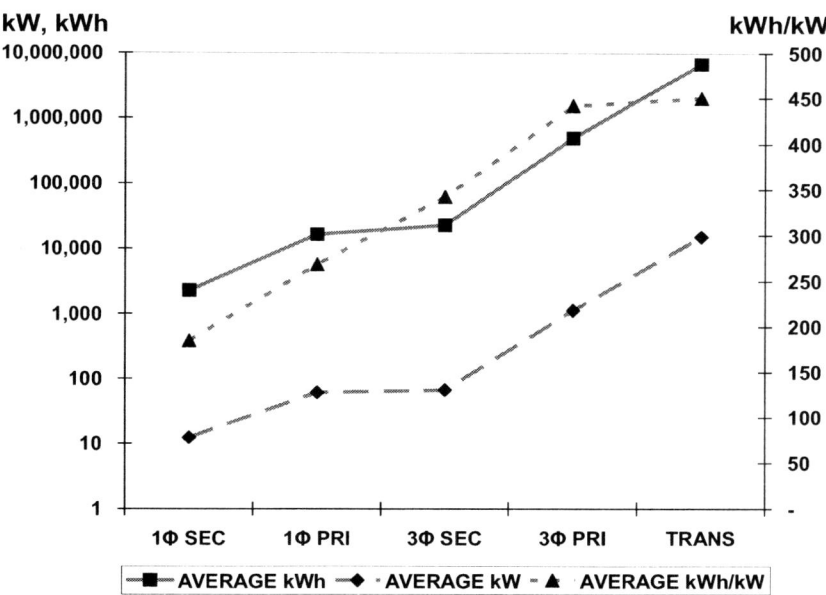

Figure 10.8 Trends in the average energy and demand usage characteristics of commercial and industrial customers as a function of delivery voltage. Average usage increases with increasing voltage.

10.3 BILL FREQUENCY ANALYSES

Tabular frequencies of electric service billing data are used for rate design, revenue forecasting, and evaluating usage characteristics for customers en masse. To produce a frequency for a group of customers requires that all bills in the group be sorted according to user specified parameters. Watt-hour metered customers bills are sorted by kWh intervals, whereas demand metered customer bills can be sorted by either kWh, demand (kW or kVA), or HUD intervals. To the extent that such features as rate minimums, demand ratchets, contract demands, or other billing determinant modifiers are used, bills can be sorted on the basis of either actual metered units or billing units.

The selection of interval size is important since the grouping of bill data by ranges of usage acts as a filter. The resulting frequency has less resolution than the original database of individual customer bill histories. The use of kWh per kW intervals by 1's (for high resolution) is generally manageable, since the average month is 730 hours. However, because of a few very high-usage customers, a residential frequency with kWh intervals by 1's would result in a substantial number of report lines (many of which would have zero entries) even when null data lines are suppressed. As a compromise, the interval widths can be varied within a frequency. Thus, for example, a residential bill frequency might be produced using kWh intervals by 1's for bills having a total usage up to 50 kWh, by 10's for bills having total usage between 50 kWh and 500 kWh, by 50's for bills having total usage between 500 kWh and 2,500 kWh, and so forth.

A group of 26 customer bills containing both kW and kWh billing determinants for a single billing month is shown below. These bills have been sorted three separate times to produce (a) the bill frequency based on kWh intervals shown in Table 10.1, (b) the bill frequency based on kW intervals shown in Table 10.2, and (c) the bill frequency based on kWh per kW intervals shown in Table 10.3.

ID	kWh	kW	ID	kWh	kW
A	30	2	N	2,255	11
B	2,275	7	O	10	1
C	1,300	20	P	1,520	19
D	2,980	4	Q	1,300	13
E	4,410	18	R	7,140	17
F	7,800	13	S	200	4
G	700	7	T	60	3
H	1,650	15	U	2,635	17
I	1,850	10	V	50	2
J	0	0	W	1,125	5
K	3,240	18	X	805	23
L	320	8	Y	3,825	17
M	660	4	Z	1,440	16

Billing Data Statistics and Applications

Table 10.1 Bill Frequency Analysis Based on Selected kWh Intervals

kWh Intervals	Bills Interval	Bills Σ	Bills %	kWh Interval	kWh Σ	kWh %	kW Interval	kW Σ	kW %
0	1	1	3.8	0	0	0.0	0	0	0.0
25	1	2	7.7	10	10	0.0	1	1	0.4
50	2	4	15.4	80	90	0.2	4	5	1.8
75	1	5	19.2	60	150	0.3	3	8	2.9
100	0	5	19.2	0	150	0.3	0	8	2.9
250	1	6	23.1	200	350	0.7	4	12	4.4
500	1	7	26.9	320	670	1.4	8	20	7.3
750	2	9	34.6	1,360	2,030	4.1	11	31	11.3
1,000	1	10	38.5	805	2,835	5.7	23	54	19.7
1,500	4	14	53.8	5,165	8,000	16.1	54	108	39.4
2,000	3	17	65.4	5,020	13,020	26.3	44	152	55.5
2,500	2	19	73.1	4,530	17,550	35.4	18	170	62.0
3,000	2	21	80.8	5,615	23,165	46.7	21	191	69.7
3,500	1	22	84.6	3,240	26,405	53.3	18	209	76.3
4,000	1	23	88.5	3,825	30,230	61.0	17	226	82.5
4,500	0	23	88.5	0	30,230	61.0	0	226	82.5
5,000	1	24	92.3	4,410	34,640	69.9	18	244	89.1
Excess	2	26	100.0	14,940	49,580	100.0	30	274	100.0

Table 10.1 actually contains nine frequency distributions based on kWh intervals: the frequency (columns labeled "Interval"), the cumulative frequency (columns labeled "Σ"), and the cumulative relative frequency (columns labeled "%") for the distribution of bills by kWh interval, the distribution of kWh associated with those bills, and the distribution of kW demands also associated with those bills. The relative frequencies for the bills, kWh, and kW have not been included in this table. Null data lines are included for this example.

The last interval indicated as "Excess" represents those bills (and their associated energy and demand usage) that have kWh that is higher than 5,000 kWh. The cumulative amounts shown on the Excess interval line represent the total bills, kWh, and kW of all customers in the group. Note that the average use of the two bills that fall within the Excess kWh interval is 7,470 kWh; thus, it can be observed from the bill frequency analysis that at least one of the two bills is considerably higher than the 5,000 kWh breakpoint. While a review of the individual bill data from above shows that the bills are 7,800 kWh (Bill F) and 7,140 kWh (Bill R), such a validation would be difficult with a database of thousands to millions of customer bills. Based on an assessment of the initial bill sort results, a revision of the original interval criteria might be necessary for the Excess kWh level (or, for that matter, any other intervals) to ensure a bill frequency resolution that meets the requirements of any subsequent analysis or application.

Similar to Table 10.1, Tables 10.2 and 10.3 each contain nine bill frequencies on the basis of kW intervals and kWh per kW intervals, respectively. Except for the interval definition, the column structures are all the same.

Table 10.2 Bill Frequency Analysis Based on Selected kW Intervals

kW Intervals	Bills Interval	Σ	%	kWh Interval	Σ	%	kW Interval	Σ	%
0	1	1	3.8	0	0	0.0	0	0	0.0
1	1	2	7.7	10	10	0.0	1	1	0.4
2	2	4	15.4	80	90	0.2	4	5	1.8
3	1	5	19.2	60	150	0.3	3	8	2.9
4	3	8	30.8	3,840	3,990	8.0	12	20	7.3
5	1	9	34.6	1,125	5,115	10.3	5	25	9.1
10	4	13	50.0	5,145	10,260	20.7	32	57	20.8
15	4	17	65.4	13,005	23,265	46.9	52	109	39.8
20	8	25	96.2	25,510	48,775	98.4	142	251	91.6
Excess	1	26	100.0	805	49,580	100.0	23	274	100.0

Table 10.3 Bill Frequency Analysis Based on Selected HUD Intervals

kWh/kW Intervals	Bills Interval	Σ	%	kWh Interval	Σ	%	kW Interval	Σ	%
0	1	1	3.8	0	0	0.0	0	0	0.0
25	4	5	19.2	150	150	0.3	8	8	2.9
50	3	8	30.8	1,325	1,475	3.0	35	43	15.7
75	1	9	34.6	1,300	2,775	5.6	20	63	23.0
100	4	13	50.0	4,960	7,735	15.6	55	118	43.1
150	1	14	53.8	1,650	9,385	18.9	15	133	48.5
200	4	18	69.2	8,385	17,770	35.8	49	182	66.4
250	4	22	84.6	11,615	29,385	59.3	51	233	85.0
300	0	22	84.6	0	29,385	59.3	0	233	85.0
350	1	23	88.5	2,275	31,660	63.9	7	240	87.6
400	0	23	88.5	0	31,660	63.9	0	240	87.6
450	1	24	92.3	7,140	38,800	78.3	17	257	93.8
500	0	24	92.3	0	38,800	78.3	0	257	93.8
550	0	24	92.3	0	38,800	78.3	0	257	93.8
600	1	25	96.2	7,800	46,600	94.0	13	270	98.5
650	0	25	96.2	0	46,600	94.0	0	270	98.5
700	0	25	96.2	0	46,600	94.0	0	270	98.5
730	0	25	96.2	0	46,600	94.0	0	270	98.5
Excess	1	26	100.0	2,980	49,580	100.0	4	274	100.0

Bill frequencies can be generated on a monthly basis, a seasonal basis, and an annual basis. For application purposes, it is generally best to exclude prorated, estimated, and other abnormal bills from a bill frequency analysis. When the exclusion of bills produces a loss of approximately five percent or less of the total billing determinants, further analysis using a *correction factor* approach will yield satisfactory results in revenue calculations.

10.4 BILL FREQUENCY REVENUE ANALYSES

The tabular form of the bill frequency provides the input data necessary to quickly and easily calculate revenue for an entire group of customers with a high degree of accuracy without having to compute bills individually. To do so, the bill frequency does not need to be highly detailed, but it must include at least those intervals that coincide with the kWh, demand, and/or HUD price blocks of the rate being applied. If an alternate rate structure is under consideration, the bill frequency should also include any potential new price blocks that might be utilized with a revised rate design.

The bill frequencies for kWh, demand, and HUD discussed previously are sufficient for evaluating many basic rate structures and attendant adjustment charges.[7] For example, the kWh bill frequency can be used to calculate revenues for declining block and inverted block structures, as well as many modified block variations. The HUD bill frequency can be used to calculate revenues for both the basic Wright and Wright-Hopkinson demand rate structures. More complex rate structures would require a more intricate bill frequency architecture. For example, a Wright rate structure having kWh blocks in one or more of its HUD steps would require a compound bill frequency having HUD intervals that are further divided by kWh subintervals. Multiple bill frequencies may be required for calculating revenues of some complex rate structures.

Revenue Calculations with Mass Billing Determinants

On an interval basis, a bill frequency compiles those billing determinants which are associated with bills having total usage that correlates to prescribed intervals. Consider the distribution of bills shown in Figure 10.9. Ten bills are sorted by total kWh usage and plotted against the breakpoints of a declining block rate having four prices. In this manner, the rate structure is being used to define four intervals of a bill frequency based on kWh usage. The billing determinants that fall within an interval are referred to as the *interval use*, which is kWh in this case.

---

[7] Adjustment charges having units of kWh or kW (kVA) demand typically can be applied to the frequency's billing determinants (e.g., fuel cost recovery or delivery voltage adjustments). However, the determination of revenues associated with power factor might require an additional frequency distribution of kVAR or kVARh usage or even an exogenous calculation.

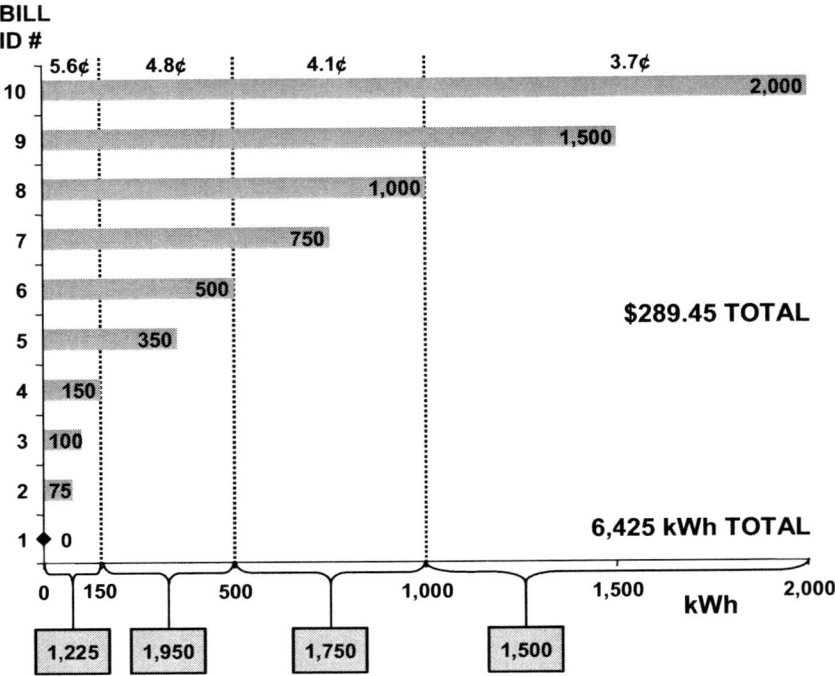

Figure 10.9 A visualization of energy use by rate block. An example of ten bills that have been sorted by using the four price blocks of a declining block rate structure to define kWh intervals of a bill frequency. The four boxes at the bottom of the chart indicate the total quantity of kWh usage that is billed at each price block. The sum of the ten individual bill amounts is equal to $289.45. Note that any applicable Customer Charge associated with the kWh prices is ignored in this example; thus, the zero kWh bill yields no revenue.

In Figure 10.9, the first interval is between 0 kWh and 150 kWh and includes all bills having a total usage from 1 kWh to 150 kWh (valid bills having no kWh usage are grouped at 0 kWh). Thus, Bill 2, Bill 3, and Bill 4 all fall within the first interval, and the interval use is observed to be 325 kWh. Likewise, the *cumulative use* for the 0 kWh and 150 kWh interval is 325 kWh. If individual revenue calculations were performed, the total kWh usage of Bill 2, Bill 3, and Bill 4 would each be billed at 5.6¢ per kWh.

The second interval is between 150 kWh and 500 kWh and includes all bills having a total usage from 151 kWh to 500 kWh. Bill 5 and Bill 6 both fall within the second interval, and the interval use is thus 850 kWh, while the cumulative use is now 1,175 kWh (325 kWh + 850 kWh). Note that Bill 5 and Bill 6 pass through the 0 kWh to 150 kWh interval. As a result, both Bill 5 and Bill 6 have a

Billing Data Statistics and Applications

portion of kWh usage that is billed at 5.6¢ per kWh (for the first 150 kWh) while the remaining kWh usage is billed at 4.8¢ per kWh. In addition, Bill 7, Bill 8, Bill 9, and Bill 10 also have a portion of kWh usage that is priced at 5.6¢ per kWh (for the first 150 kWh). For the ten bills shown, a total of 1,225 kWh is billed at 5.6¢ per kWh.

Considering further the 150 kWh to 500 kWh interval, observe that Bill 7, Bill 8, Bill 9, and Bill 10 also pass through the second interval and thus each has a portion of kWh usage that is priced at 4.8¢ per kWh (for the next 350 kWh). The total amount of energy billed at 4.8¢ per kWh, from Bills 5 through 10, is 1,950 kWh. Further observation of the graph in Figure 10.9 reveals that a total of 1,750 kWh are billed at 4.1¢ per kWh, while 1,500 kWh are billed at 3.7¢ per kWh.

By manually collecting the energy usage of all ten bills by the four price blocks of the rate, the total revenue of the group of ten bills can be easily calculated en masse:

```
1,225 kWh × $0.056 per kWh   =   $68.60
1,950 kWh × $0.048 per kWh   =   $93.60
1,750 kWh × $0.041 per kWh   =   $71.75
1,500 kWh × $0.0307 per kWh  =   $55.50

6,425 kWh        Total           $289.45
```

The collection of the energy usage by the price blocks of the rate, as used in this example, was accomplished by visual inspection. Clearly such an approach would be impractical for evaluating billing data for large groups of customers. However, the collection of energy usage by price block for groups of any number of customers can be accomplished mathematically using a bill frequency analysis.

Energy Rate Mass Revenue Calculations

A bill frequency analysis for energy-only types of rate structures, such as a typical residential block rate structure, contains distributions of bills and kWh sales by kWh intervals. Cumulative frequency data is a necessary input for the process of determining just how much of the total kWh falls within each of the price blocks of the rate. In order to calculate revenue, the bill frequency must contain at a minimum those intervals which coincide with the price blocks; however, it can contain as many additional intervals as desired.

Using the bill frequency analysis, a *consolidated factor* is calculated for each interval. The consolidated factor (at any given interval) represents the summation of (a) the cumulative use of the bills prior to and included in that interval and (b) the use through that interval for all bills which have a total usage which exceeds that interval (i.e., the pass through bills). The *use through that interval* refers to the amount of use between zero and the upper end of the interval of interest (i.e., not the use between the interval of interest, such as 500 kWh, and any previous interval, such as 150 kWh). In reference to Figure 10.9, at the 500 kWh interval, the cumulative use of the bills prior to and included in that interval is 1,175 kWh (Bills 1 through 6), and the use through that interval is 2,000 kWh (Bills 7 through 10: 4 × 500 kWh), for a total of 3,175 kWh. In other words, considering all ten bills together, there is 3,175 kWh of use between 0 and 500 kWh; thus, 3,175 kWh is the consolidated factor for the 500 kWh interval.

The consolidated factor at any interval for a kWh interval based bill frequency is determined by:

$$ECF_i = \sum_{i=1}^{T} E_i + \left[B_{Total} - \sum_{i=1}^{T} B_i \right] \times EI_i \tag{10.6}$$

where ECF_i = consolidated factor in kWh
E_i = kWh in the ith interval
B_{Total} = total number of bills
B_i = bills in the ith interval
EI_i = kWh at the upper end of the ith interval
T = total number on intervals

By applying Equation (10.6) to the billing data shown in Figure 10.9, the consolidated factor can be computed at each price breakpoint of the rate structure, as shown below. Note also that the consolidated factor formula for bills having no kWh usage (i.e., the bills identified at a kWh interval of zero) computes to zero, and that the consolidated factor for all kWh is simply equal to the total kWh usage of the ten bills (in essence the kWh amount of the largest bill in the group establishes the upper end of the last kWh interval).

Consolidated Factor = Σ kWh Use + [Total Bills – Σ Bills] × kWh Interval

ECF_0	=	0		0	10	1		0
ECF_{150}	=	1,225		325	10	4		150
ECF_{500}	=	3,175		1,175	10	6		500
$ECF_{1,000}$	=	4,925		2,925	10	8		1,000
ECF_{Total}	=	6,425		6,425	10	10		2,000

Billing Data Statistics and Applications

The revenue of all ten bills is then calculated en masse by applying the block prices to the differences between the consolidated factors which correspond to the upper and lower ends of the price blocks.

$ECF_{150} - ECF_0$	=	1,225 kWh	@ 5.6¢ per kWh		$68.60
$ECF_{500} - ECF_{150}$	=	1,950 kWh	@ 4.8¢ per kWh		$93.60
$ECF_{1,000} - ECF_{500}$	=	1,750 kWh	@ 4.1¢ per kWh		$71.75
$ECF_{Total} - ECF_{1,000}$	=	1,500 kWh	@ 3.7¢ per kWh		$55.50
				Total	$289.45

The revenue calculated by using consolidated factors equates to the revenue calculated previously by visual inspection of the ten bills.

In Table 10.4, a column of consolidated factors has been added to the frequency analysis presented before in Table 10.1. Note that the consolidated factor is a cumulative frequency of kWh, but that it differs from the cumulative use frequency since it includes kWh at each interval from the pass-through bills.

Table 10.4 Consolidated Factors for a kWh Bill Frequency

kWh Intervals	Bills Interval	Σ	kWh Interval	Σ	kW Interval	Σ	Consolidated Factor kWh	%
0	1	1	0	0	0	0	0	0.0
25	1	2	10	10	1	1	610	1.2
50	2	4	80	90	4	5	1,190	2.4
75	1	5	60	150	3	8	1,725	3.5
100	0	5	0	150	0	8	2,250	4.5
250	1	6	200	350	4	12	5,350	10.8
500	1	7	320	670	8	20	10,170	20.5
750	2	9	1,360	2,030	11	31	14,780	29.8
1,000	1	10	805	2,835	23	54	18,835	38.0
1,500	4	14	5,165	8,000	54	108	26,000	52.4
2,000	3	17	5,020	13,020	44	152	31,020	62.6
2,500	2	19	4,530	17,550	18	170	35,050	70.7
3,000	2	21	5,615	23,165	21	191	38,165	77.0
3,500	1	22	3,240	26,405	18	209	40,405	81.5
4,000	1	23	3,825	30,230	17	226	42,230	85.2
4,500	0	23	0	30,230	0	226	43,730	88.2
5,000	1	24	4,410	34,640	18	244	44,640	90.0
Excess	2	26	14,940	49,580	30	274	49,580	100.0

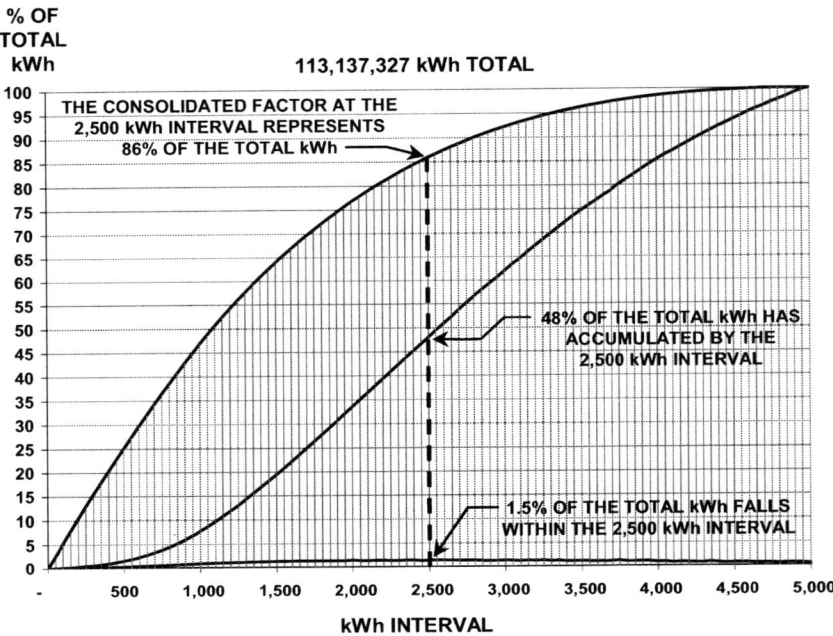

Figure 10.10 Relative frequency distributions of the interval use, the cumulative use, and the consolidated factor for a residential class of customers. The cumulative kWh use at any interval includes the kWh use in that interval, while the consolidated factor at that same interval includes the cumulative kWh use plus the portion of total kWh usage that lies between zero and the upper end of the interval for all bills having a total kWh usage that exceeds that interval.

On a larger and more practical scale, Figure 10.10 illustrates the distribution of kWh sales for a group of residential customers by means of three relative frequencies. The three curves illustrate the difference in characteristics between the interval use, the cumulative use, and the consolidated factor. The consolidated factor accumulates kWh much faster than the cumulative kWh frequency because it includes the kWh of the pass through bills.

The billing data shown in Figure 10.10 represents the monthly kWh sales of 61,900 residential bills. The revenue attained from individual bill calculations is compared to the revenue attained from applying the example block rate to the bill frequency determinants, en masse, in the computation shown below. Note the exceptional accuracy achieved with the bill frequency calculation method (the percent difference is just -0.00048%).

Bill Frequency Calculation:

Customer Charge:		61,900 Bills	@	$14.00/Bill	$866,699.00
$ECF_{150} - ECF_0$	=	8,945,351 kWh	@	5.6¢/kWh	500,939.66
$ECF_{500} - ECF_{150}$	=	19,499,813 kWh	@	4.8¢/kWh	935,991.02
$ECF_{1,000} - ECF_{500}$	=	24,546,383 kWh	@	4.1¢/kWh	1,006,401.70
$ECF_{Total} - ECF_{1,000}$	=	60,145,780 kWh	@	3.7¢/kWh	2,225,393.86
		113,137,327 kWh			

Total revenue from bill frequency calculation: $5,535,326.24

Total revenue from individual bill calculations: $5,535,352.70

Difference: Amount: ($26.46)

Percent: (0.00048%)

Such a high accuracy in results permits the utilization of bill frequency analyses for rate design, revenue forecasting, marketing evaluations, and other business applications. Efficiency is gained by the use of the bill frequency since the bills need only be sorted a single time (into carefully planned interval divisions), and no individual bill calculations are then necessary in order to determine revenue of the customer group.[8]

The mean usage for the example residential segment is 1,828 kWh per customer (113,137,327 kWh ÷ 61,900 Bills), while the average rate realization is 4.89¢ per kWh ($5,535,326.24 ÷ 113,137,327 kWh). Figure 10.11 illustrates the distribution of the kWh sales relative to the average rate realization at all levels of kWh usage for the example residential rate structure. The greatest density of kWh sales occurs where the total usage per bill exceeds 1,000 kWh (where 72% of the total bills account for 92% of the total kWh sales). From 1,000 kWh to about 5,000 kWh (the interval with the largest use bills), the average rate declines from about 6.0¢ per kWh to about 4.0¢ per kWh.

---

[8] However, individual bill calculations are important for designing rate structures for the larger commercial and industrial customer classes in order to account for potential migrations of customers from their current rate schedule to an alternate rate schedule in the event there is notable movement in the line of equality between the two rates. Not accounting for rate migrations in the design stage can result in underperforming rate structures in practice, and revenue could be lost.

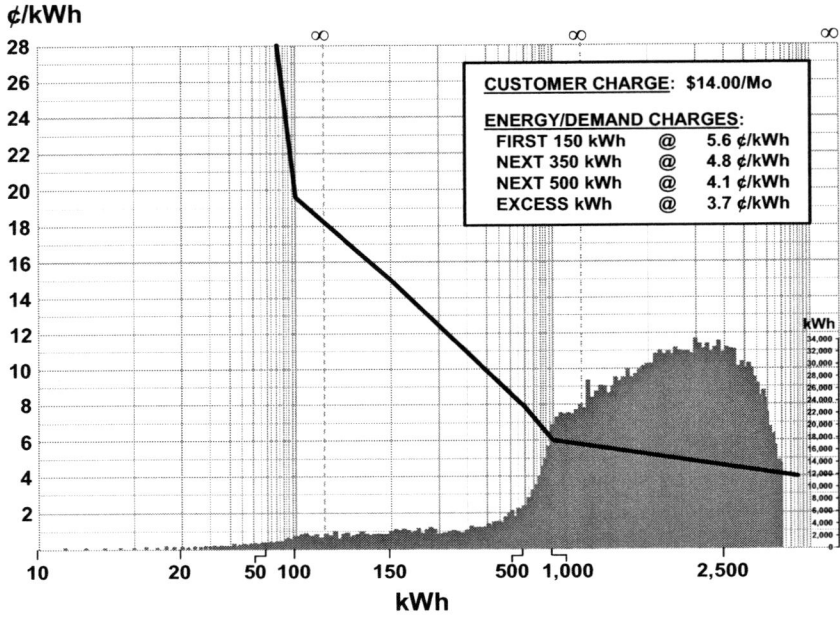

Figure 10.11 Average rate vs. sales volumes. The distribution of kWh sales by intervals is plotted against the average rate per kWh for the example residential customer group.

The percent of the total revenue generated by each price block of the example rate structure is:

Customer Charge:	@	$14.00/Bill	$866,699.00	15.7%
First 150 kWh	@	5.6¢/kWh	$500,939.66	9.0%
Next 350 kWh	@	4.8¢/kWh	$935,991.02	16.9%
Next 500 kWh	@	4.1¢/kWh	$1,006,401.70	18.2%
Over 1,000 kWh	@	3.7¢/kWh	$2,225,393.86	40.2%

The percent of total revenue yield increases significantly from the first block to the tail block, even though the unit price decreases, due to the sheer volume increase in kWh sales in the successive price blocks.

Billing Data Statistics and Applications

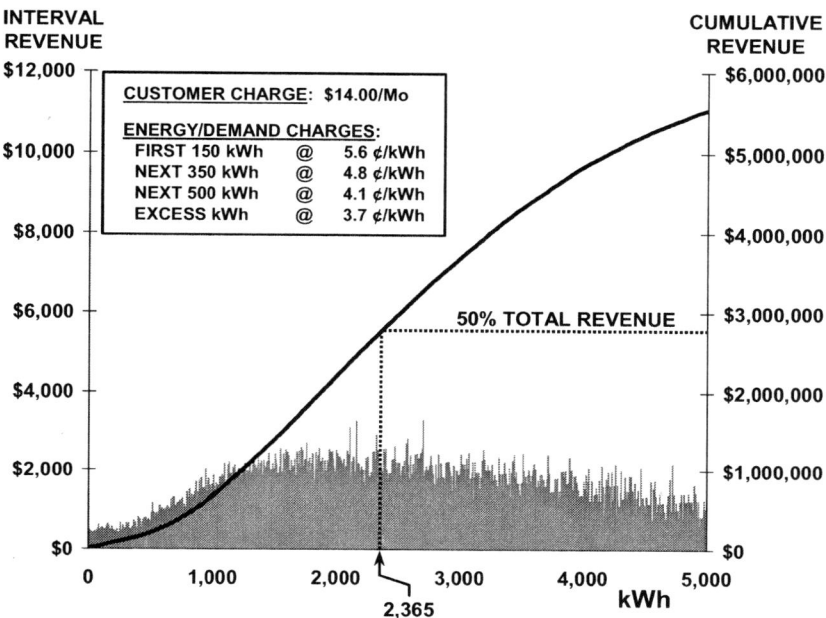

Figure 10.12 Distribution of rate revenues by kWh intervals. A histogram and cumulative frequency of the revenue yielded by application of the example rate structure to the residential class billing determinants.

Revenue for a customer group can also be shown as a frequency distribution. Bill amounts by kWh interval for the example rate and bill frequency are displayed as a histogram in Figure 10.12. The cumulative frequency of the revenue is also shown. One half of the total revenue yield is achieved at the 2,365 kWh interval. Revenue distributions can also be included in the tabular versions of a bill frequency analysis.

Demand Rate Mass Revenue Calculations

As discussed before, a bill frequency analysis can be based on intervals of kW or kVA (metered or billed units). A bill frequency based on kW demand was shown previously in Table 10.2. A bill frequency analysis that is developed on the basis of kW or kVA intervals will produce consolidated factors which also have units of demand. The methodology for calculating consolidated factors for a demand type of bill frequency is similar to that which was used in Equation (10.6) for a bill frequency based on kWh intervals. The consolidated factor at any interval for a kW or kVA interval based bill frequency is thus determined by

$$DCF_i = \sum_{i=1}^{T} D_i + \left[B_{Total} - \sum_{i=1}^{T} B_i \right] \times DI_i \qquad (10.7)$$

where DCF_i = consolidated factor in kW (or kVA)
          D_i = kW (or kVA) in the ith interval
          B_{Total} = total number of bills
          B_i = bills in the ith interval
          DI_i = kW (or kVA) at the upper end of the ith interval
          T = total number on intervals

As with the consolidated factors of the kWh interval bill frequency, the kW (or kVA) demand consolidated factors incorporate the contributions of the pass through bills throughout the bill frequency's intervals. In Table 10.5, a column of kW consolidated factors has been added to the frequency analysis originally presented in Table 10.2.

A demand bill frequency can also be based on intervals of HUD, as shown previously in Table 10.3. The methodology for determining the consolidated factor for an HUD bill frequency is similar to both the kWh and the kW (or kVA) interval expressions shown above in Equations (10.6) and (10.7). The consolidated factor at any interval for an HUD interval based bill frequency is thus determined by

$$HCF_i = \sum_{i=1}^{T} E_i + \left[D_{Total} - \sum_{i=1}^{T} D_i \right] \times HI_i \qquad (10.8)$$

where HCF_i = consolidated factor in kWh
          E_i = kWh in the ith interval
          D_{Total} = total demands in kW (or kVA)
          D_i = kW (or kVA) demand in the ith interval
          HI_i = kWh per kW (or kVA) at the upper end of the ith interval
          T = total number on intervals

Note that the consolidated factor for an HUD bill frequency has the unit of kWh and not the kWh per kW (or kVA) units of the intervals. A significant difference between the HUD consolidated factor and the kWh and kW (or kVA) consolidated factors is that the intervals throughout the HUD bill frequency contain pass-

Billing Data Statistics and Applications

through demands (as opposed to pass through bills). In Table 10.6, a column of kWh consolidated factors has been added to the HUD frequency analysis originally presented in Table 10.3.

Table 10.5 Consolidated Factors for a Demand Bill Frequency

kW Intervals	Bills Interval	Σ	kWh Interval	Σ	kW Interval	Σ	Consolidated Factor kW	%
0	1	1	0	0	0	0	0	0.0
1	1	2	10	10	1	1	25	9.1
2	2	4	80	90	4	5	49	17.9
3	1	5	60	150	3	8	71	25.9
4	3	8	3,840	3,990	12	20	92	33.6
5	1	9	1,125	5,115	5	25	110	40.1
10	4	13	5,145	10,260	32	57	187	68.2
15	4	17	13,005	23,265	52	109	244	89.1
20	8	25	25,510	48,775	142	251	271	98.9
Excess	1	26	805	49,580	23	274	274	100.0

Table 10.6 Consolidated Factors for an HUD Bill Frequency

kWh/kW Intervals	Bills Interval	Σ	kWh Interval	Σ	kW Interval	Σ	Consolidated Factor kWh	%
0	1	1	0	0	0	0	0	0.0
25	4	5	150	150	8	8	6,800	13.7
50	3	8	1,325	1,475	35	43	13,025	26.3
75	1	9	1,300	2,775	20	63	18,600	37.5
100	4	13	4,690	7,735	55	118	23,335	47.1
150	1	14	1,650	9,385	15	133	30,535	61.6
200	4	18	8,385	17,770	49	182	36,170	73.0
250	4	22	11,615	29,385	51	233	39,635	79.9
300	0	22	0	29,385	0	233	41,685	84.1
350	1	23	2,275	31,660	7	240	43,560	87.9
400	0	23	0	31,660	0	240	45,260	91.3
450	1	24	7,140	38,800	17	257	46,450	93.7
500	0	24	0	38,800	0	257	47,300	95.4
550	0	24	0	38,800	0	257	48,150	97.1
600	1	25	7,800	46,600	13	270	49,000	98.8
650	0	25	0	46,600	0	270	49,200	99.2
700	0	25	0	46,600	0	270	49,400	99.6
730	0	25	0	46,600	0	270	49,520	99.9
Excess	2	26	14,940	49,580	30	274	49,580	100.0

418   Chapter 10

En masse revenue calculation with either a kW (or kVA) or an HUD based bill frequency analysis proceeds in the same manner as with the kWh interval bill frequency. The revenue for each block is computed by applying the unit prices to the differences in consolidated factors between the upper and lower ends of the price blocks.

Compound Bill Frequencies

A compound bill frequency is required for computing revenues en masse for complex rate structures. A Wright-Hopkinson rate structure having blocks in both its Demand Charge and Energy Charge components is graphed in Figure 10.13. The 26 bills from the previous example have been superimposed as a scatter plot across the demand and energy field. The total revenue produced by applying the example rate to the 26 bills individually is $2,512.25. The rate structure graph provides a visual aid for designing a compound bill frequency that can be used to accurately calculate revenue en masse. A bill frequency analysis, having just enough detail to apply the rate correctly, is shown in Table 10.7.

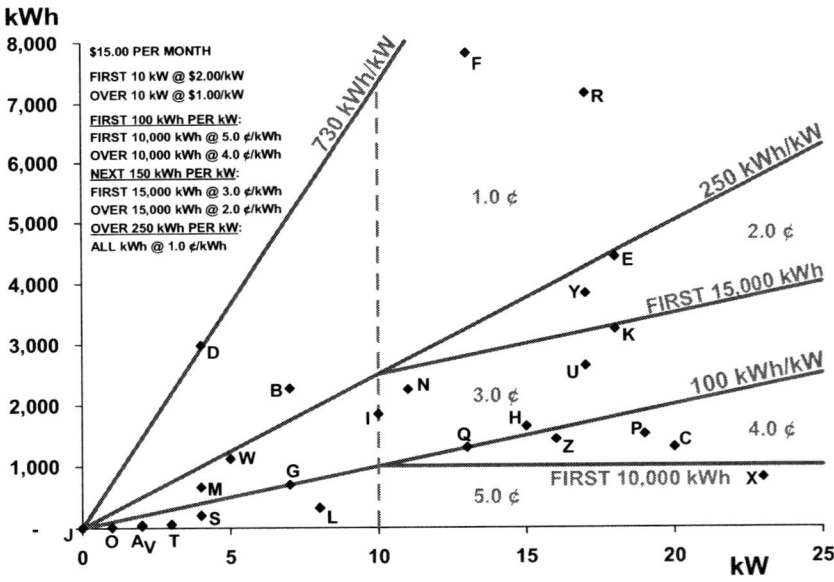

Figure 10.13 A scatter plot of billing determinants across a rate structure graph of a blocked Wright-Hopkinson rate. Bills for the example data set are distributed throughout all price sectors of the rate. The rate structure graph is useful for designing the layout and calculations needed to produce a compound bill frequency analysis.

Table 10.7 A Compound Bill Frequency Analysis

Intervals: HUD	Intervals: kWh	Bills Interval	Bills Sub Σ	kWh Interval	kWh Sub Σ	kW Interval	kW Sub Σ	Consolidated Factors: kW	Consolidated Factors: kWh	
10 kW										
100	All	7	8	1,370	1,370	27	27	--	4,370	a
250	All	3	11	3,635	5,005	19	46	--	7,755	a
745	All	2	13	5,255	**10,260**	11	57	187	10,620	a
23 kW										
100	1,000	1	1	805	805	23	23	--	12,805	b
100	Excess	4	5	5,560	6,365	68	91	--	18,965	a
250	1,500	4	9	9,780	16,145	61	152	--	28,645	c
250	Excess	2	11	8,235	24,380	35	187	--	31,880	a
600	All	2	13	14,940	**39,320**	30	217	274	39,230	a
Grand Totals:			26		49,580		274			

Notes:
a. kWh_{CF} = Sub Σ kWh + [$kW_{SUBTOTAL}$ - Sub Σ kW] × $HUD_{INTERVAL}$
b. kWh_{CF} = Sub Σ kWh + [$Bills_{SUBTOTAL}$ - Sub Σ Bills] × $kWh_{INTERVAL}$
c. kWh_{CF} = Sub Σ kWh + [($Bills_{SUBTOTAL}$ - Sub Σ Bills) × $kWh_{INTERVAL}$]
   + [($kW_{SUBTOTAL}$ - Sub Σ kW) × $HUD_{PRIOR\ INTERVAL}$]

The example complex rate structure requires the calculation of all three versions of consolidated factors. A kW demand consolidated factor, Equation (10.7), is necessary because of the two-block prices of the Demand Charge. Going from left to right with reference to the rate structure graph in Figure 10.13, the demand frequency consists of an interval from 0 to 10 kW and an interval of 10 to 23 kW (i.e., the bill with the largest value of demand). Because of the blocking of the Energy Charges above 10 kW, it is necessary to treat the analysis of the energy distribution as two subfrequencies. Going from the bottom to the top of the graph, HUD consolidated factors, Equation (10.8), are required for the kWh within the three HUD steps of the rate up to 10 kW. Above 10 kW, HUD consolidated factor are also needed except that the First 1,000 kWh block of the first HUD step (above 10 kW) requires a kWh consolidated factor, Equation (10.6). However, the First 1,500 kWh block of the second HUD step (above 10 kW) presents a more complicated situation since it rides on top of the 100 kWh per kW line. Thus, the consolidated factor for the interval defined as the First 1, 500 kWh of the Next 150 kWh per kW (greater than 10 kW) is determined by

$$CCF_i = \sum_{i=1}^{T} E_i + \left[B_{\text{Total}} - \sum_{i=1}^{T} B_i \right] \times EI_i + \left[D_{\text{Total}} - \sum_{i=1}^{T} D_i \right] \times HI_{i-1} \quad (10.9)$$

where CCF_i = compound consolidated factor in kWh
E_i = kWh in the ith interval (First x kWh of the Next y kWh/kW)
B_{Total} = total number of bills of the subfrequency
B_i = bills in the ith interval
EI_i = kWh at the upper end of the ith interval
D_{Total} = total subfrequency demands in kW (or kVA)
D_i = kW (or kVA) demand in the ith interval
HI_{i-1} = kWh per kW (or kVA) at the upper end of the ith-1 interval
T = total number on intervals

For the example bill frequency shown in Table 10.7, the ith interval (of the subfrequency where bill demands are greater than 10 kW) is defined by both the HUD step having an upper end value of 250 kWh per kW, and a kWh block having an upper end value of 1,500 kWh.[9] The first component of Equation (10.9), ΣE_i, accounts for the 16,145 kWh of the 9 bills having total usage that falls within the range between 0 and the first 1,500 kWh of the next 150 kWh per kW. The second component of Equation (10.9), $[B_{\text{TOTAL}} - \Sigma B_i] \times EI_i$, accounts for the kWh associated with those higher usage bills that pass through the 1,500 kWh block of the next 150 HUD step ($[13 - 9] \times 1,500 = 6,000$ kWh). The last component of Equation (10.12), $[D_{\text{TOTAL}} - \Sigma D_i] \times HI_{i-1}$, accounts for the kWh associated with the demands of those higher usage bills that pass through the first 100 kWh per kW interval ($[217 - 152] \times 100 = 6,500$ kWh. The consolidated factor is thus 28,645 kWh.

To apply the rate to determine total revenue, the appropriate differences in consolidated factor amounts are calculated for each associated unit price. The kWh amounts applicable to the 5.0¢, 3.0¢, and 1.0¢ kWh prices are produced from both subfrequencies. The 4.0¢ and 2.0¢ kWh prices only applies to bills greater than 10 kW. The total revenue is thus determined by

---

[9] In essence, the 1,500 kWh block in the second HUD step creates another two-interval subfrequency, i.e., a bill distribution within the 100 to 150 kWh per kW range. The same is true of the first 100 HUD step, since it contains a 1,000 kWh block. However, the complicating factor with calculating a consolidated factor for the 1,500 kWh interval is that its lower end value, the upper end of the 100 HUD step, varies with demand. In contrast, the lower end of the 1,000 kWh block in the first 100 kWh step is constant at zero for all levels of demand in excess of 10 kW, and the basic formula for calculating a kWh interval-based consolidated factor applies.

Billing Data Statistics and Applications

Bill Frequency Calculation:

Customer Charge: 26 Bills   @ $15.00/Bill $390.00

Demand Charges:

First 10 kW	187 – 0	=	187 kW	@	$2.00/kW	374.00
Over 10 kW	274 – 57	=	87 kW	@	$1.00/kW	87.00
			274 kW			

Energy Charges:

First 100 kWh per kW

First 1,000 kWh:	4,370 – 0	=	4,370 kWh			
	12,805 – 0	=	12,805 kWh			
			17,175 kWh	@	5.0¢/kWh	858.75
Over 1,000 kWh:	18,965 – 12,805	=	6,160 kWh	@	4.0¢/kWh	246.40

Next 150 kWh per kW

First 1,500 kWh:	7,755 – 4,370	=	3,385 kWh			
	28,645 – 18,965	=	9,680 kWh			
			13,065 kWh	@	3.0¢/kWh	391.95
Over 1,500 kWh:	31,880 – 28,645	=	3,235 kWh	@	2.0¢/kWh	64.70

Over 250 kWh per kW

All kWh:	10,260 – 7,755	=	2,505 kWh			
	39,320 – 31,880	=	7,440 kWh			
			9,945 kWh	@	3.0¢/kWh	391.95
			49,580 kWh			

Total revenue from bill frequency calculation: <u>$2,512.25</u>

The total revenue from the bill frequency calculation thus matches the previously stated revenue amount that results by summing the individual bill amounts.

10.5 REVENUE FORECASTING METHODOLOGIES

The projection of revenue is an essential pricing-related function as the results are used to make key business operating decisions. The comparison of projected revenues with revenue requirements provides insights for future investment and financing strategies, cost management measures, and price level adjustments.

    A revenue forecast is based in part on historical information, such as the energy and demand billing determinant records and their intrinsic relationships or trends (e.g., average kWh usage per bill or changes in load factor over time). An example of basic input billing data requirements is shown in Table 10.8. Prorated and estimated bills have been excluded, as they do not represent normal use.

Table 10.8 Historical Billing Data for a Residential Rate Class

Historical Period	Customers (Bills)	kWh Sales	Base Rate Revenue ($)	kWh Per Bill	¢ per kWh
OCT	830,736	665,265,464	39,182,225	800.815	5.8897
NOV	831,924	556,569,326	34,741,401	669.015	6.2421
DEC	826,173	661,949,859	38,851,976	801.224	5.8693
JAN	771,931	784,381,482	42,594,490	1,016.129	5.4303
FEB	833,964	723,918,662	40,988,711	868.045	5.6621
MAR	860,963	614,588,356	37,141,099	713.838	6.0432
APR	864,204	572,112,109	35,680,125	662.010	6.2366
MAY	865,079	550,811,585	34,859,163	636.718	6.3287
JUN	855,258	731,276,893	42,143,539	855.037	5.7630
JUL	825,991	905,825,300	48,492,349	1,096.653	5.3534
AUG	816,597	1,067,664,809	54,395,062	1,307.456	5.0948
SEP	821,070	914,009,830	48,761,424	1,113.194	5.3349
TOT	10,003,890	8,748,373,675	$497,831,564	874.497	5.6906

Another key input for the revenue forecast is the base forecast of the numbers of customers and the volume of kWh sales expected in future years. Such information is derived from various evaluation methods. For instance, energy usage by residential customers is often estimated by application of end-use and econometric modeling. These models operate on such factors as population projections, appliance penetration trends, end-use energy characteristics, demographics, energy price elasticity, and economic activity. Econometric modeling is also utilized for projections of the commercial and small industrial customer classes. However, the application of end-use modeling for such groups is more difficult than for residential because of the much wider diversity in end-use load devices and operating characteristics (by contrast, the entire residential class is much more homogeneous). The loads and operations of large industrial customers are far too complex to model accurately as a group. The best approach for determining future energy usage of the largest industrial customers is through informed opinion based on direct discussions with plant energy managers.

Growth in class kWh sales occurs by (a) the addition of new customers and/or (b) an increase in per capita usage. The growth rate of each of these factors is often different. Changes in the economy contribute significantly to growth rate characteristics. During national or local economic downturns, growth rates may even be negative; the number of customers in a commercial or industrial rate class may shrink due to business closings. On the other hand, a special event may create an uncharacteristic increase in kWh usage. For example, the host city of the international Olympic games will add to its infrastructure with new buildings, such as dormitories, stadiums, and event halls, thereby increasing the overall energy requirements of the area. After the occasion, at least a portion of the additional usage may continue as other tenants take over the use of the new facilities.

Over time, a natural conservation in kWh growth occurs. As the inventory of customer appliances are replaced, more efficient end-use load devices are installed. For example, the average efficiency in air conditioning units and heat pumps has increased significantly over the last several decades. The least efficient choice today has an EER specification that likely exceeds the rating of the older unit being replaced. Thus, the per capita growth in kWh usage is affected by natural efficiency improvements. An aggressive demand-side management (DSM) program can have an even greater bearing on per capita growth rates.

A projection of long-range weather conditions can not be accomplished to a degree of accuracy necessary to be of value to the forecast of kWh sales (although extreme weather effects may be modeled as a sensitivity scenario). Thus, the base case kWh sales forecast is developed on the premise of normal weather.[10] Since historical data is used as an input to the forecast, it may be necessary to make adjustments for weather variations from the norm as well as other nonrecurring aberrations (e.g., an unplanned, temporary shutdown of a large customer).

Table 10.9 shows an example of monthly projections of customers and kWh sales for a residential class. Forecasts must be specific by rate schedule or class.

Table 10.9 Base Forecast Data for a Residential Rate Class

Forecast Period	Customers (Bills)	kWh Sales	kWh Per Bill
JAN	881,096	847,614,352	962
FEB	883,528	789,874,032	894
MAR	885,967	626,378,669	707
APR	888,413	593,459,884	668
MAY	890,865	560,354,085	629
JUN	893,325	784,339,350	878
JUL	895,791	904,748,910	1,010
AUG	898,264	976,412,968	1,087
SEP	900,744	862,012,008	957
OCT	903,231	646,713,396	716
NOV	905,760	612,293,760	676
DEC	908,261	731,150,105	805
TOT	10,735,245	8,935,351,519	832

---

[10] When comparing the performance of a prior kWh sales forecast to actual results, the associated actual sales figures must be weather normalized to ensure a fair evaluation.

When demand billed rate or customer classes are considered in the forecast process, the additional estimate of individual customer maximum demands is also required in order to project revenues. Furthermore, an estimate of other related billing data, such as reactive power requirements, is likewise needed so that the associated unit prices can be applied.

Having both historical billing data and a base projection of customers and kWh sales, an analytical forecast of rate revenues can then be produced by three basic methodologies: the *Average Rate Method*, the *Ogive Method*, and the *Customer Account Method*. Each of these methods have different attributes; thus, their applications depends on such factors as the type of customer class being analyzed and the degree of forecast output information desired.

AVERAGE RATE METHOD

The *average rate method* of revenue forecasting is based on the relationship between the average rate per kWh and the average kWh per bill for a given customer or rate class. This relationship can be developed from readily available historical billing data, as exemplified previously in Table 10.8. As noted through the mathematical analysis of rates in Chapter 9, the correlation between the average rate (y) and its associated kWh usage (x) is a hyperbolic function in the form of

$$y = A + \frac{B}{x}$$

where A represents a per unit kWh charge and B represents fixed charges, i.e., Customer and/or Demand Charges (reference Equation 9.2b).[11] This relationship can be used to estimate future customer or rate class revenues given a projection of just the monthly customer (bill) counts and kWh billing determinants.

A scatter plot of the residential class monthly average rate vs. kWh usage coordinates from the data provided in Table 10.8 is illustrated in Figure 10.14. The relationship between y and x is clearly nonlinear; however, the relationship between y and the reciprocal of x (i.e., x^{-1}) is effectively linear.[12] Thus, the hyperbolic formula which relates the average rate per kWh with levels of kWh usage can be developed through a regression analysis by means of the least squares normal equations given by

---

[11] The overall relationship between the average rate and kWh usage across the whole range of usage is defined here through a single expression having one set of coefficients. In the detailed analysis of a rate structure, the coefficients A and B, as determined by rate formulization, vary when the rate is structured with multiple kWh block or HUD step prices.

[12] Linearization through the reciprocal transformation is the underlying principle for the development of the hyperbolic graph scale, as discussed in Chapter 9.

Billing Data Statistics and Applications

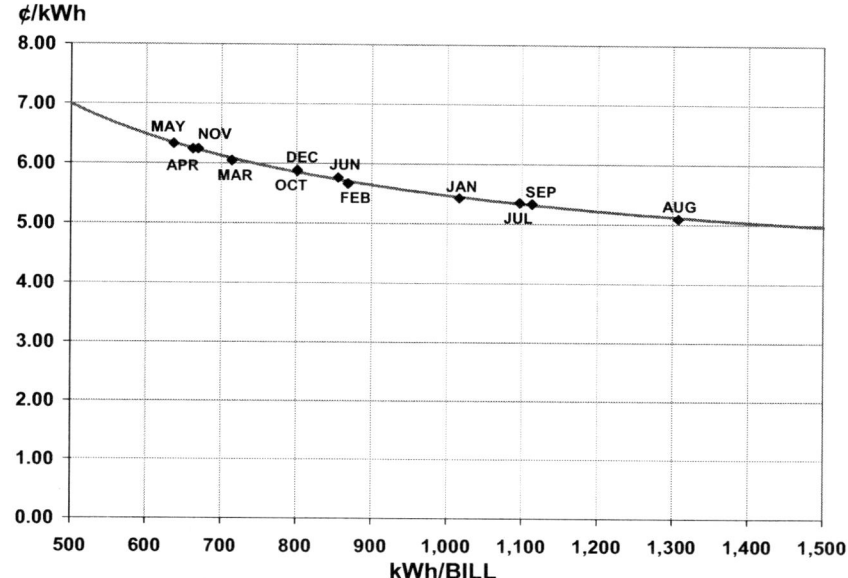

Figure 10.14 A scatter plot of the historical average monthly rate per kWh as a function of the historical average monthly kWh usage per bill. The data, from Table 10.8, is in the form of an equilateral hyperbola. The line plotted through the data is derived from a least squares regression analysis.

$$\sum_{i=1}^{N} y_i = [A \times N] + \left[B \times \sum_{i=1}^{N} \frac{1}{x_i} \right] \qquad (10.10a)$$

and

$$\sum_{i=1}^{N} \left(y_i \times \frac{1}{x_i} \right) = \left[A \times \sum_{i=1}^{N} \frac{1}{x_i} \right] + \left[B \times \sum_{i=1}^{N} \left(\frac{1}{x_i} \right)^2 \right] \qquad (10.10b)$$

where x_i = the actual (observed) value of kWh per bill in the ith month
 y_i = the actual (observed) value of ¢ per kWh in the ith month
 N = the total number of observations (months)

The two least squares normal equations have only two unknowns; therefore, the coefficients of the hyperbolic function can be determined by a simultaneous equation solution, as given by

$$A = \frac{\left[\sum_{i=1}^{N}\left(\frac{1}{x_i}\right)^2 \times \sum_{i=1}^{N} y_i\right] - \left[\sum_{i=1}^{N}\frac{1}{x_i} \times \sum_{i=1}^{N}\frac{y_i}{x_i}\right]}{\left[N \times \sum_{i=1}^{N}\left(\frac{1}{x_i}\right)^2\right] - \left[\sum_{i=1}^{N}\frac{1}{x_i}\right]^2} \quad (10.11\text{a})$$

and

$$B = \frac{\left[N \times \sum_{i=1}^{N}\frac{y_i}{x_i}\right] - \left[\sum_{i=1}^{N}\frac{1}{x_i} \times \sum_{i=1}^{N} y_i\right]}{\left[N \times \sum_{i=1}^{N}\left(\frac{1}{x_i}\right)^2\right] - \left[\sum_{i=1}^{N}\frac{1}{x_i}\right]^2} \quad (10.11\text{b})$$

For the historical billing data set presented in Table 10.8, A is found to be equal to 3.9574 and B is found to be equal to 1,515.8806. Thus, the hyperbolic function line, representing the relationship between the average rate per kWh and the average kWh usage per bill, is plotted in Figure 10.14 by means of the solution equation is given by

$$y = 3.9574 + \frac{1,515.8806}{x}$$

A visual inspection of the hyperbolic line plot in Figure 10.8 indicates a reasonable fit to the billing data set. The percent difference between the estimated y values (i.e., calculated from the solution equation) and the actual y values appears to be very minimal.

A customary method of assessing the goodness of fit of the least squares regression solution is by calculating the *coefficient of determination*, r^2, which indicates the degree to which the variations in the dependent variable are explained by variations in the independent variable. The coefficient of determination is a number in the range where $0 \leq r^2 \leq 1.0$. The square root of the coefficient of determination, which is known as the *correlation coefficient*, r, lies within the range where $-1.0 \leq r \leq +1.0$. When either r or r^2 is close to zero, the regression solution explains very little of the variation in the dependent variable, i.e., there is a very low correlation between the variables. When r^2 is close to 1.0 or when r is close to ±1.0, the regression solution explains a great deal of the variation in the dependent variable; thus, the variables are highly correlated.[13] The correlation coefficient is determined by much of the same basic data groupings used previously to determine the A and B coefficients of the hyperbolic function, as given by

$$r = \frac{\left[N \times \sum_{i=1}^{N} \frac{y_i}{x_i}\right] - \left[\sum_{i=1}^{N}\left(\frac{1}{x_i}\right) \sum_{i=1}^{N} y_i\right]}{\sqrt{\left[N \times \sum_{i=1}^{N}\left(\frac{1}{x_i}\right)^2 - \left(\sum_{i=1}^{N} \frac{1}{x_i}\right)^2\right] \times \left[N \times \sum_{i=1}^{N} y_i^2 - \left(\sum_{i=1}^{N} y_i\right)^2\right]}} \quad (10.12)$$

For the example data set, r is determined to be 0.9978 (and thus r^2 is equal to 0.9955), which is quite reasonable for applying the solution equation in order to forecast monthly revenues.

Recognizing that since y is in units of ¢ per kWh and x is in units of kWh per bill, then A (3.9574) must have the units of ¢ per kWh while B (1,515.8806) must then have the units of ¢ per bill. By dividing both A and B by 100 to convert to a basis of dollars, a simple equation can be developed for forecasting revenue for a given class j in the kth month of the forecast period, as given by

$$R_{jk} = \frac{(A_j \times E_{jk}) + (B_j \times C_{jk})}{100} \quad (10.13)$$

where R_{jk} = projected revenue for class j in the kth month
E_{jk} = projected kWh usage for class j in the kth month
C_{jk} = projected number of customers (bills) for class j in the kth month

For the example data set, the forecast of monthly revenues associated with a base projection of residential rate class monthly kWh and bills is determined by

Revenue = [0.039574 × kWh] + [15.1588 × Bills]

---

[13] A positive correlation, +r, is indicated when B is greater than zero, and a negative correlation, -r, is indicated when B is less than zero. Note that B represents the slope of the least squares linear solution of $y = A + Bx_0$, where $x_0 = x^{-1}$.

Thus, for example, the revenue for March of the forecast year, using the projected number of customers and kWh usage from Table 10.9, would be equal to: [0.039574 × 626,378,669 kWh] + [15.1588 × 885,967 Bills] = $38,218,506. All twelve months of forecasted revenues, using the average rate method, are shown in Table 10.12 (in comparison to the ogive method).

The actual monthly residential rate structure upon which the historical revenues in Table 10.8 are based is:

> Customer Charge: $7.50 per Month
>
> Energy/Demand Charges:
>   First 750 kWh   @   5.5¢/kWh
>   Next 250 kWh   @   4.0¢/kWh
>   Excess kWh   @   3.5¢/kWh

The declining block characteristic drives the overall relationship between the average rate and the amount of monthly kWh usage. Thus, the forecasted revenues are based indirectly on this same rate structure through the application of the average rate proxy. A seasonal rate structure (i.e., a combination of a peak season inverted block rate and an off-peak season declining block rate) would require an average rate determination for each season.

Since the characteristics of the rate structure are found to be an intrinsic aspect of the average rate calculation for the rate class, the average rate method can also be used to effectively forecast revenues for other customer classes having more complex rate structures. For example, the average rate calculation captures the effects of such auxiliary features as demand ratchets, power factor clauses, and minimum bill provisions. Thus, a forecast of a demand rate class requires only the same input data as for a basic nondemand rate structure, i.e., a projection of monthly customers (bills) and class kWh usage.

The average rate method can be applied quickly and with minimal input data, that is readily available from historical billing records, to forecast revenues. Other forecasting techniques are capable of yielding not only projections of revenues but also estimates of billing determinants for the forecast period.

Ogive Method

The *ogive method* of forecasting yields revenues and billing determinants for each price component of a given rate structure. Like the average rate method, the ogive Method requires a base input of projected monthly kWh sales and customers; however, the ogive method also makes use of monthly historical bill frequencies. The bill frequencies must contain at least the minimal number of intervals necessary to apply the rate structure, but the inclusion of more intervals allows for the option of testing alternative rate structure and price changes. An example bill frequency for the historical month of March is shown in Table 10.10.

Billing Data Statistics and Applications

Table 10.10 An Example Residential Bill Frequency for March

kWh Intervals	Bills Interval	Σ	kWh Interval	Σ	Consolidated kWh	Factor %
0	9,441	9,441	0	0	0	0.00
10	4,518	13,959	20,476	20,476	8,490,516	1.38
20	3,543	17,502	54,921	75,397	16,944,617	2.76
30	3,080	20,540	77,150	152,547	25,365,237	4.13
40	2,875	23,415	102,244	254,791	33,756,711	5.49
70	10,412	33,827	585,796	840,587	58,740,107	9.56
80	4,332	38,159	327,276	1,167,863	66,992,183	10.90
110	14,408	52,567	1,308,958	2,548,821	91,472,381	14.88
120	5,332	57,899	616,075	3,164,896	99,352,576	16.19
140	11,409	69,308	1,490,969	4,655,865	115,487,565	18.79
150	6,274	75,582	913,361	5,569,226	123,376,376	20.07
230	57,956	133,538	11,119,728	16,688,954	183,996,704	29.94
240	7,893	141,431	1,858,873	18,547,827	191,235,507	31.12
310	58,881	200,312	16,258,292	34,806,119	239,607,929	38.99
320	8,815	209,127	2,781,172	37,587,291	246,174,811	40.06
420	91,625	300,752	33,984,078	71,571,369	306,859,989	49.93
430	9,271	310,023	3,945,010	75,516,379	312,420,579	50.83
540	100,630	410,653	48,808,178	124,324,557	367,491,957	59.79
550	8,793	419,446	4,796,487	129,121,044	371,955,394	60.52
700	117,894	537,340	73,396,842	202,517,886	429,053,986	69.81
710	6,794	544,134	4,793,013	207,310,899	432,259,489	70.33
750	25,882	570,016	18,900,395	226,211,294	444,421,544	72.31
940	93,294	663,310	78,160,265	304,371,559	490,165,379	79.76
950	3,696	667,006	3,494,632	307,866,191	492,125,341	80.07
1,000	17,159	684,165	16,727,801	324,593,992	501,391,992	81.58
1,130	36,025	720,190	38,267,293	362,861,285	521,934,775	84.92
1,140	2,377	722,567	2,699,105	365,560,390	523,331,830	85.15
1,400	48,131	769,333	58,905,688	424,466,078	552,748,078	89.94
1,410	1,365	770,698	1,918,500	426,384,578	553,658,228	90.09
1,870	44,978	815,676	1,215,404	498,989,769	583,676,459	94.97
1,880	648	816,324	1,215,404	500,205,173	584,126,493	95.04
2,520	28,235	844,559	60,726,688	560,931,861	602,269,941	98.00
2,530	227	844,786	573,240	561,505,101	602,432,911	98.02
3,070	9,110	853,896	25,183,633	586,688,734	608,384,424	98.99
3,080	119	854,015	365,993	587,054,727	608,454,567	99.00
3,770	4,289	858,304	14,421,924	601,476,651	611,501,081	99.50
3,780	36	858,340	135,907	601,612,558	611,527,498	99.50
5,150	2,014	860,354	8,608,697	610,221,255	613,357,605	99.80
5,160	6	860,360	30,951	610,252,206	613,363,686	99.80
6,980	424	860,784	2,471,582	612,723,788	613,973,208	99.90
6,990	0	860,784	0	612,723,788	613,974,998	99.90
9,900	114	860,898	911,883	613,635,671	614,279,171	99.95
10,000	1	860,899	10,000	613,645,671	614,285,671	99.95
100,100	64	860,963	942,685	614,588,356	614,588,356	100.00

An *ogive* is a graphical representation of a cumulative frequency distribution; thus, for example, a plot of the consolidated factor vs. some function of usage would result in an ogive graph. To forecast kWh sales by the price blocks of a rate, an ogive is developed from the historical relationship between the consolidated factor, as a percentage, and the associated ratio of "use to average use." The *use to average use* is the ratio of the upper end kWh interval (corresponding to a given consolidated factor) to the monthly mean kWh usage per bill (which for the March data is 614,588,356 kWh ÷ 860,963 Bills = 713.838 kWh/Bill).

Table 10.11 shows the calculation of the ratio of use to average use for selected consolidated factors. For example, a consolidated factor of 50% is computed by multiplying the total kWh by 0.5, which is 307,294,178 kWh, and which falls within the 420 to 430 kWh interval of the bill frequency in Table 10.10. The associated use to average use is thus: 430 kWh ÷ 713.838 kWh = 0.602.

Table 10.11 Calculation of a Residential Rate Ogive for the Month of March

Consolidated Factor		kWh	Ratio of Use
Percent	kWh	Interval	To Average Use
1.00	6,145,884	10	0.014
2.00	12,291,767	20	0.028
5.00	30,729,418	40	0.056
10.00	61,458,836	80	0.112
15.00	92,188,253	120	0.168
20.00	122,917,671	150	0.210
30.00	184,376,507	240	0.336
40.00	245,835,342	320	0.448
50.00	307,294,178	430	0.602
60.00	368,753,014	550	0.770
70.00	430,211,849	710	0.995
80.00	491,670,685	950	1.331
85.00	522,400,103	1,140	1.597
90.00	553,129,520	1,410	1.975
95.00	583,858,938	1,880	2.634
98.00	602,296,589	2,530	3.544
99.00	608,442,472	3,080	4.315
99.50	611,515,414	3,780	5.295
99.80	613,359,179	5,160	7.229
99.90	613,973,768	6,990	9.792
99.95	614,281,062	10,000	14.009
99.99	614,526,897	100,100 *	140.228
100.00	614,588,356	* Largest Bill = 100,097 kWh	
Total Bills = 860,963		Average Use = 713.838 kWh/Bill	

Billing Data Statistics and Applications

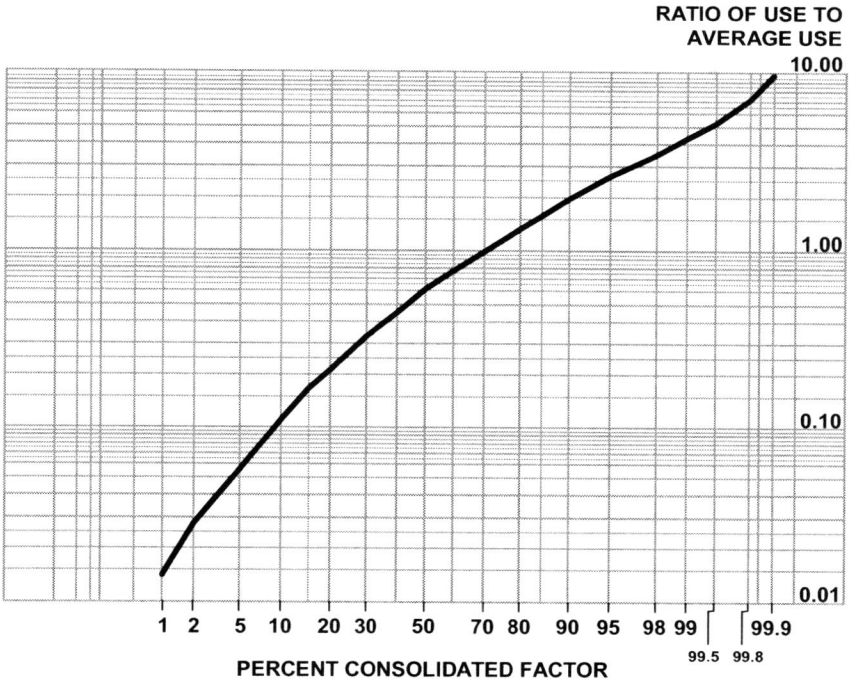

Figure 10.15 A residential class ogive. A plot of the percent consolidated factor vs. the corresponding use to average use, from the March data calculated in Table 10.11. The ogive is plotted on a Weibull graph (probability scale vs. three-cycle logarithm scale).

The percent consolidated factor vs. the ratio of the use to average use values calculated in Table 10.11 are plotted on the Weibull graph shown in Figure 10.15. The ogive provides a relationship that can be used as a basis for estimating the distribution of projected kWh sales by kWh intervals, given an input forecast of total monthly kWh and customers (bills) for the rate class. This relationship has been demonstrated to be stable as the average use per customer varies.

The ogive method of revenue forecasting is a two step process. First, the input forecast of total class kWh sales are projected into the specific blocks of the rate structure using an ogive developed from a historical bill frequency analysis. In essence, the ogive is used as an allocation mechanism to fit the total projected kWh sales into the specific price blocks of the rate structure. Specifically, the result is a projection of consolidated factors for key usage intervals. To complete the forecast, the unit prices of the rate structure are applied to the projected block billing determinants in order to calculate the revenues.

Recall that the actual monthly residential rate structure upon which the historical revenues in Table 10.8 are based is:

    Customer Charge: $7.50 per Month

    Energy/Demand Charges:
        First 750 kWh   @   5.5¢/kWh
        Next 250 kWh   @   4.0¢/kWh
        Excess kWh   @   3.5¢/kWh

Again using forecasted March sales data from Table 10.9, the total 626,378,669 kWh must be allocated between three price blocks. Considering that 885,967 customers are likewise forecasted for March, the projected average use per bill is 707 kWh.

The historical and projected data for March are shown in Figure 10.16. An ogive correlating to the price blocks of the example rate exhibits the relationship between the historical percent consolidated factor and the historical use to average use. Note that the use to average use for the tail block of the rate is based on the bill having the largest amount of kWh usage in the month. The projected use to average use can be easily calculated for each block since the projected average use per bill is known.

The projected percent consolidated factors, x_{750} and $x_{1,000}$, are then determined by proportionality using the characteristics of the historical ogive. For example, the projected use to average use for the 750 kWh interval is slightly more than its corresponding historical value (1.060820 compared to 1.050659). Thus, the projected use to average use is situated between the 750 kWh and 1,000 kWh interval boundaries, on a historical basis, as illustrated in Figure 10.16. The relationship of the projected percent consolidated factor, x_{750}, relative to the historical percent consolidated factors for the 750 kWh and 1,000 kWh interval is assumed to be equivalent to the relationship of the projected use to average use relative to the historical use to average use values for the 750 kWh and 1,000 kWh intervals, as determined by

$$\frac{1.060820 - 1.050659}{1.400878 - 1.050659} = \frac{x_{750} - 72.312067}{81.581759 - 72.312067}$$

Thus, the projected percent consolidated factor for the 750 kWh interval, x_{750}, is found to be equal to 72.581011%. The projected consolidated factor for the 750 kWh interval, y_{750}, is then determined by

$$y_{750} = 0.72581011 \times 626,378,669 \text{ kWh}_{\text{Total}} = 454,631,971 \text{ kWh}_{750}$$

Billing Data Statistics and Applications

	HISTORICAL				PROJECTED		
kWh:	614,588,356						626,378,669
BILLS:	860,963						885,967
AVERAGE USE:	713.838						707.000

kWh BILLING BLOCK	CONSOLIDATED FACTOR kWh	%	USE TO AVERAGE USE	USE TO AVERAGE USE	CONSOLIDATED FACTOR %	kWh
0	0	0	0	0	0	0
750	444,421,544	72.312067	1.050659	1.060820	x_{750}	y_{750}
1,000	501,391,992	81.581759	1.400878	1.414427	$x_{1,000}$	$y_{1,000}$
101,097 *	614,588,356	100	140.223692	141.579915	100	626,378,669

* Largest Bill

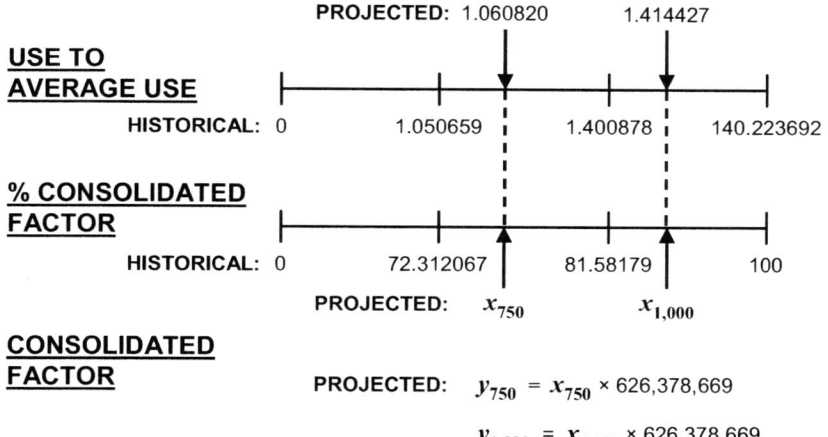

PROJECTED: $y_{750} = x_{750} \times 626{,}378{,}669$

$y_{1,000} = x_{1,000} \times 626{,}378{,}669$

Figure 10.16 A process for calculating projected consolidated factors. Calculation of projected consolidated factors for the month of March. A historical ogive (use to average use vs. percent consolidated factor) is used to estimate projected percent consolidated factors given an input of projected average use (kWh per bill). The projected consolidated factors are determined by multiplying the projected percent consolidated factors by the total projected kWh for the month.

By the same process, $x_{1,000}$ is determined to be 81.583557%, and $y_{1,000}$ is determined to be 511,021,998 kWh. The consolidated factors necessary to compute the projected revenue for March are all now estimated. The example block rate is then applied to the projected kWh consolidated factor frequency.

Bill Frequency Calculation of Projected Revenue for March:

Customer Charge:	885,967 Bills	@	$7.50/Bill	$6,644,753
$ECF_{750} - ECF_0$	= 454,631,971 kWh	@	5.5¢/kWh	25,004,758
$ECF_{1,000} - ECF_{500}$	= 53,390,027 kWh	@	4.0¢/kWh	2,255,601
$ECF_{Total} - ECF_{1,000}$	= 115,356,671 kWh	@	3.5¢/kWh	4,037,483
	626,378,669 kWh			

Total projected revenue from bill frequency calculation: $37,942,595

To accurately forecast annual revenues using the ogive method, a separate bill frequency must be projected for each billing month. A comparison of monthly forecast results from both the ogive method and the average rate method are depicted in Table 10.12. The difference in projected revenues between the two methods is found to be *de minimus*.

Table 10.12 Comparison of the Average Rate and Ogive Forecast Methods

Forecast Period	Customers (Bills)	kWh Sales	Revenue Average Rate	Revenue Ogive	Difference Amount	Difference Percent
JAN	881,096	847,614,352	$46,899,848	$46,600,202	$299,646	0.64
FEB	883,528	789,874,032	44,651,699	44,266,121	385,578	0.86
MAR	885,967	626,378,669	38,218,506	37,942,595	275,911	0.72
APR	888,413	593,459,884	36,952,856	36,879,188	73,668	0.20
MAY	890,865	560,354,085	35,679,897	35,571,973	107,924	0.30
JUN	893,325	784,339,350	44,581,180	44,770,616	-189,436	-0.42
JUL	895,791	904,748,910	49,383,650	49,378,714	4,936	0.01
AUG	898,264	976,412,968	52,257,171	51,924,389	332,782	0.64
SEP	900,744	862,012,008	47,767,461	47,699,972	67,489	0.14
OCT	903,231	646,713,396	39,284,934	39,193,936	90,998	0.23
NOV	905,760	612,293,760	37,961,148	38,062,083	-100,935	-0.27
DEC	908,261	731,150,105	42,707,681	42,838,846	-136,165	-0.32
TOT	10,735,245	8,935,351,519	$516,341,031	$515,128,635	$1,212,396	0.23

Customer Account Method

The *customer account method* is utilized for forecasting commercial and industrial rate class revenues, where projected billing determinants are also desired. While the ogive method projects monthly bill frequency determinants for a rate class (i.e., en masse), the customer account method projects the monthly billing determinants of the individual customers within a rate class.

A billing history profile for a large commercial or industrial customer account contains much more data than is recorded typically for smaller, mass market types of customers. For example, an individual commercial or industrial customer account might include such metered and calculated billing information as shown in Table 10.13. Billing detail of this nature is important to consider in the revenue forecasting process given that commercial and industrial rates are generally complex and often involve attendant billing provisions, such as demand ratchets and power factor adjustments.

Just like the average rate and ogive forecasting methods, the customer account method requires an input of projected monthly class kWh sales and bills, i.e., the base forecast. The historical monthly billing data for each customer within a particular rate class are adjusted so that the collective number of bills and kWh usage of all customers in the rate class match the input projections.

Table 10.13 Example of an Industrial Customer's Historical Billing Record

Account No.: 0123456789 Name: Grund Manufacturing
Service Voltage: 12,470/7,200 V Contract Capacity: 4,000 kW

Rate: LPS Rate Code: I-465

	Cycle Days	Actual kW Prior	Actual kW Current	Billing Demand	kWh	kVAR	kVARh	Revenue
OCT	32	4,022	4,042	4,042	1,180,800	2,383	588,508	$60,115.45
NOV	29	2,602	3,053	3,053	883,200	1,818	447,576	45,350.80
DEC	31	2,285	2,592	2,957	960,000	1,341	475,772	45,319.45
JAN	32	2,419	2,602	2,957	931,200	1,236	344,122	44,822.95
FEB	29	2,765	2,563	2,957	944,426	1,203	360,928	45,771.34
MAR	29	2,448	2,266	2,957	1,205,721	1,253	552,849	49,057.77
APR	29	2,400	3,389	3,389	1,303,285	1,934	620,093	55,095.18
MAY	32	3,302	3,725	3,725	1,735,193	2,195	859,955	65,124.90
JUN	30	3,878	4,579	4,579	2,349,784	2,832	1,249,066	83,349.91
JUL	30	4,109	4,742	4,742	2,042,517	2,893	1,063,328	80,414.46
AUG	32	4,157	4,685	4,685	1,699,200	2,843	879,917	74,658.25
SEP	30	4,224	4,262	4,262	1,392,000	2,583	720,836	74,658.25
TOT	365	38,611	42,500	44,305	16,677,326	24,514	8,162,950	$714,711.41

Historical and projected billing data for an example industrial rate class are shown in Table 10.14. Typically, revenue forecasts are determined on a calendar year basis (i.e., January through December). Due to the timing of reading meters and processing bills, historical data based on a full calendar year is not available in time to prepare a forecast for the subsequent calendar year. Thus, the twelve months of historical data must be accumulated from both the current year and the previous year. Considering a November or December forecast analysis schedule, the historical period would have to end with the current year's September or October in order to get the most recent billing data, as shown in Table 10.14. Prorated and estimated bills have been excluded.

Table 10.14 Historical and Projected Billing Units for an Industrial Class

	Historical Period					Forecast Period	
	Bills	kWh	kW	kVAR		Bills	kWh
OCT	77	47,641,140	126,990	81,364	JAN	77	47,079,000
NOV	77	44,129,240	125,218	81,863	FEB	77	42,047,000
DEC	74	40,278,760	118,182	78,496	MAR	77	41,246,000
JAN	74	43,213,500	123,248	75,734	APR	78	42,506,000
FEB	74	38,905,240	123,095	74,342	MAY	78	46,280,000
MAR	73	37,240,820	120,427	75,266	JUN	78	49,640,000
APR	75	39,111,540	123,744	78,637	JUL	78	55,828,000
MAY	75	42,697,020	123,467	79,106	AUG	78	58,046,000
JUN	72	43,681,250	116,042	77,981	SEP	78	59,286,000
JUL	75	51,027,220	126,401	84,248	OCT	79	51,078,000
AUG	77	54,469,480	131,880	86,907	NOV	79	47,403,000
SEP	77	55,792,320	130,801	87,218	DEC	79	45,107,000
TOT	900	538,187,530	1,489,495	961,198	TOT	936	585,546,000

	Historical Average Customer Characteristics				
	kWh	kW	kVAR	Load Factor	Power Factor
OCT	618,716	1,649	1,057	51.4	84.2
NOV	573,107	1,626	1,063	48.3	83.7
DEC	544,308	1,597	1,061	46.7	83.3
JAN	583,966	1,666	1,023	48.0	85.2
FEB	525,746	1,663	1,005	43.3	85.6
MAR	510,148	1,650	1,031	42.4	84.8
APR	521,487	1,650	1,048	43.3	84.4
MAY	569,294	1,646	1,055	47.4	84.2
JUN	606,684	1,612	1,083	51.6	83.0
JUL	680,363	1,685	1,124	55.3	83.2
AUG	707,396	1,713	1,129	56.6	83.5
SEP	724,576	1,699	1,133	58.4	83.2

Billing Data Statistics and Applications

As an initial step, the historical class data is reordered so that the months coincide with the forecast period, as shown for monthly bills in Table 10.15. When a growth in bills is projected, new customer accounts are synthesized to reconcile between the historical and projected numbers of customers.[14] In this example, a total of 36 bills must be added to the historical customer class database in order to reconcile with the annual base forecast. Given a monthly forecast of bills, the reconciliation is accomplished with monthly adjustments, which are noted by the column Δ Bills in Table 10.15. New customer accounts are then created as needed, and the Δ Bills are allocated to the new customer accounts. Thus, January requires three new customers to reconcile the bills. In March, a fourth customer must be added to complete that month's bill reconciliation, while two more customers must be added for the bill reconciliation in June.

Table 10.15 Reconciliation of Bills to the Forecast of Customers

	Historical		Forecast	Δ Bills	"New Customers"					
					#1	#2	#3	#4	#5	#6
OCT	77	-								
NOV	77	-								
DEC	74	-								
JAN	74	74	77	+3	1	1	1			
FEB	74	74	77	+3	1	1	1			
MAR	73	73	77	+4	1	1	1	1		
APR	75	75	78	+3	1	1	1			
MAY	75	75	78	+3	1	1	1			
JUN	72	72	78	+6	1	1	1	1	1	1
JUL	75	75	78	+3	1	1	1			
AUG	77	77	78	+1	1					
SEP	77	77	78	+1	1					
OCT	-	77	79	+2	1	1				
NOV	-	77	79	+2	1	1				
DEC	-	74	79	+5	1	1	1	1	1	
TOT		900	936	+36	12	10	8	3	2	1

Note: Historical months are reordered to coincide with the forecast period.

---

[14] In cases where negative growth may be projected (i.e., a reduction in customers), actual customer accounts are extracted from the historical database in order to match the total rate class projection of monthly customers. Such adjustments are guided generally by informed opinion in which specific customer site closings (or load reductions) are expected to occur within the forecast period.

Overall, six new customer accounts must be added to the historical database of accounts to reconcile the bills with the example base forecast. Note that New Customer #1 in Table 10.15 receives a bill allocation in every month of the forecast period, while the other five new customers receive bill allocations in less than twelve months.

Once the number of new customers required for the base forecast bill reconciliation is determined, monthly kWh, demand, and any other applicable billing determinants must be added to the historical customer database for each of the new bills.[15] Two basic approaches can be used for adding billing determinants to common groups of commercial and industrial customers. A new customer can be added by randomly cloning an account's bills from the customers that exist already in the historical database.[16] Alternatively, billing determinants representing an average customer can be added to the database for each new bill. The average customer is determined on a monthly basis simply by dividing the collective historical monthly billing determinants for the customer group by the number of monthly pre-reconciliation bills. The average customer for the example industrial class is shown in Figure 10.14. For instance, the average customer has a January energy usage of 583,966 kWh with a corresponding demand usage of 1,666 kW and 1,023 kVAR. The monthly historical billing determinants for the whole customer class are adjusted by adding an average customer's monthly billing determinants for each new bill addition, as determined by:

Historical kWh + [Δ Bills $\times$ kWh$_{AVG}$] = Adjusted Historical kWh

Historical kW + [Δ Bills $\times$ kW$_{AVG}$] = Adjusted Historical kW

Historical kVAR + [Δ Bills $\times$ kVAR$_{AVG}$] = Adjusted Historical kVAR

Thus for January, the adjusted historical billing determinants are determined to be:

43,213,500 kWh + [3 Bills $\times$ 583,966 kWh per Bill] = 44,965,398 kWh

123,248 kW + [3 Bills $\times$ 1,666 kW per Bill] = 128,246 kW

75,734 kVAR + [3 Bills $\times$ 1,023 kVAR per Bill] = 78,803 kVAR

---

[15] A base forecast that projects the addition of one or more very large commercial or industrial customers is generally guided by informed opinion in which particular new developments or existing customer expansions are expected to occur during the forecast period. In such cases, the addition of billing determinants should be specific to the expected operating characteristics of those particular customers.

[16] This approach is enhanced when the customer group is extraordinarily homogeneous (e.g., all customers in the group have the same industrial classification).

To reconcile the adjusted historical kWh with the base forecast of kWh, a monthly forecast factor is first determined by:

Forecast Factor = Base Forecast kWh ÷ Adjusted Historical kWh

Thus for January, the forecast factor is determined to be:

47,079,000 kWh ÷ 44,965,398 kWh = 1.047005

The monthly forecast factors are applied next to each customer's kWh usage in the adjusted historical database. Consequently when summed, the total kWh of the customer class will match the base forecast of kWh.

Having reconciled the kWh usage, a projection of actual demands must then be made. Monthly customer demands can be projected easily by applying the same forecast factors that are used for reconciling the kWh. In so doing, the customer class is assumed to operate at the same monthly load factors in the forecast period as in the historical period. When summed for January, the projected collective demands of the whole customer class are found to be equal to 129,041 kW (a 48% load factor assuming 730 hours in the billing month).[17] Alternatively, the average customer class load factors can be evaluated by viewing several years of historical billing data to determine if the trends in monthly load factors suggest the use of different forecast factors for determining projected actual demands.

A similar approach is used for projecting monthly kVAR demands (and/or kVARh usage). The monthly average power factors can be assumed to be constant between the historical and forecast periods, or a separate evaluation of the trend in power factors may be used when projecting the reactive power-based billing determinants.

The result of the proceeding calculations is a customer database which matches the base forecast of bills and kWh usage, and which contains projections of actual kW and kVAR demands. The historical billing data for a single example customer was shown previously in Figure 10.13. Table 10.16 shows the forecast results for that same example customer. The forecasts of monthly revenues are determined by applying the appropriate prices to each customer's projected monthly billing determinants. Note in Table 10.16 that the column indicated as the *current actual kW* is the result of the monthly demand projections for the forecast period. The column indicated as the *prior actual kW* actually contains the current year historical demands upon which the projected demands are based (reference Table 10.13). Thus, the two years of monthly demands are placed in the appropriate sequence so that any application of a demand ratchet will function appropriately. The forecast revenue for the whole customer class is determined by simply summing the projected revenues of the individual bills.

---

[17] This forecast example is based on a kW-based demand determinant. The same approach can also be used for kVA-based billing determinants.

Table 10.16 Forecast Results for the Example Industrial Customer

Account No.: 0123456789   Name: Grund Manufacturing

| | Actual kW | | Billing | | | | |
	Prior	Current	Demand	kWh	kVAR	kVARh	Revenue
JAN	2,602	2,724	3,319	974,971	1,294	360,297	$49,226.22
FEB	2,563	2,662	3,319	1,032,858	1,249	374,877	50,094.52
MAR	2,266	2,379	3,319	1,266,023	1,316	580,499	53,715.00
APR	3,389	3,541	3,541	1,361,920	2,021	647,991	57,557.65
MAY	3,725	3,882	3,882	1,808,465	2,288	896,268	67,861.68
JUN	4,579	4,803	4,803	2,464,919	2,971	1,310,268	87,418.59
JUL	4,742	4,989	4,989	2,148,734	3,043	1,118,624	84,587.41
AUG	4,685	4,929	4,929	1,787,557	2,991	925,672	78,531.51
SEP	4,262	4,471	4,471	1,460,202	2,710	756,154	68,836.63
OCT	4,042	4,224	4,224	1,223,934	2,490	614,990	62,810.41
NOV	3,053	3,196	3,492	924,702	1,903	468,608	50,528.98
DEC	2,592	2,719	3,492	1,007,033	1,407	499,081	51,565.20
TOT	42,500	44,519	47,780	14,471,318	25,683	8,553,329	$762,733.80

10.6 SUMMARY

The billing of customers for electric service is a continuous process, and it results in a massive record of time series customer usage information. The billing data of individual customers can be combined into various groupings to reveal the unique electricity consumption characteristics exhibited by a segment of similar customers viewed in concert. Knowledge of such customer group distinctions facilitates the establishment of rate schedules that correspond best to the particular attributes of those customers. Analysis of billing data yields statistics that can be used to support rate design and other pricing applications as well as additional business functions, such as marketing programs.

The historical billing data for a group of customers can be combined and ordered by sequential usage levels to produce a monthly, seasonal, or annual bill frequency. The bill frequency provides a fast and efficient means for calculating revenues from rates for a group of customers as a whole. Revenues computed from a bill frequency are highly accurate when compared to individual bill calculations, even when the group consists of a vast number of customers. As such, the bill frequency can be used as a primary analytical tool for evaluating rate structure and price level alternatives.

The bill frequency can also be used to produce a forecast of rate revenues and billing determinants. Historical billing data for customer groups, as well as

Billing Data Statistics and Applications 441

for individual customer accounts, are also used in other revenue forecasting methodologies. Revenue forecasts are used to support key business operating decisions. The forecast of billing determinants can also be used when designing rates based on a future period.

11
Cost-of-Service Methodology

11.1 INTRODUCTION

Understanding the costs required to render a particular electric service is crucial to ensuring profitability and fair pricing. Under regulation of fully integrated power companies, electricity is generally priced according to the actual costs incurred to provide service. The same is true in competitive electricity markets for those service components that continue to be regulated, such as power delivery.

There are two basic types of costs: capital and operation and maintenance (O&M) expenditures. Capital is noted to be a *fixed cost* as it represents investment in utility plant (assets), such as production, transmission, and distribution facilities and equipment. O&M is considered to be a fixed cost relative to expenditures which are constant, such as employee labor; however, O&M is considered to be a *variable cost* for expenditures which are a function of electricity consumption, such as production fuels. Identifying and differentiating the electricity cost elements between fixed and variable is a key step to developing rates which properly balance the risk shared by the service provider and the customers.

While the common costs of an electric business may be quite well known by major functions through corporate books and records, the specific costs of providing service to a particular group of customers must be determined through an extensive analysis. The main difficulty in identifying the specific costs of service lies in the circumstance that most of a utility's assets and expenses are jointly shared by the entire customer base. As a result, the stipulation of the costs required to provide service to a particular customer class must rely on a variety of cost allocation methodologies. A cost-of-service study is a model of utility accounting and financial data which relies on various engineering data and concepts to appropriately assign the detailed cost elements to the customer groups using the principle of *cost causation*.

The cost-of-service study is a detailed analysis in which all elements of cost are assigned to the customer or rate classes in the manner in which they are incurred. For example, production fuels are assigned on the basis of the customers' kWh usage since the chemical energy in the fuels are converted to electrical energy or kWh. Substations are assigned on the basis of customers' demands since the capacity provided is sized to meet the peak load requirements of the customers. Revenue meters are assigned to the classes based on the number of delivery points in each of the customer groups. All of the costs incurred to produce, transmit, and distribute power, including all of the attendant services, can be categorized into the three basic cost drivers or components: customer, demand, and energy.[1]

How and when customers utilize energy has a great bearing on the fixed and variable costs of service. The consumption of a large volume of energy over a very short period of time creates a sizeable demand, and both local and system capacity must be in place to serve such peak conditions. However, customers peak at different times due to a variety of lifestyles and operating patterns, and the production, transmission, and distribution components of the power system are designed to adequately serve the resulting diversified aggregate load of the customers. For proper cost assignment, the composite load of the system must be differentiated by the various customer or rate classes in order to determine the proportional responsibilities of these groups. In other words, the customers' load contributions to the total demand is a major cost driver.

The power system is designed, constructed, and operated in order to meet the ongoing energy and load requirements of vast numbers of diverse customers. The differences in customer electric service requirements necessitates a variety of available voltages: single phase vs. three phase, low vs. high, and regulated vs. unregulated. Both the fixed and variable costs incurred to deliver power to customers is a function of the service voltage character and differentiating costs by voltage level is a key aspect of the cost of service process.

Some customers may desire underground service for esthetics when overhead service would be less expensive. Some customers may require a significant extension of lines and equipment to receive service at a location far from the existing system facilities. Other customers may want to have redundant capacity available for enhanced service reliability. Such options represent increased fixed costs which are incurred by the utility for the benefit of specific customers and thus should be born by those customers and not by the general body of ratepayers.

The cost-of-service study is a rather complex analysis, not only because of such technical issues, but also because of the extensive amount of cost elements that must be processed. Each account in the utility's books and records must be critically reviewed, and in many cases, must be subjected to further engineering examinations in order to properly categorize the costs as being customer, energy, or demand related.

---

[1] As noted previously, rate structures are designed on the basis of these same three fundamental components since they are designed to recover the costs of service.

Cost-of-Service Methodology

Customer and system load and energy data is an essential input to the cost-of-service analysis. This information must be viewed on an hourly basis, and such information is not readily available for all customer types. Given the common use of simple watt-hour revenue metering in residential and small commercial applications, neither maximum demand nor hourly consumption data is available. As a result, the customer or rate class load shapes used for allocation of the demand cost components must be derived from load research, which is designed to yield reasonably accurate class load shapes by means of sample metering, as discussed in Chapter 3. The ability to model peak demand and annual energy losses throughout the system is another critical requirement of the cost-of-service study in order to properly allocate costs between customers at different voltage levels.

All in all, the cost-of-service study is a sophisticated model of the utility's cost structure, and its development requires a comprehensive understanding of virtually all major functional areas of utility and power system operations. This chapter focuses on the more technical underlying principles of the cost-of-service analysis process and indicates the value that can be derived from a cost-of-service study to support the rate design process.

11.2 COST-OF-SERVICE STUDY OVERVIEW

A cost-of-service study is an analysis of revenues and costs over twelve consecutive months (but not necessarily over a calendar year). This term is referred to as the *test year* or the *test period*. The test period can be fully historical, fully projected, or a hybrid which spans past and future months. Historically based studies evaluate actual investments and expenditures, while projected studies are based on forecasted data. When rates are designed on a projected basis, regulatory filing requirements may stipulate the development of both a historical (Period I) and a projected (Period II) cost analysis.

Historical test years include actual weather conditions, which can depart significantly from the average heating and cooling degree day statistics. The historical test period may comprise other anomalies such as unforeseen acute changes in the local economy, catastrophic storms, and other disasters which clearly affect customers' usage characteristics and system operating costs. A given historic period may be too stale to serve as a reasonable model of current conditions. In comparison, projected test years are based on normal weather and operations along with expected economic conditions and growth rates. Pro forma adjustments are often made to the historical period data to incorporate known and measurable changes, e.g., an increase in property tax rates effective in the subsequent (pro forma) year, in order to better identify costs relative to the rates that will be in effect.[2] Adjustments are also made to normalize volatile costs which are not representative of usual operations. The fully projected test period is more effective at

---

[2] Rates, whether designed on a historical or a projected basis, are applied to future usage.

synchronizing costs and rates, thus mitigating regulatory lag, although it is based on estimated data and is therefore subject to some degree of forecast error.

The cost-of-service study may be evaluated on an end-of period basis or an average basis. The end-of-period approach helps to mitigate regulatory lag; however, regulatory filing requirements often specify the use of a 13-month average approach. The 13-month method can be either a weighted average based on monthly cost account data or a simple average based on an account's beginning-of-year and end-of-year balances, as determined by

$$C_{AVG} = \frac{C_{EOTP} + C_{EOPP}}{13} \qquad (11.1)$$

where C_{AVG} = 13-month average balance of a given cost account
C_{EOTP} = account balance at the end of the test period
C_{EOPP} = account balance at the end of the prior period

A mix of end-of-period and average costs might also be utilized. For example, the end-of-period method could be applied to less volatile cost items, such as plant in service, whereas the 13-month average method could be applied to costs that vary substantially throughout the test period, such as materials and supplies and prepayments.

A foremost step of the cost-of-service study process is the assembly of the *rate base*, which consists primarily of tangible property in service. As investments are made, property items are booked to accounts, such as the FERC's Uniform System of Accounts, on the basis of their installed costs. These bookings represent the *first* or *original cost* of each plant account item. Regulatory requirements typically specify that the rate base is to be determined using original cost data.

The rate base can also be valued in terms of *reproduction costs*, which are the costs to "rebuild" the existing facilities at current equipment purchase prices.[3] Gathering a vast array of current prices and applying them to a detailed inventory of equipment and facilities represents a substantial effort. Furthermore, obtaining current price information for some older items of plant still in service may be literally impossible due to equipment obsolescence. A faster, simpler, and widely accepted methodology is to develop *trended original costs* on an account by account basis as an estimate of the reproduction costs for the test period. Trending can be accomplished by means of utility plant construction indices, such as the highly popular Handy-Whitman Index of Public Utility Construction Costs, which is continually updated each January and July for various geographic regions within

---

[3] The term reproduction is used in lieu of replacement as the latter term implies that in "rebuilding" the older power system, modern technologies and configurations would be employed now in order to achieve an optimal system.

Cost-of-Service Methodology

the United States.[4] The Handy-Whitman construction cost trend tables are based specifically on the FERC Plant Accounts, and the indices range back as far as 1912. Using such construction cost indicators, the reproduction cost of utility plant in service can be determined by

$$RC_{ijk} = OC_{ijl} \times \frac{CI_{jk}}{CI_{jl}} \times 100 \tag{11.2}$$

where RC_{ijk} = the estimated reproduction cost of item i in FERC Account j, in the current year k
OC_{ijl} = the original installed cost of item i in FERC Account j, booked in year l
CI_{jk} = the cost index for FERC Account j for the current year k
CI_{jl} = the cost index for FERC Account j for the original year l

As an example, the reproduction cost in 2006 (k) of the total investment in all three-phase padmount transformers (i) installed in 1997 (l) and booked to FERC Account 368 - Line Transformers (j), would be estimated by applying the ratio of the construction cost indices for those two years to the original cost of those units in 1997. Thus, the total reproduction cost basis of all items in FERC Account 368 can be easily and quickly estimated with this process.

The net rate base for either the original cost basis or the reproduction cost basis takes into account the accumulated depreciation of the property; thus, these two methods are referred to as *original cost depreciated* (OCD) and *reproduction cost new depreciated* (RCND). A third rate base valuation method is called *fair value*, and it is a composite of the OCD and RNCD methods.

Cost-of-Service Study Resolution

The cost-of-service study is a process which assigns cost responsibility to groups of customers. Utility wide, individual customers exhibit a great deal of diversity in terms of size, economic activity, operating patterns, and service requirements. To develop practical rate schedules, customers are segmented into various rate class groups, each of which has much more homogeneous customer characteristics. At a higher level, customers can be grouped in terms of basic customer classes representing residential, commercial, industrial, and lighting markets.

---

[4] The Handy-Whitman indices, published by Whitman, Requardt & Associates, LLP, specifically address cost trends by the production, transmission, and distribution plant FERC accounts. Other more general construction indices that are used for cost trending include the Turner Construction Company's Building Cost Indices, the Gross National Product price deflator, material and labor indices reported in the Engineering News-Record, and the U.S. Census Bureau Composite.

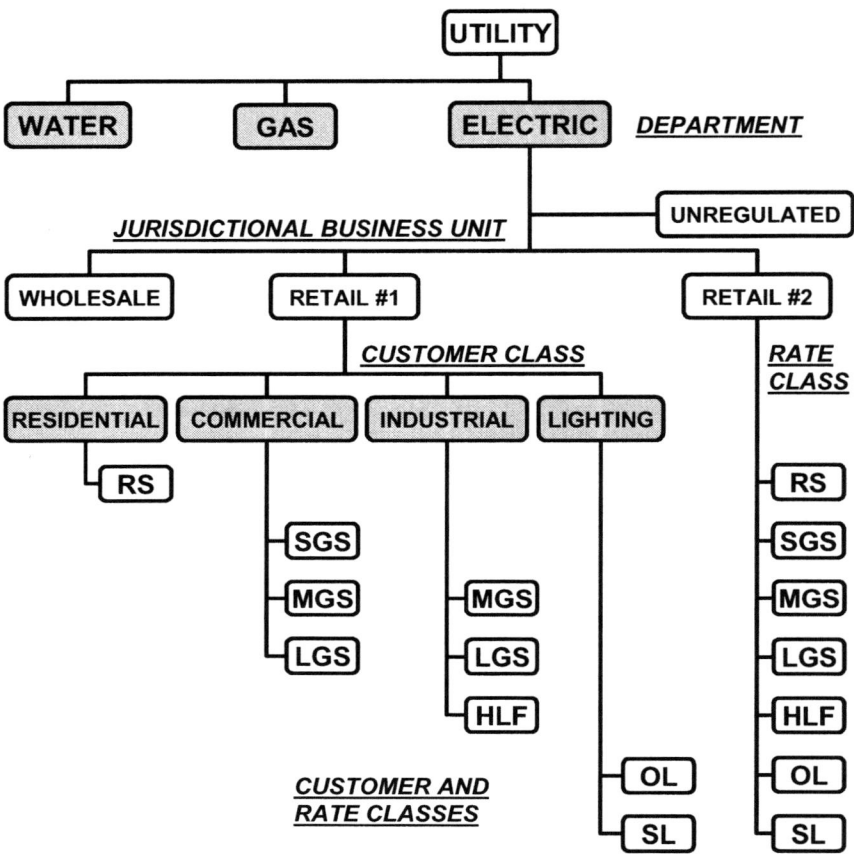

Figure 11.1 Degrees of cost separation for a multi-jurisdiction, combination utility. Cost-of-service studies are typically produced to separate costs (a) between regulatory jurisdictions, (b) between customer classes, (c) between rate classes (i.e., rate schedules), or (d) between both customer and rate classes. Under (d), general service types of rate schedule appear under multiple customer classes, as exemplified above. Combination utilities have common costs which must be allocated between the various operating departments.

Customers can also be grouped in terms of regulatory jurisdictions whereby the customers are categorized based on retail and wholesale classes of service. Utilities having service territories which extend across political boundaries, such as state lines, would have multiple retail jurisdictions under which their ultimate customers are separately categorized. The various ways in which customers can be grouped for cost separation purposes is shown in Figure 11.1.

Cost-of-Service Methodology

A cost-of-service study can be produced at various degrees of resolution in recognition of the different ways in which customers can be grouped. The minimal degree of resolution is known as a *jurisdictional separation* cost-of-service study, which assigns costs to customers grouped just according to the various regulatory agencies having ratemaking authority over the utility's services. Often the distinct jurisdictions have different philosophies, filing requirements, and procedures relative to how costs are evaluated. Costs supporting any unregulated products and services are also separated from the costs of the regulated business units to ensure that such operations would not be subsidized by the general body of ratepayers.

The retail portion of total costs can be separated to a much greater degree in a *customer class* cost-of-service study or in a *rate class* or rate schedule cost-of-service study. Similarly, the wholesale portion of total costs can be separated by customer delivery points or by different rate classes if such wholesale pricing distinctions are made. The maximum, practical degree of cost-of-service study resolution is achieved by an analysis which separates the costs of electric service both by customer class and by rate class.

As indicated previously, the cost-of-service study assigns costs to groups of customers. While it would be theoretically possible to determine the costs of service specifically for any given customer of a utility, it is generally not practical for each and every type of customer. So much of the total costs must be assigned on an allocation basis, due to the considerable joint use of the system's functional components, that apportioning all of the cost elements in a fully equitable manner to a single customer would be difficult at best. The ability to produce a cost-of-service study having a customer-by-customer level of resolution is limited further by the volumes of additional data that would be required to be gathered, processed, and modeled. Given an appropriate attention to details, the practice of using customer and/or rate class groups does yield a credible method of assigning costs for ratemaking purposes.

Despite the constraints imposed with the highest degree of study resolution, a practical single customer cost assessment can be accomplished with reasonable assurance in special cases. Very large industrial customers who are served at transmission voltages or who are served by means of a dedicated substation represent the best case for confidently identifying the specific costs of service.[5] While the production and transmission facilities are jointly used with other customers, the sophisticated metering data necessary to develop the appropriate demand cost allocators is typically available for such customers. The costs of the metering equipment itself, local facilities, such as a dedicated substation, and specific customer sales and service costs would be reasonably easy enough to quantify. As a result, some utilities model the costs of service for large, special contract customers each as a separate rate class.

---

[5] Isolating the costs of service for customers served from the distribution system with a reasonable level of confidence is much more improbable.

Cost-of-Service Study Resource Requirements

Cost-of-service studies are data intensive, and the numerous input requirements, as shown in Table 11.1, call for information from across many of the functional areas of the utility. The cost-of-service study requires both monetary and technical operating data and information. Cost and revenue figures for the study are acquired from corporate financial operating reports, which summarize data entries from various accounting systems (i.e., mass property accounting, revenue accounting, etc.). Having cost information booked in a configuration, such as the Uniform System of Accounts, facilitates the organization of the data elements within the actual cost-of-service study. Technical data is acquired from the Customer Information System (CIS) and various engineering sources.

Load research data is a crucial input to the cost-of-service study. Hourly load data must be developed for each class of service as defined by the specified study resolution. These load shapes yield demand-related allocators that are used to allocate fixed costs at each functional voltage level of the system. Although larger customers often have interval data recorder-based revenue metering, most customers are typically metered with simple watt-hour and watt-hour demand meters. Thus, the load shapes of the residential and small to medium commercial and industrial classes must be developed from interval data obtained by sampling each of the classes. The sample customers must be selected on a random basis, and each sample must be large enough to ensure a reasonable statistical accuracy.[6]

A system demand and energy loss model is also vital for development of the cost-of-service allocators. Class kW and kWh sales at the meters must be compensated for losses up through each level of the system in order to equate with the territorial input. Engineering data and analyses are used to determine the system's specific loss characteristics based on the behavior of system power flows.

Some of the cost data utilized in the study may be best analyzed exogenously with the simplified results serving as a data input to the cost-of-service model. For example, the amount of working capital to be included in the rate base is often required by regulatory agencies to be justified by an in-depth lead/lag study.[7] The separation of distribution system costs between the customer component and the demand component is a major examination in itself, and it requires information from circuit maps or an AM/FM system.[8] Single-line substation drawings must be analyzed to ensure that those investments are assigned to the proper functional level of the system in the cost-of-service study, and substations having multiple output voltages must be allocated between levels.

---

[6] An accuracy criterion of ±10% at a 90% confidence level is commonly accepted for cost-of-service study purposes.

[7] A lead/lag study evaluates the amount of annual investment funds that are required to meet a utility's monthly expense obligations based on the time difference between when expense payments are due and when revenues are received.

[8] AM/FM is a computer based Automated Mapping/Facilities Management program.

Cost-of-Service Methodology

Table 11.1 Major Cost-of-Service Study Input Data Requirements

Rate Base Items
- Electric Plant in Service
- Intangible Plant
- General Plant
- Common Plant
- Construction Work in Progress
- Plant Held for Future Use
- Fuel Stock
- Materials and Supplies
- Working Capital
- Prepayments
- Accumulated Depreciation*
- Property Insurance and Other Operating Reserves*
- Accumulated Deferred Income Taxes*
- Investment Tax Credits*
- Customer Advances*
- Customer Deposits*

*Rate Base Deductions

Expenses
- Fuel Burned
- Purchased Power
- Operations
- Maintenance
- Customer Accounts, Assistance, and Sales Expenses
- Salaries and Wages
- Depreciation Expense
- Administrative and General
- Taxes Other Than Income Taxes
- Miscellaneous Fees
- Interest on Customer Deposits
- Amortization of Nonrecurring Expenses

Income Taxes
- Federal
- State
- Tax Rates
- Deductible Interest

Financial Information
- Cost of Capital
- Capital Structure

Revenues
- Revenues from Sales
- Late Charges and Forfeited Discounts
- Miscellaneous Service Fees
- Leases and Rentals
- Nonterritorial Sales Credits
- Revenue Credits

Customer Information
- Class Designation
- Service Voltage Level
- Billing Determinants

Class Information
- Hourly Load Shapes
- Unbilled Sales

System Information
- Generation Output
- Territorial Purchases
- T&D System Loss Characteristics
- Substation Diagrams
- Distribution Circuit Characteristics
- Current Distribution Equipment Costs

The production of the cost-of-service study in a timely manner is facilitated by means of a systematic approach to gathering the vast amount of input data and information that is required in the analysis. Utilization of templates for acquiring the inputs from all responsible sources helps to format the data for entry, to ensure reporting consistency from study to study, and to fully document the information and assumptions used in each analysis. Such a methodology is particularly efficient when cost-of-service studies are produced routinely.

The cost-of-service study is a series of worksheets which can be easily constructed, implemented, and maintained in a modern PC spreadsheet application, even for high resolution studies. A summary data entry sheet is best for accommodating input information as opposed to hard coding figures that vary regularly. PC applications are also exceedingly versatile for documenting and reporting study results for many uses. The basic architecture of the cost-of-study is universal throughout the industry (including the natural gas utility business). However, because of distinctive conditions that most utilities have, such as an extraordinarily large customer with a unique service arrangement or other unusual system or contractual circumstances, and because regulatory requirements may vary significantly between jurisdictions, the basic cost-of-service model must be customized on a case-by-case basis to meet such needs properly.

Cost-of-Service Study Framework

The cost-of-service study is a key factor for establishing cost-based rates for electric service. Stated from a slightly different point of view, the cost-of-service study is a methodology for measuring the earnings position, or profitability, of the various classes of service. A fundamental result of the study is the quantification of the test period rate of return yield for each class under the existing rates for electric service. Just as in the total company view, the rate of return for each class is calculated as the percent of the class' net income to the class' rate base. To develop the class based earnings positions, the total company rate base and expenses must be apportioned to the classes using the cost causation criterion.

A rudimentary framework of the cost-of-service study process is presented in Figure 11.2. The process consists of four principal cost analysis steps:

1. **Functionalization of Costs** – Basic cost elements are organized in accordance with the major operating functions of the power system and its supporting business functions. These power system expenditures are assigned to function based on the voltage level at which the costs are incurred.

2. **Classification of Costs** – The levelized functional costs are then categorized as being energy related, demand related, or customer related. Since plant costs are associated with specific equipment and facilities, some distribution system cost elements are related to both the customer and demand cost components and must be separated accordingly.

Cost-of-Service Methodology

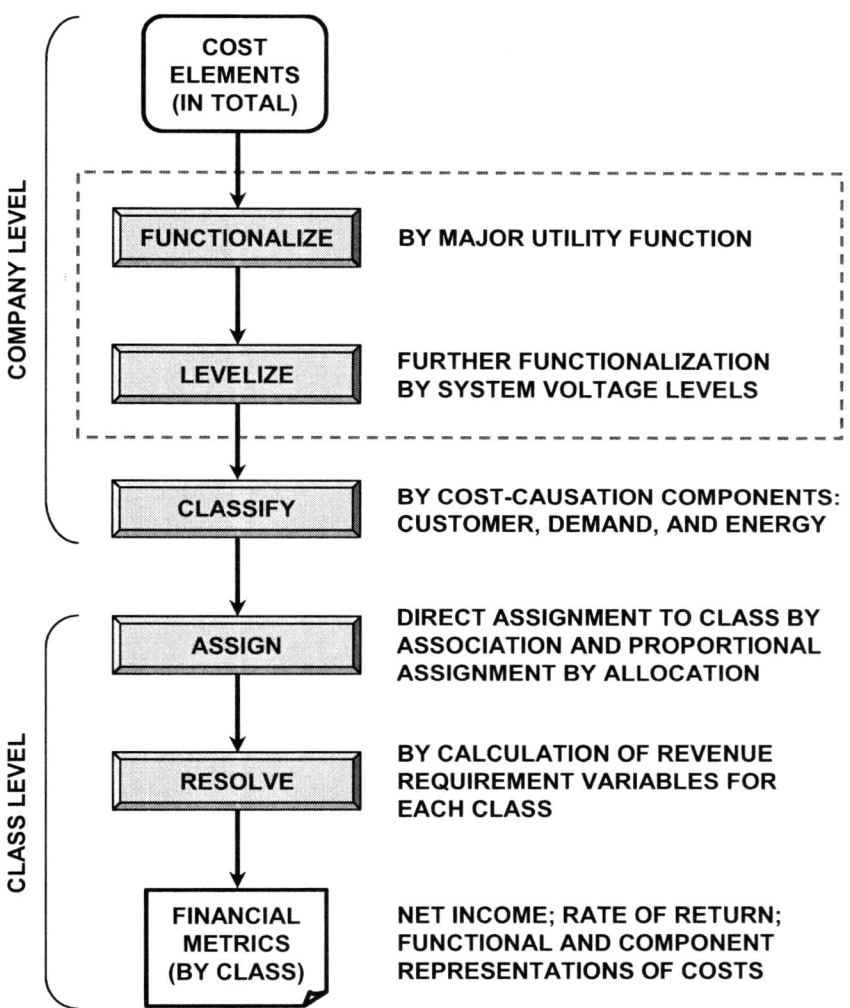

Figure 11.2 Principal steps of the cost-of-service analysis process. Functionalization and classification of rate base and expenses are executed for total company costs. The cost assignment process then apportions the total company costs to each of the classes where rates of return and other metrics can be determined on a class basis.

3. **Assignment of Costs** – The levelized functional cost components are then proportionally assigned to each customer class using (a) class kWh usage to allocate the energy cost components, (b) class load responsibilities to allo-

cate the demand cost components, and (c) the numbers of customers in each class to allocate the customer related cost components. Cost elements that are incurred exclusively for a specific class of electric service (or even a single customer) are directly assigned to that class (or to the class in which the specific customer is included).

4. **Results Quantification** – Allocated expenses and taxes are deducted from class revenues in order to determine class net income amounts, which are then divided by the allocated rate base to determine class rates of return. Other results, such as customer, demand, and energy cost component summaries, can be produced for rate design guidance and support purposes.

As noted in Figure 11.2, the functionalization and classification steps represent analyses of cost data for the company as a whole. The accounting data and the engineering inputs interact in the cost assignment step in accordance with the cost causation criterion. The completed cost-of-service study indicates results on both a class and a total company basis. The steps of the cost-of-service study process are described in further detail in the succeeding sections of this chapter.

11.3 COST FUNCTIONALIZATION

As the initial step of the cost-of-service study, functionalization recognizes the first degree of differences that exists between the various utility expenditures relative to cost causation. The result of functionalization is the assignment of plant investment and expenses to the principal utility functions, which include:

1. Production
2. Transmission
3. Distribution
4. Customer Accounts
5. Customer Assistance
6. Customer Sales

Other intermediate utility functional categories are identified during the analysis, e.g., General Plant and Administrative and General (A&G) Expenses, but their particular cost elements are ultimately functionalized in accordance with the principal utility functions indicated above.[9] Functionalization serves as an analytical process in which all of the utility cost elements are staged properly for subsequent steps of the cost-of-service study process.

---

[9] In addition, other costs, such as materials and supplies and accumulated deferred income taxes, which are typically summarized as rate base additions and deductions, ultimately can be functionalized and classified to customer, demand, and energy cost components.

Cost-of-Service Methodology

Because of their structural organization, the FERC plant accounts facilitate the functionalization of the investment-related costs. Gross plant in service, from company books and records, is assigned to function first, as illustrated in Figure 11.3. The plant accounting records provide details which expand the basic production, transmission, and distribution functions. Some items may be reclassified during this assignment process. For example, generation step-up (GSU) substations are located immediately adjacent to production facilities, although they are booked as transmission property along with other transmission-related substations. However, the GSU substations may serve two functions. The GSU transformers (TFMs) are in place to transform the medium voltage output of the generators to a transmission or subtransmission voltage for directional transport of power to the grid. The remainder of the GSU substation facilities include buswork, switches, and protective devices that may also play an integral role in the operation of network transmission, and thus those components would be retained in the transmission category along with transmission line investment.

The distribution function consists of several accounts which include distribution substations, lines, and a variety of line equipment. Utilities may or may not book distribution line conductors by operating voltage; thus, a special study might be required in order to separate the costs of primary lines and secondary lines. Such a case exemplifies the need to conduct auxiliary engineering analyses to better capture the differences in cost causation throughout the system.

Subfunctionalization of Costs

Even with the level of detail supported by the FERC plant accounts, an advanced degree of scrutiny of the system investments is required in order to build a representative model of the costs of electric service to the various customer and/or rate classes. Subfunctionalization further refines the cost elements by voltage levels, by more in-depth system operational characteristics, or even by time periods.

Levelization

The power system consists of numerous circuit configurations and voltage levels. Although most of a utility's customers are served at secondary voltages, larger customers are often served directly at various primary and transmission voltages. Service provided to a customer from an on-site, dedicated generating facility places that customer at the production level of the system. The voltage location of customers is a key aspect to determining cost causation as it defines the joint-use responsibilities of customers for each of the levels of the system. In other words, the cost of service is a function of level.

A consolidated schematic diagram of a utility's entire power system is shown in Figure 11.4. This example system consists of a large number of discrete functional voltage levels, voltage transformations, and level by-passes. Some substations fed by transmission are shown to have multiple output voltages, namely subtransmission and primary distribution (noted as T:ST/P SUBS).

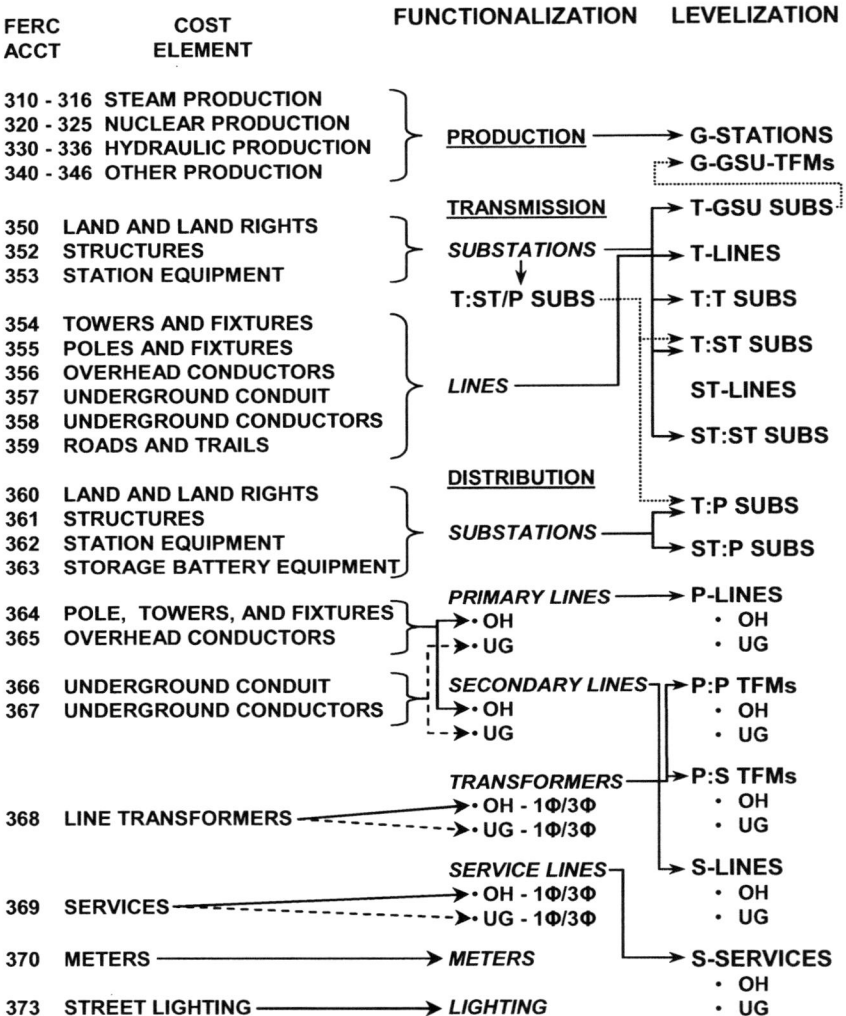

Figure 11.3 Functionalization and levelization of electric plant in service. The FERC plant accounts are generally organized by major functional divisions of the power system. However, additional analyses are typically necessary for the further refinement of the cost elements contained within each account in order to more accurately model cost causation drivers. Levelization, a subfunctionalization process, expands the principal functional categories on the basis of operational voltage levels, e.g., distribution consists of primary voltages and secondary voltages.

Cost-of-Service Methodology

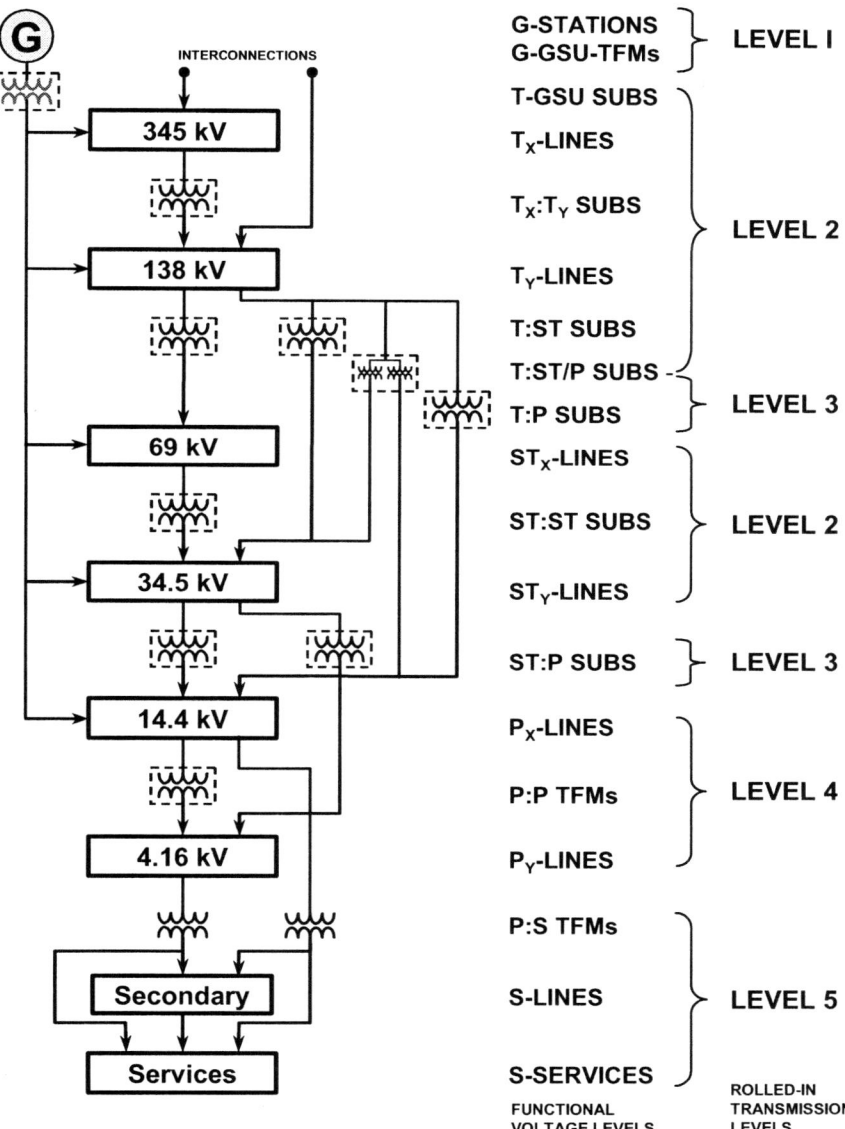

Figure 11.4 Designation of power system levels. The many discrete operating voltage levels of the power system can be condensed to a simpler structure, such as the FERC's 5-Level in which the transmission and subtransmission voltage levels are combined. Level designations: Level 1 is "Production," Level 2 is "Transmission," Level 3 is "Distribution Substations," Level 4 is "Primary Lines," and Level 5 is "Line Transformers."

Cost functionalization can be implemented on the basis of specific voltage levels, although it is a common practice to combine the specific levels by the same voltage class. Thus, for the example system, the 345 kV and the 138 kV facilities would comprise the transmission service level, the 69 kV and the 34.5 kV facilities would comprise the subtransmission service level, and the 14.4 kV and the 4.16 kV facilities would comprise the primary distribution service level. All of the various low voltage facilities are combined as the secondary distribution service level. Some jurisdictions, including the FERC, have required that transmission and subtransmission be combined ("rolled-in") for allocation as a single level.[10] As shown in Figure 11.4, this approach gives rise to a five-level system cost profile, which is mapped to the detailed functional voltage levels.[11] Levelization by discrete operating voltages requires that the associated investments have been recorded to that level of detail.

Subfunctionalization of T&D Plant

Further functionalization of the distribution plant accounts may be necessary to differentiate equipment and facilities by voltage levels. For example, in FERC Account 365 – Overhead Conductors and Devices, booking of wire and cable separately by primary and secondary voltage installations may not be the standard accounting practice. As noted previously in Table 5.4, the same conductor types are often utilized for both primary and secondary applications.[12] Analysis of the distribution circuit characteristics would then be necessary to determine the actual field inventory of conductor installations by voltage level in order to assign costs to the two levels. A comprehensive and well-maintained AM/FM (Automated Mapping/Facilities Management) system of the distribution circuits yields attributes, not only of the conductor applications but also of the other distribution equipment and facilities, in support of the cost-of-service study's voltage levelization process. Lacking a digital representation of the distribution system components, hand-drawn circuit maps or a sample-based field evaluation could be used to estimate the conductor installations and investments by level.[13]

Conversely, conductors booked in FERC Account 367 – Underground Conductors and Devices are likely to be differentiated by primary and secondary as,

---

[10] One rationale for combining subtransmission with transmission is that the subtransmission system may be operated so as to provide backup support of the transmission system.

[11] Although costs may be combined in accordance with the five elementary service levels, demand and energy loss analyses used for the development of the level allocators yield more accurate results if modeled initially on the basis of the discrete operating voltages and then simplified to the five-level system.

[12] Except that overhead duplex, triplex, and quadruplex are secondary voltage cables.

[13] An engineering assessment is crucial in the case where conductors are booked only in pounds of copper and aluminum, as is the accounting practice of some utilities.

Cost-of-Service Methodology 459

unlike overhead conductors, underground cables are particularly voltage specific. A 4/0 size underground cable is used in both low voltage (e.g., 600 volt class) and high voltage (e.g., 25 kV class) applications. However, the material cost of the high voltage version is significantly higher than the low voltage version due to greater insulation requirements. As a result, the two cable types must be booked as separate entries in order to properly account for the difference in their installed cost amounts.

Distribution poles, booked in FERC Account 364 – Poles, Towers, and Fixtures, which are used for supporting overhead conductors, might not be differentiated by voltage application in the accounting records. Poles carrying primary conductors, with or without underbuilt secondary conductors, should be assigned to the primary voltage level.[14] However, poles carrying only secondary lateral lines, and service line lift poles, should be assigned to the secondary voltage level. Conductor conduit, booked in FERC Account 366 – Underground Conduit, might not be differentiated by voltage application in the accounting records.[15] Thus, an engineering evaluation would be required to properly apportion the investment in poles and conduit between primary and secondary voltage levels.

Substations provide transformations between voltage levels throughout the T&D system. As shown in Figure 11.4, several voltage level transformations are possible, including:

- Transmission to transmission
- Transmission to subtransmission
- Transmission to primary
- Subtransmission to subtransmission
- Subtransmission to primary
- Primary to primary[16]

A *distribution substation* is so categorized by the criterion of having a primary distribution voltage as its transformation output.

Some substations may have two or more outputs of a different voltage level, such as shown in Figure 11.5 in which a transmission input voltage of 138 kV is transformed into both a subtransmission output voltage of 34.5 kV and a distribution output voltage of 14.4 kV. The total cost of such a substation would be booked in the transmission accounts. Thus, the substation must be evaluated in order to reassign its distribution-related portion of the total cost to distribution.

---

[14] In a similar manner, the FERC accounting rules state that poles and towers that carry both transmission and distribution conductors shall be booked to the transmission system (along with anchors, guys, and rights of way).

[15] The FERC accounting rules state that conduit which contains both transmission and distribution conductors shall be booked to the distribution system (along with rights of way).

[16] Lower capacity primary-to-primary transformations can be made with a line transformer.

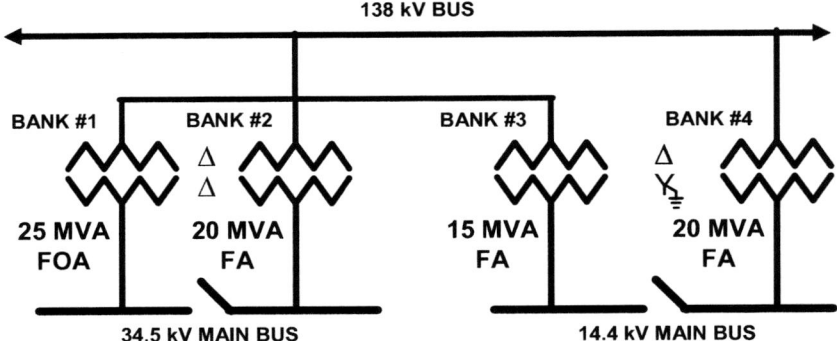

Figure 11.5 A substation having dual voltage transformations. A transmission input voltage is transformed into both subtransmission and distribution class voltages within the same yard. The entire plant investment is recorded in FERC Transmission Accounts: 350 – Land and Land Rights, 352 – Structures and Improvements, and 353 – Station Equipment. The total investment must be apportioned between the functional voltage levels of subtransmission and primary.

The substation costs for the example T:ST/P substation in Figure 11.5 would be booked into three transmission FERC Accounts: 350 – Land and Land Rights, 352 – Structures and Improvements, and 353 – Station Equipment. The substation's total booked costs can be separated between subtransmission investment and distribution investment by means of the associated transformer bank capacities along with the average costs per kVA capacity of all of the stand-alone transformations for the two output voltages, i.e., the average costs of the T:ST substations and of the T:P substations. The result of this allocation is a reclassification of a portion of the total substation's cost into three distribution FERC Accounts: 360 – Land and Land Rights, 362 – Structures and Improvements, and 363 – Station Equipment. The allocation between subtransmission and distribution is given by

Subtransmission Portion:

$$I_{35X\,T:ST} = \left[\frac{CAP_{ST} \times I_{35X\,T:ST\,AVG}}{(CAP_{ST} \times I_{35X\,T:ST\,AVG}) + (CAP_P \times I_{36X\,T:P\,AVG})} \right] \times I_{35X\,T:ST/P} \quad (11.3a)$$

Primary Distribution Portion:

$$I_{36X\,T:P} = \left[\frac{CAP_P \times I_{36X\,T:P\,AVG}}{(CAP_{ST} \times I_{35X\,T:ST\,AVG}) + (CAP_P \times I_{36X\,T:P\,AVG})} \right] \times I_{35X\,T:ST/P} \quad (11.3b)$$

Cost-of-Service Methodology   461

where $I_{35X\,T:ST}$ = the investment allocation to subtransmission
$I_{36X\,T:P}$ = the investment allocation to distribution
CAP_{ST} = subtransmission transformer capacity in kVA
CAP_P = primary distribution transformer capacity in kVA
$I_{35X\,T:ST\,AVG}$ = the average investment in \$/kVA of all transmission-to-subtransmission substations
$I_{35X\,T:P\,AVG}$ = the average investment in \$/kVA of all transmission-to-primary distribution substations
$I_{35X\,T:ST/P}$ = the T:ST/P substation's investment amount booked to the transmission FERC accounts
X = 0 for FERC accounts 350 and 360; 2 for FERC accounts 352 and 362; and 3 for FERC accounts 353 and 363

Thus, the two allocation formulas would be applied for each of the three substation FERC accounts to complete the separation of costs by output voltages. The allocated cost portions of the multiple output voltage substations are combined with the costs of the stand alone single output voltage substations to complete the levelization of substation investments.

Subfunctionalization of Production Plant

Production facilities may be subfunctionalized in the cost-of-service study under the premise that the different types of generating units have different cost-causation drivers. As discussed in Chapter 6, base generators operate nearly continuously to serve the foundation load that exists in nearly every hour of the year. Base generation typically includes nuclear units, coal-fired steam units, natural gas-fired combined cycle units, and conventional hydroelectric units. Intermediate generators operate in cyclical fashion in order to meet daily and seasonal load excursions above the base level of usage. Intermediate generators generally include fossil fuel-fired steam units, combined-cycle units, and gas and oil-fired combustion turbines. Peak generators operate more on an hourly basis to serve extreme load conditions. Peak generation generally includes gas- and oil-fired combustion turbines, internal combustion engines, and pumped-storage hydroelectric units.

The variable cost of generation per kWh at the bus bar is lowest with base generators and highest with peaking generators, with cycling generators in between. Conversely, the fixed cost of generation per kWh at the bus bar is highest with base generators and lowest with peaking generators, with cycling generators again in between. The integrated production system is designed to optimize the overall cost of producing electricity through a mix of resources that best fits the system load requirements. A network of only combustion turbines would have a low capital cost but a high operating cost. The use of high capital cost coal-fired-steam plants works within the mix to lower the average fuel cost. Thus, the fixed costs of generation can be viewed as having both demand and energy cost-causation implications. A number of production level class allocation methods used in the industry recognize an energy component of production plant.

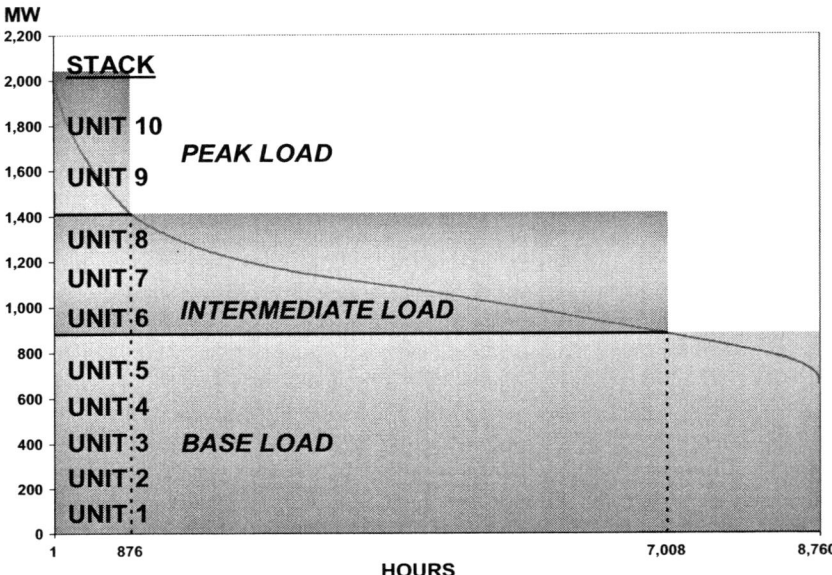

Figure 11.6 A merit order stack of generating units. In this example, base load is defined as the load up to the number of megawatts associated with 80% of the annual hours, relative to the system load duration curve. Peak load is defined as the load above the number of megawatts associated with 10% of the annual hours. Therefore, the intermediate load is the amount of megawatts between the peak load and base load. Each generating unit is evaluated to categorize its capability as base, intermediate, or peaking, and then its cost is assigned to the resulting category. The cost associated with each category is ultimately assigned to customer or rate classes using different allocation factors.

Subfunctionalization of production plant by base, intermediate, and peaking (B-I-P) categories can be accomplished by analyzing the characteristics of the generation fleet to determine a nominal merit order of unit capabilities relative to the system load categories, as shown in Figure 11.6.[17] Once categorized, each unit's specific investment can be assigned as base, intermediate, or peaking.

Production plant investment as a whole (i.e., not unit specific) can also be subfunctionalized to specific B-I-P time periods, which emulate the load duration curve to some extent. This B-I-P analysis method requires basic system load data and a definition of system seasonal and daily peak and off-peak rating periods.

---

[17] During operation some units may change order in the stack due to volatility in fuel prices. Over time, some older units may migrate upward in the stack as they are displaced by more economical capacity additions.

Cost-of-Service Methodology

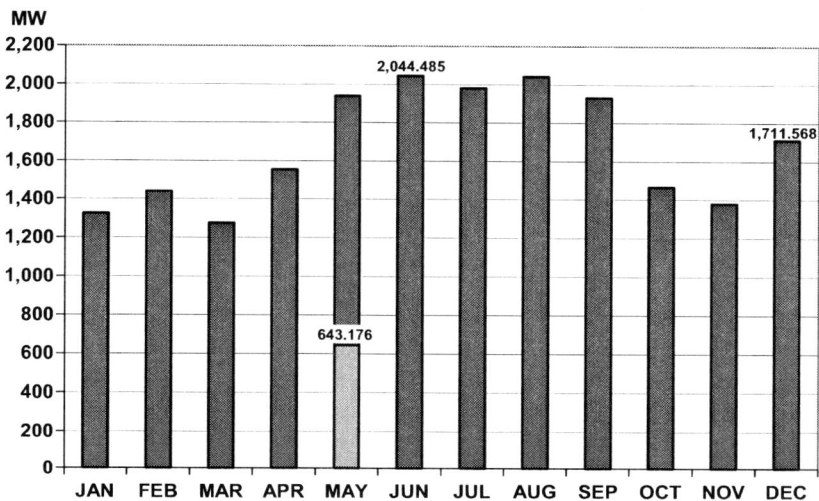

Figure 11.7 **System peak loads for each month of a test period.** An analysis of system peak loads is conducted to determine a peak season and a nonpeak season. In this example, the months of May through September are determined to define the peak season while the nonpeak season is comprised of October through April. The highest hourly load in each season is noted along with the lowest hourly load during the entire year, which in this example occurs in May.

The first step of the B-I-P allocation to rating periods is to evaluate the system monthly loads of the test period. A plot of monthly peak loads, as shown in Figure 11.7, helps to define the twelve months in terms of a peak season and a nonpeak season. In this example, May through September make up the peak season and all other months make up the nonpeak season. The highest hourly load in the peak season, the highest hourly load in the nonpeak season, and the lowest hourly load that occurs during the test period (regardless of season) is determined, as illustrated in Figure 11.7.

The next step of the B-I-P allocation to rating periods is to determine which days of the week represent system peak loading conditions. A plot of system weekly loads for January is shown in Figure 11.8. Due to routine business work schedules and typical lifestyles, the system load regularly measures higher during the weekdays than on the weekend days, and this condition is found to be generally consistent across the entire year. Thus, Monday through Friday is defined as peak days while Saturday and Sunday are defined as nonpeak days. Each month should be evaluated to ensure that weekends are consistently off peak.

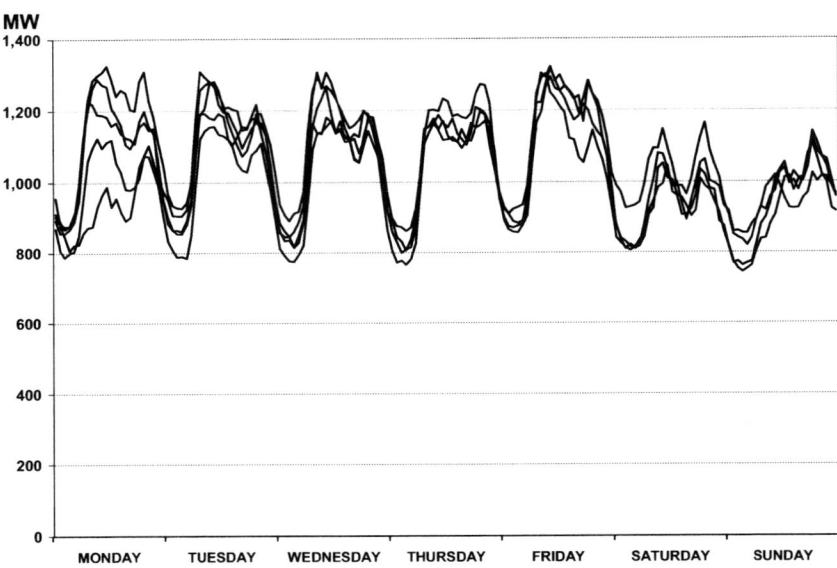

Figure 11.8 A plot of weekly loads for the month of January. Monday through Friday loads are typically higher than Saturday and Sunday loads due to customary operating cycles within the residential, commercial, and industrial sectors. The low load exhibited during one week on Monday is due to a holiday (New Year's Day). Loads on major holidays often behave similar to weekend days.

Nationally observed holidays often occur on weekdays, and their load shapes often reflect more of an off-peak characteristic due to a general shut down or reduction in industrial operations and many business activities and thus may also be categorized like weekend days. The most common U.S. holidays that are considered as being off peak include New Year's Day, Memorial Day, Independence Day, Labor Day, Thanksgiving Day (and often the day after Thanksgiving), Christmas Eve, and Christmas Day. Other national holidays or local observances may be considered as off peak if the typical load reductions tend to occur each year. Holidays falling on the weekend may be observed on the following Monday or on an adjacent day.

The next step of the B-I-P allocation to rating periods is to determine which hours of the designated peak days represent system peak loading conditions. Plots of system daily loads for all of the weekdays in February and July, excluding the July 4th holiday, of the example test period are shown in Figure 11.9. A criterion, such as when the hourly loads generally exceeds 80% of the daily peak load, could be applied as a method for defining the daily on-peak and off-peak hours. Again, the other months should be evaluated to ensure consistency of load behavior.

Cost-of-Service Methodology 465

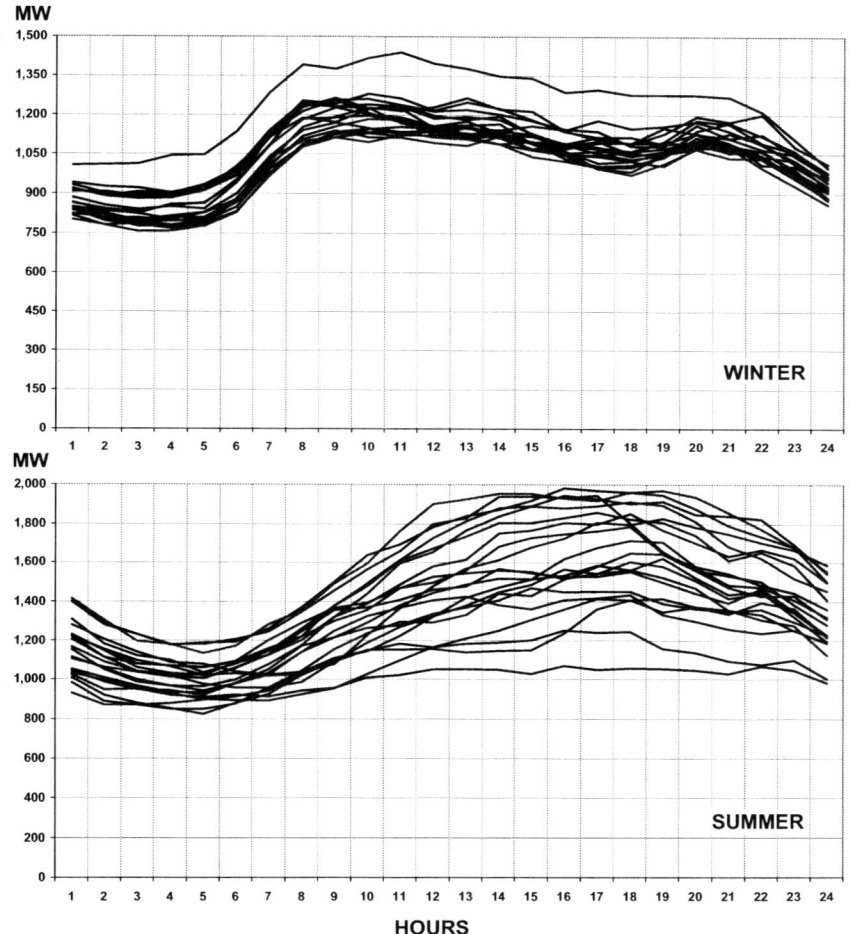

Figure 11.9 **Weekday plots of system hourly loads.** February weekdays are shown in the upper chart, while July weekdays, excluding the July 4th holiday, are shown in the lower chart. Weather-related variations can be observed in the loads (note the exceptionally cold day revealed by the winter plots). Hourly load data is useful for differentiating between daily on-peak and off-peak time periods.

In this example, the daily winter season loads are generally highest between about 7 AM to 10 PM, while the summer season daily loads are generally highest between about 10 AM and 10 PM. Thus, the definition of system rating periods by season, day of the week, and time-of day has been completed.

Table 11.2 Time-of-Use Rating Periods

Season	Days	Total Hours	On-Peak Hours	Off-Peak Hours
May – Sep			10 AM – 10 PM	10 PM – 10 AM
Weekdays	106	2,544	1,272	1,272
Weekend Days	44	1,056	--	1,056
Holidays	3	72	--	72
Season Total	153	3,672	1,272	2,400
Percent of Annual Total			36.7418%	
Oct – Apr			7 AM – 10 PM	10 PM – 7 AM
Weekdays	146	3,504	2,190	1,314
Weekend Days	60	1,440	--	1,440
Holidays	6	144	--	144
Season Total	212	5,088	2,190	2,898
Percent of Annual Total			63.2582%	
Annual Total	365	8,760	3,462	5,298

Table 11.2 is a summary of the hours in each of the time-of-use periods that have been defined for the discussion example. Approximately 37% of the annual on-peak hours occurs in the peak season (May - September), while 63% of the annual on-peak hours occurs in the nonpeak season (October - April). The off-peak hours are nearly the same amount in both seasons.

The system rating periods are illustrated against the backdrop of the system load duration curve in Figure 11.10 in relation to the (annual) maximum load which occurs in the peak season, the maximum load which occurs in the nonpeak season, and the annual minimum load. The base, intermediate, and peak period production plant allocations are determined by the three components of Equation (11.4). The allocation to the base period is based on the relationship of the minimum load (as a proportion of the annual maximum load) to the annual hours, since the minimum load increment is present in every hour; thus, the base period represents the total annual off-peak hours. Allocations to the seasonal on-peak periods are based on the relationships of the intermediate load increment (the difference between the nonpeak season maximum and the system minimum loads) to the October through April proportion of the on-peak hours, and of the peak load increment (the difference between the peak season maximum and the system minimum loads) to the May through September proportion of the on-peak hours.

Cost-of-Service Methodology

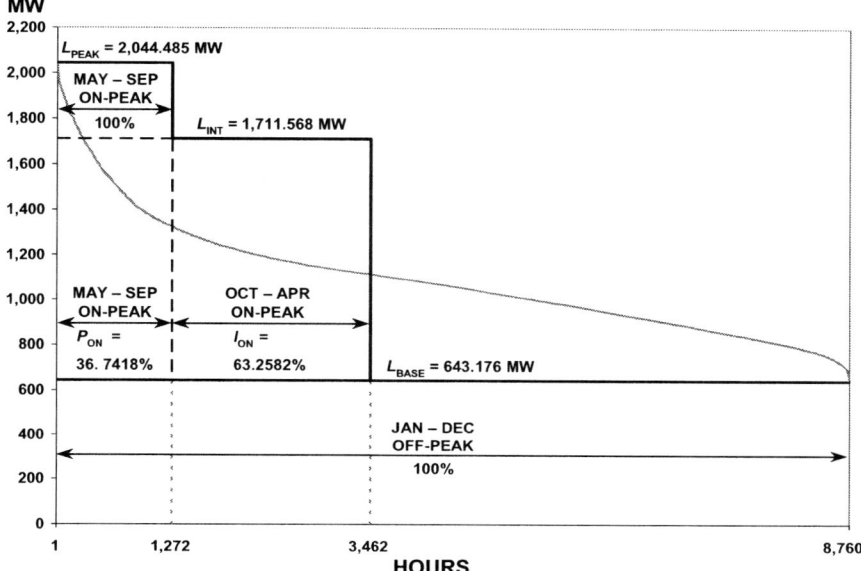

Figure 11.10 **A base, intermediate, and peak period allocation methodology for subfunctionalization of production plant.** The system load duration curve is emulated by a peak season (May through September) period of on-peak hours, an intermediate season (October through May) period of on-peak hours, and a base period of annual off-peak hours. Production plant period allocations are created by the relationships of the base, intermediate, and peak load increments to their respective rating period hours. This method allocates the investment in production plant and facilities as a whole and does not require an analysis of individual generating units.

Base Period:
$$B\% = \frac{L_{\text{Base}}}{L_{\text{Peak}}} \times 100 \qquad (11.4a)$$

Intermediate Period:
$$I\% = \frac{L_{\text{Int}} - L_{\text{Base}}}{L_{\text{Peak}}} \times I_{ON} \qquad (11.4b)$$

Peak Period:
$$P\% = \left[\frac{L_{\text{Peak}} - L_{\text{Int}}}{L_{\text{Peak}}} \times 100\right] + \left[\frac{L_{\text{Int}} - L_{\text{Base}}}{L_{\text{Peak}}} \times P_{ON}\right] \qquad (11.4c)$$

where $B\%$ = Base Period percent allocator
  $I\%$ = Intermediate Period percent allocator
  $P\%$ = Peak Period percent allocator
  L_{Base} = System hourly minimum load during the year in MW
  L_{Int} = System hourly maximum load occurring in the secondary peak season in MW
  L_{Peak} = System hourly maximum load occurring in the primary in MW (i.e., the system annual peak demand)
  I_{ON} = percent of the annual on-peak hours during the primary peak season
  P_{ON} = percent of the annual on-peak hours during the secondary peak season

Based on the example data from Table 11.2 and Equation (11.4), the B-I-P plant allocation factors would be 31.459% for the base period, 33.057% for the intermediate period, and 35.484% for the peak period.

Investment in transmission plant might also be considered for subfunctionalization into B-I-P components. The "backbone" or bulk transmission system has a particular symbiosis with the production system. Bulk transmission interconnects generating units thereby forming a production network having a greater reliability for power supply and enhanced operating flexibility compared to individual units or plants. The subfunctionalization of transmission plant and/or production plant gives rise to the use of multiple load allocation factors for the assignment of class cost responsibilities in the cost-of-service study. Such allocation methods are discussed in Section 11.5.

Functionalization of General, Intangible, and Common Plant

The plant functionalization process thus far has addressed only production, transmission, and distribution facilities. A number of general plant investments are required to support the three major functional areas, including FERC Accounts:

- 389 Land and Land Rights (not accounted for elsewhere)
- 390 Structures and Improvements (not accounted for elsewhere)
- 391 Office Furniture and Equipment
- 392 Transportation Equipment
- 393 Stores Equipment
- 394 Tools, Shop, and Garage Equipment
- 395 Laboratory Equipment
- 396 Power Operated Equipment
- 397 Communications Equipment
- 398 Miscellaneous Equipment
- 399 Other Tangible Property

Since general plant items directly support the operations of production, transmission, and distribution, the sum of the general plant investments can be functionalized simply by using *plant ratios*, i.e., the percentage of the investment amount of each major function (or even subfunction) to the total investment in production, transmission, and distribution. Alternatively, *labor ratios* based on salaries and wages associated with the major utility functions can be used to functionalize general plant. Furthermore, by including labor associated with the customer service functions (Customer Accounts, Customer Assistance, and Sales) along with production, transmission, and distribution, general plant investment is justifiably allocated across a broader spectrum of utility operations.[18]

A disadvantage of using basic plant or labor ratios to functionalize general plant is that such simple applications yield some inequities with respect to the cost causation criterion. For example, the production function would be assigned a portion of customer service offices and line trucks, and these investments clearly do not support power production operations.

A more in-depth assessment of general plant results in a greatly improved functionalization of those costs. For example, the investment in various building types can be functionalized on the basis of the labor specific to each category. Both Division level offices and Local area offices support transmission, distribution, and customer services. In general, Division level offices have relatively more resources employed in the transmission and distribution functions than in the customer service functions. In contrast, Local area offices have proportionately more resources for customer support, less resources for distribution, and minimal if any resources for transmission. On the other hand, the General Headquarters building(s) of the utility would typically house employee resources in support of all six of the functional areas. Thus, the use of functionalized labor based on the resources employed for each building type would result in a more practical allocation of those facilities and associated furnishings and equipment.

Production, transmission, distribution, and general plant items are considered to be tangible property. The utility has other costs associated with operating the business which are considered to be of an intangible nature. Intangible plant items are booked under three FERC plant accounts, which include :

301 <u>Organization</u> – Fees paid for incorporating and expenditures incidental to organizing for business readiness.

302 <u>Franchises and Consents</u> – Fees and incidental expenses paid for franchises, consents, water power licenses, or certificates of permission or approval.

303 <u>Miscellaneous Intangible Plant</u> – Costs of patent rights, licenses, privileges, and other intangible property necessary to conduct the business.

---

[18] The functionalization process incorporates a variety of internal allocations in order to place each cost element into the six major functional categories.

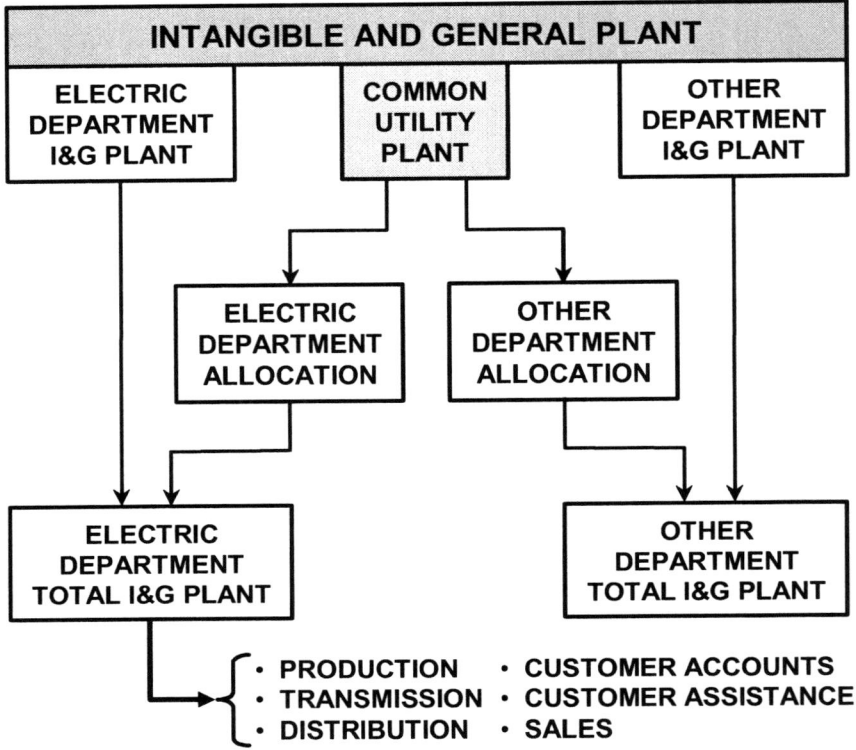

Figure 11.11 Functionalization of intangible, general, and common plant. Common intangible and general plant must be allocated between the departments for separate ratemaking treatment.

Intangible plant items can be functionalized to the major functions in a manner similar to general plant.

In the case of a combined services company, such as a gas and electric utility, portions of intangible and general plant are common to the two departments and must be separated between the different operations for ratemaking purposes, as shown in Figure 11.11. For example, the General Headquarters building might house separate groups which conduct electric system planning and gas system planning.[19] Thus, a floor space allocator could be developed and applied to sepa-

---

[19] Some groups, such as accounting and finance, support multiple services operations; thus, their total housing costs must be allocated among the different utility departments.

Cost-of-Service Methodology

rate the total cost of the floor space between the electric department and the gas department. Common Utility plant is booked under FERC Account 118 - Other Utility Plant.

Development of the Functionalized Rate Base

Production, transmission, distribution, general, and intangible plant investments (i.e., Plant In Service) are core elements of the functionalized rate base. As discussed previously, these investments can be expressed in terms of original booked costs or as reproduction costs. In either case, these plant amounts represent *gross* costs. The rate base represents *net* costs; therefore, accumulated depreciation must be applied functionally to the gross plant amounts as a deduction.

FERC Account 108 – Accumulated Provision For Depreciation of Electric Utility Plant is utilized for booking accrued depreciation of assets as a composite amount. However, subsidiary records must be maintained in order to account for depreciation by at least the major functions of: Steam Production, Nuclear Production, Hydraulic Production, Other Production, Transmission, Distribution, and General. The accumulated depreciation for each of these functions must then be allocated to the respective plant components within each function. Thus, for example, the distribution accumulated depreciation amount would be allocated to distribution substations, primary lines, line transformers, secondary lines, service lines, meters, and lighting by means of plant ratios which are based solely on the previously functionalized distribution gross investments. Once the net plant in service is determined, various additions and deductions are made to complete the rate base.

Rate Base Additions

Several other cost elements are added to net plant in service as part of rate base. These additions typically include such items as construction work in progress, electric plant held for future use, and various types of working capital.

Construction Work in Progress

Construction Work in Progress (CWIP) represents plant which is in the process of being built but which has not been placed into commercial operation (e.g., construction of a new transmission line). A large construction project, such as a generating station, requires a number of years to finish and adequate funds must be secured from investors to ensure successful completion of the facilities. Inclusion of CWIP in rate base allows for recovery of the carrying costs of construction through rates while the plant is being built. CWIP is booked in FERC Account 107 – Construction Work in Progress. Upon completion of a project, the associated cost amount is cleared from Account 107 and booked under the appropriate electric plant in service account(s). When a project having multiple components is being constructed and the in-service dates of the components will occur at different times, any facilities that are common to the whole project would be included in electric plant in service at the time in which the first component is completed (e.g., the control house for a substation having multiple transformer bays that will each be completed over the course of several years).

Regulatory treatment of CWIP varies among different jurisdictions. Some commissions allow CWIP, while many do not as a result of a strict adoption of the "used and useful" criterion. Some commissions allow CWIP to be included only for projects which are expected to be completed in the very near term, such as in the next 12 months. Some commissions have allowed only environmental-related projects under construction to be included in the rate base.

In lieu of CWIP, utilities are allowed to capitalize construction financing costs as an allowance for funds used during construction (AFUDC).[20] Upon in-service operation, these funds ultimately become part of the installed cost of plant in service and thus a part of the rate base.

Electric Plant Held For Future Use

Utilities often purchase land well in advance of a major utility project, such as for future buildings, generating facilities, and substations, to assure acquisition of appropriate sites. For example, a high growth area of new customers requires distribution line upgrades and extensions to serve the load additions in an adequate manner. Ultimately, the distribution expansion plan would require a new substation. The optimal site for new substation capacity is in the load center. Waiting too long to purchase a site would likely result in a higher purchase price as the development of adjacent land areas drive up real estate prices or even the loss of opportunity to locate the substation within close proximity of the load center. A nonoptimal substation location results in higher losses and thus higher costs.

Existing land that was utilized previously to provide electric service and is currently deferred to a later time is also considered to be *Electric Plant Held for Future Use*. Investments in other plant items which are not currently used and useful may also be considered held for future use for rate base purposes if supported by short-term installation plans.

Plant held for future use is booked in FERC Account 105 – Electric Plant Held for Future Use. Such property is classified in the same manner as for electric plant in service (i.e., FERC Accounts 301 – 399). Materials and supplies inventories, transformers, meters, and normal spare capacity of plant in service are not considered as being electric plant held for future use.

Working Capital

Working Capital represents funds that must be invested by the utility in order to meet its daily operating requirements while maintaining minimum bank balances. Working capital typically consists of materials and supplies, cash working capital, and prepayments.

---

[20] Some commissions have allowed the inclusion of CWIP in rate base with an AFUDC offset

Materials and Supplies

Materials and Supplies (M&S) consist of components of plant that are held in inventory for use in construction, operation, and maintenance of the electric system. These components consist of spare parts for production, transmission, and distribution facilities, line equipment and materials for T&D construction and repair, and other general items required to support the operation of the business. Thus M&S represents a working capital investment. M&S items are booked to FERC Account 154 – Plant Materials and Operating Supplies. Typical M&S inventory components are shown in Table 11.3.

The functionalization matrix in Table 11.3 illustrates a method for allocating the M&S components to the major functions. For example, the general inventory of poles is utilized for both T&D construction; thus, the specific application of a given pole in stock is unknown until it is placed in service. Poles and crossarms used in line construction are allocated to the T&D functions using the gross plant amounts in FERC Accounts 355 – Poles and Fixture (transmission) and 364 – Poles and Fixture (distribution).

M&S also consists of items which support all of the major functions such as Information Resources (IR) components. As with the functionalization of general plant, these components can be functionalized on the basis of labor ratios, as shown in Table 3.11.

FERC Account 163 – Stores Expense Undistributed represents the costs of labor, supplies, and other expenses incurred in general storeroom operations, and it serves as a loading charge distributed over M&S stores issues. These costs are allocated to the major functions based on the functionalized totals of the other FERC Account 154 M&S components, excluding scrubber reactant stock which is used for power plant flue gas desulphurization.[21]

Fossil fuel stock is the amount of coal and oil in inventory, i.e., the balance of fuel on hand after considering the daily burn requirements (similar to a check book balance after credits and debits). Fossil fuel inventories are booked to FERC Account 151 – Fuel Stock. FERC Account 152 – Fuel Stock Expenses Undistributed includes labor, supplies, and expenses incurred in the unloading and handling of fuel. Fuel stock costs and associated expenses which are charged initially to Accounts 151 and 152 are then cleared to FERC Account 501 – Fuel as it is utilized in production.

Nuclear fuel is accounted for over a multi-step fuel cycle, which includes fuel processing, fuel assemblies fabrication, fuel utilization within the reactor core, and fuel by-products processing.[22] Several FERC accounts are used to track the costs of nuclear fuel through its entire fuel cycle.

---

[21] Lime used for scrubbers is not a typical storeroom commodity as it is received and handled at the power plants in a manner similar to coal stock.

[22] While nuclear fuel is discussed here in the context of materials and supplies, it is often considered to be a component of plant in service.

Table 11.3 Materials and Supplies Functionalization Matrix

Item	Production	Transmission	Distribution	Customer Accounts	Customer Assistance	Sales
120 (Net) Nuclear Fuel	100%					
151/152 Fossil Fuels	100%					
154 Scrubber Reactant	100%					
Spare Parts:						
- Production	100%					
- Transmission		100%				
- Substation		352/353	361/362			
- Distribution			100%			
- Lighting			373			
Line Items:						
Poles & Arms:						
- Lines		355	364			
- Lighting			373			
Insulators		353/356	362/365			
Line Hardware		354/355	364			
Ducts & Manholes		357	366			
Wire & Cable		356/358	365/367/369/373			
Salvage Materials		356	365			
Reels & Containers		356/358	365/367/369/373			
Protection Equip.		353/356/358	362/365/367/368			
Distribution Auto.			365			
Metering Equip.			370			
Minor Materials			364/365			
General Items:						
Medical Supplies	L_{PROD}	L_{TRAN}	L_{DIST}	L_{ACCT}	L_{ASST}	L_{SALE}
Gas & Oil	L_{PROD}	L_{TRAN}	L_{DIST}	L_{ACCT}	L_{ASST}	L_{SALE}
Office Supplies	L_{PROD}	L_{TRAN}	L_{DIST}	L_{ACCT}	L_{ASST}	L_{SALE}
IR Equipment	L_{PROD}	L_{TRAN}	L_{DIST}	L_{ACCT}	L_{ASST}	L_{SALE}
163 Stores Expense Undistributed	$154*_{PROD}$	$154*_{TRAN}$	$154*_{DIST}$	$154*_{ACCT}$	$154*_{ASST}$	$154*_{SALE}$

*Totals of Account 154 excluding scrubber reactant stock.

120.1 Nuclear Fuel in Process of Refinement, Conversion, Enrichment, and Fabrication

120.2 Nuclear Fuel Materials and Assemblies- Stock Account

120.3 Nuclear Fuel Assemblies in Reactor

120.4 Spent Nuclear Fuel

120.5 Accumulated Provision for Amortization of Nuclear Fuel Assemblies

120.6 Nuclear Fuel Under Capital Leases

The accounts in this series are progressively credited and debited as the nuclear fuel moves through its fuel cycle. Net nuclear fuel investment is determined by the sum of accounts 120.1, 120.2, 120.3, 120.4, 120.6, less account 120.5. FERC Account 157 – Nuclear Materials Held For Sale includes the net salvage value of uranium, plutonium, and other nuclear materials which will not be reused in electric operations. Account 157 is debited and Account 120.5 is credited for any such net salvage value. Furthermore, in a manner similar to fossil fuel usage, FERC Account 518 – Nuclear Fuel Expense is debited (Account 120.5 is credited) as nuclear fuel is utilized in production.

Cash Working Capital
Cash Working Capital (CWC) represents the amount of money required to be on hand in order for a utility to fulfill its daily expense obligations and to preserve bank account balance minimums throughout the test period. This cash is provided by the utility investors. CWC can be determined through a comprehensive lead/lag study, which is a financial evaluation of the timing differences between when expenses must be paid (lead) and when revenues are received (lag).

Alternatively, CWC can be determined quite simply by applying a factor to the total test period O&M expenses, excluding certain expenses, to estimate the effective lag between the expenses and the revenues. One-eighth of the annual O&M expenses, excluding depreciation, taxes, and purchased power expenses, has been used frequently for determining CWC requirements in this manner.[23] Once CWC is determined, it can then be functionalized to Production, Transmission, Distribution, Customer Accounts, Customer Assistance, and Sales by means of functional ratios which represent O&M expenses (excluding purchased power, depreciation expense, and taxes).

Prepayments
Prepayments are booked in FERC Account 165 – Prepayments, and include various prepaid expenditures such as licenses, insurance, rents, taxes, interest, and

---

[23] One-eighth (or 12.5%) of a year equates to a lag of 45.6 days. Depreciation is a noncash expenditure, and taxes and purchased power costs are viewed as being paid subsequent to the receipt of revenue. Fossil fuel is also excluded in some jurisdictions.

miscellaneous items. Prepayments can be functionalized to Production, Transmission, Distribution, Customer Accounts, Customer Assistance, and Sales by means of gross plant ratios based on the previously functionalized Total Electric Plant, i.e., Gross Plant In Service (production, transmission, distribution, intangible and general plant) along with the functionalized results of Plant Held For Future Use and CWIP.

Rate Base Deductions

Several other cost elements are deducted from net plant in service as a rate base adjustment. Since the rate of return is applied to the rate base in order to determine a revenue requirement, certain deductions are necessary to recognize various accumulated capital resources that are not funded by the utility's investors. More specifically, these deductions represent deferred credits and non-current liabilities of the utility. Typical rate base deductions include such items as customer advances, contributions, and deposits, accumulated deferred income taxes, investment tax credits, and various types of operating reserves.

Customer Advances, Contributions, and Deposits

An *advance* is money paid by a customer in support of construction of lines and facilities for service that are considered to be nonstandard or which represent an excessive line extension from the existing system. Utilities often utilize a cost-to-revenue ratio measure for evaluating whether or not an advance is necessary. Refundable, or partially refundable, advances are booked in FERC Account 252 – Customer Advances for Construction. The account is credited as any applicable refunds are made to the customer.[24] Upon completion of the full amount of refund due to the customer, the balance of FERC Account 252 is credited and all applicable plant accounts are debited based on the facilities that were installed.

Customer advances are booked typically by customer class (i.e., residential, commercial, and industrial). Residential advances can be directly assigned to the distribution function since residential customers are served at low voltages, and the advances generally would support secondary and primary distribution construction requirements. The same is generally the case for commercial and small industrial customers; however, large industrial customers may also require transmission construction. Thus, industrial customer advances can be allocated between the transmission and distribution functions by means of the T&D lines components of gross plant in service and any associated CWIP.

In contrast, a *Contribution in Aid of Construction* (CIAC) represents a nonrefundable donation of money for the construction of lines and facilities, which also may be received as a result of a cost-to-revenue criterion. Nonrefundable

---

[24] For example, a line extension policy may require a customer payment for the amount of the cost of installation that is in excess of the first three years of base rate revenue, i.e., a 3 to 1 ratio. The policy may further allow for refunds to the customer as additional customers connect subsequently to the same line over some prescribed period of time.

CIAC payments are credited to all applicable plant accounts; thus, CIAC-related facilities are fully excluded from rate base since the customer has provided compensation for their construction.

Deposits are funds paid by a customer typically at the time service is established as a means to guarantee payment of the expected electric service billing.[25] An example of a criterion for deposits could require a deposit equal to twice the estimated maximum monthly electric service bill. Customer deposits typically earn interest at a rate authorized by the regulatory authority. Deposits and accumulated interest payments are typically refundable to the customer at the time service is terminated, or in some cases after a designated period of prompt payments. The deposit and interest, or a portion thereof, could be applied by the utility in order to settle the account when service is discontinued. Deposits for electric service are booked to FERC Account 235 – Customer Deposits, and deposits can be directly assigned to the Customer Accounts function.

Accumulated Deferred Income Taxes

For book and ratemaking purposes, the annual depreciation expense of a unit of property is based on a uniform depreciation rate over its expected life. Thus, the net electric plant in service is based on *straight-line depreciation.* On the other hand, an *accelerated* or *liberalized depreciation* method, such as *sum-of-the-year's digits* or *double-rate declining-balance*, is typically allowed for calculating depreciation expense deductions from income for federal and state income tax purposes. As a result of accounting for two depreciation methods, the income taxes actually paid to federal and state authorities in a particular year would be different than the income taxes calculated on the basis of book depreciation.

An example of an accelerated depreciation method in comparison with straight-line depreciation for a $100 asset over a useful life of 30 years is shown in Figure 11.12. Under straight-line depreciation, the asset depreciates as a linear function at a rate of 3.33% per year over its useful life. The rate of 3.33% per year represents the negative slope of a linear function. In contrast, under the sum-of-the-year's digits (S-Y-D) method, the asset depreciates as a quadratic function in which its slope is always negative but constantly changing over the life of the asset. In the early years, the S-Y-D depreciation rate is much greater than the straight-line depreciation rate (6.45% in year 1), and in the later years it is much less (0.22% in year 30). As a result, the accelerated depreciation yields a greater tax deduction in the early years, which in turn lowers the amount of taxable income and the subsequent amount of income taxes due, as compared to straight-line depreciation. However, in the later years, the accelerated depreciation yields a lesser tax deduction when compared to the straight-line depreciation method and thus results in a higher taxable income and higher taxes due.

---

[25] A deposit might be waved initially based on a customer's creditworthiness but might be required at a later date (or an initial deposit might be increased) in the event the customer becomes negligent in monthly bill payments.

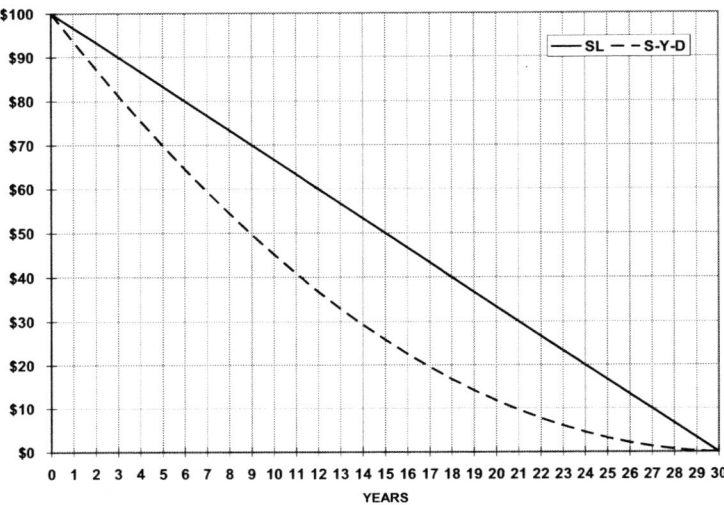

Figure 11.12 **Annual value of an asset based on straight-line vs. liberalized depreciation over its useful life.** Straight-line depreciation is used for ratemaking purposes while an accelerated depreciation method, such as sum-of-the-year's digits, is used typically for income tax deductions. The difference between the two methods gives rise to an accumulation of deferred income taxes.

Since the difference between straight-line depreciation and accelerated depreciation in the early years creates a tax obligation that will be due in the later years, this future obligation is referred to as *deferred income taxes* (DITs). The DITs accumulate over time and are thus referred to as ADITs. In the early years, electric service rates based on book depreciation would collect more tax-related revenue than is actually paid, which results in a customer-provided interest free source of funds for the utility's use until the impending time at which it must pay the deferred taxes.[26] Therefore, as with customer advances and deposits, such ADITs are deducted from the rate base so that they would not earn a rate of return through the electric service rates.

There are three deferred credit FERC accounts under which ADITs are booked:

281 ADITs – Accelerated Amortization Property - Credit income tax deferrals for restricted property for which the utility has elected to use ac-

---

[26] In the later years, the rates would collect less tax-related revenue than is actually paid.

celerated amortization over a five year period. Applicable property includes both certified defense facilities and certified pollution control facilities.

282 ADITs – Other Property - Credit income tax deferrals for all property other than accelerated amortization property includable in FERC Account 281. This account consists primarily of ADITs resulting from the use of liberalized depreciation for tax purposes, as discussed above.

283 ADITs – Other - Credit income tax deferrals for items other than property (and thus are not the result of the difference in depreciation methods as with property). For example, the loss on reacquired debt is amortized on the books over the original life of the issue being retired; however, the entire loss on the reacquired debt is deducted from income for tax purposes in the year of acquisition thereby creating a deferred income tax condition.

Because these represent deferred credits, the ADIT balances in FERC Accounts 281, 282, and 282 are deducted from rate base.

In addition, there are other ADITs created by some items that are included in income for which income taxes payable in the current year are higher because those items would not be fully reflected in determining annual net income until subsequent years. For example, the utility would record an expense on the books when adding funds to a Property Insurance Reserve (i.e., a self-insurance mechanism). However, an income tax deduction could not be claimed until such time as the utility must draw down the reserve for a qualified cause such as the repair of distribution facilities damaged by a major ice storm. Such ADITs are deferred debits, and they are recorded in FERC Account 190 – Accumulated Deferred Income Taxes. The ADIT balance in FERC Account 190 represents an addition to rate base.[27]

The record of ADITs is typically extensive and consists of fairly diverse items; thus, the functionalization process can be quite detailed with numerous unique line allocators. For instance, the Property Insurance Reserve ADITs could be functionalized using the functionalization results of the Property Insurance Reserve itself, whereas, the loss on reacquired debt ADITs may require a composite allocator incorporating several rate base components. However, given that the majority of ADITs are property related, the application of a plant in service type of allocator is commonly utilized.

Investment Tax Credits

From time to time, governments may authorize a special deduction to reduce income tax liability in order to promote business investment. For example, around

---

[27] The total amount of ADITs that is subtracted from rate base is thus the net effect of the FERC Account 281, 282, and 283 deductions and the 190 additions taken together.

1962 the U.S. government amended its tax code with a provision which allowed an *Investment Tax Credit* (ITC) of approximately 3% of the cost of qualified facilities for utility-related investments. Initially, a flow-through of ITCs to lower electric service rates was adopted by many regulatory agencies. However, this approach raised issues as to the intent of the ITC, and, by 1981, the use of flow-through accounting of ITCs was no longer permitted.

Subsequent to 1981, *normalization* was required for accounting treatment of ITCs. As with ADITs, the difference between income tax calculated on the basis of the books and the income tax actually due is treated as a deferred liability under normalization. By 1986, the ITC was completely eliminated. But given the longevity of utility plant which qualified for the ITC, a balance of unamortized ITCs may remain on the books of some utilities as of today. The deferred ITCs are recorded in FERC Account 225 – Accumulated Deferred Investment Tax Credits. Like ADITs, the deferred ITCs yield an interest free source of funds to the utility and are therefore deducted from rate base. If not functionalized at the time the amortization schedules were established for the qualified facilities, the deferred ITC amount can be functionalized by the application of a gross electric plant allocation factor.

Operating Reserves
The FERC has established four accounts designated as operating reserves which are used by utilities to accumulate funds for meeting obligations for specific expenditures that are expected to occur at some time in the future. These funds do not serve as a contingency monetary source for unpredictable events. Establishment of the annual rate of funding along with the maximum balance to be carried in an operating reserve account typically falls under the authority of the regulatory agency.[28] Furthermore, funds in these reserves could not be diverted to meet other obligations of the utility unless so directed by the regulatory agency.

Ongoing funding of the operating reserves is accomplished by the accrual of money available from receipts for electric service while taking into consideration payment requirements for O&M expenditures and the financial position of the utility. Since the funds are in essence supplied by customers, i.e., not expenditures incurred directly by the utility, the operating reserves are deducted from rate base so that they will not earn a rate of return.

Property Insurance Reserve
In general, it is difficult and often uneconomical, for utilities to procure insurance coverage for most T&D facilities, although affordable all-risk insurance is typically available for other utility electrical facilities, buildings, furnishings, vehicles, and equipment. The T&D system's level of exposure to disastrous weather events, such as tornados and ice storms, renders commercial insurance impracticable. Coastal utilities are particularly susceptible to catastrophic hurricane damage,

---

[28] FERC instructions state that "No amounts will be credited to these accounts unless authorized by a regulatory authority or authorities to be collected in a utility's rate levels."

and even a serious threat of a storm's landfall can result in significant preparation and mobilization expenditures whether or not the storm actually hits.

As a result, utilities must use self-insurance as a means to cover not just the costs of restoring the T&D system but also the rather significant deductible imposed with an all-risk insurance claim. FERC Account 228.1 – Accumulated Provision for Property Insurance is utilized to accumulate funds for such coverage. Statistics of prior storm events serve as a guide for estimating the probability of future occurrences and for predicting the costs of such events in order to establish a reasonable accrual rate as well as an accrual cap. The Property Insurance Reserve can be functionalized on the basis of a compound gross electric plant allocator which is more heavily weighted for the T&D functions.

Injuries and Damages Reserve

FERC Account 228.2 – Accumulated Provision for Injuries and Damages is an operating reserve that is utilized to accrue funds for meeting a probable liability for medical-related costs and indemnity payments, not covered by business insurance, for injuries and deaths to employees and others. In addition, the funds can be used for damages to property that is neither owned nor leased by the utility.

Thus, like the Property Insurance Reserve, Account 228.2 provides a means for a utility to provide self-insurance support for an intended purpose. The accrual is based on the best estimates available using such information as the settlements of prior claims and federal and state workers' compensation statistics and guidelines. Given the employee-oriented nature of Account 228.2, a salary and wages allocation factor is commonly utilized to functionalize the Injuries and Damages Reserve.

Miscellaneous Reserves

FERC Account 228.3 – Accumulated Provision for Pensions and Benefits is an operating reserve which includes provisions made available by the utility and amounts contributed by employees for pensions and other benefit plans, such as savings and accident and death benefits. The account includes employee pension or benefit plan funds which are included in the assets of the utility, and it does not include funds which are held by outside trustees. Given the employee-oriented nature of Account 228.3, a salary and wages allocation factor is commonly utilized to functionalize the Injuries and Damages Reserve.

FERC Account 228.4 – Accumulated Miscellaneous Operating Provisions is an operating reserve which includes funds for regulatory-authorized purposes that are not provided for elsewhere. An appropriate functionalization methodology would depend on the specific nature of the reserve.

Functionalized Rate Base Summary

An example of a functionalized original cost rate base is summarized in Table 11.4. All of the rate base elements have been assigned to the six principal utility functions of production, transmission, distribution, customer accounts, customer assistance, and sales either directly or by means of a cost causation based allocation method.

Table 11.4 Example Summary of a Functionalized Rate Base

FERC Account	Description	$ Amount (000s)
310 – 316	Steam Production Plant	812,218
320 – 325	Nuclear Production Plant	411,279
330 – 336	Hydraulic Production Plant	8,062
340 – 346	Other Production Plant	11,267
353-GSU	GSU Transformers	6,811
350 – 359	Transmission Lines	58,874
350 – 353	Transmission Substations	72,426
350 – 359	Subtransmission Lines	3,932
360 – 363	Distribution Substations	51,923
360 – 367	Primary/Secondary Distribution Lines	174,742
368	Line Transformers	107,412
369 – 370	Service Lines and Meters	59,007
373	Street Lighting and Signal Systems	20,898
301 – 303	Intangible Plant	3,503
389 – 399	General Plant	63,100
105	Plant Held for Future Use	1,046
107	Construction Work in Progress	160,216
	Total Gross Electric Plant	**2,026,717**
108	Accumulated Depreciation	744,420
	Total Net Electric Plant	**1,282,297**
120.1 – 120.6	Nuclear Fuel Stock	35,163
151 - 152	Fossil Fuel Stock	15,850
154, 163	Materials and Supplies	45,659
165	Prepayments	585
N/A	Cash Working Capital	19,153
190	Accumulated Deferred Income Taxes	5,343
	Total Rate Base Additions	**121,752**
228.1 – 228.4	Operating Reserves	9,300
235, 252	Customer Deposits and Advances	16,001
255	Accumulated Deferred Investment Tax Credits	1,530
281 – 283	Accumulated Deferred Income Taxes	237,372
	Total Rate Base Deductions	**264,204**
	TOTAL RATE BASE	**$1,139,845**

Cost-of-Service Methodology

Table 11.4—Cont. Example Summary of a Functionalized Rate Base

FERC Account	Production	Transmission	Distribution	Customer Accounts	Customer Assistance	Sales
310 – 316	812,218					
320 – 325	411,279					
330 – 336	8,062					
340 – 346	11,267					
353-GSU	6,811					
350 – 359		58,874				
350 – 353		72,426				
350 – 359		3,3932				
360 – 363			51,923			
360 – 367			174,742			
368			107,412			
369 – 370			59,007			
373			20,898			
301 – 303	2,206	172	833	246	36	10
389 – 399	39,738	3,104	15,002	4,429	643	185
105	266	349	406	20	3	1
107	147,537	9,457	2,848	316	46	13
	1,439,384	**148,314**	**433,072**	**5,010**	**727**	**209**
108	531,944	63,685	146,546	1,892	274	79
	907,440	**84,629**	**286,526**	**3,119**	**453**	**130**
120.1-6	35,163					
151 - 152	15,850					
154, 163	34,499	3,485	7,549	107	15	4
165	415	43	125	1	--	--
N/A	12,261	831	4,665	1,135	184	76
190	3,699	397	1,231	13	2	1
	101,886	**4,756**	**13,570**	**1,257**	**202**	**81**
228.1-4	3,133	957	4,854	299	43	13
235, 252		263	14,218	1,520		
255	1,219	98	213			
281 – 283	164,349	17,625	54,693	595	86	25
	168,701	**18,944**	**73,977**	**2,414**	**130**	**37**
Total	**$840,625**	**$70,441**	**$226,119**	**$1,961**	**$525**	**$174**

As part of the cost-of-service investigation, the rate base amount is used to determine:

1. The rate of return (ROR) that exists given the amount of annual net operating income (NOI), i.e., revenues realized under current rate levels less expenses and income taxes. $ROR_{Current} = NOI_{Current} \div$ Rate Base.

2. The amount of net income that would be required given a proposed ROR, i.e., the weighted average cost of capital. $NOI_{Proposed} = ROR_{Proposed} \times$ Rate Base.

These calculations are performed for the utility as a whole and on a jurisdictional, customer class, and/or rate class basis.

Table 11.5 illustrates alternative ways in which the weighted cost of capital can be calculated for use in determining net income, as described in item No. 2 above. The basic *financial view* of the weighted cost of capital includes common equity, preferred stock, and long-term debt in the capital structure. Given the cost rate assumed for each of these items, the weighted cost of capital for this example is 9.974%. The financial view method is utilized often for rate making purposes; however, some regulatory agencies may consider other sources of capital available to a utility for determining the weighted cost of capital. For example, short-term debt might also be included in the capital structure, which is designated as *Regulatory View 1* in Table 11.5. The inclusion of short-term debt yields a weighted cost of capital that is slightly lower at 9.916%.

Alternatively, some items which are typically designated as rate base deductions may be excluded from the determination of rate base but included in the calculation of the weighted cost of capital. "Regulatory View 2" in Table 11.5 incorporates ADITs, ITCs, and customer deposits in the capital structure as sources of funds available for use by the utility and are thus assigned various cost rates. ADITs are given a 0% cost rate since they represent cost free funds, but post-1970 ITCs are given a cost rate that is equal to the weighted cost of common equity, preferred stock, and long-term debt (i.e., the financial view).[29] Customer deposits are treated similar to standard savings account deposits with banks as the utility is required to accrue interest on such holdings on behalf of the customers. For this example, the cost rate applicable to customer deposits is 6%.

Compared to the basic financial view's weighted cost of capital, the effect of including ADITs, ITCs, and customer deposits as capital structure components is a measurable lowering of the weighted cost of capital to 8.799%. At the same time, however, the rate base amount would be increased since these deductible components are excluded from the total. The calculation of net income under both approaches is shown below.

---

[29] Pre-1971 ITCs were assigned a zero-cost rate.

Cost-of-Service Methodology

Table 11.5 Alternative Methods of Determining the Weighted Cost of Capital

Capital Source	Capital Amount (000s)	Financial View	Regulatory View 1	Regulatory View 2	Cost Rate
		Capital Structure			
Common Equity	$813,059	46.12%	44.44%	39.38%	12.500%
Preferred Stock	152,149	8.63%	8.32%	7.37%	8.090%
Long-Term Debt	797,769	45.25%	43.60%	38.64%	7.760%
Subtotal	$1,762,977	100.00%			
Short-Term Debt	$66,650		3.64%	3.23%	8.380%
Subtotal	$1,829,627		100.00%		
ADITs	$232,029			11.24%	0.000%
ITCs	$1,530			0.07%	9.974%
Customer Deposits	$1,520			0.07%	6.000%
Total	$2,064,706			100.00%	
		Weighted Cost of Capital			
Common Equity		5.765%	5.555%	4.922%	
Preferred Stock		0.698%	0.673%	0.596%	
Long-Term Debt		3.511%	3.384%	2.998%	
Financial View		**9.974%**			
Short-Term Debt			0.305%	0.271%	
Regulatory View 1			**9.916%**		
ADITs				0.000%	
ITCs				0.007%	
Customer Deposits				0.004%	
Regulatory View 2				**8.799%**	

	Rate Base	ROR	Net Income
Financial View:	$1,139,844,929	× 9.974% =	$113,688,133
Regulatory View 2:	$1,374,924,810	× 8.799% =	$120,979,634

Functionalization of Expenses

The FERC Uniform System of Accounts prescribes a series of accounts under which utility expenses are booked. Similar to the 300 series of plant accounts, the 500 and 900 series of expense-related accounts are already functionalized to a great degree; however, a few general cost items must be allocated between the major utility functions. The major categories of utility expense include operation and maintenance expenses, administrative and general expenses, annual depreciation expense, and taxes other than income taxes.

Operation and Maintenance Expenses

Operation and maintenance (O&M) expenses consist of labor, miscellaneous expenditures, and materials which are directly related to a major utility function.[30] Utilities follow a capitalization criterion to determine if a material item is to be treated as capital (plant) or as an expense. Generally, the cost of an item must be greater than a prescribed threshold and have a useful life of more than one year in order to be classified as a capital item. An example of an exception would be an office item, such as a desk, that is acquired to furnish a new building. The desk would be capitalized regardless of its cost amount; however, a replacement desk would be treated as an expense item. Otherwise, materials and replacement items are generally booked as an expense.

Expenses for operations and expenses for maintenance are booked separately. For example, the expenses associated with overhead distribution lines are booked as Overhead Line Expenses in Account 563 for those expenditures related to line operations and as Maintenance of Overhead Lines in Account 593 for those expenditures related to line maintenance. Patrolling distribution lines would include labor and transportation costs assignable to Account 563, whereas, stubbing a distribution pole that is currently in service would include labor, transportation, and materials costs assignable to Account 593.

As discussed previously, functionalized O&M labor is often used as an allocator for functionalizing intangible and general plant to the major functions. It is also used to allocate portions of labor and expense for which relationships to production, transmission, distribution, customer accounts, customer assistance, or sales are not easily identifiable.

Administrative and General Expenses

Administrative and general (A&G) expenses represent labor, miscellaneous expenditures, and materials that are not directly related to one of the major utility functions (e.g., the Rate Department), and therefore must be allocated. The A&G FERC Accounts consist of the following FERC Account series:

---

[30] Labor expenditures, along with various overheads, to construct a plant item are capitalized as part of the installation cost of the item.

Operation
920   Administrative and General Salaries
921   Office Supplies and Expenses
922   Administrative Expenses Transferred – Credit
923   Outside Services Employed
924   Property Insurance
925   Injuries and Damages
926   Employee Pensions and Benefits
927   Franchise Requirements
928   Regulatory Commission Expenses
929   Duplicate Charges – Credit
930.1 General Advertising Expenses
930.2 Miscellaneous General Expenses
931   Rents

Maintenance
935   Maintenance of General Plant

Functionalization of the A&G accounts is accomplished with various allocators. Employee-related items such as salaries, office supplies and expenses, injuries and damages, employee pensions and benefits, and maintenance of general plant are typically allocated to the major functions based on functionalized O&M labor. Less specific items such as outside services employed, general advertising, and miscellaneous expenses are typically allocated to function based on functionalized O&M expenses, excluding fuel and purchased power. The functionalization of rents would depend on the types of rentals involved. A labor ratio would be more appropriate for office space rentals, while a plant ratio or an O&M expense allocator might be more appropriate for other types of rentals.

If detailed sub account information is available, a more in-depth assessment of A&G expenses could be made to improve the functionalization of these costs. For example, property insurance might consist of both boiler explosion and all-risk policies. The cost of boiler explosion insurance would be directly assignable to production. The cost of all-risk insurance would then be allocated to the major functions based on a plant ratio considering the type of facilities that are covered by the policy. Miscellaneous general expenses may include costs for employee communications. That portion of the expenses could be allocated to the major functions based on employee headcounts in order to improve functionalization.

The functionalized labor portion of A&G expenses is combined with the functionalized labor portion of O&M expenses to obtain a total labor amount commonly referred to as *Salaries and Wages* (S&W). As discussed previously, an S&W allocator is often utilized for functionalizing some items of plant, such as the Accumulated Provision for Injuries and Damages (Account 228.2) and the Accumulated Provision for Pensions and Benefits (Account 228.3), because they are employee-related reserves.

Annual Depreciation Expense

The annual depreciation of an asset is recognized as an expense item, and it represents the return of capital. The depreciation expense provides a source of funds which, in part, are used to replace and expand system assets.

An item such as a distribution pole has an average life expectancy; however, some poles will fail sooner, because of exposure for instance, while some others survive well beyond their expected life. As system growth occurs and as maintenance of system components is conducted, the books and records of plant accounts reflect the changes in unit counts and monetary balances due to component additions and retirements. The mass property accounts are particularly fluidic in this respect, and the tracking of annual depreciation is thus a rather dynamic task.

Engineering-based depreciation studies are conducted periodically to develop annual depreciation rates applicable to property units which reflect current conditions of the system. The production system is evaluated on a plant or even a generating unit basis, while homogeneous or mass property items, such as line transformers, are evaluated as a group. Depreciation studies key on statistical analyses of the various plant accounts to determine such measures as average plant lives, remaining plant lives, and retirement dispersions (i.e., frequency distributions of property retirements). Iowa-type survivor curves[31] are often used in these evaluations.

As discussed previously, straight-line depreciation is utilized for book and financial reporting purposes. A basic annual depreciation rate for an item would be computed by dividing the asset's installed cost by its years of life expectancy. However, a number of assets have a salvage value upon retirement, e.g., copper conductor, which essentially reduces the magnitude of the cost to be depreciated. Traditionally, the cost to ultimately remove the asset from service, which in essence increases the original cost of the item to be depreciated, has been included as an element of annual depreciation. Thus, the annual rate of depreciation for a property unit is determined by

$$DR = \frac{100\% - S + R}{N_{AVG}} \qquad (11.5a)$$

where DR =   annual depreciation rate, in percent
  S =   salvage value as a percent of the original installed cost
  R =   cost of removal as a percent of the original installed cost
  N_{AVG} =   average service life, in years

---

[31] Around 1930, the Engineering Research Institute of Iowa State University developed three families of generalized survivor curves based on industry statistics. Each family depicts a different relationship between the mode of the retirement frequency distribution to the average life: "L" left modal, "S" symmetrical (mode and average occur at the same age), and "R" right modal. An "O" origin curve was added in the later part of the 1950s.

Equation (11.5a) is referred to as a *whole life* method as it calculates a depreciation rate at the outset of the life cycle of a property unit. Another approach, referred to as a *remaining life* method, calculates a drepreciation rate based on specific conditions at a point in time during the life cycle of a property unit. Thus, the annual rate of depreciation for a property unit is determined by

$$DR = \frac{100\% - S + R - B}{N_{REM}} \quad (11.5b)$$

where DR = annual depreciation rate, in percent
S = salvage value as a percent of the original installed cost
R = cost of removal as a percent of the original installed cost
B = current depreciation reserve balance as a percent of the current plant balance
N_{REM} = average remaining service life, in years

As discussed previously, the annual depreciation is accrued on an on-going basis, and the accumulated depreciation in FERC Account 108 is deducted from gross plant in order to determine net plant.

Depreciation expense is booked in FERC Account 403. Given that depreciation rates are calculated on a plant FERC Account basis, the annual depreciation amount is typically recorded by production, transmission, distribution, and general plant categories. The depreciation expense for general plant is then allocated to the major functions in the same manner as utilized for functionalization of the general plant assets (e.g., O&M labor).

Asset Retirement Obligations

In 2001, the Financial Accounting Standards Board (FASB) issued statement SFAS 143 which modified the accounting procedures relative to certain legal obligations of businesses. An *asset retirement obligation* (ARO) is a liability that results from a legal obligation to decommission or retire a plant asset. For example, a utility having nuclear production is required under Nuclear Regulatory Commission (NRC) regulations to dismantle and remove contaminated facilities, such as the reactor vessel, when the plant is decommissioned.

Upon recognition of an ARO liability, the utility is required to capitalize the fair value retirement cost of the asset and then depreciate that amount over the asset's useful life. As a result, the historical inclusion of the removal cost element in the determination of depreciation rates, as presented in Equations (11.5a) and (11.5b), is no longer permitted for assets having a legal obligation for retirement.

In addition, the portion of the accumulated depreciation balance in FERC Account 108 related to removal cost for existing AROs must be extracted and recategorized as an asset. The result of these adjustments is an increase in gross plant in service for those accounts containing ARO-related assets, a decrease in the amount of accumulated depreciation, and the creation of a sizeable liability. Each year, the ARO liability is credited to recognize the new fair value of the obligation. A risk-free credit adjusted discount rate is applied to compute the increase in value.

In response to the accounting changes, FERC issued Order No. 631 in 2003, which prescribed new plant accounts to be added to the Uniform System of Accounts under which asset retirement costs would be capitalized.

317	Asset Retirement Costs for Steam Production
326	Asset Retirement Costs for Nuclear Production
337	Asset Retirement Costs for Hydraulic Production
347	Asset Retirement Costs for Other Production
359.1	Asset Retirement Costs for Transmission
374	Asset Retirement Costs for Distribution
399.1	Asset Retirement Costs for General Plant

FERC Account 230 – Asset Retirement Obligations was also added to include the amount of liabilities for the recognition of AROs. In addition, FERC Account 411.10 - Accretion Expense was added to expense the liabilities associated with AROs included in FERC Account 230.

Taxes Other Than Income Taxes

Utilities are often assessed a number of taxes, other than state and federal income taxes. Such taxes are paid to various governments, including federal, province, state, county, and/or local taxing authorities. Taxes other than income taxes are recorded in FERC Account 408.1. Most such taxes are categorized as property related, payroll related, and revenue related.

Property or real estate taxes are often referred to as *ad valorem taxes* since a tax millage rate is commonly applied to the value of the assessed property.[32] Property consists of land, buildings and structures, and other property, e.g., storeroom inventory. Property taxes may be associated with the major functions when recorded; otherwise, a net plant ratio allocator could be utilized for functionalization as a reflection of depreciated property value.

Like other businesses, utilities are required to withhold and/or pay on behalf of their employees certain government taxes related to employee wages. Some employee tax payments must be matched by the employer, such as to a Social Security system, and the employer may be required to pay a state and federal unemployment tax. Payroll-related taxes are generally calculated as a percentage of an employee's gross salary, which may also be limited to a specified annual maximum for tax purposes, i.e., a wage base. Payroll-related taxes are functionalized to major function by means of a S&W allocator.

---

[32] The millage rate is applied to the assessed value of the property divided by 1,000. A mill is one-one-thousandth of a given unit of currency.

Cost-of-Service Methodology 491

Some taxes other than income taxes are assessed on the basis of the annual revenues or gross receipts of the utility. For example, state and local franchise taxes are levied for the privilege of engaging in the business of furnishing electricity within a franchised area. A franchise tax is typically a percentage applied against gross receipts less any applicable statutory deductions. CIAC payments are generally exempt from such taxes. A municipal franchise tax would apply only to the retail portion of electric sales. State public utility regulatory taxes are also typically assessed on the basis of retail revenues, which are governed by state jurisdiction.

Revenue is the natural cost causation driver for revenue-related taxes. Revenues from unbundled rates for electric services may indeed provide a reasonable means for functionalizing revenue-related taxes. However, revenues from bundled rates for electric service are not explicitly associated with the major utility functions. Demand and energy charges include a blending of the costs of production, transmission, and distribution. The customer charge includes a blending of distribution and customer services costs. Thus, at the cost functionalization stage, revenue-related taxes can be assigned directly to the customer accounts function. Later, when costs are assigned to the customer or rate classes, the revenue-related taxes are allocated on the basis of the revenues of the classes. Alternatively, the revenue-related taxes can be assigned to a seventh major function, "Other," and then allocated later to the classes in the same manner.

A utility may also be required to pay any number of miscellaneous taxes and fees. For example, a pole and wire tax is a form of a privilege tax. The tax might be assessed as a charge per distribution pole, or it might be assessed as a charge per mile of transmission line. As such, the tax would be functionalized by direct assignment to distribution or transmission as applicable. A corporate franchise tax may be based on the utility's capital value, i.e., capital stock issued and outstanding, retained earnings, deferred taxes, etc. Thus, a gross plant ratio allocator could be used for functionalization.

Some other taxes or fees may be assessed directly to the consumers, rather than the utility, while using the monthly electric service bill as a vehicle for collecting such taxes. In other words, these pass through taxes are not recorded as part of the utility's operating revenues and thus are not considered to be a part of the utility's revenue requirement. State or county sales tax is a prime example of a pass-through type of tax. The utility accounts for the monthly sales taxes collected through its revenue accounting process and then remits the taxes to the appropriate taxing authority as required.

Functionalized Expenses Summary

An example of functionalized labor and O&M expenses, A&G, depreciation expense, and taxes other than income taxes is summarized in Table 11.6. All of the expense elements have been assigned to the principal utility functions of production, transmission, distribution, customer accounts, customer assistance, and sales either directly or by means of a cost causation based allocation method.

Table 11.6 Example Summary of Functionalized Expenses

FERC Account	Description	$ Amount (000s) Labor	Expenses
500 – 509	Steam Production Operation	16,517	227,863
510 – 514	Steam Production Maintenance	14,279	26,682
517 – 525	Nuclear Production Operation	6,909	8,902
528 – 532	Nuclear Production Maintenance	2,589	2,473
535 - 540	Hydraulic Production Operation	545	1,317
541 – 545	Hydraulic Production Maintenance	307	380
546 - 550	Other Production Operation	18	334
551 – 554	Other Production Maintenance	94	121
555 - 558	Other Power Supply	626	7,034
560 – 567	Transmission Operation	2,337	3,253
568 – 573	Transmission Maintenance	934	1,309
580 – 589	Distribution Operation	11,340	16,175
590 – 598	Distribution Maintenance	4,473	11,399
901 – 905	Customer Accounts Operation	4,668	6,461
906 - 910	Customer Assistance Operation	677	1,088
911 – 916	Sales Operation	195	491
920 - 931	Administrative & General Operation	9,010	39,516
935	Administrative & General Maintenance	477	2,100
	Salaries and Wages Subtotal	75,996	
	O&M Expenses Subtotal		356,859
	Total O&M Labor and Expenses		**432,895**
403	Depreciation Expense		
	Production		40,089
	Transmission		3,764
	Distribution		13,546
	General and Common		3,001
	Total Depreciation Expense		**60,400**
408.1	Taxes Other Than Income Taxes		
	Property Taxes		6,507
	Payroll Taxes		5,957
	Revenue Taxes (refer to text discussion)		12,030
	Total Taxes Other Than Income Taxes		**24,494**
	TOTAL EXPENSES		**$517,789**

Cost-of-Service Methodology

Table 11.6–Cont. Example Summary of Functionalized Expenses

FERC Account	Production	Transmission	Distribution	Customer Accounts	Customer Assistance	Sales
500 – 509	244,380					
510 – 514	40,961					
517 – 525	15,811					
528 – 532	5,062					
535 - 540	1,862					
541 – 545	687					
546 - 550	353					
551 – 554	216					
555 - 558	7,661					
560 – 567		5,590				
568 – 573		2,243				
580 – 589			27,514			
590 – 598			15,872			
901 – 905				11,129		
906 - 910					1,765	
911 – 916						686
920 - 931	31,020	2,429	11,379	3,102	456	139
935	1,623	127	613	181	26	8
Total O&M	**349,636**	**10,389**	**55,378**	**14,412**	**2,247**	**833**
403						
Prod.	40,269					
Trans.		3,580				
Distr.			13,549			
Gen.	1,890	148	714	211	31	9
	42,159	3,728	14,263	211	31	9
408.1						
Prop.	3,709	1,041	1,672	3	--	--
Pay.	3,752	293	1,415	418	61	17
Rev.	--	--	--	*12,030*	--	--
	7,461	1,334	3,087	12,451	61	17
Total	**$399,338**	**$15,451**	**$72,728**	**$27,074**	**$2,339**	**$859**

11.4 COST CLASSIFICATION

As the second major step of the cost-of-service study, cost classification recognizes an additional degree of distinction between the various utility plant items and expenses based on cost causation. In this step, the functionalized costs are classified as fixed or variable in terms of the fundamental cost components: customer, demand, and energy.[33] Customer-related costs are fixed as they are independent of energy or load. Demand-related costs are fixed as they relate to peak load conditions, and a fixed amount of capacity is installed to meet maximum demands even though a customer's actual demand can vary considerably from hour to hour or even minute to minute. Energy-related costs are variable as they relate to the production and delivery of kWh as energy is consumed. In general, the production, transmission, and distribution substation plant functions (Levels 1, 2, and 3) are demand related. The distribution system (Levels 4 and 5) is partially demand related and partially customer related. Level 1 production fuels and some other O&M expenses are energy related. Classification further aligns the functionalized costs in preparation for assignment to jurisdiction or classes using customer, demand, and energy allocation factors.

The classification process yields additional plant functionalization. The FERC Uniform System of Accounts represents a fairly detailed view of the cost elements of which a power system is composed. Subaccounts provide even more detail; however, the complexity of the varieties, sizes, and ratings of power system components is extensive. For practical purposes, utility accounting records often reflect a roll-up of similar items which filters out some of the details that would be desirable for the cost-of-service study. For example, OH distribution conductors are not necessarily differentiated by voltage level when recorded in FERC Account 365. If so, the classification analysis should, at a minimum, differentiate conductors (and other associated line equipment) by primary distribution and secondary distribution in order to recognize cost differences between Levels 4 and 5. An auxiliary analysis using engineering circuit diagrams or GIS maps could be conducted to complete a separation of conductor costs to voltage levels. Similar analyses are often required to recapture details lost when new line equipment and facilities are booked.

Classification of Distribution Plant

The distribution system has a number of sub functions of its own. A distribution substation transforms transmission or subtransmission voltages to primary distribution voltages and provides capacity to meet customer demand on an area-wide basis. Primary distribution feeders extend out from a substation in order to reach customer sites. Line transformers transform primary distribution voltages to sec-

---

[33] As noted previously, the underlying cost driver for a limited number of expenses is revenue which in essence serves as a fourth cost component.

Cost-of-Service Methodology

ondary voltages and provide capacity to meet the demand of a single or a few adjacent customers. Secondary distribution lines and service lines extend from line transformers and connect to a customer's low voltage wiring system. The distribution system provides reliability through its switching and protection schemes. It also provides power quality by means of voltage regulation equipment. Meters record the energy utilization characteristics of customers.

On the principle of cost causation, some distribution facilities can be classified as purely demand related, while other distribution facilities can be classified as purely customer related. A substantial amount of distribution plant investment has cost causation attributes of both demand and customer cost components. For instance, a line transformer, is both customer related and demand related as it provides both a load and a no-load function. This distinction is derived through a *minimum distribution system* analysis of poles, conductors, and a variety of line equipment, which is discussed below.

Distribution line equipment and facilities investment costs are recorded in FERC Accounts 364 – 370. Outdoor area and roadway lighting facilities investment costs are recorded in FERC Account 373. The details of items recorded in these accounts, as specified by the Uniform System of Accounts, is shown in Table 11.7. To properly classify distribution costs, the cost-of-service study must examine the detailed cost items which are booked in each account.

Meters and Service Lines

With some exceptions, such as street lighting, each customer has a revenue meter which has a mission to register kW or kVA demand and/or kWh usage for billing purposes. The same inexpensive, basic watt-hour meter can be used to measure the kWh usage of a small efficiency apartment or a 3,500 ft^2 residence or a small commercial office or shop. A particular small customer's meter reading in any given billing period could be 400 kWh for example, while in another billing period, the customer's meter reading could be 4,000 kWh. The cost of the meter is independent of the customer's load and energy utilization; thus, meters are deemed to be a customer-related fixed cost.

More expensive, sophisticated meters are used for medium to large commercial and industrial customers, although these higher costs represent only a small fraction of the total bill amount. The application of watt-hour revenue meters would be woefully inadequate for larger customers as the measurement of complex demand units is necessary to design and administer equitable rates for customers having such diverse energy and demand requirements. In addition, instrument transformers are necessary for metering of high current and/or high voltage delivery points. As a result, the cost of metering generally increases with the size of the customer, but this relationship does not imply that such metering is demand related. As before, the cost of the demand meter is independent of the customer's load and energy utilization as even large industrial customers, having the same type of metering equipment, can have significant variations in their usage characteristics from one billing period to another.

Table 11.7 Distribution System Plant FERC Account Items

364 Poles, Towers, and Fixtures
1. Anchors, head arm, and other guys, including guy guards, guy clamps, strain insulators, pole plates, etc.
2. Brackets.
3. Crossarms and braces.
4. Excavation and backfill, including disposal of excess excavated material.
5. Extension arms.
6. Foundations.
7. Guards.
8. Insulator pins and suspension bolts.
9. Paving.
10. Permits for construction.
11. Pole steps and ladders.
12. Poles, wood, steel, concrete, or other material.
13. Racks complete with insulators.
14. Railings.
15. Reinforcing and stubbing.
16. Settings.
17. Shaving, painting, gaining, roofing, stenciling, and tagging.
18. Towers.
19. Transformer racks and platforms.

365 Overhead Conductors and Devices
1. Circuit breakers.
2. Conductors, including insulated and bare wires and cables.
3. Ground rods, clamps, etc.
4. Insulators, including pin, suspension, and other types, and tie wire or clamps.
5. Lightning arrestors.
6. Railroad and highway crossing guards.
7. Splices.
8. Switches.
9. Tree trimming, initial cost including the cost of permits therefore.
10. Other line devices.

366 Underground Conduit
1. Conduit, concrete, brick, and tile, including iron pipe, fiber pipe, Murray duct, and standpipe on pole or tower.
2. Excavation, including shoring, bracing, bridging, backfill, and disposal of excess excavated material.
3. Foundations and settings specially constructed for and not expected to outlast the apparatus for which constructed.
4. Lighting systems.
5. Manholes, concrete or brick, including iron or steel frames and covers, hatchways, gratings, ladders, cable racks and hangers, etc., permanently attached to manholes.
6. Municipal inspection.
7. Pavement disturbed, including cutting and replacing pavement, pavement base, and sidewalks.
8. Permits.
9. Protection of street openings.
10. Removal and relocation of subsurface obstructions.
11. Sewer connections, including drains, traps, tide valves, check valves, etc.
12. Sumps, including pumps.
13. Ventilating equipment.

367 Underground Conductors and Devices
1. Armored conductors, buried, including insulators, insulating materials, splices, potheads, trenching, etc.
2. Armored conductors, submarine, including insulators, insulating materials, splices in terminal chamber, potheads, etc.
3. Cables in standpipe, including pothead and connection from terminal chamber or manhole to insulators on pole.
4. Circuit breakers.
5. Fireproofing in connection with any items listed herein.
6. Hollow-core oil filled cable, including straight or stop joints, pressure tanks, auxiliary air tanks, feeding tanks, terminals, potheads and connections, etc.
7. Lead and fabric covered conductors, including insulators, compound-filled, oil-filled or vacuum splices, potheads, etc.
8. Lightning arrestors.
9. Municipal inspection.
10. Permits.
11. Protection of street openings.
12. Racking of cables.
13. Switches.
14. Other line devices.

Cost-of-Service Methodology

Table 11.7–Cont. Distribution System Plant FERC Account Items

368 Line Transformers
1. Installation, labor of (first installation only).
2. Transformer cut-out boxes.
3. Transformer lightning arrestors.
4. Transformers, line and network.
5. Capacitors.
6. Network Protectors.

369 Services
1. Brackets.
2. Cables and wires.
3. Conduit.
4. Insulators.
5. Municipal inspection.
6. Overhead to underground, including conduit or standpipe and conductor from last splice on pole to connection with customer's wiring.
7. Pavement disturbed, including cutting and replacing pavement, pavement base, and sidewalks.
8. Permits.
9. Protection of street openings.
10. Service switch.
11. Suspension wire.

370 Meters
1. Alternating current, watt-hour meters.
2. Current limiting devices.
3. Demand indicators.
4. Demand meters.
5. Direct current watt-hour meters.
6. Graphic demand meters.
7. Installation, labor of (first installation only).
8. Instrument transformers.
9. Maximum demand meters.
10. Meter badges and their attachments.
11. Meter boards and boxes.
12. Meter fittings, connections, and shelves (first set).
13. Meter switches and cut-outs.
14. Prepayment meters.
15. Protective devices.
16. Testing new meters.

373 Street Lighting and Signal Systems
1. Armored conductors, buried or submarine, including insulators, insulating materials, splices, trenching, etc.
2. Automatic control equipment.
3. Conductors, overhead or underground, including lead or fabric covered, parkway cables, etc., including splices, insulators, etc.
4. Lamps, arc, incandescent, or other types, including glassware, suspension fixtures, brackets, etc.
5. Municipal inspection.
6. Ornamental lamp posts.
7. Pavement disturbed, including cutting and replacing pavement, pavement base, and sidewalks.
8. Permits.
9. Posts and standards.
10. Protection of street openings.
11. Relays or time clocks.
12. Series contactors.
13. Switches.
14. Transformers, pole or underground.

Since metering is a customer-related component, the associated costs will ultimately be allocated based on rate schedule counts of the numbers of customers. To improve the cost causation basis for allocation, metering equipment costs recorded in FERC Account 370 should be subfunctionalized to capture key differences in electric service methodology, including:

- Watt-hour meters: single phase and three phase
- Demand meters: single phase and three phase
- Low voltage CTs and PTs
- High voltage CTs and PTs

Service lines are the low voltage conductors that extend from line transformers and secondary distribution lines and make a connection to the customer's wiring system at the service entrance point. Both overhead and underground services are utilized. Services may be configured as open wire or bundled cable, i.e., n-plex. In addition, service lines are either single phase and three phase.

Service lines are often deemed to be entirely customer related. However, the argument can be made that services may also have a demand-related attribute. If one customer's maximum load is expected to be greater than a similar customer's maximum load, then service lines having different capacity ratings could be sized to meet the two different loads (even though the same rating of meter could be used for both services). When a utility carries in stock a variety of sizes of service lines for use with similar types of customers (e.g., the residential class), then the cost of service lines could be separated between customer and demand components in the same manner used for cost separation of primary and secondary conductors. However, utilities often minimize size varieties of service lines that it stocks as a matter of practicality and economics. Thus, nearly all customers in a particular class of service could have the same size service line, particularly with smaller customer rates, and extracting a demand-related cost component would be more or less inconsequential. Given also that each service line is fully dedicated to a single customer, a customer-related classification is a reasonable judgment.

Similar to metering, the associated costs of service lines will ultimately be allocated based on rate schedule counts of the numbers of customers. To improve the cost causation basis for allocation, service line costs recorded in FERC Account 369 should be subfunctionalized to capture key differences in electric service methodology, including:

- Overhead service lines: single phase and three phase
- Underground service lines: single phase and three phase

Minimum Distribution System Analysis

The concept of a minimum distribution system recognizes that the primary and secondary distribution system has both customer-related and demand-related attributes. As discussed previously, the customer cost component is associated with no-load conditions, whereas the demand cost component is associated with load conditions, i.e., equipment and facilities have the capacity to meet peak loads. Some devices can be categorized as being either customer related or demand related, while some individual devices are related to both components.

A direct approach to identifying and quantifying the two cost components is to evaluate each major distribution item in terms of its mission. For example, a distribution system protection scheme is a mix of circuit breakers, reclosers, and fused cutouts that are coordinated to minimize customer outages in the event of faults on the system. A simple feeder schematic including protective devices is shown in Figure 11.13.

Cost-of-Service Methodology 499

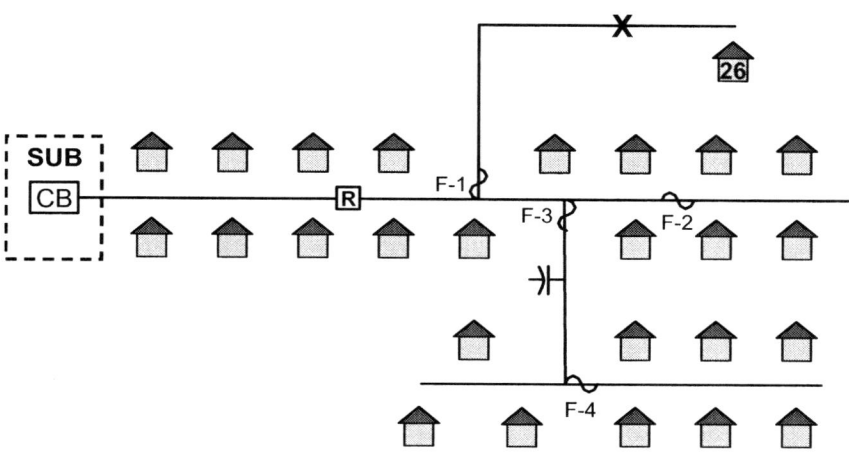

Figure 11.13 **A schematic of a primary feeder system.** A fault occurs on the primary tap line which connects customer #26 to the main feeder. Under a temporary fault condition, customers downstream from the recloser will experience a momentary outage, but service is restored to all customers. Under a permanent fault condition, the same momentary outage will occur to all customers, but only customer #26 will lose service as fuse F-1 blows to clear the fault from the main feeder. The capacitor located between fuses F-3 and F-4 provide feeder voltage support under heavy loading conditions.

The protective devices work together to preserve service to as many customers as possible under various fault conditions. In essence, the protection scheme safeguards the "voltage path" which connects each customer on the circuit to the source, i.e., the substation. Consider a case where all customer load is removed from the circuit, and a fault occurs as before, the protection scheme will operate exactly as it did under load in an effort to maintain the voltage path to as many customers as possible. The protection scheme's mission is independent of load or demand and thus is a customer-related function. Furthermore, other facilities which provide a connection of a customer to the source have a customer-related attribute.

Consider again the circuit in Figure 11.13 during peak demand periods. Under light loading conditions, the capacitor is switched off. However, once high load currents create an unacceptable voltage drop, the capacitor connects to the circuit to raise the voltage along the feeder profile, thereby releasing capacity. The capacitor's mission is to support voltage under load and is thus dependent on load. Absent customer load, the capacitor is not required at all, even as the circuit is energized to create a voltage path. The capacitor is independent of the voltage path. As a result, the capacitor is a demand-related component.

The mission of a number of other distribution system devices have both customer-related and demand-related attributes. Primary and secondary conductors clearly create a path interconnecting customers to a source, but they also have capacity to carry load. Line transformers are integral to the path as they connect the secondary conductor system to the primary conductor system, but they also provide capacity to serve customers' loads.

When a single device has both customer-related and demand-related attributes, its total cost must be allocated. The *minimum intercept* or *zero-intercept methodology* provides a rational basis for separating the cost of a device between its customer and demand components. The zero-intercept methodology is a weighted linear regression of the unit costs of standard ratings or sizes of a specific device, such as a single-phase overhead line transformer, plotted as a function of its capacity-related characteristic, which would be kVA for a line transformer. The objective of the regression analysis is to determine the *y*-intercept. The *y*-intercept represents that portion of a device's total cost that is associated with zero capacity and thus the customer-related component.[34] The unit costs must be weighted by the numbers of devices because of the uneven distribution of the various ratings or sizes of the devices in service. The slope and the *y*-intercept of the weighted linear regression equation is given by

$$m = \frac{\left[N \times \sum_{i=1}^{\tau} n_i X_i Y_i\right] - \left[\sum_{i=1}^{\tau} n_i X_i \times \sum_{i=1}^{\tau} n_i Y_i\right]}{\left[N \times \sum_{i=1}^{\tau} n_i X_i^2\right] - \left[\sum_{i=1}^{\tau} n_i X_i\right]^2} \quad (11.6a)$$

$$b = \frac{\sum_{i=1}^{\tau} n_i Y_i}{N} - m \times \left[\frac{\sum_{i=1}^{\tau} n_i X_i}{N}\right] \quad (11.6b)$$

where m = slope
  b = *y*-intercept
  X_i = unit size
  Y_i = unit cost
  n_i = number of units of a particular size or rating
  N = total number of all units (i.e., $\Sigma\ n_i$)
  τ = total number of sizes or ratings

---

[34] The slope of the regression equation signifies a demand component of cost.

Cost-of-Service Methodology

Once the y-intercept unit cost is calculated for a given category of devices, the total cost of all such devices (units) in that category is apportioned between the customer cost component and the demand cost component by:

$$\text{Customer Cost} = y\text{-Intercept} \times \text{Total Number of Units} \quad (11.7a)$$

$$\text{Demand Cost} = \text{Total Category Cost} - \text{Customer Cost} \quad (11.7b)$$

There are two approaches for applying the zero-intercept analysis. The first approach utilizes embedded cost data obtained directly from plant account records. The unit costs for the different sizes or ratings of the subject device are determined on the basis of the booked investment and the numbers of units, and the results are then used in the regression analysis. A problem arises often with the embedded cost approach because of wide variations in equipment vintage. Over time, equipment ratings have generally trended upward due to higher distribution voltages and increased customer loads. However, the book records may reflect a significant number of small sized devices in service that have relatively low average unit costs due to the older age of their installations. Concurrently, larger size devices have much higher average unit costs as they were more recently installed. This temporal disparity in unit costs can distort the regression results and even produce a negative y-intercept.

An alternative approach overcomes such incongruities as it is based on an estimate of the cost required to "rebuild" all of the devices in service all at once using current unit construction costs applied to the full inventory of devices. The resulting total rebuild cost will be somewhat greater than the total embedded cost due to the increases in materials and construction costs that have occurred between the vintage periods and the current time. The total cost of the rebuilt devices now represents the "Total Category Cost" in Equation (11.7b). These same current unit installation costs are used in the regression to determine the y-intercept for use in Equation (11.7a).[35] The rebuild y-intercept will also be greater than the y-intercept determined from average book costs. Once the allocation of the total rebuild cost is made to the customer cost component and the demand cost component, the total embedded or book cost is allocated proportionately by

$$\% \text{ Rebuild Customer Cost} \times \text{Total Book Cost} = \text{Customer Cost} \quad (11.8a)$$

$$\% \text{ Rebuild Demand Cost} \times \text{Total Book Cost} = \text{Demand Cost} \quad (11.8b)$$

A comparison of zero-intercept regression analyses results based on both embedded costs and rebuild costs is shown in Figure 11.14.

---

[35] The regression equation can be used to interpolate and extrapolate unit installation costs for devices which are in service (and therefore must be "rebuilt"), but which are neither no longer purchased as standard inventory nor no longer manufactured.

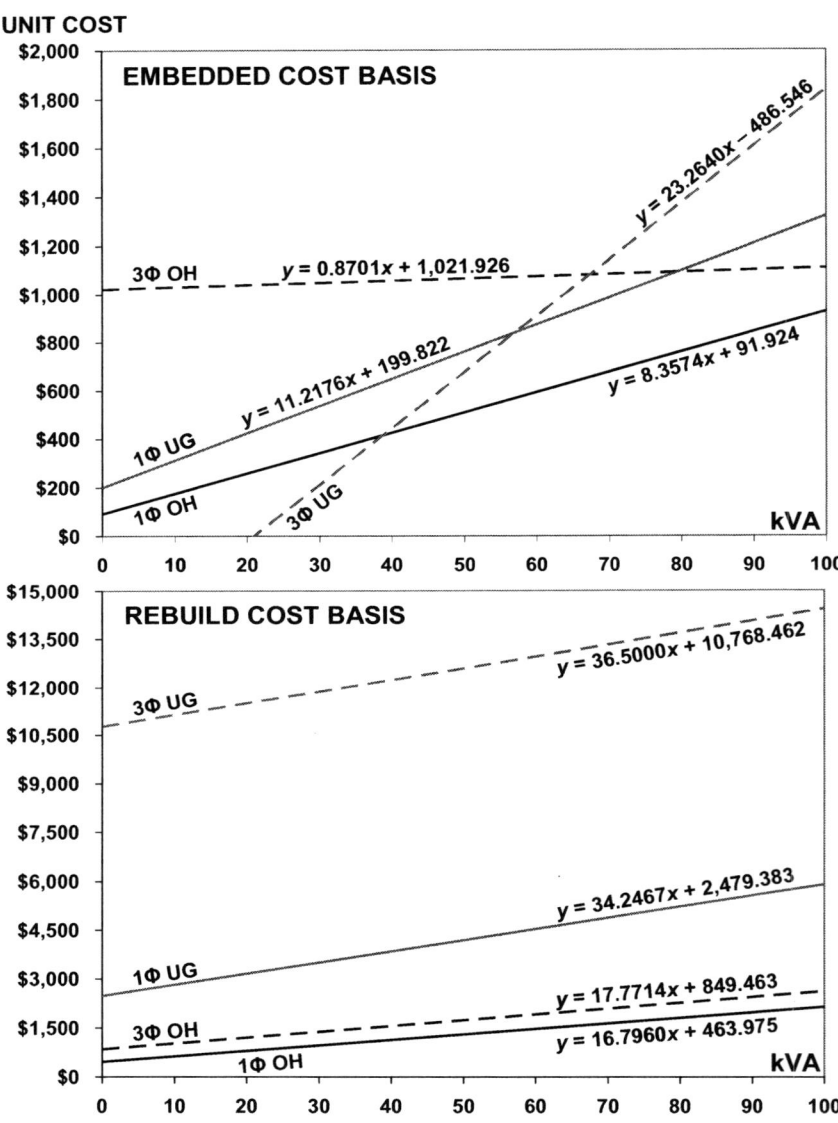

Figure 11.14 Zero-intercept regression analysis: embedded cost basis vs. rebuild cost basis for line transformers. Problematic relationships which often result from variations in vintage costs (upper chart) can be corrected by "rebuilding" the system on a current cost basis and then applying a weighted regression using current unit costs (lower chart).

Line Transformers

As discussed above, line transformers, which are located at Level 5 in the system structure, have attributes of both the customer and demand cost components; therefore, their total cost should be apportioned accordingly using the zero-intercept analysis. Line transformer gross plant, which is booked in FERC Account 368, is typically itemized by kVA ratings, input and output voltages, and phases. In addition, transformers are further differentiated between pole mount (overhead) and pad mount (underground). Independent weighted regressions should be applied for each of the four major transformer categories:

- Pole mount line transformers: single phase and three phase
- Pad mount line transformers: single phase and three phase

Typical standard kVA ratings of line transformers are shown in Table 5.2. It is not critical to have unit cost data points for all transformers in service; however, a range of the smaller size units in each category is important to ensure data linearity and reasonable y-intercept results. In addition, the data points of a particular regression should represent the same input and output voltages (e.g., 15 kV, 208Y/120 volt pad mount transformers). Given the wide range of characteristics (voltages, one vs. two bushings, etc.) within a major transformer category, the regression analysis can be conducted using the most common or representative transformer type in each major category with the results then applied to the population of transformers in each respective category. An example of the zero-intercept analysis results obtained for the four major categories of line transformers, using the system rebuild approach, is shown in Figure 11.14.

As shown in Table 11.8a, the system rebuild-based y-intercepts from the regression examples shown in Figure 11.14 are multiplied by the number of total line transformers, by major category, to compute the rebuild customer-related costs. The customer-related amounts are then expressed as a percentage of the total system rebuild costs.

Table 11.8a Results of Level 5 Line Transformer Classification Example

Transformer Type	Number of Units	y-Intercept Unit Costs $/Unit	Current Cost Rebuild Basis		
			Customer Cost $(000s)	Total Rebuild Cost $(000s)	Percent of Total
1φ Pole Mount	98,728	463.975	45,807	109,961	41.66%
3φ Pole Mount	1,135	849.463	964	6,339	15.21%
1φ Pad Mount	6,188	2,479.383	15,342	24,946	61.50%
3φ Pad Mount	1,310	10,768.462	14,107	41,220	34.22%
Total	107,361		$76,220	$182,466	

Table 11.8b Results of Level 5 Line Transformer Classification Example

Transformer Type	Percent of Total Rebuild Costs	$ Embedded Costs (000s)		
		Total Book Cost	Customer Component	Demand Component
1φ Pole Mount*	41.66%	41,412	17,251	24,161
3φ Pole Mount*	15.21%	3,550	540	3,010
1φ Pad Mount	61.50%	4,386	2,697	1,689
3φ Pad Mount	34.22%	10,701	3,662	7,039
Total		$60,049	$24,150	$35,899

* Includes vault-type transformers.

As shown in Table 11.8b, the customer-related cost percentages determined on the basis of rebuilding the line transformer system at current costs are applied to the total transformer book costs, by major category, in order to compute the embedded customer-related cost components. Thus, the associated embedded demand-related cost components are the differences between total embedded costs and the customer-related components. The result of this analysis is the classification of distribution line transformer gross plant.

These transformer classification results can be used also to classify the costs of associated transformer installation facilities. For instance, concrete and fiberglass pads, also booked in FERC Account 368, serve as foundations for underground system line transformers. As they are an integral part of line transformer installations, the costs of the transformer pads should be separated between the customer and demand cost components at Level 5 in the same proportions as for the pad-mount transformers. Different transformer pads are used for single-phase transformers than for three-phase transformers; thus, both voltage categories of pad-mount transformers would be used for classifying the transformer pads.

Vault-type transformers are used typically in urban underground systems located in high density load areas where the installation of above ground facilities is impractical. A vault transformer is essentially an overhead type of transformer, mounted in a below grade concrete manhole or vault, to which primary and secondary underground cables are attached. A vault may contain one to several transformers and may include ventilation equipment for dissipating heat and a sump pump for discharging any water that may accumulate in the vault. Vaults and associated equipment are booked in FERC Account 366. Like transformer pads, vaults are a crucial part of the underground transformer installation. Thus, transformer vaults should be separated between the customer and demand components at Level 5 in the same proportions as for pole mounted transformers.

Cost-of-Service Methodology

Distribution Conductor Systems

The distribution system consists of single-phase and three-phase lines, which are operated at primary and secondary voltages. Aerial conductors are supported by poles, while underground cables are buried directly in earth as well as placed in conduits where extra physical protection is necessary. Primary voltage submarine cables are used for underwater crossings.

Conductors form a voltage path connecting customers' loads to a source: secondary conductors link customer service lines to line transformers while primary lines link line transformers to distribution substations. As with a line transformer, the connectivity attribute of a distribution conductor is independent of load or energy usage and therefore has a customer-related component of cost. In addition, a conductor transports electric current and thus has a load-bearing capability, referred to as ampacity, which indicates a demand-related component of cost. As a result, the zero-intercept analysis is applicable to the primary and secondary conductor systems for separating out the associated customer and demand-related components of cost.

As exhibited in Table 5.4, a large array of conductor types and sizes is typically in service within a utilitywide distribution system. The costs of overhead conductors are booked in FERC Account 365, while underground conductor costs are booked in FERC Account 367. However, the level of detail needed to successfully conduct a system rebuild analysis may not be present in the plant accounting records. For instance, the book costs of conductors are not always differentiated between primary and secondary voltage applications, and conductor sizes are sometimes grouped by ranges, e.g., 1/0 to 4/0. If available, a detailed inventory of in-service conductors can be produced from an AM/FM system query. Otherwise, field samples of the distribution system can be performed to estimate conductor types, sizes, and lengths on a utility-wide basis. Current costs are applied to the conductor inventory either as a cost per unit of length for material plus a cost per span for labor or as a combined cost per unit of length for both material and labor (i.e., an average span length is assumed). The zero-intercept analysis can be simplified by selecting representative conductor types for:

- Overhead conductors: primary and secondary
- Underground conductors: primary and secondary

Overhead Conductors

For the example regression analysis of the aerial conductor system, bare ACSR has been selected on behalf of the primary system, weatherproof (covered) aluminum has been selected on behalf of the open-wire secondary system, and aluminum duplex cable has been selected on behalf of the n-plex secondary system. Conductor MCM was utilized in place of conductor ampacity as the independent variable for the regression examples because it yielded the better r^2 result From a perspective of load bearing capability, conductor size is a reasonably linear proxy for conductor ampacity. The regression results are shown in Figure 11.15.

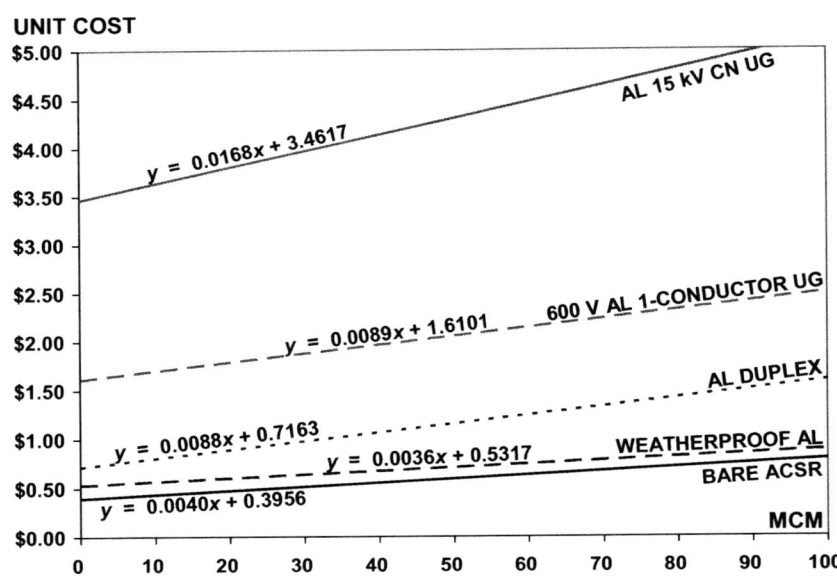

Figure 11.15 A Zero-intercept analysis for primary and secondary overhead and underground conductors. The analysis is based on current unit costs in $ per foot of selected conductor types. The installed feet of each conductor size is used for the regression weighting factors. Note that duplex consists of both a phase wire and a neutral wire.

In order to determine the customer component of cost for overhead conductors, the y-intercept is applied to the distribution line length on a completed circuit basis, i.e., a single-phase route extending from the substations (line transformers) to the end of the primary (secondary) lines along with a neutral return. Simply stated, the distance measurements of all of the primary lines and all of the open-wire secondary lines in terms of "pole miles" are each doubled to determine the minimum distribution primary and secondary circuit lengths. However, twice the pole mile route would overstate the lengths of the minimum distribution circuits in the case where line construction practice allows the primary and secondary systems to share a common neutral.

A simplified single-line schematic diagram of the entire system of overhead primary and secondary distribution circuits, which incorporates a partial sharing of the neutral, is shown in Figure 11.16. As illustrated, not all of the pole miles are comprised of both primary and secondary lines. With a few AM/FM system measurements, or estimates based on a sufficient field sampling of the conductor system, the primary and secondary minimum distribution circuits lengths can be determined.

Cost-of-Service Methodology

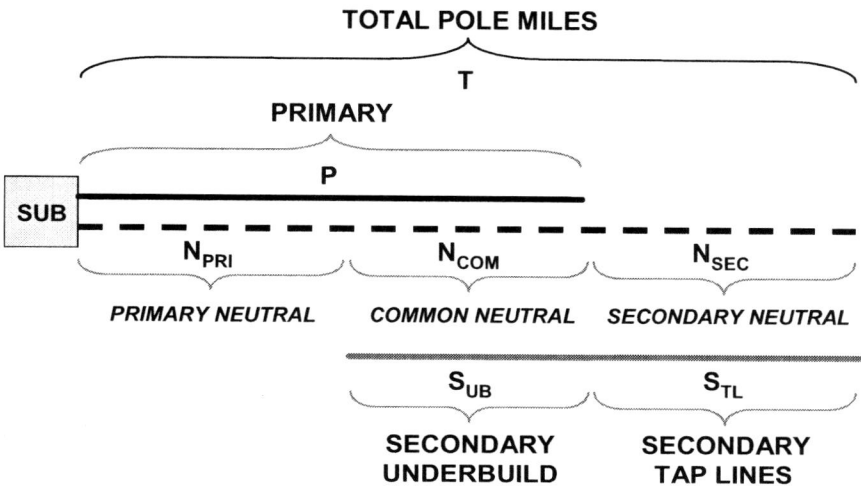

Figure 11.16 A simplified equivalent circuit diagram of the overhead distribution system. Shown are the primary, secondary, and common segments of the neutral conductor which runs the full length of the primary and secondary circuits. The primary and secondary lines of this example are assumed to be of a grounded wye configuration throughout.

Given: Total pole miles of lines = T = 4,950 miles
Total primary pole miles = P = 2,975 miles
Total secondary pole miles = S = 2,550 miles
Categorization of secondary:
Secondary open wire pole miles = S_{OW} = 190 miles (7.45%)
Secondary n-PLEX pole miles = $S_{N\text{-}PLEX}$ = 2,360 miles (92.55%)

Thus: Secondary tap lines = S_{TL} = $T - P$
= 4,950 − 2,975 = 1,975 miles
Secondary underbuild = S_{UB} = $S - S_{TL}$
= 2,550 − 1,975 = 575 miles

The total length of the neutral conductor (N) is equal to the total pole miles (T), which is 4,950 miles. The length of the secondary segment of the neutral (N_{SEC}) is equal to the pole miles of secondary tap lines (S_{TL}), which is 1,950 miles, while the length of the common neutral segment (N_{COM}) is equal to the miles of secondary underbuild (S_{UB}), which is 575 miles. Thus, the length of the primary segment of the neutral is determined to be:

Primary neutral = N_{PRI} = $N - N_{SEC} - N_{COM}$
= 4,950 − 1,975 − 575 = 2,400 miles

The minimum distribution circuit length for the overhead primary system, considering a common neutral with the overhead secondary, is determined by:

$$MDSL_{PRI} = P + [N_{PRI} + \tfrac{1}{2}N_{COM}] \qquad (11.9a)$$

For this example, $MDSL_{PRI}$ = 2,975 + 2,400 + 0.5(575) = 5,662.50 miles or 29,898,000 feet.

The determination of the minimum distribution for the secondary system is slightly more complicated since the conductors consist of both open-wire and n-plex cable construction. Two spans of open wire are necessary to complete the minimum distribution circuit, whereas only one span of duplex is required given that the cable incorporates both a covered phase wire and a bare neutral wire. The per unit costs of duplex, and thus the y-intercept, are for both wires together.

The minimum distribution circuit length for the overhead open-wire, secondary system, considering a common neutral with the overhead primary, is determined by:

$$MDSL_{SEC\text{-}OW} = S_{OW} + [N_{SEC} + \tfrac{1}{2}N_{COM}] \times \%S_{OW} \qquad (11.9b)$$

For this example, $MDSL_{SEC\text{-}OW}$ = 190 + (1,975 × 0.0745) + 0.5(575 × 0.0745) = 358.56 miles or 1,893,177 feet.

The minimum distribution circuit length for the overhead n-plex, secondary system, considering a common neutral with the overhead primary, is determined by:

$$MDSL_{SEC\text{-}N\text{-}PLEX} = S_{N\text{-}PLEX} + [N_{SEC} + \tfrac{1}{2}N_{COM}] \times \%S_{N\text{-}PLEX} \qquad (11.9c)$$

For this example, $MDSL_{SEC\text{-}N\text{-}PLEX}$ = 2,360 + (1,975 × 0.9255) + 0.5(575 × 0.9255) = 4,453.94 miles or 23,516,823 feet. Note that half of $MDSL_{SEC\text{-}N\text{-}PLEX}$ represents the actual length of the n-plex system, and this amount must be used when applying the y-intercept cost of the duplex.[36] As a check, the sum of $MDSL_{PRI}$, $MDSL_{SEC\text{-}OW}$, and $MDSL_{SEC\text{-}N\text{-}PLEX}$ should be equal to the sum of P, S, and N, which is 10,475 miles in this example.

Underground Conductors
Underground cables are designed and manufactured to a voltage class specification. Common distribution ratings include 600 volt, 5 kV/8 kV, 15 kV, 25 kV, and 35 kV. Accounting records typically differentiate underground cable costs and lengths by voltage class, thus providing a level of detail adequate for distin-

---

[36] Unlike open-wire secondary conductors, the y-intercept of the duplex conductor regression represents the zero-intercept cost of two conductors per unit of length.

Cost-of-Service Methodology

guishing between primary and secondary distribution applications. For the regression analyses of the underground conductors system, unjacketed, 15 kV aluminum URD concentric neutral (CN) type cable has been selected on behalf of primary, and 600 volt single-conductor aluminum UD type cable has been selected on behalf of secondary. The analysis results are shown in Figure 11.15. As indicated in the chart, underground conductors are generally more costly than their overhead counterparts, particularly with respect to higher voltage applications. Cable installation also includes the costs of boring under paved surfaces and trenching and backfilling.

Unlike aerial conductor systems, underground primary and secondary systems do not share a neutral even though the two conductor systems may at times be placed in a common trench or duct bank. Often a neutral is integral to the cable design as with the primary URD cable used in the example. With such cables, the minimum distribution system distance is simply equal to the trench length of the cables. The low-voltage, single-conductor cable used in the example requires a separate neutral conductor; thus, the associated minimum distribution system distance is twice the trench length of the secondary underground system.

Conductor Classification

The calculations for the classification of overhead conductors in FERC Account 365 and underground conductors in FERC Account 367 are shown in Table 11.9.

Table 11.9a Results of the Distribution Conductor Classification Example

Conductor Type	Number of MDS Units (000s of ft)	y-Intercept Unit Costs ($ per ft)	Current Cost Rebuild Basis		
			Customer Cost $(000s)	Total Rebuild Cost $(000s)	Percent of Total
Primary					
Bare ACSR OH	29,898	0.3956	11,828	56,416	20.97%
15 kV CN UG	4,213	3.4617	14,584	25,396	57.43%
Secondary					
WP OH	1,893	0.5317	1,007	2,623	38.39%
Duplex OH	11,758	0.7163	8,422	26,837	31.38%
1-Conductor UG	6,422	1.6101	10,340	17,044	60.67%
Voltage Level Allocators					
OH Primary				56,416	95.56%
OH Secondary*				2,623	4.44%
Total OH				$59,039	

* Excluding n-plex cables.

Table 11.9b Results of the Distribution Conductor Classification Example

Conductor Type	Percent of Total Rebuild Costs	$ Embedded Costs (000s)		
		Total Book Cost	Customer Component	Demand Component
FERC Acct. 365		$62,087		
OH Primary	20.97%	41,302	8,661	32,641
OH Secondary*	38.39%	1,919	737	1,182
OH n-plex	31.38%	18,866	5,920	12,946
FERC Acct. 367		$32,169		
UG Primary	57.43%	17,877	10,267	7,610
UG Secondary	60.67%	14,292	8,670	5,622

* Excluding n-plex cables.

Note that the FERC 365 cost for overhead, open-wire conductors (i.e., net of n-plex) has been allocated to voltage level. OH Primary = ($62,087 - $18,866) × 95.56% = $41,302. OH Secondary = ($62,087 - $18,866) × 4.44% = $1,919.

The primary voltage-related costs are assigned to Level 4, while the secondary voltage-related costs are assigned to Level 5. The total book cost amounts shown in Table 11.9b represent only the conductor portions of FERC Accounts 365 and 367. These FERC Accounts also contain the costs of other line equipment, including switches and switchgear, reclosers, and sectionalizers (reference Table 11.7).[37] As discussed previously, these particular devices are classified 100% as customer cost related, and they would be assigned to Level 4 as they are operated at a primary voltage.

Other items in FERC Accounts 365 and 367 are classified on the basis of the conductor classifications. For example, potheads would be classified based on primary underground cable as they are used for primary level terminations, and secondary pedestals would be classified based on secondary underground cable.

FERC Account 366 contains a variety of conduit and duct lines that are used to protect primary and secondary underground conductors. Thus, the primary and secondary underground conductor classifications are applied to the book costs in order to classify the conduit and duct line portion of costs in FERC Account 366 to the customer and demand related costs at Levels 4 and 5.

---

[37] The items specified in FERC Accounts 365 and 367 include lightning arrestors, which are utilized for line protection. However, utility plant accounting practice may consolidate these costs in FERC Account 368 – Line Transformers along with the lightning arrestors used for equipment protection.

Poles and Aerial Conductor Supports

Poles of various heights and classes are used to support overhead conductors, transformers, and other distribution equipment. As discussed previously, conductors create a voltage path to customers; thus, a major function of the poles is to support the voltage path. The placement of poles is based on the geographical dispersion of customers. While distribution poles are spaced generally a few hundred feet apart along an express corridor, poles are also placed typically at customer lot lines so that a service line of a customer does not cross over the property of an adjacent customer. Thus, more poles are required to reach customers in rural settings due to long distances (low customer density), but more poles are also needed in higher density urban and suburban settings.

The height of a pole relates primarily to safety requirements. A pole line follows the terrain; thus, taller poles are required when a line traverses a dip in the terrain, as shown in Figure 11.17. Otherwise, the uplift forces on the insulators could cause a conductor to break free of its attachment point. In addition, pole height is critical to maintaining electrical safety clearances, such as those prescribed in the National Electrical Safety Code (NESC). To maintain such clearances, taller poles are required for crossing elevated roadways and railroad tracks. Taller poles may also be required for maintaining clearances from other electrical cables attached to the same pole, including telephone and cable television lines. The standard urban pole may need to be a size larger than its rural counterpart because of these additional utility lines and other attachments, such as outdoor lighting fixtures and fiber optics lines.[38] All such height increases are related to safety and are independent of load or energy usage.

The standard distribution pole may need to be upsized to the next available dimension of height (+ 5 feet) in order to maintain proper clearances when line equipment is to be installed. Based on the particular type of equipment, the increase in height requirement is driven by a customer-related device (recloser or sectionalizer), a demand-related device (capacitor or voltage regulator), or a device that is both customer related and demand related in nature (line transformer). Most upsizing of pole height for equipment involves a few pole sizes above the standard size pole, and a number of those increases are due to clearance purposes, as discussed above. The larger range of distribution pole heights are needed primarily for clearance purposes and most often can accommodate equipment installations without additional height increases. Overall, a portion of the cost of a pole is related to the customer component.

Pole circumference on the other hand must be increased from the standard specification with heavy equipment installations. As shown in Figure 11.18, pole class requirement is a function of transformer capacity. Transformer weight is directly proportional to capacity due to greater core mass, a greater volume of insulating material, and a larger tank.

---

[38] "Standard pole" as used here is in reference to the utility's specification for normal primary voltage construction, e.g., a 40 foot, class 5 pole.

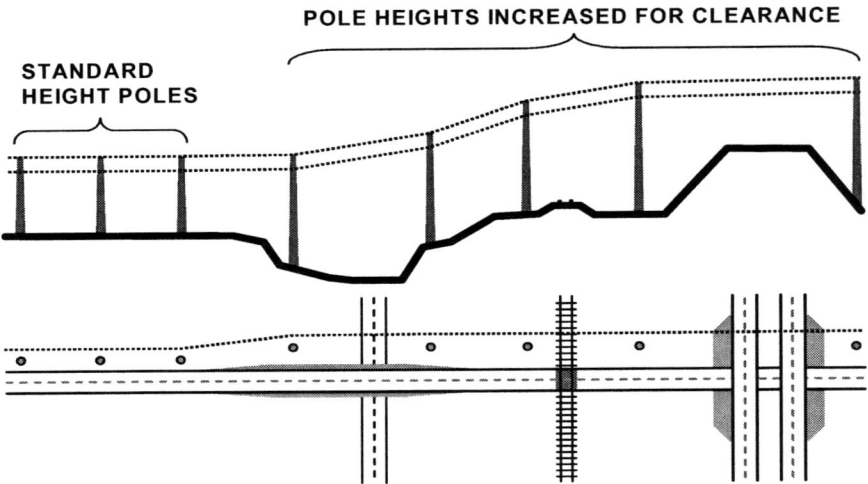

Figure 11.17 A plan and profile drawing of a distribution line. Lines are graded as to the terrain to prevent uplift stresses on insulators and to maintain proper clearances. Such pole height increases are independent of load and are thus predominantly customer related.

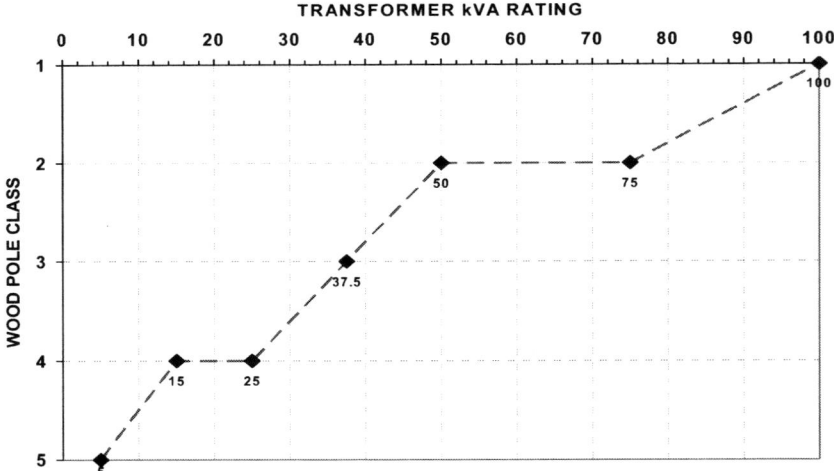

Figure 11.18 A plot relating wood pole class requirements to transformer kVA ratings. Transformer weight increases as a function of capacity; thus, a larger pole class is required to support a greater weight. Source: *Power Distribution Systems*, U.S. Department of Defense, UFC 3-550-03N, 16 January 2004.

Cost-of-Service Methodology

Greater pole classes are also required for supporting larger, heavier conductors, higher tension conductors, and a greater number of conductors (i.e., multiple primary voltage circuits or three-phase vs. single-phase circuits). Such circumstances are indicative of the load bearing function of aerial conductors. Thus, a portion of the cost of a pole is related to the demand component.[39]

Taken as a whole, pole height is found to be predominantly correlated with nonload-bearing conditions, while pole class is found to be predominantly correlated with load-bearing conditions. Extraction of a minimum distribution pole cost component can be accomplished in a method comparable to conductors. With conductors, the regression variables are cost per unit of length as the y-variable and MCM as the x-variable. MCM represents the cross-sectional area of a conductor, which is indicative of its load bearing capability. The length of the conductor is independent of load. In a similar manner, the cost per unit of pole height serves as the y-variable. However, while the cross-sectional area of wood poles varies with pole class, the cross-sectional area of concrete and steel poles may not vary with pole class. For example, a 40-foot pre-cast concrete pole could have the same exact tip and butt dimensions for classes 1-4, which then eliminates cross-sectional area as the x-variable for the linear regression. However, each of these concrete pole classes have different *ground-line moment capacities* (GLMC) based on a transverse load applied two feet from the tip of the pole.[40] The GMLC is indicative of mechanical load bearing capacity, which in turn is indicative of electrical load bearing capability, e.g., higher capacity electrical equipment.[41]

While there are numerous sizes of poles, it is not essential to develop a regression analysis for each and every pole size. Calculations of the zero-intercepts for the most frequently utilized wood, concrete, and/or steel poles produce reasonable results which can then be applied to the whole populations of wood, concrete, and/or steel poles. Poles used strictly for supporting secondary voltage conductors should be identified, if possible, as their costs would be assigned to Level 5, while poles supporting primary voltage conductors, regardless of any secondary underbuild conductors, would be assigned to Level 4. The smaller poles booked in FERC Account 364 would generally be used in secondary voltage applications (as they lack sufficient heights to meet primary voltage clearance requirements).

---

[39] Taller poles require greater circumference dimensions just to support themselves safely (without consideration of conductors, equipment, or other attachments). As a result, the minimum pole class available for a given pole height increases with taller poles. This aspect of circumference or class is independent of load.

[40] ANSI 05.1 specifies horizontal loadings for each standard wood pole class. For concrete and steel poles, the NESC prescribes an ultimate strength of 62.5% for the comparable height and class of wood poles. The moment is equal to the product of force and distance.

[41] Since poles are not electrical components like conductors a mechanical proxy for electrical capacity is needed to identify the customer- and demand-related costs.

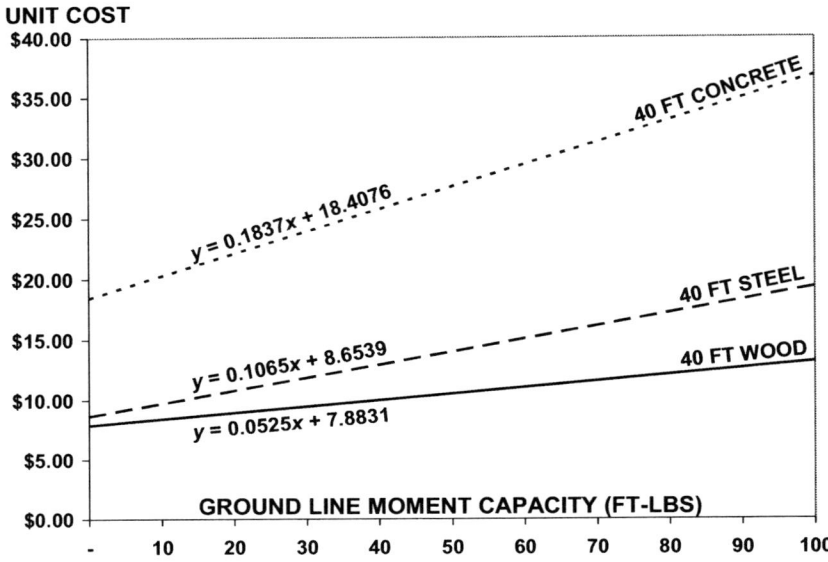

Figure 11.19 Zero-Intercept analysis for forty foot wood, concrete, and steel poles. The analysis is based on ground-line moment capacity as a proxy for the electrical load-bearing capability of the poles. The ground-line moment capacity is calculated for each of the various pole classes.

The results of the zero-intercept analyses for example wood, concrete, and steel poles, based on rebuild costs, are shown in Figure 11.19. The unit costs represent the $ per foot of pole height, and the regression weighting factors are based on the pole inventory for each class of 40 foot pole. The y-intercept corresponds to the per unit cost of a forty foot pole having no mechanical strength and thus, by proxy, no electrical load bearing capability. The y-intercept values are applicable to both primary and secondary voltage system poles. The y-intercepts are multiplied by the total lengths of poles in order to determine the customer components of cost, as shown in Table 11.10. The total pole length is determined by summing the products of the total inventory of poles for each pole height, regardless of class, and the respective heights, e.g., the total length for forty foot wood poles: 40 feet × 55,643 forty foot poles = 2,225,720 feet. The total length of the population of wood poles is 5,579,390 feet.[42]

---

[42] Both standard and ornamental poles, dedicated for street and outdoor lighting purposes, are booked in FERC Account 373 and are excluded from the minimum distribution system analysis.

Cost-of-Service Methodology

Table 11.10a Results of Distribution Pole Classification Example

Pole Type	Number of MDS Units (ft)	Current Cost Rebuild Basis			
		y-Intercept Unit Costs ($ per ft)	Customer Cost $(000s)	Total Rebuild Cost $(000s)	Percent of Total
Wood	5,579,390	7.8831	43,983	66,255	66.38%
Concrete	19,405	18.4076	357	755	47.28%
Steel	10,875	8.6539	94	164	57.32%
Total			**$44,434**	**$67,174**	

Table 11.10b Results of Distribution Pole Classification Example

Pole Type	Percent of Total Rebuild Costs	$ Embedded Costs (000s)		
		Total Book Cost	Customer Component	Demand Component
Primary				
Wood	66.38%	31,599	20,975	10,624
Concrete	47.28%	467	221	246
Steel	57.32%	112	64	48
Total Primary		**32,178**	**21,260**	**10,918**
Secondary				
Wood	66.38%	10,551	7,004	3,547
Concrete	47.28%	4	2	2
Steel	57.32%	12	7	5
Total Secondary		**10,567**	**7,013**	**3,554**
Total Poles		**$42,745**	**$28,273**	**$14,472**

A variety of pole hardware and facilities are used to attach line conductors to distribution poles. Major pole attachments recorded in FERC Account 364 are shown by voltage level application in Table 11.11. Plant account records may group these items into a single *fixtures* category. As the majority of pole fixture investment is related to the support of primary voltage conductors, the customer and demand components previously determined for overhead primary conductors can be used to classify pole fixture investment at Level 4.

Table 11.11 Typical Pole Fixtures Recorded in FERC Accounts 364

	Primary – Level 4	Secondary – Level 5
Pole Top Pins	X	
Crossarms	X	
Standoff Brackets	X	
Clevis (for neutral & n-plex)	X	X
Clevis Racks (for open wire)		X
Down Guys	X	X
Anchors	X	X

Insulators and ground rods are booked in FERC Account 365. Insulators are used predominantly for supporting high voltage conductors, and thus the associated investment can be classified on the basis of overhead primary conductors. A copper ground rod driven into the earth at the base of a pole is used for grounding the neutral, transformers, and other line equipment. The ground rod is connected to a copper wire that is stapled to the side of the pole, and in essence becomes an integral part of the pole. As the majority of the investment in ground rods is related to the primary distribution system, ground rods can be classified on the same basis as primary poles.

Line Equipment and Facilities
As discussed previously, the mission or function of a particular distribution system device provides a guide as to its classification as a demand and/or customer component of cost. The classification of common distribution line devices is shown in Table 11.12. Classification of these devices is accomplished either by direct designation as customer or demand related or on the basis of a previously classified distribution component to which the device is associated.

Regulators and capacitors support line voltage magnitude under heavy loading conditions and thus are designated as demand-related investments. System protection devices, including reclosers, sectionalizers, and inline fused cutouts, function to maintain line voltage continuity throughout the primary distribution circuit during a fault, regardless of the system's loading conditions at the time, and thus are designated as customer-related investments. Likewise, lightning arresters protect lines during over-voltage surges and thus are designated as customer-related investments.

Line switches are operated to reconfigure distribution circuits to better balance substation loading, which is a demand-related function. In addition, line switches are operated to maintain voltage continuity to customers from an alternate substation and/or feeder source when the normal service requires an outage for maintenance or repair. As a result, line switches can be classified in the same manner as primary conductors.

Cost-of-Service Methodology

Table 11.12 Classification of Line Equipment and Facilities

	Primary – Level 4		Secondary – Level 5	
	Demand	**Customer**	**Demand**	**Customer**
Regulators	X			
Capacitors	X			
Reclosers & Sectionalizers		X		
Cutouts for:				
• OH Line Transformers			X	X
• Line Protection		X		
• Capacitors	X			
Arresters for:				
• OH Line Transformers			X	X
• Line Protection		X		
• Capacitors	X			
• Regulators	X			
• Reclosers & Sectionalizers		X		
OH & UG Line Switches	X	X		
Bypass Switches for:				
• Regulators	X			
• Reclosers & Sectionalizers		X		

Cutouts and arresters are also used for line equipment protection and therefore should be classified in the same manner, e.g., arresters installed for capacitor protection are demand related, while arresters installed for recloser protection are customer related. Cutouts and arresters become an integral part of an overhead transformer and are thus both customer and demand related.[43] Bypass switches are installed with regulators, reclosers, and sectionalizers; thus, the classification of these devices follows accordingly.

Street and Outdoor Lighting Facilities

Unlike the other power classes of electric service, lighting is a specific end-use based electric service. Lighting service is provided to governmental entities for the illumination of streets and highways, major thoroughfare interchanges, bridges, parks, and athletic fields. Lighting service is also provided to individual customers for special applications, such as billboards, and for area lighting, such as around a building, a private drive, a parking lot, and other outdoor venues.

---

[43] This approach treats the ensemble of the transformer and its protection devices like a CSP transformer. Note however that a CSP transformer located in a high fault current area may require an auxiliary current limiting fuse.

Street and outdoor lighting is generally an unmetered service.[44] The energy usage is specific to the particular lighting technology and lamp rating and the load of the ballast, and monthly kWh consumption can be calculated based on the connected load and the hours of darkness for the location. Customers may have the option to install and own their own lighting systems and purchase only the maintenance and/or energy services from the utility. Otherwise the monthly rate also includes pole and/or fixture charges.

Street and outdoor lighting facilities consist of luminaries, mounting brackets, and poles.[45] Luminaries are available in a variety of technologies, and they may be simple or highly ornamental. Poles may also be relatively plain, like basic distribution poles, or highly ornamental. Lighting poles are made of wood, concrete, galvanized steel, aluminum, and fiberglass. Multiple lighting fixtures can be mounted on a common pole. Street lighting applications often require a minimal number of poles as the luminaries can be mounted on existing power service poles that are typically located along the right-of-way of a road. In addition, lighting fixtures may be connected to an underground distribution system.

Street and outdoor lighting luminaries and poles are booked in FERC Account 373 – Street Lighting and Signal Systems. Other equipment and facilities used specifically for providing lighting service, including conductors and transformers, are also booked in FERC Account 373. From a cost functionalization perspective, lighting facilities are a part of distribution. From a cost allocation view, the investment in lighting facilities represents a distinctive direct assignment. For cost classification purpose, lighting fixtures and poles represent a fixed cost and thus a customer-related component.

Classification of Substations, Transmission, and Production Plant

Distribution substations provide a voltage transformation between the transmission or subtransmission system and the primary distribution system. This connection between different voltage classes is conceptually the same as the connection provided by line transformers between the primary system and the secondary system. As discussed previously, a line transformer is installed to serve either a single customer or just a few adjacent customers at a specific site, and thus its connectivity or voltage path characteristic is recognized as being customer related. The line transformer's demand-related characteristic, capacity, is sized on the basis of the specific load requirements of the one to a few customers.

In contrast, a common distribution substation is installed to aggregate and connect a large number of customer loads to the transmission system, and its trans-

---

[44] Special lighting applications, such as billboards and traffic signals, are often metered and served under either a small general service rate schedule or a rate designed on the basis of the particular end-use load and its operating characteristics.

[45] Street and outdoor lights are controlled by photocells which sense the transitions between daytime and nighttime; thus, lighting is a natural off-peak service.

Cost-of-Service Methodology   519

former is sized to serve the loads en masse on an area-wide basis. The collective diversified peak load characteristics of the customers, not the number of customers connected, is the cause for substation transformer capacity sizing. Compared to a line transformer from which a small number of similarly sized customers are served, the individual loads of the huge amount of customers served by the substation will vary by several orders of magnitude, i.e., from 1 kW to 10 MW or more. Two substations each having the same total count of customers of customers could have significantly different loads due to the difference in mix of customers served. Siting of a line transformer is based more on the location of one or a few adjacent customers, whereas siting of substation capacity is based more on the location of the load center as opposed to where individual customers are situated. Consequently, the notion of a customer-related component becomes overwhelmed when applied to a distribution substation, and thus distribution substations are recognized as being totally demand-related investments.

In a similar manner, transmission system capacity is sized to transport bulk power to the distribution substation points, and the sheer number of customers which comprise the load is irrelevant to meeting the overall capacity requirement. Therefore, the transmission system investment, including subtransmission lines and transmission class substations, are recognized as being totally demand-related investments.

Production plant facilities are generally deemed to be demand related while fossil and nuclear fuel stocks are classified as energy related. However, some cost assignment methodologies treat production plant as having an element of cost that is a function of energy. The previously discussed B-I-P production subfunctionalization process allocates a portion of total plant investment cost to a base period which traverses all hours of the year. The base period cost amount is then allocated to the customer or rate classes on the basis of the class average loads, which is the same as allocating on the basis of class energy (MW_{AVG} = MWh ÷ annual number of hours).

Another example is the *equivalent peaker* allocation methodology, which is based on a premise that recognizes that the lowest cost of generation capacity is attained through combustion turbines (CTs). However, CT fuel costs are much higher than other conventional generation methods. A more economical solution is found by substituting base load generation for a portion of the CT generation to realize lower fuel costs in order to optimize the total fixed and variable costs of production. With the equivalent peaker methodology, only the CTs and the CT-equivalent capacity cost portion of all other units are treated as demand related. The remainder of the total plant investment cost is thus treated as energy related.

Classification of Other Plant Items and Accumulated Depreciation

Within the functionalization process, general and intangible plant costs are allocated to the major functions of Production, Transmission, Distribution, Customer Accounts, Customer Assistance, and Sales typically by means of a plant or labor ratio. The subsequent classification of general and intangible plant for each of the six major functions is a direct extension of the functionalization logic. For example, if the production function's allocated share of general and intangible plant is determined by means of a functional O&M labor allocator, then the demand-related and energy-related portions of production O&M labor are then utilized to classify the production-related portion of general and intangible plant. The classification of functional O&M labor and expenses is discussed below.

Transmission is generally deemed to be a demand cost function. Thus, in accordance with transmission gross plant, the transmission function's share of the allocated general and intangible plant is generally considered to be demand related and can be directly assigned as such.

If the distribution system's share of general and intangible plant is determined also by means of a functional O&M labor allocator, then that functionalized amount can be classified by means of the composite results of the distribution O&M labor classification, as derived from the minimum distribution system analysis, in order to carry forward a consistent cost treatment. The Customer Accounts, Customer Assistance, and Sales functions are each designated completely as a customer-related function; thus, their respective shares of the allocated general and intangible plant are classified directly to the customer cost component.

The separation of Plant Held for Future Use and CWIP into the classification cost components is also achieved in a fairly straightforward manner. For the production and transmission functions, these other plant items are treated in a manner similar to the classification of the associated gross electric plant. The distribution function's component of Plant Held for Future Use and CWIP identified for distribution substations is classified entirely as a demand component of cost, which follows specifically from the direct assignment of distribution substation gross plant to demand. On the other hand, the classification of the distribution function's share of Plant Held for Future Use and CWIP for distribution lines is made between the customer and demand cost components. A compound allocator fashioned from the minimum distribution system analysis results of FERC Accounts 364 - 367 could be used to separate these costs between the customer and demand components. The Customer Accounts, Customer Assistance, and Sales functions' shares of Plant Held for Future Use and CWIP would be classified simply as customer component costs.

If applicable, the make up of Plant Held for Future Use and CWIP may also include some cost elements associated with common or general plant. As such, each of the six major functions could be allocated a portion through the functionalization process. Similar to above, the production function's share would be separated between the demand and energy components based on the functionalization logic, e.g., an O&M labor ratio. The transmission, and distribution substation functions' shares would be designated as demand components. The distribution line's share would be separated between customer and demand components on a basis compatible with the logic used for functionalization, while the three customer functions' shares would be designated as customer components of cost.

Cost-of-Service Methodology

Accumulated depreciation is apportioned on the basis of the gross plant classification results. Once again, the production function's share would be separated between the demand and energy components, the transmission and distribution substation functions' shares would be designated as demand components, the distribution line's share would be separated between customer and demand, and the three customer functions' shares would be designated as customer components of cost. Minimum distribution system analysis results could be used for allocating accumulated depreciation between the customer and demand components for each level of detail, e.g., single-phase, overhead line transformers.

Classification of Rate Base Adjustments

Rate base addition and deduction items are classified to cost components in general accordance with the functionalization process. For example, if cash working capital (CWC) is functionalized on the basis of O&M expenses less fuel and purchased power, then an O&M expense ratio developed for the cost components in each function would be a consistent application for classification. Thus, CWC would be classified between demand and energy for the production function, between customer and demand for the distribution function, and simply as demand for transmission given that customer and energy components are not typically defined for that function. The three customer functions would also treat their associated shares of CWC as being entirely customer cost components.

Production fuel stocks and scrubber reactant are classified directly to the energy cost component; however, classification of several other plant materials and supplies requires cost separation. For example, if salaries and wages (S&W) are used to functionalize general inventory items such as office supplies, then O&M and A&G labor is classified first for each of the major functions, and the results are then used to develop functional S&W ratios. The application would produce cost components in the same categories as the CWC classification example discussed above. Classification ratios developed from the minimum distribution system analysis can be used to classify various distribution materials. For example, insulator inventory could be classified on the same basis as determined for the overhead conductor classification.

A functional S&W allocator is a natural extension for classifying employee related operating reserves, such as the provisions for Injuries and Damages and Pensions and Benefits. Associated plant ratios would be used for classifying rate base adjustments which are functionalized on the basis of plant, such as the Provision for Property Insurance reserve, Accumulated Deferred Investment Tax Credits and Accumulated Deferred Income Taxes.

Classified Rate Base Summary

An example summary of a classified rate base is shown in Table 11.13. The amounts shown correlate to the amounts included in the functionalized rate base summary shown in Table 11.4

Table 11.13 Example Summary of Rate Base Classification

FERC		Total $ (000s)	Cost Components	
Production			Demand	Energy
310 - 346	Production Plant	907,440	896,264	11,176
	Fuel Stock (Energy Component)	51,012	--	51,072
	Production Rate Base Additions	47,174	27,223	19,952
	Production Rate Base Deductions	165,002	161,844	3,157
Transmission			Demand	
350 - 359	Transmission Lines and Subs	84,629	84,629	--
	Transmission Rate Base Additions	4,359	4,359	--
	Transmission Rate Base Deductions	70,441	70,441	--
Distribution			Demand	Customer
360 - 363	Distribution Substations	36,698	36,698	--
	Primary Lines and Equipment			
364	Poles	124,915	64,821	60,094
	Pole Hardware and Fixtures	20,556	16,267	4,288
365	1φ and 3φ OH Conductors	27,230	21,549	5,680
	OH Switches	1,459	1,155	304
	Reclosers and Sectionalizers	2,080	--	2,080
366	Conduit and Cable Manholes	4,423	1,890	2,533
367	1φ and 3φ UG Conductors	12,511	5,345	7,166
	UG Switchgear	303	130	174
368	Voltage Regulators	489	489	--
	Capacitors	2,062	2,062	--
	Cutouts – Pri. Equipment Protection	267	267	--
	Cutouts – Pri. Line Protection	4,323	--	4,323
	Arrestors – Pri. Equipment Protection	502	294	208
	Arrestors – Pri. Line Protection	3,637	--	3,637
	Line Transformers			
368	1φ OH Transformers	28,271	16,582	11,689
	3φ OH Transformers	2,432	2,066	366
	Cutouts - Transformers	10,280	6,248	4,032
	Arrestors - Transformers	9,713	5,904	3,810
	1φ UG Transformers	2,987	1,159	1,828
	3φ UG Transformers	7,312	4,831	2,482
	1φ UG Transformer Pads	1,035	607	428
	3φ UH Transformer Pads	292	249	44

Note: Amounts represent total net electric plant.

Cost-of-Service Methodology

Table 11.13–Cont. Example Summary of Rate Base Classification

FERC		Total $ (000s)	Cost Components	
Distribution – Continued.			Demand	Customer
	Line Transformers - Continued			
366	1φ Transformer Vaults	140	82	58
	3φ Transformer Vaults	1,159	931	228
	Secondary Lines and Equipment			
364	Poles and Hardware	7,238	2,462	4,776
365	1φ and 3φ OH Conductors	11,175	7,671	3,504
366	Conduit and Cable Pedestals	1,175	467	708
367	1φ and 3φ UG Conductors	9,780	3,895	5,905
	Service Lines			
369	1φ OH Services	18,727	--	18,727
	3φ OH Services	341	--	341
	1φ UG Services	10,108	--	10,108
	3φ UG Services	305	--	305
	Revenue Meters			
370	1φ Watt-hour Meters	6,866	--	6,866
	3φ Watt-hour Meters	5	--	5
	1φ Demand Meters	2,104	--	2,104
	3φ Demand Meters	574	--	574
	Low Voltage CTs	806	--	806
	High Voltage CTs	79	--	79
	Low Voltage PTs	118	--	118
	High Voltage PTs	154	--	154
	Street Lighting			Specific
373	Fixtures and Standards			14,233
	Distribution Rate Base Additions	12,339	6,684	5,654
	Distribution Rate Base Deductions	72,746	36,083	36,663
Customer Services				Customer
	Customer Accounts	3,119	--	3,119
	Customer Assistance	453	--	453
	Customer Sales	130	--	130
	Customer Svcs. Rate Base Additions	1,524	--	1,524
	Customer Svcs. Rate Base Deductions	2,566	--	2,566
TOTAL RATE BASE		**$1,139,845**		

Classification of Expenses

Like plant and other investment items, operating expenses are classified by the customer, demand, and energy cost components. Furthermore, the classification process for expenses is a general extension of the associated functionalization logic. Thus, if an expense item is functionalized on the basis of O&M labor, then the application of a cost component-based O&M labor ratio, specific to each of the six major functions, provides a consistent means for classifying each function's share of that expense item.

As with functionalization, a certain classification order is required. For instance, classification of O&M labor precedes classification of some A&G FERC Accounts. Both of these steps must be accomplished before a salaries and wages ratio can be developed for use in classifying various employee-related expense items.

Classification of O&M Labor and Expenses

Classification of O&M labor and expenses, booked in FERC Accounts 901 through 916, for the Customer Accounts, Customer Assistance, and Sales functions is straightforward since all such expenses are independent of the amount of demand or energy utilization by customers and thus are entirely customer cost related. For instance, the cost to read a residential customer's meter, render a monthly bill, and account for payment is the same given a registration of 1 kWh or 10,000 kWh.

The classification of distribution O&M labor and expenses is more involved as much of the O&M labor and expenses must be apportioned between the customer and demand-related cost components. Results from the minimum distribution system analysis can be utilized for this allocation. For example, the O&M labor and expenses for overhead lines (FERC Accounts 583 and 593) can be classified by means of a compound allocation factor derived from the classification results of poles (FERC Account 364) and conductors (FERC Account 365). O&M labor and expenses related to distribution substations are classified completely as demand-related costs, while O&M labor and expenses related to meters are classified completely as customer-related costs.

Distribution supervision and engineering O&M labor is classified as customer and demand-related costs based on the classification of the underlying O&M FERC accounts. For example, the labor booked in FERC Account 590 is allocated between customer and demand using the sums of the customer and demand labor costs determined for FERC Accounts 591 through 598. These O&M labor subtotals can also be used to classify the supervision and engineering related expenses, which are also booked in FERC Account 590. An example of the classification of distribution O&M labor and expenses is shown in Table 11.14.

Transmission O&M labor and expenses are booked in FERC Accounts 560 through 563. As with transmission plant investment, transmission O&M labor and expenses are generally considered to be demand-related costs.

Cost-of-Service Methodology

Table 11.14 Example Classification of Distribution O&M Labor and Expenses

FERC Account	Description	Total	$ Amount (000s) Customer	Demand
	Distribution Operation			
580	Operation Supervision and Engineering	2,506	1,138	1,368
581	Load Dispatching	312		312
582	Station Expenses	3,277		3,277
583	Overhead Line Expenses	8,937	3,456	5,481
584	Underground Line Expenses	3,712	2,119	1,593
585	Street Lighting and Signal Systems	578	(Excluded from total)	
586	Meter Expenses	2,266	2,266	--
587	Customer Installation Expenses	349	349	--
588	Miscellaneous Distribution Expenses	5,428	2,465	2,963
589	Rents	149		149
	Distribution Maintenance			
590	Maintenance Supervision and Engineering	397	174	224
591	Maintenance of Structures	958		958
592	Maintenance of Station Equipment	982		982
593	Maintenance of Overhead Lines	8,832	3,416	5,416
594	Maintenance of Underground Lines	1,568	895	673
595	Maintenance of Line Transformers	1,326	525	801
596	Maintenance of Street Lighting Systems	1,063	(Excluded from total)	
597	Maintenance of Meters	618	618	
598	Maintenance of Misc. Distribution Plant	129	129	
	Total Distribution O&M Expenses	**41,746**	**17,549**	**24,197**

Production O&M labor and expenses are classified as demand-related and energy-related cost components. While variations may exist, two basic methods have been utilized typically for classifying production O&M labor and expenses. These methods are referenced here as the "NARUC Method" and the "FERC Method."[46] In general, the NARUC Method treats many of the labor cost elements as being demand-related fixed costs, while treating expense cost elements (e.g., materials) as being energy-related variable costs. The FERC Method is an all-or-none predominance approach to classification. Thus, if more than half of a given production O&M FERC account is related to demand (energy) cost, then the whole account is considered to be a demand (energy) cost. A comparison of these two methods is exemplified in Table 11.15.

---

[46] The "NARUC Method" is summarized in the National Association of Regulatory Utility Commissioners's *Electric Utility Cost Allocation Manual*. The "FERC Method" is represented in various utility case decisions.

Table 11.15 Methodologies of Production O&M Classification

FERC Account		Total $ (000s)	FERC Demand	FERC Energy	NARUC Demand	NARUC Energy
	Steam Production O&M					
	Operation					
500	Supervision and Engineering	2,463	All	--	Labor*	Labor*
501	Fuel and Fuel Handling	198,817	--	All	--	All
502	Steam Expenses	28,077	All	--	Labor	Materials
503	Steam From Other Sources	0	--	All	--	All
504	Steam Transferred - Credit	0	--	All	--	All
505	Electric Expenses	7,036	All	--	Labor	Materials
506	Miscellaneous Steam Expenses	7,887	All	--	All	--
507	Rents	100	All	--	All	--
509	Allowances	0	--	All	--	All
	Maintenance					
510	Supervision and Engineering	3,203	--	All	Labor*	Labor*
511	Structures	3,581	All	--	All	--
512	Boilers	22,934	--	All	--	All
513	Electric Plant	6,908	--	All	--	All
514	Miscellaneous Steam Plant	4,336	All	--	--	All
	Nuclear Production O&M					
	Operation					
517	Supervision and Engineering	1,598	All	--	Labor*	Labor*
518	Fuel	5,553	--	All	--	All
519	Coolants and Water	565	All	--	Labor	Materials
520	Steam Expenses	2,909	All	--	Labor	Materials
521	Steam From Other Sources	0	All	--	--	All
522	Steam Transferred - Credit	0	All	--	--	All
523	Electric Expenses	227	All	--	Labor	Materials
524	Miscellaneous Nuclear Expenses	4,959	All	--	All	--
525	Rents	0	All	--	All	--
	Maintenance					
528	Supervision and Engineering	1,528	--	All	Labor*	Labor*
529	Structures	521	All	--	All	--
530	Reactor Plant Equipment	1,364	--	All	--	All
531	Electric Plant	1,074	--	All	--	All
532	Miscellaneous Nuclear Plant	576	--	All	--	All

* The Supervision and Engineering Account total is allocated between Demand and Energy based on the labor subtotal of the other FERC Accounts within the group, e.g., Account 500 Labor and Materials are allocated based on the total division of labor to Demand and Energy for Accounts 501 – 509.

Note: All = O&M Labor + O&M Materials

Table 11.15–Cont. Methodologies of Production O&M Classification

FERC Account		Total $ (000s)	FERC Demand	FERC Energy	NARUC Demand	NARUC Energy
	Hydraulic Production O&M					
	Operation					
535	Supervision and Engineering	126	All	--	Labor*	Labor*
536	Water for Power	0	All	--	All	--
537	Hydraulic Expenses	66	All	--	All	--
538	Electric Expenses	891	All	--	Labor	Materials
539	Miscellaneous Hydraulic Expenses	73	All	--	All	--
540	Rents	706	All	--	All	--
	Maintenance					
541	Supervision and Engineering	86	All	--	Labor*	Labor*
542	Structures	31	All	--	All	--
543	Reservoirs/Dams/Waterways	23	All	--	All	--
544	Electric Plant	118	--	All	--	All
545	Miscellaneous Hydraulic Plant	429	All	--	--	All
	Other Production O&M					
	Operation					
546	Supervision and Engineering	22	All	--	All	--
547	Fuel	229	--	All	--	All
548	Generation Expenses	16	All	--	All	--
549	Miscellaneous Other Expenses	57	All	--	All	--
550	Rents	30	All	--	All	--
	Maintenance					
551	Supervision and Engineering	6	All	--	All	--
552	Structures	3	All	--	All	--
553	Generating and Electric Plant	206	All	--	All	--
554	Miscellaneous Other Plant	1	All	--	All	--
	Other Power Supply O&M					
555	Purchased Power - Demand	1,261	All	--	All	--
	Purchased Power – Energy	5,029	--	All	--	All
556	System Control/Load Dispatching	1,313	All	--	All	--
557	Other Expenses	57	All	--	All	--
	Total O&M	$316,993	$39,312	$277,681	$69,660	$247,333
	Fuel Component			209,628		204,598
	Variable O&M Component			68,053		42,734

Classification of A&G Labor and Expenses

A&G labor and expenses are classified after the classifications of plant and O&M labor and expenses have been completed. As discussed previously, the different A&G cost elements are first functionalized to production, transmission, distribution, customer accounts, customer assistance, and sales typically using various functional O&M labor costs, O&M expenses, and plant ratios. Once the costs associated with these ratios are subsequently classified, then function-specific O&M labor, O&M expense, and plant ratios can be developed in order to complete the classification of the A&G cost elements for each major function.

For example, through the functionalization process, the distribution function is allocated a portion of total A&G salaries based on the ratio of distribution O&M labor to total company labor. Distribution O&M labor and O&M expenses are then each classified between demand and customer cost components (total distribution O&M classification results are shown in Table 11.14). The distribution-related A&G salaries can then be classified proportionally between demand and customer costs components using the distribution O&M labor cost classification results.

Classification of Depreciation Expense and Taxes Other Than Income Taxes

As discussed previously, annual depreciation expense typically is calculated and booked on a FERC plant account basis. As a result, the classification of the plant accounts can be used to classify the annual depreciation expense. For example, the depreciation expense associated with FERC Account 368 could be classified to the customer and demand cost components based on the classification results of all plant items booked in FERC Account 368, i.e., the minimum distribution system analysis results of line transformers, plus 100% of capacitors to demand, plus 100% of line protection cutouts to customer, etc.

Taxes other than income taxes are categorized generally as payroll related, property related, and revenue related. Functional payroll taxes can be classified using the classification results of functionalized salaries and wages. Functional property taxes can be classified using an appropriate classified plant basis, e.g., net total electric plant. Functional revenue taxes can not be accurately allocated to the customer, demand, and energy cost components within the classification process.

Classified O&M Expenses Summary

An example summary of classified O&M labor and expenses is shown in Table 11.16. The totals shown correlate to the totals included in the functionalized O&M labor and expenses summary shown in Table 11.6. Note that O&M labor and expenses for FERC Accounts 585 and 596 for Street Lighting and Signal Systems are included only in the total amount and not by cost component. These costs are directly assigned to the outdoor lighting class of customers and are not allocated by means of a customer, demand, or energy allocator as is the case of all other costs shown, except revenue taxes, which are allocated using revenues.

Table 11.16 Example Summary of Classified O&M Labor and Expenses

FERC Account	Description	Total	Energy	Demand	Customer
		\$ Amount (000s)			
500 – 509	Steam Production Operation	244,380	223,671	20,709	
510 – 514	Steam Production Maintenance	40,961	37,046	3,915	
517 – 525	Nuclear Production Operation	15,811	6,500	9,311	
528 – 532	Nuclear Production Maintenance	5,062	4,106	956	
535 - 540	Hydraulic Production Operation	1,862	447	1,415	
541 – 545	Hydraulic Production Maintenance	687	653	34	
546 - 550	Other Production Operation	353	229	124	
551 – 554	Other Production Maintenance	216	--	216	
555 - 558	Other Power Supply	7,661	5,029	2,631	
560 – 567	Transmission Operation	5,590		5,590	
568 – 573	Transmission Maintenance	2,243		2,243	
580 – 589	Distribution Operation	26,936		15,143	11,793
590 – 598	Distribution Maintenance	14,809		9,054	5,756
585 & 596	Street Lighting (Total only)	1,641			
901 – 905	Customer Accounts Operation	11,129			11,129
906 - 910	Customer Assistance Operation	1,765			1,765
911 – 916	Sales Operation	686			686
920 - 931	A&G Operation	48,525	12,365	27,343	8,817
935	A&G Maintenance	2,578	648	1,439	490
	Total O&M Labor and Expense	**432,895**			
403	Depreciation Expense				
	Production	42,159	755	41,405	
	Transmission	3,728		3,728	
	Distribution	14,263		7,790	6,473
	Customer Accounts	211			211
	Customer Assistance	31			31
	Sales	9			9
	Total Depreciation Expense	**60,400**			
408.1	Taxes Other Than Income Taxes				
	Property Taxes	6,507	11	5,643	853
	Payroll Taxes	5,957	1,498	3,327	1,132
	Revenue Taxes (Total only)	12,030			
	Total Other Taxes	**24,494**			
	TOTAL O&M	**\$517,789**			

11.5 COST ASSIGNMENT

The third major step of the cost-of-service study assigns the functionalized and classified cost elements to customer groups. The number of groups can be minimal or substantial depending on the degree of study resolution desired (reference Figure 11.1). Some of the costs can be directly assigned to a customer group by association. For example, outdoor lighting facilities would be directly assigned to a lighting rate group, and a substation dedicated to a particular customer would be assigned directly to the precise rate schedule and/or class of service under which that specific customer is served.

Overall, the preponderance of the costs of electric service arise from joint-use facilities that are designed, constructed, and operated to serve the general body of customers and, consequently, must be allocated on a cost-causation criterion. These joint-use cost elements have been identified as being customer related, demand related, and energy related at the various functional levels of the system. To ensure customer equity, a variety of cost allocators is necessary to apportion all of these diverse electric service cost elements to the customer groups.

Customer-related costs are identified largely within the primary and secondary distribution systems and are thus allocated to only those customer groups served at Levels 4 and 5. Some customer costs, such as meter reading expenses, would be applicable to all metered customer groups served at all system levels and should be apportioned accordingly. Since customer-related costs are independent of load and energy usage, customer cost allocators are developed on the basis of the numbers of customers served at each level. Note that secondary voltage customers served at Level 5 also make use of primary distribution facilities at Level 4; therefore, both primary and secondary voltage customers share the customer-related costs at Level 4.

Demand-related costs are identified at all functional levels of the system. All customer groups make use of production and transmission facilities and thus share in the associated demand-related costs at Levels 1 and 2. However, transmission customers served at Level 2 would not share in the distribution demand-related costs at Levels 3, 4, and 5. Furthermore, primary distribution customers served at Level 4 would not share in the secondary distribution demand-related costs at Level 5; however, secondary distribution customers served at Level 5 would share in the primary distribution demand-related costs at Level 4. The demand-related cost at each level of the system is a general function of the peak load at each level; thus, customer group MW demands, considering coincidence of load characteristics, form the basis for the development of demand allocators at each system level. *Demand losses* must be taken into account in order to reflect all demand meter readings at lower system levels to each successive higher system level for proper development and application of the demand-related cost allocator at each functional level.

Energy-related costs of fuel and variable O&M are identified typically only at the production function (Level 1), where all customer groups would share these

Cost-of-Service Methodology

costs proportionally. Similar to the development of demand allocators, *energy losses* must also be taken into account in order to reflect all kWh meter readings at lower system levels to the system production level.

Since customer, demand, and energy cost allocators are developed and applied at all functional levels of the system, a count of customers, along with their associated energy and demand usage, must be identified by functional level on the basis of their electric service characteristics. An example of a customer count levelization analysis summary, by major customer class, is shown in Table 11.17. As discussed later, this analysis of customer counts will serve as the basis for development of the customer cost allocation factors. The associated levelization analysis of customer group MWh usage and MW loads will serve as the basis for development of the energy and demand cost allocation factors.

Table 11.17 Example of a Customer Levelization Analysis – Customer Counts

Level/Function	Residential	Commercial	Industrial	Lighting	Total
1 Production					
2 Transmission			8		8
3 Distribution Subs		--	2		2
4 Pri. Distribution:					
3φ Primary		50	83		133
1φ Primary		7	5		12
5 Sec. Distribution:					
3φ Secondary	55	10,900	1,100		12,055
1φ Secondary	224,945	41,000	--	15,250	281,195
Total Customers	225,000	51,957	1,198	15,250	293,405
Lighting Fixtures				87,200	

Key Load and Energy Characteristics of Cost Allocation

Allocation factors are based on cost-causation criteria which are driven by various technical characteristics of the power system. All customers do not peak at the same time due to load diversity throughout the system. Thus, coincidence of load is a major issue which must be considered in the development of rational demand allocation factors. In addition, demand and energy losses are key issues to be considered in the formulation of the ultimate demand and energy allocators to ensure cost allocation equity among the customer groups.

Consideration of Load Diversity

The cost-causation effect of an individual customer's load varies when viewed at different levels of the system. This effect occurs as a result of the interaction of the customer's load with other customer loads at that particular service level and at the higher levels of the system.

Monthly load shapes for an example commercial customer, an entire commercial class of customers, and a total system are illustrated in Figure 11.20. As shown in the upper chart of Figure 11.20, the system peak demand is established on the third Friday of the month at a magnitude of about 1,600 MW. In the center chart of Figure 11.20, the commercial class peak for the month is established on Friday of the following week at a magnitude of about 270 kW. Because of the time difference between the occurrence of the system peak and the commercial class peak, the value of the class maximum load is referred to as the class *non-coincident peak* (NCP). The magnitude of the commercial class demand at the time of the system peak is about 200 kW, and this value is referred to as the class *coincident peak* (CP).[47] As shown in the lower chart of Figure 11.20, an individual customer within the commercial class is found to establish a monthly maximum demand of approximately 37 kW during the third Tuesday of the month. However, the customer's demand at the time of the commercial class peak is found to be less than 5 kW. Furthermore, the customer's demand at the time of the system peak is also less than 5 kW as its operation is observed to have shut down to a minimum load level just prior to the system peak hour. A comparison of the three load shapes illustrates the nature of load diversity throughout the system. The customer's peak demand is significantly diverse with respect to both the peak load of the commercial class and the peak load of the system.

Demand-related cost drivers are functions of load diversity. The cost to serve a customer at the local level is based on sizing the capacity of local facilities to meet the customer's peak demand, which could occur at any hour of the day and on any day of the year. In this example, line transformer capacity that can properly handle the kVA load associated with the customer's 37 kW peak demand would be required. As observed in Figure 11.20, the customer's load at the time of either the commercial class peak or the system peak would not provide an accurate representation of the customer's service transformer capacity requirements.

In contrast, a customer's peak demand does not provide a true representation of that customer's production level capacity requirements. The example commercial customer actually requires less than 5 kW of production capacity, which is a small fraction of its peak demand of 37 kW. While some customers may establish their peak demands simultaneously with the system peak, many other customers' peaks occur in other hours. Production capacity is sized to meet the diversified loads of all customers as a whole, i.e., where load diversity is greatest. Since costs are typically allocated to rate or customer classes, class CP load values are highly indicative of production cost responsibilities assignable to the classes.

---

[47] The CP value will always be less than or equal to the NCP value. Generally, a CP value will approach the NCP value as a function of load factor. At 100% load factor, the class hourly loads would be equal throughout the month. Thus, with no load diversity, the CP and NCP values would be the same.

Cost-of-Service Methodology

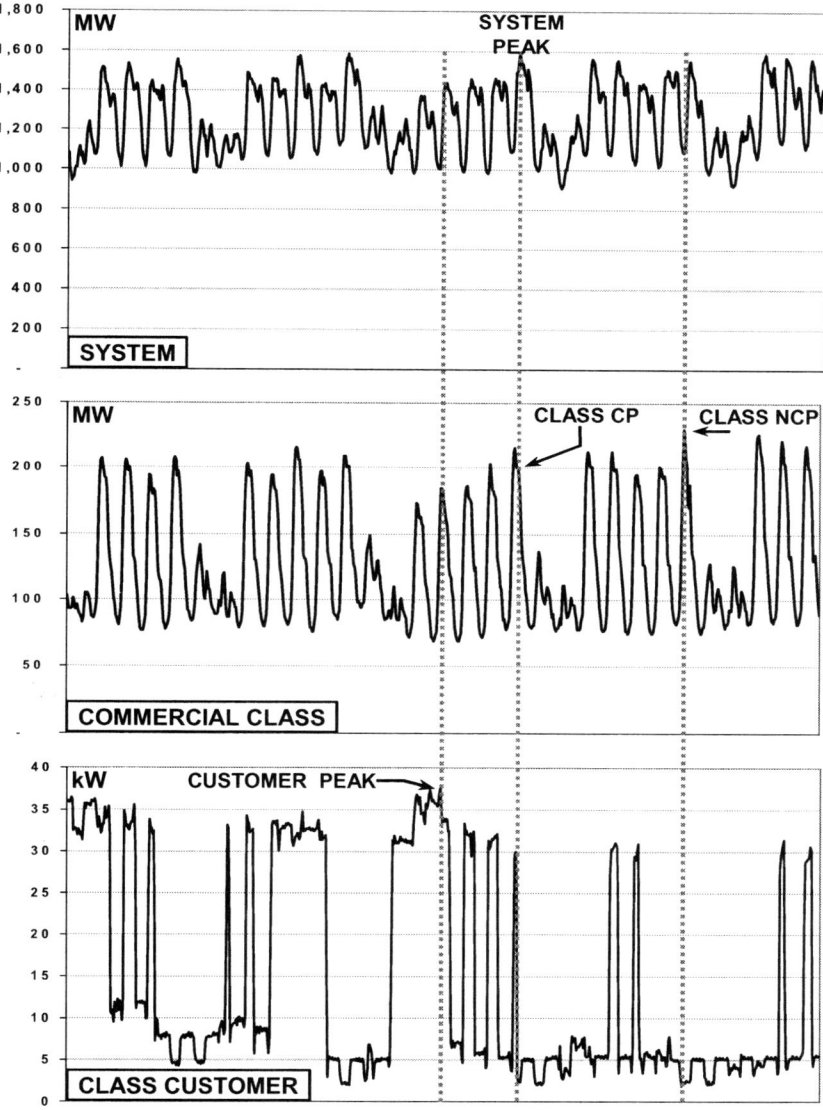

Figure 11.20 An example of load diversity during the course of a month. A single customer's peak demand (lower chart) does not occur at the same time as either the class peak (center chart) or the system peak (upper chart). Load diversity characteristics are key to providing an adequate amount of capacity at each functional level of the power system.

Building production capacity to serve an extreme load circumstance that may happen only occasionally would not be the most economical decision when other operational options exist for providing capacity.[48] For example, supplemental capacity might be acquired from tie-line interchange with neighboring utility systems and/or interruptible service contracts with large customers might be invoked during system critical loading conditions. An understanding of the typical system load diversity characteristics is a key factor in not only providing adequate capacity to serve load but also in determining the capacity-related cost responsibilities of the customer or rate classes.

Figure 11.20 provides just a single month's view of load diversity. During another month, the example customer may in deed peak at the same time as the class peak, system peak, or both as the class peak might also occur at the time of system peak in any given month. A more comprehensive view of load diversity at the upper functional levels of the system is provided by viewing the interactions of load over multiple months or a whole year. An average of multiple monthly class peaks relative to the average of multiple monthly system peaks is more appropriate for determining the typical amount of load diversity at the production and transmission levels of the system.

A "12-CP" methodology is utilized frequently for the allocation of Level 1 demand–related production plant costs as it yields a very high load diversity characteristic to the demand allocation factors. The 12-CP allocation factor for each customer or rate class is determined by

$$\text{DAF}_{\text{CP}_j} = \frac{\sum_{i=1}^{12} \text{CP}_{ij}}{\sum_{i=1}^{12} \text{SP}_i} \times 100 \qquad (11.10)$$

where DAF_{CP_j} = 12-CP allocation factor for the jth class, as a percent
CP_{ij} = coincident load of the jth class in the ith month, in MW
SP_i = system monthly peak load in the ith month in, MW

The result of Equation (11.10) is a class allocation factor representing an average month of the test year. Averaging across the twelve monthly system and class peaks captures the diversity in weather-sensitive end-use load devices in utility systems which experience appreciable seasonal peak load cycles.

---

[48] Building capacity to serve the connected load of the system at any functional level would be wholly irrational as significant excess capacity would result even during the most elevated peak load conditions. Some amount of system load diversity is always evident.

Like the production level, the transmission system's capacity requirements are based on highly diversified customer loads. As a result, demand allocators which reflect a system CP load character of diversity represent a rational means for apportioning demand-related transmission costs A 12-CP allocation factor is also used frequently for assigning transmission costs to the rate or customer classes, particularly for the transmission "backbone" facilities, as these lines unify the central production stations. On the other hand, subtransmission may exhibit aspects of both network transmission and radial or localized distribution, and the application of a 12-CP allocation factor might overstate the inherent load diversity. Taken together, a Level 2 allocation factor that exhibits a bit less diversity than a Level 1 production allocation factor may be more representative of the associated load characteristics. Given a system that experiences a prominent peak loading season (summer air conditioning vs. winter electric heating), an average of the CP values based on just the associated peak months may yield a more fitting allocation factor for transmission and/or subtransmission costs with respect to the load diversity of Level 2.

The system of distribution substations experiences a significant amount of load diversity although somewhat less than the load diversity realized within the transmission system. The load on some distribution substations may follow closely throughout the year the peak load characteristics of the overall system, while other substation peak loading conditions may be totally divergent from the system peak load in several or all months of the year. Figure 11.21 is a plot of the days and hours of the peak summer month and the peak winter month at which a system of distribution substation transformer banks experienced their maximum monthly loading conditions relative to the system peak in that same month. In the summer month example, not one substation bank is observed to peak at the exact hour of the system peak, although the influence of air conditioning load in the band between the hours of noon and six P.M. is quite evident. This pattern provides an indication of those substations which have fairly high concentrations of residential and commercial customers since air conditioning load is a prominent driver of summer peaks. Substation banks outside of this band are likely to be more concentrated with higher load factor commercial and/or industrial customers as a number of the associated peak hours occur well into the late night and early morning hours of the day all across the month.

In the winter month chart of Figure 11.21, approximately ¼ of the substation transformer banks peak during the exact hour of the system peak. Given that the winter system peak hour occurs at six to seven A.M. on day 24, those substation banks clustered tightly around the time of the system peak are indicative of electric heating load as colder winter nights typically give rise to system peak load conditions right around dawn. The pattern of this data constellation would also suggest that the temperature on that morning was much lower than other surrounding mornings. The remaining substation bank peaks are widely dispersed across the hours and days of the month, likely again due to industrial load and nonelectric heating customers.

536   Chapter 11

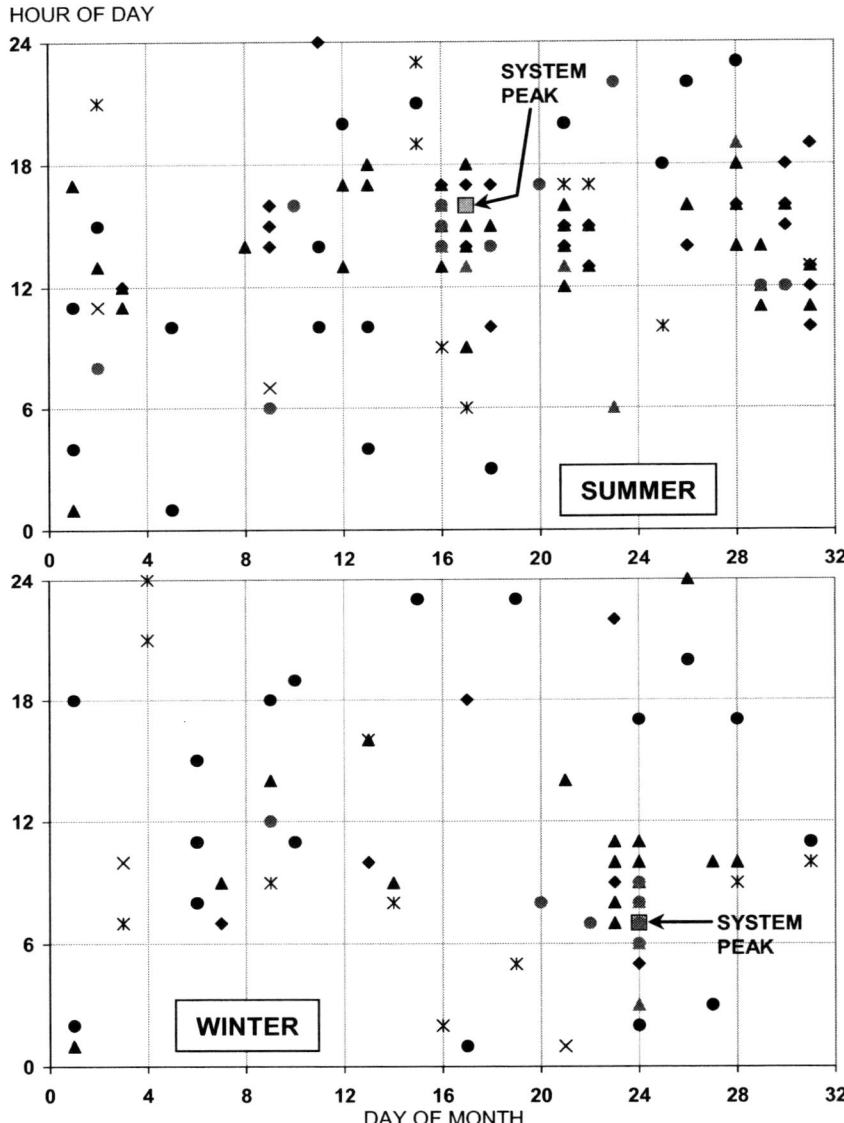

Figure 11.21 Distribution substation transformer bank peaks. Shown are the days and hours that transformer banks establish peak demands relative to the day and hour of the system peak demands for a summer and a winter month. The graphs illustrates a significant level of load diversity at the distribution substation level of the power system.

Cost-of-Service Methodology

In many cases, the annual diversified load shape composed of the hourly loads of all of the distribution substations would be found to peak in the same hour as the overall system since a majority of the total load typically is served from the distribution system. It is possible, however, for the distribution substation system's annual peak to occur at a different time if a sizeable amount of customer load is served directly from the transmission system and if the diversified load shape of the transmission customers is also found to peak off of the system hour, e.g., diversity between retail and wholesale customer loads. For distribution substation cost assignment purposes, a single CP allocator based on the distribution level customer or rate class load contributions to the distribution substation system's annual peak hour is a method that acknowledges an additional loss of load diversity as expected when moving from Level 2 to Level 3.[49]

Primary distribution feeder systems emanate from the distribution substations into the surrounding areas serving both primary and secondary delivery voltage customers. Each distribution feeder is unique in terms of its mix of customer types. A given feeder may serve a very homogeneous class of load, such as an industrial park of light manufacturing customers or a highly dense area of office complexes. Another feeder, even from the same substation, may serve a predominance of residential customers as its route takes it through neighborhoods of homes while encountering an occasional small commercial load. Overall, few distribution feeders would likely match the exact proportional class load mix of the system as a whole. As discussed above, the aggregate load of all feeders at the substation level is fairly diverse due to the interactions of the class loads; however, a single feeder that is more or less class predominant realizes a lower level of load diversity.

Given that primary distribution feeder capacity is closely associated with customer or rate class peak loads, i.e., NCP demands, a system CP type of allocator may not be truly representative of the Level 4 load diversity. As a result, class NCP demands are often utilized to develop allocation factors for primary distribution demand-related costs. The NCP allocation factor for each primary and secondary voltage customer or rate class is determined by

$$DAF_{NCP_j} = \frac{NCP_j}{\sum_{j=1}^{N} NCP_j} \times 100 \qquad (11.11)$$

where DAF_{NCP_j} = NCP allocation factor for the jth class, as a percent
NCP_j = annual noncoincident load of the jth class, in MW
N = the total number of customer or rate classes

---

[49] Depending on the particular mix of customers an NCP-based allocator or even a CP-NCP weighted allocator might be more representative of distribution substation load diversity.

The line transformers and secondary voltage distribution lines realize the least amount of load diversity. Consider three comparable homes connected to a common line transformer. Perhaps these homes were constructed by the same building contractor at the same time and include similar appliances and other end-use load devices as well as comparable envelope insulation levels. The load patterns of the three homes will be guided mostly by the lifestyles of the occupants. One home is occupied by a family of five and only one of the adults works outside of the home during weekdays. Another home is occupied by two adults who both work outside the home through the week but who are generally at home on weekends. The other home is owned by a single adult who not only works outside of the home but also travels frequently for business and on weekends, thereby leaving the home unoccupied a significant amount of time.

The same end-use load devices will be used at different times and with different frequencies in each of the homes, and monthly kWh usage will likely be different. However, unless the air conditioning thermostats are controlled and not just set and left by the occupants, the three air conditioning units will respond to hot summer afternoon temperatures in the same manner. In other words, the three air conditioning units have a fairly high probability of simultaneous operation at some times, i.e., there would be some periods of overlap as the three units go through their independent thermostat-controlled operating cycles. While load diversity does exist between these three homes, it is much less than the load diversity of all homes when viewed at the other functional levels of the power system. Since air conditioning is a major domestic load, the capacity requirement of the line transformer in particular is significantly impacted by the size of the air conditioning units and their expected coincidence of operation.

Similar to the class NCP allocation factor, a noncoincident customer peak allocation factor is often used for assigning the demand-related Level 5 costs of line transformers and secondary conductors to the secondary voltage classes. The allocation factor is based on the maximum annual demands of customers for each secondary voltage customer or rate class and is determined by

$$DAF_{MAX_j} = \frac{\sum_{k=1}^{n} D_{jk}}{\sum_{j=1}^{N} \sum_{k=1}^{n} D_{jk}} \times 100 \qquad (11.12)$$

where DAF_{MAX_j} = maximum load allocation factor for the jth class, as a percent
D_{jk} = maximum annual demand of the kth secondary voltage customer of the jth customer or rate class, in MW
n = the total number of secondary voltage customers in each class
N = the total number of secondary voltage customer or rate classes

Cost-of-Service Methodology

Table 11.18 Example of Coincidence Factors for Different Customer Classes

	Residential		Commercial		Industrial	
	kW$_{AVG}$	% of MAX	kW$_{AVG}$	% of MAX	kW$_{AVG}$	% of MAX
MAX	6.68	100	156.0	100	4,597	100
NCP	4.08	61	137.2	88	4,174	91
1-CP	3.67	55	116.8	75	3,701	81
5-CP	3.48	52	114.3	73	3,589	78
12-CP	2.59	39	106.6	68	3,499	76

The variations in load diversity across the power system can be illustrated by calculating and plotting coincidence factor relationships for various customer or rate classes.[50] As shown in the Table 11.18 example, the average customer demands for the residential, commercial, and industrial customer classes have been calculated on the basis of the total class-specific NCP and CP loads. For example, the average residential customer maximum demand at the meter is determined to be 6.68 kW per customer. The average residential customer demand at the time of the residential class peak, i.e., the NCP kW, is determined to be 4.08 kW per customer. The reduction in per customer demand is directly due to the diversity of load that exists between all of the residential customers taken together compared to either one typical customer at the meter or even just a few typical customers at the line transformer. In other words, the resulting increase in load diversity is a function of not only the difference in time at which the loads are observed to peak but also the number of customer demands being considered in the calculation of the average load per customer. As discussed above, the class NCP demand is indicative of the magnitude of load diversity at the primary feeder level.

The average demand per customer continues to drop progressively on the basis of a 1-CP, 5-CP, and 12 CP view of the residential class load. Note that the 12-CP view of load diversity is based on a number of observations which essentially represents twelve times the total number of class customers. Therefore, the average residential customer maximum demand at the system production level becomes less than half as much as at the meter level. Note also that neither the commercial class nor the industrial class exhibit a reduction in kW per customer as pronounced as the residential class example. This disproportionate change in average demand across the system is due to the different degrees of class load factor. As shown in Table 11.18, a system coincidence factor profile can be developed for each of the three customer classes by dividing each classes' average kW per customer result at the NCP, 1-CP, 5-CP, and 12-CP levels by its associated average customer demand at the meter, i.e., kW MAX.

---

[50] As discussed in Chapter 2, the coincidence factor is the reciprocal of the diversity factor.

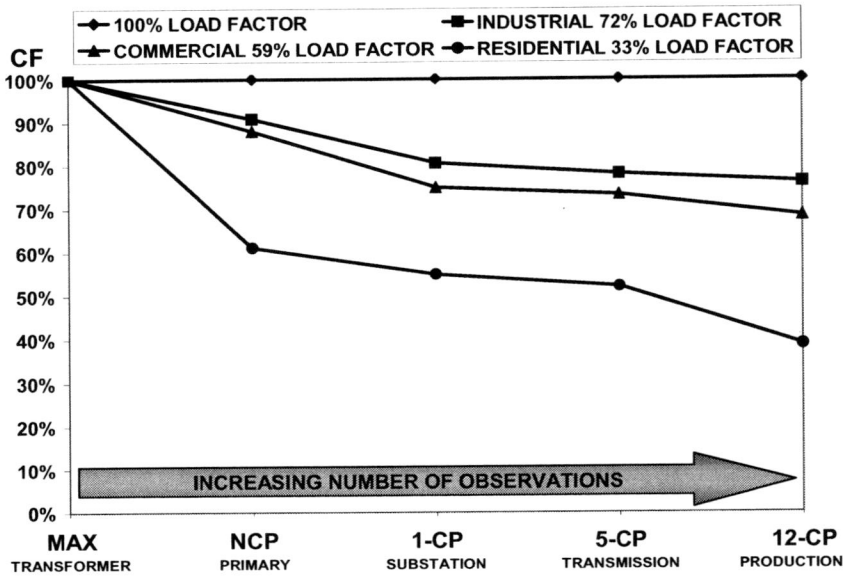

Figure 11.22 A plot showing the effect of load factor on the coincidence of load at various power system levels. The number of observations (customer demands), upon which the calculations of the class coincidence factors are based, is a key aspect of the level of load diversity realized. The coincidence factors for the residential, commercial, and industrial classes decrease at different rates when moving from the service transformer to the generator due to differences in class load factors. The decrease is more pronounced with lower load factors; however, at 100% load factor, a load is unconditionally coincident with the peak at every level of the system regardless of the number of observations considered.

Coincidence factor values, which correspond to the major functional levels of the power system, are plotted for the example residential, commercial, and industrial customer classes in Figure 11.22. A substantial difference in load factor exists among the three classes. The correlation between load factor and load diversity is evidenced by the difference in shapes of the class coincidence factor profiles. Higher load factor class loads are more coincident with system peaks than are lower load factor class loads. With the exception of a 100% load factor customer, this relationship holds for groups (classes) of customers as opposed to individual customers since even a solitary low load factor customer could peak simultaneously with the total system load.

The NCP and the various CP coincidence factor results are highly representative of the varying degrees of interclass load diversity throughout the power system; thus, they are effective in characterizing the demand-related cost causation principle. As noted above, CP-based allocation factors are commonly used to

Cost-of-Service Methodology

assign the demand cost components of production, transmission, and distribution substations to the customer or rate classes.

A concern with utilizing a CP-based allocation factor may be introduced if a particular customer or rate class is found to be prominently or completely off peak in nature. For example, outdoor lighting could avoid some of the demand cost assignment as system peaks generally occur during daylight hours.[51] Since lighting requires production, transmission, and substation facilities, albeit mostly in the off-peak period, it should receive some portion of those costs, otherwise other customers would subsidize the lighting services. A direct assignment of these costs could be made to ensure equity between the classes. Employment of an NCP allocation factor for demand-related primary distribution costs would not be challenged by an off-peak load since the maximum demand of each class, regardless of the time of occurrence, is used in the formulation of the allocator.

An *Average and Excess Demand* (AED) cost allocation methodology[52] has also been used as a means of addressing the issue of off-peak loads relative to a CP-based allocation. The AED allocation is based on both NCP and average demands. The system load factor is first applied to apportion the demand-related cost component into two portions, i.e., an Average Demand portion = LF × Demand Cost and an Excess Demand portion = (1 − LF) × Demand Cost.

The resulting average demand portion of cost is then allocated to the customer or rate classes by means of the ratios of the class average demands to the average demand of all of the classes taken together. The average demands are determined by dividing annual energy usage by the number of hours in the year. Note that an allocation on the basis of average demands will yield the same cost assignment results as simply allocating on the basis of just energy. In this manner, the AED allocation methodology presupposes that a portion of the total cost, which by nature is classified as a demand-related (fixed) cost component, is more appropriately allocated as if it were more of a variable cost component. Thus, the AED allocation methodology is generally utilized for assigning production plant under the viewpoint that such facilities are designed and built to provide both capacity and energy. In theory, the total system load could be met by an extensive fleet of gas-fired combustion turbines, which are relatively low in capital cost but high in fuel cost. However, in practice, a significant amount of production plant consists of high capital cost, large-scale steam and hydroelectric base-load units, which are operated nearly continuously in order to capture fuel cost savings thereby yielding the lowest total cost.

---

[51] Photocell-controlled outdoor lighting would be turned off during most summer peak periods when days are long and air conditioning drives mid to late afternoon peaks. Lighting is more likely to be on peak in a winter month when a cold snap causes electric heating to drive an evening or early morning peak when longer hours of darkness occur. With the variability in weather, lighting could be turned on at the time of the system peak during a few winter months in one year and off peak in every month of the following year.

[52] Developed by W. J. Greene, circa 1925.

The excess demand portion of the demand-related cost component is allocated to the customer or rate classes based on the differential amount by which the NCP values exceed the associated average demand values. The ratios of the class excess demands to the excess demands of all of the classes taken together provide the allocation factors.

Instead of the two step allocation process described above, the average demand and the excess demand calculations can be combined into a single allocation factor equation that would be applicable to the total demand cost amount, as determined by

$$DAF_{AED_j} = \left[\frac{LF_{SYS} \times E_j}{\sum_{j=1}^{N} E_j} + \frac{(1 - LF_{SYS}) \times \left(NCP_j - \frac{E_j}{T} \right)}{\sum_{j=1}^{N} NCP_j - \frac{1}{T}\sum_{j=1}^{N} E_j} \right] \times 100 \quad (11.13)$$

where DAF_{AED_j} = average and excess load allocation factor for the jth class, as a percent
LF_{SYS} = system annual load factor
E_j = annual energy of the jth class, in MWh
NCP_j = annual noncoincident demand of the jth class, in MW
N = the total number of customer or rate classes
T = the number of hours in the year

Another compound demand cost assignment methodology that recognizes energy as a subcomponent cost allocation constituent follows from the BIP subfunctionalization process previously described. The total cost of production plant is apportioned to Base, Intermediate, and Peaking periods using Equations (11.4a, 11.4b, and 11.4c). The Base portion of cost is then allocated to the customer or rate classes on the basis of energy (or average demand). The Intermediate portion of cost is allocated to class by means of the off-peak season CP values. The Peak portion of cost is allocated to class by means of the peak season CP values. A benefit of the BIP method is that it establishes a framework of costs from which time-differentiated rates can be designed.

A number of other allocation methodologies have been utilized for assigning the demand-related costs of electric service to customer groups and are described throughout the literature. A good many of these methods are variations of the load diversity-based concepts and practices discussed herein. Different methods applied to the same data can often result in a wide variation in allocation results. Selection of a demand cost allocation scheme should be based on the characteristics of the power system, considering such factors as load mix and behavior at the

Cost-of-Service Methodology

different functional levels. For instance, an NCP or an AED allocator may be more representative of the load characteristics of the distribution substation level in one system, while a 1-CP allocator may be more fitting in another system. A logical demand cost allocation design that works well for one utility may not be optimal for another utility due to unique distinctions in power system and customer characteristics.

All demand allocation methods require knowing the magnitude of each customer or rate classes' load at a certain hour or hours of the year. Large commercial and industrial customers are often metered by consecutive 15 or 30 minutes interval recorders for billing purposes; thus, hourly class load profiles can be developed from the revenue meters. Most of the other customers' revenue meters often have either no demand reading, as typical with residential and small commercial watt-hour meters, or a single monthly demand reading based on the customer's demand interval of maximum usage. As a result, quality load research data is required to build the necessary class load profiles of a substantial portion of the customer base. Adherence to industry accepted sample development and load expansion standards is critical to ensure a reasonable and dependable level of accuracy. A considerable amount of costs are at stake in the allocation process, and inaccurate load survey data will result in an inequitable representation of the class costs of service.

Consideration of Electrical Losses

Electrical losses throughout the T&D system represent a major factor in the development of energy and demand cost allocators. Energy sales are recorded by revenue meters located at each customer's point of service. Since customers receive service at different voltage levels, the adjustment of the metered kWh units to incorporate total system line and transformation losses ensures that allocations of the fuel and variable O&M costs incurred at the production level of the system are accomplished in a fair and equitable manner. In other words, prior to allocation, class sales at the meters are transformed to their equivalent share of energy production at the Level 1 generator buses. For cost allocation purposes, energy-related losses are assessed on an annual basis.

Demand-related losses are assessed on a system peak load basis, since they are related to system capacity requirements. Class demands at the service points are adjusted for losses up through the system to the level or levels at which they will be applied as demand-related cost allocators. For example, a CP type of allocator used for assigning production plant costs would require that the associated class demands be adjusted for total system line and transformation losses in order to reflect their equivalent share of production capacity at Level 1. In contrast, secondary voltage class NCP demands would first be adjusted for secondary line and line transformer losses to make them equivalent with primary voltage class NCP demands. The loss adjusted secondary voltage class demands would then be further adjusted, along with the primary voltage class demands, to incorporate primary feeder losses for development of a Level 4 NCP set of allocation factors.

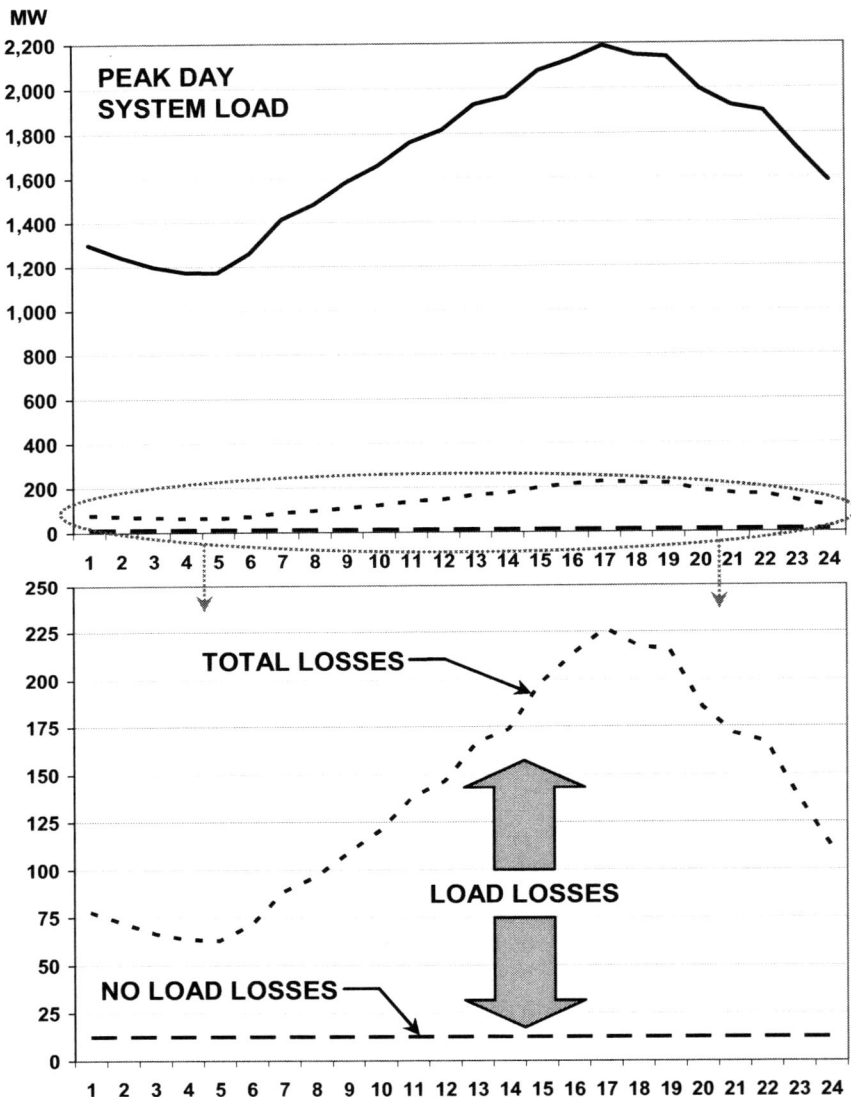

Figure 11.23 Peak day profile of system load and the associated T&D losses.

As discussed in Chapter 5, there are two types of electrical losses: transformer core losses and load losses. Core, or no-load, losses are constant in time as

Cost-of-Service Methodology

they represent a steady-state condition caused simply by energizing the transformers. Core losses are independent of a transformer's loading conditions. On the other hand, load losses are a function of load current, i.e., i^2R, and are present in both the windings of transformers and in line conductors.

Figure 11.23 illustrates the i^2R aspect of losses. The load-related losses in a given hour are proportional to the square of the load in that hour. As a result, the losses at peak loading conditions are proportionately higher relative to the system load than during off-peak loading conditions. At 5:00 AM, when system load is at its minimum for the day, the associated load losses are slightly more than 4% of the load in that hour. However, at the 5:00 PM system peak hour, losses represent nearly 10% of the system load. The magnitude of the no-load losses is constant in every hour.

A plot comparing a unitized system load with its associated load-related losses, on an annual load duration basis, is shown in Figure 11.24. Compared to a system load factor of 55% (i.e., the unitized average load), the i^2R nature of losses results in a 29% load factor for the losses.

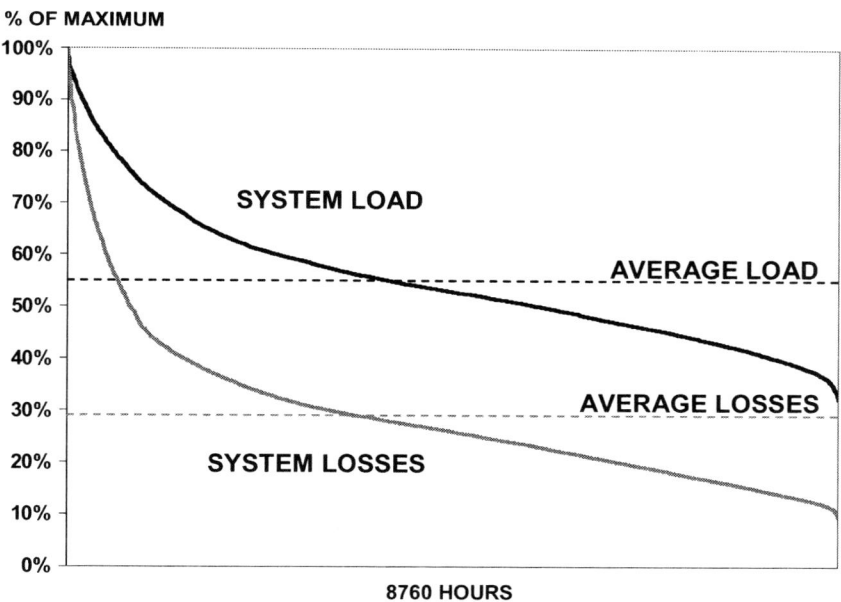

Figure 11.24 Annual unitized load duration curves of system load and the associated load-related losses. The load factor of the losses is much less than the load factor of the load itself as the average losses relative to the peak system losses are much lower. The load factors are equal to the unitized average values.

The calculation of load factor is provided by equation (2.19). The load factor of the load-related losses is a function of the square of the load, as determined by

$$LF_{Loss} = \frac{1}{T}\sum_{i=1}^{T}\left(\frac{L_i}{L_{Max}}\right)^2 \qquad (11.14)$$

where LF_{Loss} = the annual load factor of the load losses, in percent
L_i = the load in the ith hour, in MW
L_{Max} = the maximum hourly load occurring during the year, in MW
T = the number of hours in the year

The load factor of the losses is a key component for modeling line and transformer load-related energy and demand losses.

For the cost-of-service study, system losses are modeled by voltage level. A simplified schematic diagram for an entire power system, such as shown in Figure 11.4, is used as a framework for modeling peak demand and annual energy flows from the territorial inputs (generators and intersystem tie lines) to the customer delivery points. Transformer and line losses are accounted for at each voltage level. Each voltage level is analyzed as a node with power flowing in from various sources, including the higher voltage subsystems, and with power flowing out in terms of level sales, level losses, and outputs to lower voltage subsystems. The sum of the input and output power flows for each node is equal to zero.

A detailed schematic diagram of a 34.5 kV subsystem is shown in Figure 11.25. Power flows into this 34.5 kV node, which is indicated by the encircling dashed line, from the 138 kV and 69 kV subsystems at points "A" and "B." In addition, one or more generators are connected to the 34.5 kV system by means of 13 kV to 34.5 kV GSU transformers. Note that the GSU and the input substations (SUB-1 and SUB-2) represent the total of all such voltage transformations. For instance, SUB-2 may represent the composite of just a few or even dozens of 69 kV to 34.5 kV substations that are scattered throughout the power system territory. The input substations are metered at the low voltage buses of the transformers (M-1 and M-2), while the generators are metered at their output terminals (M-G).

Sales to customers taking service at 34.5 kV are metered for billing purposes (M-R). In reality, customer delivery points are distributed along the lines; however, for modeling losses for the cost-of-service study, sales can be treated as a single nodal output from the composite 34.5 kV line represented by "x" to "y". The 34.5 kV level outputs to the two lower voltage subsystems are indicated as C" and "D;" however, these outputs are metered at the low voltage buses of SUB-3 and SUB-4 (M-3 and M-4), which are located a voltage transformation beyond the edge of the 34.5 kV node.

Cost-of-Service Methodology

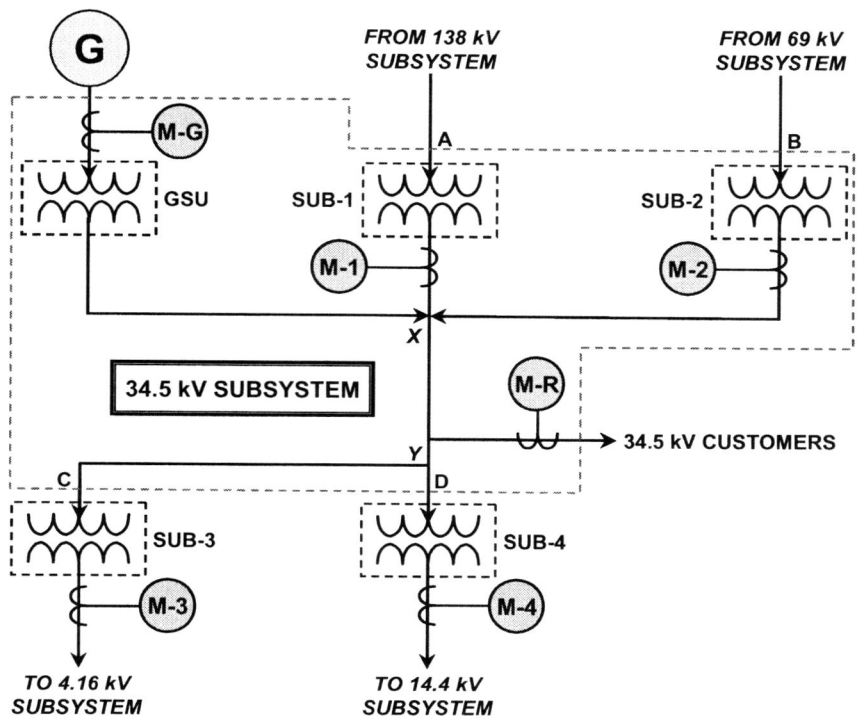

Figure 11.25 A nodal model of a 34.5 kV subsystem. Power inputs to the node are shown at "A," "B," and the generator "G." Power outputs are shown at "C," "D," and the 34.5 kV customers. Implicit outputs to be determined include transformer losses and line losses occurring along the 34.5 kV lines, which are indicated by the line segment "x" to "y."

As indicated in Figure 11.25, metered load data is essential to determining the power flows into and out of the node. Differences in power flows are key to quantifying the energy and demand losses. Basic watt-hour metering adequately captures the annual energy flows, but interval metering is necessary to determine the demands during the time of system peak. While large commercial and industrial customers typically are metered with interval recorders for billing purposes, peak demands for residential and smaller commercial and industrial customers might have to be estimated from load research sample data.

Transformer data is also required for the loss analysis, including factory results of full load loss and no-load loss tests. Individual test results are best for substation power transformers, but typical results may be sufficient for line transformers. Installed transformer MVA capacities are needed as well, including substation OA/FA/FOA ratings.

Transformer demand and energy losses are determined from the factory test data. Since the no-load related transformer loss data is constant under any transformer loading condition, the associated demand and energy no-load losses for a given transformer are determined simply by

$$D_{TNL} = NLL \tag{11.15a}$$

$$E_{TNL} = NLL \times T \tag{11.15b}$$

where D_{TNL} = peak hour demand-related transformer no-load losses in MW
NLL = the no-load loss rating of the transformer in MW
E_{TNL} = annual energy-related transformer no-load losses in MWh
T = the number of hours in the year

The no-load related demand and energy losses of individual transformers are summed to get the total for each of the different voltage transformations. For the 34.5 kV subsystem shown in Figure 11.25, total no-load loss values are calculated for each of the four substation transformers and for the GSU transformers.

Load-related transformer demand and energy losses are functions of transformer loading conditions. The demand load losses of a transformer at the time of system peak are determined by

$$D_{TLL} = FLL \times \left(\frac{L_{@Peak}}{T_{CAP} \times PF}\right)^2 \tag{11.16a}$$

where D_{TLL} = peak hour demand-related transformer load losses in MW
FLL = full-load loss rating of the transformer in MW
T_{CAP} = transformer capacity in MVA
PF = power factor of the transformer load
$L_{@Peak}$ = transformer load coincident with the system peak in MW

The quantity contained within the brackets of Equation (11.15a) represents the percent loading of the transformer at the time of system peak.

A transformer's energy load loss across the year is a function of its demand load loss as determined by

$$E_{TLL} = \frac{D_{TLL}}{\left(\dfrac{L_{@Peak}}{L_{Max}}\right)^2} \times T \times LF_{Loss} \tag{11.16b}$$

Cost-of-Service Methodology

where
- E_{TLL} = annual energy-related transformer load losses in MWh
- D_{TLL} = peak hour demand-related transformer load losses in MW
- $L_{@Peak}$ = transformer load coincident with the system peak in MW
- L_{MAX} = transformer maximum annual load
- T = the number of hours in the year
- LF_{Loss} = the annual load factor of the losses in percent

The denominator of Equation (11.16b) adjusts the transformer's demand load losses at the time of system peak to the transformer's demand losses at the time of its peak loading condition. In other words, this adjustment takes into account load diversity between the system peak and the transformer peak. Without this adjustment, the annual energy load losses would be understated.

A Methodology for Evaluating Losses

The loss characteristics of the power system are evaluated by modeling the power flows from the generators to the loads. Generally, the analysis begins at the highest voltage levels and ends with the lowest voltage levels. However, each voltage level (node) can be analyzed as a subsystem and then consolidated with the other voltage subsystems. The 34.5 kV subsystem schematic diagram in Figure 11.25, along with the input data provided below, will serve as the basis for this example. Note that even at three decimal places, some rounding difference will be observed.

Transformer Data	Capacity (MVA)	NLL (MW)	FLL (MW)
GSU	19.412	0.008	0.038
SUB-1	70.185	0.100	0.301
SUB-2	10.910	1.015	0.042
SUB-3	20.605	0.029	0.091
SUB-4	7.035	0.002	0.008

Meter Data	Annual MWh	MW @ Peak	LF*	LF_{LOSS}*	CF*
M-G	24,900	18.261	15.57%	15.44%	100.00%
M-1	251,170	29.772	77.63%	58.24%	78.53%
M-2	37,850	4.498	75.63%	58.24%	78.53%
M-3	89,405	18.913	53.96%	30.52%	100.00%
M-4	27,438	6.662	47.00%	23.94%	100.00%
M-R	192,000	25.710			

*Load Factor, Loss Factor, and Coincidence Factors determined from hourly loads.

Other Data

Annual Hours = 8,760
Power Factor @ Peak = 98%

Calculation of 34.5 kV Subsystem Inflow @ "A"	MW	MWh
SUB-1 Metered Demand and Energy (M-1)	29.772	251,170
SUB-1 NL Demand Losses (Eq. 11.14a)	0.100	
SUB-1 NL Energy Losses (Eq. 11.14b)		879
SUB-1 Demand Load Losses (Eq. 11.15a)	0.056	
SUB-1 Energy Load Losses (Eq. 11.15b)		467
Total Inflow From 138 kV Subsystem	<u>29.929</u>	<u>252,515</u>

In the same manner, the demand and energy inflows from the 69 kV subsystem to SUB-2 @ "B," the demand and energy outflows to the 4.16 kV subsystem (SUB-3) @ "C," and the demand and energy outflows to the 14.4 kV subsystem (SUB-4) @ "D" are determined to be:

	MW	MWh
34.5 kV Subsystem Inflow @ "B"	4.522	38,153
34.5 kV Subsystem Outflow @ "C"	6.672	27,472
34.5 kV Subsystem Outflow @ "D"	19.027	89,886

Note that the sum of the demands and energies for "C' and "D" represents the 34.5 kV subsystem outflows at "Y," which are: 25.699 MW and 117,358 MWh.

Generation is also connected to the 34.5 kV subsystem. The output of the GSU is determined by subtracting the calculated transformer no-load and load-related demand and energy losses from the generator terminal demand and energy meter readings (M-G), which are determined to be: 18.218 MW and 24,781 MWh. The total input to the 34.5 kV lines @ "X" can now be determined:

	MW	MWh
SUB-1 Metered Demand and Energy (M-1)	29.772	251,170
SUB-2 Metered Demand and Energy (M-2)	4.498	37,950
GSU Output:	18.218	24,781
Total Input to the 34.5 kV Lines @ "X"	<u>52.488</u>	<u>313,901</u>

While line losses are distributed along the lengths of the conductors and demand and energy sales are also distributed based on customer delivery points, both the load-related losses and sales can be represented as single nodal outputs. The line losses between "X' and "Y" can be determined, as residual values, given the known inputs and outputs:

	MW	MWh
Total Input to the 34.5 kV Lines @ "X"	52.488	313,901
Less: Total Output of the 34.5 kV Lines @ "Y"	25.699	117,358
Less: Metered Demand and Energy Sales (M-R)	25.710	192,000
Total Line Losses	<u>1.079</u>	<u>4,543</u>

Cost-of-Service Methodology

Table 11.19a Total Power Flow through an Example 34.5 kV Subsystem

	MW @ Peak	Loss Factor	Branch Factor	Annual MWh	Loss Factor	Branch Factor
34.5 kV Subsystem Outflows						
34.5 kV to 4.16 kV	6.672			27,472		
34.5 kV to 14.4 kV	19.027			89,886		
34.5 kV Sales	25.710			192,000		
Sub-Total Outflow	**51.409**			**309,358**		
34.5 kV Line Losses	1.079	2.100%		4,542	1.468%	
	52.488			313,900		
Transformer Outflows	**52.488**			**313,900**		
69 kV to 34.5 kV Subs	4.498		8.570%	37,950		12.090%
138 kV to 34.5 kV Subs	29.772		56.722%	251,170		80.016%
34.5 kV GSU Subs	18.218		34.708%	24,781		7.895%
Transformer Load Losses	**0.100**			**584**		
69 kV to 34.5 kV Subs	0.009	0.190%		71	0.186%	
138 kV to 34.5 kV Subs	0.056	0.189%		467	0.186%	
34.5 kV GSU Subs	0.035	0.191%		47	0.190%	
Transformer No-Load Losses	**0.124**			**1,084**		
69 kV to 34.5 kV Subs	0.015			133		
138 kV to 34.5 kV Subs	0.100			879		
34.5 kV GSU Subs	0.008			72		
Total 34.5 kV Losses	1.303	2.472%		6,210	1.968%	
Total 34.5 kV Outflow	**52.712**			**315,568**		
34.5 kV Subsystem Inflows						
From 69 kV Subsystem	4.522		8.579%	38,153		12.090%
From 138 kV Subsystem	29.929		56.779%	252,515		80.019%
From 34.5 kV Generation	18.261		34.642%	24,900		7.891%
Total 34.5 kV Inflow	**52.712**			**315,568**		

The results of the calculated losses and power flows through the example 34.5 kV subsystem are organized in Table 11.19a. While the initial analysis of system losses proceeded by evaluating power flows from the generators and the upper level voltage subsystems to the loads and to the lower voltage subsystems (a downward direction), the loss analysis model developed for the cost-of-service study works in reverse (an upward direction).

To complete the model framework, (a) load-related loss factors for lines and transformers and (b) power flow branching factors are determined. For example, the 34.5 kV line losses were calculated to be 1.079 MW at peak load and 4,542

MWh across the year. Thus, the respective line loss factors as a percentage of the subsystem demand and energy outflows are 2.100% (1.079 MW ÷ 51.409 MW = 0.020988) and 1.468% (4,542 MWh ÷ 309,358 MWh = 0.014683). Similarly, transformer load-related loss factors are determined by the ratio of transformer loss to transformer outflow. For example, the load-related loss factor for the 138 kV to 34.5 kV substation transformers is 0.189% (0.056 MW ÷ 29.772 MW = 0.001894).

Branching factors indicate how the system power flows split between different voltage levels. For example, the total power inflow to the 34.5 kV subsystem from 34.5 kV generation and the higher voltage subsystems is 52.712 MW at peak load. The inflow from the 138 kV subsystem alone is 29.992 MW or 56.7789% of the total subsystem inflow. Due to no-load and load-related losses, the power flow out of the 138 kV transformers to the 34.5 kV lines is 29.722 MW. Since the total 34.5 kV subsystem transformer outflow to the 34.5 kV lines is 52.488 MW at peak load, the 138 kV to 34.5 kV transformer output represents 56.7219% of the total transformer outflow.

All information needed to model power flows and calculate losses for each of the customer or rate classes has now been determined. The model integrates all of the voltage subsystems together into a single, uniform analysis. Given the determination of the various system loss and branching factors, the only input required to determine the results for a given customer or rate class is that class's annual energy sales and load at the system peak.

An example of the power flows through the 34.5 kV subsystem due to strictly the sales delivered to residential customers served at secondary voltage, which are 824.893 MW and 2,644,799 MWh, is shown in Table 11.19b. The outflows from the 34.5 kV subsystem to the 4.16 kV and 14.4 kV subsystems are the results of the power flows through those two systems that are required to serve the residential customers in the secondary voltage subsystem. The MW and MWh sales to 34.5 kV customers have been zeroed out since this analysis is only considering residential sales. The branching factors calculated in Table 11.19a on the basis of all customer sales are now applied to determine the 34.5 kV subsystem power flow trifurcations associated with the residential sales.

The loss factors are applied respectively to the subtotal outflows in order to determine the line losses and to the substation outflows to determine load–related transformer losses. Transformer no-load losses, however, must be allocated proportionately to each of the classes since no-load loss values are fixed amounts. This allocation can be made on the basis of the transformer outflow for each class relative to the transformer outflow for all classes. For residential, the allocation of the demand-related no-load transformer loss would be determined by

<u>138 kV Substation Transformer Outflows to 34.5 kV Lines</u>:
Outflow due to residential class = 7.656 MW
Outflow due to all classes = 29.772 MW
Outflow ratio = 0.25715

Cost-of-Service Methodology

Table 11.19b Residential Power Flow through an Example 34.5 kV Subsystem

	MW @ Peak	Loss Factor	Branch Factor	Annual MWh	Loss Factor	Branch Factor
34.5 kV Level Outflows						
34.5 kV to 4.16 kV	3.838			11,738		
34.5 kV to 14.4 kV	9.382			30,089		
34.5 kV Sales	------			------		
Subtotal Outflow	13.219			41.827		
34.5 kV Line Losses	0.278	2.100%		614	1.468%	
	13.497			42,442		
Transformer Outflows	13.497			42,442		
69 kV to 34.5 kV Subs	1.157		8.570%	5,131		12.090%
138 kV to 34.5 kV Subs	7.656		56.722%	33,960		80.016%
34.5 kV GSU Subs	4.685		34.708%	3,351		7.895%
Transformer Load Losses	0.026			79		
69 kV to 34.5 kV Subs	0.002	0.190%		10	0.186%	
138 kV to 34.5 kV Subs	0.015	0.189%		63	0.186%	
34.5 kV GSU Subs	0.009	0.191%		6	0.190%	
Transformer No-Load Losses	0.032			147		
69 kV to 34.5 kV Subs	0.004			18		
138 kV to 34.5 kV Subs	0.026			119		
34.5 kV GSU Subs	0.002			10		
Total 34.5 kV Losses	0.335	2.472%		840	1.968%	
Total 34.5 kV Outflow	**13.554**			**42,667**		
34.5 kV Level Inflows						
From 69 kV Subsystem	1.163		8.579%	5,159		12.090%
From 138 kV Subsystem	7.696		56.779%	34,142		80.019%
From 34.5 kV Generation	4.696		34.642%	3,367		7.891%
Total 34.5 kV Inflow	**13.554**			**42,667**		

Residential class portion of 138 kV to 34.5 kV transformer no-load losses = 0.25715×0.100 MW = 0.026 MW

The energy-related no-load loss amounts must also be allocated, or the allocated demand-related no-load loss amounts can be multiplied by the number of hours in the year.

Since the demand loss portion of the model discussed above is calibrated to the system annual peak, i.e., a 1-CP allocator, adjustments must be incorporated when applying the model for development of other load allocators. For example, the residential load at the time of system peak was noted above as being 824.893 MW. The corresponding residential loads are 782.376 MW for 5-CP, 582.255 MW for 12-CP, and 918.466 MW for NCP. The adjustments are required because of the i^2R relationship between losses and load, and it is made to the load-related loss factors. Considering the 12-CP allocator as an example, the adjustment of the loss factor associated with the line losses, in each voltage subsystem, is made by multiplying the factor by the ratio of the 12-CP case subtotal level outflow to the 1-CP base case subtotal level outflow. Thus, the line loss factors would decrease from the 1-CP base case for the multiple CP allocators and increase from the 1-CP base case for the NCP allocator. Similar adjustments are made to the transformer load loss factors based on transformer outflow ratios.

The results of the analysis of electrical losses are key to the development of equitable class demand and energy allocation factors. Loss adjusted MWh sales are used to allocate production energy costs. Loss adjusted demands are used to allocate demand-related costs at all five levels of the system.

Allocation of Rate Base

Allocation of the functionalized and classified rate base cost elements is rather straightforward. Loss adjusted customer or rate class MWh sales are used to allocate fuel stocks, scrubber reactant stock, and other energy-related production rate base items. Various loss adjusted customer or rate class loads are used to allocate the demand-related cost components of the production, transmission, distribution, customer accounts, customer assistance, and sales functions. Recognizing that the selection of demand allocators is based on a case-by-case basis considering such issues as system load characteristics and regulatory governance, the example presented below represents only one view as it is based on the various diversified load concepts previously discussed. The customer-related cost allocations are based on numbers of customers, some of which may include weighting factors to capture various cost-causation differentials.

Examples of customer, demand, and energy allocation factors used to assign costs to customer classes are shown in Table 11.20. A given allocator is indicated by a series of percentage-based factors which represent the cost-causation share of the total cost element for each applicable class. The allocators take into account customer service voltage characteristics (magnitude and phase).[53] Commercial and industrial customer or rate classes often contain secondary, primary, and even transmission voltage level customers. Thus, in allocation factor development, it is important to include only those customers meeting the functional requirements of a particular cost element. For example, primary and transmission customers do not share in the cost of line transformers.

---

[53] Cost allocations for this case study were actually conducted on the basis of customer voltage levels in order to achieve a high degree of resolution. The detailed results were then consolidated at the major customer class level.

Cost-of-Service Methodology

Table 11.20 Example Customer, Demand, and Energy Allocation Factors

Allocation Factor	Total	Customer Class			
		Residential	Commercial	Industrial	Lighting
Customer					
Level 4:					
▪ R/C/I	278,145	225,000 80.8931%	51,957 18.6798%	1,188 0.4271%	
▪ Total	293,397	225,000 76.6853%	51,957 17.7083%	1,190 0.4083%	15,250 5.1976%
Level 5:					
▪ 3φ OH/UG-LT	12,055	55 0.4562%	10,900 90.4189%	1,100 9.1248%	
▪ 1φ UG-LT	265,945	224,945 84.5833%	41,000 15.4167%		
▪ 1φ OH-LT (Weighted)	317,360	225,110 70.9321%	73,700 23.2228%	3,300 1.0398%	15,250 4.8053%
▪ Total	293,250	225,000 76.6879%	51,900 17.6982%	1,100 0.3751%	15,250 5.2003%
Levels 1-5:	293,405	225,000 76.6858%	51,957 17.7083%	1,198 0.4083%	15,250 5.196%
Lighting Fixtures:					87,200
Demand					
▪ Level 1 MW (12-CP)	1,646	652 39.6254%	588 34.4077%	427 25.9669%	-- --
▪ Level 2 MW (5-CP)	1,857	842 45.3550%	588 31.6674%	427 22.9777%	-- --
▪ Level 3 MW (1-CP)	1,757	874 49.7345%	572 32.5479%	311 17.7176%	-- --
▪ Level 4 MW (NCP)	2,085	985 47.2216%	717 34.3832%	368 17.6537%	15 0.7415%
▪ Level 5 MW (MAX-A)	1,738	1,738 55.2195%	1,228 39.0272%	164 5.2168%	17 0.5364%
▪ Level 5 MW (MAX-B)	3,050	1,689 55.3877%	1,184 38.849%	160 5.2403%	16 0.5370%
Energy					
▪ Level 1 MWh	9,346	2,896 30.8193%	3,236 34.4368%	3,199 34.0447%	66 0.6992%

Allocation of customer-related costs requires a number of unique allocators to accurately address the varied cost driver aspects of the assorted cost elements. For example, the allocation of the customer component of primary distribution poles is complicated by the fact that a reasonable count of street lighting "customers" is difficult to define in the same sense as the residential, commercial, and industrial classes. To use the number of lighting fixtures (which can be considered as individual service delivery points) as an equal customer count along with the power service classes would result in assigning an overabundance of pole cost to the lighting class (reference customer counts in Table 11.17). Lighting does not drive the need for and construction of the primary feeder system. The primary system is built to serve the power classes, and lighting, in essence, takes advantage of the space available on the existing poles. Furthermore the use of lighting accounts would not be a representative customer account as the number of individual lighting fixtures served under each account could vary significantly.

A compromise position that would assess a more reasonable amount to the lighting class is to assign a fraction, such as one foot, of the customer-related cost of a typical primary system pole (e.g., 40 foot, Class 5 wood pole) to each lighting fixture. The remainder of the wood pole customer-related cost would then be allocated to just the residential, commercial, and industrial classes, served at primary and secondary voltage levels, using the Level 4 R/C/I customer allocator shown in Table 11.20.

Line transformers add a level of complexity to the allocation process regarding their associated customer cost components. A different customer allocation factor is needed for each type of transformer. Overhead and padmount 3ϕ units are used for 3ϕ service customers, so lighting and other 1ϕ service customers would typically be excluded (reference the Level 5 3ϕ OH/UG-LT allocator in Table 11.20). Padmount 1ϕ units would be minimally applicable (if at all) to the industrial customer class (reference the Level 5 3ϕ UG-LT allocator in Table 11.20). On the other hand, overhead 1ϕ units are applicable to both 1ϕ and 3ϕ service customers where three 1ϕ units are joined in a wye or delta configuration in order to form a 3ϕ transformer bank. A weighted allocation factor is appropriate for such a case; thus, a weighting factor of three is applied to the 3ϕ customers so that they will be assigned additional costs commensurate with having a three transformer bank.[54]

	Residential	Commercial	Industrial	Lighting
1ϕ	224,945	41,000	--	15,250
3ϕ	$55 \times 3 = 165$	$10,900 \times 3 = 32,700$	$1,100 \times 3 = 3,300$	--
Total	225,110	73,700	3,300	15,250

[54] Open-delta banks are sometimes used for small 3ϕ services, such as lift stations. Consideration of these two transformer services would result in a weighting factor of less than 3.

Cost-of-Service Methodology

These weighted customer counts are used to developed the Level 5 1φ OH-LT allocation factor shown in Table 11.20. A similarly weighted allocator would be appropriate for assigning cost differentials for instrument transformers.

Demand allocation factors are required for each level of the system since demand-related costs are defined at each level. Demand sales at the customer meters are adjusted for losses as described previously. Table 11.20 indicates two demand allocation factors at Level 5. The Level 5 (MAX-B) allocation factor incorporates secondary voltage system losses, i.e., the nondiversified demands at the secondary customer meters are loss adjusted to the low-voltage side of the line transformers. Thus, Level 5 (MAX-B) would be used for assigning the demand-related costs of the secondary lines, poles, and equipment to the low voltage customers. Further adjusting these demands to the high-voltage side of the line transformers incorporates the transformer load and no-load losses thereby yielding the Level 5 (MAX-A) allocation factor, which is then used for allocating not only the line transformer costs but also the associated cutout and arrester costs.

The Level 4 (NCP) allocator further adjusts the secondary service loads, along with the primary service loads, to the low-side buses of the distribution substation transformers. Therefore, the Level 4 (NCP) allocation factor is used for assigning the demand-related feeder system costs, including the lines, poles, and equipment, to the secondary and primary customers.

The Level 3 (1-CP) allocation factor further adjusts for distribution substation transformer load and no-load losses, and it is therefore used to assign the costs of *common* distribution substations. Common substations serve multiple customers as they are sized and built to serve area load. On the other hand, a substation may be *dedicated*, i.e., designed to serve a specific customer. The cost of dedicated substations should be directly assigned to the customer or rate class under which that particular customer is served. The development of the Level 3 allocation factor for common substations should exclude the demands of the dedicated substation customers in order to prevent an over-allocation of costs to that class.

The Level 2 (5-CP) allocation factor continues to adjust the lower voltage customer loads, along with subtransmission and transmission system customer loads for transmission level losses up to the output buses of the GSU transformers. Note that a similar situation as described above could occur at Level 4 whereby a transmission customer could have a dedicated substation, e.g., a 345 kV to 138 kV transformation. As noted above, the dedicated substation cost should be directly assigned to that customer. The Level 1 (12-CP) allocator further adjusts all of the customers' loads for GSU load and no-load losses and thus serves as the factor for assigning production plant costs to the customer or rate classes.

Energy sales at the customers' meters are loss adjusted in the same fashion as the demand sales in order to determine the equivalent class sales at the Level 1 generator buses (the intermediate energy value solutions at the lower system levels are of no consequence). Results of the customer, demand, and energy allocations for the example rate base components are summarized in Table 11.21. Note that FERC Account 373 investment is directly assigned to the lighting class.

Table 11.21 Example Summary of Rate Base Allocation

Function [Level]	Total $ (000s)	Customer Class			
		Residential	Commercial	Industrial	Lighting
Production					
Demand [1]	761,642	301,804	262,064	197,775	--
Energy [1]	78,982	24,342	27,199	26,889	552
Transmission					
Demand [2]	70,441	31,805	22,206	16,430	--
Distribution					
Demand:					
• Substations [3]	32,184	15,890	10,252	6,042	--
• Primary [4]	45,825	22,522	14,577	8,341	385
• Transformers [5]	27,869	13,887	12,229	1,589	165
• Secondary [5]	11,145	5,620	4,823	636	66
Customer:					
• Primary [4]	35,850	31,734	4,906	(600)	(190)
• Transformers [5]	17,479	9,871	6,784	429	395
• Secondary [5]	10,798	8,897	2,043	8	(150)
• Service Lines [5]	23,995	20,198	3,794	4	--
• Meters	8,974	4,403	3,754	826	--
Specific:					
• Lighting [5]	12,000				12,000
Customer Svcs					
Customer:					
• Accounts	1,961	1,730	206	(156)	181
• Assistance	525	402	93	2	27
• Sales	174	134	31	712	9
Total	**$1,139,845**	**$493,162**	**$374,767**	**$257,280**	**$14,636**
		43.27%	32.88%	22.57%	1.28%

As noted in Table 11.21, it is possible to calculate a negative rate base component in some cases. These negative values are caused by the further netting effect of the rate base additions and deductions being applied to the net electric plant in service. For instance, the negative amount observed for the industrial Customer Accounts function is caused primarily by the difference in magnitude between its smaller allocated portion of general plant (the major component of net

Cost-of-Service Methodology 559

plant in the three customer services functions) and its significantly larger allocated portion of the industrial class customer deposits, which is a rate base deduction.

Allocation is a necessary, yet viable process for development of the cost of service study. Taken together however, cost assignments based on a logical and consistent cost-causation principle are reasonable, and they yield useful results. Drilling down too deeply in an effort to single out the costs associated with a small component can prove to be misleading

Allocation of Expenses: O&M and A&G Labor and Expenses, Depreciation Expense, and Taxes Other Than Income Taxes

Assignment of O&M and A&G labor and expenses to the customer or rate classes for the production function is quite similar to the assignment of production plant. The Level 1 Energy Allocator is utilized for allocation of those production expenditures which are classified as being energy related, while the Level 1 Demand Allocator is utilized for allocation of those production expenditures which are classified as being demand related. In a similar fashion, the Level 2 Demand Allocator is utilized for allocation of the transmission expenditures, which are generally classified as being demand related.

In contrast, assignment of distribution O&M and A&G expenditures is more involved. Both distribution O&M and A&G demand and customer cost assignments require multiple allocation factors. Unlike the situation with the five other major system functions, the span of expenditures across Levels 3, 4, and 5 prohibits the use of a single Level allocator.

As noted in Table 11.14, the FERC O&M accounts are spread across the distribution functions of substations, lines, line transformers, meters, and lighting. The O&M amounts in the lighting accounts (585 and 596) are directly assignable to the Lighting customer class. The O&M amounts in the Meter Expense and Maintenance of Meters accounts (586 and 597) are allocated to the customer or rate classes using the results of the allocation to class of the meter gross plant (FERC Account 370). The Maintenance of Line Transformers (595) is allocated using the appropriate Level 5 demand and customer allocation factors.

The O&M amounts in FERC Accounts 582, 591, and 592 relate to distribution substation expenditures. Station Expenses expenditures (582) are allocated to the customer or rate classes based on the combined results of the gross plant allocation of FERC Account 361 – Structures and Improvements and FERC Account 362 – Station Equipment. Maintenance of Structures (591) would be assigned only on the basis of the FERC Account 361 allocation results, while Maintenance of Station Equipment (592) would be assigned only on the basis of the FERC Account 362 allocation results

FERC Account 581 – Load Dispatching is a function of both substations and distribution primary lines. Thus, a representative portion of the O&M expenditures would be allocated on the basis of the Level 3 demand allocation factor and the remaining portion would be allocated on the basis of the Level 4 demand allocation factor.

The O&M FERC Accounts for overhead and underground line expenses (583, 584, 593, and 594) apply to a diverse mix of poles, conductors, and line equipment and facilities. Like line transformers, these expenditures are classified into both demand-related and customer related cost components as a result of the minimum distribution classification process. These expenditures can be allocated between primary and secondary voltages based on previous distribution system gross plant allocations and then allocated to the customer or rate classes using the appropriate Level 4 and 5 customer and demand allocation factors.

Supervision and engineering operations and maintenance expenditures (580 and 590) can be allocated based on the assignment results of the underlying O&M FERC accounts. These same results can be used to allocate the miscellaneous O&M account expenditures contained in FERC Accounts 588 and 598. This approach parallels the method discussed previously to classify these four accounts by customer and demand components.

The distribution A&G cost elements are assigned on the basis of other allocated costs continuing the basic cost-causation logic used for functionalization and classification. For example, A&G labor in FERC Account 920 is assigned to the customer or rate classes on the basis of the allocation results of distribution O&M labor.

The Customer Accounts, Customer Assistance, and Sales O&M and A&G accounts are classified totally as customer-related costs and are thus allocated by means of customer count based allocators. Weighting factors may be appropriate in some cases. For example, FERC Account 902 – Meter Reading expenses apply to all metered service rates. However, the cost to read an industrial meter may be greater than the cost to read a residential or small commercial meter, as would be the case if different labor skill categories are used for these functions. Thus, a cost differential factor could be developed in order to expand the large industrial rate customer counts in order to improve the allocation accuracy.

A departure from the basic customer, demand, and energy cost allocation factor approach is appropriate for some instances. The cost-causation aspect of FERC Account 928 - Regulatory Commission Expenses is more closely related to revenue from sales than to customer counts, demands, or energy. Thus, the revenues of just the retail customer or rate classes are used as the means of assigning the costs of associated regulatory expenditures. FERC Account 930.1 – General Advertising Expenses is applicable to all classes of service, including native load wholesale customers; thus, total territorial sales is used for cost assignment.

Functional depreciation expense is generally assigned on the basis of the standard Level allocation factors, except with distribution (again because of the multiple level aspect). Assignments to the customer or rate classes are made on the basis of the corresponding gross plant allocation results.

There are three major categories of Taxes Other Than Income Taxes (FERC Account – 408.1). Property taxes are assigned on the basis of allocated plant. Payroll taxes are assigned on the basis of allocated labor. Revenue Taxes are assigned on the basis of the revenues of the customer or rate classes.

Cost-of-Service Methodology

Table 11.22 Example Summary of Expenses Allocation

Function [Level]	Total $ (000s)	Customer Class			
		Residential	Commercial	Industrial	Lighting
Production					
Demand [1]	106,380	42,146	36,603	27,622	10
Energy [1]	292,957	30,299	100,885	99,724	2,049
Transmission					
Demand [2]	15,451	6,959	4,865	3,626	1
Distribution					
Demand:					
• Substations [3]	10,325	5,038	3,554	1,733	--
• Primary [4]	19,883	9,650	6,808	3,322	104
• Transformers [5]	5,000	2,849	1,834	286	31
• Secondary [5]	4,978	2,842	1,820	286	31
Customer:					
• Primary [4]	13,774	9,439	2,793	181	1,360
• Transformers [5]	2,932	2,014	588	28	301
• Secondary [5]	4,626	3,183	936	48	459
• Service Lines [5]	2,702	2,070	604	28	--
• Meters	5,282	4,000	1,188	93	--
Specific:					
• Lighting [5]	3,226				3,226
Customer Svcs					
Customer:					
• Accounts	27,074	16,301	6,834	3,151	788
• Assistance	2,340	1,793	415	10	121
• Sales	859	658	152	4	45
Total	**$517,789**	**$199,914**	**$170,093**	**$141,328**	**$6,454**
		38.61%	32.85%	27.29%	1.25%

Note: Amounts include O&M, A&G, depreciation expense and taxes other than income taxes.

The example allocation of expenses to the customer classes is summarized in Table 11.22. Each entry combines the allocated amounts for O&M, A&G, depreciation expense, and taxes other than income taxes for that function.

11.6 COST-OF-SERVICE STUDY RESULTS

Key goals of the cost-of-service analysis are (1) to separate the total utility costs by regulatory jurisdictions and/or by customer or rate classes within those jurisdictions through the cost functionalization, classification, and assignment processes and (2) to determine the financial performance of those designated customer segments under present rates over the course of an annual test period. The financial performance is a measure of the return on investment (ROI) and the return on equity (ROE) under present or proposed rates. These financial measures are determined by means of the basic revenue requirement formula.[55] Equations (2.21) and (2.22) can be reformulated to define the ROI for the total utility or for a given customer group j, as determined by

$$ROI_j = \frac{REV_{PRES_j} - EXP_j - IT_j}{RB_j} \qquad (11.17)$$

where ROI_j = ROI for group j
REV_{PRES_j} = group j annual revenue under present rates
EXP_j = O&M expenses assigned to group j
IT_j = income taxes assigned to group j
RB_j = rate base (investment) assigned to group j

The numerator in Equation (11.17) is referred to as the *net operating income* (NOI), while the result of $[REV_{PRES_j} - EXP_j]$ is referred to as the *NOI before income tax* (NOI$_{BIT}$).

Rate Base and O&M expenses are assigned to the customer or rate classes as discussed previously. To complete the cost-of-service study and determine the financial performance of the classes, the operating revenues of the customer groups must be analyzed, and income taxes must be assigned.

Analysis of Revenues

For a vertically integrated utility, the largest component of annual revenue comes from the sales of electricity to native load customers. Since retail and territorial wholesale revenues are known by rate schedule, no allocation to the customer or rate classes is required. Given a favorable production capacity and energy price position, the opportunity for nonterritorial or off-system sales transactions may represent a noteworthy source of annual revenue to the utility. Sales of electricity to the major customer classes are booked under the following FERC Accounts:

---

[55] The cost-of-service study does not calculate a revenue requirement. Rather it determines the return component given an annual amount of revenue produced under present rates.

Cost-of-Service Methodology

- 440 – Residential Sales
- 442 – Commercial and Industrial Sales
- 444 – Public Street and Highway Lighting
- 446 – Sales to Railroads and Railways
- 447 – Sales for Resale
- 448 – Interdepartmental Sales
- 449 – Other Sales

FERC Account 447 – Sales for Resale includes both territorial wholesale revenues and nonterritorial (off-system) sales. The territorial wholesale revenues are directly assigned to the native load wholesale class of service. Nonterritorial sales are made from production resources whose costs are supported by all native load customers; thus, the off-system sales revenues are credited, in whole or in part, to the territorial customer or rate classes.[56] Off-system sales revenues are comprised of a demand component and an energy component. The production and transmission demand components of off-system sales revenues are allocated to the customer or rate classes by means of the respective Level 1 and Level 2 demand allocators. The production energy component of off-system sales is allocated to class by means of the Level 1 energy allocator.

Other operating revenue sources include a variety of charges and credits. FERC Account 450 – Forfeited Discounts includes additional charges from customers who neglect to remit monthly payment for electric service by a specified date. A rate schedule's Term of Payment provision may be structured either as a loss of early payment discount or as a late payment fee should the bill not be paid by the date indicated on the monthly billing statement. Forfeited discounts are directly assignable to the customer or rate classes since those occurrences are known by rate schedule.

A number of Miscellaneous Service Revenues are booked under FERC Account 451. A variety of auxiliary service fees are used by utilities to recover extraordinary costs associated with specific customer situations. These separate fees prevent the subsidization by other customers if such costs were otherwise included in base rates. Some of the more commonly assessed miscellaneous fees include:

- Service Connection/Reconnection
- Meter Reread
- Meter Test
- Returned Checks
- Collection Fees
- Current Diversion Case Expenses
- Temporary Service

---

[56] Some regulatory jurisdictions allow for a specified level of sharing of the off-system sales revenues between the customers and the utility as an incentive for the utility to seek opportunity sales that provide benefits to customers.

Miscellaneous service charges are typically prescribed in the general terms and conditions section of the electric tariff. If booked by rate schedule, miscellaneous service charge revenues would be directly assignable to the customer or rate classes. Other revenues may be derived from sales of water and water power, which are booked in FERC Account 453. These sales relate to water used for domestic, industrial, irrigational, and other uses as well as the development of water power by others.

Rent from electric property is booked in FERC Account 454. A common example of rental income is pole attachment revenues, which are assessed to other utilities having aerial facilities, such as telephone and cable television services, for the use of space on the electric utility's distribution poles. Thus, these revenues are credited to the customer or rate classes on the basis of the gross plant amounts for primary and secondary distribution poles, as previously allocated.

Income from other departments of the utility, such as gas and water services, for rents that are credited to the electric department is booked in FERC Account 455. An example would be the use by another department of any floor space or other property of the electric department, which has not been identified and separated as common facilities. Related general plant allocation factors could be developed to credit such rentals to the customer or rate classes.

Other electric revenues are booked in FERC Account 456. One example would be the sale of excess inventory of materials and supplies. Allocation of income from these property sales would be allocated as a revenue credit to the customer or rate classes using the results of the class rate base assignments for the related materials and supplies items.

Other income includes amounts booked to FERC Accounts: 456.1 – Revenue from Transmission of Electricity of Others (typically referred to as wheeling), 457.1 – Regional Transmission Service Revenues, and 457.2 – Miscellaneous Revenues (related to RTO operations). As these FERC Accounts relate to transmission operations, a Level 2 demand allocator would be appropriate for class assignment purposes.

Customer class revenues from sales, along with allocated revenue credits, are summarized in Table 11.23 for the example cost-of-service analysis. As discussed previously, revenues from sales are utilized to allocate some A&G costs as well as revenue taxes to the customer or rate classes.

Assignment of Income Taxes

Income tax is a principal component of the revenue requirement determination. Income taxes are based on an adjusted income amount. Taxable income for a business is calculated by subtracting deductibles, O&M expenses and interest paid, from revenues. A tax rate is then applied to the adjusted income in order to determine the tax liability. State and federal income taxes are calculated and paid annually by the utility as a whole. Consequently, a cost-of-service study allocation is necessary in order to compute an income tax liability for each of the customer or rate classes.

Cost-of-Service Methodology

Table 11.23 Example Summary of Revenue and Revenue Credit Allocation

	Total $ (000s)	Customer Class			
		Residential	Commercial	Industrial	Lighting
Rev. from Sales	719,311	277,697	247,368	184,701	9,545
Late Payments	2,519	2,247	269	2	--
Misc. Services	1,154	1,057	93	3	--
Elec. Prop. Rent	646	420	163	52	11
Other Elec. Rev.	14	6	5	4	--
Wheeling Rev.	433	196	137	99	--
Off-System Sales:					
▪ Demand Rev.	8,826	3,498	3,037	2,292	--
▪ Energy Rev.	22,112	6,815	7,615	7,528	155
Total	**$755,015**	**$291,936**	**$258,687**	**$194,681**	**$155**
		38.67%	34.26%	25.79%	1.29%

Since O&M labor and expenses have already been allocated to the customer or rate classes, the subtotal taxable income of each class is easily calculated. The interest paid must then be allocated to complete the calculation of the class taxable income amounts. Interest on long-term debt is booked in FERC Account 427, and other interest expense, including short-term notes and interest on customer's deposits, is booked in FERC Account 431. The total deductible interest can be allocated on the basis of the assigned gross plant in service in order to determine the taxable income amount for each of the customer or rate classes. The resulting class taxable income amounts can then be used as an allocation method to assign the total income taxes paid to the customer or rate classes.

Alternatively a formula can be developed on the basis of total utility figures that can then be applied to the comparable amounts for each customer or rate class in order to assign an income tax liability. In addition, this approach utilizes the combined state and federal income tax rate. The combined tax rate takes into account that state taxes are deductible for federal tax purposes, as given by

$$T_C = T_S + (1 - T_S) \times T_F \qquad (11.18)$$

where T_C = combined state and federal income tax rate in percent
T_S = state income tax rate in percent
T_F = federal income tax rate in percent

As an example, if the state income tax rate is 8.25% and the federal income tax rate is 34%, the combined income tax rate would be 8.25% + (1 - 0.0825) × 34% = 39.445%.

The formulary income tax approach first requires the calculation of an income tax deduction factor, as given by

$$K = \frac{(T_C \times NOI_{BIT}) - IT}{RB'} \quad (11.19)$$

where K = an income tax deduction factor
T_C = the combined state and federal income tax rate in percent (from Equation 11.18)
NOI_{BIT} = total net operating income before income tax
IT = total income taxes paid
RB' = total rate base, excluding CWIP and CWC

The state and federal tax assignment to each class is the given by

$$T_j = (T_C \times NOI_{BIT_j}) - (K \times RB'_j) \quad (11.20)$$

where T_j = income taxes assigned to class j
T_C = the combined state and federal income tax rate in percent (from Equation 11.18)
NOI_{BIT_j} = total net operating income before income tax of class j
K = the income tax deduction factor (from Equation 11.19)
RB'_j = rate base, excluding CWIP and CWC, of class j

The assigned income taxes are then subtracted from the class NOI_{BIT} amounts in order to determine the NOI result for each class.

Calculation of ROI and ROE

Rate of return (ROR) was introduced previously as a key financial metric that is calculated from the cost-of-service study. More specifically, ROR is calculated as a *return on investment* (ROI) in which investment is actually the rate base, as given by

$$ROI = \frac{NOI}{RB} \times 100 \quad (11.21a)$$

The ROI is a percentage relates to the weighted cost of capital (refer to Table 11.5). The ROI is a financial performance measure that indicates to what degree the current rates produce revenue relative to the costs of providing service during the test period. In other words, the ROI is compared to the weighted cost of capital as one means of evaluating the rates. Since rate base, expenses, and income taxes are allocated, the cost-of-service study provides the means to determine the ROI performance on a customer or rate class basis.

Cost-of-Service Methodology

Common equity is a component of the cost of capital that is of particular interest for an investor-owned business, and the *return on equity* (ROE) can also be measured on both a total and a customer or rate class basis using the results of the cost-of-service study. Various elements of the capital structure are used to calculate the ROE, as given by

$$ROE = \frac{NOI - [RB \times (D + PS)]}{RB \times EQ_{CAP}} \times 100 \qquad (11.21b)$$

where ROE = the return on equity in percent
  NOI = the net operating income
  RB = rate base
  D = cost rate for long-term debt (bonds) in percent
  PS = cost rate for preferred stock in percent
  EQ_{CAP} = common equity capitalization in percent

The numerator in Equation (11.21b), $NOI - [RB \times (D + PS)]$, is referred to as the *Return Available for Equity*. Note that the calculation of ROE assumes that debt and preferred stock are fully compensated. As a result, it is possible to compute a negative ROE for a customer or rate class when its present rate revenue is low compared to its allocated portion of rate base.

The ROI and ROE results for the cost-of-service study example are shown in Table 11.24. The returns of the customer classes are found to vary considerably compared to the total.

Table 11.24 Example Summary of ROI and ROE Results

	Total $ (000s)	Residential	Commercial	Industrial	Lighting
Revenue	755,015	291,936	258,687	194,681	9,710
Expenses	517,789	199,914	170,093	141,328	6,454
NOI_{BIT}	237,226	92,021	88,594	53,354	3,257
Income Tax	148,229	60,204	52,825	33,104	2,096
NOI	88,997	31,818	35,769	20,250	1,160
Rate Base	1,139,845	493,163	374,767	257,280	14,636
ROI	7.81%	6.45%	9.54%	7.87%	7.93%
ROE	7.80%	4.86%	11.57%	7.94%	8.06%

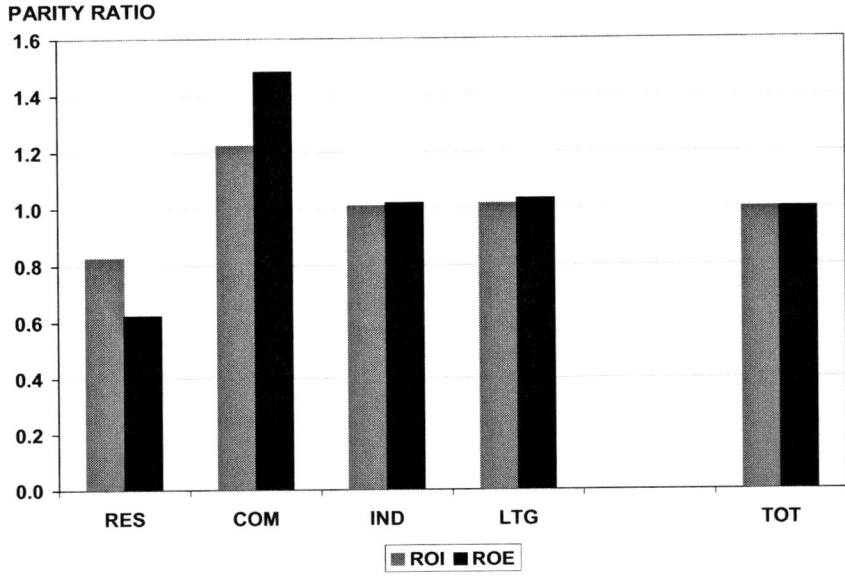

Figure 11.26 Analysis of parity based on customer class ROI and ROE results. The ROI (ROE) of each class is divided by the total ROI (ROE) in order to compute a ratio. These results indicate that the commercial class is subsidizing the residential class on a cost-of-service basis.

The class ROI and ROE results can be unitized in order to determine *a parity ratio*. In other words, the return of each class is divided by the total return. The parity ratio provides a simple means for comparing the financial performance of the classes, as shown by the graph in Figure 11.26. In this example, both the industrial and lighting classes are yielding an ROI and ROE that is only slightly above the average. However, the commercial class is observed to subsidize the residential class by a significant amount.

Cost-of-Service Study Applications

The cost-of-service study serves as a resource for supporting regulatory activities. The quantification of financial results provides guidance for pricing strategy development. The cost-of-service study helps to answer key questions such as:

- Should rate revenues be increased?
- How should a revenue increase amount be allocated to the customer or rate classes?
- How should rate structure charges be modified?

Cost-of-Service Methodology

With capital expansion and rising costs, rate increases are inevitable from time to time to prevent earnings erosion. The example cost-of-service study indicates that an increase in rates would be needed to improve the ROI to the target (weighted cost of capital) of 9.974%. Achieving this level of ROI would result in a 12.75% ROE. Thus, the overall revenue requirement to produce these results is determined by a series of calculations:

Rate Base	$1,139,844,929
Weighted Cost of Capital	9.974%
Target NOI	$113,688,133
Less Current NOI	- 88,997,033
Incremental NOI	24,691,100
Income Tax Expansion Factor [Factor = $1/(1 - T_C)$]	1.651391
Incremental Operating Income	40,774,668
Plus Current Operating Income	237,225,807
Target Operating Income (NOI$_{BIT}$)	278,000,474
Expenses	517,788,787
Target Revenue Requirement	$795,789,261
Current Revenue	755,014,593
Revenue Increase: Amount	$40,774,668
Percent	5.40%

Most of the revenue increase would be derived from revisions in rate schedule pricing, i.e., revenue from sales. The remainder would then need to be acquired from the other revenue sources, such as miscellaneous service fees.

In order to develop new pricing, the revenue increase to sales must first be allocated to the rate schedules. This allocation is based on base rate revenues (excluding any embedded fuel costs). This allocation results in a uniform percent increase to each rate schedule. With respect to class parity findings, the consideration of adjusting specific customer or rate class increase amounts is warranted. As discussed above, the commercial class is found to be subsidizing the residential class by a significant margin. Thus, it would be appropriate to give less than the average increase to the commercial class while giving the residential class more, so as to move the two classes closer to parity. However, achieving parity may take time. Reaching parity all at once in this example would result in a significant increase to residential rates with offsetting decreases to the commercial class rates. Consideration of customer impacts is also a part of revenue allocation.

The cost-of-service study also provides information which is vital for the pricing function. In particular, the customer, demand, and energy cost components can be cast as component revenue requirements. These data serve as inputs to the rate design process by providing a starting point for establishing rates that are intended to recover costs on a cost causation basis as will be discussed in Chapter 12.

11.7 SUMMARY

The cost-of-service study is a meticulous investigation of the capital investment and expenses required to render electric service to customers. The cost-of-service study is composed of accounting and financial data which are organized, evaluated, and assigned to regulatory jurisdictions and to their respective customer or rate classes. Various analyses based on utility engineering principles and applications are key to capturing the cost causation aspects of providing electric service. Thus, the cost-of-service study is a business model which not only provides an indication of financial performance but also provides guidance for ensuring the development of effective rate structures.

The three classifications of the costs of electric service are customer, demand, and energy. The key drivers of these cost components are numbers of customers, peak loads, and kWh consumption. Allocation factors are developed for each of these cost components so that their root costs are apportioned to customers in a fair and equitable manner. For instance, customer end-use loads, operated to meet lifestyle and business needs, interact at all levels of the power system, since loads at the meters are ultimately presented to the generators by means of the transmission and distribution system. The varying degrees of load diversity throughout the power system result in a dynamic configuration of cost-causative loads. As such, a variety of load-based allocation factors are required to properly model customers' demand-related costs at each system level. Load research is key to acquiring quality load data. Incorporation of electrical losses is essential to ensure equitable load and energy-based allocations.

The example cost-of-service allocations presented herein represent a number of the commonly applied methods and, perhaps, some unique approaches; however, they are not meant to signify an exclusive methodology. While both the goals and the procedural steps of a cost-of-service study are universal, the exact cost allocation logic employed depends on the specific conditions of the system being evaluated. Customer mix, system load characteristics, and regulatory precedents are but a few of the considerations which support the development of a particular cost allocation algorithm. The key lesson is to understand the underlying principles that give rise to why a certain cost element is incurred to provide electric service so that it may be assigned to customers in a sensible and justifiable manner.

12

Translating Costs to Prices

12.1 INTRODUCTION

Rate design is the process of establishing a rate structure for a given class of electric service, along with its constituent price components, that is capable of achieving a specified revenue requirement at a commensurate level of risk. As discussed in Chapter 8, rate structure risk is shared both between the electric service provider and the customer and between the customers served together under a given rate schedule.

The cost-of-service study is a key resource for the rate design function. The cost-of-service study provides electric service costs on a customer, demand, and energy component basis for the major power system functions of production, transmission, distribution (primary and secondary), customer accounts, customer assistance, and sales. Such cost information, together with the associated customer, demand, and energy billing units of the rate schedule classes, provides a basis from which to develop elementary rates for electric service. For example, consider the component cost data for a residential class of service:

Residential Electric Service	**Revenue Requirements**
Customer Component	
▪ Distribution	$35,630,382
▪ Customer Accounts	15,441,501
▪ Customer Assistance	1,853,112
▪ Sales	677,552
Customer Total	$53,602,547

Demand Component	
• Production	$94,201,326
• Transmission	12,594,916
• Distribution	29,978,186
Demand Total	$136,774,428
Energy Component	
• Nonfuel	$23,232,828
• Fuel/Purchased Power	64,087,162
Energy Total	$87,319,990
Total Revenue from Sales	$277,696,965
Nonfuel Target Revenue	$213,609,803

For this example, the total cost of O&M production fuel is considered to be recovered through an auxiliary rate schedule. Thus, the remaining fixed and variable costs are recovered through a rate structure that represents base rate billing. The residential class consists of 225,000 customers with an annual sales volume of 2,644,798,967 kWh. Monthly service is measured by means of watt-hour meters; thus, the demand cost component of revenue must be recovered on the basis of kWh consumption. A two-part rate structure with a monthly (or daily) customer charge and a basic straight-line (flat) kWh rate would be calculated by:

Customer Charge = $53,602,547 ÷ (225,000 Customers × 12 Months)
= $19.8528 per Bill → $19.85 per Bill

Energy/Demand Charge = ($23,232,828 + $136,774,428) ÷ 2,644,798,967 kWh
= $0.060498835 per kWh → 6.050¢ per kWh

Due to rounding of the charges for practical application, the rate yields slightly less than the target revenue:

$19.85 per Bill × 2,700,000 Bills =	$53,595,000	
$0.06050 per kWh × 2,644,798,967 kWh =	160,010,338	
	213,605,338	
	4,465	Under Target

The shortfall is 0.002% of the target. Because of the volume of kWh, and with only a single kWh charge to adjust, it is difficult to fine-tune the rate to acquire the exact target. Changing the rate to 6.0501¢ per kWh would yield an additional

$2,645 thereby reducing the shortfall to $1,820. Changing the rate to 6.0502¢ per kWh would yield an additional $5,290, which exceeds the target revenue by $825. A rate of 6.05017¢ per kWh would result in revenue that is just $31 over the target. However, carrying out the rate to too many decimal places results in *phantom revenue*. In other words, the extra decimal places have no effect on lower volume bills because of rounding to dollars and cents. In general, not exceeding 3 to 4 decimal places on a kWh-based rate is preferable. Generally, a rate is designed to produce no more than the target revenue.

Alternatively, the rate could be designed to achieve other cost recovery aspects. For example, the rate could be seasonally differentiated and/or kWh blocks could be incorporated to recognize cost variations based on monthly usage levels. As shown in Figure 12.1, a plot of monthly system peak loads reveals any seasonality that may exist throughout the year. In this example, May through September costs would be higher than the other months of the year. A major driver of this load pattern is the residential class, which uses a greater amount of energy in the summer months as a result of air conditioning. A dual block rate design would be effective in recovering the system's seasonal cost variations.

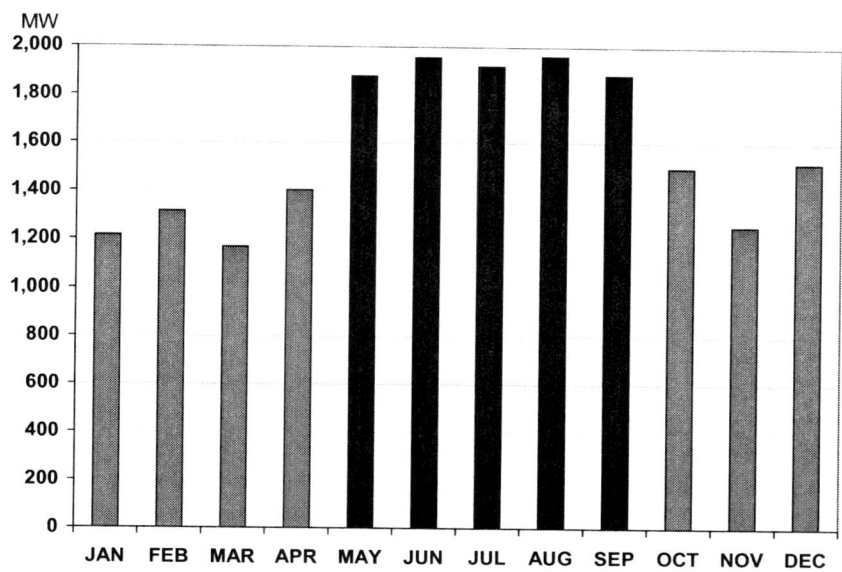

Figure 12.1 Determination of peak and off-peak seasons. A plot of monthly system peak demands helps to define seasonal rating periods. Compared to a year round flat rate, a seasonal rate design would produce revenues more on a cost-causation basis. Blocking of the kWh charges would provide additional price signals to customers.

As shown in Table 3.3, monthly residential base use for nonweather sensitive appliances is generally on the order of 500 to 600 kWh per month. In this example, base use will be defined as 600 kWh per month, and thus a kWh block will be established at 600 kWh for both the summer month and the nonsummer month rate structures. The first 600 kWh block will be priced at the same rate per kWh since it represents year round usage of nonweather sensitive end-use devices. The monthly Customer Charge calculated previously will also be used in this example. The concept rate structure is given by

Customer Charge: $19.85 per Bill

Demand/Energy Charges:

May – September Rate Structure
1st 600 kWh @ $Z\cent$ per kWh
Excess kWh @ $X\cent$ per kWh

October – March Rate Structure
1st 600 kWh @ $Z\cent$ per kWh
Excess kWh @ $Y\cent$ per kWh

Seasonal block rate design requires a monthly kWh bill frequency analysis of the rate class, and the bill frequency must calculate the consolidated factors for the 600 kWh intervals.

	Total kWh	1st 600 kWh	Excess kWh
May – September	1,281,758,967	471,843,107	809,915,860
October – March	1,363,040,000	823,379,257	539,660,743
Total	2,644,798,967	1,295,222,364	1,349,576,603

In addition, differential block pricing guidance is needed. One approach is to use system lambda in order to spread the prices among the seasonal kWh blocks. An analysis of system lambda indicates that the ratio of the average May through September lambda to the annual average lambda is 1.107169, and the comparable ratio for October – March is 0.992661. By applying these ratios, the tail step kWh prices in the summer and non-summer periods will reflect the marginal costs of generation in those seasons. A two-part seasonal rate structure with a monthly (or daily) customer charge and a combined inverted block—declining block kWh rate would be calculated by:

Total Nonfuel Target Revenue	$213,609,803
Less Customer Charge Revenue (based on $19.85/Bill)	53,595,000
Demand/Energy Charge Revenue Target	$160,014,803

Translating Cost to Prices

The seasonal lambda ratios can be used to differentiate the summer and non-summer tail block prices as a function of the year-round base use block:

Base Use Block = Z
Summer Tail Block = $X = 1.107169 \times Z$
Nonsummer Tail Block = $Y = 0.992661 \times Z$

The Demand/Energy Charge revenue target is set equal to the sum of the revenue produced by the application of the three unit prices to the base use kWh, the excess summer kWh, and the excess nonsummer kWh, as given by

$[Z \times 1{,}295{,}222{,}364] + [1.107169 \times Z \times 809{,}915{,}860] + [0.992661 \times Z \times 539{,}660{,}743] = \$160{,}014{,}803$

Thus, the base use block price, Z, is calculated by

$$Z = \frac{\$160{,}014{,}803}{2{,}689{,}860{,}079 \text{ kWh}} = \$0.059488 \text{ per kWh}$$

The summer tail block price, X, is thus found to be equal to $0.065863 per kWh, and the nonsummer tail block price, Y, is found to be equal to $0.054887 per kWh.

The abridged block prices selected for the revenue yield test are: $Z = 5.949¢/\text{kWh}$, $X = 6.586¢/\text{kWh}$, and $Y = 5.489¢/\text{kWh}$. These rates are found to produce $160,015,815, which is $1,012 over the revenue target. Adjustments of the third significant decimal place in each rate are made to reduce this amount so as to not exceed the target. The final rates below result in a revenue production that is only $91 under the target revenue.

Customer Charge: $19.85 per Bill

Demand/Energy Charges:

<u>May – September Rate Structure</u>
1st 600 kWh @ 5.951¢ per kWh
Excess kWh @ 6.580¢ per kWh

<u>October – March Rate Structure</u>
1st 600 kWh @ 5.951¢ per kWh
Excess kWh @ 5.489¢ per kWh

Nondemand rate design is relatively straightforward and can be accomplished effectively with minimal analysis. On the other hand, demand rate architecture is much more complex as the metering of customers' monthly maximum demands adds another dimension to the design problem. Capturing the cost-causation aspects of demand-metered customers requires a special analysis that is essentially an extension of inter-class cost-of-service study principles within a given rate class. While the conventional cost-of-service study apportions total costs between customer or rate classes, these assigned costs must then be further allocated to customers within the class on the basis of the customers' load factors. The key to this intra-rate cost-of-service analysis is an understanding of how the coincidence of load relates to load factor.

12.2 COINCIDENCE FACTOR—LOAD FACTOR RELATIONSHIP

The relationship between customer group coincidence factors and individual customer's load factors was first investigated and developed in the late 1930s by Constantine W. Bary of the Philadelphia Electric Company. This empirical relationship was explained from a theoretical sense in a mathematical exposition by H. E. Eisenmenger of the Consolidated Edison Company of New York, Inc.[1] Their results were presented to the Load Research Committee of the Association of Edison Illuminating Companies (AEIC). Over the next several decades, the Load Research Committee conducted repetitive studies of the coincidence factor—load factor (CF—LF) relationship using load research data supplied by AEIC member companies.

In essence, the relationship shows that coincidence factor has a nonlinear relationship to load factor, as illustrated by the CF—LF curve plotted in Figure 12.2. Specifically, the relationship is expressed in the form of a third-order polynomial, as given by

$$CF = \mathbf{A} + \mathbf{B}\, LF + \mathbf{C}\, LF^2 + \mathbf{D}\, LF^3 \qquad (12.1)$$

The x-axis represents the load factors of individual customers based on monthly demand and energy usage. The y-axis represents the coincidence factors for groups of customers. These groups are composed of individual customers having the same monthly load factors.

The plot shows that as load factor starts to increase from zero, the coincidence factor increases at a very fast rate, i.e., the slope of the curve approaches vertical. In other words, there is a quick loss of load diversity as load factor rises. As load factor continues to increase, the slope of the curve decreases substantially, and, at mid-range load factors, the slope of the curve approaches zero. At higher load factors, the slope of the curve increases so that at 100% load factor, the group loads are 100% coincident.

---

[1] In his treatise, Eisenmenger referred to the relationship between coincidence factor and load factor as the *Bary curve*.

Translating Cost to Prices

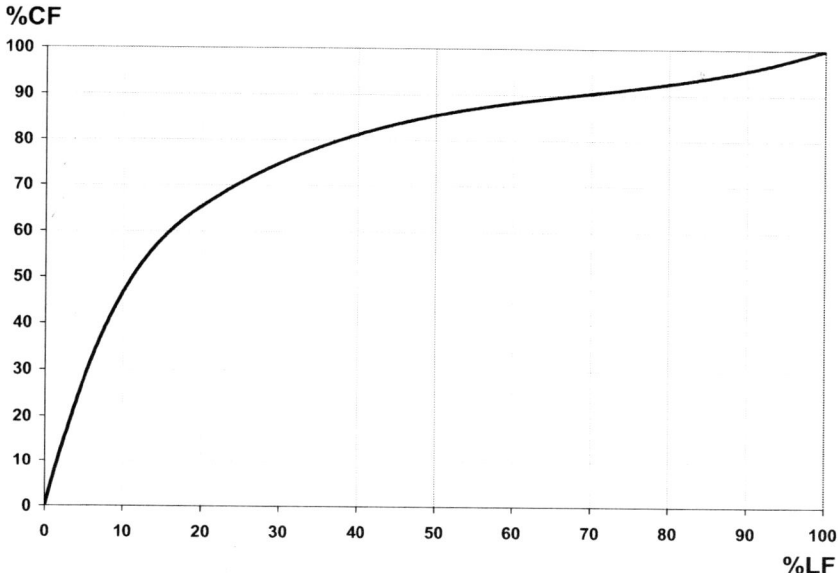

Figure 12.2 An example of the nonlinear relationship between coincidence factor and load factor. The *x*-axis represents monthly load factors of individual customers. The *y*-axis represents the coincidence factors of groups of customers having the same monthly load factors. The CF—LF relationship was first identified by Constantine W. Bary, and it is commonly referred to as the Bary Curve. The Bary curve is a key input to the development of cost curves for demand metered rate class.

The CF—LF curve can be viewed as representing the probability that customer loads will, or will not, peak simultaneously. As shown in Figure 12.3a, the probability of coincidence is low with groups of low load factor customers. Conversely, the probability of coincidence is high with groups of high load factor customers, as shown in Figure 12.3b. The crux of the CF—LF relationship is that it is based upon groups of customers.[2] While the loads of any two 100% load factor customers are obviously 100% coincident in every hour, the individual peak loads of a group of numerous low load factor customers will be distributed about the month based on variations in end-use device types and operating patterns. With some exceptions, low load factor customers are typically small and highly diverse commercial operations.

---

[2] Typically, thousands of customers are served under general service rate schedules. The CF—LF relationship is not based on any single rate schedule but on the general body of commercial and industrial customers.

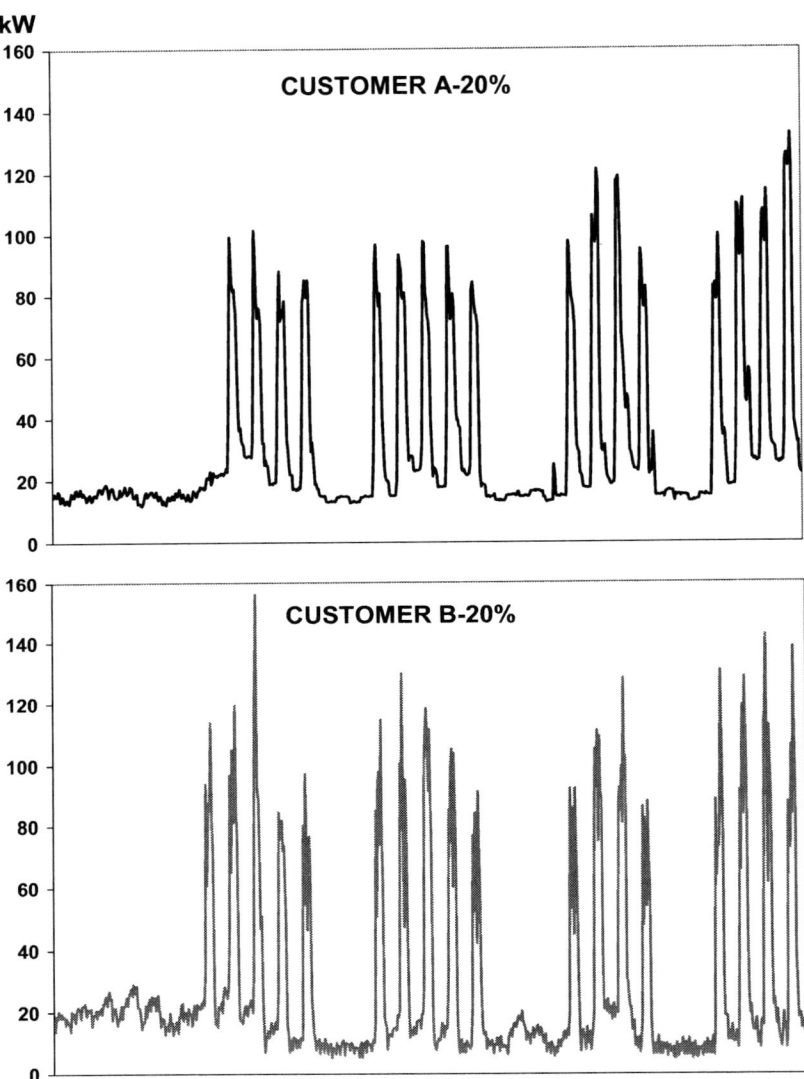

Figure 12.3a Load diversity as a function of load factor. Shown are example hourly load curves of two 20% load factor commercial customers for a common month. With the highly cyclical characteristics of these load curves, there is a fairly low probability of coincidence. While any two such load shapes could result in peaking simultaneously, a group of several such low load factor customers is not highly likely to all peak at the same time.

Translating Cost to Prices

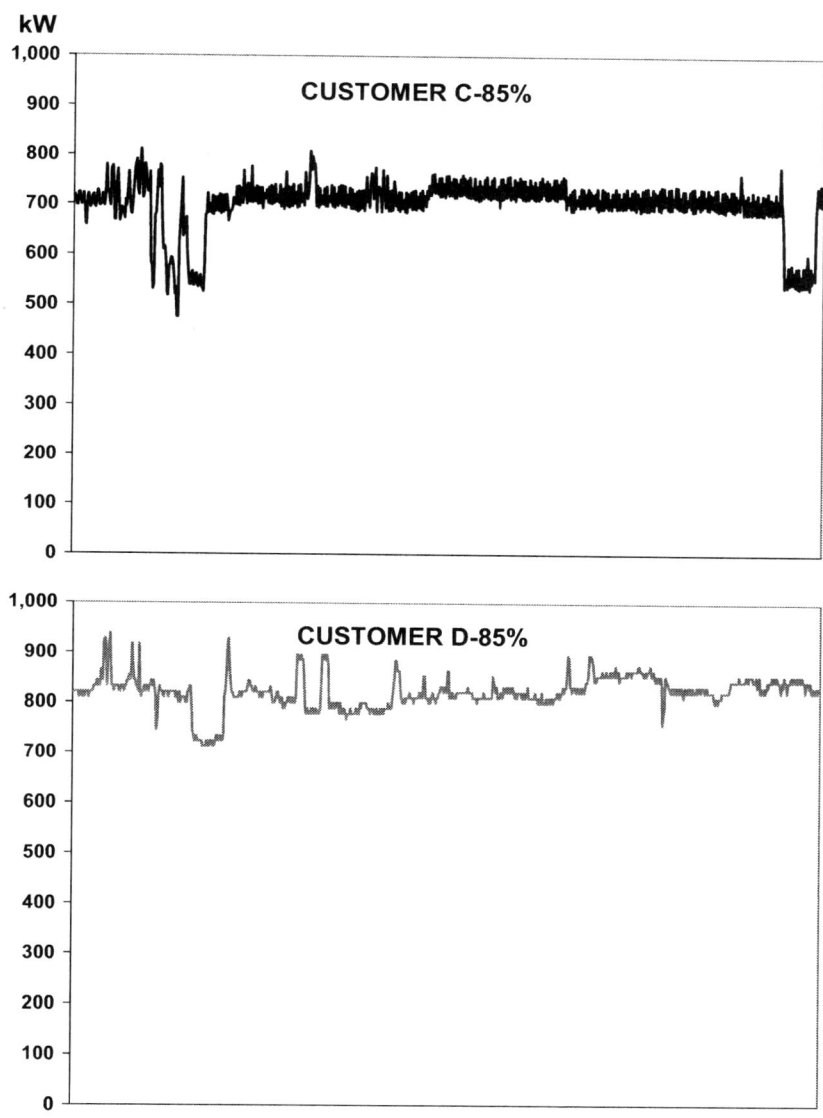

Figure 12.3b Load diversity as a function of load factor. Shown are example monthly load curves of two 85% load factor industrial customers. At such a fairly high load factor, there is a high probability that the two loads would be highly coincident across the month. The rather flat load profile offers minimal opportunity for load diversity. At a 100% load factor, just two loads alone would be 100% coincident.

Development of the Coincidence Factor—Load Factor Relationship

The CF—LF curve is an estimate of system-wide customer load interactions and thus is not specific to any particular rate schedule class. Typically, the relationship is determined using commercial and industrial customer loads. The development of the CF—LF relationship requires hourly customer load data, which is acquired from interval revenue metering and load research survey recorders.

Since the independent variable of the CF—LF relationship is represented by individual customer load factors, data from customers having a common load factor are joined to form a load factor group from which its particular coincidence factor can be calculated. The customer input data is sorted and aggregated by load factor in order to form the groups. Since the calculation of the coincidence factor is dependent on a compilation of loads, the sorting of customers is better accomplished by load factor ranges as opposed to discrete load factors.[3] To provide for a more robust set of groups, the customers can be sorted by load factor ranges. Thus, the 10% load factor group could be composed of those customers having monthly load factors of 6% to 15%. The 20% load factor group would then consist of customers having monthly load factors of 16% to 25%. The result would be ten load factor groups, with 96% to 100% representing the last group.[4]

The coincidence factor for a group of loads is defined in Equation (2.16b). While the sorting and grouping of customers is based on load factor, any particular group could have a wide distribution of customer maximum demands. A very large industrial load could be grouped with several smaller commercial and industrial customers. The result of calculation of the coincidence factor, as defined in Equation (2.16b), could be skewed by the dominance of the single large customer. Since the goal is to determine typical coincidence factors across the population of customers, the loads of the customers in each of the load factor groups are unitized prior to the calculation of the coincidence factor. This approach gives each observation an equal weight and focuses the result on load interaction behavior as opposed to load magnitude.

Once the load factor group coincidence factors are calculated, they can be plotted as a function of load factor, as shown in Figure 12.4. A continuous function that relates coincidence factor to load factor is then developed by applying polynomial regression, in the form of Equation (12.1), to the (CF,LF) data points. The regression line should not be forced through the origin.

---

[3] Even if interval metered data were available for the entire population of customers, sorting of the customers by discrete load factor values (i.e., 1%, 2%, 3% . . . 100%) would result in 100 load factor groups in which some may have only a single or just a few observations. Thus, the calculation of a group coincidence factor would be either impossible or unreliable.

[4] As a result of the load diversity in each group, the load factor of a group can be higher than the range of its individual customer load factors. This effect is more pronounced with the lower ranges of load factor where the load diversity is greatest.

Translating Cost to Prices

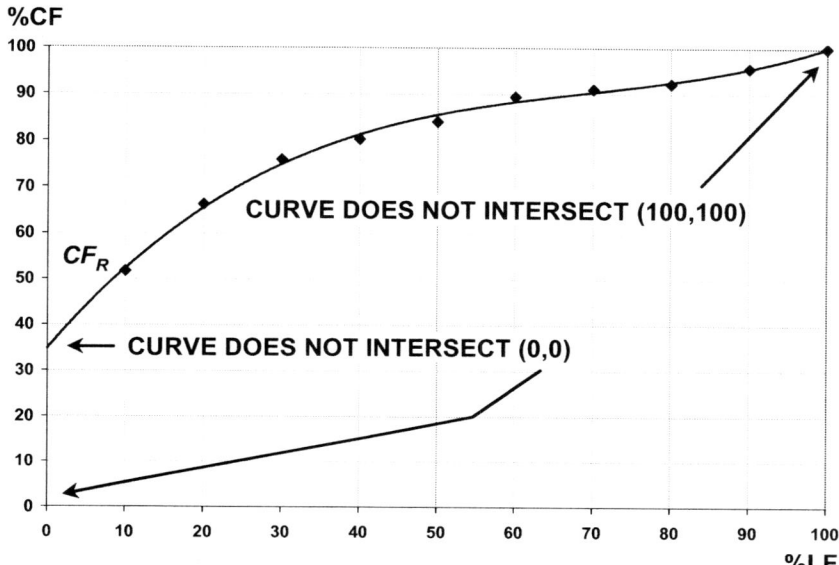

Figure 12.4 Results of a CF—LF regression analysis. The regression solution indicated by CF_R fits the data well with an r^2 near unity. However, the regression line has a nonzero y-intercept, and it yields a coincidence factor that slightly exceeds 100% at 100% load factor. Both of these issues must be addressed before the CF—LF relationship can be used as a model for intra-rate demand cost allocation.

The polynomial curve fit is a cubic least squares solution. The problem is defined by a system of four simultaneous linear equations given by

$$\sum_{i=1}^{n} y_i = An + B\sum_{i=1}^{n} x_i + C\sum_{i=1}^{n} x_i^2 + D\sum_{i=1}^{n} x_i^3 \quad (12.2a)$$

$$\sum_{i=1}^{n} x_i y_i = A\sum_{i=1}^{n} x_i + B\sum_{i=1}^{n} x_i^2 + C\sum_{i=1}^{n} x_i^3 + D\sum_{i=1}^{n} x_i^4 \quad (12.2b)$$

$$\sum_{i=1}^{n} x_i^2 y_i = A\sum_{i=1}^{n} x_i^2 + B\sum_{i=1}^{n} x_i^3 + C\sum_{i=1}^{n} x_i^4 + D\sum_{i=1}^{n} x_i^5 \quad (12.2c)$$

$$\sum_{i=1}^{n} x_i^3 y_i = A\sum_{i=1}^{n} x_i^3 + B\sum_{i=1}^{n} x_i^4 + C\sum_{i=1}^{n} x_i^5 + D\sum_{i=1}^{n} x_i^6 \quad (12.2d)$$

where x_i = observations of load factor, as a percentage
       y_i = observations of coincidence factor, as a percentage
       n = number of observations

The objective of the curve fit analysis is to determine the regression coefficients A, B, C, and D. Equation (12.2) can be restated in matrix form, as given by

$$[Y] = [X] \times \begin{bmatrix} A \\ B \\ C \\ D \end{bmatrix} \quad (12.3a)$$

and solved by inverting $[X]$, as given by

$$\begin{bmatrix} A \\ B \\ C \\ D \end{bmatrix} = [X]^{-1} \times [Y] \quad (12.3b)$$

Given the input data:

Load Factor Group		Coincidence Factor	
$x_i =$	10	$y_i =$	51.76
	20		66.21
	30		75.90
	40		80.28
	50		84.03
	60		89.43
	70		91.07
	80		92.23
	90		95.63
	100		100.00

the solution to the regression problem is found to be:

$$CF_R = 0.000125 \times LF^3 - 0.02599 \times LF^2 + 2.002\ LF + 34.8$$

The subscript R merely represents that the equation is a result of regression. The solution line is plotted in Figure 12.4.

CF—LF Model Calibration

As noted in Figure 12.4, there are two significant issues that must be addressed in order for the CF—LF relationship to be in a format which can be used as a demand cost allocation model. The curve must fit exactly between the origin and 100% (as supported by the empirical evidence).

The first issue is the positive intercept of the regression solution, which, in this example, is 34.8%. A nonzero intercept is a common occurrence in the development of the CF—LF relationship, in part, because of the lack of observations available at low load factors.[5] Additional metering could improve the curve fit at the low end; however, the associated effort would not be cost effective. Forcing the regression solution through the origin could be done (with great distortion), but the resulting r^2 would indicate a poor fit to the data. Application of a higher order polynomial is not a reasonable approach.

A workable solution is to synthesize a third-order polynomial, to be noted as CF_S, which intersects the origin, and splice it to the regression solution at a point where its results appear to be sound. Given the underlying data set used for the example, a 20% load factor is a reasonable point in which to splice a synthesized curve to the regression curve. To ensure a smooth transition from the synthesized curve to the regression curve, both equations must yield the same coincidence factor at 20% load factor. In addition, the slopes of the two curves must be equal at the 20% load factor intersection. Thus, constraints at 20% load factor are represented by:

$$CF_S = CF_R$$

and

$$dCF_S = dCF_R$$

The first constraint provides a data input for the synthesized curve as $CF = 65.44\%$. The origin (0,0) also serves as a data input. A third-order regression requires at least three data points, so an approximation by inspection is made at 10% load factor. A final solution is found by iteratively adjusting the 10% load factor value and testing the resulting regression solution against the two constraints identified above until they are met. A solution which meets the 20% load factor constraints is found to be:

$$CF_S = 0.003157 \times LF^3 - 0.2356 \times LF^2 + 6.721\ LF$$

---

[5] Another reason for the nonzero intercept is due to the somewhat erratic nature of load diversity of low load factor loads. Calculation of coincidence factors for each month of the year for the 10% load factor group would result in a fairly wide distribution of results above and below the regression trend line. Similar calculations for the other load factor groups would also result in distributions above and below the trend line; however, the distributions become tighter as 100% is approached.

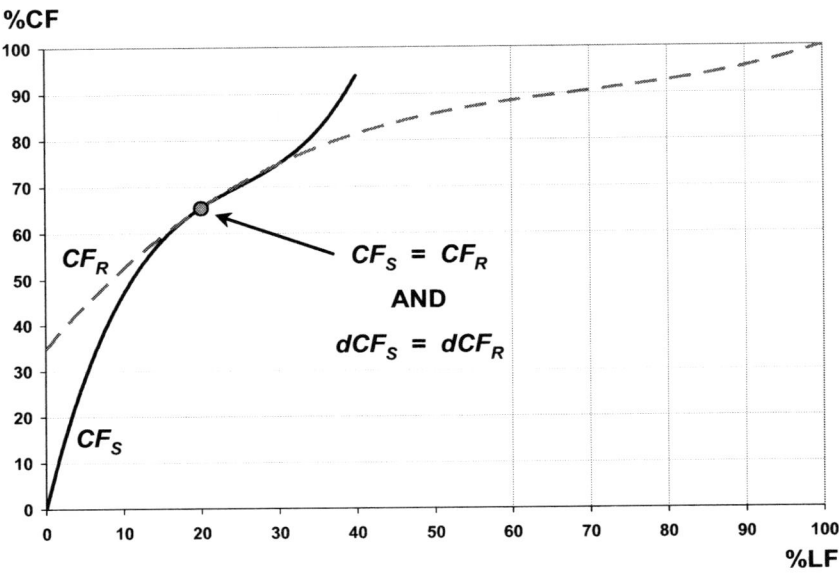

Figure 12.5 **Synthetic CF—LF curve splice.** The synthetic third-order polynomial, CF_S, makes up for a nonzero y-intercept deficiency which occurs commonly in CF—LF regression analyses. The coincidence factor value and slope constraints at the 20% load factor splice point yield a smooth transition from one curve to the other. The synthetic curve solution is required in order to develop the CF—LF relationship for application as an intra-rate demand cost allocator.

The synthetic curve is plotted in Figure 12.5, and it is observed to tie in smoothly with the original regression curve at the 20% load factor constraint. For application purposes, CF_S is used for load factors between 0 and 20%, and CF_R is used for load factors between 20% and 100%. However, both curves require a slight adjustment before the overall CF—LF relationship is fully calibrated between 0 and 100%. As indicated in Figure 12.4, when the regression equation CF_R is tested at $LF = 100$, the solution does yield a coincidence factor that slightly exceeds 100.0. In this example, the result is found to be 100.1. Calibration is completed by multiplying both CF_R and CF_S by the ratio of 100 to 100.1. Thus, the final CF—LF solution is described by:

0 – 20% LF: $0.003153846 \times LF^3 - 0.235364635 \times LF^2 + 6.7143\, LF$

20 – 100% LF: $0.000124875 \times LF^3 - 0.025964036 \times LF^2 + 2.0\, LF + 34.76523$

Translating Cost to Prices

When plotted together over their applicable load factor ranges, the calibrated CF_S and CF_R equations yield the CF—LF curve illustrated in Figure 12.2. When developing a rate class cost curve as a function of load factor, or the hours use of demand, the CF—LF curve serves as a key allocator for certain demand-related cost components.

Variations of the CF—LF Relationship

The development of the CF—LF curve discussed above is based on the conventional calculation of the coincidence factor, which is the ratio of the maximum demand of a group of individual loads to the sum total of the maximum demands of the individual loads, as defined by Equation (2.16b). In this context, the individual loads are customers having a common value of load factor. When considering the load factor group's position with respect to the total system load, the resulting maximum demand of the customer group represents a non-coincident peak (NCP) type of factor. In other words, the group peak may or may not occur simultaneously with the system peak demand.[6] The cost-of-service study regards the NCP type of factor as being highly indicative of the cost causation of the demand-related component of the primary distribution system. Thus, the conventional CF—LF relationship can be utilized as an intra-rate cost allocator for the primary distribution demand costs assigned to a particular rate class during the cost-of-service analysis.

The conventional CF—LF relationship is not a plausible indicator of cost causation for production and transmission system demand-related costs since load diversity is much greater at these levels of the system than at the distribution system level. As discussed in Chapter 11, a strong indicator of the cost causation driver of production and transmission is a coincident peak (CP) type of demand cost allocation factor. Consequently, an adapted coincidence factor is required for development of a modified CF—LF relationship in order to properly allocate these costs in an intra-rate cost curve development.

Figure 12.6 illustrates a monthly load shape for an example load factor group of customers. This particular group's maximum demand occurs one week after the system maximum demand is established for that month. The group's load at the time of the system peak is notably less that the group's maximum peak due to the innate diversity characteristics associated with that group's load factor. A modified coincidence factor type of calculation can be formulated by substituting the group's load at the time of system peak in place of the group peak load as the numerator of the conventional coincidence factor equation. Furthermore, the modified coincidence factor can be expected to be less than the conventional coincidence factor because the numerator has decreased while the denominator, the sum of customer maximum demands, remains the same.

---

[6] The higher load factor groups are more likely to peak at the same time as the system since load diversity is minimal.

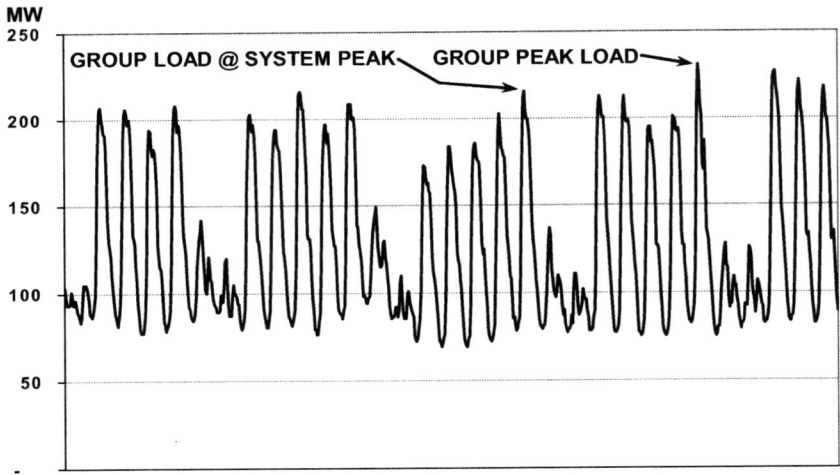

Figure 12.6 Relationship of coincidence factor to the cost of service study demand allocators. An example load factor group's peak demand occurs at a time different from the system peak demand. A coincident factor calculation based on the group peak relates to an NCP type of cost-of-service allocation factor, while a coincidence factor based on the group's load at the time of system peak relates to a CP type of allocation factor.

Given two different coincidence factors for the same load factor group, the conventional coincidence factor will be designated as CF_{NCP}, and the modified coincidence factor will be designated as CF_{CP}.[7] To summarize, the load factor group coincidence factors are determined by

$$CF_{NCP} = \frac{D_{MAX}}{\sum_{i=1}^{n} d_i} \qquad (12.4a)$$

$$CF_{CP} = \frac{D_{SYS}}{\sum_{i=1}^{n} d_i} \qquad (12.4b)$$

---

[7] A coincidence factor can be derived consistent with a number of typical cost-of-service study allocators, e.g., 12-CP, 5-CP, 1-CP, and NCP. Development of a CF—LF relationship for each of these conditions would reveal a family of Bary type curves.

where CF_{NCP} = a load factor group's NCP type of coincidence factor
CF_{CP} = a load factor group's CP type of coincidence factor
D_{MAX} = a load factor group's maximum demand
D_{SYS} = a load factor group's demand at the time of system peak
d_i = maximum demand of the ith customer in a load factor group
n = number of customers in the load factor group

The conventional CF—LF relationship is also not a plausible indicator of cost causation for other demand-related cost components, such as line transformers and secondary distribution. As discussed in Chapter 11, the significant demand-related cost causation driver of local facilities is customer maximum demands. In essence, a single customer's maximum demand is fully coincident with itself under all conditions (i.e., 1 kW ÷ 1 kW = 1 kW), since diversity of a single load is undefined. Thus, the coincidence factor based on customer maximum demands is represented as 100% at all levels of load factor, and it can be designated as CF_{MAX}.

A chart comparing CF_{CP}, CF_{NCP}, and CF_{MAX} curves is shown in Figure 12.7. The CF_{CP} curve is based on the average of twelve monthly coincidence factor values; thus, it relates to the 12-CP demand cost allocation factor developed for the example cost-of-service study. The CF_{NCP} curve is the same curve that is plotted in Figure 12.2. Furthermore, this CF_{NCP} curve is actually based on the average of twelve monthly coincidence factor values. A pure CF_{NCP} curve based on just the highest peak demand for the year for each load factor group would lie above the CF_{NCP} curve shown in the two figures. This average-based CF_{NCP} curve was developed to help simplify the cost curve example presented in the next section and will be discussed further there. The CF_{MAX} curve is represented as a straight line across the top of the chart.

A kWh per kW scale has been added to the x-axis of the CF—LF chart in Figure 12.7. A cost curve for a rate class is in a form similar to a multi-step hours use of demand rate structure; thus, it is more practical to work with kWh per kW units than with load factor percentages.

12.3 COST CURVE DEVELOPMENT

A cost curve bridges the gap between the cost-of-service study and rate design. Unit costs are translated into prices within a rate structure. Cost curve development can be thought of as an intra-rate cost allocation methodology. Like the cost-of-service study, the more intricate part of the cost curve allocation process revolves around the demand cost components. The cost curve serves as an essential model for designing and evaluating alternative Hopkinson and Wright rate structures, particularly when displayed in graphical form. In addition, cost curves can be used to evaluate the operation of current rate structures and price levels.

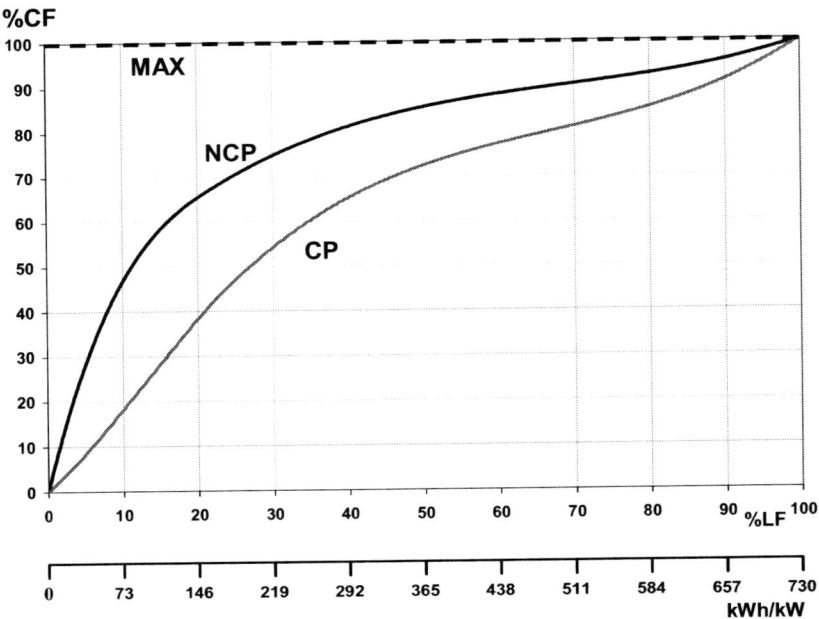

Figure 12.7 Comparison of the coincident peak and non-coincident peak CF—LF curves. The CP curve lies below the NCP curve due to greater customer load diversity at the time of the system peak as opposed to the time of a load factor group's peak. The MAX curve represents a customer's demand at the local facilities level where diversity is non-existent for all practical purposes. Also shown is a second x-axis scale having units of kWh per kW, which facilitates the chart's use for development of cost curves.

A flowchart describing the input data and analytical processes needed to develop a cost curve for a given demand rate class is shown in Figure 12.8. Input data includes the cost-of-service study component revenue requirements for the given rate. These revenue requirements are comparable to the cost components used previously in this chapter to illustrate a nondemand rate design example.

Another input is an hours use of demand bill frequency, which is based on the subject rate's kWh usage and actual kW maximum demands.[8] The frequency needs to be at full resolution, i.e., based on single kWh per kW intervals.

---

[8] It is important to utilize actual demands rather than billing demands, since actual demands relate to cost causation, while billing demands relate to cost recovery assurance. Billing demands vary from actual demands due to the effects of demand ratchets and rate and contract demand minimums.

Translating Cost to Prices

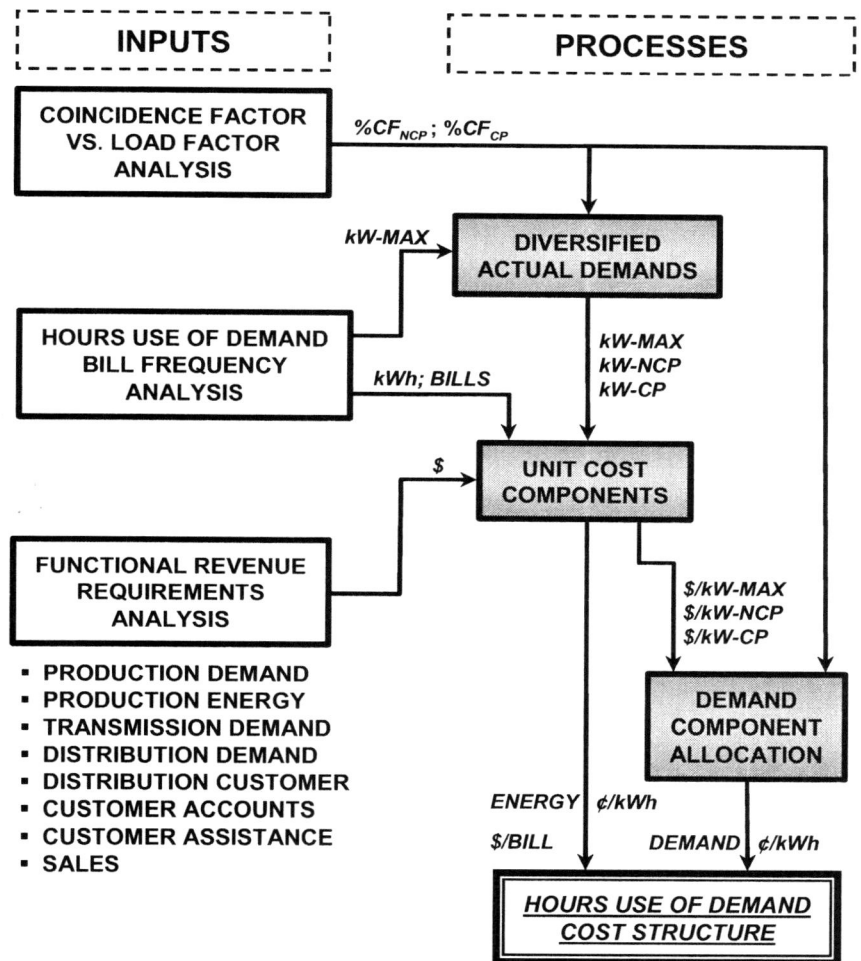

Figure 12.8 Cost curve development process. Three major analytical steps are implemented to produce a cost curve for a given rate class. The cost curve is of a form similar to an hours use of demand rate structure, and it can be easily plotted and compared to the associated rate structure that is currently used for billing.

The final input data to the cost curve development process are the CF—LF curves. The example cost-of-service study discussed in Chapter 11 was based on five demand cost allocation factors: 12-CP for production, 5-CP for transmission, 1-CP for distribution substations, 1-NCP for primary distribution, and customer

MAX for line transformer and secondary distribution. For simplicity, CF—LF curves based on 12-CP, average NCP, and MAX are used for the cost curve example below. The 12-CP coincidence factors will be applied to both production and transmission. Average NCP coincidence factors will be applied to both distribution substations and primary distribution as a compromise for the difference in load diversity of the two functional levels. Consistent with the cost-of-service study, the MAX 100% coincidence factors are used for line transformers and secondary lines. Development and application of CF—LF curves for all five of the cost-of-service demand allocation factors would provide a more granular cost curve solution. However, the use of only three CF—LF curves will yield a cost curve solution which has a reasonable level of resolution.

Diversification of Actual Demands

The first step of the cost curve development process is to diversify the actual demands of the rate class by applying the CF—LF curves to the hours use histogram, as shown in the abridged Table 12.1. The coincidence factors are applied as a function of hours use as a proxy for load factor since the histogram is based on kWh per kW intervals. The diversification effects on the distribution of actual demands are plotted in Figure 12.9.

Table 12.1 Hours Use of Demand Histogram – Diversification of Demands

HUD (kWh/kW)	LF (%)	Actual kW	CF_{NCP} (%)	NCP kW	CF_{CP} (%)	CP kW
0	0.00	0	0.00	0	0.00	0
20	2.74	5,026	16.69	839	4.42	222
50	6.85	7,097	35.96	2,552	11.93	847
100	13.70	11,298	55.92	6,318	25.71	2,904
150	20.55	14,952	65.98	9,865	39.32	5,880
200	27.40	23,450	72.64	17,034	50.61	11,867
250	34.25	18,102	77.83	14,088	59.36	10,476
300	41.10	14,731	81.77	12,046	66.03	9,727
350	47.95	11,705	84.73	9,918	71.04	8,315
400	54.79	11,945	86.94	10,385	74.82	8,937
450	61.64	10,265	88.64	9,099	77.80	7,986
500	68.49	9,882	90.07	8,901	80.42	7,947
550	75.34	8,867	91.47	8,111	83.10	7,369
600	82.19	2,308	93.09	2,148	86.29	1,992
650	89.04	520	95.15	495	90.40	470
700	95.89	821	97.91	804	95.88	787
≥ 730	100.00	46,080	100.00	46,080	100.00	46,080
Total		6,888,228		5,273,832		4,099,734

Translating Cost to Prices

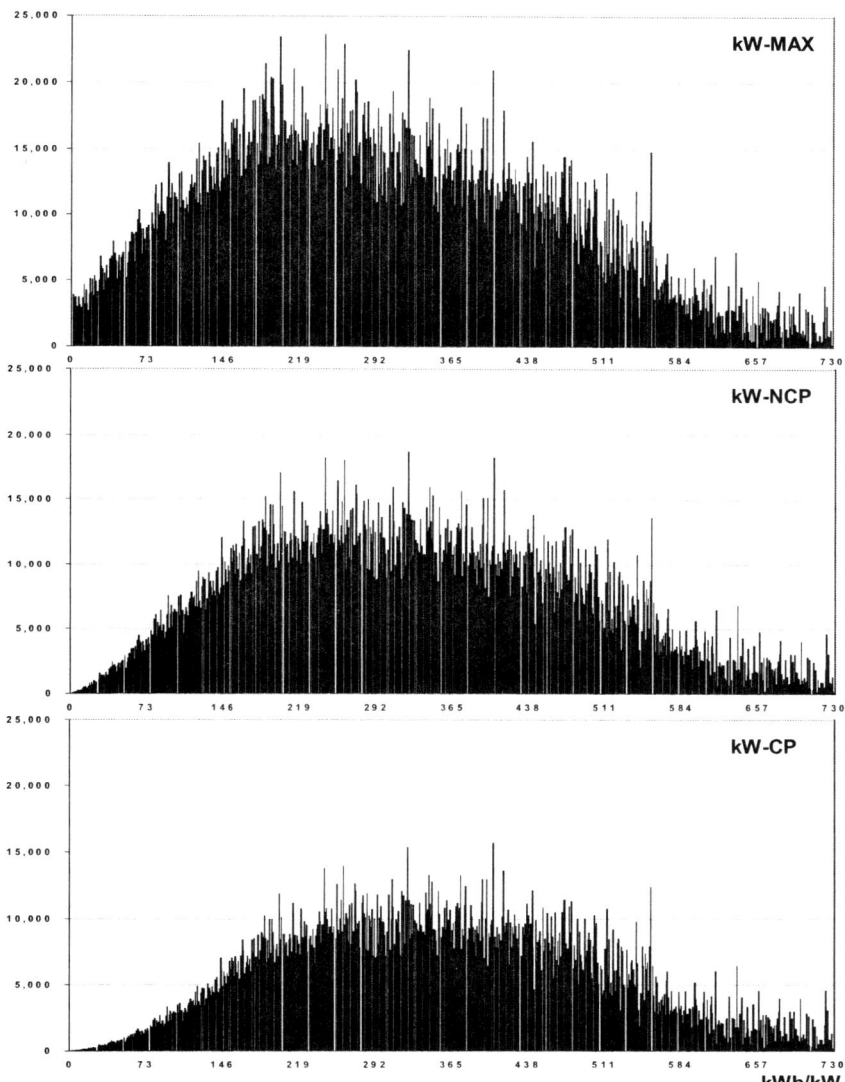

Figure 12.9 Diversification effects on the hours use of demand bill frequency. Shown are histograms of rate class maximum demands and NCP and CP diversified demands. The effect of diversifying the maximum demands across the range of hours use of demand is not uniform since the NCP and CP CF—LF curves are not linear. Diversification is observed to have a great impact on the magnitude of demands at the lower kWh/kW range and a minimal impact on the magnitude of demands at the higher kWh/kW range.

Development of Unit Cost Components

The second step of the cost curve development process is to compute unit cost components using the results of the demand diversification analysis and functional revenue requirements from the cost-of-service study. For example, consider the component cost data for a medium commercial/industrial class of service:

Secondary 3ϕ General Service	**Revenue Requirements**
Customer Component	
▪ Distribution Primary	$902,889
▪ Line Transformers	1,395,329
▪ Secondary Distribution	317,872
▪ Service Line	246,754
▪ Revenue Meter	647,684
▪ Customer Accounts	3,236,323
▪ Customer Assistance	101,462
▪ Sales	37,104
Customer Total	$6,885,417
Demand Component	
▪ Production	$61,233,793
▪ Transmission	6,326,692
Demand Subtotal	$67,560,485
▪ Distribution:	
▪ Substation	$3,606,378
▪ Primary Lines	5,845,192
Demand Subtotal	$9,451,570
▪ Line Transformers	$2,639,972
▪ Secondary Lines	1,732,945
Demand Subtotal	$4,372,917
Demand Total	$81,384,973
Energy Component	
▪ Nonfuel	$19,637,444
▪ Fuel/Purchased Power	51,292,168
Energy Total	$70,929,612
Total Revenue from Sales	$159,200,002

Translating Cost to Prices

The example rate class consists of 12,000 commercial and industrial secondary three-phase customers having an annual energy use of 2,100,000 MWh. The maximum and diversified actual demands are provided in Table 12.1. The unit cost components are calculated below:

Unit Customer Cost   = $6,885,417 ÷ (12,000 Customers × 12 Months)
                                    = $47.8154 per Bill

Unit Demand Costs:
- Production/Transmission   = $67,560,485 ÷ 4,099,734 kW_{CP}
                                    = $16.4792 per kW_{CP}

- Substation/Primary Lines   = $9,451,570 ÷ 5,273,832 kW_{NCP}
                                    = $1.7922 per kW_{NCP}

- Transformers/Sec. Lines   = $4,372,917 ÷ 6,888,228 kW_{MAX}
                                    = $0.6348 per kW_{MAX}

Unit Energy Costs:
- Nonfuel   = $19,637,444 ÷ 2,100,000,000 kWh
                                    = $0.009351 per kWh

- Fuel/Purchased Power   = $51,292,168 ÷ 2,100,000,000 kWh
                                    = $0.024425 per kWh

Allocation of Unit Demand Cost Components

The functional unit demand costs developed above represent the cost assignments to a 100% load factor customer. A 100% load factor customer is totally coincident with both the system peak and its group peak. Thus, all of the high load factor customer peak demands would be assigned $15.4792 per kW of peak demand for production and transmission capacity and $1.7922 per kW of peak demand for the distribution substations and primary lines.

In contrast, lower load factor customers' peak demands are not all coincident with either the system peak or the group peak, as indicated by the CP and NCP variations of the CF—LF curves. As noted previously, any given customer at any load factor may peak at the exact same time as the system peak and/or the group peak. But with a group of similar loads, the average customer's peak load at the meter translates to a lesser average load at the distribution level and a lesser average load still at the production and transmission levels. Given a group of customers, customers in the lower load factor groups should be assessed less than the full unit costs of demand (excluding the line transformer and secondary lines as previously discussed) in recognition of their load diversity characteristics.

Table 12.2 Allocation of Unit Demand Costs

HUD (kWh/kW)	CF_{MAX} (%)	CF_{NCP} (%)	CF_{CP} (%)	$ Per kW_{MAX}	$ Per kW_{NCP}	$ Per kW_{CP}	$ Per kW_{TOTAL}
0	100.00	0.00	0.00	0.0000	0.0000	0.0000	0.0000
20	100.00	16.69	4.42	0.6348	0.2992	0.7288	1.6628
50	100.00	35.96	11.93	0.6348	0.6445	1.9661	3.2454
100	100.00	55.92	25.71	0.6348	1.0021	4.2361	5.8731
150	100.00	65.98	39.32	0.6348	1.1825	6.4804	8.2978
200	100.00	72.64	50.61	0.6348	1.3018	8.3395	10.2762
250	100.00	77.83	59.36	0.6348	1.3947	9.7828	11.8123
300	100.00	81.77	66.03	0.6348	1.4655	10.8814	12.9817
350	100.00	84.73	71.04	0.6348	1.5186	11.7065	13.8599
400	100.00	86.94	74.82	0.6348	1.5582	12.3292	14.5222
450	100.00	88.64	77.80	0.6348	1.5886	12.8207	15.0441
500	100.00	90.07	80.42	0.6348	1.6142	13.2521	15.5012
550	100.00	91.47	83.10	0.6348	1.6393	13.6946	15.9688
600	100.00	93.09	86.29	0.6348	1.6682	14.2194	16.5225
650	100.00	95.15	90.40	0.6348	1.7053	14.8975	17.2376
700	100.00	97.91	95.88	0.6348	1.7547	15.8002	18.1898
≥ 730	100.00	100.00	100.00	**0.6348**	**1.7922**	**16.4792**	18.9062
Unit Costs:				**0.6348**	**1.7922**	**16.4792**	

The CF—LF curves are used to prorate the full unit costs for production and transmission and for distribution substations and primary distribution lines, as shown in Table 12.2.[9] The prorated unit demand costs are keyed to various hours use of demand levels, as shown in Table 12.2. The prorated unit costs are applicable to a customer's peak demand, based on that customer's hours use of demand, since the unit costs themselves account for the characteristics of load diversity between a customer's peak demand and its effective contribution to the system peak and the load factor group peak loads. The three unit demand cost components are summed to obtain a total unit cost assessment as a function of hours use of demand. The total unit demand cost per kW is plotted in Figure 12.10.

The plot in Figure 12.10 is a form of a rate class cost curve, but it only addresses the demand components of cost. By converting this cost per kW curve to an equivalent ¢ per kWh curve, the non-fuel energy costs, which are expressed in units of ¢ per kWh, can then be added in. This conversion is accomplished by approximating the cost per kW curve with straight-line segments, as given by

---

[9] As discussed previously, the coincidence factor associated with a customer's maximum demand is 100% at any load factor; thus, the unit cost for line transformers and secondary lines is not prorated.

Translating Cost to Prices

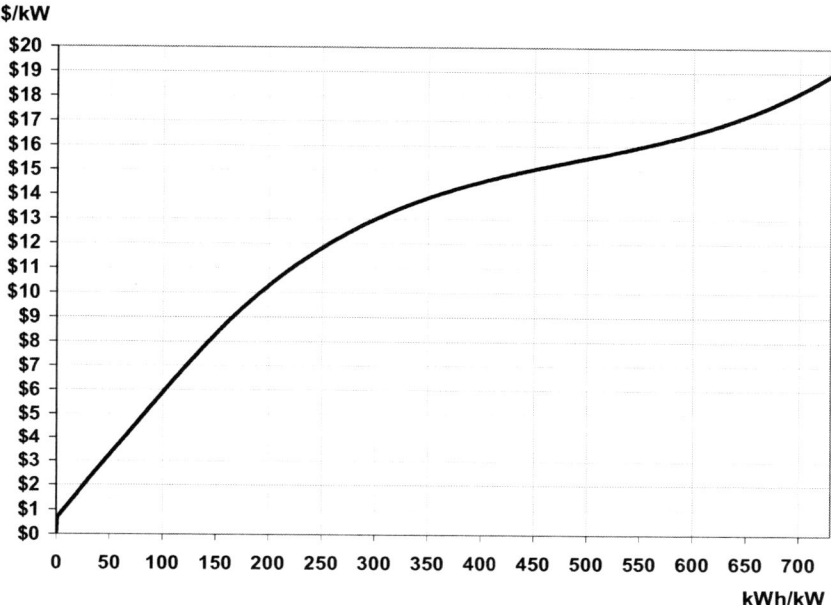

Figure 12.10 Unit demand cost allocation results. The curve represents the sum of 100% of the unit demand cost per kW of line transformers and secondary lines plus prorated amounts of the unit demand costs per kW of production, transmission, distribution substations, and primary distribution as a function of hours use of demand (as a proxy for load factor).

$$E_{AVG} = \frac{\Delta D_{TOT}}{\Delta HUD} \times 100 \qquad (12.5)$$

where E_{AVG} = average demand cost in ¢ per kWh
ΔD_{TOT} = difference in unit demand cost in $ per kW
ΔHUD = difference in hours use of demand in kWh per kW

Equation (12.5) is used to develop the average demand cost on a ¢ per kWh basis between two points on the cost curve. Generally, increments of 50 kWh per kW are sufficient enough to graph the average cost curve, particularly on a hyperbolic scale of hours use of demand. For example, the average cost between 100 kWh per kW and 150 kWh per kW, based on Table 12.2 data, is calculated as

[($8.2978/kW - $5.8731/kW) ÷ (150 kWh/kW − 100 kWh/kW)] × 100 = 4.8494¢ per kWh

Table 12.3 Development of the Hours Use of Demand Cost Structure

HUD (kWh/kW)	Demand		Energy	Total
	$ Per kW	¢ Per kWh	¢ Per kWh	¢ Per kWh
0	0.0000	0.0000	0.0000	0.0000
20	1.6628	8.3142	0.9351	9.2493
50	3.2454	5.2752	0.9351	6.2103
100	5.8731	5.2554	0.9351	6.1905
150	8.2978	4.8494	0.9351	5.7845
200	10.2762	3.5957	0.9351	4.5308
250	11.8123	3.0723	0.9351	4.0074
300	12.9817	2.3388	0.9351	3.2739
350	13.8599	1.7562	0.9351	2.6913
400	14.5222	1.3246	0.9351	2.2597
450	15.0441	1.0439	0.9351	1.9790
500	15.5012	0.9141	0.9351	1.8492
550	15.9688	0.9353	0.9351	1.8704
600	16.5225	1.1073	0.9351	2.0424
650	17.2376	1.4304	0.9351	2.3655
700	18.1898	1.9043	0.9351	2.8394
≥ 730	18.9062	2.3881	0.9351	3.3232

Note that the resulting value of 4.8494¢ per kWh represents the slope of the straight-line segment between 100 kWh per kW and 150 kWh per kW.

The results of applying Equation (12.5) to the demand cost curve is shown in Table 12.3. The nonfuel energy cost component is added to the average demand cost to obtain the total average demand and energy cost as a function of kWh per kW. Expressed in this manner, the average cost values resemble a multistep Wright rate structure.

The hours use-based cost structure is plotted in the same manner as a Wright rate structure, as shown by the chart in Figure 12.11. The combined demand and energy cost structure is plotted both with and without the customer cost component (determined previously as $47.8154 per bill). Since the customer cost component is truly fixed, i.e., not a function of either kW or kWh, plotting of the cost curve with the customer cost component requires an assumption of demand so that its average cost per kWh can be calculated as a function of kWh per kW. A 50 kW value was assumed for this plot as it is approximately equal to the average monthly kW per customer for this example rate class. The y-intercept at $x = 20$ kWh per kW is approximately 14¢ per kWh. A cost curve based on a demand higher than 50 kW would have a comparable intercept that is somewhat less than 14¢ per kWh because of the relatively higher amount of kWh that exists, even at low hours use. The opposite would occur with demands that are less than 50 kW.

Translating Cost to Prices

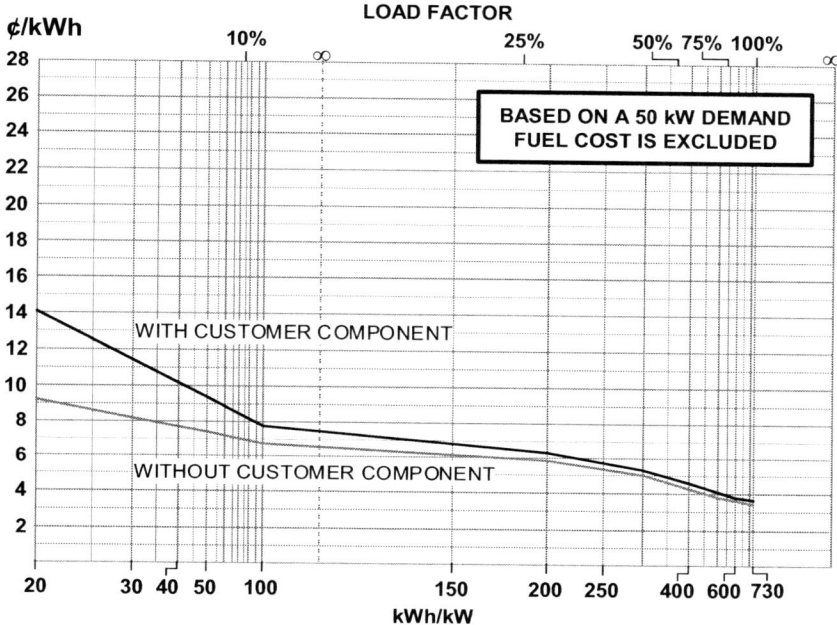

Figure 12.11 **Demand rate cost curves.** A cost curve is plotted in a fashion similar to an average rate graph. The demand and energy cost structure is plotted both with and without the customer cost component. A demand of 50 kW is assumed for plotting the curve with the customer component. The selection of different demand assumptions would result in a family of curves.

12.4 RATE DESIGN METHODOLOGY

The intra-rate cost analysis techniques discussed in the previous section yields a cost curve as a function of hours use of demand. The curve incorporates the variation in load diversity from low to high load factors as measured by groups of common customers. The continuous function cost curve can be simplified for plotting purposes by means of straight-line segments over short ranges of hours use. The slopes of these line segments have the units of ¢ per kWh. In essence, transforming the cost function in this manner is rate design. Even though it could serve as a billing mechanism for the example rate class, the 50 kWh per kW price increment structure is too complex for practical applications. Averaging the demand cost curve over longer line segments will produce a more reasonable design without sacrificing its inherent cost causation aspect.

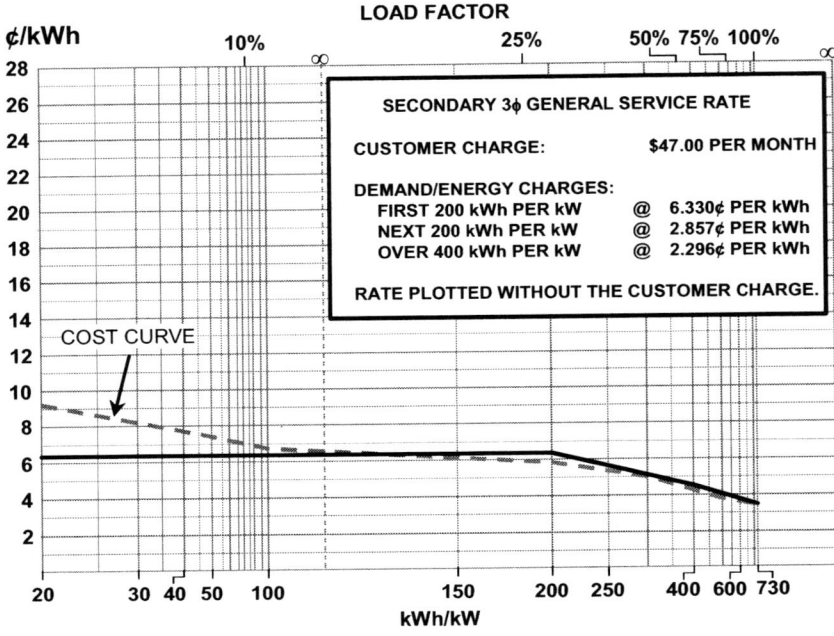

Figure 12.12 Wright rate design. A three-step hours use of demand rate designed to track the cost curve closely between 100 kWh per kW and 730 kWh per kW.

Wright Rate Structure Design

The design of a practical Wright rate structure, which tracks the cost curve, could be approached in different ways. For this example, a three-step rate is designed for the hours use ranges of 0 to 200 kWh per kW, 200 to 400 kWh per kW, and over 400 kWh per kW. To simplify the rate design process, the rate design initially addresses only the kWh charges (the finalization of the Customer Charge is considered during the fine tuning process). Thus, the combined demand and energy revenue target is $101,022,417 (the total base rate revenue target with the customer-related cost is $107,907,834). The demand- and energy-related cost curve, without the customer cost component, is shown in Figure 12.12.

The 0 to 200 kWh per kW price step is considered first. The average cost is found to be 6.808¢ per kWh at 100 kWh per kW and 5.850¢ per kWh at 200 kWh per kW. By inspection of the cost curve, it is reasonable to set the initial unit price of the first 200 kWh per kW step in the middle of these values, i.e., 6.33¢ per

Translating Cost to Prices

kWh. Without inclusion of the Customer Charge, the first hours use of demand price of a Wright rate structure will display as a flat line when graphed, as observed in Figure 12.12.

Now consider the tail step of the concept Wright rate structure. By inspection, a straight-line extrapolation of the cost curve based on the range of 400 to 730 kWh per kW indicates an average rate at infinity of approximately 2.3¢ per kWh. This rate at infinity would also represents the value of the tail step price. Thus, initial values for the first and last steps of the rate structure have been selected. The revenue produced by each of these two hours use steps can then be determined by applying the two prices to the appropriate kWh quantities (i.e., the differences in interval consolidated factors), which are derived from the hours use of demand bill frequency and summarized below.

kWh/kW Interval	Consolidate Factor	Δ Consolidated Factor
0	0	
		1,219,371,207
200	1,219,371,207	
		665,625,796
400	1,884,997,003	
		215,002,997
Total	2,100,000,000	

		Revenue
First HUD step: 1,219,371,207 kWh × $0.0633/kWh =		$77,186,197
Last HUD step: 215,002,997 kWh × $0.0230/kWh =		4,945,069
		$82,131,266

The initial price of the penultimate step is then determined by dividing the second step kWh into the difference between the demand and energy revenue target and the revenue produced by the first and last step prices:

($101,022,417 - $82,131,266) ÷ 665,625,796 kWh = $0.02838 per kWh
→ 2.84¢ per kWh

When the rounded initial kWh prices are applied back to the billing determinants, a revenue target overcollection of $12,622 is calculated. The customer-related cost component is now incorporated for fine tuning to the total base rate revenue target. Recall that the full cost-based customer component is $47.8154 per bill. The final base rate design for the secondary three-phase general service rate class, which is only $239 below the revenue target, is determined to be:

Secondary Three-Phase General Service Rate

Customer Charge: $47.00 per Month

Energy/Demand Charges:

First 200 kWh/kW @ 6.330¢ per kWh
Next 200 kWh/kW @ 2.857¢ per kWh
Over 400 kWh/kW @ 2.296¢ per kWh

The final rate, excluding fuel and the Customer Charge, is plotted in Figure 12.12 along with the cost curve.

Another rate design consideration would be to include an explicit Demand Charge to the structure, i.e., a Wright-Hopkinson design. Reducing the first 200 kWh per kW price by ½¢, would free up revenue, in the amount of $6,096,856, for the Demand Charge. The Demand Charge is then determined by dividing this amount by the sum of the customer maximum demands:

$$\$6,096,856 \div 6,888,228 \text{ kW} = \$0.8851/\text{kW}$$

After fine-tuning (to only $179 below the revenue target), the equivalent rate design would be:

Secondary Three-Phase General Service Rate

Customer Charge: $47.00 per Month

Demand Charge: $0.89 per kW

Energy/Demand Charges:

First 200 kWh/kW @ 5.827¢ per kWh
Next 200 kWh/kW @ 2.861¢ per kWh
Over 400 kWh/kW @ 2.285¢ per kWh

The Wright-Hopkinson rate design is plotted in Figure 12.13. The rate curve now tracks the cost curve closely across all hours use of demand. The introduction of the explicit Demand Charge causes the average rate to increase notably at low hours use of demand. This elevation in the curve occurs because the Demand Charge behaves like a fixed charge, and a fixed charge contributes greatly to the average rate per kWh at low hours use of demand because of the associated low kWh usage. This effect is similar to the impact of a truly fixed Customer Charge. However, a Demand Charge's impact on the average rate per kWh is independent of the kW magnitude because the associated kWh is self-scaling.

Translating Cost to Prices

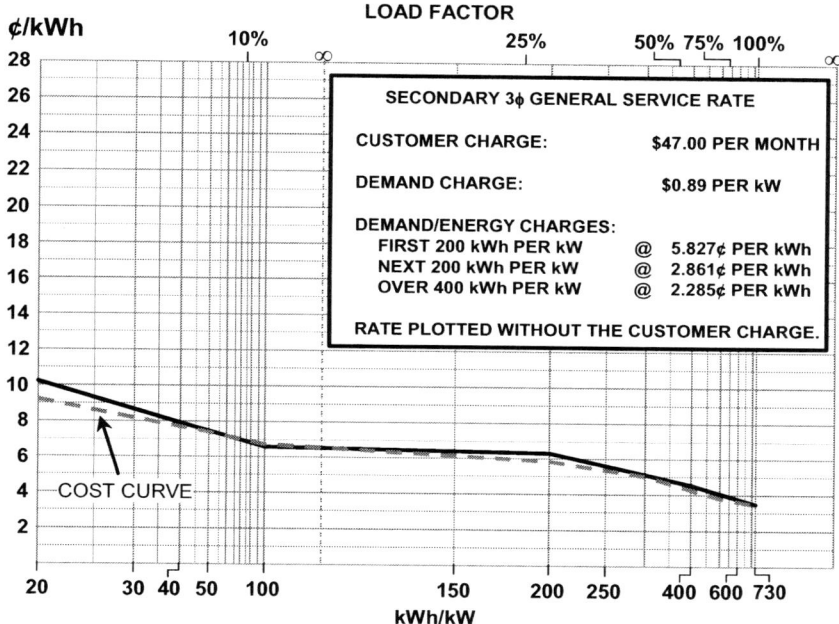

Figure 12.13 Wright-Hopkinson rate design. By adding an explicit Demand Charge to a basic Wright rate structure, the average rate is elevated at low hours use of demand since the Demand Charge is a fixed charge and thus provides the same declining average rate effect as caused by a Customer Charge.

Hopkinson Rate Structure Design

The CF—LF relationship was utilized previously to develop a cost curve based on diversified demands. An alternative cost curve, which is not based on diversified demands, can also be developed from the cost-of-service study and demand unit data for the example secondary three-phase general service rate. In this case, the total demand-related revenue requirements for production, transmission, distribution substations, primary distribution lines, line transformers, and secondary distribution lines are summed and then divided by the total of the customer maximum demands. Thus, the nondiversified cost curve, plotted in Figure 12.14, is represented by just two cost components:

Demand: $81,384,973 ÷ 6,888,228 kW = $11.8151 per kW

Energy: $19,637,444 ÷ 2,100,000,000 kWh = $0.009351 per kWh

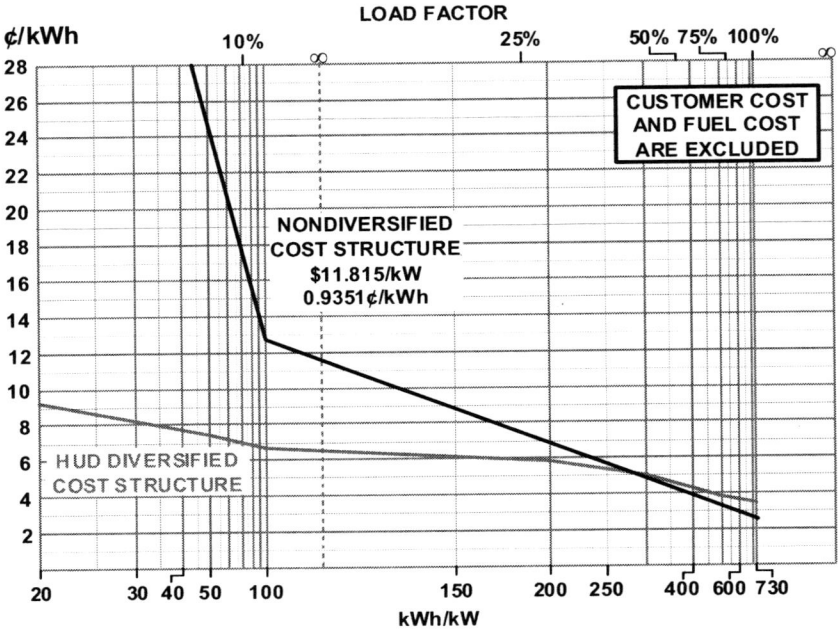

Figure 12.14 Nondiversified vs. diversified cost curves. A nondiversified cost curve is developed from the cost-of-service study functional demand and energy revenue requirement components using nondiversified maximum demands. The curve is then plotted in comparison to the cost curve that is based on diversified demand units.

The nondiversified cost curve intersects the diversified cost curve at approximately 285 kWh per kW (the average hours use of the class is 305 kWh per kW). At low hours use of demand, the nondiversified cost curve greatly exceeds the comparable diversified cost curve in terms of the average cost per kWh. However, at high hours use of demand, the nondiversified cost curve is observed to be less than the diversified cost curve in terms of the average cost per kWh.

With minor fine tuning (to only $566 under target), the nondiversified cost curve could be implemented as a billable rate:

Secondary Three-Phase General Service Rate

Customer Charge: $47.50 per Month

Demand Charge: $11.70 per kW

Energy Charge: 0.975¢ per kWh

Translating Cost to Prices

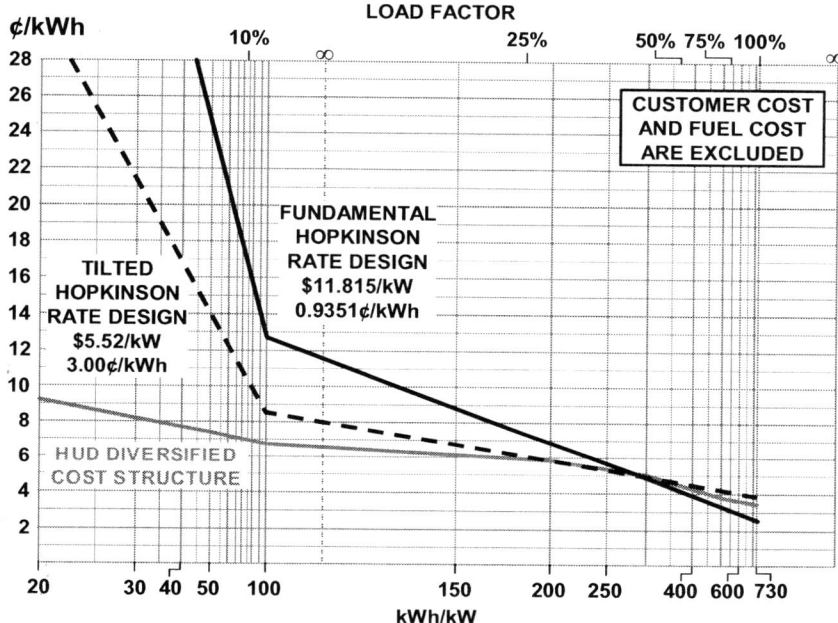

Figure 12.15 Hopkinson rate design. The nondiversified cost curve is designated as the "Fundamental Hopkinson" rate design as it represents a rate structure in which all demand-related costs are recovered through an explicit Demand Charge and all energy-related costs are recovered through the Energy Charge. By tilting the Fundamental Hopkinson rate, a portion of the demand-cost recovery is transferred to the kWh price. In so doing, the average rate per kWh of the "Tilted Hopkinson" rate structure rate shifts downward on the left and upward on the right. The pivot point is 305 hours use of demand, i.e., the average hours use of the rate class.

As illustrated in Figure 12.15, the nondiversified cost structure (plotted in Figure 12.14) has been designated as the "Fundamental Hopkinson" rate design as its Demand Charge consists of only demand-related costs while its Energy Charge consists of only the nonfuel energy-related costs. By shifting a portion of demand cost recovery responsibility from the Demand Charge to the Energy Charge, i.e., by applying the principle of rate tilt, the average rate per kWh is observed to decrease on the left side of the chart and to increase on the right side of the chart. The point of pivot is located at 305 kWh per kW, which is also the hours use of demand at which the Fundamental Hopkinson rate and the Tilted Hopkinson rate intersect. Furthermore, 305 kWh per kW represents the average hours use of demand for the example rate class. Any degree of applied rate tilt will cause the average rate curve to pivot at this point. With full rate tilt, the average rate would be plotted as a straight line across the entire chart, i.e., a uniform rate per kWh; thus, the Hopkinson rate structure would be transformed into a nondemand, flat energy rate structure.

Note in Figure 12.15 how the Tilted Hopkinson rate tracks the HUD diversified cost curve generally over the range of 175 to 500 kWh per kW. Further rate tilt could be applied so as to make the Hopkinson rate coincide with the HUD diversified cost curve at 100 kWh per kW. Then the Hopkinson rate would track the HUD diversified cost curve generally over the range of 100 to 275 kWh per kW. However, at hours use above about 275 kWh per kW, the average rate would rise well above the HUD diversified cost curve.

Choosing a Base Rate Structure Design

As demonstrated above, cost curves provide a guide for the design of practical rate structures. Additional guidance is provided by assessing the distribution of customer maximum demands, as a function of hours use of demand, for the subject class of service. Examples of histograms of customer demands by kWh per kW intervals are illustrated for three different rate classes in Figure 12.16.

The top chart shows a histogram of fairly small commercial customers having an average class load factor of just less than 20%. The distribution of demands generally ranges from zero to about 50% load factor. The middle chart shows a histogram of medium commercial and industrial customers having an average class load factor of 37%. In this case, the distribution of demands ranges soundly from zero to 100% load factor. The bottom chart shows a histogram of large industrial customers having an average class load factor of nearly 62%. There are few customers in this class; however, the distribution of demands are contained in a fairly tight range from about 50% load factor to about 75% load factor.

Considering the impact of rate tilt on the average rate, as a function of the hours use of demand, the low load factor rate class in the top chart would be served effectively under a highly tilted Hopkinson design. A nondemand rate might even be of practical consideration. On the other hand, the fairly high load factor rate class in the bottom chart would be served effectively under just a slightly tilted Hopkinson design. In contrast, the wide distribution of demands across the whole range of load factor of the rate class in the middle chart calls for a more complex solution, as a simple Hopkinson design, at any degree of rate tilt, would be problematic for many customers. Two approaches could be utilized.

First, the medium customer rate class could be re-segmented by creating three new rate classes in which the new groups would have more homogeneous characteristics than the original single class. Then, a uniquely tilted Hopkinson rate could be designed for each group. Alternatively, a multi-step Wright rate could be designed to adequately span the wide range of load factors. Recall from Chapter 9 that when a Wright rate structure is formulized, each of its price steps yields an equivalent Hopkinson rate structure whose implicit rate tilt is commensurate with load factor.

Translating Cost to Prices

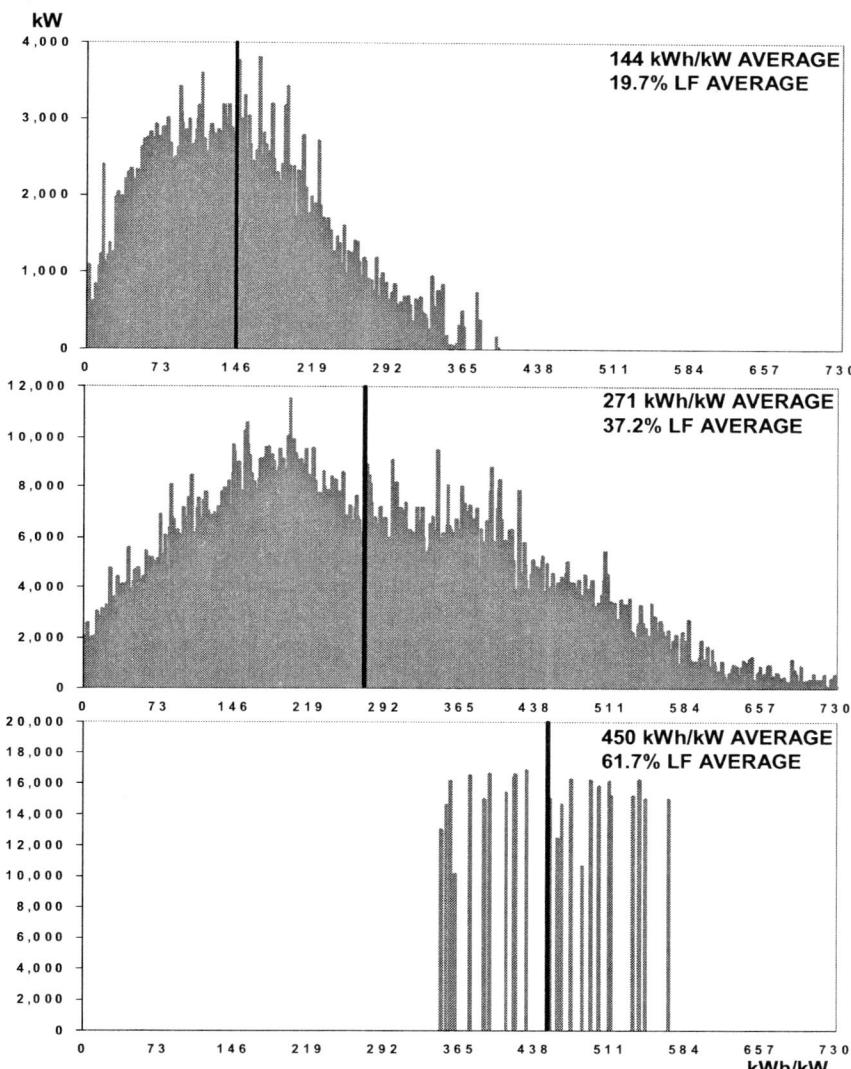

Figure 12.16 Rate class load characteristics as a guide for rate structure selection. Shown are plots of customer maximum demand histograms for three commercial and industrial rate classes. The average hours use of demand is shown by the black reference lines on each chart. The top chart indicates the applicability of a highly tilted Hopkinson rate design, while the bottom chart indicates the applicability of a slightly tilted Hopkinson rate design. The middle chart suggests the applicability of a multi-step Wright rate which operates effectively as a series of Hopkinson rates having various degrees of rate tilt.

Fuel Cost Recovery Charges

As discussed in Chapter 8, the recovery of generator fuel costs is often addressed outside of the base rates for electric service. However, the kWh charges of the base rates may include an embedded amount of fuel cost. Generally, this base fuel price is established at the time of a general rate case. An elementary form of a fuel cost recovery, or fuel adjustment charge (FAC), factor is given by

$$FAC = \frac{F}{S} - \frac{F_{BASE}}{S_{BASE}} \qquad (12.6)$$

where FAC = current fuel adjustment charge per kWh
  F = total cost of fuel in the current period
  F_{BASE} = total cost of fuel established in the base period
  S = total kWh sales in the current period
  S_{BASE} = total kWh sales in the base period

The term [$F_{BASE} \div S_{BASE}$] would be equal to zero if no fuel cost recovery is embedded in the base rates.

In general, the FAC is based on the actual cost of fuel, which includes the invoice price of fuel, less any cash or other discounts, and the charges necessary to transport the fuel from the acquisition point(s) to the unloading point(s). However, the FAC generally excludes charges for unloading fuel from the shipping medium as well as handling of the fuel up to the point where the fuel enters the first boiler plant bunker, hopper, tank, or other boiler house structure.

In addition to direct fuel costs, the FAC generally includes the net energy cost component of purchased power, when such energy is purchased on an economic dispatch basis or to displace higher cost production in order to serve native load. However, the FAC generally excludes the capacity or demand cost component of purchased power, which is then recovered through base rates or a special rate rider. Other costs typically recovered through the FAC include emissions allowances for SO_2 and NO_X and emissions reagents, such as limestone for flue gas desulphurization (FGD) and ammonia for selective catalytic reduction (SCR), as discussed in Chapter 6. The FERC accounts in which FAC expenses are booked include:

501 Fuel – Fossil fuels used in the production of steam for the generation of electricity;

509 Allowances – Allowances purchased for SO_2 and NO_X emissions that are directly related to generator output;

518 <u>Nuclear Fuel Expense</u> – Nuclear fuel used in the production of steam for the generation of electricity;[10]

547 <u>Fuel</u> – Natural gas and petroleum-based fuel products used in other power generation; and

555 <u>Purchased Power</u> – The energy-related cost component of purchased power, including energy purchased from qualifying facilities (QFs) under an approved rate schedule or special contract for purchase.

Other fuel-related expenses that may be included in the FAC are the costs, including gains and losses, associated with options, futures, and financial derivatives used for the purpose of hedging the costs of energy and fuel. In addition, carrying charges for all or a portion of the over/underrecovery fuel balances may be allowed.

When revising the FAC for current costs and sales, a true-up adjustment is incorporated in order to reconcile the over/underrecovery balance. FAC revisions which operate monthly are based on historical costs. Semiannual and annual FAC revisions are typically based on projected fuel costs and sales. In either case, the total fuel costs must be allocated appropriately between the retail and wholesale jurisdictions.

Voltage Level Differentiation of the FAC Factor

Calculation of the FAC by means of Equation (12.6) results in an average billing factor per kWh. To better recognize the cost-causation of energy delivery to customers, given T&D system energy losses, the average FAC factor can be adjusted for service at the different delivery voltage levels at which customers are served.[11] The energy loss model developed for energy-related cost allocation to the customer or rate classes can be used to delineate the average FAC factor by standard voltage classifications, as exemplified below.

---

[10] Fossil fuels, which may be used for ancillary steam facilities and booked to FERC Account 518, are excluded if they have already been included in the cost of fuel booked elsewhere.

[11] Given a typical mix of customers, the overall average delivery voltage level is generally somewhere between secondary and primary. Rate schedules are sometimes used as a proxy to differentiate the FAC in recognition of system losses. Residential is clearly a secondary service. Large industrial rates are often designed for high voltage service since larger customers generally tend to take service at primary or even transmission voltages. Medium-size commercial and industrial customers represent a blend of secondary and primary services; thus, associated general service rate schedules align closely with the average FAC factor.

Table 12.4 Voltage Differentiated Energy Loss Expansion Factors

Voltage Class	Retail Sales MWh	Supply MWh	Energy Loss Factor	Energy Loss Multiplier
Secondary				
< 600 V	5,874,800	6,429,293	1.09438500	1.01901562
Primary				
4.16 kV	48,200	50,742	1.05273859	
14.4 kV	2,081,800	2,158,766	1.03697089	
Total	2,130,000	2,209,508	1.03732770	0.96588781
Subtransmission				
34.5 kV	192,000	198,076	1.03164583	
69 kV	67,900	69,246	1.01982327	
Total	259,000	267,322	1.02855714	0.95772127
Transmission				
138 kV	480,100	485,468	1.01118100	0.94154182
Total	8,744,800	9,391,591	1.07396293	

The energy loss model, used to allocate energy-related costs to customer or rate classes in a cost-of-service study, is utilized to develop FAC factor multipliers which differentiate the average fuel cost per kWh by functional delivery voltage levels. As shown in Table 12.4 for retail sales, an energy loss expansion factor is calculated for each voltage level by dividing supply, or territorial input, energy by sales, or delivered energy. For example, the energy loss factor for secondary voltage sales is calculated by:

$$6{,}429{,}293 \text{ MWh} \div 5{,}874{,}800 \text{ MWh} = 1.09438500$$

In addition, an overall retail sales expansion factor is calculated.

Next an energy loss multiplier is then determined for each voltage level as the ratio of the voltage level energy loss factor to the total energy loss factor. For example, the energy loss multiplier for sales delivered at the secondary voltage level is calculated by:

$$1.09438500 \div 1.07396293 = 1.01901562$$

Translating Cost to Prices

The fact that the secondary energy loss multiplier is greater than unity is an indication of the greater amount of losses incurred to deliver a MWh at low voltage as compared to the losses incurred to deliver a MWh at a higher voltage.

The average retail FAC rate per kWh is calculated from the retail portion of projected fuel and purchased costs and projected retail sales:

$$\$352,561,937 \div 8,977,895 \text{ MWh} = \$39.27/\text{MWh} \rightarrow 3.9270\cent \text{ per kWh}$$

The average FAC rate is then adjusted by the voltage level energy loss multipliers:

Secondary: 3.9270¢ per kWh × 1.01901562 = 4.0017¢ per kWh
Primary: 3.9270¢ per kWh × 0.96588781 = 3.7930¢ per kWh
Subtransmission: 3.9270¢ per kWh × 0.95772127 = 3.7610¢ per kWh
Transmission: 3.9270¢ per kWh × 0.94154182 = 3.6974¢ per kWh

The initial rates are tested for revenue yield by applying them to their associated projected sales by voltage level:

Secondary: $40.017 per MWh × 6,021,670 MWh = $240,967,624
Primary: $37.930 per MWh × 2,203,345 MWh =   83,573,789
Subtransmission: $37.610 per MWh ×   263,178 MWh =   9,730,835
Transmission: $36.974¢ per MWh ×   489,702 MWh =   18,106,412
Total $352,378,659

Due to rounding of the initial rates, the revenue produced is found to be $183,278 below the target. Fine tuning of the rates results in revenue yield which is just $177 below target. The final FAC rates for the projected test period are thus determined to be:

Secondary: 4.0019¢ per kWh
Primary: 3.7931¢ per kWh
Subtransmission: 3.7609¢ per kWh
Transmission: 3.6975¢ per kWh

These rates are applied to the customers' metered monthly sales based on the specific voltage level at which each customer receives electric service.

12.5 GUIDING PRINCIPLES OF RATEMAKING

To a great extent, rate design is about recovering the costs of electric service in a manner which balances the risk between the service provider and the customer, and between customers themselves. Rates must recover both fixed and variable costs in a timely manner. While cost recovery is a crucial function of electricity pricing, other factors are important for sound ratemaking.

The design of a rate does not occur in a vacuum. A typical electric utility tariff contains a variety of rate schedules under which a given customer may be qualified to take service and from which the customer may select for service. End use-oriented rate and TOU-based rates are two examples of alternatives to a standard rate. In addition, customer choice commonly exists between standard offer rates, e.g., small-, medium-, and large-general service and, perhaps, high load factor service. The designs of these rates are interrelated. Changing only one rate or changing multiple rates disproportionately can seriously impact the relationship between two or more rates within the tariff. Even with careful attention to this issue, rate modifications often cause a shift in the equilibrium between rates. Tariff modifications occur on a grand scale as part of the rate case process and maintaining, or improving, the relationships between the rates is critical for managing customer rate shock and ensuring a degree of stability under change. The differential chart analysis discussed in Chapter 9 is a key tool to assure a sound tariff.

The positions of prices, rate schedules, and the overall tariff are not always optimal. Variations over time in the needs of the electricity pricing stakeholders can result in a tariff structure that may satisfy short-term concerns but which does not provide a sound position for meeting impending industry issues. Changes in economic conditions and environmental concerns, along with new legislative and regulatory initiatives, may challenge the tariff position. Changes in technologies, such as metering, may create new opportunities that likewise challenge the tariff position. Developing and implementing a pricing strategy which recognizes such contingencies helps to maintain a sound and progressive tariff. Even so, tactical changes in rate structures and pricing levels must be applied gradually to avoid customer rate shock.

To ensure practical rates, both cost and non-cost issues must be brought into balance when pricing electricity. In 1960, Dr. James C. Bonbright, an economics professor at Columbia University, summarized a criteria for sound ratemaking. Endearingly referred to as Bonbright's "Big Eight," these key guiding principles serve as a timeless doctrine for the ratemaking professional:

1. The related "practical" attributes of simplicity, understandability, public acceptability, and feasibility of application;

2. Freedom from controversies as to proper interpretation;

3. Effectiveness in yielding total revenue requirements under the fair-return standard;

Translating Cost to Prices

4. Revenue stability from year to year;
5. Stability of the rates themselves, with a minimum of unexpected changes seriously adverse to existing customers;
6. Fairness of the specific rates in the apportionment of total costs of service among the different consumers;
7. Avoidance of "undue discrimination" in rate relationships; and
8. Efficiency of the rate classes and rate blocks in discouraging wasteful use of service while promoting all justified types and amounts of use:
   a. In the control of the total amounts of service supplied by the company;
   b. In the control of the relative uses of alternative types of service.[12]

12.6 SUMMARY

A considerable amount of resources are available for determining rate structures and price levels for electric service. The cost-of-service study provides in-depth functional cost data and cost component revenue requirements, which help to frame the rate design problem. Historical billing data provides insight to customer and rate class characteristics and offers a quantification of statistical attributes that are important to know while attacking the rate design problem. Load research data provides a level of detail that allows for the development of crucial relationships between key factors. All things considered, an understanding of the relationship between coincidence of load and load factor is perhaps the most important underlying principle of electricity pricing. Well proven analytical techniques and mathematical routines form the muscle that provides a sound basis for a rate design solution.

But electricity pricing incorporates both art and science. Rate design is not simply a recipe process. Experience in rate design develops a judgment skill and an intuition about the way in which things respond when prices and/or rate structures are modified. Understanding the stakeholder perspectives is also important to the development of successful rates. Together, the art and science provide the comprehensive support necessary to develop and implement practical rate design solutions.

---

[12] Bonbright, James C. *Principles of Public Utility Rates.* New York: Columbia University Press, 1961, p. 291.

Appendix A

Unit Conversions

TIME UNITS

1 Yr	=	8,760 Hr (8,784 Hr for a leap year)		
1 Mo (Avg)	=	730 Hr	=	30.42 Day
1 Mo (Avg)	=	1,460 ½ Hr Intervals	=	2,920 ¼ Hr Intervals

ELECTRICAL UNITS

Power

1 W	=			0.001 kW	=	0.000001 MW
1 kW	=	1,000 W	=			0.001 MW
1 MW	=	1,000,000 W	=	1,000 kW		
1 hp	=	0.7457 kW (@100% Efficiency)				

Energy

1 Wh	=			0.001 kWh	=	0.000001 MWh
1 kWh	=	1,000 Wh	=			0.001 MWh
1 MWh	=	1,000,000 Wh	=	1,000 kWh		
1 kWh	=	3,413 Btu (@100% Efficiency)				

LOAD CHARACTERISTICS UNITS

Hours Use of Demand	Monthly Load Factor*	Power Factor	kVA Per kW	kVAR Per kW
0	0.00%	100%	1.000	0
50	6.85%	99%	1.010	0.14249
100	13.70%	98%	1.020	0.20306
150	20.55%	97%	1.031	0.25062
182.5	**25.00%**	96%	1.042	0.29167
200	27.40%	95%	1.053	0.32868
250	34.25%	94%	1.064	0.36295
300	41.10%	93%	1.075	0.39523
350	47.95%	92%	1.087	0.42600
365	**50.00%**	91%	1.099	0.45561
400	54.79%	90%	1.111	0.48432
450	61.64%	89%	1.124	0.51232
500	68.49%	88%	1.136	0.53974
547.5	**75.00%**	87%	1.149	0.56673
550	75.34%	86%	1.163	0.59337
584	80.00%	85%	1.176	0.61974
600	82.19%	84%	1.190	0.64594
620.5	85.00%	83%	1.205	0.67200
650	89.04%	82%	1.220	0.69800
657	90.00%	81%	1.235	0.72399
700	95.89%	80%	1.250	0.75000
730	**100.00%**	75%	1.333	0.88192
		70%	1.429	1.02020
		65%	1.538	1.16913
		60%	1.667	1.33333

*Based on an average month of 730 hours.

Unit Conversions

PRICING UNITS

1 mill	=			0.1 ¢	=	$ 0.001
1 ¢	=	10 mills	=			$ 0.01
1 $	=	1,000 mills	=	100 ¢		

$1.00/MWh = 0.1 ¢/kWh
$10/MWh = 1.0 ¢/kWh
$100/MWh = 10.0 ¢/kWh
$1000/MWh = $1.00/kWh

$100/kW-Yr. = $8.33/kW-Mo. = 27.4 ¢/kW-Day
1.0 ¢/Day = 30.42 ¢/Mo. = $3.65/Yr.

Appendix B

Uniform System of Electric Accounts

THE FERC'S UNIFORM SYSTEM OF ACCOUNTS

Utility accountants utilize the FERC's Uniform System of Accounts (or a similar accounting system) for booking, tracking, and reporting expenditures and financial information. The accounting System is referenced in a variety of utility applications, including cost-of-service studies.

The FERC's Uniform System of Accounts is applied to both *Major* and *Nonmajor* classifications of utilities. A Major utility is one that in each of the last three consecutive years exceeded any one or more of the following criteria for sales or transmission service: (a) 1,000,000 MWh of total sales, (b) 100 MWh of sales for resale, (c) 500 MWh of power exchanges delivered, or (d) 500 MWh of wheeling for others (deliveries plus losses). Nonmajor utilities are those not classified according to the above criteria, but which had total sales in the last three consecutive years of 10,000 MWh or more. Some of the accounts in the System are indicated as being applicable to "(Major Only)" or "(Nonmajor Only)."

Some accounts are indicated as being "[Reserved]." In several instances, sequential account numbers within the System have been skipped to allow for possible later expansion or to permit better coordination with the numbering systems of other utility departments. These skipped account numbers have been included in this appendix and are indicated as "[*Not Used*]" (author's notation) in order to provide continuity and to clearly illustrate the layout of the System's numbering scheme. Out-of-sequence numbers are cross-referenced to facilitate the location of such specific accounts.

Utility accountants often establish one or more *subaccounts* under the main accounts of the System as a means of identifying and booking cost information in greater detail. An example of a subaccount structure is provided at the end of this appendix.

UNIFORM SYSTEM OF ACCOUNTS – ELECTRIC UTILITIES

BALANCE SHEET CHART OF ACCOUNTS

Assets and Other Debits

1. UTILITY PLANT

101	Electric Plant in Service (Major Only)
101.1	Property under Capital Leases
102	Electric Plant Purchased or Sold
103	Experimental Electric Plant Unclassified (Major Only)
103.1	Electric Plant in Process of Reclassification (Nonmajor Only)
104	Electric Plant Leased to Others
105	Electric Plant Held for Future Use
106	Completed Construction Not Classified – Electric (Major Only)
107	Construction Work in Progress – Electric
108	Accumulated Provision for Depreciation of Electric Utility Plant (Major Only)
109	[Reserved]
110	Accumulated Provision for Depreciation and Amortization of Electric Utility Plant (Nonmajor Only)
111	Accumulated Provision for Amortization of Electric Utility Plant (Major Only)
112	[Reserved]
113	[Reserved]
114	Electric Plant Acquisition Adjustments
115	Accumulated Provision for Amortization of Electric Plant Acquisition Adjustments (Major Only)
116	Other Electric Plant Adjustments
117	[Not Used]
118	Other Utility Plant
119	Accumulated Provision for Depreciation and Amortization of Other Utility Plant (Nonmajor Only)
120	[Not Used]
120.1	Nuclear Fuel in Process of Refinement, Conversion, Enrichment, and Fabrication (Major Only)
120.2	Nuclear Fuel Materials and Assemblies – Stock Account (Major Only)
120.3	Nuclear Fuel Assemblies in Reactor (Major Only)
120.4	Spent Nuclear Fuel (Major Only)
120.5	Accumulated Provision for Amortization of Nuclear Fuel Assemblies (Major Only)

Uniform System of Electric Accounts

120.6 Nuclear Fuel under Capital Leases (Major Only)

2. OTHER PROPERTY AND INVESTMENTS

121	Nonutility Property
122	Accumulated Provision for Depreciation and Amortization of Nonutility Property
123	Investment in Associated Companies (Major Only)
123.1	Investment in Subsidiary Companies (Major Only)
124	Other Investments
125	Sinking Funds (Major Only)
126	Depreciation Fund (Major Only)
127	Amortization Fund – Federal (Major Only)
128	Other Special Funds (Major Only)
129	Special Funds (Nonmajor Only)

3. CURRENT AND ACCRUED ASSETS

130	Cash and Working Funds (Nonmajor Only)
131	Cash (Major Only)
132	Interest Special Deposits (Major Only)
133	Dividend Special Deposits (Major Only)
134	Other Special Deposits (Major Only)
135	Working Funds (Major Only)
136	Temporary Cash Investments
137	[Not Used]
138	[Not Used]
139	[Not Used]
140	[Not Used]
141	Notes Receivable
142	Customer Accounts Receivable
143	Other Accounts Receivable
144	Accumulated Provision for Uncollectible Accounts – Credit
145	Notes Receivable from Associated Companies
146	Accounts Receivable from Associated Companies
147	[Not Used]
148	[Not Used]
149	[Not Used]
150	[Not Used]
151	Fuel Stock (Major Only)
152	Fuel Stock Expenses Undistributed (Major Only)
153	Residuals (Major Only)

154	Plant Materials and Operating Supplies
155	Merchandise (Major Only)
156	Other Materials and Supplies (Major Only)
157	Nuclear Materials Held for Sale (Major Only)
158	[Not Used]
158.1	Allowance Inventory
158.2	Allowances Withheld
159	[Not Used]
160	[Not Used]
161	[Not Used]
162	[Not Used]
163	Stores Expense Undistributed (Major Only)
164	[Not Used]
165	Prepayments
166	[Not Used]
167	[Not Used]
168	[Not Used]
169	[Not Used]
170	[Not Used]
171	Interest and Dividends Receivable (Major Only)
172	Rents Receivable (Major Only)
173	Accrued Utility Revenues (Major Only)
174	Miscellaneous Current and Accrued Assets
175	Derivative Instrument Assets
176	Derivative Instrument Assets - Assets
177	[Not Used]
178	[Not Used]
179	[Not Used]
180	[Not Used]

4. DEFERRED DEBITS

181	Unamortized Debt Expense
182	[Not Used]
182.1	Extraordinary Property Losses
182.2	Unrecovered Plant and Regulatory Study Costs
182.3	Other Regulatory Assets
183	Preliminary Survey and Investigation Charges (Major Only)
184	Clearing Accounts (Major Only)
185	Temporary Facilities (Major Only)
186	Miscellaneous Deferred Debits
187	Deferred Losses from Disposition of Utility Plant

Uniform System of Electric Accounts

188   Research, Development, and Demonstration Expenditures (Major Only)
189   Unamortized Loss on Reacquired Debt
190   Accumulated Deferred Income Taxes
191   [*Not Used*]
192   [*Not Used*]
193   [*Not Used*]
194   [*Not Used*]
195   [*Not Used*]
196   [*Not Used*]
197   [*Not Used*]
198   [*Not Used*]
199   [*Not Used*]
200   [*Not Used*]

Liabilities and Other Credits

5. PROPRIETARY CAPITAL

201   Common Stock Issued
202   Common Stock Subscribed (Major Only)
203   Common Stock Liability for Conversion (Major Only)
204   Preferred Stock Issued
205   Preferred Stock Subscribed (Major Only)
206   Preferred Stock Liability for Conversion (Major Only)
207   Premium on Capital Stock (Major Only)
208   Donations Received from Stockholders (Major Only)
209   Reduction in Par or Stated Value of Capital Stock (Major Only)
210   Gain on Resale or Cancellation of Reacquired Capital Stock (Major Only)
211   Miscellaneous Paid-In Capital
212   Installments Received on Capital Stock
213   Discount on Capital Stock
214   Capital Stock Expense
215   Appropriated Retained Earnings
215.1   Appropriated Retained Earnings – Amortization Reserve, Federal
216   Unappropriated Retained Earnings
216.1   Unappropriated Undistributed Subsidiary Earnings (Major Only)
217   Reacquired Capital Stock
218   Noncorporate Proprietorship (Nonmajor Only)
219   Accumulated Other Comprehensive Income
220   [*Not Used*]

6. LONG-TERM DEBT

221	Bonds
222	Reacquired Bonds (Major Only)
223	Advances from Associated Companies
224	Other Long-Term Debt
225	Unamortized Premium on Long-Term Debt
226	Unamortized Discount on Long-Term Bebt – Debit

7. OTHER NONCURRENT LIABILITIES

227	Obligations under Capital Lease – Noncurrent
228	[Not Used]
228.1	Accumulated Provision for Property Insurance
228.2	Accumulated Provision for Injuries and Damages
228.3	Accumulated Provision for Pensions and Benefits
228.4	Accumulated Miscellaneous Operating Provisions
229	Accumulated Provision for Rate Refunds
230	Asset Retirement Obligations

8. CURRENT AND ACCRUED LIABILITIES

231	Notes Payable
232	Accounts Payable
233	Notes Payable to Associated Companies
234	Accounts Payable to Associated Companies
235	Customer Deposits
236	Taxes Accrued
237	Interest Accrued
238	Dividends Declared (Major Only)
239	Matured Long-Term Debt (Major Only)
240	Matured Interest (Major Only)
241	Tax Collections Payable (Major Only)
242	Miscellaneous Current and Accrued Liabilities
243	Obligations under Capital Leases – Current
244	Derivatives Instrument Liabilities
245	Derivatives Instrument Liabilities – Hedges
246	[Not Used]
247	[Not Used]
248	[Not Used]
249	[Not Used]
250	[Not Used]

Uniform System of Electric Accounts

9. DEFERRED CREDITS

251	[Reserved]
252	Customer Advances for Construction
253	Other Deferred Credits
254	Other Regulatory Liabilities
255	Accumulated Deferred Investment Tax Credits
256	Deferred Gains from Disposition of Utility Plant
257	Unamortized Gain on Reacquired Debt
258	[Not Used]
259	[Not Used]
260	[Not Used]
261	[Not Used]
262	[Not Used]
263	[Not Used]
264	[Not Used]
265	[Not Used]
266	[Not Used]
267	[Not Used]
268	[Not Used]
269	[Not Used]
270	[Not Used]
271	[Not Used]
272	[Not Used]
273	[Not Used]
274	[Not Used]
275	[Not Used]
276	[Not Used]
277	[Not Used]
278	[Not Used]
279	[Not Used]
280	[Not Used]
281	Accumulated Deferred Income Taxes – Accelerated Amortization Property
282	Accumulated Deferred Income Taxes – Other Property
283	Accumulated Deferred Income Taxes – Other

284 - 300 [Not Used]

Appendix B

ELECTRIC PLANT CHART OF ACCOUNTS

1. INTANGIBLE PLANT

301	Organization
302	Franchises and Consents
303	Miscellaneous Intangible Plant
304	[*Not Used*]
305	[*Not Used*]
306	[*Not Used*]
307	[*Not Used*]
308	[*Not Used*]
309	[*Not Used*]

2. PRODUCTION PLANT

A. Steam Production

310	Land and Land Rights
311	Structures and Improvements
312	Boiler Plant Equipment
313	Engines and Engine-Driven Generators
314	Turbogenerator Units
315	Accessory Electric Equipment
316	Miscellaneous Power Plant Equipment
317	Asset Retirement Costs for Steam Production Plant
318	[*Not Used*]
319	[*Not Used*]

B. Nuclear Production

320	Land and Land Rights (Major Only)
321	Structures and Improvements (Major Only)
322	Reactor Plant Equipment (Major Only)
323	Turbogenerator Units (Major Only)
324	Accessory Electric Equipment (Major Only)
325	Miscellaneous Power Plant Equipment (Major Only)
326	Asset Retirement Costs for Nuclear Production Plant
327	[*Not Used*]
328	[*Not Used*]
329	[*Not Used*]

C. Hydraulic Production

330	Land and Land Rights
331	Structures and Improvements
332	Reservoirs, Dams, and Waterways
333	Water Wheels, Turbines, and Generators
334	Accessory Electric Equipment
335	Miscellaneous Power Plant Equipment
336	Roads, Railroads, and Bridges
337	Asset Retirement Costs for Hydraulic Production Plant
338	[*Not Used*]
339	[*Not Used*]

D. Other Production

340	Land and Land Rights
341	Structures and Improvements
342	Fuel Holders, Producers, and Accessories
343	Prime Movers
344	Generators
345	Accessory Electric Equipment
346	Miscellaneous Power Plant Equipment
347	Asset Retirement Costs for Other Production Plant
348	[*Not Used*]
349	[*Not Used*]

3. TRANSMISSION PLANT

350	Land and Land Rights
351	[Reserved]
352	Structures and Improvements
353	Station Equipment
354	Towers and Fixtures
355	Poles and Fixtures
356	Overhead Conductors and Devices
357	Underground Conduit
358	Underground Conductors and Devices
359	Roads and Trails
359.1	Asset Retirement Costs for Transmission Plant

4. DISTRIBUTION PLANT

360	Land and Land Rights
361	Structures and Improvements
362	Station Equipment
363	Storage Battery Equipment
364	Poles, Towers, and Fixtures
365	Overhead Conductors and Devices
366	Underground Conduit
367	Underground Conductors and Devices
368	Line Transformers
369	Services
370	Meters
371	Installations on Customers' Premises
372	Leased Property on Customers' Premises
373	Street Lighting and Signal Systems
374	Asset Retirement Costs for Distribution Plant Production Plant
375	[*Not Used*]
376	[*Not Used*]
377	[*Not Used*]
378	[*Not Used*]
379	[*Not Used*]

5. REGIONAL TRANSMISSION AND MARKET OPERATION PLANT

380	Land and Land Rights
381	Structures and Improvements
382	Computer Hardware
383	Computer Software
384	Communication Equipment
385	Miscellaneous Regional Transmission and Market Operation Plant
386	Asset Retirement Costs for Regional Transmission and Market Operation Plant
387	[Reserved]
388	[*Not Used*]

6. GENERAL PLANT

389	Land and Land Rights
390	Structures and Improvements
391	Office Furniture and Equipment
392	Transportation Equipment

Uniform System of Electric Accounts

393	Stores Equipment
394	Tools, Shop, and Garage Equipment
395	Laboratory Equipment
396	Power Operated Equipment
397	Communication Equipment
398	Miscellaneous Equipment
399	Other Tangible Property
399.1	Asset Retirement Costs for General Plant

INCOME CHART OF ACCOUNTS

1. UTILITY OPERATING INCOME

400	Operating Revenues
401	Operation Expense
402	Maintenance Expense
403	Depreciation Expense
403.1	Depreciation Expense for Asset Retirement Costs
404	Amortization of Limited-Term Electric Plant
405	Amortization of Other Electric Plant
406	Amortization of Electric Plant Acquisitions and Adjustments
407	Amortization of Property Losses, Unrecovered Plant, and Regulatory Study Costs
407.1	[Not Used]
407.2	[Not Used]
407.3	Regulatory Debits
407.4	Regulatory Credits
408	[Reserved]
408.1	Taxes Other Than Income Taxes, Utility Operating Income
408.2	[See "2. C. Taxes Applicable to Other Income and Deductions]
409	[Reserved]
409.1	Income Taxes, Utility Operating Income
409.2	[See "2. C. Taxes Applicable to Other Income and Deductions]
409.3	[See "2. C. Taxes Applicable to Other Income and Deductions]
410	[Reserved]
410.1	Provisions for Deferred Income Taxes, Utility Operating Income
410.2	[See "2. C. Taxes Applicable to Other Income and Deductions]

411	[Reserved]
411.1	Provision for Deferred Income Taxes – Credit, Utility Operating Income
411.2	[See "2. C. Taxes Applicable to Other Income and Deductions]
411.3	[Reserved]
411.4	Investment Tax Credit Adjustments, Utility Operations
411.5	[See "2. C. Taxes Applicable to Other Income and Deductions]
411.6	Gains from Disposition of Utility Plant
411.7	Losses from Disposition of Utility Plant
411.8	Gains from Disposition of Allowances
411.9	Losses from Disposition of Allowances
411.10	Accretion Expense
412	Revenues from Electric Plant Leased to Others
413	Expenses of Electric Plant Leased to Others
414	Other Utility Operating Income

2. OTHER INCOME AND DEDUCTIONS

A. Other Income

415	Revenues from Merchandising, Jobbing, and Contract Work
416	Costs and Expenses of Merchandising, Jobbing, and Contract Work
417	Revenues from Nonutiltiy Operations
417.1	Expenses of Nonutiltiy Operations
418	Nonoperating Rental Income
418.1	Equity in Earnings of Subsidiary Companies (Major Only)
419	Interest and Dividend Income
419.1	Allowance for Other Funds Used During Construction
420	Investment Tax Credits [See also "2. C. Taxes Applicable to Other Income and Deductions]
421	Miscellaneous Nonoperating Income
421.1	Gain on Disposition of Property

B. Other Income Deductions

421.2	Loss on Disposition of Property
422	[Not Used]
423	[Not Used]
424	[Not Used]
425	Miscellaneous Amortization
426	[Reserved]
426.1	Donations

Uniform System of Electric Accounts

426.2 Life Insurance
426.3 Penalties
426.4 Expenditures for Certain Civic, Political, and Related Activities
426.5 Other Deductions
    Total Other Income Deductions
    Total Other Income and Deductions

C. Taxes Applicable to Other Income and Deductions

408.2 Taxes Other Than Income Taxes, Other Income and Deductions
409.2 Income Tax, Other Income and Deductions
409.3 Income Taxes, Extraordinary Items
410.2 Provision for Deferred Income Taxes – Other Income and Deductions
411.2 Provision for Deferred Income Taxes – Credit, Other Income, and deductions
411.5 Investment Tax Credit Adjustments, Nonutility Operations
420 Investment Tax Credits [*See also "2. A. Other Income*]
    Total Taxes on Other Income and Deductions
    Net Other Income and Deductions

3. INTEREST CHARGES

427 Interest on Long-Term Debt
428 Amortization of Debt Discount and Expense
428.1 Amortization of Loss on Reacquired Debt
429 Amortization of Premium on Debt – Credit
429.1 Amortization of Gain on Reacquired Debt – Credit
430 Interest on Debt to Associated Companies
431 Other Interest Expense
432 Allowance for Borrowed Funds Used During Construction – Credit
433 [*See "Retained Earnings Chart of Accounts"*]

4. EXTRAORDINARY ITEMS

434 Extraordinary Income
435 Extraordinary Deductions

RETAINED EARNINGS CHART OF ACCOUNTS

433	Balance Transferred from Income
434	[*See "Income Chart of Accounts: 4. Extraordinary Items"*]
435	[*See "Income Chart of Accounts: 4. Extraordinary Items"*]
436	Appropriations of Retained Earnings
437	Dividends Declared – Preferred Stock
438	Dividends Declared – Common Stock
439	Adjustments to Retained Earnings

OPERATING REVENUE CHART OF ACCOUNTS

1. SALES OF ELECTRICITY

440	Residential Sales
441	[*Not Used*]
442	Commercial and Industrial Sales
443	[*Not Used*]
444	Public Street and Highway Lighting
445	Other Sales to Public Authorities (Major Only)
446	Sales to Railroads and Railways (Major Only)
447	Sales for Resale
448	Interdepartmental Sales
449	Other Sales (Nonmajor Only)
449.1	Provision for Rate Refunds

2. OTHER OPERATING REVENUES

450	Forfeited Discounts
451	Miscellaneous Service Revenues
452	[*Not Used*]
453	Sales of Water and Water Power
454	Rent from Electric Property
455	Interdepartmental Rents
456	Other Electric Revenues

Uniform System of Electric Accounts

456.1 Revenue from Transmission of Electricity of Others
457 [Not Used]
457.1 Regional Transmission Services Revenues
457.2 Miscellaneous Revenues

458 - 499 [Not Used]

OPERATION AND MAINTENANCE EXPENSE CHART OF ACCOUNTS

1. POWER PRODUCTION EXPENSES

A. Steam Power Generation

Operation

500 Operation Supervision and Engineering
501 Fuel
502 Steam Expenses (Major Only)
503 Steam from Other Sources
504 Steam Transferred – Credit
505 Electric Expenses (Major Only)
506 Miscellaneous Steam Power Expenses (Major Only)
507 Rents
508 Operation Supplies and Expenses (Nonmajor Only)
509 Allowances

Maintenance

510 Maintenance Supervision and Engineering (Major Only)
511 Maintenance of Structures (Major Only)
512 Maintenance of Boiler Plant (Major Only)
513 Maintenance of Electric Plant (Major Only)
514 Maintenance of Miscellaneous Steam Plant (Major Only)
515 Maintenance of Steam Production Plant (Nonmajor Only)
516 [Not Used]

B. Nuclear Power Generation

Operation

517	Operation Supervision and Engineering (Major Only)
518	Nuclear Fuel Expense (Major Only)
519	Coolants and Water (Major Only)
520	Steam Expenses (Major Only)
521	Steam from Other Sources (Major Only)
522	Steam Transferred – Credit (Major Only)
523	Electric Expenses (Major Only)
524	Miscellaneous Nuclear Power Expenses (Major Only)
525	Rents (Major Only)
526	[*Not Used*]
527	[*Not Used*]

Maintenance

528	Maintenance Supervision and Engineering (Major Only)
529	Maintenance of Structures (Major Only)
530	Maintenance of Reactor Plant Equipment (Major Only)
531	Maintenance of Electric Plant (Major Only)
532	Maintenance of Miscellaneous Nuclear Plant (Major Only)
533	[*Not Used*]
534	[*Not Used*]

C. Hydraulic Power Generation

Operation

535	Operation Supervision and Engineering
536	Water for Power
537	Hydraulic Expenses (Major Only)
538	Electric Expenses (Major Only)
539	Miscellaneous Hydraulic Power Generation Expenses (Major Only)
540	Rents
540.1	Operation Supplies and Expenses (Nonmajor Only)

Maintenance

541	Maintenance Supervision and Engineering (Major Only)
542	Maintenance of Structures (Major Only)
543	Maintenance of Reservoirs, Dams, and Waterways (Major Only)

Uniform System of Electric Accounts

544   Maintenance of Electric Plant (Major Only)
545   Maintenance of Miscellaneous Hydraulic Plant (Major Only)
545.1   Maintenance of Hydraulic Production Plant (Nonmajor Only)

D. Other Power Generation

Operation

546   Operation Supervision and Engineering
547   Fuel
548   Generation Expenses (Major Only)
549   Miscellaneous Other Power Generation Expenses (Major Only)
550   Rents
550.1   Operation Supplies and Expenses (Nonmajor Only)

Maintenance

551   Maintenance Supervision and Engineering (Major Only)
552   Maintenance of Structures (Major Only)
553   Maintenance of Generating and Electric Plant (Major Only)
554   Maintenance of Miscellaneous Other Power Generation Plant (Major Only)
554.1   Maintenance of Other Power Production Plant (Nonmajor Only)

E. Other Power Supply Expenses

555   Purchased Power
556   System Control and Load Dispatching (Major Only)
557   Other Expenses
558   [*Not Used*]
559   [*Not Used*]

2. TRANSMISSION EXPENSES

Operation

560   Operation Supervision and Engineering
561   [*Not Used*]
561.1   Load Dispatch – Reliability
561.2   Load Dispatch – Monitor and Operate Transmission System
561.3   Load Dispatch – Transmission Service and Scheduling

561.4   Scheduling, System Control, and Dispatch Services
561.5   Reliability, Planning, and Standards Development
561.6   Transmission Services Studies
561.7   Generator Interconnection Studies
561.8   Reliability, Planning, and Standards Development Series
562   Station Expenses (Major Only)
563   Overhead Line Expenses (Major Only)
564   Underground Line Expenses (Major Only)
565   Transmission of Electricity by Others (Major Only)
566   Miscellaneous Transmission Expenses (Major Only)
567   Rents
567.1   Operation Supplies and Expenses (Nonmajor Only)

Maintenance

568   Maintenance Supervision and Engineering (Major Only)
569   Maintenance of Structures (Major Only)
569.1   Maintenance of Computer Hardware
569.2   Maintenance of Computer Software
569.3   Maintenance of Communication Equipment
569.4   Maintenance of Miscellaneous Regional Transmission Plant
570   Maintenance of Station Equipment (Major Only)
571   Maintenance of Overhead Lines (Major Only)
572   Maintenance of Underground Lines (Major Only)
573   Maintenance of Miscellaneous Transmission Plant (Major Only)
574   Maintenance of Transmission Plant (Nonmajor Only)
575   [*Not Used*]

3. REGIONAL MARKET EXPENSES

Operation

575.1   Operation Supervision
575.2   Day-Ahead and Real-Time Market Administration
575.3   Transmission Rights Market Administration
575.4   Capacity Market Administration
575.5   Ancillary Services Market Administration
575.6   Market Monitoring and Compliance
575.7   Market Administration, Monitoring, and Compliance Services
575.8   Rents
576   [*Not Used*]

Uniform System of Electric Accounts

Maintenance

576.1	Maintenance of Structures and Improvements
576.2	Maintenance of Computer Hardware
576.3	Maintenance of Computer Software
576.4	Maintenance of Communication Equipment
576.5	Maintenance of Miscellaneous Market Operation Plant
577	[Not Used]
578	[Not Used]
579	[Not Used]

4. DISTRIBUTION EXPENSES

Operation

580	Operation Supervision and Engineering
581	Load Dispatching (Major Only)
581.1	Line and Station Supplies and Expenses (Nonmajor only)
582	Station Expenses (Major only)
583	Overhead Line Expenses (Major only)
584	Underground Line Expenses (Major only)
585	Street Lighting and Signal System Expenses
586	Meter Expenses
587	Customer Installations Expenses
588	Miscellaneous Distribution Expenses
589	Rents

Maintenance

590	Maintenance Supervision and Engineering (Major Only)
591	Maintenance of Structures (Major Only)
592	Maintenance of Station Equipment (Major Only)
592.1	Maintenance of Structures and Equipment (Nonmajor only)
593	Maintenance of Overhead Lines (Major Only)
594	Maintenance of Underground Lines (Major Only)
594.1	Maintenance of Lines (Nonmajor Only)
595	Maintenance of Line Transformers
596	Maintenance of Street Lighting and Signal Systems
597	Maintenance of Meters
598	Maintenance of Miscellaneous Distribution Plant

599 - 900 [Not Used]

5. CUSTOMER ACCOUNTS EXPENSES

Operation

- 901 Supervision (Major Only)
- 902 Meter Reading Expenses
- 903 Customer Records and Collection Expenses
- 904 Uncollectible Accounts
- 905 Miscellaneous Customer Account Expenses (Major Only)

6. CUSTOMER SERVICE AND INFORMATIONAL EXPENSES

Operation

- 906 Customer Service and Informational Expenses (Nonmajor Only)
- 907 Supervision (Major Only)
- 908 Customer Assistance Expenses (Major Only)
- 909 Informational and Instructional Advertising Expenses (Major Only)
- 910 Miscellaneous Customer Service and Informational Expenses (Major Only)

7. SALES EXPENSES

Operation

- 911 Supervision (Major Only)
- 912 Demonstrating and Selling Expenses (Major Only)
- 913 Advertising Expenses (Major Only)
- 914 [*Not Used*]
- 915 [*Not Used*]
- 916 Miscellaneous Sales Expenses (Major Only)
- 917 Sales Expenses (Nonmajor Only)
- 918 [*Not Used*]
- 919 [*Not Used*]

8. ADMINISTRATIVE AND GENERAL EXPENSES

Operation

- 920 Administrative and General Salaries
- 921 Office Supplies and Expenses

922	Administrative Expenses Transferred – Credit
923	Outside Services Employed
924	Property Insurance
925	Injuries and Damages
926	Employee Pensions and Benefits
927	Franchise Requirements
928	Regulatory Commission Expenses
929	Duplicate Charges – Credit
930	[*Not Used*]
930.1	General Advertising Expenses
930.2	Miscellaneous General Expenses
931	Rents
932	[*Not Used*]
933	Transportation Expenses
934	[*Not Used*]

Maintenance

935	Maintenance of General Plant

Example of SubAccounts for Electric Meter Plant and Associated O&M

Account 370 – Meters

370.01	Electric Meters
370.02	Electric Meter Installations
370.03	Electric Meter Accessories

Account 586 – Meter Expenses

586.01	Supervision of Electric Meter Operations
586.02	Installing, Removing, and Resetting Electric Meters
586.03	Installing, Removing, and Resetting Electric Load Survey Recorders
586.04	Routine Inspection of Metering Equipment
586.05	Field Testing of Single-Phase Electric Metering
586.06	Field Testing of Poly-Phase Electric Metering
586.07	Electric Service Turn On/Turn Off
586.08	Maintaining Electric Meter Records

Account 597 – Maintenance of Meters

597.01	Repair of Electric Meters
597.02	Repair of Electric Meter Socket Bases
597.03	Repair of Instrument Transformers, Load Survey Recorders, etc.
597.04	Shop Testing of Single-Phase Electric Metering
597.05	Shop Testing of Poly-Phase Electric Metering
597.06	Instrument Repair and Testing

Appendix C
Industry Classification Systems

NAIC AND SIC CODES

Industry classification systems are used to congregate economic units (production and services sectors) by similar processes, outputs, or activities. Utilities use industrial classification systems to group their customers into common categories so that the customer base can be segmented readily in order to conduct financial, statistical, and other technical analyses. The ability to group customers with similar operating and load characteristics can be of great value for pricing purposes, and an accurate classification of customers is necessary to ensure the development of homogeneous customer groups.

An industry classification system long used by utilities and other organizations in the United States is the *Standard Industrial Classification (SIC) System*. SIC codes were first introduced in 1938 when a list of manufacturing industries was compiled. A list of nonmanufacturing industries was compiled and released in the following year. Since that time, the SIC system has undergone periodic revisions in conjunction with changes in the economy.[1] The most recent SIC code manual update occurred in 1987.

In 1997, the new *North American Industry Classification (NAIC) System* was compiled and released with the objective of consolidating the industry classification systems of Canada (Canadian SIC, 1980), Mexico (Mexican Classification of Activities and Products, 1994), and the United States into a common set of codes. Since 1997, the NAIC system has undergone periodic revisions. Both the SIC and the NAIC code systems will likely coexist in the foreseeable future until utilities and other organizations are able to convert to the newer system.

---

[1] In 1948, the United Nations adopted an International Standard Industrial Classification (ISIC) System and made subsequent revisions in 1958, 1968, and 1989.

Both classification systems have major group subdivisions: A through K in the SIC system and 11 through 92 in the NAIC system (although the numbering is not fully inclusive). The SIC code system is based on a 2-3-4 digit hierarchical structure while the NAIC system is based on a 3-4-5-6 digit hierarchical structure. Classification systems users can work with a particular level of detail depending on the specific application and the desired resolution. An example comparing the two system structures is made for the classification of firms which manufacture portable electric space heaters.

SIC Classification

D.	Manufacturing
36	Electronic and Other Electrical Equipment, Except Computer Equipment
363	Household Appliances
3634	Electric Housewares and Fans

NAIC Classification

33	Manufacturing
335	Electrical Equipment, Appliance, and Component Manufacturing
3352	Household Appliance Manufacturing
33521	Small Electrical Appliance Manufacturing
335211	Electric Housewares and Household Fan Manufacturing

The listing of codes in this Appendix include only two digit SIC and three-digit NAIC classifications. Descriptions of the higher detailed codes can be found in the 1987 Standard Industrial Classification Manual and the 2007 North American Industry Classification Manual. In addition, the NAIC manual provides a cross-reference listing that maps SIC codes to NAIC codes.

Note: NAICS is intended to provide an industry structure and hierarchy that ensures comparability of statistics between countries. Some code detail necessary for U.S. industry cannot be recognized by Canada and/or Mexico because of size, specialization, or organization of the industry. The NAICS United States publication was used for this Appendix. Thus, in the listing of NAIC codes, some categories may be denoted with the symbol [US] or [CAN]. If marked with [US], the code applies only to United States industry. If marked with [CAN], only United States and Canadian industries are comparable. If blank, Canadian, Mexican, and United States industries are comparable. NAICS Canada and NAICS Mexico publications contain details relative to the two countries individually.

Industry Classification Systems

North American Industry Classification System - 2007

11	Agriculture, Forestry, Fishing, and Hunting
21	Mining, Quarrying, and Oil and Gas Extraction
22	Utilities
23	Construction
31	Manufacturing
32	Manufacturing
33	Manufacturing
42	Wholesale Trade
44	Retail Trade
45	Retail Trade
48	Transportation and Warehousing
49	Transportation and Warehousing
51	Information
52	Finance and Insurance
53	Real Estate and Rental and Leasing
54	Professional, Scientific, and Technical Services
55	Management of Companies and Enterprises
56	Administration and Support and Waste Management and Remediation Services
61	Educational Services
62	Health Care and Social Assistance
71	Arts, Entertainment, and Recreation
72	Accommodation and Food Services
81	Other Services (Except Public Administration)
92	Public Administration

Standard Industrial Classification System - 1987

A.	Agriculture, Forestry, and Fishing
B.	Mining
C.	Construction
D.	Manufacturing
E.	Transportation, Communications, Electric, Gas, and Sanitary Services
F.	Wholesale Trade
G.	Retail Trade
H.	Finance, Insurance, and Real Estate
I.	Services
J.	Public Administration
K.	Nonclassifiable Establishments

North American Industry Classification System - 2007

NAIC Code	NAIC Title
11	**AGRICULTURE, FORESTRY, FISHING, AND HUNTING**
111	Crop Production
112	Animal Production
113	Forestry and Logging
114	Fishing, Hunting, and Trapping
115	Support Activities for Agriculture and Forestry
21	**MINING**
211	Oil and Gas Extraction
212	Mining (Except Oil and Gas)
213	Support Activities for Mining
22	**UTILITIES**
221	Utilities [CAN]
23	**CONSTRUCTION**
236	Construction of Buildings
237	Heavy and Civil Engineering Construction
238	Special Trade Contractors
31-33	**MANUFACTURING**
311	Food Manufacturing
312	Beverage and Tobacco Product Manufacturing
313	Textile Mills
314	Textile Product Mills
315	Apparel Manufacturing
316	Leather and Allied Product Manufacturing
317	[*Not Used*]
318	[*Not Used*]
319	[*Not Used*]
320	[*Not Used*]
321	Wood Product Manufacturing
322	Paper Manufacturing
323	Printing and Related Support Activities
324	Petroleum and Coal Products Manufacturing
325	Chemical Manufacturing
326	Plastics and Rubber Products Manufacturing
327	Nonmetallic Mineral Product Manufacturing
328	[*Not Used*]
329	[*Not Used*]

North American Industry Classification System - 2007

NAIC Code	NAIC Title
330	[Not Used]
331	Primary Metal Manufacturing
332	Fabricated Metal Product Manufacturing
333	Machinery Manufacturing
334	Computer and Electronic Product Manufacturing
335	Electrical Equipment, Appliance, and Component Manufacturing
336	Transportation Equipment Manufacturing
337	Furniture and Related Product Manufacturing
338	[Not Used]
339	Miscellaneous Manufacturing
42	**WHOLESALE TRADE**
423	Merchant Wholesalers, Durable Goods [US]
424	Merchant Wholesalers, Nondurable Goods [US]
425	Wholesale Electronic Markets and Agents and Brokers [US]
44-45	**RETAIL TRADE**
441	Motor Vehicle and Parts Dealers [CAN]
442	Furniture and Home Furnishings Stores [CAN]
443	Electronics and Appliance Stores [CAN]
444	Building Material and Garden Equipment and Supplies Dealers [CAN]
445	Food and Beverage Stores [CAN]
446	Health and Personal Care Stores [CAN]
447	Gasoline Stations [CAN]
448	Clothing and Clothing Accessories Stores [CAN]
449	[Not Used]
450	[Not Used]
451	Sporting Goods, Hobby, Book, and Music Stores [CAN]
452	General Merchandise Stores [CAN]
453	Miscellaneous Store Retailers [CAN]
454	Nonstore Retailers [CAN]
48-49	**TRANSPORTATION AND WAREHOUSING**
481	Air Transportation
482	Rail Transportation
483	Water Transportation
484	Truck Transportation
485	Transit and Ground Passenger Transportation

North American Industry Classification System - 2007

NAIC Code	NAIC Title
486	Pipeline Transportation
487	Scenic and Sightseeing Transportation
488	Support Activities for Transportation
489	[*Not Used*]
490	[*Not Used*]
491	Postal Service
492	Couriers and Messengers
493	Warehousing and Storage
51	**INFORMATION**
511	Publishing Industries (except Internet)
512	Motion Picture and Sound Recording Industries
513	[*Not Used*]
514	[*Not Used*]
515	Broadcasting (except Internet)
516	[*Not Used*]
517	Telecommunications
518	Data Processing, Hosting, and Related Services
519	Other Information Services
52	**FINANCE AND INSURANCE**
521	Monetary Authorities – Central Bank
522	Credit Intermediation and Related Activities
523	Securities, Commodity Contracts, and Other Financial Investments and Related Activities
524	Insurance Carriers and Related Activities
525	Funds, Trusts, and Other Financial Vehicles [US]
53	**REAL ESTATE AND RENTAL AND LEASING**
531	Real Estate
532	Rental and Leasing Services
533	Lessors of Nonfinancial Intangible Assets (except Copyrighted Works)
54	**PROFESSIONAL, SCIENTIFIC, AND TECHNICAL SERVICES**
541	Professional, Scientific, and Technical Services
55	**MANAGEMENT OF COMPANIES AND ENTERPRISES**
551	Management of Companies and Enterprises

North American Industry Classification System - 2007

NAIC Code	NAIC Title
56	**ADMINISTRATION AND SUPPORT AND WASTE MANAGEMENT AND REMEDIATION SERVICES**
561	Administrative and Support Services
562	Waste Management and Remediation Services
61	**EDUCATIONAL SERVICES**
611	Educational Services
62	**HEALTH CARE AND SOCIAL ASSISTANCE**
621	Ambulatory Health Care Services
622	Hospitals
623	Nursing and Residential Care Facilities
624	Social Assistance
71	**ARTS, ENTERTAINMENT, AND RECREATION**
711	Performing Arts, Spectator Sports, and Related Industries
712	Museums, Historical Sites, and Similar Institutions
713	Amusement, Gambling, and Recreation Industries
72	**ACCOMMODATION AND FOOD SERVICES**
721	Accommodation
722	Food Services and Drinking Places
81	**OTHER SERVICES (EXCEPT PUBLIC ADMINISTRATION)**
811	Repair and Maintenance
812	Personal and Laundry Services
813	Religious, Grantmaking, Civic, Professional, and Similar Organizations
814	Private Households
92	**PUBLIC ADMINISTRATION**
921	Executive, Legislative, and Other General Government Support [US]
922	Justice, Public Order, and Safety Activities [US]
923	Administration of Human Resource Programs [US]
924	Administration of Environmental Quality Programs [US]
925	Administration of Housing Programs, Urban Planning, and Community Development [US]
926	Administration of Economic Programs [US]

North American Industry Classification System - 2007

NAIC Code	NAIC Title
927	Space Research and Technology [US]
928	National Security and International Affairs [US]

Industry Classification Systems

Standard Industrial Classification System - 1987

SIC Code	SIC Title
A.	**AGRICULTURE, FORESTRY, AND FISHING**
01	Agricultural Production – Crops
02	Agricultural Production – Livestock and Animal Specialties
03	[Not Used]
04	[Not Used]
05	[Not Used]
06	[Not Used]
07	Agricultural Services
08	Forestry
09	Fishing, Hunting, and Trapping
B.	**MINING**
10	Metal Mining
11	*[Not Used]*
12	Coal Mining
13	Oil and Gas Extraction
14	Mining and Quarrying of Nonmetallic Minerals, Except Fuels
C.	**CONSTRUCTION**
15	Building Construction - General Contractors and Operative Builders
16	Heavy Construction Other Than Building Construction – Contractors
17	Construction - Special Trade Contractors
18	*[Not Used]*
19	*[Not Used]*
D.	**MANUFACTURING**
20	Food and Kindred Products
21	Tobacco Products
22	Textile Mill Products
23	Apparel and Other Finished Products Made from Fabrics and Similar Materials
24	Lumber and Wood Products, Except Furniture
25	Furniture and Fixtures
26	Paper and Allied Products
27	Printing, Publishing, and Allied Industries
28	Chemicals and Allied Products
29	Petroleum Refining and Related Industries
30	Rubber and Miscellaneous Plastics Products

Standard Industrial Classification System - 1987

SIC Code	SIC Title
31	Leather and Leather Products
32	Stone, Clay, Glass, and Concrete Products
33	Primary Metal Industries
34	Fabricated Metal Products, Except Machinery and Transportation Equipment
35	Industrial and Commercial Machinery and Computer Equipment
36	Electronic and Other Electrical Equipment and Components, Except Computer Equipment
37	Transportation Equipment
38	Measuring, Analyzing, and Controlling Instruments; Photographic, Medical and Optical Goods; Watches and Clocks
39	Miscellaneous Manufacturing Industries
E.	**TRANSPORTATION, COMMUNICATIONS, ELECTRIC, GAS, AND SANITARY SERVICES**
40	Railroad Transportation
41	Local and Suburban Transit and Interurban Highway Passenger Transportation
42	Motor Freight Transportation and Warehousing
43	United States Postal Service
44	Water Transportation
45	Transportation by Air
46	Pipelines, Except Natural Gas
47	Transportation Services
48	Communications
49	Electric, Gas, and Sanitary Services
F.	**WHOLESALE TRADE**
50	Wholesale Trade – Durable Goods
51	Wholesale Trade – Nondurable Goods
G.	**RETAIL TRADE**
52	Building Materials, Hardware, Garden Supply, and Mobile Home Dealers
53	General Merchandise Stores
54	Food Stores
55	Automotive Dealers and Gasoline Service Stations
56	Apparel and Accessory Stores
57	Home Furniture, Furnishings, and Equipment Stores

Industry Classification Systems

Standard Industrial Classification System - 1987

SIC Code	SIC Title
58	Eating and Drinking Places
59	Miscellaneous Retail
H.	**FINANCE, INSURANCE, AND REAL ESTATE**
60	Depository Institutions
61	Nondepository Credit Institutions
62	Security and Commodity Brokers, Dealers, Exchanges, and Services
63	Insurance Carriers
64	Insurance Agents, Brokers, and Service
65	Real Estate
66	[Not Used]
67	Holding and Other Investment Offices
68	[Not Used]
69	[Not Used]
I.	**SERVICES**
70	Hotels, Rooming Houses, Camps, and Other Lodging Places
71	[Not Used]
72	Personal Services
73	Business Services
74	Not Used]
75	Automotive Repair, Services, and Parking
76	Miscellaneous Repair Services
77	[Not Used]
78	Motion Pictures
79	Amusement and Recreation Services
80	Health Services
81	Legal Services
82	Educational Services
83	Social Services
84	Museums, Art Galleries, and Botanical and Zoological Gardens
85	[Not Used]
86	Membership Organizations
87	Engineering, Accounting, Research, Management, and Related Services
88	Private Households
89	Services, Not Elsewhere Classified
90	[Not Used]

Standard Industrial Classification System - 1987

SIC Code	SIC Title
J.	**PUBLIC ADMINISTRATION**
91	Executive, Legislative, and General Government, Except Finance
92	Justice, Public Order, and Safety
93	Public Finance, Taxation, and Monetary Policy
94	Administration of Human Resource Programs
95	Administration of Environmental Quality and Housing Programs
96	Administration of Economic Programs
97	National Security and International Affairs
98	[*Not Used*]
K.	**NONCLASSIFIABLE ESTABLISHMENTS**
99	Nonclassifiable Establishments

Appendix D
Internet Resources

UTILITIES AND TARIFFS

Utility Company Links
http://www.utilityconnection.com/

Utility Tariff Links
http://www.tariffbooks.com/

INDUSTRY TRADE ASSOCIATIONS

American Council for an Energy-Efficient Economy (ACE[3])
http://aceee.org/

American Gas Association (AGA)
http://www.aga.org/

American Public Power Association (APPA)
http://www.appanet.org/

American Wind Energy Association (AWEA)
http://www.awea.org/

Association of Edison Illuminating Companies (AEIC)
http://www.aeic.org/

Association of Energy Engineers (AEE)

http://www.aeecenter.org/

Canadian Electricity Association (CEA)

http://www.canelect.ca/

Canadian Energy Research Institute (CERI)

http://www.ceri.ca/

Edison Electric Institute (EEI)

http://www.eei.org/

Electric Power Research Institute (EPRI)

http://www.epri.com/

Electricity Consumers Council (ELCON)

http://www.elcon.org/

Energy Networks Association (ENA)

http://2009.energynetworks.org/

Institute of Electrical and Electronics Engineers (IEEE)

http://www.ieee.org/

Institute of Engineering and Technology (IET)

http://www.theiet.org/

International Association for Energy Economics (IAEE)

http://www.iaee.org/

National Rural Electric Cooperative Association (NRECA)

http://www.nreca.org/

North American Electric Reliability Council (NERC)

http://www.nerc.com/

Rural Utility Services (RUS)

http://www.usda.gov/rus/

Solar Electric Power Association (SEPA)

http://www.solarelectricpower.org/

Internet Resources

REGULATORY ORGANIZATIONS

Energy Regulators Regional Association (ERRA)
http://www.erranet.org/

Federal Energy Regulatory Commission (FERC)
http://www.ferc.gov/

National Association of Regulatory Utility Commissions (NARUC)
http://www.naruc.org/

National Regulatory Research Institute (NRRI)
http://www.nrri.org/

ENERGY INFORMATION

Dow Jones Indexes
http://averages.dowjones.com/mdsidx/?event=energyUSDaily
http://averages.dowjones.com/mdsidx/?event=energyEuroPower

Energy Information Administration (EIA)
http://www.eia.doe.gov/

International Energy Agency (IEA)
http://www.iea.org/

New York Mercantile Exchange (NYMEX)
http://www.nymex.com/

U.S. Department of Energy (DOE)
http://www.energy.gov/

UTILITY ACCOUNTING SYSTEMS

FERC Electric Accounts
http://www.ferc.gov/legal/acct-matts/usofa.asp

RUS Electric Accounts
http:www.usda.gov/rus/electric/pubs/a/1767b-1.pdf

INDUSTRY CLASSIFICATION SYSTEMS

International Standard Industrial Classification (ISIC) System
http://unstats.un.org/unsd/cr/registry/regcst.asp?Cl=27

North American Industry Classification (NAIC) System
http://www.census.gov/eos/www/naics

Standard Industrial Classification (SIC) System
http://www.osha.gov/pls/imis/sic_manual.html

OTHER

Financial Accounting Standards Board (FASB)
http://www.fasb.org/

Note: Web site URL addresses may change from time to time.

Bibliography

POWER SYSTEMS ENGINEERING REFERENCES

ABB Power Systems Inc. *Electrical Transmission and Distribution Reference Book*. Pittsburgh, PA: ABB Power Systems Inc., 1989.

_____. *Electric Utility Engineering Reference Book: Distribution Systems*. Pittsburgh, PA: ABB Power Systems Inc., 1989.

Beatty, H. Wayne. *Handbook of Electric Power Calculations*. 3rd ed. New York, NY: McGraw Hill, 2001.

Bureau of Naval Personnel. *Basic Electricity*. 2nd ed. New York, NY: Dover Publications, Inc., 1970.

Dillard, Clyde R., and David E. Goldberg. *Chemistry*. New York, NY: Macmillian Company, 1971.

Dugan, Roger C., Mark F. McGranaghan, and H. Wayne Beaty. *Electrical Power Systems Quality*. New York, NY: McGraw Hill, 1996.

Edison Electric Institute. *Handbook for Electricity Metering*. 10th ed. Washington DC: Edison Electric Institute, 2002.

Elgerd, Olle I., and Patrick D. van der Puije. *Electric Power Engineering*. 2nd ed. New York, NY: Chapman and Hall, 1998.

El-Hawary, Mohamed E. *Electric Power Systems: Design and Analysis*. Reston, VA: Reston Publishing Inc., 1983.

Energy Information Administration. *Household Energy Consumption and Expenditures 1993*. Washington D.C.: U.S. Government Printing Office, 1995.

_____. *Manufacturing Consumption of Energy 1994*. Washington D.C.: U.S. Government Printing Office, 1997.

_____. *A Look at Commercial Buildings in 1995: Characteristics, Energy Consumption, and Energy Expenditures*. Washington D.C.: U.S. Government Printing Office, 1998.

Faulkenberry, Luces M., and Walter Coffee. *Electrical Power Distribution and Transmission*. Upper Saddle River, NJ: Prentice-Hall, 1996.

Fitzgerald, A. E., Charles Kingsley, Jr., and Alexander Kusko. *Electric Machinery*. 3rd ed. New York, NY: McGraw Hill, 1971.

Gönen, Turan. *Electric Power Distribution Systems Engineering*. New York, NY: McGraw Hill, 1986.

Grainger, John J., and William D. Stevenson, Jr. *Power Systems Analysis*. New York, NY: McGraw-Hill, Inc., 1994.

Grisbey, L. L., ed., *The Electric Power Engineering Handbook*. Boca Raton, FL: CRC Press LLC, 2001.

Hayt, Jr., William H., and Jack E. Kemmerly. *Engineering Circuit Analysis*. 2nd ed. New York, NY: McGraw Hill, 1971.

Hingorani, Norain G., and Laszlo Gyugyi. *Understanding FACTS: Concepts and Technology of Flexible AC Transmission Systems*. New York, NY: IEEE Press, 2000.

Institute of Electrical and Electronics Engineers. *Electric Power Distribution for Industrial Plants: IEEE Std. 141-1993*. New York, NY: Institute of Electrical and Electronics Engineers, 1994.

Johansson, Thomas B., et al. *Renewable Energy Sources for Fuels and Electricity*. New York, NY: Island Press, 1993.

Kurtz, Edward B., Thomas M. Shoemaker, and James E. Mack. *Lineman's and Cableman's Handbook*. 9th ed. New York, NY: McGraw Hill, 1997.

Miller, Robert H., and James H. Malinowski. *Power System Operation*. 3rd ed. New York, NY: McGraw Hill, 1994.

Orfeuil, Maurice. *Electric Process Heating*. Columbus, OH: Battelle Press, 1987.

Pansini, Anthony J. *Electrical Distribution Engineering*. New York, NY: McGraw Hill, 1983.

Patrick, Dale R., and Stephen W. Fardo. *Electrical Distribution Systems*. Lilburn, GA: The Fairmont Press, 1999.

Stein, Benjamin, John S. Reynolds, and William J. McGuiness. *Mechanical and Electrical Equipment for Buildings*. 7th ed. New York, NY: John Wiley & Sons, 1986.

Turner, Wayne C. *Energy Management Handbook*. 3rd ed. Lilburn, GA: Fairmont Press, 1997.

University of Oklahoma. *Energy Alternatives: A Comparative Analysis*. Washington D.C.: U.S. Government Printing Office, 1975.

Vanek, Francis M., and Louis D. Albright, *Energy Systems Engineering: Evaluation and Implementation*. New York, NY: McGraw Hill, 2001.

Vogt, Lawrence J., and David A. Conner, *Electrical Energy Management*. Lexington, MA: D. C. Heath, Lexington Books, 1977.

Wiebe, Michael. *A Guide to Utility Automation: AMR, SCADA, & IT Systems for Electric Power*. Tulsa, OK: PennWell Books, 1999.

Wood, Allen J., and Bruce F. Wollenberg. *Power Generation, Operation, and Control*. 2nd ed. New York, NY: John Wiley & Sons, Inc., 1996.

PRICING AND REGULATORY REFERENCES

AEIC Load Research Committee. *Load Research Manual.* 2nd ed. Birmingham, AL: Association of Edison Illuminating Companies, 2001.

Alt Jr., Lowell E. "Energy Utility Rate Setting." 2006.

Ajello, Julian et al. *Electric Utility Cost Allocation Manual.* Washington, DC: National Association of Regulatory Utility Commissioners, 1992.

American Gas Association. *Gas Rate Fundamentals.* 4th ed., Arlington, VA: American Gas Association, 1987.

_____. *Introduction to Public Utility Accounting.* Written and Compiled by Marvin P. Reeser. Washington DC: American Gas Association, 1976.

Argonne National Laboratory and National Economic Research Associates, Inc. *Load Research Manual Volume 1: Load Research Procedures.* Washington DC: U.S. Department of Energy, 1980.

_____. *Load Research Manual Volume 2: Fundamentals of Implementing Load Research Procedures.* Washington DC: U.S. Department of Energy, 1980.

Argonne National Laboratory and Gordian Associates, Inc. *Load Research Manual Volume 3: Load Research for Advanced Technologies.* Washington DC: U.S. Department of Energy, 1980.

Bary, Constantine W. *Operational Economics of Electric Utilities.* New York, NY: Columbia University Press, 1963.

Bonbright, James C. *Principles of Public Utility Rates.* New York, NY: Columbia University Press, 1961.

Bonbright, James C., Albert L. Danielsen, and David Kamerschen. *Principles of Public Utility Rates.* 2nd ed., Arlington, VA: Public Utility Reports, 1988.

Brown, Steven J. and David S. Sibley. *The Theory of Public Utility Pricing.* New York, NY: Cambridge University Press, 1986.

Bryant, J. M., and R. R. Herrman. *Elements of Utility Rate Determination.* New York, NY: McGraw-Hill Book Company, Inc., 1940.

Caywood, Russell E. *Electric Utility Rate Economics.* New York, NY: McGraw-Hill, 1972.

Charles Rivers Associates for Edison Electric Institute. *A Guide To Electricity Forecasting Methodology*, Washington DC: Edison Electric Institute, 1986.

Christensen and Associates Energy Consulting, LLC for Edison Electric Institute, *Retail Electricity Pricing and Rate Design in Evolving Markets.* Washington DC: Edison Electric Institute Institute, 2007.

Cicchetti, Charles J., William J. Gillen, and Paul Smolensky. *The Marginal Cost and Pricing of Electricity: An Applied Approach.* Cambridge, MA: Ballinger Publishing Company, 1977.

Cravath, R. J. [1910] "Demand and Diversity Factors and Their Influence on Rates." In *The Development of Scientific Rates for Electricity Supply.* Detroit, MI: The Edison Illuminating Company of Detroit, 1915.

Doane, John. [1910] "High Efficiency Lamps—Their Effect on the Cost of Light to the Central Station." In *The Development of Scientific Rates for Electricity Supply*. Detroit, MI: The Edison Illuminating Company of Detroit, 1915.

Doherty, Henry L. [1900] "Equitable, Uniform and Competitive Rates." In *The Development of Scientific Rates for Electricity Supply*. Detroit, MI: The Edison Illuminating Company of Detroit, 1915.

Doran, John J., Frederick M. Hoppe, Robert Koger, and William W. Lindsay. *Electric Utility Cost Allocation Manual.* Washington, DC: National Association of Regulatory Utility Commissioners, 1973.

Edison Electric Institute and American Gas Association. *Introduction to Depreciation and Net Salvage of Public Utility Plant and Plant of Other Industries.* Washington DC: Edison Electric Institute, 2003.

_____. *Introduction to Public Utility Accounting.* 5th ed. Washington DC: Edison Electric Institute, 2001.

Electric Power Research Institute, Edison Electric Institute, American Public Power Association, and National Rural Electric Cooperatives for the National Association of Regulatory Utility Commissioners. 1976 – 1981. *Electric Utility Rate Design Study*, 94 vols. Palo Alto, CA: Electric Power Research Institute.

Ferguson, Louis A. [1911] "Effect of Width of Maximum Demand on Rate Making." In *The Development of Scientific Rates for Electricity Supply*. Detroit, MI: The Edison Illuminating Company of Detroit, 1915.

Garfield, Paul J., and Wallace F. Lovejoy. *Public Utility Economics.* Englewood Cliffs, NJ: Prentice Hall, Inc., 1964.

George, Stephen S. *Competitive Metering, Billing and Customer Services: An Analysis of Operational Issues, Part I.* Washington DC: Edison Electric Institute, 1997.

_____. *Competitive Metering, Billing and Customer Services: An Analysis of Operational Issues, Part II.* Washington DC: Edison Electric Institute, 1998.

Greene, W. J. [1896] "A Method of Calculating the Cost of Furnishing Electric Current and a Way of Selling It." In *The Development of Scientific Rates for Electricity Supply*. Detroit, MI: The Edison Illuminating Company of Detroit, 1915.

Hopkinson, John. [1892] "On the Cost of Electric Supply." In *The Development of Scientific Rates for Electricity Supply*. Detroit, MI: The Edison Illuminating Company of Detroit, 1915.

ICF Incorporated for Electric Power Research Institute and American Public Power Association. *The Design of Alternative Rates for Public Power Systems: Issues and Procedures.* Palo Alto, CA: Electric Power Research Institute, 1984.

Jeynes, Paul H. *Profitability and Economic Choice.* Ames, IA: The Iowa State University Press, 1968.

Kahn, Alfred E. *The Economics of Regulation: Principles and Institutions*, Cambridge, MA: The MIT Press, 1988.

Laurits R. Christensen Associates, Inc., for Edison Electric Institute. *Redesigning Distribution Tariffs for Restructured Electric Power Markets.* Washington, DC: Edison Electric Institute, 2000.

Lyndon, Lamar. *Rate-Making for Public Utilities.* New York, NY: McGraw-Hill Book Company, Inc., 1923.

NARUC Staff Subcommittee on Accounting and Finance. *Rate Case and Audit Manual.* Washington, DC: National Association of Regulatory Utility Commissioners, 2003.

Nash, L. R. *Public Utility Rate Structures.* New York, NY: McGraw-Hill, 1933.

Office of the Federal Register. "Part 101 – Uniform System of Accounts Prescribed for Public Utilities and Licensees Subject to the Provisions of the Federal Power Act." In *Code of Federal Regulations: Title 18 – Conservation of Power and Water Resources.* Washington D.C.: U.S. Government Printing Office, 2007.

Oxtoby, James V. [1910] "Reasonable Profit: Its Definition, Collection and Distribution." In *The Development of Scientific Rates for Electricity Supply.* Detroit, MI: The Edison Illuminating Company of Detroit, 1915.

Page, Roy. *Primer of Electric Service Costs.* New York, NY: Harper & Brothers Publishers, 1937.

Philipson, Lorrin, and H. Lee Willis. *Understanding Electric Utilities and De-Regulation.* New York, NY: Marcel Dekker, Inc., 1999.

Phillips Jr., Charles F. *The Regulation of Public Utilities.* 3rd ed. Arlington, VA: Public Utility Reports, 1993.

Schmidt, Michael. *Automatic Adjustment Clauses: Theory and Application.* East Lansing, MI: The Institute of Public Utilities, Michigan State University, 1980.

Schrock, Derek W. *Load Shape Development.* Tulsa, OK: PennWell Books, 1997.

Small, Michael E. *A Guide to FERC Regulation and Ratemaking of Electric Utilities and Other Power Suppliers.* 3rd ed. Washington, DC: Edison Electric Institute, 1994.

Suelflow, James E. *Public Utility Accounting: Theory and Application.* East Lansing, MI: The Institute of Public Utilities, Michigan State University, 1973.

The Brattle Group for Edison Electric Institute. *Electric Utility Automatic Adjustment Clauses: Benefits and Design Considerations.* Washington, DC: Edison Electric Institute, 2006.

Walters, Frank S. *The Art of Rate Design.* Washington, DC: Edison Electric Institute, 1984.

Wilson, Robert B. *Nonlinear Pricing.* New York, NY: Oxford University Press, Published in association with the Electric Power Research Institute, 1993.

Wright, Arthur. [1896] "Cost of Electricity Supply." In *The Development of Scientific Rates for Electricity Supply.* Detroit, MI: The Edison Illuminating Company of Detroit, 1915.

Index

AC power transmission, 173-184
  congestion management, 190
  contract path, 190
  loop flows, 190
  surge impedance loading, 180
accelerated depreciation, 477-478
accounts. *See* Uniform System of Accounts
accumulated deferred income taxes (ADITs), 451, 477-479, 482, 521
administrative and general (A&G) expenses, 451, 486-487
  assignment of, 559-560
  classification of, 528
  functionalization of, 487
ad valorem taxes. *See* property taxes
Advanced Metering Infrastructure (AMI), 254
allocation factors
  compound, 542
  development of, 554
  weighted, 556-557
allowance for funds used during construction (AFUDC), 472
alternative fuels. *See* geothermal production, hydroelectric production, solar power, wind generator
AM/FM system, 450, 458, 505-506

ancillary services, 3, 17, 116
Applicability Clause, 268-269, 332
as available power, 132
asset retirement obligation (ARO), 489-490
Association of Edison Illuminating Companies (AEIC) Load Research Committee, 576
attendant billing provisions, 309-333
  delivery voltage adjustment, 319-320
  determination of billing demand, 310-318
  metering voltage adjustments, 320-321
  minimum bill provision, 326-327
  power factor adjustments, 321-325
  special billing adjustments, 327-330
  taxes and tax adjustments, 330
  term of payments, 331
attendant service provisions, 267-271
  Applicability Clause, 268-269, 332
  Availability Clause, 267-268, 332
  Character of Service Clause, 269

Term of Contract Clause, 270, 333
Availability Clause, 267-268, 332
average rate chart, 340, 349, 377-378. *See also* hyperbolic chart
avoided costs
    avoided capacity costs, 132
    avoided energy costs, 132, 223-224

Bary, Constantine W., 576-577
Bary curve. *See* CF—LF curve
bill chart, 337-339, 377-378
    three-dimensional, 363
bill classifications
    estimated bill, 258-259
    normal bill, 259
    prorated bill, 258-259
bill comparisons
    graphical, 369-380, 382-387
    tabular, 365-366
bill equations
    development of, 352-361
    geometric interpretation of, 362-363.
See also billing formulas
bill frequency, 407
    compound frequency, 418-421
    demand frequency, 415-417
    HUD frequency, 416-417
    kWh frequency, 405
bill frequency analysis, 404-407, 431
    consolidated factor, 409, 413
    cumulative use, 408
    interval use, 407
    interval size, 404
    use through that interval, 409-410
bill proration, 258-259
billing demand
    determination of, 310-318
    metering of, 237-241
billing determinant., 299, 302, 389
    en masse, 407
    forecasting of, 424, 428, 431, 435, 438-440
    scatter plot of, 418
    of single bills, 404
billing formulas, 335-336
    See also bill equations, miminum bill provision, revenue calculations
billing month, 257
billing of electric service, 256-261
    conjunctional billing, 251
billing options
    levelized billing, 118
    summary billing, 119
block energy rate
    block discounts, 290-291
    block extender, 301-302
    block inserts, 291-294
    declining block, 288-289
    design of, 571-575
    floating tail block, 295
    inverted block, 288-289
    penultimate block insert, 293
    percentage-of-use price blocks, 295-297, 299-301
    quick-break declining block, 288
    split block insert, 293-294
    stopper, 289-290
    stretcher block. *See* block extender
    tail block insert, 292
block Hopkinson rate, 277
Bonbright, James C. 18, 610
bundled rate, 263-265, 277, 316, 332-333
cable
    concentric neutral cable, 147-148
    duplex cable, 147
    n-plex cable, 147, 505
    quadruplex cable, 147
    submarine cable, 155
    tree wire, 154
    triplex cable, 147
    underground cable, 155, 508-509

Index

See also conductors
capacitors, 26-28, 112-113, 124, 157-159, 516-517, 522
capacity, 106-107
capacity factor, 216, 218
capital structure, 54-55, 484-485
capitalization criterion, 486
cash working capital (CWC), 475
   classification of, 521
   functionalization of, 475
central tendency, 391, 394, 397
CF—LF curve, 576-577
   CF—LF model calibration, 583-585
   CF—LF model variations, 585-587
   synthetic CF—LF curve splice, 583-584
Character of Service Clause, 269
chemical end uses
   battery charging, 73-74
   electrolysis, 73
   electroplating, 73
cogeneration, 11, 128-132
   Backup Electric Service, 132
   bottoming cycle, 129
   combined cycle, 129
   Maintenance Electric Service, 132
   parallel operation, 130
   Qualifying Facility (QF), 13, 131-132, 223-224
   Small Power Production (SPP), 131
   Supplementary Electric Service, 132
   topping cycle, 129-130
coincidence factor, 44, 576
   system profile of, 539-540
coincidence factor—load factor (CF—LF) relationship, 576-577
Columbia University, 610-611
common plant, 470-471
   functionalization of, 468-469

competition, 11-12
Competitive Transition Charge (CTC), 17, 264, 330
conductors
   aerial conductors, 152, 505-508
   aluminum wire, 152
   American Wire Gauge (AWG), 153
   ampacity, 148, 505
   circular mils, 153
   conductor resistance, 139
   copper wire, 152
   underground conductors, 154-155, 508-509
   See also cable
conduit, 148, 510
confidence level, 100, 102, 251
Consolidated Edison Company of New York, Inc., 576
construction cost indicators, 446-447
construction work in progress (CWIP)
   classification of, 520
   functionalization of, 471-472
contribution in aid of construction (CIAC), 275, 476-477
correction factor, 407
cost basis
   first cost. *See* original cost
   original cost, 446-447, 488
   rebuild cost, 501-503, 509
   reproduction cost, 446-447, 471
   trended original cost, 446-447
cost causation, 57, 443, 452, 481, 491, 494
cost classification
   of expense, 524-529
   of plant, 494-523
cost functionalization, 454-493
   subfunctionalization, 455-468
cost levelization, 455, 458
cost of capital. *See* weighted average cost of capital (WACC)

cost-of-service study, 9, 19, 266, 443-570, 592
  applications, 568-570
  framework, 452-454
  jurisdictional separation, 449
  pro forma adjustments in, 445
  resolution of, 447-448
  resource requirements for, 450-451
cost-to-revenue criterion, 476
cost, types of
  embedded cost, 55
  fixed cost, 57, 443, 461, 495, 518
  installed cost, 50
  marginal cost, 55-57
  stranded cost, 17
  variable cost, 57, 443, 461, 541
cost curve development, 587, 590
critical peak pricing (CPP), 298
cross-subsidization
  inter-class, 567-568
  intra-rate, 272
current diversion, 252
customer
  as energy manager, 120-124
  classes, 3, 8-9
  responsibilities of, 126-127
  See also obligation to serve
customer advances for construction, 476-477
Customer Charge Structures, 273-276
  based on customer density, 275
  based on monthly demand, 274
  daily fee approach for, 273
  voltage differentiated, 274
customer density, 6, 150, 165-166, 275
customer information system (CIS), 389-390, 450
customer-related costs, 273-276
customer service, 119-120, 451, 486, 560
customer, types of, 3-4

chain accounts, 122
compound retail, 133
municipal utilities, 10, 133
native customers, 3
retail, 3
rural electric cooperatives, 10, 133
seasonal usage, 276
territorial, 3
wholesale, 3, 133
cycle billing, 256-259

DC power transmission, 184-187
  bipolar DC line, 185-186
  converter stations, 184-186
  homopolar DC line, 185
  monopolar DC line, 185
degree day (DD), 75-76, 83
delivery point, 106
delivery voltage
  adjustments, 319-320, 358
  characteristics, 402-403
demand, 46-47
  contract demand, 317
  counted demand, 311
  hybrid kVA, 311, 387
Demand Charge Structures
  bifurcated, 277, 310, 316
  explicit, 277, 326
  hours use of, 281-287
  implicit, 286, 326
  time-differentiated, 278-281
demand control, 123
demand cost allocation
  methodologies
    average and excess demand (AED), 541-543
    B-I-P, 462-468, 542
    coincident peak (CP), 532-537
    equivalent peaker, 519
    noncoincident customer peak, 538
    noncoincident peak (NCP), 532, 537
    12-CP, 534-535

Index

demand interval, 46-47, 238, 241
  rolling demand, 239-241
  sliding window. *See* rolling demand
  synchronous interval, 240-241
demand ratchet, 312-317
demand-related costs, 276-277
Demand-Side Management Cost Recovery Rider, 329-330
deposits, 451, 477, 482, 485
  as a component of capital structure, 484-485
depreciation
  accelerated, 477-478
  accumulated, 471, 489, 521
  double-rate declining-balance, 477
  Iowa-type survivor curves, 488
  remaining life method, 489
  straight-line, 477-478, 488
  studies, 488-489
  sum-of-the-year's digits, 477-478
  whole life method, 489
depreciation expense, 488-489
  assignment of, 560
  classification of, 528
  functionalization of, 489
differential charts, 367-383
  applications of, 382-383
  line of equality in, 367-368
  power factor consideration in, 380-381
  three-dimensional, 368
  topological view of, 378-380
  TOU rates, 384-386
  See also rate comparisons
distribution
  FERC indicators of, 173
  network configuration, 172
  performance, 168-172
  spatial attributes of, 165-168
  voltage levels, 146, 152
distribution equipment
  bypass switches, 517

capacitors, 157-159, 516-517
circuit breakers, 160-161
cutouts, 150-151, 155, 160, 517
insulators, 151, 154, 516
lightning arresters, 150-151, 159, 517
line switches, 516-517
reclosers, 160, 516-517
regulators, 157, 516-517
relays, 161
sectionalizers, 160, 516
substations, 161-164
transformers. *See* line transformers
distribution plant
  assignment of, 555-558
  classification of, 494-518
diversity factor, 44
Doherty, Henry L., 271
Doherty rate, 271, 341, 335

economic development, 12
economic dispatch, 219-222, 606
  merit order, 222, 328
  transmission penalty factor, 222
Edison, Thomas A., 67, 73, 228
eddy currents, 138, 230
efficacy, 68
efficiency, 36
  coefficient of performance (COP), 63-64
  efficacy, 68
  energy efficiency ratio (EER), 64
  seasonal energy efficiency ratio (SEER), 64
Eisenmenger, H. E., 576
electric current, 24
  alternating current (AC), 25
  direct current (DC), 25
  displacement current, 27
  exciting current, 138
  fault current, 159-160
  magnetizing current, 138
  skin effect, 139

electric plant held for future use
  classification of, 520
  functionalization of, 472
electric service requirements
  energy and capacity, 106-107
  power quality, 112-115
  reliability, 111-112
  voltage and frequency, 107-110
electric tariff, 7-9, 17
  general terms and conditions, 7, 265-266
  rate schedule, 7-8, 15-17, 263-331
electrical loads, 36-49
  individual loads, 37-40
  systems of electrical loads, 40-49
electrical losses, 138-141, 164, 543-554
  copper losses. *See* load losses
  core losses. *See* no-load losses
  demand losses, 141, 530, 543, 548
  eddy-current losses, 138
  energy losses, 141, 531, 543, 548-554
  full load losses, 547-549
  hysteresis losses, 138
  load losses, 138, 247-248, 544-545, 548-554
  no-load losses, 138, 247-248, 545, 547-554
  system modelling of, 546-554
electricity pricing
  concepts of, 4-7
  guiding principles of, 610
  role of engineering in, 18-19
  stakeholders, 7-9
electricity pricing considerations
  configuration of local facilities, 5-6
  cost recovery, 58-59
  spatial cost variations, 5
  temporal cost variations, 5
  type of construction, 6

electronic end uses, 74
emissions, 198, 200, 204, 214
  allowances, 606
  reagents, 198, 606
end-use applications, 62-89
  commercial sector, 84-85
  industrial sector, 86-89
  process heating, 66
  residential sector, 75-83
end-use load devices
  adjustable speed drives, 72
  air conditioner, 39-40, 79
  heat pump, 63-64
  lamps, 67-68
  microwave oven, 77
  motors, 69-72
  multi-stage load devices, 38
  operating patterns, 17
  refrigerator, 81
  resistance heater, 63
  water heater, 39, 65
end-use trends, 91-96
  life cycles, 91-92
  market penetration, 92-95
  obsolescence, 92, 95
end-use value, 89-91
energy audit, 124
Energy Charge Structures
  block, 288-297
  time-differentiated, 297-298
  See also block energy rate
energy consumption, 36, 45, 48
  base use, 75-78, 82, 89-90
  end-use requirements, 74-88
  seasonal use, 75-78, 89-90
energy conversion, 61, 195, 225
Energy Policy Act (EPAct), 13
energy procurement, 121-122
energy-related costs, 287
energy service, 106
Energy Supply Cost Recovery, 328-329. *See also* fuel and purchased power
Environmental Cost Recovery, 329

Index

Federal Energy Regulatory Commission (FERC), 10, 15, 116, 131, 133
   asset retirement obligation (ARO), 490
   5-level system for cost of service, 457
   organization of, 10
   variable O&M predominance method, 525-527
Financial Accounting Standards Board (FASB), 489
firm power, 132
flat rate, 288
Flexible AC Transmission Systems (FACTS), 183-184, 190
   Static Synchronous Compensator (STATCOM), 183
   Static Synchronous Series Compensator (SSSC), 183
   Unified Power Flow Controller (UPFC), 183
flue gas desulfurization (FGD), 198, 606
forecast. *See* revenue forecast
frequency (system), 107
   harmonics, 114-115
   motor speed, 71
frequency distribution, 391, 393-394, 401
   bill analysis, 404-407
   cumulative relative frequency distribution, 395-396
   relative frequency distribution, 391, 393
   revenue analysis, 407-421
   skew in a, 392, 394, 397, 399, 401
fuel and purchased power cost recovery, 287-288, 328-329, 606
fuel cells, 204-205

gas-fired production, 197-198
   combined cycle (CC), 203
   combustion turbine (CT), 202-203
   microturbines, 204
general plant
   classification of, 519-520
   functionalization of, 468-471
general terms and conditions, 7
generation
   backup generation, 112, 127-128
   base load generation, 216, 462
   combined cycle (CC), 203
   cycling generation, *See* intermediate generation
   distributed generation (DG), 141, 193, 204, 210
   efficiency, 219
   generation step-up (GSU), 137, 193, 455, 546, 557
   intermediate generation, 216, 462
   load following, 131
   operations, 219-225
   peaking generation, 216, 462
   screening curves, 216-217
generation reserves, 218-219
   cold reserves, 219
   hot reserves, 219
   operating reserves. *See* spinning reserves
   reserve capacity, 218-219
   reserve margin, 218
   spinning reserve, 219, 222
generator
   armature, 194-196
   capability curves, 196
   cooling, 195
   exciter, 194, 208
   field, 194-196
   generation step-up (GSU). *See* transformer
   heat rate, 219-220
   induction generator, 208
   input-output curve, 220
   loading. *See* capability curves, 196

merit order stack, 462
nameplate capacity, 211, 214
overexcited state, 194-196
prime mover, 193-195
synchronous generator, 194-196
transfer switch, 127-128, 131
underexcited state, 194-196
generator turbines
   aeroturbines, 207-208
   Francis-type reaction turbine, 206
   gas-fired turbines, 202-203
   Kaplan-type reaction turbine, 206
   microturbine, 204
   Pelton wheel-type impulse turbine, 206
   steam turbines, 197, 199-201
geothermal production, 200
   binary cycle, 200
   flash tank, 200
Greene, W. J., 541
gross receipts. *See* revenue
gross receipts tax. *See* revenue taxes

Handy-Whitman Index of Public Utility Construction Costs, 446-447
harmonics, 114-115
heat rate, 219-220
heating end uses, 63-66
   degree day (DD), 75-76, 83
   process heat, 66
   water heating, 65
high load factor rate, 268, 308, 326
   minimum bill provision of, 326
high voltage metering. *See* instrument transformers
histogram, 391, 415
Hopkinson, John 18, 277, 281
Hopkinson rate structure, 277
   block Hopkinson structure, 277
   design of, 603-604
hours use of demand (HUD), 282-287
   effect on average rate, 284
   relation to load factor, 282
   temporal aspects of, 287
   visualization of, 283
hydraulic production
   conventional, 205-206
   pumped storage, 206
   run-of-the-river, 205
   tidal power, 206-207
hydroelectric production. *See* hydraulic production
hyperbolic chart, 343-349. *See also* average rate chart

impedance, 28-30
   capacitive reactance, 25, 28-30
   inductive reactance, 25, 28-30
   in parallel, 29
   resistance, 24
income taxes
   assignment of, 564-566
   combined state and federal income tax rate, 565
   deferred income taxes (DITs), 477-479
   expansion factor for, 569
   flow-through accounting, 480
   normalization, 480
industry classifications, 268, 390
   NAICS codes, 86-87, 113, 214, 639-646
   SIC codes, 86-88, 639-640, 647-650
industry restructuring, 12-15
infinite bus, 130
initial charge rate, 276
injuries and damages reserve
   classification of, 521
   functionalization of, 481
instrument transformers, 228, 245-247, 495, 557
   current transformers (CTs), 245-247
   potential transformers (PTs), 245-247

Index

voltage transformers (VTs). *See* potential transformers
intangible plant
   classification of, 519-520
   functionalization of, 469-470
intercept. *See* zero intercept
interruptible service, 123-124, 534
inverse scale. *See* hyperbolic chart
investment, 50
investment tax credit (ITC), 479-480
   flow-through accounting, 480
   normalization, 480
Iowa State University, 488

labor
   assignment of, 559-560
   classification of, 524-528
   functionalization of, 469
   labor ratios, 469
   See also salaries and wages
lambda (λ), 220-223
   real-time pricing (RTP), 224
   use in rate design, 574-575
late or deferred payment provision, 331
lead/lag study, 450, 475
   See also cash working capital
liberalized depreciation. *See* accelerated depreciation
lifeline rate, 289
lighting efficiency. *See* efficacy
lighting end uses, 67-68
   fluorescent lamps, 67
   high intensity discharge (HID) lamps, 67
   incandescent lamps, 67
   lumen, 68
line extension policy, 476
line transformer, 137, 148-151, 162
   assignment of, 538, 556-557
   classification of, 502-504
   completely self-protected (CSP), 151
load characteristics, 36-49

   coincidence factor, 44, 539-540
   connected load, 40
   demand, 46-47
   diversity factor, 44
   energy consumption, 45
   instantaneous load, 45-47
   load diversity, 44, 123, 531-543, 549
   load factor, 48-49, 282
   nameplate rating, 37, 124
load control, 117, 122-123
load density, 166
load diversity, 44, 123, 531-543, 549
   cost-of-service considerations, 531-543
   feeder load diversity, 165
load duration curve, 214-217, 224, 462, 467
   of system losses, 545
load factor, 48-49, 268, 272
   of load-related losses, 546
   in mimimum bills, 326-327
   relationship to hours use of demand (HUD), 282
   See also CF—LF relationship
load growth, 167-168, 422-423
   spatial variations in, 165-168
   S-shape characteristic of, 168
load research, 99-103, 251
   confidence interval, 100, 102, 251
   for cost of service, 450
   load shapes, 96-99
   normal distribution curve, 102-103
   sample design, 100
   sampling accuracy, 100
   standard deviation, 102
   standard error, 85, 88, 102
   translation, 100, 236, 240, 250, 255-256
loss compensation, 247-248, 320-321
loss correction factors, 321

loss expansion factor, 608
loss load factor, 546
losses. *See* electrical losses

materials and supplies (M&S), 473-475
   classification of, 521
   functionalization of, 474
mean, 394, 397-398
mechanical end uses, 69-72
median, 394-395
   point estimate of, 395
meet or release clause, 271
meter
   A-Base type, 232-233
   bidirectional electronic meter, 230
   detents, 230
   disk, 229-232, 249
   disk constant (Kh), 231-232
   Edison meter, 73
   indicating demand type, 238-239, 310
   integrating demand meters, 238-244
   pulse initiators, 236, 249
   Q-hour meter, 243-244
   S-Base socket type, 232-233
   self-contained meters, 228, 245
   time-of-use metering, 235-237, 240-241
   totalizing relays, 249
   watt-hour meter, 229, 232
   Wright meter, 228
metering
   demand, 238-241
   high voltage, 245-247
   interval, 236-237, 239-241
   load research, 100-101
   loss compensation, 247-248
   pre-pay, 254
   reactive demand, 242-244
   reactive energy, 234-235
   revenue, 228
   submetering, 133
   totalization, 248-251, 255
   volt-ampere metering, 244
   watt-hour, 232-234, 495
Metering Voltage Adjustments, 320-321, 358
meter reading, 252-254
   Automatic Meter Reading (AMR), 254
minimum bill provision, 326-327
minimum demand requirement. *See* rate minimum
minimum distribution system, 495, 498-517, 520-521, 524, 528
   conductors, 505-510
   line transformers, 502-503
   pole, 511-515
   voltage path concept, 498-499, 505, 511, 518
minimum intercept. *See* zero intercept
mode, 394
   point estimate of, 394
motors, 69-72, 87-88, 269
   adjustable speed drives, 72
   commutator, 69-70
   direct current (DC), 69-70
   induction motor, 71-72
   rotor, 69-72
   squirrel-cage rotors, 72
   stator, 69-72
   synchronous motor, 71-72
   wound rotors, 72

National Association of Regulatory Utility Commissioners (NARUC), 525
   variable O&M classification method, 525-527
National Electric Code (NEC), 126
National Electrical Safety Code (NESC), 142, 511
National Grid system, 190
natural conservation, 423

Index

net operating income (NOI), 484, 562
neutral conductor, 144-145, 147, 154, 159
   common neutral, 144, 154, 506-508
nitrogen oxides (NO$_X$), 198
NO$_X$. *See* nitrogen oxides
North American Industrial Classification System (NAICS), 86-87, 113, 214, 639-646
North American Electric Reliability Council (NERC), 188-189
nuclear production
   boiling water reactors, 199
   Chernobyl, 199
   light water reactors, 199
   liquid metal fast breeder reactor, 210
   pressurized water reactors, 199
   steam generator, 199
   Three Mile Island, 199
Nuclear Regulatory Commission (NRC), 489

obligation to connect, 14
obligation to serve, 9
ogive, 428-434
   use to average use, 430
Ohm's Law, 24, 29
Open Access Same-Time Information System (OASIS), 13
operating reserves, 451, 480-481
   injuries and damages, 481
   pensions and benefits, 481
   property insurance, 480-481
operation and maintenance (O&M)
   expenses, 50, 486
   assignment of, 559-561
   classification of, 524-527
   functionalization of, 486
   supervision and engineering O&M, 524
outages, 168-170
   See also reliability
outdoor lighting. *See* street lighting
overcurrent, 159-160

parallel operation, 130
parity, 568-569
   parity ratio, 568
payroll-related taxes, 490
peak shaving, 132
Pearl Street generating station, 67
Philadelphia Electric Company, 576
plant, 50
   plant ratios, 469
plant held for future use
   classification of, 520
   functionalization of, 472
pole, 151-152, 154
   circumference. *See* pole class
   class, 151, 511-513
   fixtures, 515-516
   ground line moment capacities (GLMC), 513-514
   height of, 511-513
polyphase service, 109
power, 30-34
   apparent power (VA), 32
   hybrid kVA, 242, 311, 321
   reactive power (VAR), 34
   real power (Watts), 33
   three-phase power, 144
power delivery service, 106
power exchange (PX), 225
power factor, 34-35
   average power factor, 235, 311-312
   base power factor, 312, 322-324
   correction, 124, 325
   hybrid power factor, 242
   improvement. *See* correction
   lagging power factor, 35
   leading power factor, 35
   price signal, 324-325
power factor rate adjustments, 321-325

billing demand adjustment
  approach, 312
  excess kVA approach, 322-323
  excess kVAR approach, 322-324
  formulization of, 358-359
power flow
  flow of reactive power, 178-179
  flow of real power, 140, 178-183
  inadvertent flows, 188-189
  interconnected utility systems, 187-190
  loop flows, 190
  parallel paths, 181-182, 190
  receiving-end voltage, 177
  reference bus voltage, 177, 179
  sending-end voltage, 177
power line carrier (PLC), 253
power management, 122-124
power pool, 187, 224-225
power quality, 112-114
  dips. *See* sags
  flicker, 113
  harmonics, 114-115
  nonlinear load, 115
  overvoltage, 113-114, 159-160, 168,
  sags, 113, 127
  surges, 112-113
  swells, 113
  transients, 112
  undervoltage, 113-114
power system, 1-3
  levels, 457
prepayments, 451, 475, 482
  functionalization of, 475-476
price regulation, 10-12
pricing strategy, 610
pricing structures. *See* rate structures
primary distribution, 151-156
  network configuration, 172
  radial system, 155
production
  central station, 135, 173, 193
  distributed generation (DG), 204
  See also generators
production cost modeling, 223
production facilities
  air preheater, 197
  axial compressor, 203
  combustor, 203
  condenser, 197, 199, 202
  cooling ponds, 202
  cooling towers, 202
  economizer, 197
  flash tank, 200
  heat recovery steam generator (HRSG), 203
  heliostats, 201
  linear trough, 200
  penstock, 205-206
  powerhouse, 206
  precipitators, 198
  receiver pipe, 200-201
  reformer, 205
  regenerator, 203
  reheater, 197
  scrubber, 198
  selective catalytic reduction (SCR), 198
  steam generator, 199
  superheater, 197
  tailrace, 205
  turbine inlet cooling, 203
  windmill, 207-208
production fuels, 211-213
  biomass and wood wastes, 210, 214
  fossil fuels, 195, 197-198, 201, 210
  landfill gas, 203-204, 210, 214
  methane, 202-205
  natural gas, 197, 202, 205
  nuclear, 198-199, 210
  waste materials, 210
production O&M costs
  assignment of, 559
  classification of, 525-527
production plant

Index

assignment of, 534
classification of, 518-519
subfunctionalization of, 461-468
prompt or deferred payment provision. *See* late payment provision
property insurance reserve
  ADIT associated with, 479
  functionalization of, 480-481
property taxes, 490
Public Service Commission (PSC), 10
Public Utility Holding Company Act (PUCHA), 13
Public Utility Regulatory Policies Act (PURPA), 13, 131-132
purchased power, 214, 222, 328, 606-607

Q-Metering, 243-244
Qualifying Facility (QF), 13, 131-132, 223-224

ratchet. *See* demand ratchet
rate base, 52, 446
  additions, 471-476
  allocation factors, 555
  assignment of, 554-559
  classification of, 494-523
  deductions, 476-481
  development of, 471
  end-of period basis, 446
  fair value, 447
  functionalization of, 454-483
  original cost depreciated (OCD), 447
  reproduction cost new depreciated (RCND), 447
  13-month average, 446
rate class cost curve, 587-597
rate comparisons
  bill tables, 364-365, 383
  differential charts, 367-383
  for TOU rates, 383-387

rate design
  choosing a base rate structure design, 604-605
  demand rates, 598-605
  for fuel cost recovery, 606-609
  nondemand rates, 571-576
rate formulization, 352-363
  of complex billing structures, 358-361
  equation matrix, 354
  geometric interpretation of, 362-363
  inherent rate tilt, 356
ratemaking principles, 610-611
rate minimum, 317-318
  in differential charts, 372
  in rate formulization, 354
  in rate structure charts, 350, 352-353
rate riders, 266, 310, 327, 330
rate schedule, 15-17, 331-333
  anatomy of, 266-267
  applications of, 8
rate structure
  combinations, 298-302
  risk, 302-309
  three-part rate, 271
  two-part rate, 272
  See also block energy rate, Doherty, Hopkinson, time-of-use, unbundled, Wright, Wright-Hopkinson
rate structure chart, 350-351
  in bill frequency analysis, 418
  in differential chart analysis, 369-371
  for rate formulization, 352-353, 355
  relationship to bills, 362-363
rate tilt, 272, 299, 306, 308
  in rate design, 603-605
  in Wright rates, 355-356
reactive power. *See* power
real estate taxes. *See* property taxes

real-time pricing (RTP), 224, 237, 254, 298
reciprocal scale. *See* hyperbolic chart
reciprocating engines, 210
Regional Transmission Organizations (RTOs), 190-191, 265
regulator (voltage), 157, 516-517
regulatory agency. *See* Federal Energy Regulatory Commission, Public Service Commission
reliability, 111-112, 142, 171-172, 176, 188
  customer requirements, 111-112
  outages, 142, 168-172, 176
reliability indices, 171-172
  Customer Average Interruption Duration Index (CAIDI), 171
  Customer Average Interruption Frequency Index (CAIFI), 171
  Momentary Average Interruption Frequency Index (MAIFI), 172
  momentary interruption indices, 172
  System Average Interruption Duration Index (SAIDI), 171-172
  System Average Interruption Frequency Index (SAIFI), 171
restructuring. *See* industry restructuring
return
  rate of return (ROR), 52-53, 484, 566
  return available for equity, 567
return on equity (ROE), 54, 562, 567-569
return on investment (ROI), 54, 562, 566-569
revenue
  analysis of, 562-565
  credits, 564-565
  from miscellaneous services, 563-564
  phantom, 573
  from sales, 562-563
revenue calculations
  for individual bills, 261, 412-413
  with mass billing determinants, 407-413, 415-421
revenue forecast, 421-440
  average rate method, 424-428, 434
  customer account method, 435-440
  ogive method, 428-434
revenue requirement, 52-53, 562, 569
revenue taxes, 491
rolled-in transmission, 457
rules and regulations. *See* terms and conditions

salaries and wages (S&W), 469, 487, 521, 524, 528
  See also labor
sales
  electricity, 562-563
  for resale, 563
  See also unbilled sales
sample
  mean, 102, 398
  standard deviation, 102, 398
SCADA. *See* Supervisory Control and Data Acquisition
scrubber. *See* flue gas desulfurization
secondary distribution, 146-148
  network configuration, 172
selective catalytic reduction (SCR), 198, 606
service. *See* interruptible service, single-phase service, polyphase service, three-phase service, wholesale service
service drop. *See* service line

Index

service line, 146, 148, 482, 495, 523, 558, 561
service reliability. *See* reliability
single-phase service, 109
SO$_2$. *See* sulfur dioxide
solar power, 200-201, 208-210
    central station (grid-support) PV, 209
    insolation, 209
    linear trough, 200
    photovoltaics, 208
    receiver pipe, 200
standard deviation
    of a sample, 106, 398
    of the population, 397
Standard Industrial Classification (SIC) codes, 86-88, 639-640, 647-650
straight-line rate, *See* flat rate
steam production, 197-202
stopper rate mechanism, 289-290
street and outdoor lighting, 517-518
subfunctionalization, 456-468
    of production plant, 461-468
    of T&D plant, 458-461
subsidy. *See* cross-subsidization
substation, 161-164, 174, 176, 459
    assignment of, 543, 557
    classification of, 518-519
    dual voltage, 459-460
    load diversity aspects of, 535-537
    service boundaries, 163-164
subtransmission, 141, 161, 174-175, 535
    distribution acting as, 164
    levelization of, 456-457
sulfur dioxide (SO$_2$), 198
Supervisory Control and Data Acquisition (SCADA), 252
system lambda. *See* lambda

take-or-pay provision, 317
tariff, 7
    organization of, 265-266

taxes and tax adjustments, 330
taxes other than income taxes, 451
    assignment of, 560
    classification of, 528
    franchise tax, 491
    functionalization of, 490
    pass through taxes, 491
    payroll taxes, 490
    property taxes, 490
    revenue taxes, 491
Term of Contract Clause, 270, 333
Term of Payment, 331
terms and conditions, 265-266
test period. *See* test year
test year, 53
    historical, 445
    projected, 445
thermodynamic cycles
    Brayton cycle, 202
    Rankine cycle, 197
three-phase service, 109
time-of-day. *See* time-of-use
time-of-use (TOU)
    rate structures, 278-281, 297-298
    rating periods, 463-466
transformer
    generation step-up (GSU), 137, 193, 455, 546, 557
    load tap changing (LTC), 157
    losses, 138-139, 548-549
    pad, 504
    pad-mounted, 148-150
    phase-shifting, 183, 235
    pole-mounted, 148-150
    power class, 137, 163
    vault-type, 504
    See also line transformers
transformer connections
    delta, 143-145
    open delta, 145
    wye, 143-145
translation system, 236, 240, 250, 255-256
transmission

backbone. *See* bulk transmission
bulk transmission, 174-175, 468
  insulators, 175
  rolled-in, 457
  penalty factor, 222
  See also subtransmission
transmission interconnection, 187-190
  inadvertent energy, 188-189
  power interchange, 187
  See also Regional Transmission Organizations, wheeling
transmission plant
  assignment of, 535, 557
  classification of, 518-519
transmission tariff (OATT), 17
  ancillary services, 3, 17, 116
turbine. *See* generator
Turner Construction Company's Building Cost Indices, 447

unbilled sales, 257
unbundled rates, 15-17, 264-265, 316
Uniform System of Accounts, 50-51, 446, 494, 617-637
  distribution plant accounts, 496-497
  in functionalization, 455-456, 486
  subaccounts, 494, 638
Uninterruptible Power Supply (UPS), 112, 128
unit commitment, 222-223
unit cost components, 592
unit train, 197
used and useful criterion, 472

variance, 397-398
VAR. *See* power
voltage, 2-3, 24-27
  effective voltage, 31-32
  voltage classes, 151-152
  voltage level bypasses, 140-141
  voltage regulators, 157

voltage support, 161, 179-180
voltage level adjustments, 318-321
voltage path concept, 499-500, 505, 511

weighted average cost of capital (WACC), 32, 484, 566, 569
  financial view of, 484
  regulatory view of, 484-485
wheeling, 10, 188, 564-565
Whitman, Requardt & Associates, LLP, 447
wholesale service
  full requirements service, 3
  partial requirements service, 3
wind generator, 207-208
  aeroturbine, 207
  cut-in speed, 207
working capital, 451, 472, 475, 482
  cash working capital (CWC), 475
  lead/lag study, 450, 475
Wright, Arthur 18, 228
Wright-Hopkinson rate design, 600
Wright rate structure, 281, 284-287
  design of, 589-600
  forumlarization of, 352-357
  graph of, 350-352

zero-intercept methodology, 500-501
  for conductors, 505-510
  for poles, 511-515
  for line transformers, 502-503
  See also minimum distribution system